Cockroaches as Models for Neurobiology: Applications in Biomedical Research

Volume I

Editors

Ivan Huber, Ph.D.
Professor
Department of Biological and Allied Health Sciences
Fairleigh Dickinson University
Florham Park - Madison Campus
Madison, New Jersey

Edward P. Masler, Ph.D.
Research Physiologist
Insect Reproduction Laboratory
Agricultural Research Service
U.S. Department of Agriculture
Beltsville, Maryland

B. R. Rao, Ph.D.
Professor
Department of Biological Sciences
East Stroudsburg University
East Stroudsburg, Pennsylvania

CRC Press, Inc.
Boca Raton, Florida

Library of Congress Cataloging-in-Publication Data

Cockroaches as models for neurobiology: applications in biomedical research/editors, Ivan Huber, Edward P. Masler, B. R. Rao.
p. cm.
Includes bibliographical references.
ISBN 0-8493-4838-2 (v. 1) — ISBN 0-8493-4839-0 (v. 2)
1. Neurobiology—Research—Methodology. 2. Plectoptera (Cockroach)—Physiology. I. Huber, Ivan, 1931- . II. Masler, Edward P., 1948- . III. Rao, B. R., 1936- .
QP357.C63 1990
595.7′22′0724—dc20

89-22149
CIP

This book represents information obtained from authentic and highly regarded sources. Reprinted material is quoted with permission, and sources are indicated. A wide variety of references are listed. Every reasonable effort has been made to give reliable data and information, but the author and the publisher cannot assume responsibility for the validity of all materials or for the consequences of their use.

Direct all inquiries to CRC Press, Inc., 2000 Corporate Blvd., N.W., Boca Raton, Florida 33431.

International Standard Book Number 0-8493-4838-2
International Standard Book Number 0-8493-4839-0

Library of Congress Card Number 89-22149
Printed in the United States

THE EDITORS

Ivan Huber, Ph.D., is Professor of Biology at Fairleigh Dickinson University, Madison, New Jersey.

Dr. Huber received his A.B. degree in Zoology from Cornell University, Ithaca, New York in 1954 and his Doctorate in Entomology from the University of Kansas, Lawrence in 1968.

Dr. Huber is a member of the Entomological Society of America, The New York Entomological Society, the Society for the Study of Evolution, the Society of Systematic Zoology, and the honorary society Sigma Xi. He has been the recipient of research grants from the Merck Institute for Therapeutic Research.

Dr. Huber has written on allomones in chigger mites and the ecological genetics of flour beetles, but most of his work has been on cockroaches. He has published articles on the taxonomy, population biology, and oviposition behavior of this group. He currently is studying learning and memory in cockroaches.

Edward P. Masler, Ph.D., is a Research Physiologist in the Insect Reproduction Laboratory, Agricultural Research Service, U.S. Department of Agriculture, Beltsville, Maryland. He received his A.B. in Biology from St. Anselm College, Manchester, New Hampshire (1970) and an M.S. in Genetics from the University of New Hampshire, Durham (1973). He was awarded a Ph.D. in Biology from the University of Notre Dame, South Bend, Indiana (1978), where he became interested in the physiology of insect development and reproduction. He was awarded a postdoctoral fellowship to the Roche Institute of Molecular Biology, where he spent 2 years in the Department of Biochemistry studying vitellogenic protein structure and modification in the developing oocyte. A postdoctoral appointment to the Department of Entomology, Cornell University, Ithaca, New York followed, where he investigated neuropeptides responsible for ecdysteroid and vitellogenin production. Dr. Masler has been a Research Physiologist with the U.S. Department of Agriculture since 1982 and currently is interested in the biochemistry and molecular biology of development and reproduction in selected dipteran and lepidopteran species. Special emphasis is placed on the identities and roles of neuropeptides in metamorphosis and ovarian maturation. Memberships include the International Brain Research Organization, the International Society of Invertebrate Reproduction and Development, The Society for Developmental Biology, the Society for Neuroscience, and the Society of Sigma Xi.

Balakrishna R. Rao, Ph.D., is Professor of Biology at East Stroudsburg University, East Stroudsburg, Pennsylvania. He received his B.S. in Agriculture from Banaras Hindu University, Varanasi, India in 1957, his M.S. in Agriculture from Karnatak University, Dharwad, Karnataka, India in 1959, and a Ph.D. in Entomology from The Ohio State University, Columbus in 1964. His doctoral work in insect physiology culminated in his dissertation, entitled "Trypsin Activity Associated with the Reproductive Development in the Female Tampa Cockroach, *Nauphoeta cinerea* Oliv.". Dr. Rao had been a Postdoctoral Fellow at The Johns Hopkins University, the University of Connecticut, and The Marine Biological Laboratories, Woods Hole, Massachusetts prior to becoming Associate Professor of Biology at East Stroudsburg University. He is a member of the Entomological Society of Pennsylvania, the Entomological Society of America, Sigma Xi, the the Commonwealth of Pennsylvania University Biologists. Dr. Rao's current interests include reproductive physiology and digestive enzymes in cockroaches as well as mosquito reproductive physiology.

CONTRIBUTORS

Volume I

David W. Alsop, Ph.D.
Department of Biology
Queens College, CUNY
Flushing, New York

Moray Anderson, Ph.D.
Department of Zoology and Comparative Physiology
University of Birmingham
Birmingham, England

David J. Beadle, Ph.D.
School of Biological and Molecular Sciences
Oxford Polytechnic
Headington, Oxford
England

William J. Bell, Ph.D.
Department of Entomology
University of Kansas
Lawrence, Kansas

Isabel Bermudez, Ph.D.
School of Biological and Molecular Sciences
Oxford Polytechnic
Headington, Oxford
England

Jonathan M. Blagburn, Ph.D.
Institute of Neurobiology
University of Puerto Rico
San Juan, Puerto Rico

Jean-Jacques Callec, Ph.D.
Laboratory of Animal Physiology
University of Rennes I
Rennes, France

Yesu T. Das, Ph.D.
Innovative Scientific Services, Inc.
Piscataway, New Jersey

Charles R. Fourtner, Ph.D.
Department of Biological Sciences
State University of New York
Buffalo, New York

Ayodhya P. Gupta, Ph.D.
Department of Entomology
Rutgers University
New Brunswick, New Jersey

Ivan Huber, Ph.D.
Department of Biological and Allied Health Sciences
Fairleigh Dickinson University
Florham Park — Madison Campus
Madison, New Jersey

Bernard Hue, Ph.D.
Laboratory of Physiology
URA CNRS 611
University of Angers
Angers, France

Charles J. Kaars, Ph.D.
Department of Biological Sciences
State University of New York
Buffalo, New York

E. P. Masler, Ph.D.
Agricultural Research Service
U.S. Department of Agriculture
Beltsville, Maryland

Thomas A. Miller, Ph.D.
Department of Entomology
University of California
Riverside, California

Marcel Pelhate, Ph.D.
Laboratory of Physiology
URA CNRS 611
University of Angers
Angers, France

Yves Pichon, Ph.D.
Laboratory of Cellular Neurobiology
Department of Biophysics
C.N.R.S.
Gif-sur-Yvette, France

Robert M. Pitman, Ph.D.
Department of Biology and
Preclinical Medicine
Gatty Marine Laboratory
St. Andrews, Fife
Scotland

B. R. Rao, Ph.D.
Department of Biological Sciences
East Stroudsburg University
East Stroudsburg, Pennsylvania

David B. Sattelle, Ph.D.
Unit of Insect Neurophysiology
and Pharmacology
Department of Zoology
University of Cambridge
Cambridge, England

Karel Sláma, Ph.D.
Insect Chemical Ecology Unit
Institute of Organic Chemistry
Czechoslovak Academy of Sciences
Prague, Czechoslovakia

C. S. Thompson, Ph.D.
Department of Zoology
University of Toronto
Toronto, Ontario
Canada

S. S. Tobe, Ph.D.
Department of Zoology
University of Toronto
Toronto, Ontario
Canada

Hiroshi Washio, Ph.D.
Laboratory of Neurophysiology
Mitsubishi-Kasei Institute
of Life Sciences
Machida, Tokyo
Japan

CONTRIBUTORS

Volume II

Michael E. Adams, Ph.D.
Division of Toxicology and Physiology
Department of Entomology
University of California
Riverside, California

Cynthia A. Bishop, Ph.D.
Department of Psychology
Stanford University
Stanford, California

Benjamin J. Cook, Ph.D.
Veterinary Toxicology and Entomology Research Laboratory
Agricultural Research Service
U.S. Department of Agriculture
College Station, Texas

Roger G. H. Downer, Ph.D.
Department of Biology
University of Waterloo
Waterloo, Ontario
Canada

Franz Engelmann, Ph.D.
Department of Biology
University of California
Los Angeles, California

Derek W. Gammon, Ph.D.
Agricultural Research Division
American Cyanamid Co.
Princeton, New Jersey

Ivan Huber, Ph.D.
Department of Biological and Allied Health Sciences
Fairleigh Dickinson University
Florham Park — Madison Campus
Madison, New Jersey

Larry L. Keeley, Ph.D.
Laboratories for Invertebrate Neuroendocrine Research
Department of Entomology
Texas Agricultural Experiment Station
Texas A&M University
College Station, Texas

Manfred J. Kern, Ph.D.
Pflanzenschutzforschung-Biologie
Hoechst Aktiengesellschaft
Frankfurt am Main, Federal Republic of Germany

Michael K. Leung, Ph.D.
Multidisciplinary Center for the Study of Aging and Chemistry/Physics Program
SUNY/College at Old Westbury
Old Westbury, New York

Michael I. Mote, Ph.D.
Department of Biology
Temple University
Philadelphia, Pennsylvania

Michael O'Shea, Ph.D.
Laboratory of Neurobiology
University of Geneva
Geneva, Switzerland

Terry L. Page, Ph.D.
Department of Biology
Vanderbilt University
Nashville, Tennessee

Susan M. Rankin, Ph.D.
Department of Entomology
Texas A&M University
College Station, Texas

Coby Schal, Ph.D.
Department of Entomology
Cook College
Rutgers University
New Brunswick, New Jersey

Berta Scharrer, Ph.D.
Department of Anatomy and Structural Biology
Albert Einstein College of Medicine
Bronx, New York

Günter Seelinger, Ph.D.
Institute for Zoology
University of Regensburg
Regensburg, Federal Republic of Germany

Alan F. Smith, Ph.D.
Department of Entomology
Cook College
Rutgers University
New Brunswick, New Jersey

Rajindar S. Sohal, Ph.D.
Department of Biological Sciences
Southern Methodist University
Dallas, Texas

George B. Stefano, Ph.D.
Multidisciplinary Center for the Study of Aging and Biological Sciences Program
SUNY/College at Old Westbury
Old Westbury, New York

Renée M. Wagner, Ph.D.
Veterinary Toxicology and Entomology Research Laboratory
Agricultural Research Service
U.S. Department of Agriculture
College Station, Texas

Jane L. Witten, Ph.D.
Laboratory of Neurobiology
University of Geneva
Geneva, Switzerland

Stephen Zawistowski, Ph.D.
Division of Social Sciences
St. John's University
Staten Island, New York

Sasha N. Zill, Ph.D.
Department of Anatomy
Marshall University School of Medicine
Huntington, West Virginia

TABLE OF CONTENTS

Volume I

TABLE OF CONTENTS

Volume II

IV. Neurohormones and Neurotransmitters

V. Metabolic Control

VI. Sense Organs, Plasticity, and Behavior

PROLOGUE

Ivan Huber

Some years ago, I toured the animal rooms at the College of Pharmacy, St. John's University, Staten Island, NY. In casual conversation, the curator commented that in a few years he expected his rats and mice to be replaced by insects or other invertebrates. The prime reasons for this were the ever rising cost of maintaining facilities for mammals and the increasingly stringent government regulations because of agitation by animal rights activists. It occurred to me then that my favorite research subject, the cockroach, would make an admirable substitute for mammalian models in many biomedical research programs.[1] There are a number of very good reasons to consider using cockroaches for these purposes, and it is appropriate at this point to mention them. Cockroaches are large enough to handle, yet do not require much room. Large numbers of them can be raised very inexpensively. Furthermore, many species with a variety of reproductive strategies are already available in culture. Cockroaches are particularly resistant to disease and trauma. A circumstance which may pacify antivivisectionists is that insects are said to be insensitive to pain.[2,3] Prospective users should be encouraged by the fact that considerable literature already exists dealing with cockroach physiology and anatomy (for references, see Introduction), so it should not be necessary to establish parameters for a completely new organism. Finally, but most significantly, in neurobiology, at least, many functional analogies can be made between cockroach and mammalian systems.[4]

In the fall of 1984, a small workshop entitled "Cockroaches as Models in Biomedical Research" was organized for an Eastern Branch Meeting of the Entomological Society of America. My intention was to explore, in a public forum, the potential of cockroaches in biomedicine. Response was positive and gratifying. Karl Maramorosch of Rutgers University later suggested that the presentations could form the nucleus of a book, and for this I thank him. The other organizers of the workshop (E. P. Masler, Agricultural Research Service, U.S. Department of Agriculture, Beltsville, MD and B. R. Rao, East Stroudsburg University, East Stroudsburg, PA) agreed with me that such a volume would be of benefit to illustrate the use of insects as alternative experimental animal systems. We realized that such an undertaking rapidly could become unmanageable if some limitations were not set. As we examined the subjects addressed by the workshop, it became clear that all the papers dealt in some way with neurobiology. We therefore chose this as the topic of our book, inviting the speakers and other contributors to participate.

The chapters in this book are concerned partly, and in some cases, solely, with methods. We believe this will be especially useful to those for whom this book is primarily intended, workers in the biomedical sciences. We hope, through this volume, to persuade more researchers to adopt the cockroach as their experimental animal. Since many of these investigators may be more familiar with mammals than with insects, it is our intention not only to review neurobiological areas in which cockroaches could serve as models, but also to provide a source of information on many aspects of work with cockroaches. For those progressive readers who already are using insects in their research, we trust that they, too, will find something of value in these pages.

During the 16th century, animals, including insects (a source of remedies), were studied as an essential part of the pharmacopeia. As a result, a close connection between entomology and medicine developed and persisted for the next 200 years. Marcello Malpighi (1628—1694), the Italian physician and professor of medicine, was the first to separate the two disciplines because he considered the aims of animal anatomists to be different from those of the medical profession.[5] It is my hope that after reading this book researchers will reassociate the two fields to their mutual benefit.

REFERENCES

1. Committee on Models for Biomedical Research, *Models for Biomedical Research: A New Perspective,* Board on Basic Biology, Commission on Life Sciences, National Research Council, National Academy Press, Washington, D.C., 1985.
2. **Wigglesworth, V. B.,** Do insects feel pain?, *Antenna,* No. 1, 8, 1980.
3. **Eisemann, C. H., Jorgensen, W. K., Merritt, D. J., Rice, M. J., and Cribb, B. W.,** Do insects feel pain? A biological review, *Experientia,* 40, 164, 1984.
4. **Scharrer, B.,** Insects as models in neuroendocrine research, *Annu. Rev. Entomol.,* 32, 1, 1987.
5. **Beier, M.,** The early naturalists and anatomists during the Renaissance and seventeenth century, in *History of Entomology,* Smith, R. F., Mittler, T. E., and Smith, C. N., Eds., Annual Reviews, Palo Alto, CA, 1973, 81.

INTRODUCTION

E. P. MASLER

Insects present an extremely diverse and rich source of material for numerous aspects of biomedical research. It was precisely this diversity that forced us to limit our focus to a single suborder, the cockroaches, and to a single theme, neurobiology. As indicated in the Prologue, cockroaches possess a number of attractive features which make them suitable for laboratory experimentation. To recapitulate somewhat, they are a convenient size, inexpensive to rear and maintain, produce a number of generations per year, and are easy to sex, even at hatching. There is also a substantial basic literature[1-6] which emphasizes their physiology.

Our concentration on neurobiology results from a number of observations, not least of which is the explosive growth of this field as a research discipline over the last two decades. Registration figures for the annual meetings of the Society for Neuroscience show an increase from 1,400 in 1971 to 8,610 in 1983 and to over 12,000 in 1987. This increase cannot avoid placing demands upon available resources. It is prudent, therefore, to consider alternative biological systems for some research efforts.

Cockroaches have been used for numerous studies in insect physiology. Their size facilitates electrophysiological recording through the anatomical localization of single neurons and the location, excision, and manipulation of other organs such as the brain, hemolymph, gonads, Malpighian tubules, and fat body (analogous to the vertebrate liver). In addition, the amenability of insects, including cockroaches, to experimental manipulation makes them suitable for a number of approaches, ranging from neuroendocrine and metabolic studies to perception and behavior.

It was our intention with these volumes to introduce the reader to the use of cockroaches in biomedical research, to provide an overview of the anatomy and biology of the cockroach, and to present current research results and applications. Authors were free to construct their chapters according to individual expertise and judgment. Each author was aware that the volume was intended to provide the impetus for the adoption of insect systems by researchers in other disciplines. To this end, each section includes discussion and review of individual areas of research and the treatment of relevant practical methods and techniques. When appropriate, authors cite specific instances where biomedical research may be served by studies with cockroaches. In other cases, the author may draw on research reports with other insects in an effort to complete discussion of a topic, illustrate with examples an experimental approach which may be possible with cockroaches but has not yet been reported, or identify limitations of the cockroach as an experimental tool.

Two features of these volumes should be noted. First, the six sections address three distinct but interrelated considerations of the cockroach as an experimental tool. Basic biology, anatomy, and rearing are featured in Section I. Sections II through IV deal with specific aspects of individual systems as they relate to neurobiology. Neural structure and neurohemal organs are described here, along with signal transmission and cellular communication. Chemical and electrical signals, including neurotransmitters and peptide hormones, are also discussed. In Sections V and VI, the integration of neural and neuroendocrine signals in controlling higher level processes, the regulation of internal processes of metabolism and reproduction, and the transduction of external signals are considered.

The second feature to note is the integration between individual chapters. Authors were encouraged to cross-reference other chapters as much as possible, allowing the reader who selects a chapter title of particular interest to be referred to other chapters where related subjects can be found. This should be of value to those already somewhat familiar with the subject as well as those approaching insect neurobiology for the first time. Although the volume is arranged in logical sequence from general to specific and integrative aspects of the cockroach neurobiological system, it nevertheless can be entered from almost any section or chapter. Above all, it should be regarded as a resource manual as well as a review of the literature. Although various analogies between insect and vertebrate systems are discussed (e.g., the corpus allatum-corpus cadiacum complex vs. the vertebrate hypothalamo-hypophyseal system, fat body vs. liver, neuroendocrine systems, sensory and motor innervation and proprioception, and so forth) and authors speculate freely on potential uses for cockroach systems, it is hoped that the reader will recognize new opportunities for adaptation of insect systems to the unique problems of his or her research interests.

REFERENCES

1. **Beier, M.,** Blattariae (Schaben), in *Handbuch der Zoologie,* Vol. 4, Walter de Gruyter, Berlin, 1974, 1.
2. **Bell, W. J.,** *The Laboratory Cockroach,* Methuen, New York, 1982.
3. **Bell, W. J. and Adiyodi, K. G., Eds.,** *The American Cockroach,* Chapman and Hall, New York, 1981.
4. **Cornwell, P. B.,** *The Cockroach,* Vol. 1, Hutchinson and Co., London, 1968.
5. **Cornwell, P. B.,** *The Cockroach,* Vol. 2, Associated Business Programmes, London, 1976.
6. **Guthrie, D. M. and Tindall, A. R.,** *The Biology of the Cockroach,* St. Martin's Press, New York, 1968.

Section I. Biology

Chapter 1

BIOLOGY OF THE COCKROACH

William J. Bell

There is at present an emphasis on finding and deploying invertebrates to serve as models for vertebrate systems. In her recent paper on this topic, Scharrer[1] states that "the current focus is on lower forms for several reasons, including relative simplicity, greater accessibility, and applicability for certain experimental procedures. But first and foremost, the relevance of information gained from the use of such models has been established by the demonstration of remarkable structural and functional parallels" between insects and vertebrates. The research presented in this volume certainly attests to the parallelisms between cockroach and vertebrate biology and suggests that the cockroach will continue to reign high among potential candidates as an invertebrate model system. What is it about the cockroach that has led to its prevailing acceptability?

First, we should narrow the subject to include only those cockroach species that are used in biomedical research. Unfortunately, cockroaches are often referred to as though they were a totally homogeneous grouping, when actually there are more than 4000 species of cockroaches comprising the suborder Dictyoptera, and they are quite diverse. Cockroaches are placed in five families which belong to two major phyletic lines (Blattoidea and Blaberoidea) separated on the basis of reproductive strategies and morphology[2] (Figure 1). Females of the Blattoidea (families Cryptocercidae and Blattidae) are oviparous, producing hard, rigid egg cases which are dropped shortly after their formation; embryogenesis proceeds in the deposited ootheca. Common examples of the Blattidae are in the genera *Periplaneta* (the American cockroach, *P. americana*) and *Blatta* (the Oriental cockroach, *B. orientalis*). Several different oviposition mechanisms characterize the Blaberoidea (families Blaberidae, Polyphagidae, and Blattellidae). Females in the Polyphagidae rotate the ootheca and then carry it attached externally at the genital vestibulum. Species in the family Blaberidae, typified by *Blaberus craniifer, Blaberus discoidalis, Nauphoeta cinerea,* and *Leucophaea maderae,* are ovoviviparous. They rotate the egg case and retract it internally into the uterus, where embryogenesis occurs. *Diploptera punctata* represents the most highly evolved reproductive mode within the Blaberidae because a highly nutritious "milk", rich in protein, is transferred from the mother to the embryo. This transfer occurs from the wall of the uterus via pleuropodia (embryonic first abdominal segment appendages).[2a] Thus, this species is a truly viviparous one. Family Blattellidae includes only oviparous species, but some (e.g., the German cockroach, *Blattella germanica*) carry their oothecae externally throughout embryogenesis in an apparent intermediate condition between oviparity and ovoviviparity.

Cockroaches can be found in almost all habitat types where insects occur, but they are most diverse in warm, humid tropical regions.[3] Many of the Polyphagidae (*Arenivaga, Polyphaga*) inhabit desert regions of North Africa, Asia, and the southwestern U.S., and many have adopted unique behavioral and physiological strategies to maintain water balance, such as vertical movements through sand.[4-6] Aquatic species are the least studied of all cockroaches, although they have been observed inhabiting streams, pools, and ephemeral water-filled bromeliads.[7-8]

Of more than 50 species recorded in man-made habitats, four *Periplaneta,* two *Blatta, Blattella,* and several blaberid genera (*Blaberus, Eublaberus, Leucophaea,* and *Nauphoeta*) are most commonly given a domiciliary status.[9-10] Most domiciliary cockroaches are of African or Indo-Malayan origin,[11] and they have spread through trade routes (especially by air and sea) as commerce with tropical regions has proliferated.

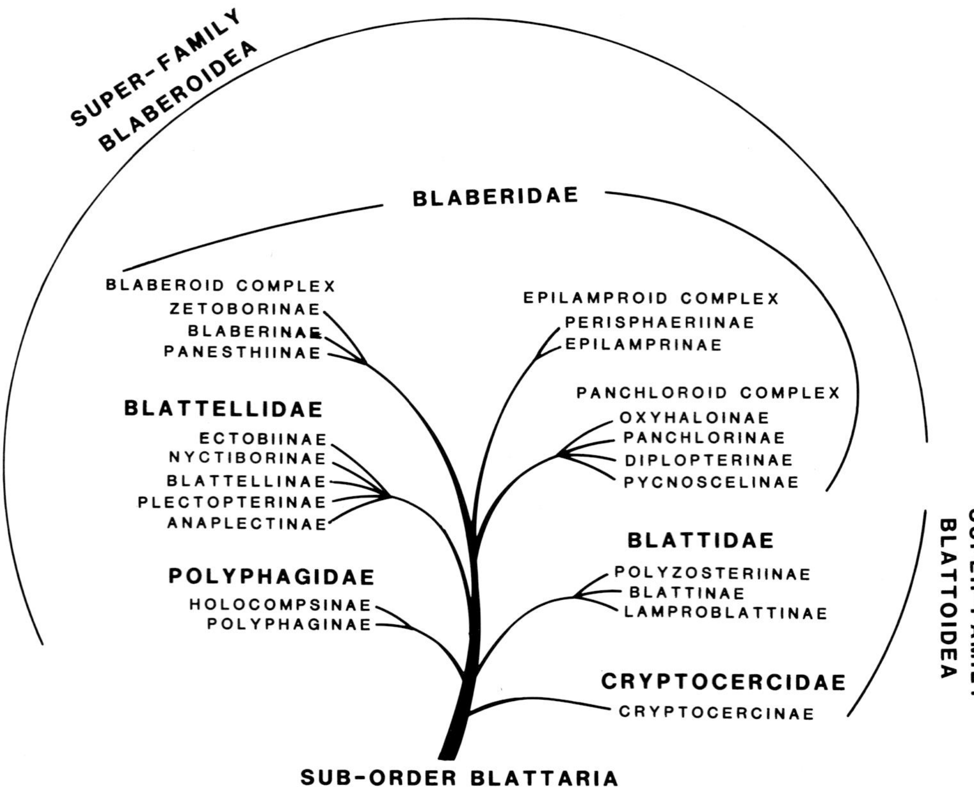

FIGURE 1. Phyletic relationships of the families and subfamilies of the cockroaches according to McKittrick.[2] The Blattaria are regarded by the author as a suborder of the Dictyoptera, the other two suborders being the termites and mantises.

Cavernicolous cockroach species have been studied with respect to social interactions, mating systems, and natural history; the degree of specialization to cave habitats varies in cockroaches from accidental habitation of caves by outdoor or domiciliary species to restriction to such habitats accompanied by adaptations that may be morphological (loss of vision and wings), behavioral (no circadian rhythmicity), or physiological (narrow range of temperature or humidity tolerance).[12]

The majority of studies on outdoor cockroaches are the result of systematic collection, thus providing excellent distribution data, but little biological information. Our knowledge of the behavioral ecology of cockroaches in forests and grasslands has increased vastly over the past decade, however. Studies of tropical habitats where cockroaches are most diverse and abundant indicate a large variety of habitats where cockroaches are found, including leaf litter, dead leaves trapped above ground, caves, hollow trees, rotting logs, pools and streams, nests of other animals, flowers, at various altitudes, and, of course, in homes.

It is not necessary to visit the tropics to observe cockroaches in their natural habitat. In temperate regions there are cockroaches living in woods and grasslands, *Parcoblatta* in North America and *Ectobius* in northern Europe and Asia. Although these species are successful in winter diapause and other mechanisms for coping with temperate climates, they cannot live in our houses. They may fly in occasionally in the late spring, but they do not colonize the bathroom or the kitchen. In fact, most of the species used in research are very good fliers, although they do not usually fly unless the temperature is quite high. *Blaberus,* for example, causes considerable commotion when it flies into poolside lights at tropical resorts.

With the exception of a few species that inhabit caves, nests of ants or birds, or burrows in the desert, most adult cockroaches spend the day hidden under bark or in subterranean or aerial leaf litter, and then at dusk they begin to move onto the vegetation and, in many species, to migrate vertically to a preferred height.[13] They are found more often on leaves or stems than on large trunks, probably because vibrations of predators can be perceived more easily on a leaf or stem than on a surface that does not transmit vibrations as well. The sensory mechanisms for detecting vibrations from the substratum and the air have been studied extensively. When disturbed, cockroaches jump off the leaf or run quickly to a hiding place.

Cockroaches play major roles in their niche. In the tropics they are preyed upon by army ants which periodically raid hollow trees inhabited by *Blaberus* species;[12] they comprise 92% of the diet of Australian net-casting spiders,[14] and they are captured by scorpions in deserts.[5] There is also evidence of cockroaches being preyed upon by toads, birds, geckos, bats, and opossums.[9,12] In turn, cockroaches turn over considerable quantities of nitrogen-rich foods and also may be important in transporting fungus spores that act upon leaf litter. They consume huge quantities of leaf litter, contributing to the efficient return of the nutrients locked up in leaves back into the soil.

Anyone who has watched a cockroach closely would have to admit that they are intriguing beasts. We are impressed by their stealthy walk and their boldness in scurrying across the kitchen table. However, few scientists select the cockroach as their experimental animal because of its charms. The cockroach most often is selected because it is relatively large and hardy and is easy to obtain and culture — and because of its prolific breeding qualities. Hardiness is certainly a major advantage for the cockroach as compared to many other invertebrates; this hardiness can be manifested in various ways, all of which contribute to the usefulness of the beast in experimental work. One way that cockroaches show their resiliency is in their response to experimental manipulations, such that after injections or transplantation therapy the wound heals quickly and there is no immune response to intra-specific tissues. Many commonly used species also show broad tolerance to environmental conditions. Most species used in research are of tropical origin, and they must be kept warm and humid. However, they can tolerate a relatively wide range of temperature and humidity conditions, especially compared to many tropical insects that have rather narrow tolerances for temperature and humidity. For example, *Periplaneta americana* adults are active between 15 and 37°C, although they prefer the middle of this range. Perhaps the most astounding example of cockroach hardiness is their ability to survive without food and water. Female *P. americana* at 27°C and 30 to 40% relative humidity survive for up to 42 d with no food or water and up to 90 d with water, but with no food.[15] It is not surprising, then, that Alsop writes in Chapter 2 of this work that cockroaches are easier to rear than many other invertebrates.

Cockroaches are primitive on the family tree of insects, and as such most of the functional systems of cockroaches are fairly unspecialized compared to representatives of higher insect orders. Lack of specialization may be a good trait for a laboratory animal in that general rules can be tested rather than highly specific ones. Thus, the nervous system of the cockroach is less centralized and cephalized than those of Hymenoptera, Diptera, Coleoptera, and Lepidoptera, and the digestive system is designed to handle a variety of food types rather than specific types such as fluids in mosquitoes and aphids or leaves in grasshoppers and lepidopterous larvae. In addition to being unspecialized, some species of cockroaches, such as *Blaberus* and *Leucophaea,* are quite large. This attribute is a distinct advantage in research because collecting tissue or cells for extraction can be accomplished efficiently and quickly. Since individual organs such as the gut, brain, and antenna are large enough to penetrate with electrodes, there are excellent possibilities for electrophysiological recording and for sampling of fluids. Size often correlates with accessibility of physiological systems. The cockroach has been used extensively to study endocrine and ovarian physiology because the corpus allatum and the ovary are exposed and operated upon so easily.

A major advantage of cockroaches is the diversity within this closely related taxon. It is no accident, for example, that among the five most common laboratory species we find representatives characteristic of oviparous, ovoviviparous, and viviparous reproduction. Early workers studying various aspects of cockroach reproduction chose representatives of these three reproductive strategies in order to better understand how control mechanisms with the same components could bring about very different types of reproduction.[16] In addition, because the reproductive cycle of cockroaches is controlled by the neuroendocrine system, which includes feedback components, it is analogous in many ways to the mammalian reproductive cycle.

Many of the characteristics of the four or five species commonly maintained in the laboratory relate to preadaptations of certain species for surviving in domestic habitats and to natural selection which has occurred since cockroaches joined human households when cavemen set up housekeeping. Unfortunately, we know nothing about the extent of natural selection that has operated on cockroaches over the past several thousand years. This is because we have not been able to locate natural populations of *Blattella germanica, Blatta orientalis,* or *Supella longipalpa,* and, except for cave habitats, most outdoor populations of *P. americana* are still associated with man. Some of the nondomestic species, such as *Blaberus craniifer* and *Blaberus discoidalis,* have been studied extensively in natural habitats, while others such as *L. maderae* have not. Nevertheless, we can speculate about the preadaptations that may have allowed some species of cockroaches to survive in our houses and/or in the laboratory. Like rats and mice, most cockroach species are nocturnal, a characteristic that offers considerable protection since natural cockroach predators are largely absent in the domestic environment and because humans are active during the day. To be sure, fewer cockroaches would survive if they had diurnal habits. Again, we must be careful not to generalize because there are some diurnal species in the tropics; in fact, *Euphyllodromia angustata,* a strikingly beautiful bronze, red, and blue insect, is only active during the day.

Because we now find some species of cockroaches in caves and because the conditions under which they live probably have been the same for millions of years, we assume that caves have provided habitats for cockroaches for a long time. Cockroaches such as *Periplaneta* and *Blaberus* feed on bat guano and on pieces of fruit dropped by bats onto the floor of a cave or hollow tree.[17] Caves, which in some ways are similar to man-made structures, are ecologically stable habitats. In fact, it has been hypothesized that the association between cockroaches and man originated when cockroaches followed man's food into caves.[18] Because cavernicolous cockroaches (and domiciliary species) associate with man's shelter and food even when man is not present, it seems more likely that caves were already inhabited by cockroaches when man abandoned his arboreal habitats.[12] Associations with other cavernicolous animals (e.g., bats, porcupines, opossums) may have preadapted cockroaches to later associations with man. Thus, it seems likely that cockroaches were already living in caves when humans took up similar residences and that the two species were sufficiently compatible that the travels of man allowed cockroaches to exploit new areas by depending on humans rather than on bats for food and safe housing.

Another reason why a few select species are able to survive in homes and in the laboratory is related to their general nutritional requirements. Most cockroach species have special nutritional requirements that would be impossible to supply on a large scale in the laboratory. For example, *Cryptocercus punctulatus* lives in social family groups within rotting logs, and adults ingest wood and digest cellulose using enzymes produced by protozoans within their guts.[19] Nymphs are fed by trophallaxis.[20] Many of the species used for experimentation, however, are largely omnivorous; at least they can survive on laboratory chow, which is supremely nutritious, but certainly quite different from the diets of most cockroach species.

Anecdotal observations of cockroaches in our houses add to the notion that they are omnivorous. *P. americana,* for example, has been observed (in India) to ingest paper, boots, hair, bread, fruit, clothing, and book bindings, as well as (in Taipei) banana, fish, peanuts, bean cake, old rice, putrid sake, oil paper, animal hides, and crepe de Chine. Cockroaches are really not omnivores, however, in the sense that they "eat everything". As with most so-called omnivores, cockroaches actually balance their diet by selecting among available types of foods[21] rather than trying to obtain all required nutrients from one type of food. Honeybees, for example, obtain everything they need from nectar and pollen, and they never eat leaves or seeds. On the other hand, female *Xestoblatta hamata* select among various types of food according to vitellogenic requirements.[22] Attempts to rear species such as *X. hamata* in the laboratory generally have been unsuccessful.

An important characteristic of laboratory cockroaches is that they can withstand and in many cases thrive in crowded conditions. Often cockroaches in containers are literally on top of one another, and they seem quite able to carry on feeding, mating, and production of offspring under these circumstances. Interestingly, under very crowded conditions the mean size of individuals decreases, allowing more individuals to be packed into a small space. If and when National Institutes of Health guidelines are finally issued for cockroach husbandry, they surely will allow a measure of crowding because some species are unhealthy when living in small groups and/or when they are continually transferred to clean cages. As with many other characteristics of laboratory-reared cockroaches, the ability to withstand crowding and/or the natural tendency to aggregate are not typical of cockroaches in general. Although species that inhabit caves often live in somewhat crowded conditions, individuals of most species are not at all crowded in nature. Researchers who study agonistic or mating behaviors of cockroaches in the laboratory are invariably amazed when they are unable to observe these activities in nature. In one 3-year study of cockroach behavior only four incidences of agonistic behavior were recorded, while in laboratory cages agonistic behavior occurs nearly continuously among males. The low frequency of agonistic behavior in nature seems to result from the low frequency of interactions between adults. Most of the time we find individual cockroaches when we search for them at night in the rain forest. Small groups are sometimes observed feeding and two individuals may be copulating, but they are never crowded.

There are a few minor disadvantages to using cockroaches in research. Some species secrete foul-smelling odors (although most do not), but this does not suggest that cockroaches are dirty. In fact, cockroaches often spend more than 50% of their time cleaning their antennae, legs, feet, and wings. Another disadvantage is the ease with which susceptible people become allergic to cockroach (usually cuticular) antigens. It is quite possible, however, that the incidence of allergies to cockroaches is no greater than that to rats, mice, or moths.

In addition to learning more about vertebrates through the use of cockroaches, studying model systems leads to a better understanding of the evolutionary history of biological phenomena and elucidates the general applicability of basic principles. These objectives are important ones, since research goals should not always be so anthropocentric. At least as interesting is knowledge about how man is related to other organisms on earth, and, whereas the comparison is quite complete at a morphological level, it is severely lacking at the physiological and biochemical levels.

One of the most remarkable conclusions we can derive from this book is that studies of cockroaches as biomedical models, driven to a large extent by grant funding possibilities, have inadvertently produced a large body of knowledge about insect physiology. This information has opened new vistas in entomology and will in the future provide ideas for the development of new pest management schemes. Whenever we gain information about how insects work, we expose vulnerable points in their biochemistry or physiology which can be exploited by imaginative entomologists. The results have gone far in dismissing the myth that insects are "simple"; indeed, if we continue to find all of the biochemical complexity in the cockroach that is typical of the vertebrate, we will soon have to say that insects are "complex". If that happens we may have to seek a simple vertebrate as a model system to better understand complex animals such as cockroaches!

REFERENCES

1. **Scharrer, B.,** Insects as models in neuroendocrine research, *Annu. Rev. Entomol.,* 32, 1, 1987.
2. **McKittrick, F. A.,** Evolutionary study of cockroaches, *Mem. Cornell Agric. Exp. Stat.,* 389, 1, 1964.

2a. **Stay, B. and Coop, A.,** Developmental stages and chemical composition in embryos of the cockroach, *Diploptera punctata,* with observations on the effect of diet, *J. Insect Physiol.,* 19, 147, 1973.

3. **Guthrie, D. M. and Tindall, A. R.,** *The Biology of the Cockroach,* Edward Arnold, London, 1968.
4. **Edney, E. B., Haynes, S., and Gibo, D.,** Distribution and activity of the desert cockroach *Arenivaga investigata* (Polyphagidae) in relation to microclimate, *Ecology,* 55, 420, 1974.
5. **Hawke, S. D. and Farley, R. D.,** Ecology and behavior of the desert burrowing cockroach, *Arenivaga* sp., *Oecologia,* 11, 263, 1973.
6. **Cohen, A. C. and Cohen, J. L.,** Microclimate, temperature and water relations of two species of desert cockroaches, *Comp. Biochem. Physiol.,* 69, 165, 1981.
7. **Takahashi, R.,** Observations on the aquatic cockroach, *Opistoplatia maculata, Dobutsugaku Zasshi,* 38, 89, 1926.
8. **Albuquerque, R. S. I., Tibana, R., Jerberg, J., and Rebordões, A. M. P.,** Contribuição para o conhecimento ecològico de *Poeciloderrhis cribosa* (Burmeister) e *Poeciloderrhis verticalis* (Burmeister) com um estudo sobre a genitalia externa (Dictyoptera: Blattariae), *Rev. Bras. Biol.,* 36, 239, 1976.
9. **Roth, L. M. and Willis, E. R.,** The biotic associations of cockroaches, *Smithson. Misc. Collect.,* 141, 1, 1960.
10. **Vargas, M. and Fisk, F. W.,** Two new records of roaches invading houses in Costa Rica, *J. Med. Entomol.,* 10, 411, 1973.
11. **Rehn, J. A. G.,** Man's uninvited fellow traveller — the cockroach, *Sci. Mon.,* 61, 265, 1945.
12. **Schal, C., Gauthier, J.-Y., and Bell, W. J.,** Behavioural ecology of cockroaches, *Biol. Rev. Cambridge Philos. Soc.,* 59, 209, 1984.
13. **Schal, C. and Bell, W. J.,** Vertical community structure and resource utilization in neotropical forest cockroaches, *Ecol. Entomol.,* 11, 411, 1986.
14. **Austin, A. D. and Blest, A. D.,** The biology of two Australian species of dinopid spider, *J. Zool. (London),* 189, 145, 1979.
15. **Willis, E. R. and Lewis, N.,** The longevity of starved cockroaches, *J. Econ. Entomol.,* 50, 432, 1957.
16. **Roth, L. M. and Willis, E. R.,** An analysis of oviparity and viviparity in the Blattaria, *Trans. Am. Entomol. Soc. (Philadelphia),* 83, 221, 1958.
17. **Gautier, J.-Y.,** Etude comparée de la distribution spatiale et temporelle des adultes de *Blaberus atropos* et *B. colosseus* (Dictyoptères) dans cinq grottes de l'ile de Trinidad, *Biol. Behav.,* 9, 237, 1974.
18. **Chopard, L.,** La biologie des Orthoptères, *Encycl. Entomol.,* A20, 1, 1938.
19. **Cleveland, L. R., Hall, S. R., Sanders, E. P., and Collier, J.,** The wood-feeding roach, *Cryptocercus,* its protozoa, and the symbiosis between protozoa and roach, *Mem. Am. Acad. Arts Sci.,* 17, 185, 1934.
20. **Seelinger, G. and Seelinger, U.,** On the social organisation, alarm and fighting in the primitive cockroach *Cryptocercus punctulatus* Scudder, *Z. Tierpsychol.,* 61, 315, 1983.
21. **Cohen, R. W., Heydon, S. L., Waldbauer, G. P., and Friedman, S.,** Nutrient self-selection by the omnivorous cockroach *Supella longipalpa, J. Insect Physiol.,* 33, 77, 1987.
22. **Schal, C.,** Behavioral and Physiological Ecology and Community Structure of Tropical Cockroaches (Dictyoptera: Blattaria), Ph.D. thesis, University of Kansas, Lawrence, 1983.

Chapter 2

COCKROACH CULTURING

David W. Alsop

TABLE OF CONTENTS

I. INTRODUCTION

The primary aim in culturing cockroaches is to achieve a steady, high output of healthy specimens with a minimal expenditure of time, space, and money. Such conditions have rarely been met, primarily because of a lack of information about most species' requirements and behavior, and about optimal culturing methods and conditions. As a result, in most places the culturing of these insects has been on a purely ad hoc basis, with little attempt being made to accurately assess the cost and health of the specimens, or of the actual output of the cultures. The reader interested in insect rearing is referred to the volume edited by King and Leppla.[1] The contributions deal with such topics as the genetics of reared insects, especially the decline of variability and performance and ways to guard against both; actual rearing systems for a diverse selection of insects intended for a variety of uses; and the management of insect-rearing programs, including data-processing techniques, systems management, and quality control.

Most of the literature on the culturing of cockroaches concerns only the three main pest species, *Periplaneta americana, Blatta orientalis,* and *Blattella germanica* (see, for example, References 2 and 3). There are, however, nearly 4000 other species, some of which may well prove to be even better subjects for a variety of investigations than these pests. Thus, it is fortunate that over the years a few investigators have devoted some effort to devising methods for raising many different species under standardized conditions at low cost in limited spaces and with relatively little expenditure of time. The chief purpose of this chapter, then, is to present the culturing methods and materials which these researchers have found to be most useful, with the hope that investigators will not only be able to better their culturing of the three domiciliary pest species, but will also be drawn to investigate others.

II. OBTAINING SPECIES FOR CULTURING

A. GOVERNMENT REQUIREMENTS

The importation and interstate movement of cockroaches is regulated by the United States Department of Agriculture, Animal and Plant Health Inspection Service, Plant Pest and Quarantine Section (USDA-APHIS-PPQ). Persons wishing to obtain permits for interstate movement and foreign importation must apply to USDA-APHIS, Federal Center Building, Hyattsville, MD 20782. An application must be filled out and acted on affirmatively by APHIS-PPQ and a pertinent state regulatory agency before permits are issued.

B. SOURCES AND THEIR REQUIREMENTS

Periplaneta americana can be purchased from many of the larger biological suppliers (e.g., Carolina Biological Supply Company, Connecticut Valley Biological Company, Nebraska Scientific, Wards Natural Science). Of the above companies, only Carolina Biological also sells *Blattella germanica* and one of the *Blaberus* spp. (probably *B. craniifer*). *P.*

americana, Blattella germanica, and *Blatta orientalis* can also be obtained from the USDA-ARS Live Insects Laboratory, Building 476, Beltsville Agricultural Center, Beltsville, MD 20705. Delays of 4 weeks to 6 months may be expected when obtaining species from this source.[3] Other species are available from a number of university sources (those of the author included), and here, also, considerable delays are often to be expected.

All sources require permits and as a general rule will not ship during the cold winter months or very hot summer months. Only commercial suppliers can be expected to provide immediate shipment, and even they have limitations. Most noncommercial culturers have tightly controlled cultures that provide only the specimens they need. Thus, considerable lead times are often required to build up the excess numbers needed to provide additional specimens to other parties. Finally, unless cultures are being exchanged between individuals, the requestor should, at a minimum, expect to pay for the postage and handling of the shipment, and as a matter of courtesy large numbers of specimens should neither be requested nor expected.

III. PHYSICAL REQUIREMENTS FOR CULTURING

A. INCUBATORS AND ASSOCIATED FACILITIES

Ideally, cockroaches should be cultured in a sealed constant-temperature incubator with light and humidity controls and containing both a sink and a work table. USDA-APHIS-PPQ requires that incubators have an insect-proof door and that they be sealed, negative-pressure containments. Thus, air must be actively, rather than passively, exhausted from the room. The racks for holding cages should have shelf spacing allowing 2 to 3 in. above and around the cage for air circulation. Open-back metal racks are best for this purpose, with those made of stainless steel being preferred because they do not rust and are easily cleaned. Traps and/or poison baits should be placed under the racks so that cockroaches escaping from cages are prevented from infesting other cages and the surrounding area. Besides the incubator, other requirements for culturing are (1) a large outside sink for washing waterers and cages, (2) a freezer for disposing of cultures and waterer plugs, and (3) storage space for extra cages, waterers, and animal food.

B. LIGHT, TEMPERATURE, AND HUMIDITY REGIMES

The great majority of cockroaches are subtropical or tropical in origin and therefore are best raised at temperatures of 26 ± 2°C, 40 to 60% relative humidity, and day lengths approximating 12 h light:12 h darkness.

1. Temperature

Temperatures over 29°C are near the upper thermal limits of many species, particularly cavernicolous ones, and many species do not do well below 21°C. In all cases, growth is considerably slowed at lower temperatures and at temperatures >30°C.

2. Humidity Control

In general, cockroaches require large amounts of water to prevent death by dessication. Maintaining 40 to 60% relative humidity requires that humidification be provided in the winter and dehumidification in the summer. At <40% relative humidity, waterers dry out rapidly and the nymphs of many species die of dessication. On the other hand, while most species do very well and actually prefer >60% relative humidity, two problems become dominant at higher humidities. First, the acaroid mites associated with virtually all species rarely become numerous enough to cause discernable damage at <60% relative humidity, but can rapidly build to numbers great enough to annihilate a culture at higher humidities.[4] Mold growth on food, waterers, harborage materials, and carcasses also rapidly becomes rampant at high humidities. Here it is important to note that, in addition to controlling the humidity in the incubator, it is also necessary to provide adequate ventilation within the cages so that their internal humidity does not reach higher values.

3. Lighting

All species do well on a 12 h light/12 h dark cycle. The vast majority of species are most active at night, and a light-dark cycle is necessary to entrain their many circadian rhythms. Considerable variation in light intensity is permissible, and it can be held at a comfortable level for the investigator if the cages are provided with enough harborage material so that the cockroaches can hide when they wish.

IV. NUTRITIONAL REQUIREMENTS

The best food for raising cockroaches is dried (pelleted) dog food. For most species this is all that need be provided. It is one of the most completely fortified of all animal feeds, and many species eat it entirely, while others (particularly many of the blattellids) honeycomb it in a variety of patterns, an indication that they apparently do not require or are avoiding certain of its components. On the other hand, most species do not grow well on any of the many laboratory chows (rabbit, mouse, rat, etc.), presumably because these are apparently lacking one or more substances required for their optimal growth. Also, while most species are detritivores that will eat almost anything (including each other) when hungry, it does not necessarily mean that they will grow well on whatever is presented to them (i.e., bread, lettuce, etc.) or that it is necessary to provide them with a variety of different foods. Indeed, "minimal" diets such as bread and/or lettuce are merely a means of slowly starving them to death.

Two precautions are necessary when using dried dog food: (1) preventing it from becoming a route for the introduction of pests into the cultures and (2) keeping it from getting wet. Pest introduction (mainly beetles) can be prevented by freezing the food or storing it in a low-humidity cold room. Keeping it dry is necessary because (1) it molds rapidly when wet, and certain species will not eat it in this condition; and (2) when wet it attracts and becomes a favored breeding ground for fungus flies (Phoridae) which often become a major pest problem in cultures.

A few species, all live-bearing blaberids, seemingly do better when additionally provided with certain fruit supplements (e.g., banana peels and melon rinds) and/or brewer's yeast. The matter of additional supplementation is one of the greatest areas of uncertainty in cockroach culturing, since it is not at all clear which species require various types of dietary additives (beyond those present in dog food) for optimal growth. This lack of knowledge is one of the reasons that certain species have never been cultured successfully.

V. WATERERS

All cockroaches require water provided *ad lib.* While nearly all of them can survive for relatively long periods without food, virtually none can do so without being given fairly constant access to water. Since the majority of time spent in caring for cultures is devoted to watering the insects, a major aim has been to develop low-cost, high-capacity waterers that are easy to prepare, are of long-term usage, and which prevent water from spilling or siphoning into the cage.

Because of the great variety of cage sizes (see Section VI below), two general types of long-term waterers have been developed for culturing: *horizontal* ones for smaller cages and *vertical* ones for the larger cages (Figures 1 and 2). There are a number of variations on these, including a special type of vertical waterer developed for culturing species that produce

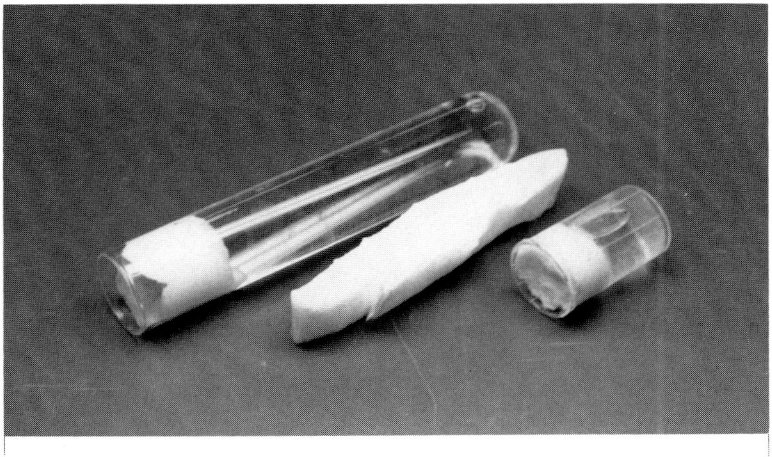

FIGURE 1. Horizontal waterers: 200-mm culture tube, left; 15-dram vial, right; folded strip of absorbent paper wadding, center.

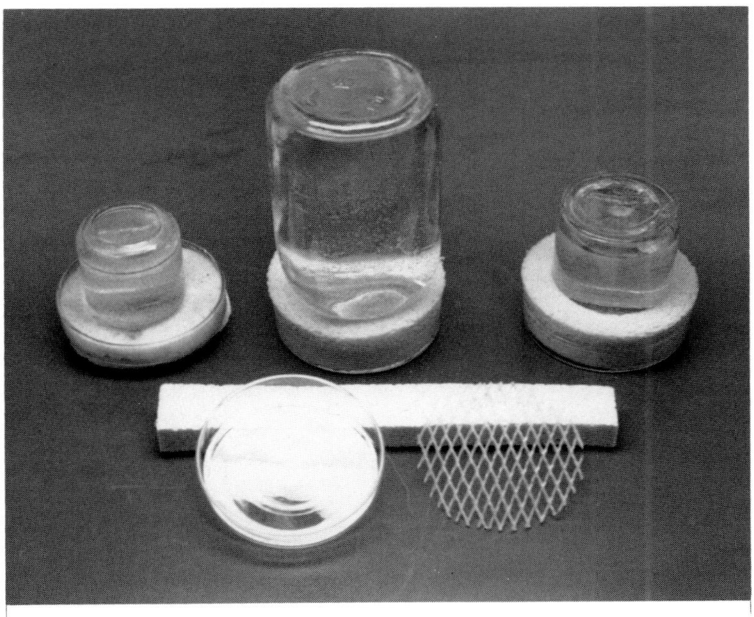

FIGURE 2. Vertical waterers. Background: assembled waterers — special waterer for oothecal moisture uptake, left; quart jar, center; pint jar, right. Foreground: additional materials — deep petri dish cover and disk of plastic mesh leaning against an expanded compressed sponge strip.

oothecae which must acquire water from a substrate in order for development to occur (Figure 3).

A. HORIZONTAL WATERERS

Materials include

- Plastic vials, 15 dram, and/or 35 × 200 mm culture tubes

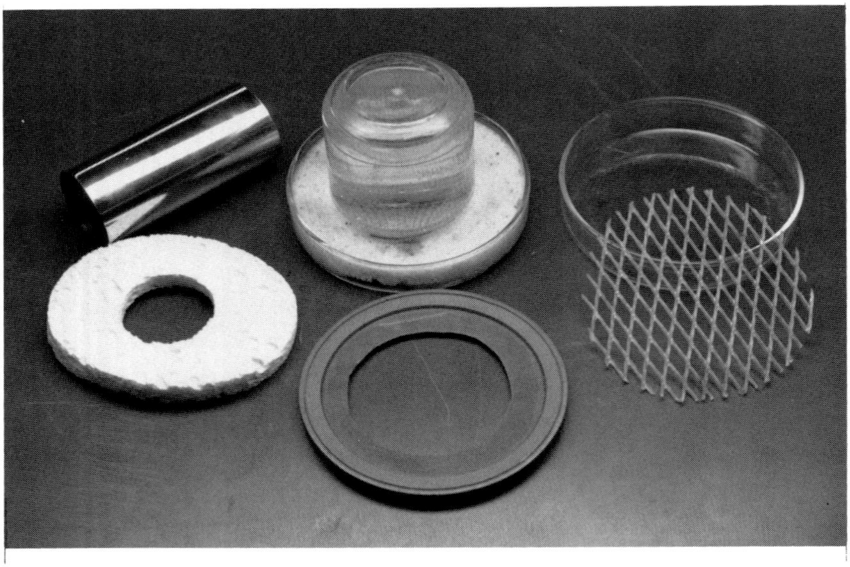

FIGURE 3. Special waterer for oothecal moisture uptake. Clockwise from center top: assembled waterer; petri dish and disk of plastic mesh; cutout coffee can lid that fits over waterer so that it can be inverted for refilling without losing oothecae; oval sponge with hole cut in it; piece of sink drain tailstock used to cut holes in dried sponges.

- Nonsterile absorbant paper wadding
- Racks for the tubes
- Paper cutter

Waterers of this type are used mainly in small or low cages, but also should be employed in addition to large vertical waterers (Section V.B below) in large cages if the nymphs are very small or have difficulty climbing up the edge of the vertical waterers. Two basically different types of horizontal waterers have been employed: (1) pieces of sponge or wads of cotton in a plastic or glass petri dish and (2) large-diameter tubular containers fitted with absorbent plugs. While the first type offers a large access surface from which the cockroaches may drink, it requires fairly constant attention because the amount of water held is only that which can be taken up by the sponge or cotton. The second type is greatly preferred because the tubes can hold substantial amounts of water and, depending upon their volume and the number of individuals in a cage, will last for as long as 3 weeks before having to be replaced.

Tubular waterers can be made from any of a variety of plastic vials or culture tubes (Figure 1). The best ones are those having mouth diameters of 30 to 35 mm because they hold much greater amounts of water than waterers with smaller diameters and they offer a plug face accessible to all but the largest of the cockroaches. Plastic snap-cap vials (15 dram short, 33-mm opening and 15 dram long, 29-mm opening) make useful waterers for very small cages. These can be bought in gross quantities from most medical supply houses and a variety of other sources, and they will last for a long time if cared for properly. In larger cages it is best to use large glass culture tubes (200 × 35 mm) because of their far greater capacities. These tubes, together with the stainless steel racks necessary to hold them, can be obtained from Bellco Glass, Inc., P. O. Box B, 340 Erudo Road, Vineland, NJ 08360.

In terms of cost, preparation time, and quality control (i.e., not getting "leakers"), nonsterile absorbent paper wadding is by far the best material to use to make plugs for large-diameter horizontal waterers. It is much less expensive than using wads of adsorbent cotton for plugs, and it yields far fewer leakers. Also, while sponge plugs can be used safely in vials with small mouth openings (e.g., 10 to 15 mm), they become expensive and time-consuming to make, and they almost invariably leak when placed in large-diameter tubes.

Nonsterile absorbent paper wadding (formerly called Cellu-Cotton) is a multilayered product that can be obtained from certain paper companies and medical supply houses, usually in the form of 24-in.-high (5-lb) rolls. The rolls are first cut in half with a cross-cut saw or band saw to yield two 12-in.-high rolls. These half rolls are then unwound, and the wadding is cut into either 14-in. (for 29- to 30-mm-opening tubes) or 16-in. (for 35-mm-opening tubes) lengths with a paper cutter. The resulting lengths are folded in thirds and cut crosswise with the paper cutter into ca. 1-in.-wide strips (Figure 1). The yield from each folded length should be about 12 to 13 strips for tubes with 29- to 30-mm openings and 10 to 11 strips for those with 33- to 35-mm openings.

Plugs are formed by rolling the strips to approximately the same diameter as the waterer opening. These are then pushed firmly with a slight twisting motion into already filled waterers, and any exuded water is shaken off. They can be inserted dry into large tubes, but because of their great absorbency should be wetted thoroughly before being inserted into the shorter vials. If the insertion is done properly, the surface of the plug will remain uniformly moist as long as there is any water in the tube, but will not permit it to leak into the cage. Leakage is most often caused by (1) having insufficiently long lengths of wadding in the roll, (2) winding the wadding rolls so tightly that they cannot fully expand against the walls of the waterers, and (3) culturing species that actively burrow into the surface of the plugs (e.g., *Panchlora* and *Pycnoscelus* spp.). When culturing burrowing species, the best way to avoid leakage is to cut the wadding strips into ca. $1^1/_2$-in. widths so that the plugs are sufficiently deep not to be broached before the water in the tube is exhausted.

Horizontal waterers of this type can be prepared very quickly. A total of 20 to 30 lengths of wadding are cut and folded at a time, then stored for use as needed. In a few minutes, 40 strips can be cut with a paper cutter and then rolled and inserted into filled waterers at the rate of 10 to 15 per minute. As well as being much less expensive than using cotton to make the plugs, this is far faster, neater, and yields virtually no leakers.

Plugs are removed from spent waterers most easily with large (ca. 12-in.) forceps which are forced down into the tubes with their points open on either side of the plugs, then squeezed and withdrawn. Dried wadding is very hard and should be moistened before being removed. Spent plugs should always be frozen to kill any roaches, oothecae, and/or pests present in them before discarding them.

B. VERTICAL WATERERS

Materials include

- Half pint, pint, or quart mason jars, or the equivalent
- Lids of deep (200 × 20 mm) glass petri dishes
- Compressed sponge
- Open-mesh plastic laboratory matting
- Unglazed tiles (6-in. square)
- Paper cutter

Large waterers (Figure 2) are best made from any of a variety of jars or bottles having approximately 65-mm outside neck diameters (e.g., mason jars) because strips of compressed sponge can then be used as the water distributors and the lids of deep petri dishes (glass, 100 × 20 mm) as the waterer bases. These lids are readily available and can be obtained separately from the bottoms through most laboratory suppliers. Bottles having mouth openings with other diameters require bases of different (usually more difficult to obtain) sizes.

The best water distributor is a strip of sponge wrapped around the bottle neck and held in place by the petri dish. Using pads of cotton or paper wadding alone in the dish directly leads to an unstable situation in which the slightest movement of the cage can tip the bottle over. Sheets of compressed sponge ($^1/_8$ in. thick, expanding to ca. 1 in.) are obtainable from a number of suppliers. When dry they are easily cut with a paper cutter. Strips ($10^1/_2 \times 1^3/_{16}$ in.) are cut for mason jars (other jars or dishes require different sizes), then wetted and squeezed dry several times before being used. It is also necessary to insert a disk of open-mesh plastic padding (of the type widely sold for use on laboratory sinks and drainboards) between the mouth of the bottle and the petri dish to ensure a free flow of water between the bottle and the sponge.

Waterers of this type usually do not have to be refilled for several weeks. They take little time and effort to set up: the jar is filled to the top with water, a strip of thoroughly wetted sponge is wrapped around its neck, a plastic mesh disk is laid across the mouth of the jar atop the sponge, and a petri dish is forced down over the mesh and sponge. Almost no water is lost when inverting the jar if the petri dish has been pushed all the way down to the mouth of the jar and the sponge packed up against it.

Large waterers (1 qt or more) used in large cages should be placed on 6-in. squares of unglazed tile rather than directly on the cage floor or harborage material. The tiles provide a stable base for the waterers, serve to elevate them above the debris that inevitably builds up in the cages, and help to keep the cage and harborage materials from getting wet.

C. SPECIAL WATERER FOR OOTHECAL WATER UPTAKE

Many species produce oothecae having such an initially low water content that an exogenous source of moisture must be provided if development is to occur. This is a widespread occurrence in the cockroaches, occurring in some or all members of the Cryptocercidae, Lamproblattinae (Blattidae), Polyphagidae, and Blattellidae.[5] Species with this requirement are among the most difficult of all cockroaches to culture.

Many different moistened materials have been used in attempts to meet this need, including tubes and dishes of sand, sphagnum moss, filter paper, cotton, absorbant paper wadding, and sponges. The major problems they have presented have been (1) the almost constant attention required to keep the substrates moist and (2) the growth of molds on the substrate and oothecae. While the waterer described below is not a final answer to culturing all species with this requirement, it has been used to successfully culture a fair number of blattellids that actively select a wet substrate for oothecal deposition. Its main advantages are that it requires only weekly attention and only rarely supports mold growth.

This waterer has a wide, moist surface. It consists of (1) a ca. 5-oz baby food (or similar) bottle ($2^1/_4$ in. diameter, $2^3/_4$ in. high) as the reservoir; (2) a deep petri dish cover (see Section V.B above) as the base plate; (3) a $^1/_2$-in.-thick, $3^1/_4 \times 4^1/_4$ in. oval cellulose sponge as the water distributor; and (4) a disk of woven plastic window screening to be placed beneath the sponge (Figure 3). In addition, it is also necessary to have a plastic cover with an overall diameter slightly greater than that of the petri dish, having a central hole slightly larger than that of the bottle, for use when recharging the waterer.

The oval sponges are widely available. They have the distinct advantage of having many large surface pores into which many species will insert their oothecae. Before use they must be washed thoroughly to remove the humifactants in them, and then dried. Holes are bored through the middle of the dried sponges with a $1^3/_4$-in.-diameter metal cutter with notched edges. (Note: it is almost impossible to cut holes in sponges which are wet!) Assembling the waterers is straightforward: the plastic mesh is placed in the petri dish and a wet sponge pushed down on top of it; this is then inverted and pushed down over an already filled bottle, and the waterer is then turned upright. There is virtually no water loss while doing this, particularly if the sponge has been saturated thoroughly.

If mold growth becomes a problem, as happens with only certain species, the sponge should be soaked initially in a solution of 0.1% methyl *para*-hydroxybenzoate (sold under the trade names Nipagin M, Methylparaben, and Tegosept M). To make up the 0.1% solution, a 20% stock solution of methyl *para*-hydroxybenzoate in 70% ethyl alcohol is prepared first. The stock solution is then added to distilled water to produce the final concentration. Furthermore, if the oothecae show signs of supporting mold growth, they should be sprayed lightly with the solution at 1- to 3-d intervals.

These waterers last for 1 week or longer before having to be refilled. They are never taken completely apart until the sponge degenerates. Instead, to refill them, the plastic cover is first slipped over the bottle down onto the petri dish, and the entire assembly is inverted. The water bottle is then pulled out of the hole in the sponge, filled, and reinserted, and the waterer is then turned upright. This sequence prevents the loss of oothecae loosely deposited atop and along the sides of the sponge, and it permits moisture to be supplied continuously to oothecae which require more than 1 week to develop.

D. WATERER CLEANING

Other than the paper wadding, all of the waterer materials discussed above can be used repeatedly. Vials, culture tubes, jars, petri dishes, and plastic disks are soaked in a strong detergent solution for a few days and then washed. Sponge strips are soaked in hot water, rinsed and squeezed dry a few times, and then either immediately reinserted into waterers or dried out for reuse at a later time. Because the sponge deteriorates slowly, the strips usually have to be replaced after four to six uses. The plastic mesh disks deteriorate more slowly and should be disposed of only when they become brittle. Thus, the ongoing expense of making waterers is quite low since, if properly taken care of, all of the other materials last for many years.

VI. CAGES FOR SMALL- AND LARGE-SCALE CULTURING

The type and size of the cage chosen to house a particular species are dependent on the size of the individuals, their initial numbers, their spacing requirements, whether they prefer to rest on horizontal or vertical surfaces, and the final colony size desired. While many different types of cages have been used, ranging in size from small and large plastic boxes, through various types and sizes of glass jars and aquaria, to large metal containers and plastic or metal garbage cans, the types which have been found to be of overall greatest use are

1. Round acrylic containers (6 in. diameter, $2^1/_2$ in. high) with tightly fitting lids (Figure 4)
2. Rigid rectangular acrylic "refrigerator chests" ($12^3/_8 \times 10^1/_{16} \times 3^7/_8$ in.) (Figure 4)
3. Wide-mouth gallon bottles, battery jars of a number of sizes (including ones that can be made by removing the necks from ca. 1-gal glass chemical bottles), and tall cylindrical plastic containers (Figure 4)
4. Aluminum or galvanized steel cages, usually of a large size (e.g., ca. $10 \times 14 \times 24$ in.)
5. Plastic garbage cans (20 to 30 gal) having semirigid lids that can be clamped tightly against the rim of the can by heavy steel handles

These containers have the following features in common: (1) long-term durability, (2) ease of cleaning, (3) resistance to corrosion or being eaten by the cockroaches, (4) tightly fitting lids or the ability to have cheese cloth stretched tightly across their openings, and (5) all can be provided with a means of permitting airflow into the cage so that elevated humidity does not cause the food to become moldy or mites to become a serious problem in the colony.

FIGURE 4. Plastic cages: type 1 (right, front), type 2 (left), type 3 (right, rear). See Section VI of text for details.

Type 1 cages are mainly used for starting colonies of small- to medium-sized species from a few individuals, particularly when they are of the sort which does best under crowded or semicrowded conditions. Type 2 cages are ideal for long-term culturing of small- to medium-sized species that favor horizontal spacing and do not otherwise require special care. Type 3 containers are the best for most blattellids (this also includes large metal canisters), while type 4 are best for culturing medium- to large-sized blaberids and very large blattellids (e.g., *Nyctibora* spp.). The large garbage cans (type 5) are primarily used for mass rearing of small- to medium-sized species that tolerate or do best on vertical surfaces.

Durable acrylic containers of a variety of sizes and shapes that are ideal for making type 1, 2, and 3 cages are available from Tri-State Plastics (Crown Hill Industries, Inc., 4779 Upper Valley Pike, Urbana, OH 43078). Battery jars of various sizes can be obtained from most glass suppliers or can be made from large ($\geq$2 qt) bottles if a proper glass cutoff wheel is available. Various sizes of rectangular aluminum or galvanized metal cages must be fabricated, while plastic garbage cans are relatively inexpensive and widely available. Very light plastic containers (such as the widely available shoe boxes, etc.) should not be used because they are not durable, they crack easily, and they usually do not have tightly fitting lids. Similarly, while glass aquaria can be used, they are not recommended for large-scale culturing because of their expense, weight, and high cost of repair.

Provision for air circulation within the cage is a necessity no matter what type of cage is used. Battery jars and other type 3 containers that have large openings are best covered with several layers of cheese cloth held in place with bands made from cloth-covered elastic tape. Although the elastic tape is more expensive than rubber bands, its use is highly recommended because it lasts far longer than rubber bands, which age rapidly and have a disconcerting tendency to break when least expected, allowing the cockroaches to escape. If plastic cage tops are used, merely punching holes in them with a heated needle or rod is not sufficient. Instead, sizable openings equivalent to one half or more of the cage top must be made and then covered with fairly fine-mesh metal screening (Figure 4). Holes can be easily cut in rigid plastics with a high-wattage (300-W) soldering iron whose tip has been cut and ground to a knife-like blade, and in soft plastics with tin shears. The screening, which should be stainless steel to withstand repeated washing, is then sealed around the edge of the openings by being melted into the plastic with a lower-wattage (150-W) soldering iron. When metal containers are used, the screening, usually with a heavier $^3/_8$-in.-mesh hardware cloth supporting layer, must be attached to the lids by welding or riveting. Breaks in plastic cages are best repaired by fusing screen across the splits with a hot soldering iron.

All cages must have some provision made for the placement of food and waterers, and they must also have some form of bedding material placed in them (see Section VII below). In addition, a band of petroleum jelly should always be applied around the inside of the upper part of the cage to keep the cockroaches from crawling onto the lid or cheese cloth. This can be obtained cheaply in 1- to 5-lb containers and is applied most easily and neatly to the cages with a 1-in.-wide paint brush.

VII. HARBORAGE MATERIALS

Nearly all cockroaches thrive when provided with hiding spaces that approximate their body height. Additionally, most have a decided preference for resting on either horizontal or vertical surfaces. The harborage materials used must take these necessities into account and in addition (1) should be low in cost or extremely durable; (2) should facilitate, rather than hinder, the transfer of cultures; (3) should not interfere with or be harmed by the waterers; and (4) for many blattids, must be of a material that they can excavate and to which they can then cement and cover their oothecae.

The materials which have proved to be the best for meeting these requirements are

1. Large-sized (8 × 11 in.) and medium-sized (5 × 11 in.) papier-mâché separators from fluorescent light tube cartons
2. Tempered sheet Masonite® in several configurations
3. Hardward cloth ($^1/_4$-in. mesh)
4. Paper toweling and wadding, both of which are always used as adjuncts to other materials
5. Pelleted laboratory bedding ($^1/_8$-in. size; a corncob product)

The materials *not* recommended for use are wood shavings, shredded paper, and rolls of corrugated cardboard. Many species find the oils from the wood shavings objectionable. Furthermore, shavings and shredded paper tend to get knocked into the waterers and act as wicks, turning the entire bedding into a wet mass that molds rapidly. Additionally, it is very difficult to remove nymphs from them when the cultures have to be transferred. Rolls of corrugated cardboard tend to be eaten by the cockroaches and then collapse; also, because of the air circulation problems they present, they tend to mold rapidly.

Large papier-mâché separators are ideal harborage materials for raising most blattids, blaberids, and many blattellids, particularly larger ones such as *Nyctibora* and *Xestoblatta* species. These separators cost nothing and can be obtained in large quantities wherever there is extensive use of fluorescent lighting. They provide a multitude of spaces and of horizontal and vertical surfaces, and they also admirably meet requirement 4 above. Most species do well when provided with nothing more than horizontal stacks formed either by having alternating layers running crosswise to each other or by having them all run in the same direction, but offset so that the grooves of one layer sit atop the flat crests of the one beneath it.

Some of the smaller burrowing blaberids such as the *Panchlora* and *Pycnoscelus* spp. do best when provided with much closer spacing, a requirement that is met most easily by placing sheets of multilayered paper wadding (see Section V above) between the separator layers. On the other hand, some of the very large blaberids that inhabit caves and hollow trees (e.g., the two *Archimandrita* spp. and *Blaberus* spp. such as *B. colosseus, B. giganteus,* and *B. parabolicus*) do well only when provided with extensive vertical resting surfaces. This requirement is met most easily by stapling paired separators to the sides of short lengths of 1 × 2 in. furring strips and then standing them upright in large type 4 cages. A 2- to 3-in. layer of the pelleted laboratory bedding should be placed in these cages (into which the small nymphs can burrow), and a block of wood or a brick must be used to elevate the waterer above this bedding because it molds rapidly when damp.

Because of their smaller size, most blattellids require much closer spacing than can be provided by the papier-mâché separators. The ideal harborage material for raising them is a Masonite® spacer with staple legs. The Masonite® spacers are cut into squares or octagons that will fit loosely into battery jars or similar type 3 containers and which have four or more $^{3}/_{8}$-in. staples driven into them as spacing legs. These are then stacked in the containers, with a piece of paper toweling between each layer to provide the very tight spaces favored by the tiny nymphs.

Cornwell[6] recommended the use of rolls of corrugated paper in battery jars or large can containers for blattelids and others. Their use presents many problems. The rolls must be prepared by inserting in them some form of spacer material such as a strip of the corrugated paper (this requires unwinding and rewinding the roll) so that the resulting spiral has its layers held about $^{3}/_{8}$ in. apart. They also must be supported above the floor of the cage by wood or metal spacers so that the cockroaches can move easily from one layer to the next. Additionally, care must be taken with waterer placement to ensure that no water reaches the rolls, since they collapse and mold rapidly when wet. For these reasons, even though the initial cost and time expended in making the Masonite® separators is greater than that involved in using paper rolls, the former are best for long-term culturing in battery jars because of their great durability and ease of cleaning and because they also permit a much easier transfer of cultures from one cage to another.

When mass rearing blattids such as *Blatta orientalis* and the various *Periplaneta* spp., the required spacing must be between $^{3}/_{4}$ and 1 in., something that is difficult to maintain if corrugated paper is used. It also does not last long because both of these genera excavate the paper for oothecal attachment. A much better material for culturing such species is standing lengths of $^{1}/_{4}$- or $^{3}/_{8}$-in.-mesh hardware cloth that have been folded back and forth, leaving about 1 in. between the layers (see Section VIII.E). The screens must be supported above the floor of the cage (usually a plastic garbage can), like the paper rolls, but when using them there is no danger of collapse due to a waterer spill or of the paper being eaten by the cockroaches.

VIII. SETTING UP CAGES

The primary requirements for setting up cages are to provide adequate harborage and a means for keeping both harborage and food away from the waterers. Each type of cage (see Section VI) has its own requirements. Since most can be used for both continuing and age-class culturing, the methods for doing the latter are discussed separately in Section VIII.F below.

A. TYPE 1 CAGES

A type 1 cage makes use of a 15-dram vial waterer and a papier-mâché separator cut two "U"s wide and then in half crosswise. The separator will hold the vial firmly against the side of the cage. Dog food can be placed alongside or on top of the separator, but always must be kept away from the waterer face.

B. TYPE 2 CAGES

These cages use 35 × 200 mm culture tubes as waterers. While a large papier-mâché separator can be employed to hold the culture tube against the long wall of the cage, this is satisfactory only for culturing small blattellids such as *Supella* and *Shawella* spp. For the larger blattid genera (*Blatta, Neostylopyga,* and *Periplaneta*) and, in general, all but the largest blattellid (*Nyctibora*) and blaberid (*Archimandrita, Blaberus,* etc.) genera, a much better solution is to use a Masonite® base provided with wood block legs, which firmly pins the culture tube against the side of the cage, keeps it from sliding back and forth, and also permits the cockroaches to pass freely beneath it (Figures 5 to 7).

These bases are very durable, with some in the author's laboratory having been used for more than 10 years. (They are simply cleaned by first being frozen and then having the oothecae and other detritus scraped from them with a spatula and scrub brush.) The dimensions of the base and the positioning and lengths of the leg blocks are given in Figure 7. Wood glue and short flathead nails should be used to attach the blocks to the base. If many separators are to be made, it is best to first make up a jig that holds the base and permits accurate positioning of the legs. This allows completed bases to be prepared in 1 min or less and thus, in the long run, amply makes up for whatever time was spent making the jig.

The cage is prepared by first putting a base into it, then a filled waterer, and then some food down the back so that it slides under the base. A large papier-mâché separator is then placed on top of the base and shoved as far forward as possible. More food is added at the back, and another separator is added on top of it with its "V" bases sitting on the flattened top of the first separator. One great advantage of this type of arrangement is that waterers are easy to remove and replace; another is that colonies can be enlarged or transferred merely by taking the top separator off and transferring it to another freshly prepared cage.

C. TYPE 3 CAGES

Cages of this type are most useful for raising the smaller blattellids (e.g., *Blattella*). Stacks of Masonite® disks with staple legs are layered in the containers. If vials are used as waterers, they should be supported in "M"-shaped saddles made from hardware cloth to keep them from shifting. The food should be placed in a plastic or aluminum cup or in a small open-top box of hardware cloth to keep it from coming into contact with the waterer.

The special vertical waterer (Section V.C) should be employed if a longer time between watering is desired or if blattellids, which require a moist substrate for oothecal water uptake, are being cultured. Because these waterers are heavy and tend to slide around on the Masonite® surface, they must be confined in position. One way is to use three short wood blocks attached to the uppermost Masonite® spacer. An alternative is to attach to a spacer (with wood glue) a $^1/_2$-in.-thick layer of polyurethane foam with a hole in it large enough to hold the waterer. The hole is made most neatly with the rim of a 1-lb coffee can, which easily cuts through the plastic when heated, leaving an opening only slightly larger than the petri dish base of the waterer. Because of the space taken up by such waterers, it is best to use rectangular hardware cloth holders for the food (Figure 8).

D. TYPE 4 CAGES

Larger cages such as these are ideal for raising the large blaberids, and overall they are easier to handle and require much less space than the type 5 cages (see Section E below). They are sufficiently large to allow the use of the 1-qt mason jar waterers and provide sufficient space to raise even the largest cockroaches. The waterer is placed on a ceramic tile and surrounded with 7- to 8-in.-high, square, large-mesh wire fence which keeps the papier-mâché harborage material from falling onto the waterer surface. Fences are best created from the vinyl-coated fencing rolls made for garden and lawn edges, widely available in hardware stores. This material has approximately a $2^1/_2$-in. spacing between horizontals

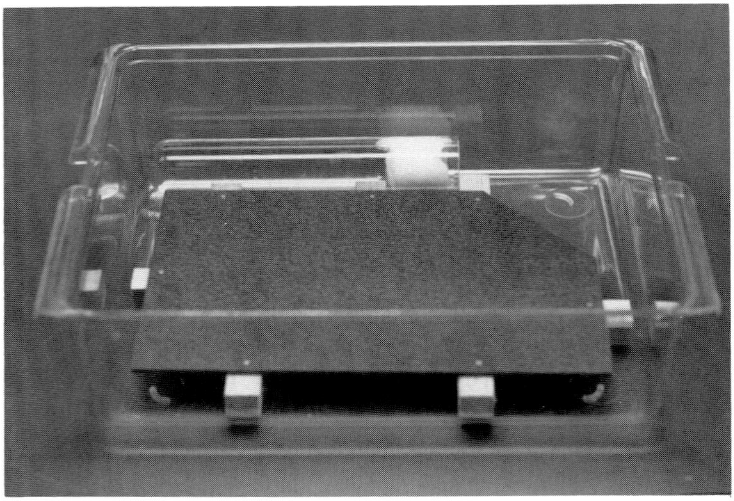

FIGURE 5. Type 2 cage with Masonite® base and 200-mm culture tube in place.

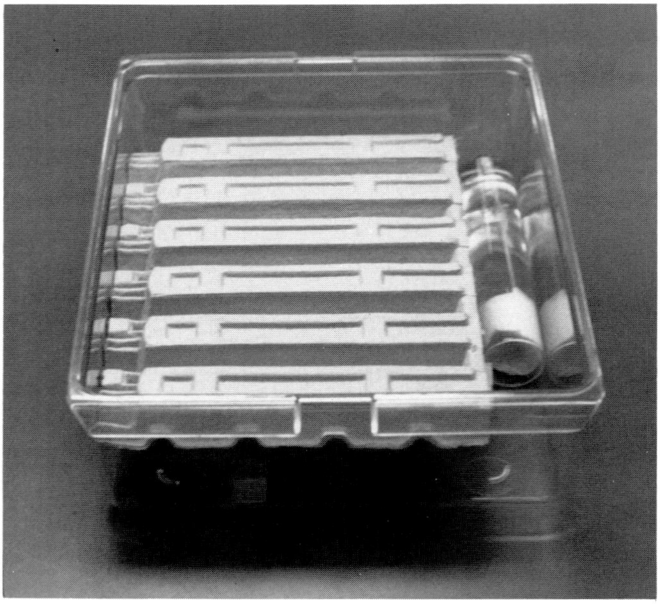

FIGURE 6. Same as Figure 5, with large papier-mâché separator placed on top of the Masonite® base.

and verticals so that it can be bent easily to form a $7^1/_2$-in. square and can be cut to the required height. As with the type 2 cages, the best harborage material is the large or small papier-mâché separator. In these larger cages, however, the layers of separators should be placed at a 90° angle to each other because they tend to be shifted greatly by the large-bodied cockroaches. An alternative arrangement (still with a 1-qt waterer within a fence) used for the very largest species (*Archimandrita* spp., *Blaberus colosseus,* and *B. giganteus*) has been given in Section VII.

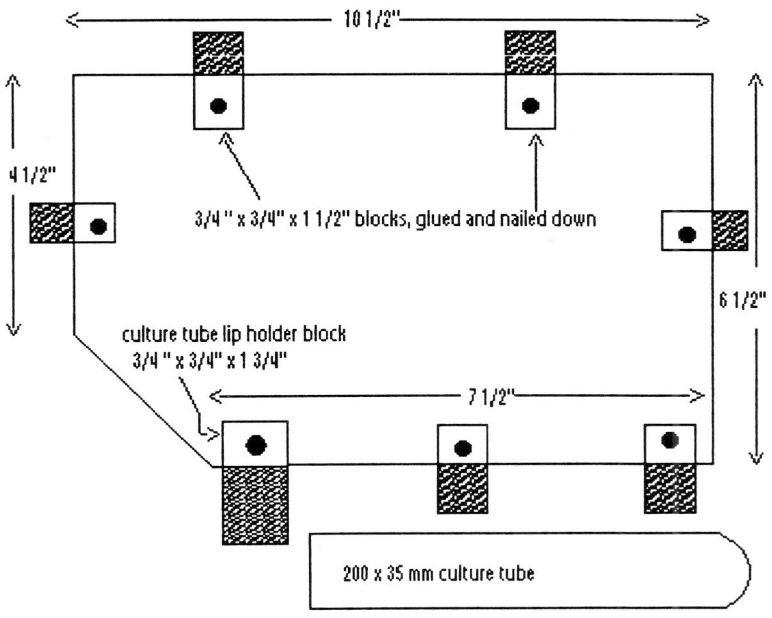

FIGURE 7. Masonite® base for use with $12^3/_8 \times 10^1/_{16} \times 3^7/_8$ in. acrylic refrigerator chests.

FIGURE 8. Waterer holder for type 3 cages. Hole was cut in polyurethane with a heated 1-lb coffee can. Food holder is made from $^1/_2$-in.-mesh hardware cloth.

E. TYPE 5 CAGES

Large garbage can containers are best for the mass rearing of blattids (*Blatta, Eurycotis, Periplaneta*) which do well on vertical surfaces. Other than waterers, the only materials employed in these cages are $^1/_4$- or $^3/_8$-in.-mesh hardware cloth and either two bricks or two 8-in. lengths of 2 × 4 in. wood. First, a circular fence of $^1/_4$- or $^3/_8$-in.-mesh hardware cloth is made that just fits within the container and extends to within about 6 in. of the top. This should have a number of holes cut into it so that roaches can pass from one side to the other. Next, a length of hardware cloth whose width is 1 in. less than the height of the container is folded back and forth into a zig-zag that fills the space within the outer circular fence, with the folds being made around a $1^1/_2$-in. wood form to give proper spacing between the layers. This inner ziz-zag portion is supported on the bricks or pieces of wood to allow the passage of roaches between the layers. A 5-in.-wide piece of hardware cloth is then placed on top of the zig-zag portion and firmly attached to both it and the outer circle. This piece will support the waterer, which should be placed on a tile. No special feeders are necessary as the roaches are fed by pouring dog food down between the vertical leaves of the zig-zag.

F. AGE-CLASS CULTURING

Setting up cultures to produce large numbers of adults that are all approximately the same age requires either very large cultures or considerable lead time. The reason, very simply, is that one has only two alternatives: either the cultures must be of such a size that a sufficient number of freshly molted adults can be collected within a very short period of time, or it must begin with obtaining fresh oothecae (blattids), ootheca-bearing females (*Blattella* spp.), pregnant females (blaberids), or recently eclosed nymphs (all groups). In any event, it always requires building the cultures to a large size in holding cages (types 2 to 4) so that sufficient numbers become available.

If the age classes are to be generated in terms of birthdate, the best methods are to have a good ootheca collector or a means of collecting freshly eclosed nymphs. For blattids which actively attach their oothecae to surfaces, the best way to do this is to have a number of standing population cages with papier-mâché separator harborage (this must be added to the type 4 cages; see Section VIII.E above). New separators are added to the cages. When a sufficient number of oothecae have been collected, adults and nymphs are shaken off the separator, which is then moved to a newly prepared cage (usually type 2) with a waterer. The ootheca-bearing separators are moved to new cages at 3- to 5-d intervals, with all nymphs appearing during that interval being left together within a cage. If this is done with five to ten standing-population cages as the sources, then outputs of several hundred nymphs per interval can be sustained for some time.

Blattella spp. must be handled differently because the females bear their oothecae until just prior to eclosion of the nymphs. A very successful method for collecting age-classed nymphs is to place either only ootheca-bearing females or a mixture of males and females in an aluminum-screened cage that is within a larger cage. The nymphs eclosing from the oothecae are very small and easily slip through the screening into the larger cage, where they can be collected at whatever interval desired. With such a system, depending on the number of females and holder cages employed, up to several hundred nymphs born within a short time of each other can be collected daily.

As all blaberids are live bearers, the best method for obtaining nymphs is to confine adults within a cage with a raised hardware mesh floor through which the adults cannot squeeze. In this circumstance, it is best to have nothing more than food and water in the upper cage so that the nymphs, instead of finding hiding places, will tend to slip through the mesh to the floor below. Unless nymphs are collected daily, a means of providing them with water must be employed, as they tend to dessicate rapidly.

No matter which of the nymph-collecting methods is used, a central problem remains: not all nymphs will grow at the same rate. Thus, there should be no expectation that they will all become adults at the same time. For this reason, if large numbers of adults of known age are required, it is best to start with a number of large age-classed nymphal cultures in order to be able to remove on a daily basis a sufficient number of adults that have emerged within a short time of each other.

IX. TRANSFERRING CULTURES

All cultures must be transferred at intervals into fresh cages, both as a sanitation measure and because the population density within them becomes too high. For most species, the best way to judge when this should be done is to note the rate of water usage: in general, it is when a 300-mm horizontal waterer or a vertical waterer becomes depleted in 1 week or less.

For all but the large blaberids and species in garbage cans, the best way to set up a new culture is to take the top papier-mâché separator or several Masonite® separators from the old cage and place it or them in a new cage. If the population densities have become extremely high, or if the species has started to eat the papier-mâché or it has become moldy, then the uppermost layer should be discarded and a new separator placed in the old cage, left for a period of time, and then transferred to the new cage. For species in garbage cans, placing separators in the can and allowing the roaches to crawl onto them is the best method for easily obtaining a sufficient number to transfer to a new cage.

Large blaberids (*Archimandrita* spp., *Blaberus* spp., etc.) are handled differently. Here, 10 to 15 pregnant females, an equal number of males, and ca. 30 nymphs of assorted sizes are picked by hand from the old cage and transferred to a new, already set up cage. If pelleted bedding has been used in the cage, then it is almost always necessary to use a $^1/_4$-in.-mesh sieve to obtain a sufficient number of small nymphs from the bedding.

The timing for the transfer of large *Gromphadorhina* spp. and large blaberids given vertical surfaces on which to rest (*Archimandrita* spp., *Blaberus giganteus,* etc.) is slightly different from the above. For species resting on vertical surfaces, the colony should be split or transferred when the adults occupy approximately half of the vertical surfaces because larger nymphs require a vertical surface on which to molt successfully and are forced to attempt to do so on horizontal surfaces when the adult density is too high. The large *Gromphadorhina* spp. seemingly represent a unique case, in that a colony can become senescent if the adult density becomes too high. A sure sign that this is occurring is to have a culture containing a large number of adults and small nymphs, but virtually no mid-sized ones. There are two reasons for this:

1. The adults are large and heavy bodied and have long, thick spines on their legs. Under crowded conditions, while walking around they inadvertently puncture the cuticle of young nymphs, particularly freshly molted ones. When the adult density is very high the nymphal death rate is so great as to almost eliminate any new recruitment to adulthood.
2. At high densities the females either self-abort their oothecae or have them partially or entirely eaten by other adults while they are being extruded prior to rotation and withdrawal into the internal incubating uterus. As a result, with this genus the colony must be divided or a new one started as soon as the relative number of mid-sized nymphs starts to fall.

X. CULTURE AND SPECIMEN DISPOSAL

Deep-freezing is the best method for disposal of specimens and cultures. It avoids the use of noxious chemicals and is far better than autoclaving, which always results in a wet, malodorous sludge that is difficult to dispose of properly. Freezing at $-30°C$ for 24 to 48 h kills everything and yields a relatively dry mass that easily can be scraped into a plastic bag and disposed of with the regular garbage. Before washing, the petroleum jelly should be scraped from the rim of the cage with a spatula, and any that remains is removed with a paper laboratory wipe (tissue). This eliminates leaving jelly films on the cages during washing and subsequent fouling (and blockage) of sinks and drain lines. The cages are then either filled with a strong, hot detergent solution or submerged in a tank of detergent solution, allowed to soak, and then washed. If metal cages are being used, they should then be wiped dry to prevent rusting or other corrosion.

Masonite® separators should not be wetted. Instead, after being frozen they should be scraped clean with a spatula and then placed in a 70°C oven and baked for 1 d or more to kill any mold spores. All waterers, bricks, unglazed tiles, etc. are soaked in detergent solutions and then washed and rinsed. Paper waterer plugs are disposed of after being frozen. If sponges are used, they should be washed thoroughly and then baked dry in a 45°C oven.

XI. PESTS AND PEST CONTROL

The major "pest" problems that can occur in cockroach cultures are mold growth, phorid flies, acaroid mites, spiders, beetles, and, in multispecies cultures, other cockroaches.

A. MOLDS

The occurrence of molds in cultures is primarily a manifestation of high-humidity conditions, either in the incubator or locally within a cage. Dog food that becomes moist or wetted (e.g., by a waterer spill) rapidly becomes moldy. While many species will eat moldy food, others will not. If mold growth becomes prolific, small nymphs may become entrapped in it and die. Thus, any cage in which there has been a waterer spill must be cleaned out immediately.

The only cure for mold growth is to lower the overall humidity and/or to thoroughly clean out the cages as outlined above. This is not merely a matter of proper sanitation because molds are direct attractants of phorid flies.

B. PHORID FLIES

These are small, hump-backed flies with an absolute avidity for wet food and fresh cockroach carcasses. While primarily attracted by molding food, their larvae require a semiliquid medium for growth, so that in addition to carcasses they will start to grow on waterers and any other source of moisture in an incubator when they become numerous. Although they seemingly present no problem to the cockroaches, these flies can rapidly become so numerous that it becomes difficult to work in an incubator without inhaling them.

Preventing invasion by phorids is a matter of eternal vigilance. Fly paper and *Drosophila* traps charged with molding dog food can be used as early-warning systems, but with traps of these types the absence of flies does not mean that the cultures have not been invaded. All cultures should be examined weekly, particularly during the summer months. If flies are seen in a cage, it should be removed from the incubator *immediately*; then either the cage should be disposed of directly by freezing, or some individuals should be removed from it in a hood (or outdoors) prior to freezing of the cage.

It takes prodigious efforts to exterminate these flies once they become established in an incubator. For all intents and purposes, most or all cultures must be restarted. Individuals and/or oothecae must be put in clean, fly-proof cages and then moved to a new location where they can be observed to ensure that no flies have been carried over. This must be done for every cage in the incubator, with all infested cages and their waterers being frozen. Finally, when everything has been removed from the incubator, its temperature should be raised to 40°C or more for 1 week to ensure that all adults, larvae, and pupae of the flies have been killed. This general sanitation procedure also kills any other pests in the incubator.

C. ACAROID MITES

Most cockroaches have or can become infested with acaroid mites. Small as they are, if they become numerous enough, they can seriously debilitate or annihilate a culture. Mite populations increase prodigiously at temperatures >28°C and at >60% relative humidity. While it is probable that they can never be totally eliminated, they usually can be controlled to a low level merely by keeping temperature and humidity levels in the incubator and within the cages below these values and by practicing proper cage sanitation. This is yet another reason why the insides of the cages must be kept dry and waterer spills avoided. It also is why cages must be cleaned at intervals to prevent a buildup of carcasses, since these are humidity traps and can greatly elevate the relative humidity, and thus mite production, within a cage.

D. SPIDERS

Spiders are adventives that become established in incubators only if there are many escaping cockroach nymphs or phorid flies. They only become a real problem if they become established within a cage, since most will eat very small (nymphal) cockroaches. Fortunately, they are easily controlled by direct killing or by sucking them up with a vacuum cleaner. If a vacuum cleaner is used, its bag should be frozen after each use to kill whatever living material it contains.

E. BEETLES

Most of the beetles occurring in cultures have been brought in with and are living on dog food. Prevention of infestation is therefore a matter of ensuring that they are not furnished with this means of entry. When they become exceedingly numerous, the only way to get rid of them is to follow the procedure for the phorid flies, outlined above in Section XI.C.

Dermestids are the only other type of beetle likely to infest an incubator. In general, this only occurs when someone else in the same building or insectory is raising them. Their larvae are voracious, avidly devouring most dry organic materials in a cage, paper and carcasses included. If very numerous, they will even begin to eat freshly molted roaches. As is the case with other beetles, they cannot be controlled to unobjectionable levels; it is necessary to eliminate them from the incubator as outlined above.

F. COCKROACHES

Several cockroach species are escape artists, particularly as nymphs, and have a propensity for establishing themselves in the cages of other species. Included in this category are *Blattella germanica, Blaberus discoidalis, Blaberus craniifer, Gromphadorhina portentosa (?),* the *Periplaneta* spp., *Symploce capitata,* and, among the exotics, *Phoetalia pallida.* If left unattended, they actually can outcompete certain other species.

To a large extent, preventing escapees is primarily a matter of ensuring that the Vaseline® bands around the cage rims remain intact, that all cage tops are closed and their openings covered, and that poison traps are set around the incubator. Other measures are required when working with blattellids such as *Blattella germanica* and *S. capitata,* since the adults of both of these will fly short distances under incubator conditions. When working with these cockroaches, it is necessary to set their cages on blocks in the middle of fairly large moats filled with water to which a small amount of detergent has been added. This assures that any individual jumping from the cage will be wetted and drowned.

REFERENCES

1. **King, E. G. and Leppla, N. C., Eds.,** Advances and Challenges in Insect Rearing, Agricultural Research Service (Southern Region), U.S. Department of Agriculture, New Orleans, LA, 1984.
2. **Cornwell, P. B.,** *The Cockroach,* Vol. 1, Hutchison and Co., London, 1968.

3. **Morgan, N. O.,** Blattidae, in *Handbook of Insect Rearing,* Vol. 1, Singh, P. and Moore, R. R., Eds., Elsevier, Amsterdam, 1985, 321.
4. **Piquett, P. G. and Fales, J. H.,** Rearing Cockroaches for Experimental Purposes, ET-301, U.S. Department of Agriculture, 1952.
5. **Roth, L. M.,** Water changes in cockroach oöthecae in relation to the evolution of ovoviviparity and viviparity, *Ann. Entomol. Soc. Am.,* 60, 928, 1967.
6. **Cornwell, P. B.,** *The Cockroach,* Vol. 2, Associated Business Programmes, London, 1976, chap. 12.

Chapter 3

ANATOMY OF THE COCKROACH

A. P. Gupta, Y. T. Das, and B. R. Rao

TABLE OF CONTENTS

I. INTRODUCTION

This chapter is intended to provide basic information on the external and internal anatomy of the cockroach to a scientist in the biomedical and veterinary sciences who may not be familiar with insect anatomy and who may wish to venture into dissecting one, either to satisfy his/her scientific curiosity or with a view to using it as a model for his/her research. An entomologist or neurobiologist already familiar with insect anatomy, therefore, may find most of the contents of the chapter elementary, but then its purpose is to inform the novice rather than one with expertise in entomology.

The presentation on each organ system includes a brief description of its basic features, presented in a jargon-free style, although the important technical terms have been included. Should further details be necessary, the reader may consult the many reviews and original papers referred to in the section on each organ system. Note that sections on the nervous system and sense organs are not included because they are described in detail in other chapters in this work.

The chapter is organized into three major sections: external anatomy, including integument; dissection instructions; and internal anatomy.

II. EXTERNAL ANATOMY

The body of a cockroach, or for that matter any other insect, is composed of some 20 segments that are grouped into three easily recognizable regions: *head, thorax,* and *abdomen.* It also has a neck (cervix) which is not often visible.

A. HEAD

The head (Figure 1) is composed of six or seven segments and bears a pair of segmented *antennae*, two *compound eyes,* two simple eyes (*ocelli*), and the chewing type of *mouthparts.* Because the latter point downward, the head in cockroaches is termed *hypognathous.* In a frontal view, the head shows several grooves (*sulci*) which define various parts that form the head capsule.

The *antennae* represent the first set of appendages on the head and consist of three main parts: the basal part (*scape*), the middle or the second part (*pedicel*), and the remaining multisegmented part (*flagellum*). The pedicel has a sensory structure (*Johnston's organ*) in most insects, and the flagellum is the longest part in cockroaches, with approximately 88 segments in *Blattella germanica.*[1] The flagellum is also the site of various sensory receptors. Each antenna has an antennal sulcus (*circumantennal sulcus*) around its base.

Each *compound eye* is composed of a group of separate units, or *ommatidia* (2000 in *Periplaneta americana*); each ommatidium consists of an outer *cornea* or corneal lens, a *crystalline cone* (produced by four cone cells) behind the cornea, and several elongated *retinal or retinula cells* which are grouped together to form an optic rod (*rhabdom*). In addition, some pigment cells are present. A nerve connects the rhabdom to the optic nerve. For anatomical and ultrastructural details of the compound eye in *P. americana,* see the papers of Butler[2,3] and Chapter 23 in this work.

The *ocelli* (simple eyes) of cockroaches also often have a corneal lens and a crystalline cone, but are not involved in image formation. They principally act as photoreceptor cells that are capable of little more than distinguishing between light and darkness. They are also regarded as having a stimulatory effect on the compound eyes. For details, see the article by Guthrie and Tindall[4] and Chapter 23 of this text.

The *mouthparts* (Figure 2) are a set of structures that surround the mouth opening in such a manner that they enclose an extraoral or preoral space (*cibarium*). They consist of the following parts:

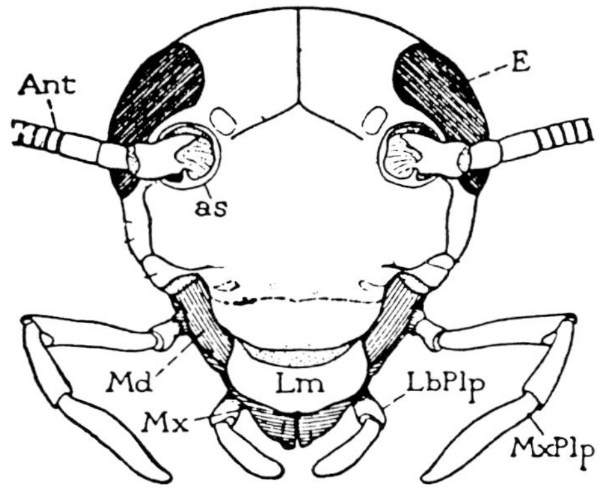

FIGURE 1. Frontal view of head of *Blatta orientalis.*[15] Ant = antenna; as = antennal sulcus; LbPlp = labial palp; Lm = labrum; Md = mandible; Mx = maxilla; MxPlp = maxillary palp; E = eye.

1. *Mandibles* (the main pair of jaws) are located on each side and are used for chewing.
2. *Maxillae* (the second pair of jaws) are located just behind the mandibles and are used chiefly for picking up and holding food; each maxilla consists of a cardo (basal part), a stipes (middle part), two distal pieces, an inner lacinia and an outer galea, and a maxillary palp.
3. The *labrum* is the upper lip; a median lobe on its internal or ventral surface is called the *epipharynx,* which serves as a taste organ.
4. The *labium,* or lower lip, represents the two fused second maxillae and closes the rear of the circle of structures around the mouth; it helps to hold food in the cibarium and serves as a tasting organ. The labium consists of a basal part (submentum), followed by a middle portion (mentum) and a distal prementum; from the latter arise a pair of inner glossae and a pair of outer paraglossae. Like the maxilla, the labium has a labial palp. Internally, there is another structure (*hypopharynx*) on the upper surface of the labium; it also serves as a taste organ or tongue.

Such primitive and generalized type of mouthparts are also found in centipedes and symphylans; the latter are considered the progenitors of insects.

B. NECK

The neck (*cervix*) in insects is considered a reduced body region (*"microthorax"*) and is strengthened on each side by lateral sclerites (cervical sclerites).

C. THORAX

A characteristic part of the insect body, the thorax (Figure 3) is the locomotor center, with three pairs of *legs* and two pairs of *wings* in the winged (pterygote) insects. It has three segments, the *pro-, meso-,* and *metathorax,* each one bearing a pair of legs. The wings, when present, are located only on the meso- and metathorax (pterothorax).

The thorax is like a box, consisting of a dorsal side (*notum* or *tergum*), two lateral sides (*pleura*), and a ventral or underside (*sternum*). Any hardened part or subdivision of any of the four sides of the thorax is called a *sclerite,* those of the notum, pleuron, and sternum being termed *tergite, pleurite,* and *sternite,* respectively. By the same token, a tergite belonging to the metathorax may be called a *metatergite* and its sternite, the *metasternite,* and so on. Each notum, pleuron, and sternum is often subdivided into parts, each part

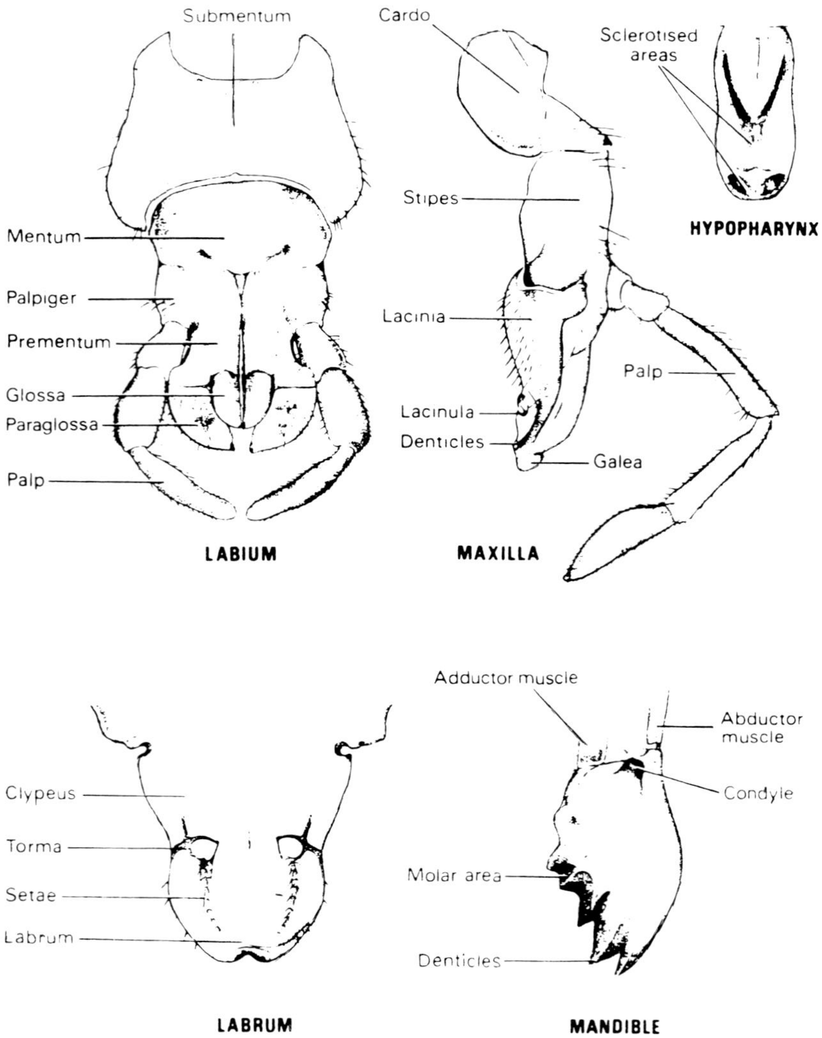

FIGURE 2. Mouthparts of *Periplaneta americana*. (From Bell, W. S. and Adiyodi, K. G., Eds., *The American Cockroach,* Chapman and Hall, New York, 1981. With permission.)

(sclerite) bearing a different name (e.g., *scutum, scutellum,* etc.). The thorax has two pairs of spiracles (see Section V.C).

1. Legs

All three pairs of legs have a common pattern, and each leg consists of the following five parts (Figure 4):

1. The *coxa* (or limb base) acts as the ball of a ball-and-socket joint by which the leg is attached to the body.
2. The *trochanter* is a small leg segment which connects with the coxa by a dicondylic articulation and is fused with the next segment, the femur.
3. The *femur* is the largest and stoutest part of the leg; it is packed with muscles.
4. The *tibia* follows the femur; it has a dicondylic articulation with the latter and is often

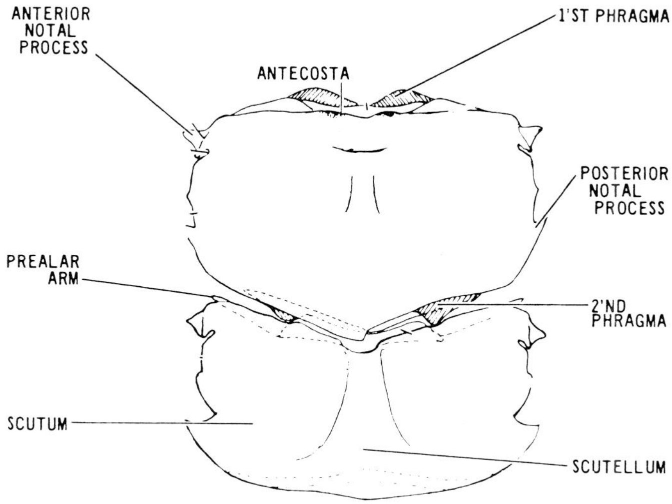

FIGURE 3. Dorsal view of the pterothorax (meso- and metathorax combined) of *Arenivaga aegyptica.*[58]

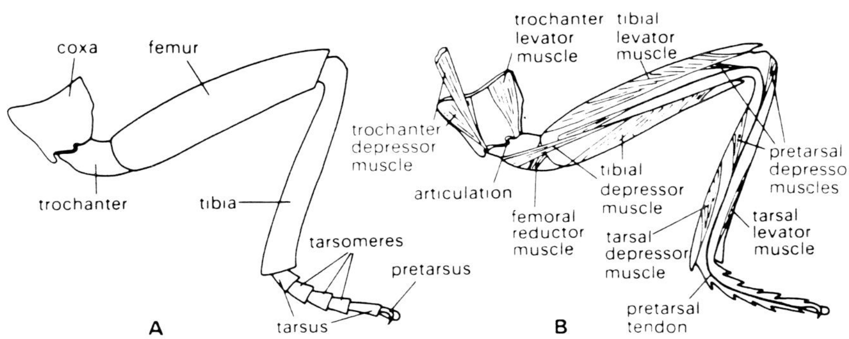

FIGURE 4. A typical insect leg and its muscles.[59]

longer and more slender than the femur. It often bears spines and (at or near its outer end) movable spurs, which provide traction during locomotion.

5. The *tarsus,* or foot, is typically a five-jointed part of the leg.
6. The *pretarsus* bears a pair of claws (*ungues*) at its tip and also various pads or glands.

The cockroach legs are of the running (cursorial) type.

2. Wings

Insects are the only animals that have wings (Figure 5) that evolved for flight; bird and bat wings are modified, preexisting limbs. Many believe that there were originally three pairs of wings, one on each of the three thoracic segments. However, during the course of evolution, those on the prothorax were lost.

In cockroaches, the first pair, or forewings (located on the mesothorax), are rather hardened and leathery in texture and are called *tegmina,* while the second pair (located on the metathorax) are membranous; both pairs of wings contain many veins that strengthen them.

The bases of all wings are connected to the body by membranous hinges in which are

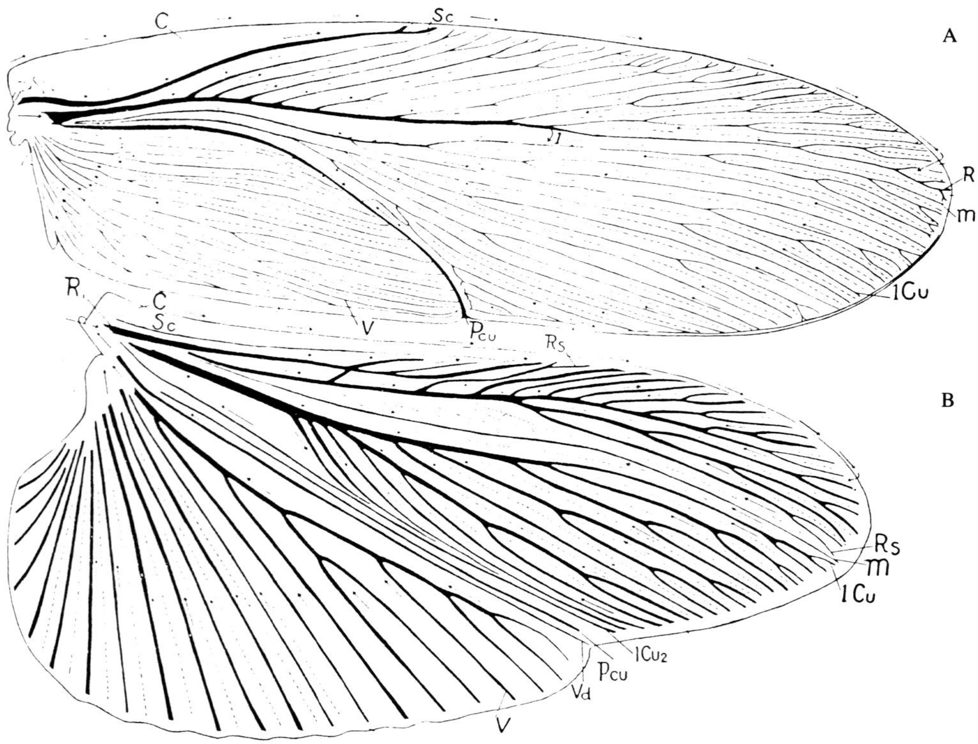

FIGURE 5. Fore- (A) and hindwings (B) of *P. americana*.[21] Letter abbreviations and arrows refer to various wing veins and directions of blood flow, respectively. Axillary sclerites at wing bases are not shown.

located a group of small sclerites (*axillary sclerites*) that articulate with the edge of the notum.

D. ABDOMEN

The abdomen typically consists of 11 segments and bears several appendages, including the *external genitalia* (see Section V.G). The jointed and prominent *cerci*, which are covered with tactile sense organs, are located on the 11th segment.

The segmental nature of the insect body is more evident in the abdomen than in the head and thorax. As in the thorax, each abdominal segment has a tergum, a sternum, and a pleuron, as well as eight pairs of spiracles (see Section V.C).

It is particularly easy to determine the sex of an adult cockroach from the terminal abdominal segments. In the male, nine sterna can be seen, while in the female, only seven are visible. In addition, females of Blattidae (e.g., *Blatta* and *Periplaneta*) have a keel-shaped subgenital plate attached to the seventh sternum (see Section V.G.1.b).

III. INTEGUMENT

The insect integument consists of a layer of epidermal cells which secrete and maintain a rigid wall, called the *cuticle*. The integument maintains the shape and rigidity of the body, acts as a barrier to toxic materials, prevents loss of body water by evaporation, and provides camouflage for the cockroach.

The integument is made up of three major layers: (1) the *epidermis,* (2) the *procuticle,* and (3) the *epicuticle*. The epicuticle is approximately 2 μm thick and can be divided into

FIGURE 6. Diagram of cockroach epicuticle.[6] (1) Cement layer; (2) wax layer; (3) outer epicuticle; (4) inner epicuticle.

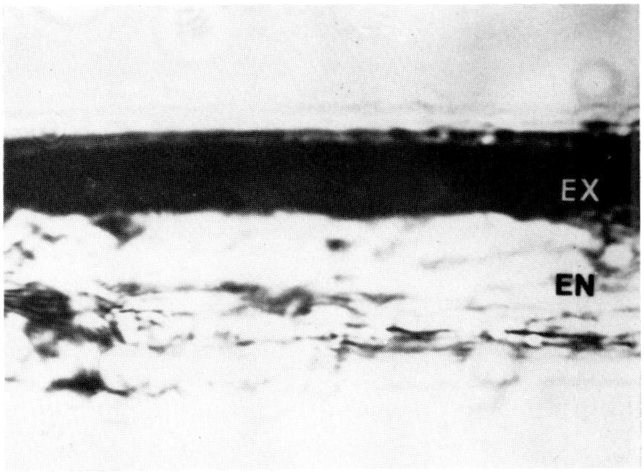

FIGURE 7. Cross section of the pronotum of *Blattella germanica*,[69] showing the nonsclerotized endocuticle (EN) and sclerotized exocuticle (EX). Sections were 10 μm thick and unstained.

an inner *epicuticle,* an outer *epicuticle,* a *wax layer,* and an outermost *cement layer* (Figure 6). The epicuticle is composed of proteins, lipids, and phenols.

The *procuticle,* which is the predominant region of the cuticle, can be separated into a sclerotized region (*exocuticle*) and a nonsclerotized region (*endocuticle*; see Figure 7). Sclerotization is a process by which the cuticular proteins are hardened and darkened. The process involves a variety of transformations and requires *N*-acetyldopamine.

Chemically, the cuticle is essentially made up of proteins, lipids, and carbohydrates. While the chemical characterization of individual molecules is still largely unknown, at least one important constituent, *chitin,* has been characterized. It is an *N*-acetylglucosamine polymer with a β-1,4 linkage (Figure 8). For details on the insect integument, one may refer to the works of Lipke et al.,[5] Mills,[6] and Hepburn.[7]

IV. DISSECTION INSTRUCTIONS

A. GENERAL INSTRUCTIONS

To study the *external anatomy* of any insect, one should take a particular part of the body, boil it for 10 to 15 min in 10% KOH to remove all soft parts, and then wash the material several times in water to remove all traces of the KOH. If necessary, the material to be examined may be stained in borax carmine or eosin and then studied in saline or 70% alcohol.

For an examination of the components of the *integument,* either a small portion of the body wall or the entire pronotum should be first dehydrated in an ethanol series and then embedded in a suitable medium and sectioned on a microtome. It is preferable to use unstained sections because it is easier to study the natural colorations (tanning and melanization) of

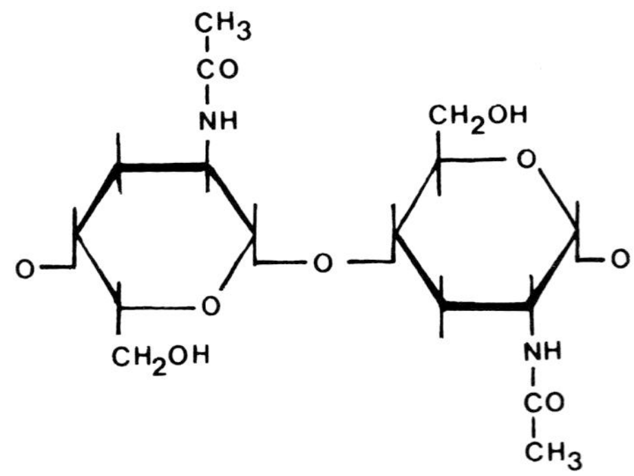

FIGURE 8. Chemical structure of *N*-acetylglucosamine.

the various layers. Very thin sections (<10 μm) are usually difficult to obtain because of the brittleness of the cuticle.

To study any of the *internal organ systems,* one has to open the body cavity (*hemocoel*), clear the fat body and the tracheae, and expose the various organs (viscera). Depending on the size of the cockroach, one would need a shallow or deep wax-layered dissection dish (DD) and physiological saline. (To prepare the latter, dissolve 9.32 g NaCl, 0.77 g KCl, 0.18 g $NaHCO_3$, and 0.01 g NaH_2PO_4 in 800 ml distilled water; after complete dissolution, add 0.5 g anhydrous $CaCl_2$, mix well, and bring the volume to the 1-l mark.) Alternately, 50 or 70% ethyl alcohol may be used.

The following protocol is recommended for cockroach dissection:

1. First, anesthetize the cockroach with CO_2 gas or immobilize it by placing it in the freezer for 20 to 30 min. Take a pair of fine dissecting scissors and, while holding the cockroach upside down in one hand, carefully make a continuous incision along its lateral sides between the terga and the sterna. One may find it convenient to remove all legs and wings, and even the head, before making the incision, either while holding the animal or after placing it in the DD. If you prefer to place it in the DD, make sure that it remains submerged in the saline at all times, especially after exposing the viscera.
2. After making the incision, place the cockroach on its ventral surface if you wish to look at all the internal organs except the dorsal vessel. Place it on its back for the latter structure.
3. It is necessary at this stage to anchor the roach to the wax layer of the DD using insect pins. Insert pins, slanting away from the cockroach, all along the periphery of the terga or sterna (whichever are in contact with the wax), stretching them slightly.
4. With fine forceps, gently peel away the unpinned body wall, exposing the viscera or the dorsal vessel. Note that all the internal organs are covered with fat body, tracheae, and tracheoles. Carefully remove them so that all the organs are clearly visible. For the study of a specific system, follow the instructions below for that particular system.

B. INSTRUCTIONS FOR VARIOUS SYSTEMS

To expose the *digestive and excretory systems* (Sections V.A and B; Figure 9), simply remove the fat body, tracheae, and tracheoles and separate the various digestive organs and the Malpighian tubules.

It is possible to study the *trachea* (Section V.C; Figure 10) in the above dissection by carefully separating the main branches of the tracheae and following some of them to the spiracles. The latter can be studied according to the methods for any part of the external anatomy.

Dissecting *muscles* (Section V.D; Figure 11) requires more care and patience, simply because most of them are very minute. Each muscle should be traced from its site of origin to that of its insertion, or vice versa.

The *dorsal vessel* (Section V.E; Figure 12) is located on the dorsal diaphragm; therefore, to study this structure, one should place the anesthetized insect on its back and remove the sterna and all internal organs to expose the heart and other associated structures (if present). The *hemocytes* can be studied and identified using Gupta's identification keys.[8,9]

The *reproductive system* (Section V.G.2; Figures 20 and 21) can be exposed by separating and removing the entire digestive system, fat body, and other tissues. Dissections of the spermatheca, its associated structures (the accessory glands), and the testes require more care and patience.

The *brain* (Section V.H; Figures 22 to 25), its associated structures, and the *ventral nerve chord* can be exposed by placing the cockroach on its ventral side and removing all overlying viscera. This will expose the entire length of the ventral nerve chord except its anterior-most part, which connects with the brain.

To dissect the brain, secure the head by inserting a small pin vertically down through the anterior-most region of the head. Starting from the prothorax, cut open the dorsal wall of the neck and the head along an arbitrary median line. To fully expose the brain, carefully separate the integument from the brain tissue and pin down the split integument into the DD. The integument may also be removed in small pieces and then discarded. Because of the small size of the brain, extreme care should be taken in removing the integument so that the brain tissue is not damaged in the process. This can be accomplished by gently pushing down the brain tissue as the integument is lifted up.

As in other animals, the nerve tissue in insects is slightly opaque, and care should be taken to decipher the location of the ganglia with proper illumination and magnification. Sometimes it is useful to stain the entire system by injecting 20 to 30 μl of 2% methylene blue in physiological saline 1 h prior to dissection.[10]

Dissection of the *endocrine system* (Section V.H; Figures 22 to 24) is more difficult than for most of the above systems. We suggest the following method:

1. After anchoring the roach to the DD, insert a small pin vertically down through the anterior-most region of the head. By holding the pin with forceps and pulling the head slightly forward (to stretch the neck region), secure the pin to the DD. A desirable posture of the head should have the mouthparts facing as far posteriorly as possible so that the neck region is fully exposed.
2. Using a pair of fine scissors, cut open the dorsal body wall along an arbitrary median line, starting from the pronotum and continuing through the head. Remove the thoracic body wall, tracheae, fat body, muscles, and digestive organs.
3. Pin down the split ends of the head integument with minute pins. The brain area, the corpora cardiaca, and the corpora allata should now be visible.
4. Under the dissecting microscope, arrange the light source so that these parts are seen clearly under minimum light intensity. In the German cockroach, the corpora allata are somewhat spindle shaped and have a maximum mean diameter of 75 μm in the male and 121 μm in the female.[10a]
5. The paired corpora cardiaca are part of the retrocerebral complex and are located dorsal to the hypocerebral ganglion. The anterior portions of the corpora cardiaca are intimately associated with the lateroventral wall of the aorta.

Since the ecdysial gland, which is juvenile tissue, persists in the adults in several cockroach species,[11] it is possible to dissect this gland in both adults and nymphs. These authors[11] have used the following procedures:

1. Using a 100-μl Hamilton® syringe fitted with a 30-gauge needle, inject into the sixth-instar nymph or adult 20 to 30 μl of 0.25% methylene blue in double-distilled water 2 to 4 h prior to dissection.
2. Cut open the prothorax and the neck region ventrally in ice-cold physiological saline and examine the glands.

C. DISSECTING TOOLS AND MATERIALS

Small dissecting dishes or trays can be prepared by pouring blackened (or plain) melted wax onto the bottom half of a suitable petri dish to a depth of one half to one third of the height. The thickness should be such that the dissections remain submerged without spilling the saline or alcohol.

For useful tips on microdissection, one may refer to Reference 66.

Dissecting pins, watchmaker's forceps with fine tips, iris microdissecting scissors, and other paraphernalia can be purchased from biological supply houses.

V. INTERNAL ANATOMY

A. DIGESTIVE SYSTEM

The alimentary canal (Figure 9) is divided into foregut (*stomodeum*), midgut (*mesenteron*), and hindgut (*proctodeum*). The fore- and hindgut are said to be of ectodermal origin; they have a cuticular lining on the inside, whereas the midgut has a specially synthesized peritrophic membrane that envelops the food contents.

The foregut consists of the undifferentiated *pharynx* and *esophagus,* the latter leading to a distensible *crop* that terminates in the *proventriculus* (gastric mill). Internally, the latter is lined with six hard cuticular teeth. The *stomodeal valve* is located at the end of the foregut.

The paired salivary glands, appearing as grape-like clusters (acini) with their paired reservoirs, are seen wrapped around the esophagus with delicate connective tissues. After dislodging them, the ducts can be traced passing through the neck, finally terminating at the base of the hypopharynx. The common reservoir ducts join the common salivary ducts to form the ultimate salivary duct. Salivary glands are whitish and translucent, and the reservoirs are filled with fluid.

Eight finger-like blind pouches (*gastric caeca*) are located at the junction of the fore- and midgut. The midgut is a simple tube and is the primary digestive and absorptive region. Numerous coiled Malpighian tubules are seen at the junction of the mid- and hindgut. These tubules function as excretory organs and are involved in water regulation and ionic balance.

The hindgut can be conveniently divided into a narrower *ileum*, a dilated, sac-like *colon,* and a thicker, oval-shaped *rectum.* The ileum accommodates the *pyloric valve,* consisting of six tiny spicules on the inner wall where the *peritrophic membrane* might be torn off. The rectum is provided with six longitudinal rectal pads that aid in resorption of water from the fecal matter.

Cockroaches are considered to be omnivorous, with mouthparts adapted for solid foods. However, water is an important dietary item. The ingested food passes through the alimentary canal, assisted by rhythmic peristalsis, at a specific rate, depending on the nature of the food.

The food is well moistened by the salivary secretions in the buccal cavity (cibarium) during mastication and is then passed into the esophagus via the pharynx. The crop mixes the partially digested food with the midgut digestive juices that pass forward from the midgut via antiperistaltic movement. The food is further triturated by the proventricular denticles before passing into the midgut through the stomodeal valve.

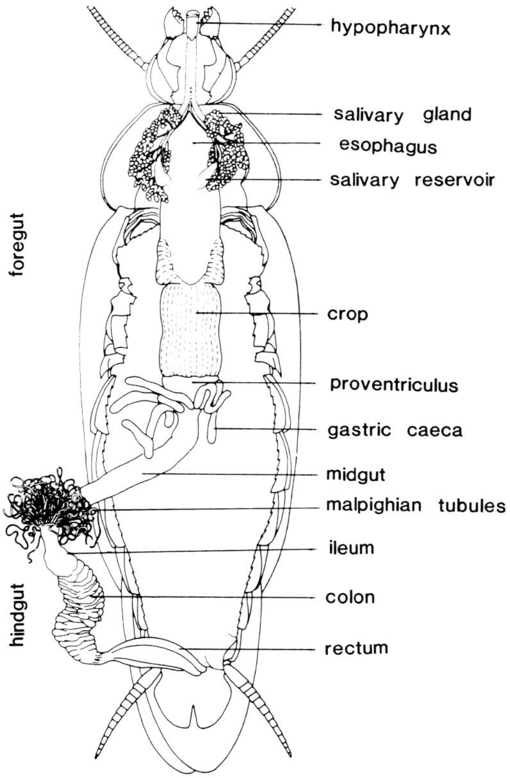

FIGURE 9. Digestive and excretory systems of *P. americana*. (From MacLeod, E. G., *The Biology of Insects: Laboratory Manual,* Burgess Publishing, Minneapolis, 1972. With permission.)

The gastric caeca are thought to have secretory and absorptive functions. The midgut is lined with a very thin peritrophic membrane which is formed by endodermal epithelial cells and is replaced continuously. This tubular sheath, in addition to its protective function for the delicate epithelial lining, is permeable to enzymes.

Much of the digestion takes place in the midgut. Several enzymes play a major role in the digestion of ingested food. For example, the salivary gland in *P. americana* contains amylase and some cellulase, whereas the midgut and caeca produce amylase, maltase, invertase, lactase, lipase, glycosidase, and β-glucosidase. Lipase has been found in the crop, midgut, and caeca. The enzymes detected in the crop may have been derived from either the salivary glands or the midgut. Enzymes with a lower pH aid in carbohydrate digestion in the crop, while trypsin and other proteinases, such as dipeptidases, aminopeptidases, and carboxypeptidases (but not pepsin), are found in the crop and the midgut. Some midgut enzymes are derived from intestinal microorganisms. Very little digestion takes place in the hindgut; however, resorption of water and ions occurs in this region. For a complete list of digestive enzymes one may refer to the chapter by Bignell in Reference 12.

B. EXCRETORY SYSTEM

The toxicity of the end product of nitrogenous metabolism appears to be correlated to the nitrogen:hydrogen (N:H) ratio. Insects, with their high rate of metabolic activity, must get rid of the nitrogenous waste safely and quickly. Being primarily terrestrial animals, insects (except for aquatic species) cannot afford to have large amounts of water to excrete this product in a soluble form. Hence, uric acid, an insoluble compound with the least toxic effect (1:1 N:H ratio), is the chief excretory product in insects.

The *Malpighian tubules* (Figure 9) are the primary excretory organs, located at the junction of mid- and hindgut in the form of coiled tubular structures. They appear to be vaguely arranged in six bundles of 12 to 15 each (or ranging from 60 to 150 in number) and are slightly yellowish in color. The tubules are of endodermal origin, with a one-cell-thick wall surrounding a lumen, and are closed distally. They are enveloped by strands of intrinsic muscles which produce the characteristic spiraling movement. A peritoneal sheath made of tracheoblasts covers the outside.

The uptake of waste material from the hemolymph can be studied by injecting dyes such as indigo carmine, neutral red, or trypan blue. For example, about 90% of the indigo carmine (2.5% aqueous) injected into the pericardial cavity of the American cockroach nymph can be excreted in about 5 h.

The probable mechanism of filtration is that soluble sodium and potassium urates, entering the tubules across the cell membrane, precipitate as insoluble uric acid and pass into the hindgut. The uric acid is then excreted along with the fecal matter. The ions and water are resorbed into the hemolymph, either at the base of the tubules or in the hindgut.

In addition to the Malpighian tubules, cockroaches are known to utilize other structures for excretion. The *fat body* can perform "storage excretion". *Uricose glands* in certain males can discharge urates by incorporating them in the *spermatophore,* which is then transferred into the female's bursa during copulation. Some waste material can be deposited in the cuticle and later expelled during the molting process. For more details on the excretory system of insects, see Reference 13.

C. RESPIRATORY SYSTEM

The high metabolic rate in insects is possible because of the availability of oxygen to the internal organs directly from the air rather than through an intermediate tissue or a fluid such as blood. This is accomplished by an elaborate network of ducts (*tracheae*) with openings on the lateral aspects of the body through which air enters and ultimately reaches the tissues and cells; this network is referred to as the *tracheal system.*

The noncellular part of the respiratory system (*ventilatory system*; see Figure 10) can be described in the following manner. Air enters the body by way of ten pairs (eight abdominal and two thoracic) of openings (*spiracles*) that regulate the flow of air by valves which are controlled by a set of muscles. Abdominal spiracles open inward and are recessed. Air then passes through the tracheal system, finally reaching the tissues such as muscles.

Tracheal structures are of ectodermal origin and are characterized by a spiral cuticular lining (*taenidia*) which prevents the collapse of the tubes. In dissection they appear as flexible silvery tubes because they contain air. Larger tubes, forming the main ducts, are called the tracheal trunks, whereas smaller to minute tubules, which are formed by repeated branchings, are known as *tracheoles.* These terminate as the tracheolar end cells that have intimate contact with the tissues.

1. Tracheal Arrangements

Using a blaberid cockroach, the following tracheal arrangement can be outlined. The dorsal view shows a pair of lateral longitudinal trunks that connect the spiracles on each side. A pair of dorsal trunks, located on either side of the middorsal heart and running parallel to the heart, are connected to each other by short commissures. These dorsal trunks are connected to the lateral trunks by stout anterior and posterior transverse connectives. In the thorax, however, the longitudinal dorsal trunks are quite enlarged, with shortened connectives.

The ventral view shows a pair of thin ventral longitudinal trunks that run on either side

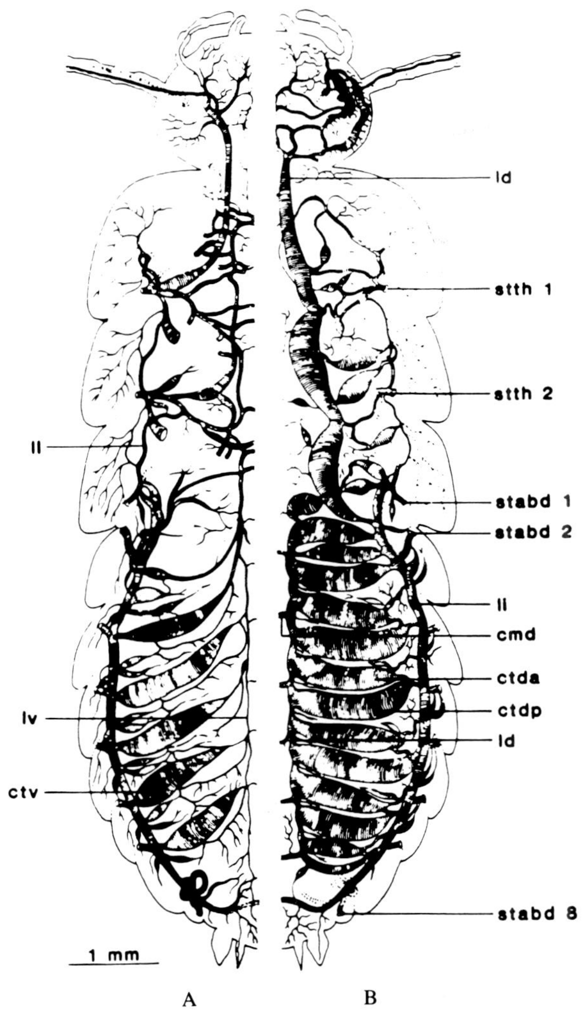

FIGURE 10. Tracheal morphology of first-instar nymph of *Gromphadorhina portentosa*. (A) Ventral view, (B) dorsal view; cmd = dorsal commissure; ctda, ctdp = anterior and posterior dorsal transverse connectives, respectively; ctv = ventral transverse connective; ld, ll, lv = dorsal, lateral, and ventral longitudinal trunks, respectively; stabd = abdominal spiracles 1 to 8; stth = thoracic spiracles 1, 2. (From Miller, T. A., in *The American Cockroach*, Bell, W. S. and Adiyodi, K. G., Eds., Chapman and Hall, New York, 1981, chap. 8. With permission.)

of the central nervous system. These are connected to the ventral lateral trunks by single transverse connectives. The thoracic ventral longitudinals are not as enlarged on the dorsal side. In addition, there are large visceral trunks, giving out branches that anastomose into various internal structures, such as parts of the alimentary canal, Malpighian tubules, reproductive organs, and fat body. Appendicular tracheae are also seen embedded in muscles. For more details on the respiratory system of insects, refer to Reference 14.

D. MUSCULATURE

The unique feature of an insect form is its ability to accommodate large numbers of muscles (Figure 11) because of the increased surface area of the exoskeleton. This enables

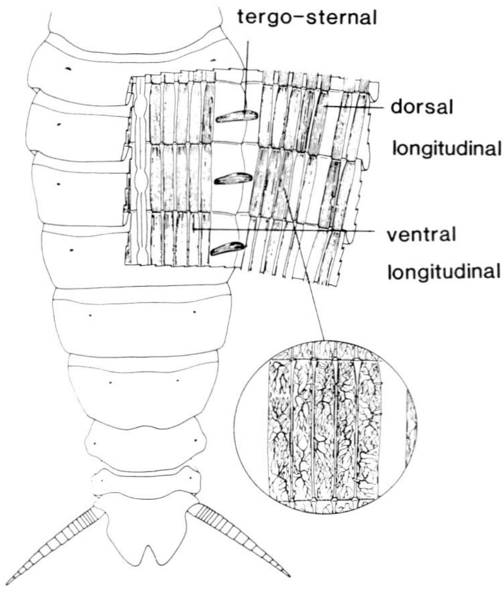

FIGURE 11. Dorsal and ventral longitudinal and tergo-sternal muscles of the abdomen of *P. americana*. (From McLeod, E. G., *The Biology of Insects: Laboratory Manual,* Burgess Publishing, Minneapolis, 1972. With permission.)

the body to perform intricate movements despite its segmented nature. Humans with their endoskeletal system have about 700 muscles, while the American cockroach has as many, if not more. All insect muscles are of the striated type, even though functionally they can be differentiated into skeletal and visceral muscles.

The gross skeletal muscles of the American cockroach have been summarized from various sources by Guthrie and Tindall.[4] They compiled the names, numbers, and locations, along with their insertion and origin from each region of the body. For further details, refer to References 15 to 19.

Naming muscles involves functional aspects, which include many complex terms. For example, Hughes,[20] in describing leg movement, uses terminology such as protraction, promotion, retraction, remotion, adduction, abduction, levation, depression, extension, and flexion. The muscles that participate in this movement are grouped into two categories: extrinsic (arising from outside the leg) and intrinsic (arising wholly within the leg, running from one segment to the next). The coxa is moved by a set of extrinsic muscles that arise in the thorax. Intrinsic muscles are simpler and consist of pairs of antagonistic muscles on each segment. For instance, there are three levator muscles within the trochanter, arising from the coxa, and three depressor muscles (two from a coxal origin and a third from the pleural ridge). The femur is immovably attached to the trochanter, and the tibia is moved by extensor and flexor muscles arising in the femur and inserted into tendons from the membrane at the base of the tibia. The tarsal levator and depressor muscles arise from the tibia and are inserted into the top of the base of the tarsus. There are no intrinsic tarsal muscles. Characteristically, the pretarsus has only depressor muscles, whose fibers, in the form of a long tendon, are inserted into the femur and tibia and thus reach the pretarsus (Figure 4).

In addition to the large surface area created by the exoskeleton, there are modified internal portions of the skeleton, such as the *tentorium* (in the head), pleural and sternal *apophyses, phragmata* and *spinae* (in the thorax), and the *antecostal ridges* of the abdomen, which provide more areas to anchor muscles.

1. Basic Muscle Structure

Each muscle is made up of long, multinucleated fibers bounded by the *sarcolemma*, or plasma membrane. The cytoplasm and endoplasmic reticulum are referred to as *sarcoplasm* and *sarcoplasmic reticulum,* respectively. The plasma membrane is deeply invaginated into the fibers to form a regular radial canal. *Myofibrils* (filaments), a characteristic feature of muscle cells, are embedded in the sarcoplasm and extend from end to end along the fiber. The stouter fibrils are composed of myosin and the thinner ones of actin. The myosin filaments are surrounded by a ring of 10 to 12 actin filaments. Both actin and myosin filaments are linked by cross bridges, which provide structural and mechanical continuity. Units of 10 to 20 fibers are separated by tracheolated membranes. Tracheoles are in close contact with the outside of the muscle fiber, providing an adequate oxygen supply. Each fiber unit may also have its own nerve supply, or several muscle units may have common innervation. Furthermore, each unit is innervated by both a fast and a slow axon (sometimes inhibitory axons, too). The fibers are amply supplied with mitochondria.

Visceral muscles (found in the gut and heart) differ from skeletal muscles in having uninucleate cells, and the fibers are found individually rather than in groups. There are fewer mitochondria. The visceral muscles contract slowly; however, they are capable of supercontraction.

E. CIRCULATORY SYSTEM

The circulatory system of the cockroach consists of a simple tubular *dorsal vessel* (Figure 12), the blood (*hemolymph*), and the blood cells (*hemocytes*; Figure 13). Because in insects there are no real arteries or veins that carry blood directly to the tissues, the circulatory system in insects is called an *open system,* as opposed to the *closed system* in vertebrates. The body cavity (*hemocoel*) is divided longitudinally into three chambers (*sinuses*) by two longitudinal muscular partitions (*septa*; Figure 14). The dorsal septum (*diaphragm*) lies just below the heart and is attached to the terga laterally. It is made up of sheets of striated muscle fibers (*alary or aliform muscles*) and defines the *pericardial sinus* dorsally. The dorsal vessel virtually rests on it. The *ventral diaphragm,* when present, lies over the ventral nerve cord and defines the *perineural sinus*. The sinus lying between the pericardial and perineural sinuses is called the *perivisceral sinus*. The dorsal septum assists in the *diastole* of the heartbeat, while the ventral one directs blood flow (by its undulations) backward and laterally to ensure hemolymph irrigation of the nerve cord.

The *dorsal vessel* is a simple tube that runs along the dorsal midline below the terga. It is open anteriorly, but closed posteriorly, and it has two recognizable regions: the *aorta* and the *heart,* the latter being the distended, muscular posterior part. The heart is highly contractile, often confined to the abdomen, and generally consists of 10 to 12 chambers that are held in place by alary (fan-shaped) or aliform muscles. The anterior portion of the aorta, located in the head, is called the *cephalic aorta*. Blood flows into the heart through minute segmental openings (*ostia*) located on the dorsolateral sides of the heart, the cephalic aorta and the aorta being devoid of both alary muscles and ostia. The latter have valves that prevent expulsion of blood from the heart into the hemocoel during contraction (systole). Anteriorly, the cephalic aorta discharges blood near the base of the brain.

Some cockroaches (e.g., *Blaberus* and *P. americana*) also have *segmental vessels,* or "arteries" (Figure 12). In addition, some cockroaches (e.g., *P. americana*) possess the so-called *pulsatile organs,* or "hearts", located at the base of the antennae in the head and the wings.[21]

Each heartbeat consists of a *diastole* and *systole,* which together constitute a single

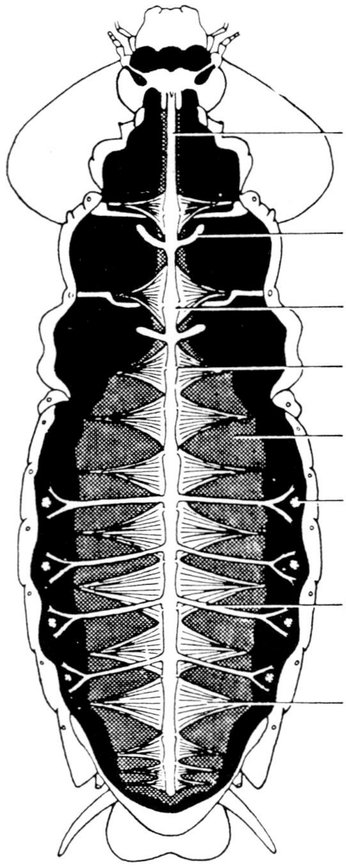

FIGURE 12. Dissection showing the dorsal vessel, lateral arteries (segmental vessels), dorsal diaphragm, ostia, and alary (aliform) muscles of *Blaberus trapezoides*.[62]

cardiac cycle. It is controversial whether the insect heart is neurogenic or myogenic. Certainly, the heart in *Periplaneta* is heavily innervated. The heartbeat in insects originates in the posterior part of the heart and proceeds in an anterior direction. However, it is reversed under certain conditions.

The *hemocoel* is rarely full of hemolymph, which in insects is a clear liquid, serves as a medium of transport for its constituents, and constantly irrigates all the internal organs. In insects, blood does not transport oxygen to tissues and cells. It comprises about 16 to 20% of the total body weight, and it consists of *hemocytes* and all kinds of organic and inorganic ions, enzymes, and humoral immune factors (see Section V.F). Its pH ranges from 6.00 to 8.00.

Insect blood coagulates owing to hemocyte agglutination, hemolymph gelation, or a combination of both. In most insects, the granules of the granulocytes or coagulocytes cause coagulation. For details of the circulatory system and the hemolymph in insects, see the reviews of Miller[22] and Mullins,[23] respectively.

1. Hemocytes

The most commonly found hemocytes in insects are the *prohemocytes, plasmatocytes, granulocytes, spherulocytes, adipohemocytes, coagulocytes,* and *oenocytoids* (Figure 13).

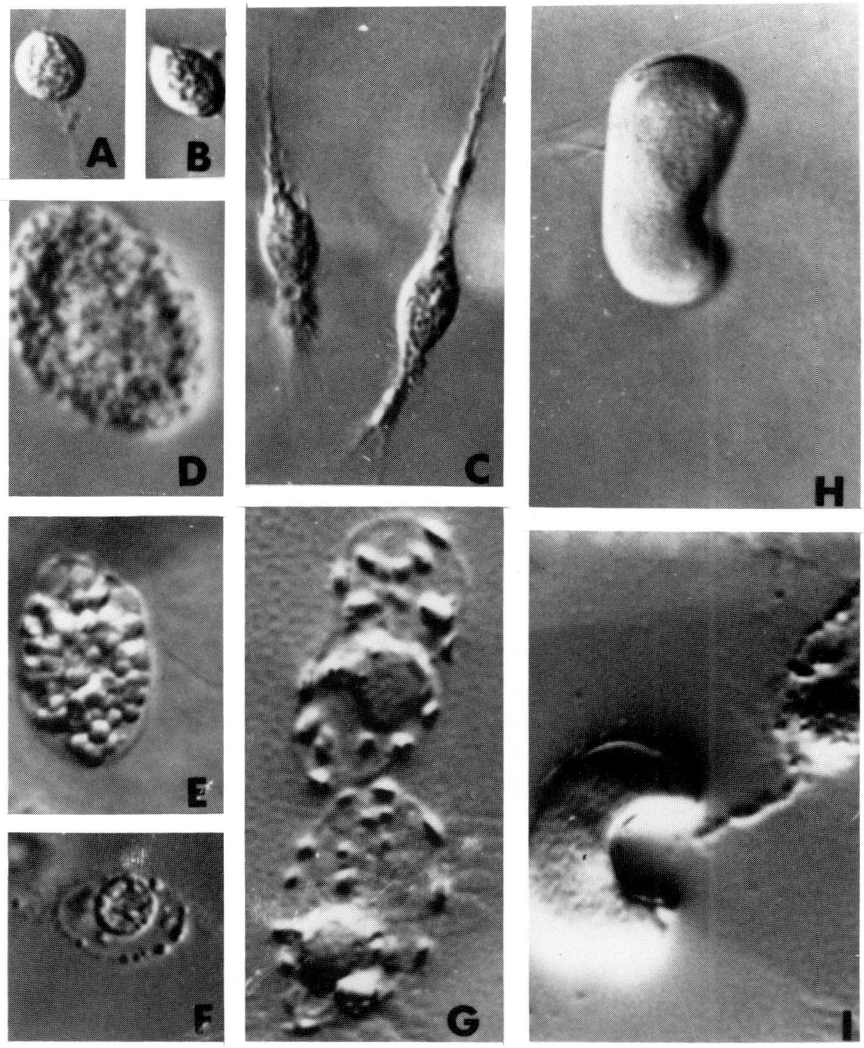

FIGURE 13. Major hemocyte types of *Gromphadorhina portentosa*. (A) Prohemocyte (magnification × 4333); (B) prohemocyte (magnification × 5633); (C) plasmatocytes (magnification × 6500); (D) granulocyte (magnification × 5687); (E) spherulocyte (magnification × 5687); (F) coagulocyte (magnification × 6314); (H) enucleate oenocytoid (magnification × 5525); (I) oenocytoid with extruded nucleus (magnification × 7150); Nomarski phase-contrast. (Reprinted with permission from Gupta, A. P., in *Comprehensive Insect Physiology, Biochemistry and Pharmacology*, Vol. 3, Kerkut, G. A. and Gilbert, L. I., Eds., Copyright 1981, Pergamon Books Ltd.)

Most of these can be identified easily using Gupta's identification key for hemocytes.[8,9] In the German cockroach, *Blattella germanica,* all seven types are present,[24,25] while in the hissing cockroach, *Gromphadorhina portentosa,* there are only six types, the adipohemocytes being absent.[26] For various details of hemocytes and their functions, see the reviews of Gupta.[8,9,27]

F. IMMUNE SYSTEM

The immune system is essentially a part of the circulatory system and consists of two types of blood cells as well as several humoral factors found in the blood. Insects have

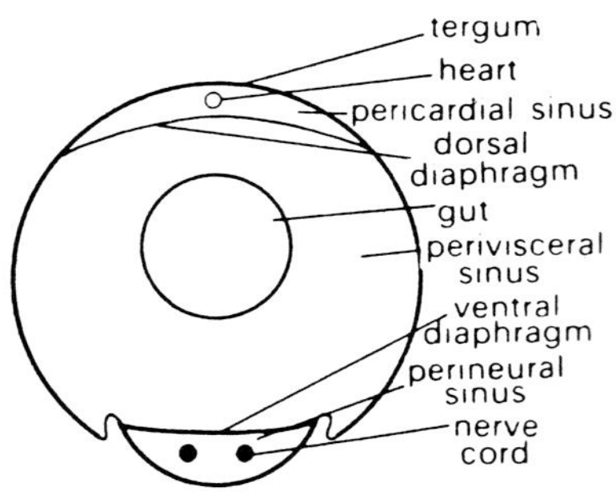

FIGURE 14. Cross section of abdomen of a generalized insect, showing the three sinuses.[63]

several kinds of blood cells (see Section V.G), two of which (the *granulocytes* [GRs] and *plasmatocytes*; Figures 13 and 15) were called *immunocytes* by Gupta[26,27] because of their ability to discriminate between self and nonself tissues. These two immunocytes function in two important ways: (1) they serve as the first line of defense (*cell-mediated immunity*) and (2) they synthesize and secrete the following immunologic factors, some of which provide the second line of defense (analogous to *antibody-mediated immunity*): *hemagglutinins* (lectins),* *lysozymes* (also produced by the fat body), *complement-like factors, antibacterial* and *antiviral factors,* the *prophenoloxidase system, coagulogens,* and the *clotting enzyme*. For details of the insect immune system, see References 27 and 28.

GRs are very reactive to foreign antigens and even chemicals. When exposed to them, GRs become activated and are characterized by the presence of numerous plasma membrane extensions (pseudopods; see Figure 15) and pronounced development of peripheral microtubules and nuclear pores.[29,30]

G. REPRODUCTIVE SYSTEM

1. External Genitalia

External genitalia represent the greatly modified appendages of the eighth and ninth abdominal segments. In males, they are used to clasp the female during mating; in the female, they are concerned with oviposition and are therefore called *ovipositors*. The stingers of wasps, bees, and ants are ovipositors that have become adapted for defense and offense.

a. Male Genitalia

In insects, the external genitalia of the male develop from a pair of phallic lobes (*phallomeres*), which divide to form a pair of inner (*mesomeres*) and a pair of outer (*parameres*) structures. In *Blatta,* there is also an additional ventral phallomere (*hypophallus*).[31]

The *male genitalia* in insects are located on the ninth segment (Figure 16) and consist of a pair of claspers (*parameres*) and the penis (*aedeagus*). The inner chamber of the aedeagus, which continues with the *ejaculatory duct* (see below), is called the *endophallus,* and its opening at the tip or end of the aedeagus is the *phallotreme*. Note that the true *gonopore* is at the junction of the ejaculatory duct and the endophallus; hence, it is internal. However, in many insects, the endophallic duct is reversible so that the gonopore assumes a terminal position during copulation (Figure 17). A well-formed endophallus does not occur in *Blatta*.[31] The claspers and the aedeagus may be mounted on a common base, the *phallobase*. Snodgrass[15] uses the term *phallus* to include the phallobase and its various processes,

* Functionally analogous to antibodies; it must be mentioned that insects do not possess antibodies similar to those in vertebrates, but have molecules that perform antibody functions.

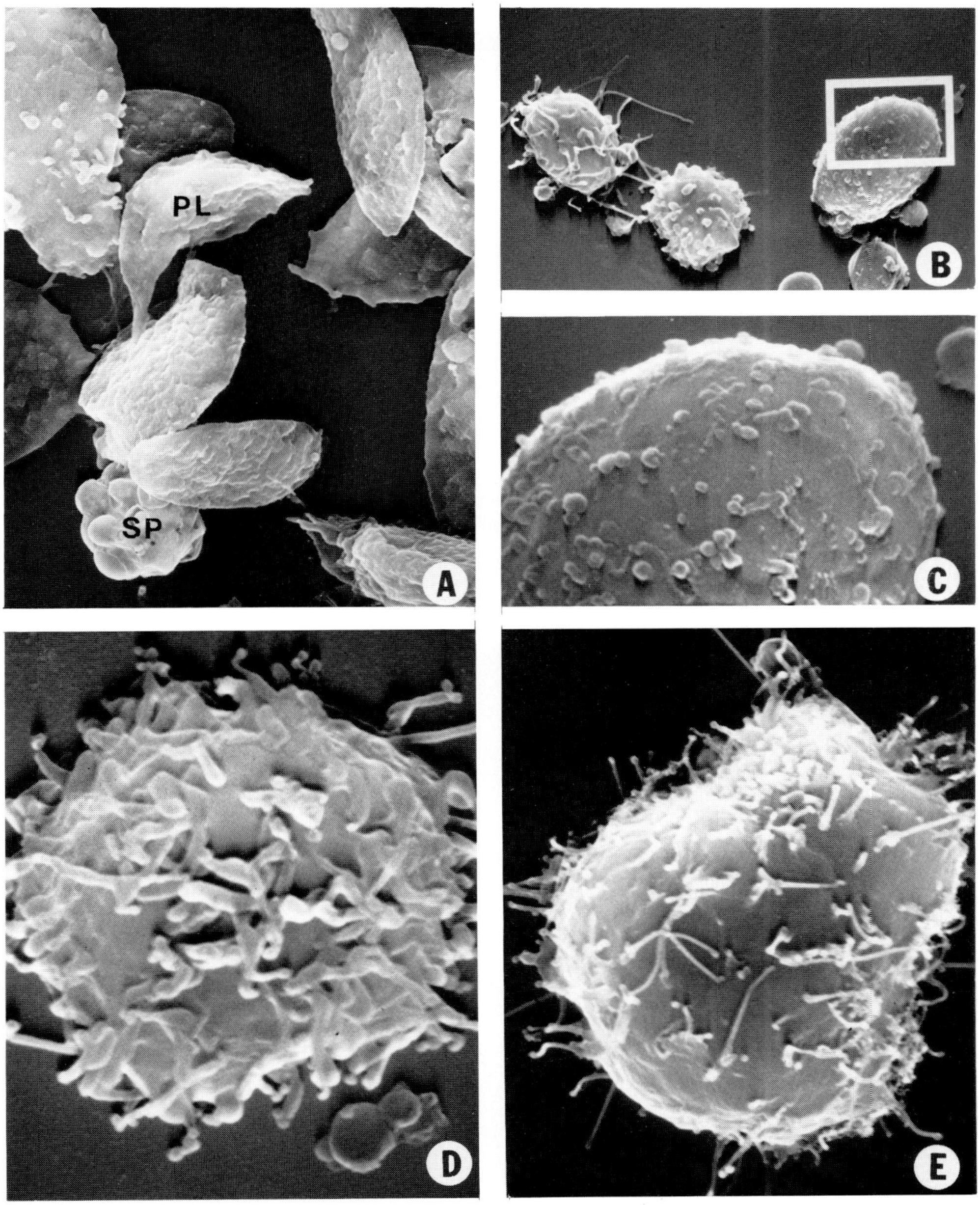

FIGURE 15. Granulocytes in various stages of activation. (A) Cluster of hemocytes (plasmatocyte, granulocytes, and spherulocyte) of *B. germanica*; PL = plasmatocyte, SP = spherulocyte. All unlabeled cells are granulocytes (GRs). Note that the GR is the major immunocyte in this cockroach. (Magnification × 4,000.) (B) Three GRs of *B. germanica*. (Magnification × 3,000.) (C) Magnified view of portion of one GR (in Figure 15B, right). (Magnification × 12,000.) (D) Activated GR of *B. germanica*. Note that this immunocyte, when reacting to foreign material (i.e., in its activated state), including chemicals in a saline solution, gives rise to numerous extensions of its plasma membrane (also called pseudopods). This type of cellular reaction of the GR is analogous to similar reaction of the vertebrate T-lymphocyte. (Magnification × 10,000.) (E) Activated GR of *G. portentosa*.

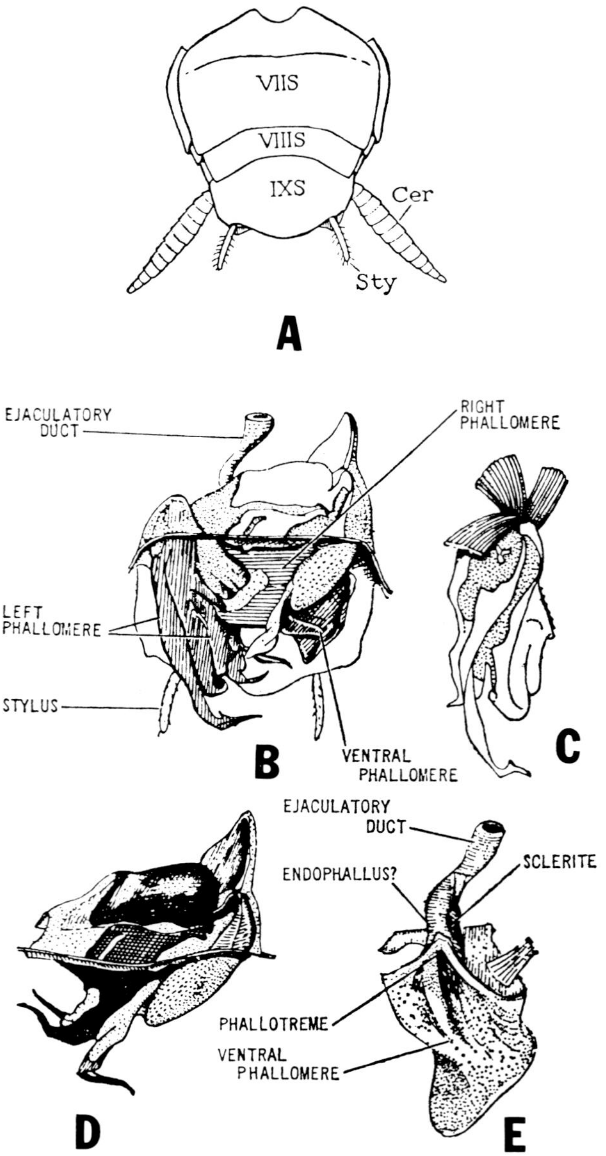

FIGURE 16. External male genitalia of *B. germanica*. (A) End of abdomen showing three abdominal segments (7 to 9), the cerci, and stylus;[15] (B) genitalia within the genital chamber; (C) left phallomere; (D) right phallomere; (E) ventral phallomere, with phallotreme from endophallic pouch continuous with ejaculatory duct.[31]

the aedeagus, and the endophallus. Note that in Dictyoptera not only are the typical claspers and aedeagus absent (because the phallomeres do not form these typical structures; see article by Khalifa[32]), but there is bilateral asymmetry in the genitalia (e.g., in *Blattella*). According to Matsuda,[31] the male genitalia in Blattaria are concealed by the ninth sternum (*subgenital plate*). For more details of the external genitalia in Blattaria see References 31, 33, and 34, and for further reading on genitalic muscles and nerves see References 35 and 36.

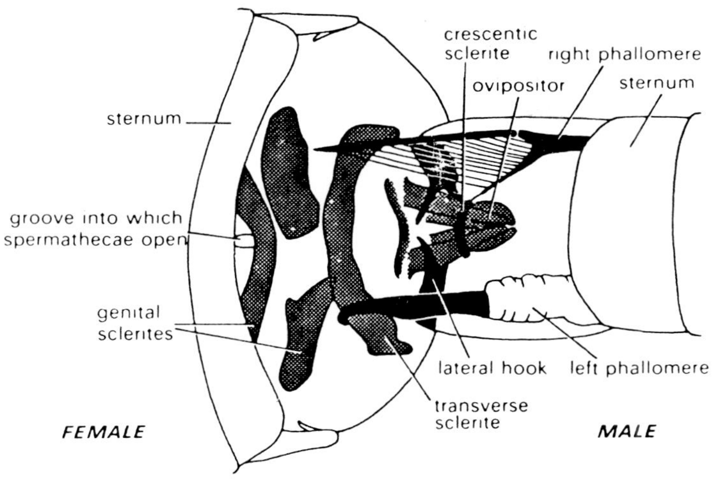

FIGURE 17. Ventral view of male and female genitalia of *Blattella* in copula (end-to-end position).[32] Subgenital plate and endophallus have been removed.

FIGURE 18. Ootheca of *Blatta orientalis,* split open to show arrangement of eggs. (From *The Oriental Cockroach,* Vol. 4, Neil A. Maclean Co., Belmont, CA, 1969.)

b. Female Genitalia

When present, the *female genitalia* in insects are located on the eighth and ninth segments. In Blattaria, the subgenital plate (seventh sternum) together with the dorsal sclerites of the eighth and the ninth sterna form a large chamber, which has a smaller anterior chamber and a larger posterior one (vestibule); the egg case (ootheca; see Figure 18) is formed in the latter chamber, and the anterior chamber supports the opening of the common oviduct (Figure 19). The floor of this chamber forms the *genital chamber,* which in cockroaches represents the forerunner of the highly sclerotized *bursa copulatrix* found in other insects.[31]

The ovipositor is often hidden by the subgenital plate in most cockroaches, and the stylus of the ninth segment is absent. Typically, an ovipositor (Figure 19B) has a base (*gonocoxa* or *valvifer*) that is articulated to an elongated, slender process (*gonapophysis* or *valvula*) that forms the shaft of the ovipositor. At the base of the ovipositor is a small sclerite (*gonangulum*) which, in generalized insects, is attached to the base of the first gonapophysis

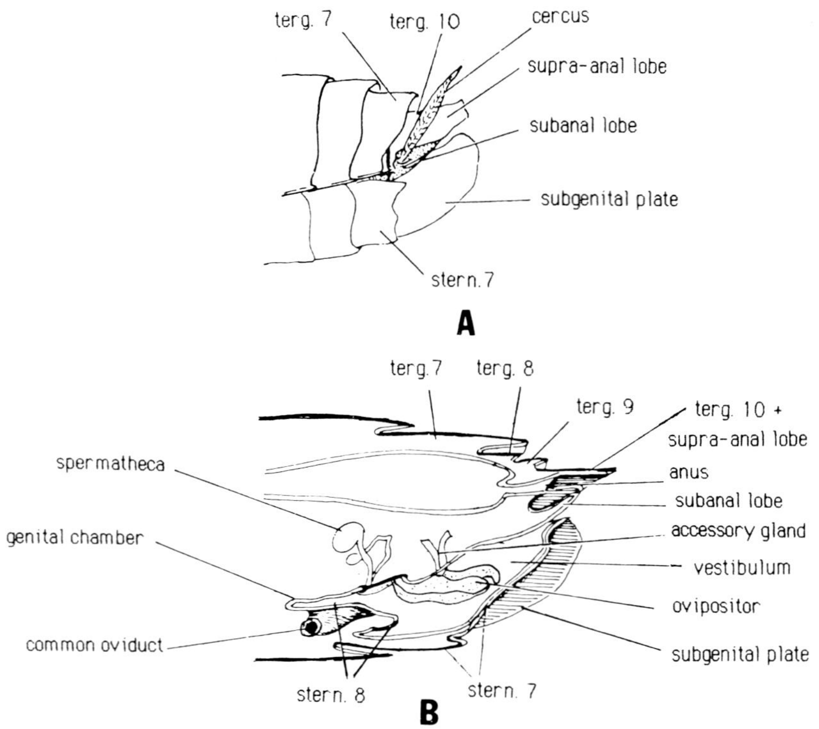

FIGURE 19. Female genitalia of *Blatta orientalis*.[31] (A) Lateral view of posterior abdominal segments (eighth and ninth segments concealed). (B) Median section through posterior abdominal segments.

and articulates with the second gonocoxa (of the ninth segment) and the tergum of this segment. It probably represents the coxa of the ninth segment. In some insects, the second gonocoxa has a process (the *gonoplac* or third valvula).

In Blattaria there are three pairs of reduced valvulae that form the ovipositors, and both the second (inner) and third (lateral) valvulae are supported by the second gonocoxa (valvifer).[31]

2. Internal Reproductive Organs

a. Male Organs

The male reproductive organs (Figure 20) in insects consist of a pair of *testes* (made up of sperm tubes) that are connected to a median duct (*ejaculatory duct*) by a pair of lateral ducts (*vasa deferentia*). The posterior part of each vas deferens is distended and is called a *seminal vesicle*. In addition, there are accessory glands whose secretion (together with the semen) produces capsules containing sperm (*spermatophores*) in the ejaculatory duct. This duct opens to the exterior through an opening (*gonopore*).

According to Matsuda,[31] in Blattaria the number of sperm tubes in the testes of various cockroaches varies: *Blattella* spp. have four, while *Cryptocercus* and *Leucophaea* have many. Furthermore, their shapes also vary: globular or fusiform (*Blatta*), rounded or oval (*Blattella*), rounded (*Cryptocercus*), and mulberry like (*Blaberus*).

The *vasa deferentia* in cockroaches terminate in an *ampulla,* which is an anterior mesodermal part of the definitive ejaculatory gland.[31]

The *accessory glands* in cockroaches may be of several types. Louis and Kumar[37] found three types in *P. americana* and *Blaberus,* on the basis of their lengths. However, such

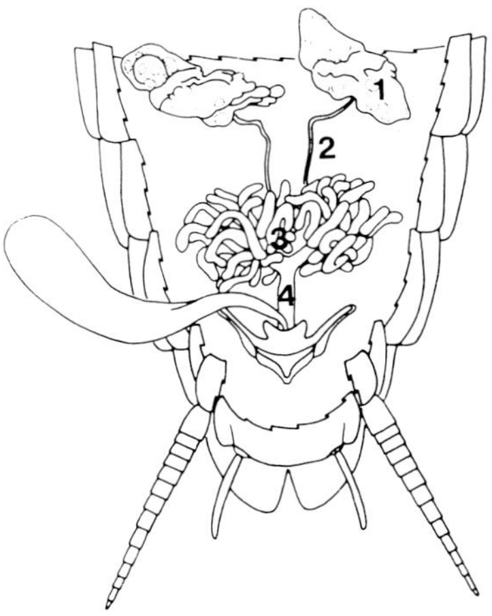

FIGURE 20. Internal reproductive organs of male *P. americana;* 1 = testis, 2 = vas deferens, 3 = accessory gland, 4 = ejaculatory duct. (From MacLeod, E. G., *The Biology of Insects: Laboratory Manual,* Burgess Publishing, Minneapolis, 1972. With permission.)

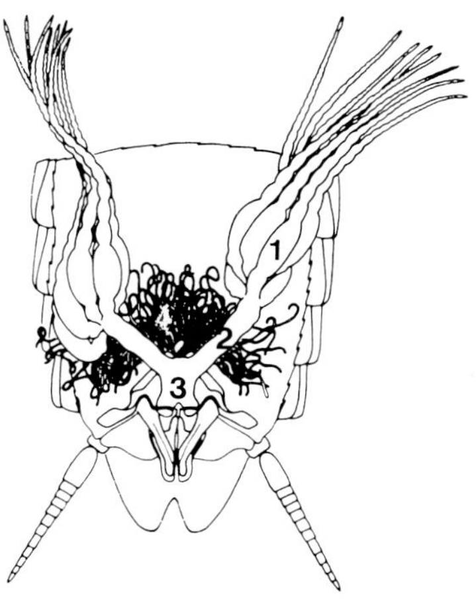

FIGURE 21. Internal reproductive organs of female *P. americana*; 1 = ovary, 2 = lateral oviduct, 3 = common oviduct, 4 = colleterial or accessory gland. (From MacLeod, E. G., *The Biology of Insects: Laboratory Manual,* Burgess Publishing, Minneapolis, 1972. With permission.)

differences do not occur in *Leucophaea, Nauphoeta,* and *Blattella*. Large amounts of uric acid are often found in the accessory glands in Blattellidae and Blaberidae; most of this is poured over the spermatophore during copulation. For details of spermatophore formation in cockroaches, consult References 4, 32, and 38 to 43.

b. Female Organs

The female internal reproductive organs (Figure 21) in insects consist of a pair of *ovaries* that are connected to a pair of *lateral oviducts.* These, in turn, open posteriorly into a common oviduct, which opens into the *genital chamber* formed by an inflection of the body wall behind the eighth sternum. Because the genital chamber serves as a copulatory pouch during mating, it is also called the *bursa copulatrix.*[15] When the genital chamber becomes elongated and tubular and is continuous with the median oviduct, it is called the *vagina.* In addition, there are accessory or *colleterial glands, spermathecae,* and a *spermathecal gland.*

Each ovary is composed of several egg tubes (*ovarioles*), which are of the primitive *panoistic* type (i.e., without specialized nurse cells) in cockroaches. The number of ovarioles varies: *Diploptera* have 6; *Blatta,* 6; *Leucophaea,* 15 to 20; and *Periplaneta,* 8. The paired lateral oviducts are short. The number of spermathecae also varies. In *Blatta* and *Periplaneta* (Blattidae), it is a two-branched structure. *Blattella* and *Diploptera* have two pairs and *Leucophaea,* one pair. Whenever there are branched spermathecae, one branch is considered to function as a spermathecal gland.[31] For more details on spermathecae, see References 33 and 44.

The accessory or colleterial glands in cockroaches also open into the genital chamber via a single basal tube, and they consist of a number of long tubules whose secretion forms the egg case (ootheca). For details on accessory glands and the formation of the egg case, see References 4, 34, 43, and 45 to 51.

H. NEUROENDOCRINE SYSTEM

Essentially, the insect pars intercerebralis-corpora cardiaca complex is functionally analogous to the vertebrate hypothalamic-hypophyseal complex in terms of coordinating the activities of the nervous and endocrine systems.

Historically, insects have fascinated scientists for their ability to metamorphose, and they have provided some basic insight into the science of endocrinology. As early as 1922, Kopeć[52] found that a chemical factor released from the brain was necessary for pupation in the gypsy moth larva, thus implicating, for the first time in any animal, the nervous system in endocrine control of growth and development. Subsequently, Clever and Karlson[53] provided the first evidence on the mode of action of steroid hormones at the cellular and molecular levels in a fly.

There are four main components in the insect neuroendocrine system: (1) neurosecretory cells in the brain and ganglia of the ventral nerve cord (Figure 22); (2) the corpora cardiaca (paired neurohemal organs that store and release neurosecretions; Figure 23); (3) the ecdysial, molting, or prothoracic glands (Figure 24); and (4) the corpora allata (paired lateral glands; Figure 23). The latter two are nonneural epithelial endocrine glands.[54]

Growth, differentiation, and morphogenesis are regulated by three classes of developmental hormones, viz., the neurohormones, the ecdysones, and the juvenile hormones, produced by the brain and/or the neurohemal organs, the ecdysial glands, and the corpora allata, respectively. In a typical situation, the pars intercerebralis neurosecretory cells of the protocerebrum of the brain secrete a tropic hormone *(ecdysiotropin)* which passes into the hemolymph through the corpora cardiaca. The ecdysial glands, upon stimulation by ecdysiotropin, secrete a morphogenetic hormone (*β-ecdysone*;[54,64] Figure 25) which initiates the molting process. The other morphogenetic hormone, *juvenile hormone* (Figure 26), determines the characteristics of the molt (nymphal or adult). During the early nymphal stages the titer of the juvenile hormone is high, and its gradual decline and eventual "absence" results in the nymphal-adult molt.

The juvenile hormone, however, reappears in the adult cockroach for a variety of different roles (i.e., egg maturation, activity of the reproductive accessory glands, and sex pheromone production and release). The chapters in Part V of this work consider some of these topics in more detail. For details of neurohemal organs and insect endocrinology, one may refer to References 55 to 57.

VI. SUMMARY

This chapter consists of basic information on the external anatomy of the cockroach

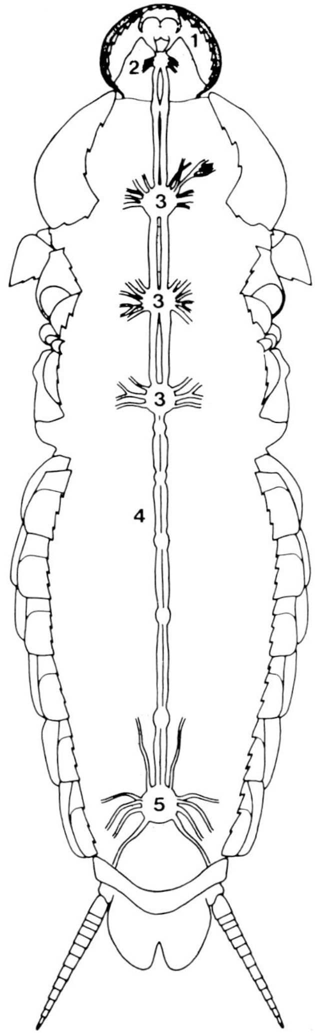

FIGURE 22. Diagram of central nervous system of *P. americana*;[61] 1 = head (brain) region, 2 = subesophageal ganglion, 3 = thoracic ganglia, 4 = ventral nerve chord, 5 = terminal abdominal ganglion.

(including the integument), general and specific instructions for dissecting cockroaches and their various systems, and the internal anatomy, including the digestive, excretory, respiratory, musculature, circulatory, immune, reproductive, and endocrine systems. It is mainly intended for scientists in the biomedical and veterinary sciences who may not be familiar with insect anatomy.

ACKNOWLEDGMENTS

This is the New Jersey Agricultural Experiment Station Publication No. F08125-02-88, supported by state funds and U.S. Hatch funds.

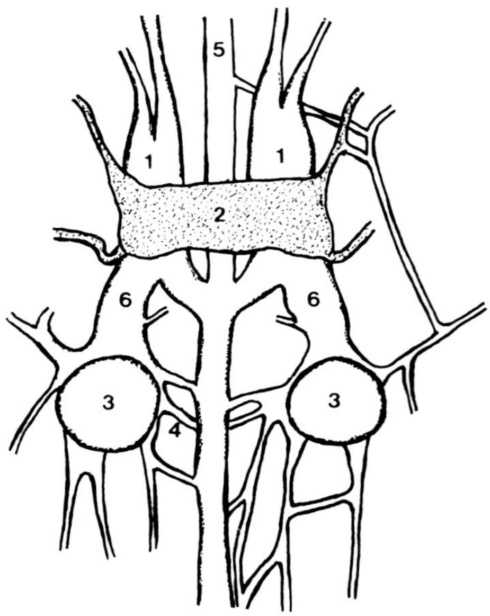

FIGURE 23. Diagram of partial retrocerebral complex of *P. americana*;[12] 1 = corpus cardiacum, 2 = cardiaca-commissural organ (posterior portion removed), 3 = corpus allatum, 4 = allatal commissure, 5 = recurrent nerve, 6 = nervi corporis cardiaci.

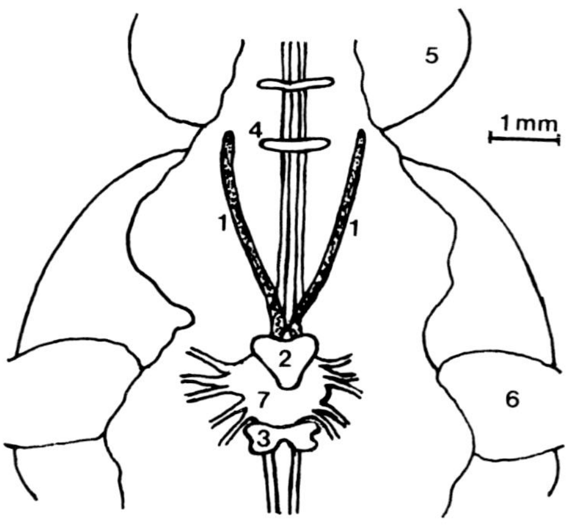

FIGURE 24. Ventral view of partial dissection of cervical and prothoracic region of *P. americana* nymph;[12] 1 = ecdysial glands, 2 = sternum, 3 = poststernum, 4 = cervical sclerite, 5 = head, 6 = coxa, 7 = thoracic ganglion.

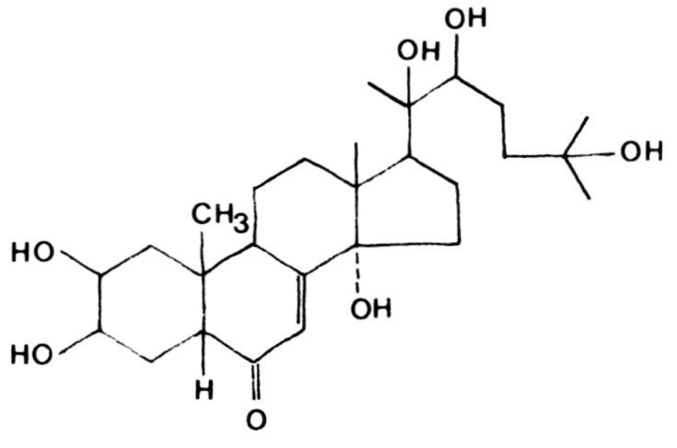

FIGURE 25. Structure of β-ecdysone.

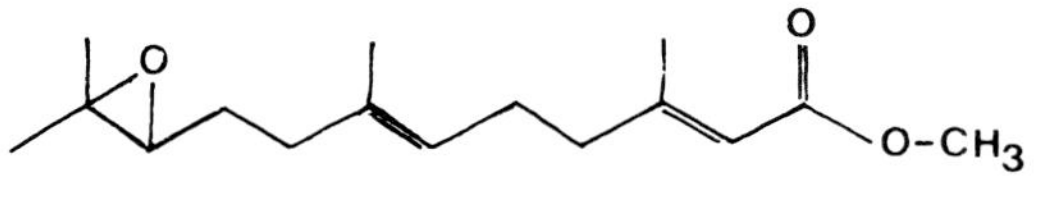

FIGURE 26. Structure of cockroach juvenile hormone: methyl *cis*-10,11-epoxy-3,7,11-trimethyl-*trans*-2,*trans*-6-dodecadienoate.[65]

REFERENCES

1. **Ramaswamy, S. B. and Gupta, A. P.,** Sensilla of the antennae and the labial and maxillary palps of *Blattella germanica* (L.) (Dictyoptera: Blattellidae): their classification, *J. Morphol.,* 168, 269, 1981.
2. **Butler, R.,** The anatomy of the compound eye of *Periplaneta americana* (L.). I. General features, *J. Comp. Physiol.,* 83, 223, 1973.
3. **Butler, R.,** The anatomy of the compound eye of *Periplaneta americana* (L.). II. Fine structure, *J. Comp. Physiol.,* 83, 239, 1973.
4. **Guthrie, D. M. and Tindall, A. P.,** *The Biology of the Cockroach,* Edward Arnold, London, 1968.
5. **Lipke, H., Graves, B., and Leto, S.,** Polysaccharides and glycoprotein formation in the cockroach. II. Incorporation of D-glucose-^{14}C into bound carbohydrate, *J. Biol. Chem.,* 240, 601, 1965.
6. **Mills, R. R.,** Integument, in *The American Cockroach,* Bell, W. S. and Adiyodi, K. G., Eds., Chapman and Hall, New York, 1981, chap. 2.
7. **Hepburn, H. R.,** Structure of the integument, in *Comprehensive Insect Physiology, Biochemistry and Pharmacology,* Vol. 3, Kerkut, G. A. and Gilbert, L. I., Eds., Pergamon Press, Oxford, 1985, chap. 1.
8. **Gupta, A. P.,** Identification key for hemocyte types in hanging drop preparations, in *Insect Hemocytes,* Gupta, A. P., Ed., Cambridge University Press, London, 1979, chap. 17.
9. **Gupta, A. P.,** Cellular elements in the hemolymph, in *Comprehensive Insect Physiology, Biochemistry and Pharmacology,* Vol. 3, Kerkut, G. A. and Gilbert, L. I., Eds., Pergamon Press, Oxford, 1985, chap. 10.
10. **Hazarika, L. K. and Gupta, A. P.,** Anatomy of the retrocerebral complex of *Blattella germanica* (L.) (Dictyoptera: Blattellidae), *Zool. Anz.,* 219(5/6), 257, 1987.

10a. **Das, Y. T. and Gupta, A. P.,** unpublished data, 1988.

11. **Hazarika, L. K. and Gupta, A. P.,** Structure, innervation, persistence, and effects of juvenile hormone on the prothoracic glands in adult *Blattella germanica* (L.) (Dictyoptera: Blattellidae), *Zool. Sci.,* 3, 145, 1986.
12. **Bell, W. S. and Adiyodi, K. G., Eds.,** *The American Cockroach,* Chapman and Hall, New York, 1981.
13. **Bradley, T. J.,** The excretory system: structure and physiology, in *Comprehensive Insect Physiology, Biochemistry and Pharmacology,* Vol. 4, Kerkut, G. A. and Gilbert, L. I., Eds., Pergamon Press, Oxford, 1985, chap. 10.
14. **Mill, P. J.,** Structure and physiology of the respiratory system, in *Comprehensive Insect Physiology, Biochemistry and Pharmacology,* Vol. 3, Kerkut, G. A. and Gilbert, L. I., Eds., Pergamon Press, Oxford, 1985, chap. 13.

15. **Snodgrass, R. E.,** *Principles of Insect Morphology,* McGraw-Hill, New York, 1935.
16. **Carbonell, C. S.,** The thoracic muscles of the cockroach *Periplaneta americana, Smithson. Misc. Collect.,* 107, 1, 1947.
17. **Chadwick, L. E.,** The ventral intersegmental muscles of cockroaches, *Smithson. Misc. Collect.,* 131, 1, 1957.
18. **Shankland, D. L.,** Nerves and muscles of the pregenital abdominal segments of the American cockroach, *Periplaneta americana* (L.), *J. Morphol.,* 117, 353, 1965.
19. **Alsop, D. W.,** Comparative analysis of the intrinsic leg musculature of the American cockroach, *Periplaneta americana, J. Morphol.,* 158, 199, 1978.
20. **Hughes, G. M.,** The coordination of insect movements. I. The walking movements of insects, *J. Exp. Biol.,* 29, 267, 1952.
21. **Yeager, J. F. and Hendrickson, G. O.,** Circulation of blood cells in wings and wing pads of the cockroach, *Periplaneta americana* (Linn.), *Ann. Entomol. Soc. Am.,* 27, 739, 1934.
22. **Miller, T. A.,** Structure and physiology of the circulatory system, in *Comprehensive Insect Physiology, Biochemistry and Pharmacology,* Vol. 3, Kerkut, G. A. and Gilbert, L. I., Eds., Pergamon Press, Oxford, 1985, chap. 8.
23. **Mullins, D. E.,** Chemistry and physiology of the hemolymph, in *Comprehensive Insect Physiology, Biochemistry and Pharmacology,* Vol. 3, Kerkut, G. A. and Gilbert, L. I., Eds., Pergamon Press, Oxford, 1985, chap. 9.
24. **Hazarika, L. K. and Gupta, A. P.,** Variations in hemocyte populations during various developmental stages of *Blattella germanica* (L.) (Dictyoptera: Blattellidae), *Zool. Sci.,* 4, 307, 1987.
25. **Chiang, A. S., Gupta, A. P., and Han, S. S.,** Arthropod immune system. I. Comparative light and electron microscopic accounts of immunocytes and other hemocytes of *Blattella germanica* (Dictyoptera: Blattellidae), *J. Morphol.,* 198, 257, 1988.
26. **Gupta, A. P.,** The identity of the so-called crescent cell in the hemolymph of the cockroach, *Gromphadorhina portentosa* (Schaum) (Dictyoptera: Blaberidae), *Cytologia,* 50, 739, 1985.
27. **Gupta, A. P.,** Arthropod immunocytes: identification, structure, functions, and analogies to the functions of vertebrate B- and T-lymphocytes, in *Hemocytic and Humoral Immunity in Arthropods,* Gupta, A. P., Ed., John Wiley & Sons, New York, 1986, chap. 1.
28. **Gupta, A. P., Ed.,** *Hemocytic and Humoral Immunity in Arthropods,* John Wiley & Sons, New York, 1986.
29. **Han, S. S. and Gupta, A. P.,** Arthropod immune system. V. Activated immunocytes (granulocytes) of the German cockroach, *Blattella germanica* (L.) (Dictyoptera: Blattellidae) show increased number of microtubules and nuclear pores during immune reaction to foreign tissue, *Cell Struct. Funct.,* 13, 333, 1988.
30. **Han, S. S. and Gupta, A. P.,** Arthropod immune system. II. Encapsulation of implanted nerve cord and "plain gut" surgical suture by granulocytes of *Blattella germanica* (L.) (Dictyoptera: Blattellidae), *Zool. Sci.,* 6, 303, 1989.
31. **Matsuda, R.,** *Morphology and Evolution of the Insect Abdomen,* Pergamon Press, Oxford, 1976.
32. **Khalifa, A.,** Spermatophore production in *Blattella germanica* (L.) (Orthoptera: Blattidae), *Proc. R. Entomol. Soc. London,* A25, 53, 1950.
33. **McKittrick, F. A.,** Evolutionary Studies of Cockroaches, Cornell Univ. Exp. Stn. Mem. No. 389, Cornell University, Ithaca, NY, 1964.
34. **Roth, L. M.,** Evolution and taxonomic significance of reproduction in Blattaria, *Annu. Rev. Entomol.,* 15, 75, 1970.
35. **Pipa, R. L.,** Muscles and innervation of the genital and postgenital abdominal segments of the female American cockroach *Periplaneta americana* (L.) (Dictyoptera: Blattidae), *Int. J. Insect Morphol. Embryol.,* 16, 17, 1987.
36. **Pipa, R. L.,** Muscles and nerves of the posterior abdomen and genitalia of the male *Periplaneta americana* (L.) (Dictyoptera: Blattidae), *Int. J. Insect Morphol. Embryol.,* 17(6), 455, 1988.
37. **Louis, D. and Kumar, R.,** Morphology and histology of the mushroom-shaped gland in some Dictyoptera, *Ann. Entomol. Soc. Am.,* 64, 977, 1971.
38. **Ito, H.,** Contribution histologique et physiologique à l'étude des annexes des organes genitaux des Orthoptérès, *Arch. Anat. Microsc.,* 20, 343, 1924.
39. **Zabinski, J.,** Copulations extérieure chez les Blattes, *C.R. Soc. Biol.,* 112, 596, 1933.
40. **Gupta, B. L.,** On the structure and function of spermatophore in the cockroach, *Periplaneta americana* (Linn.), *Indian J. Entomol.,* 8, 79, 1947.
41. **Van Wyk, L. E.,** The morphology and histology of the genital organs of *Leucophaea maderae* (L.), *Entomol. Soc. S. Afr.,* 15, 3, 1952.
42. **Davey, K. G.,** The evolution of the spermatophore in insects, *Proc. R. Entomol. Soc. London,* A35, 107, 1960.

43. **Cornwell, P. B.,** *The Cockroach,* Vol. 1, Hutchinson Press, London, 1968.
44. **Gupta, B. L. and Smith, D. S.,** Fine structural organization of the spermatheca in the cockroach, *Periplaneta americana, Tissue Cell,* 1(2), 295, 1969.
45. **Brunet, P. C. J.,** The formation of the ootheca by *Periplaneta americana, Q. J. Microsc. Sci.,* 92, 113, 1951.
46. **Hagan, H. R.,** *Embryology of the Viviparous Insect,* Ronald Press, New York, 1951.
47. **Roth, L. M.,** The evolutionary significance of rotation of the ootheca in Blattaria, *Psyche,* 74, 85, 1967.
48. **Roth, L. M.,** Water changes in cockroach oothecae in relation to the evolution of ovoviviparity and viviparity, *Ann. Entomol. Soc. Am.,* 60, 828, 1967.
49. **Roth, L. M.,** Uricose glands in accessory sex gland complex of male Blattaria, *Ann. Entomol. Soc. Am.,* 60, 1203, 1967.
50. **Roth, L. M.,** Oothecae of Blattaria, *Ann. Entomol. Soc. Am.,* 61(1), 83, 1968.
51. **Roth, L. M.,** Additions to the oothecae, uricose glands of Blattaria, *Ann. Entomol. Soc. Am.,* 64, 127, 1971.
52. **Kopeć, S.,** Studies on the necessity of the brain for the inception of insect metamorphosis, *Biol. Bull.,* 42, 322, 1922.
53. **Clever, U. and Karlson, P.,** Induktion von Puff-veranderungen in den Speicheldrusen Chromosomen von *Chironomus tentans* durch Ecdyson, *Exp. Cell Res.,* 20, 623, 1960.
54. **Doane, W. W.,** Role of hormones in insect development, in *Developmental Systems: Insects,* Vol. 2, Counce, S. J. and Waddington, C. H., Eds., Academic Press, New York, 1973, chap. 4.
55. **Gupta, A. P., Ed.,** *Neurohemal Organs of Arthropods,* Charles C Thomas, Springfield, IL, 1983.
56. **Kerkut, G. A. and Gilbert, L. A., Eds.,** *Comprehensive Insect Physiology, Biochemistry and Pharmacology,* Vol. 7 and 8, Pergamon Press, Oxford, 1985.
57. **Gupta, A. P., Ed.,** *Morphogenetic Hormones in Arthropods,* Vol. 1 to 3, Rutgers University Press, New Brunswick, NJ, 1989.
58. **Matsuda, R.,** Morphology and evolution of the insect thorax, *Mem. Entomol. Soc. Can.,* No. 76, 1, 1970.
59. **Snodgrass, R. E.,** Morphology and mechanism of the insect thorax, *Smithson. Misc. Collect.,* 80(1), 1, 1927.
60. **Das, Y. T. and Gupta, A. P.,** Nature and precursors of juvenile hormone-induced excessive cuticular melanization in German cockroach, *Nature (London),* 268(5616), 139, 1977.
61. **MacLeod, E. G.,** *The Biology of Insects: Laboratory Manual,* Burgess Publishing, Minneapolis, 1972.
62. **Nutting, W. L.,** A comparative anatomical study of the heart and accessory structures of the orthopteroid insects, *J. Morphol.,* 89, 501, 1951.
63. **Richards, A. G.,** Ventral diaphragms of insects, *J. Morphol.,* 113, 17, 1963.
64. **King, D. S. and Marks, E. P.,** The secretion and metabolism of α-ecdysone by cockroach (*Leucophaea maderae*) tissue *in vitro, Life Sci.,* 15, 147, 1974.
65. **Muller, P. J., Masner, P., Trautman, K. H., Suchy, M., and Wipf, H. K.,** The isolation and identification of juvenile hormone from cockroach corpora allata *in vitro, Life Sci.,* 15, 915, 1974.
66. **Wyman, R. J.,** A beginners guide to microdissection: *Drosophila* style, *Fine Sci. Points,* 1, 1, 1987.

Section II. Nervous System

Chapter 4

ANATOMY OF THE CENTRAL NERVOUS SYSTEM AND ITS USEFULNESS AS A MODEL FOR NEUROBIOLOGY

Charles R. Fourtner and Charles J. Kaars

TABLE OF CONTENTS

I. INTRODUCTION

The nervous system of the cockroach has been and continues to be an excellent experimental model for a variety of physiological, pharmacological, toxicological, and behavioral investigations. The chapters of this work provide ample testimony to the usefulness of the cockroach nervous system as a model for the discovery and verification of phenomena of general biological significance. It is our intent to provide a brief introduction to the overall gross morphology of the nervous system in the cockroach, to discuss some of the neuronal components of the cockroach nervous system, to describe the neuronal circuits underlying some behavioral models, and, finally, to describe the glial sheath around the central nervous system (CNS) and some of the interesting changes that occur in the glial elements during peripheral regeneration. Where gross and cellular morphologies are presented elsewhere in this work, we will direct the reader to the appropriate figures.

II. GROSS MORPHOLOGY OF THE CENTRAL NERVOUS SYSTEM

The cockroach CNS is comprised of two basic gross morphological elements: ganglia and connectives. Ganglia are aggregations of neurons. The neuronal cell bodies form a peripheral rind (cortex). The local processes of these neurons and the processes of neurons whose somata are located in other ganglia and in the periphery form a central neuropil. Neuronal processes are found within two morphologically distinct regions: tracts and neuropil. Tracts are regions in which all the neuronal processes have the same general orientation; they are virtually free of synaptic contacts. Neuropil is characterized by processes which have no general orientation, but in which synaptic contacts are made between processes. Connectives are the central neural structures which join adjacent ganglia, and tracts are the neural pathways within a ganglion which are continuous with anterior and posterior connectives. The laterally running neural pathways within a ganglion are known as commissures and are the major communications link between the segmental (right and left) hemiganglia. Except for regions of synaptic contacts, neurons and their processes are invested with nonneural glial cells. The entire CNS is enveloped in a sheath composed of glia and noncellular elements.

Embryologically, each segment develops a protoganglion. During embryonic development there occurs considerable fusion of protoganglia, with the result that the postembryonic nervous system has 11 principal ganglia: 2 in the head (the supra- and subesophageal ganglia, which are joined by the circumesophageal connectives), 3 in the thorax (the pro-, meso-, and metathoracic ganglia), and 6 in the abdomen (for review see References 1 to 3).

A. THE CEPHALIC NERVOUS SYSTEM

The supraesophageal ganglion is primarily a region for integration of sensory information, as well as sensory-to-motor synaptic interactions, initiation of locomotory commands, and the integration and initiation of neuroendocrine reflexes. The supraesophageal ganglion is the product of fusion of an uncertain number of segmental ganglia (most likely three) during embryonic development. Three distinct regions of the supraesophageal ganglion can be recognized. Figure 1 illustrates the gross morphology and the nerves arising from the cephalic nervous system.

1. Protocerebrum

The protocerebrum, the largest and most anterior region of the supraesophageal ganglion, receives input from the ocelli and compound eyes. The neural pathways and projections from the ocelli and compound eyes to this region are discussed in more detail in Chapter

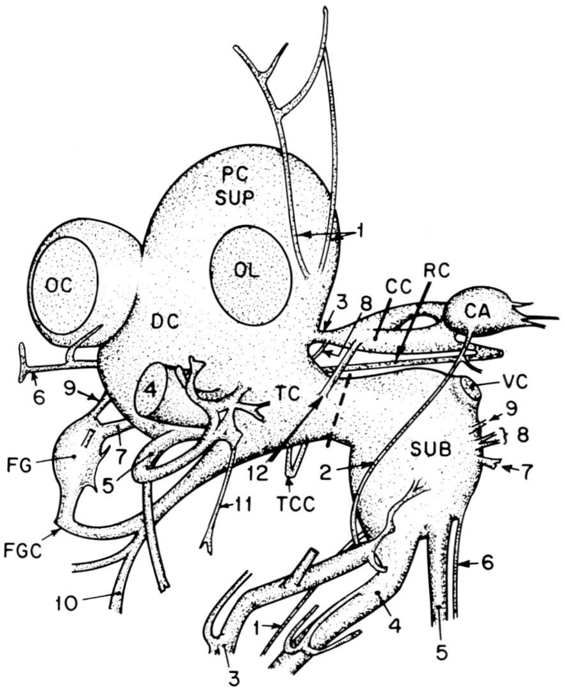

FIGURE 1. A schematic illustration of the left lateral aspect of the cephalic region of the nervous system of *Periplaneta americana,* with the peripheral projections of the optic lobe (OL) and the ocellar nerve (OC) severed. SUP, supraesophageal ganglion; SUB, subesophageal ganglion; CA, corpus allatum; CC, corpus cardiacum; FG, frontal ganglion; FGC, frontal ganglion connective; RC, recurrent nerve; TCC, tritocerebral commissure; VC, ventral connective; PC, protocerebral region; DC, deutocerebral region; TC, tritocerebral region. The dashed line designates the circumesophageal connectives joining the supra- and subesophageal ganglia. For a description of the distribution of the peripheral nerves, see Tables 1 and 2. (From Pipa, R. and Delcomyn, F., in *The American Cockroach,* Bell, W. J. and Adiyodi, K. G., Eds., Chapman and Hall, London, 1981, 175. With permission.)

23 of this work. Also within the protocerebrum are the pars intercerebralis and the pars lateralis, which play an extremely important role in neuroendocrine function; from these regions emerge the major neuronal processes that innervate the corpora cardiaca (CC), an organ which serves as a neurosecretory store analogous to the anterior pituitary in vertebrates.

Using a variety of histological stains and cobalt sulfide backfilling procedures it is clear that a number of cells along the medial border of the pars intercerebralis send axonal processes into the CC nerves 1 and 2. For further discussion on this literature see the review of Pipa and Delcomyn,[2] and for a general description of the gross morphology of the supraesophageal/neuroendocrine pathways see Chapter 5 in this work. Krauthamer,[4] using electrophysiological procedures and injection of horseradish peroxidase (HRP), recently demonstrated that there are two types of cells within the pars intercerebralis. Type 1 is clearly a neurosecretory type of neuron and sends axonal processes to the CC nerves. Type 2, a nonneurosecretory type cell, is larger in diameter, with its axonal process projecting to the circumesophageal connectives. Since these cells can be clearly defined by their physiological properties, they represent a good model for the study of neurosecretory cells having their somata within the

CNS. Thompson and Tobe (see Chapter 5 of this work) have reviewed the cellular morphology of the neural elements associated with the supraesophageal/neuroendocrine pathways in the cockroach *Diploptera punctata*.

The largest structures within the supraesophageal ganglion are the mushroom bodies (the corpora pedunculata). Each mushroom body is subdivided into two lobes (a and b) and two calyces, a lateral and a medial (see Figure 3 in Chapter 26 of this work). Although there have been a substantial number of studies regarding the structure of the mushroom bodies, of the neuronal projections into and out of these highly organized structures, and of the neuronal types within the pedunculata, there is still considerable debate as to the function of these dominant structures. Numerous hypotheses regarding their function have been proposed, from serving a purely olfactory purpose (e.g., olfactory memory[5]) to being a highly complex integrating region for parallel processing of visual and olfactory information.[6]

There have been, as yet, few studies on the physiology of thse structures in the cockroach. Due to the projections from the antennal afferents, Weiss[7] has argued that the mushroom bodies may be a region of olfactory integration.

2. Deutocerebrum

Posterior to the protocerebrum is the deutocerebrum, the principal center for the antennal olfactory input which is processed in the many neuropilar regions or glomeruli. Output from the glomeruli projects to the mushroom bodies of the protocerebrum. The morphology of the olfactory pathways and the neuronal projections is summarized and discussed by Seelinger in Chapter 26 of this work.

3. Tritocerebrum

The most posterior region of the supraesophageal ganglion is the tritocerebrum, which is intimately associated with the frontal ganglion, the primary ganglion of the stomatogastric (also termed the stomodaeal or vegetative) nervous system. In addition to receiving input directly from the frontal ganglion, it also receives input from the hypocerebral ganglion via the frontal ganglion. The hypocerebral ganglion is a small ganglion of the stomatogastric system through which other, more caudal stomodaeal inputs/outputs pass and is connected to the frontal ganglion by the recurrent nerve of the stomodaeal nervous system.[8] Stefano et al., in Chapter 16 of this work, further consider some of the pharmacological studies involving the stomatogastric nervous system, particularly the posterior recurrent nerve.

There is a large neuropilar area of the tritocerebrum which may function as an area for integration of sensory input from the mouth parts[9] and the antennal neuropil of the deutocerebrum. It also appears to be another important neuroendocrine pathway, since the third nerve (NCC III) projecting to the CC arises from the tritocerebrum; however, it is not clear if any of the axons exiting NCC III originate in the tritocerebrum. For a general discussion of the anatomical arrangements of the neuroendocrine and stomodaeal projections to the supraesophageal ganglion, see Chapter 5 in this work and the review of Pipa and Delcomyn.[2]

4. Subesophageal Ganglion

The other principal component of the cephalic nervous system is the subesophageal ganglion, the primary motor control system for the muscles operating the mouth parts (the mandibles, maxillae, and labium). While it most likely results from the fusion of three embryonic segments, its origins remain uncertain due to the complexity of its neuropilar structure and the few detailed studies of its structure. The subesophageal ganglion may also function as a site of origin for command elements responsible for controlling locomotory behavior, since there is behavioral evidence that some inhibitory influences on reflexes are removed when the thoracic ganglia are isolated from the subesophageal ganglion.[10] The gross anatomy of the subesophageal ganglion is depicted in Figure 1.

The subesophageal ganglion of the locust has been studied extensively during the past decade. In a recent review of that literature, Altman and Kien[11] describe the gross and cellular morphology and discuss the role of the subesophageal ganglion in maintaining motor excitability, but more interestingly as a cross (left-right) comparator aiding motor coordination. Although the cockroach CNS would be amenable to extensive studies of the subesophageal ganglion, only a few of the cellular components of this structure have been identified.[12]

Figure 1 also illustrates the major nerve branches arising from the supra- and subesophageal ganglia, and the legend provides information concerning the primary sites innervated by the cephalic nerves.

A very recent compendium of papers on the arthropod brain[13] should serve as a useful description for those wishing to compare the cockroach supra- and subesophageal ganglia to those of other arthropods.

B. THE THORACIC NERVOUS SYSTEM

The thoracic CNS of the cockroach is composed of three ganglia (pro-, meso-, and metathoracic) joined by their paired connectives. Each of these ganglia is responsible for processing and integrating sensory information from and controlling motor output to their appropriate thoracic segments. The pro- and mesothoracic ganglia each gives rise to eight bilaterally symmetrical pairs of nerves. The metathoracic ganglion, a composite of the embryonic ganglia of the third thoracic and first abdominal segments, gives rise to eight pairs of segmental nerves, with nerve 8 innervating the spiracle of the first abdominal segment. Figure 2 illustrates the locations and branching patterns of the nerves arising from the mesothoracic ganglia.

The nomenclature for the peripheral nervous system is a simple alphanumeric system in which each branch (ramus) arising from a segmental nerve is numbered in sequential order, ramus number 1 being the most centrally located. For example, the first ramus branching from segmental nerve 5 is labeled 5r1. In those cases in which a nerve decussates into two or three branches of similar diameter an alpha designation is assigned; for example, segmental nerve 3 decussates into two nerves of similar size which are labeled nerve 3A (anterior) and nerve 3B (posterior). Branching patterns of the individual rami in the periphery are noted by lower-case alpha designations. For example, the nerve innervating one of the muscles located in the coxa of a walking leg is designated 5r1a.[14,15]

For a brief description of the segmental nerves see Tables 1 to 3, and for a more complete description of the peripheral distribution of the nerves and the approximate number of axonal profiles contained in each nerve and subsequent branch see the review by Guthrie and Tindall[1] and the series of papers by Dresden and Nijenhuis[14-16] and by Pipa et al.[17,18]

C. THE ABDOMINAL NERVOUS SYSTEM

Within the abdomen there are six ganglia, identified as abdominal ganglia 1 through 6. Abdominal ganglion 1 is the ganglion of the second embryonic segment, and ganglion 6 results from the fusion of ganglia originating in embryonic segments 7 to 11. The gross anatomy of the five anterior ganglia is similar in a given species of cockroach, with two bilateral segmental pairs, termed Nerve 1 (dorsal) and Nerve 2 (ventral). In some species there is a sharp distinction between Nerves 1 and 2, but in others fusion can occur at the ganglion or at various distances from the ganglion. In addition, in some species there is a very small transverse nerve which projects to the periphery and primarily contains neurosecretory axons. Nerve 1 innervates the spiracles, with Nerve 2 innervating the abdominal musculature.[18a]

The description of the gross anatomy of the sixth abdominal ganglion is presented by Hue and Callec in Chapter 8 of this work, which is on synaptic mechanisms in the CNS.

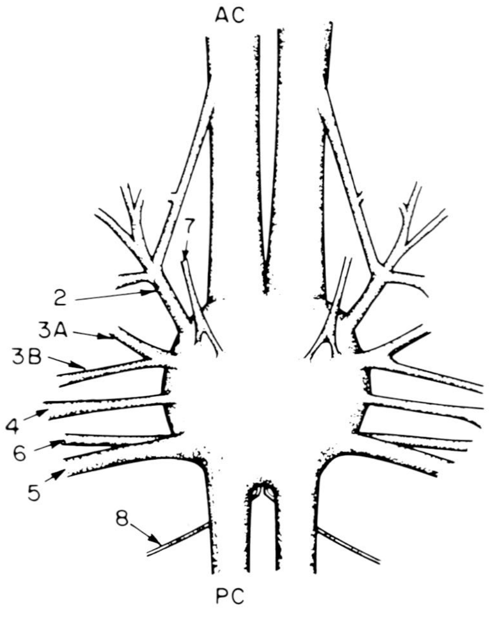

FIGURE 2. A schematic illustration of the mesothoracic ganglion of the cockroach *Periplaneta americana*, in ventral view. AC, anterior connective; PC, posterior connective. The numbers 2 through 8 designate the segmental nerves from anterior to posterior consecutively. The anterior connective is considered the first nerve emanating from the thoracic ganglia. For a distribution of the peripheral nerves, see Table 3.

TABLE 1
Peripheral Nerves Arising from Supraesophageal Ganglion

Nerve #	Name	Type[a]	Region
1	Dorsal integumentary	Sen	Vertex and frons
2	Ocellar	Sen	Ocelli
3	Dorsal cardiac	Mot	Corpus cardiacum
4	Antennal	Sen/mot	Antenna
5	Dorsal deutocerebral	Mot/sen	Extrinsic dorsal muscles
6	Ventral	Mot/sen	Antenna
7	Median frontal	Connective	Frontal ganglion
8	Main cardiac (NCC II)	Mot	Corpus cardiacum
9	Lateral frontal	Connective	Frontal ganglion
10	Labral	Sen/mot	Clypeus/labrum
11	Lateral integumentary	Sen	Genal region
12	Ventral cardiac (NCC III)	Mot	Corpus cardiacum

[a] Sen = sensory, mot = motor.

III. MICROSCOPIC ORGANIZATION OF THE CENTRAL NERVOUS SYSTEM

We will limit our discussion here primarily to the thoracic ganglia. A comprehensive study on the microscopic anatomy of the brain, comparable to that of the honeybee,[6] has

TABLE 2
Peripheral Nerves Arising from Subesophageal Ganglion

Nerve #	Name	Type[a]	Region
1	Hypopharyngeal	Mot/sen	Hypopharynx
2	Corpus allatum	Mot	Corpus allatum
3	Mandibular	Mot/sen	Mandibular muscles, hypopharynx
4	Maxillary	Mot/sen	Maxilla
5	Labial	Mot/sen	Labrum
6	Cervical-1	Mot	Ligula neck muscles
7	Cervical-2	Mot/sen	Ligula neck muscles
8	Cervical-3	Mot	Ligula neck muscles
9	Prothoracic gland	Mot	Prothoracic gland

[a] Mot = motor, sen = sensory.

TABLE 3
Peripheral Distribution of Nerves Arising from Thoracic Ganglia

Nerve #	Type[a]	Innervation
2A	Sen	Ventral thoracic musculature
2B	Sen/mot	Wings, meso and meta; neck muscles, pro
2C	Mot	Dorsal thoracic and neck musculature
3A	Mot	Coxal and thoracic musculature
3B	Sen/mot	Primary sensory nerve to coxa; coxal and femoral musculature
4	Mot	Bifunctional muscles which originate in thorax and insert on trochantin
5	Sen/mot	Coxal, femoral, and distal leg musculature and sense organs
5r8	Sen	Large sensory nerve to distal leg segments
6A	Mot	Tergal and coxal musculature
6B	Mot/sen	Coxal; muscles/meron
7	Sen	Basisternum
8	Sen/mot	Spiracle

Note: For a more complete description of specific rami, see References 12, 14—18, 35, and 78.

[a] Sen = sensory, mot = motor.

not been published. There are descriptions of a few cellular pathways in the supraesophageal ganglion and some of the neurons associated with those pathways elsewhere in this work (see Chapter 23). One recent study on the morphology of third-order ocellar interneurons in the protocerebrum is a particularly excellent example of this type of neuronal-anatomical research in the cockroach.[19,20] Figure 3 illustrates one of those interneurons (OL-II) which projects to the optic lobes on both sides of the brain and is included as an example of the morphological complexity and of the distance over which an interneuron in the supraesophageal ganglion can project.

The pathways in the thoracic ganglia originally described by Pipa et al.[18] and recently in more detail by the superb studies of Gregory[21,22] have proven to be exceptionally useful for those studying the central neural components responsible for a variety of behaviors.

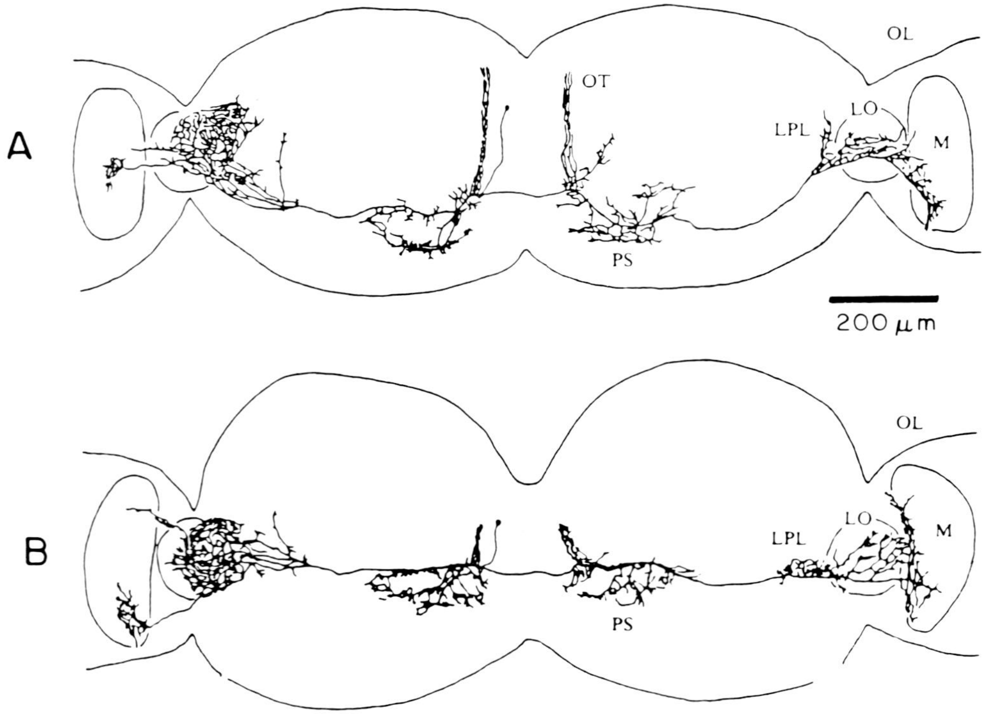

FIGURE 3. The microscopic anatomy of an interneuron (OL-II, or optic lobe neuron II) located in the protocerebrum. The cell body is located in the pars intercerebralis, and its processes form an extensively branching arborization in both optic lobes and in the area of both ocellar tracts (OT). PS, posterior slope of protocerebrum; M, medulla; LO, lobula; LPL, lateral protocerebral lobe. (A) Dorsal view; (B) posterior view. (From Mizunami, M. and Tateda, H., *J. Exp. Biol.*, 125, 57, 1986. With permission.)

Using a variety of silver staining procedures, Gregory has clearly defined the regions of longitudinal tracts, commissures, and synaptic interaction within the neuropil of the thoracic ganglia. Thoracic ganglia consist of a rind two to seven cells in thickness surrounding the neuropilar areas, along with the tracts and commissures connecting the neuropilar areas.

The thoracic ganglia contain seven longitudinal tracts connecting each anterior connective to its corresponding posterior connective. These are illustrated in Figure 4 and are labeled as follows: (1) the dorsal tract, which consists of medial and lateral regions (the MDT and LDT, respectively); (2) the dorsal intermediate tract (DIT); (3) the dorsal medial tract (DMT); (4) the ventral lateral tract (VLT); (5) the ventral medial tract (VMT); (6) the ventral intermediate tract (VIT); and (7) the ventral tract, which consists of five (inner and outer) lateral bundles and three medial bundles (o.LVT, i.LVT, MVT1, MVT2, and MVT3).

There are also ten commissures which connect the right and left hemiganglia; six are located dorsally and are labeled DCI to DCVI (anterior to posterior), while four are ventral: (1) ventral commissure I (VCI), (2) ventral commissural loop II (VCLII), (3) posterior ventral commissure (PVC), and (4) supramedian commissure (SMC).

Gregory[21] also charted the projections of the segmental nerves as they enter the ganglion and invade the appropriate neuropilar and commissural regions. For example, the fibers in nerve 2 divide into two roots, dorsal and ventral, after they enter the ganglion; nerve 3 divides into seven roots, 4 into two, 5 (which is the largest) into eleven identifiable roots, 6 into four, and 7 into three. Nerve 8 is formed from a single root. From these data he was able to identify the specific projections for a number of identified motoneurons (see below). Whereas the commissures and the connective tracts are primarily areas of nonsynaptic

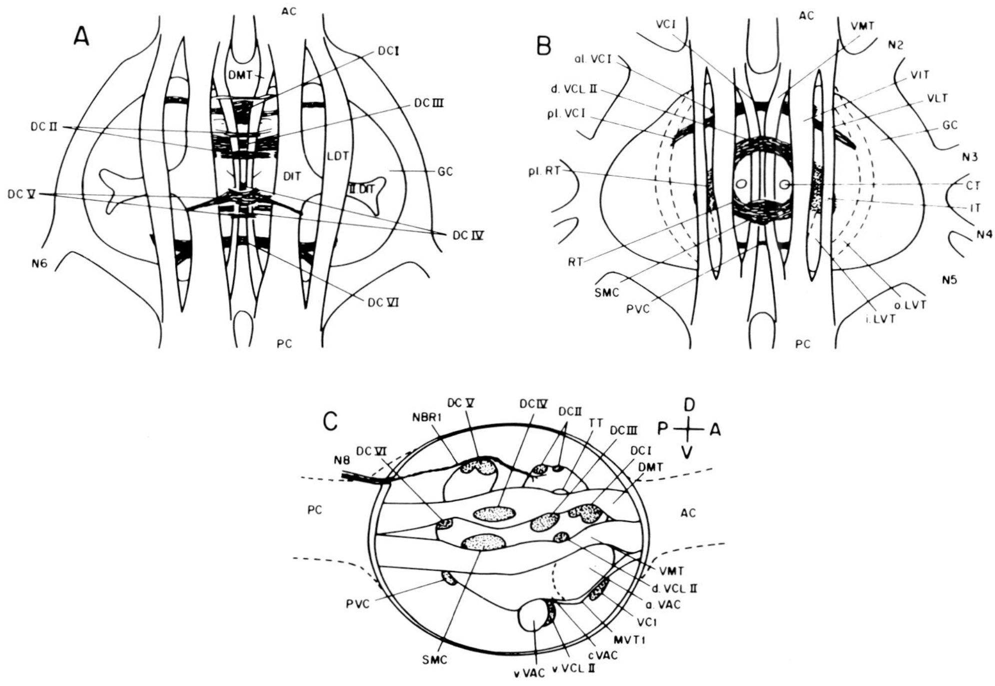

FIGURE 4. Schematic illustration of the tracts and commissures in the mesothoracic ganglion. (A) frontal section, dorsal view; (B) frontal section, ventral view; (C) midsaggital view. See text (Section III) for description of the tracts and commissures and the corresponding abbreviations not listed here. AC, anterior connective; PC, posterior connective; GC, ganglionic core; RT, ring tract; pl.RT, posterior limb of ring tract; IT, I-tract; CT, C-tract; TT, T-shaped tract; ll.DIT, lateral limb of dorsal intermediate tract; al.VCI and pl.VCI, anterior and posterior limbs of ventral commissure I, respectively; d.VCLII and a.VCII, dorsal and anterior regions of ventral commissural loop II, respectively; N8R1, nerve 8 root 1; a.VAC, c.VAC, and v.VAC, anterior, cylindrical, and posterior region of the ventral association centers, respectively. (From Gregory, G. E., *Philos. Trans. R. Soc. London,* 267, 421, 1974.With permission.)

interaction, the nerve roots are very probably the regions of significant synaptic input to motoneurons.[21]

Gregory[21] has also extended the previous morphological work on the location and distribution of somata. By using tracts, commissures, and glial invaginations into the somatal rind, he has grouped somata by their location in the ganglion. He has arranged the somata of the mesothoracic ganglion into approximately 25 groups and numerous subgroups. There are two anterior lateral groups (with a possible third), four anterior ventral, one anterior dorsolateral, four anterior ventrolateral, two posterior lateral, two posterior ventrolateral, and three posterior ventral groups. The remaining somata are grouped in regions on the dorsal and ventral midline and are described in detail with their neuritic projections into the various tracts and commissures of the ganglion.

IV. STRUCTURE AND FUNCTION OF IDENTIFIED NEURONS IN VENTRAL NERVE CORD

A. TECHNICAL CONSTRAINTS IN PHYSIOLOGICAL STUDIES OF THE NERVOUS SYSTEM

Anderson et al. (Chapter 10 of this work) have described the various techniques for electrophysiological recording that are currently in use for investigating neural activity in the central and peripheral nervous systems of the cockroach. Some additional procedures which are generally appropriate follow.

Animals should first be anesthetized with cold or carbon dioxide (a 5-s exposure to CO_2 is usually sufficient to keep animals quiet for a minute or two). Regardless of whether a relatively intact or highly dissected preparation is used as an experimental model, the dorsal surface of the animal should first be opened and the gut removed. If the gut is left in place, any perforation of the midgut will lead to the release of enzymes which will cause rapid deterioration of the centrally generated neural responses. The gut can be extracted by gently pulling it with two pairs of forceps, thus separating the gut at both the oral and anal ends. When the gut is removed carefully, the entire reproductive system will also be extracted.

Removal of the exoskeleton must be done with care, particularly when exposing the ventral surface of the animal, since it is extremely important to keep the tracheal system and oxygen flow intact. In fact, Krauthamer[23] devised an interesting experimental system in his investigations of the neurosecretory cells of the pars intercerebralis. To study the cells *in situ*, he used an isolated head in which the main neck tracheae were cannulated and the cephalic tracheal system was perfused with humidified O_2. This procedure maintained the supra- and subesophageal ganglia in viable condition for up to 30 h.

One further constraint on the experimental protocols associated with investigation of the CNS *in situ* is use of the proper saline. For experiments in which the neurolemma surrounding the nervous system is left intact, any one of a variety of ionic compositions can be utilized. For example, for *Periplaneta americana* we suggest a MOPS (morpholinopropane-sulfonic acid)-buffered saline (124 m*M* NaCl, 10 m*M* KCl, 5 m*M* $CaCl_2$, 1 m*M* $MgCl_2$, 3 m*M* MOPS, 3.9 m*M* MOPS sodium salt [pH 7.21], and 40 m*M* sucrose[27]). The MOPS buffer allows a higher calcium concentration than the standard phosphate buffers and also permits oxygenation of the saline with little change in pH. The sucrose is added for proper osmotic balance; it should be added immediately preceding the experiment to reduce the possibility of the growth of microorganisms, which often produce toxins.

For experiments in which the neurolemma must be removed, we suggest a saline with a composition similar to the ionic environment surrounding the neural elements of the CNS;[25-27] for example, 124 m*M* NaCl, 10 m*M* KCl, 8 m*M* $MgCl_2$, 20 m*M* $CaCl_2$, 3 m*M* MOPS, 3.9 m*M* MOPS sodium salt, and 40 m*M* sucrose.

B. MOTONEURONS

The cockroach nervous system has proven to be a superb experimental model for analysis of the morphology of single, physiologically identified neurons. The original attempts at mapping the motoneurons of the meso- and metathoracic ganglia in cockroaches were performed using various staining procedures,[28,29] the most useful being that of Young,[29] which involved staining the perinuclear RNA. Figure 5 is an example of such a map for the mesothoracic ganglia. With the advent of the use of Procion dyes,[30] either pressure injected or iontophoresed into somata, several investigators began to study not just location of the motoneuronal somata, but the three-dimensional structure of the neuropilar arborization of the physiologically identified motoneurons. The study of the morphology of the processes of identified neurons was advanced tremendously through the work of Pitman et al.,[31] who demonstrated the utility of iontophoresing cobaltous ions directly into the cell body of a physiologically identified motoneuron, D_f, the motoneuron which produces the very rapid twitch of the femur. This procedure proved more useful because the cobaltous ion moved throughout the cell much more quickly than the less mobile Procion dyes. Standard microscopy could then be utilized for viewing the cellular morphology, thus permitting a more complete description of the entire neuronal arbor of the injected motoneurons (see also Chapter 11 in this work).

A typical motoneuron consists of a rather large unipolar cell body, generally between

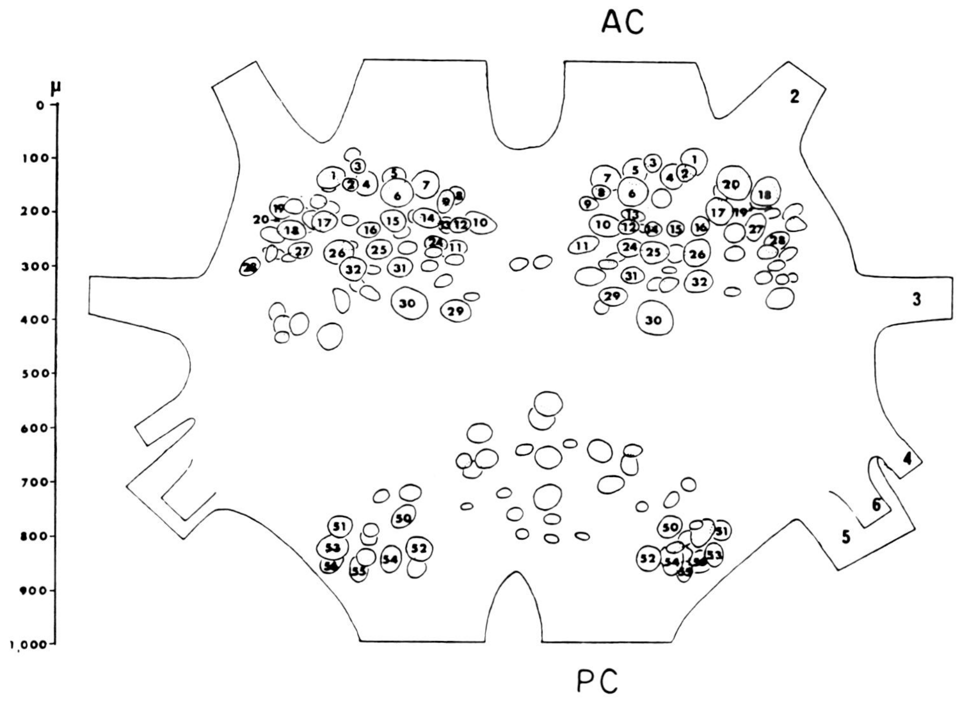

FIGURE 5. A map of the position of neural somata in the mesothoracic ganglion of the cockroach. The somata placement was sufficiently consistent from one preparation to another to enable some of the profiles to be identified and numbered. This type of map may prove beneficial for further studies of identified neurons, particularly motoneurons, in the CNS. AC, anterior connective; PC, posterior connective. (From Young, D., *J. Insect Physiol.*, 15, 1175, 1969. With permission.)

20 and 50 μm in diameter. Arising from the cell body is a very thin fiber which may run for several tens of microns and may expand to as large as 5 μm in diameter. This expanded fiber, the neuropilar segment,[32] is generally located in one of the nerve roots described by Gregory;[21] from it originate a large number of smaller processes which may extend to many regions of the neuropil ipsilateral to the soma. For an exquisite view of the complexity of the processes of one motoneuron, D_s, see Figure 9A in Chapter 14 of this work. The peripheral axon also arises from this expanded process. Near the edge of the ganglion there is generally a slight reduction in the diameter of the axon where the axon exits the neuropilar region; this constricture may be the low-threshold initial segment of the motor axon.

The expanded region (neuropilar segment) of the motoneuronal neuritic field is a neurophysiologist's dream, since it is possible to penetrate this region of an identified neuron and record synaptic events, particularly those which influence the nearby spike-initiating zone. In original experiments on insect central nervous elements, recordings were obtained primarily from the somata which were spatially and electrically isolated from the extensive neuronal arbor.[33]

The cell bodies of the motoneurons are generally located on the ventral and ventral lateral surfaces of the ganglion, ipsilateral to the nerve root of origin for the motor axon. There are two notable exceptions to this general rule. First, the cell body of motoneuron D_s, innervating the slower muscles which extend the femur, is located on the ipsilateral dorsoposterior surface of the thoracic ganglia[34] in the posterior lateral group. Second, the cell body of the motoneuron innervating the dorsal longitudinal flight muscle is located on the contralateral ventroanterior surface of the ganglion.[12,35]

In addition to serving as a model for the analysis of the morphology of identified motoneurons, the cockroach CNS has been an experimental model for the study of populations of motoneurons innervating individual muscles and even subgroups of muscles of known fiber type.[36] Cobalt backfilling is the technique most usefully applied for such studies. This technique was originally used to identify the number and distribution of the common inhibitory motoneurons which innervate several of the muscles of the cockroach leg.[37] Two types of inhibitory neurons were found, one termed the widespread common inhibitory neuron since its axon projects into the segmental nerves 3, 4, 5 (via 4), and 6. Its soma is located in the posterior ventral medial group. The other type of inhibitor is the local inhibitory neuron, the axon of which exits through nerve 5, but bifurcates almost immediately after leaving the ganglion, sending an axonal branch to nerve root 4 through the small anastomosis connecting segmental nerves 4 and 5. The somata of the local common inhibitors are located within the posterior ventral groups.

Several studies have revealed the location and structure of the somata and neuronal processes of several motoneurons innervating the thoracic musculature of the cockroach.[35-36,38] Pearson and Fourtner[37] mapped the location of several motoneuronal somata which innervate the musculature of the walking legs. Using the dye Lucifer yellow both to backfill motor nerves and to iontophorese into physiologically identified motoneurons, Ritzmann et al.[35] and Davis[12] have described the locations and the major neuropilar projections of 16 motoneurons which innervate flight muscles and are active during flight behavior. As in the case of limb motoneurons, the neuronal processes of the flight motoneurons are extensive and radiate into many of the neuropilar regions of the thoracic ganglia.

Using the cobalt backfilling techniques, Iles[38] studied the motoneurons of the prothoracic ganglion and found with few exceptions that the location of somata and the distribution of neurites were similar to those previously described in the meso- and metathoracic ganglia.

Backfilling motoneuronal axons in concert with recording from individual motoneurons within the CNS has allowed the physiological characterization of over 50 identified motoneurons within the meso- and metathoracic ganglia of the cockroach which are associated with various leg and wing movements. Although the neuronal maps of the meso- and metathoracic ganglia described above would certainly contain the somata of these identified motoneurons, the specific locations of most of the physiologically identified motoneuron somata have not been charted on those maps. Although the ability to identify a specific motoneuronal soma from preparation to preparation would allow analysis comparable to that on identified neurons in molluscs,[39] with currently available information it appears that it is only possible to identify a given soma as belonging to one of Gregory's groups prior to penetration of the neuron.

The cockroach nervous system has also proven to be an excellent model for studying the role of motoneurons in a variety of behavior patterns. Since the gross morphology and the action of most thoracic skeletal muscles have been well described and the distribution of motor axons to individual muscles and muscle groups has been studied, the neural basis of behaviors ranging from simple reflexes to complex motor patterns such as locomotion, walking, or flight can be and has been investigated extensively. Using the type of electrophysiological techniques described by Anderson et al. in Chapter 10 of this work, several investigators have correlated muscle activity occurring during well-described behavioral patterns with centrally generated motor output recorded extracellularly from peripheral motor axons or intracellularly from motoneurons or other central neurons (for review, see References 2, 31, and 40). In addition, the cockroach has served as an excellent model for the investigation of sensory-to-motor reflexes and of sensory inputs to interneurons. For a comprehensive discussion of sensory-motor interactions controlling reflexes of the limb, see Chapter 25 of this work.

For a number of years, the cockroach has been the insect of choice for investigating walking behavior and interlimb coordination (for review of the earlier literature in this field, see Reference 41). During the past 20 years, several investigators, using high-speed cinematography, electrophysiological recording techniques, and the staining procedures described above, have refined and advanced models proposed by earlier workers. It is now clear that insect walking is controlled by a circuit of neurons in the CNS, the central pattern generator (CPG) of which is activated by some type of central command and can be modulated by a variety of sensory inputs. This literature has been reviewed on numerous occasions during the past decade[42-44] and will not be discussed except in reference to specific interneurons.

C. DORSAL UNPAIRED MEDIAL (DUM) NEURONS

A group of neurons having somata located along the dorsal midline were originally described and were found, using electrophysiological data, to send axonal branches to both sides of the body.[45,46] In addition, Crossman et al.[45] suggested that members of this group might send axonal projections out several of these bilateral nerve roots. The physiological importance of these cells has been established only recently, with most of the research concentrated on the locust rather than the cockroach.[47-49] One DUM neuron which innervates the extensor tibiae muscle in the locust releases octopamine in the vicinity of the muscle and can modulate muscular contraction.[50] For a review of this literature, see the article by Evans and Meyers[50] and Chapter 17 of this work. Studies on the DUM cells in the cockroach and the effects of transmitters such as octopamine and proctolin (see Chapter 14 of this work) eventually may explain the manner in which cockroaches maintain and modulate the high level of resting tension in skeletal muscles.

Gregory[51] has identified three groups of dorsal median cells, with the DUM cells probably being located in the posterior medial group. It is evident that not all medial cells have the characteristic bilateral symmetry of DUM cells. Gregory has identified one neuron, located in the middorsal medial group, which appears to innervate only one side of the animal and which has also been identified by Davis[12] as being a common neuron innervating all three of the dorsal longitudinal flight muscles. This strongly suggests that not all dorsal medial neurons are unpaired.

D. INTRAGANGLIONIC NEURONS

Investigations of the cockroach nervous system and its role in the control of locomotion led to the discovery of another group of neurons, each member of which is confined entirely within a single thoracic ganglion. Pearson and colleagues,[52,53] using cobalt iontophoresis, described a pair of interneurons that played a significant role in the neuronal circuit, the CPG responsible for producing rhythmic leg movements. These intraganglionic interneurons had small cell bodies (15 to 20 μm in diameter) located on the ventral surface contralateral to their extensively branching processes (Figure 6). Experimental manipulation of the membrane potential of these interneurons demonstrated that they influenced a broad pool of motoneurons innervating the leg contralateral to their somata, particularly those motoneurons responsible for movement of the femur.

One other important physiological characteristic of these local intraganglionic interneurons is that they are not capable of producing all-or-nothing action potentials or even the graded spikes which have been observed in other neuronal types (e.g., amacrine cells). Therefore, these cells apparently produce their postsynaptic effects through graded release of transmitter. The studies on these intraganglionic neurons established a new class of physiologically and morphologically identified, local, nonspiking neurons which (1) were clearly part of a CPG and (2) could control activity in a specific motoneuronal pool through the graded release of neurotransmitters. This model system has led to the discovery of similar neuronal types in other arthropods[44,54-58] and may prove to be a useful model for nonspiking neurons in vertebrate CNSs.

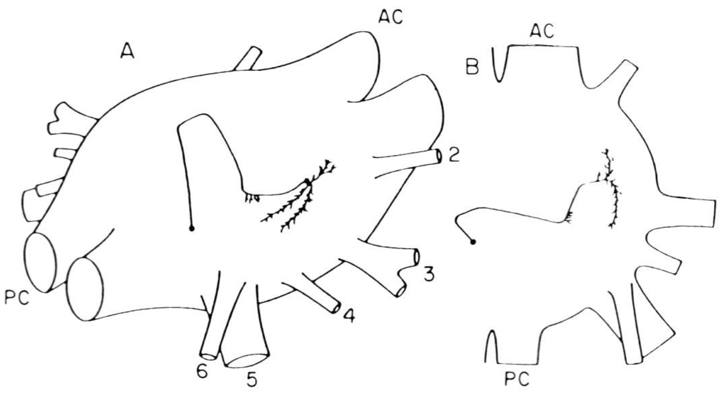

FIGURE 6. Drawing of a cobalt sulfide-stained interneuron (Interneuron I) in the metathoracic ganglion of the cockroach.[37] This neuron is an essential element of the CPG which produces rhythmic leg movements. It has no obvious axon and does not produce action potentials. (A) A dorsal three-dimensional view; (B) a dorsal two-dimensional view. AC, anterior connective; PC, posterior connective.

E. INTERGANGLIONIC NEURONS

In addition to the intraganglionic interneurons, a large number of interganglionic interneurons have also been characterized morphologically and physiologically in the thoracic nervous system of the cockroach. The most obvious of these are the giant axons to be discussed in Section IV.F below. However, Ritzmann et al.[59,60] have recently described a series of interganglionic interneurons which are intercalated between the giant fibers and the motoneurons controlling the behavioral escape responses in the cockroach. The processes of these interganglionic interneurons branch extensively in the neuropil of both sides of the ganglion, with the cell bodies located contralateral to the connectives through which their axons exit the ganglion. Five of these interneurons were identified and named on the basis of the characteristic shape of their neuritic processes: "Lambda", "Cross", "J", "Reversed J", and "T". Figure 7 illustrates several examples of these identified interganglionic interneurons.

One population of these interganglionic interneurons, the dorsal posterior group (DPG) neurons, is found in the dorsal posterior region of the meso- and metathoracic ganglia. The morphology of these cells has been elucidated using iontophoresis of the dye Lucifer yellow,[61] a powerful staining technique, since even after injection of the dye the cell continues its normal physiological activity unless selectively inactivated by directing UV light at the region containing the dye.[62] Using the mapping nomenclature of Gregory,[21] Ritzmann and Pollack[59] have described the intraganglionic projections of each of these interganglionic neurons and their axonal projections from the ganglion. For example, "Lambda" projects anteriorly and posteriorly via contralateral DIT; "J", anteriorly via LDT; "Reversed J", posteriorly via LDT; and "Cross", anteriorly and posteriorly via VIT (see Figure 7). Another population of neurons called "T" cells have their somata in the ventral region at the level of nerve 3 and have their axonal projection through the contralateral anterior DIT.

DPG neurons excite the motoneuron D_s, which innervates a slow extensor of the femur. All of the DPGs except the "Cross" cell initially excite the widespread common inhibitor. "Lambda" activates it with a very short latency. This explains the rapid activation of the common inhibitor observed following direct stimulation of the abdominal ventral nerve cord.[63] These motor responses always occur in the motor units contralateral to the DPG somata. One of the fascinating riddles yet to be answered is the purpose of the common inhibitors, particularly with reference to their early activation by the DPG interneurons.

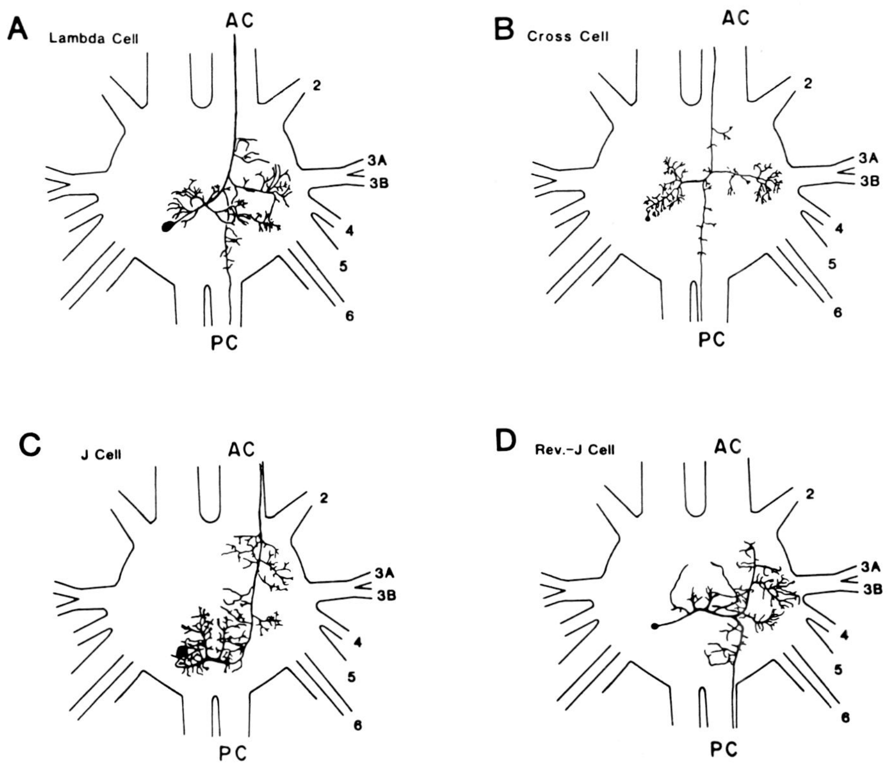

FIGURE 7. Drawings of four identified interganglionic interneurons associated with specific motoneuronal activity in the meta- and mesothoracic ganglia of the cockroach. The somata of these neurons are all located within the dorsal posterior group of somata. (A) "Lambda"; (B) "Cross"; (C) "J"; (D) "Reversed J". AC, anterior connective; PC, posterior connective. See Section IV.E of text for a complete description of the physiology of these cells. (From Ritzmann, R. E. and Pollack, A. J., *J. Comp. Physiol. A*, 159, 639, 1986. With permission.)

The "T" neuron evoked a more complex motor response, with an initial contralateral excitation of both D_s and the widespread common inhibitor followed by a long-lasting burst of activity in the leg flexor (levator) motor neuronal pool.

F. GIANT AXONS

One of the most extensively studied systems in the cockroach CNS is the system of giant axons, which run the length of the ventral nerve cord.[64-66] Ever since the description of the rapid escape response in the cockroach, there has been continuing discussion on the role which these giant axons might play in escape behavior. The giant axons have been divided into two morphological groups: a ventral group of four, three extremely large axons (50 to 60 μm in diameter) and a smaller one (30 μm), and a dorsal group of three axons ranging from 20 to 30 μm in diameter.[67]

Each of the giant axons has a soma which is located in the sixth abdominal ganglion. The soma is contralateral to the connective containing its axon and its major dendritic arbor. There are few small processes ipsilateral to the cell body. There are also processes arising from the giant axons as they pass through the meta- and mesothoracic ganglia. These are small and are generally located in the ventral neuropilar regions. Figure 8 illustrates the morphology of the seven giant axons (see also Figure 1 in Chapter 8 of this work for a photograph of the morphology of a cobalt-stained ventral giant 2).

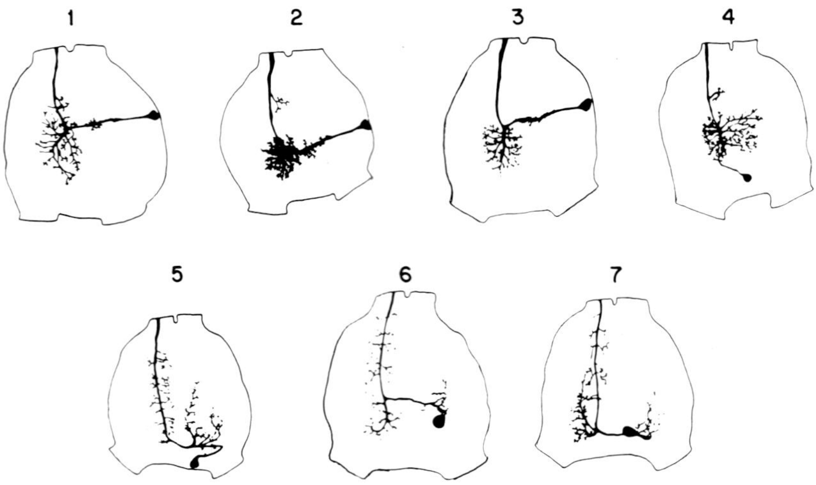

FIGURE 8. Drawings of cobalt-filled giant fiber processes and somata in the sixth abdominal ganglion. The numbers above each drawing correspond to the number of the giant cell depicted; 1 through 4 represent the ventral giants and 5 through 7 the dorsal giants. The soma is always contralateral to the process projecting to the anterior connective. (From Daley, D. L., Vardi, N., Appignani, B., and Camhi, J. M., *J. Comp. Neurol.*, 196, 41, 1981. With permission.)

Each of the seven giant axons is relatively easy to penetrate with microelectrodes in the abdominal nerve cord without disrupting sensory input via the cercal afferents and motor output from the thoracic ganglia. Westin et al.[68,69] demonstrated that each of the giants is responsive to air streams directed at the cerci and has unique receptive fields, thus permitting a selective activation of giants with well-controlled noninvasive inputs.

Studies of the motor output from thoracic ganglia resulting from stimulation of giant fibers (either singly or in pairs) have revealed that the giants, other than ventral giants 2 and 3, can influence the motor output to the population of motoneurons used in walking.[70-72] Recent physiological studies have demonstrated separate functions for the dorsal and ventral giants. It is now fairly well established that the giant axons (at least the dorsal ones) can be utilized to initiate escape by activating the motoneuronal pool used for running and may directly activate the CPG for running. The ventral giants may be extremely important in determining the directionality of escape.[71,71a] Daley and Delcomyn,[73-75] using a tightly tethered cockroach with the abdominal ventral nerve cord held in place on an immovable platform, were able to record from the giant axons intracellularly. While their data strongly suggested that during walking there was simultaneous activity in the dorsal giants, but little or no activity in the ventral giants, there may in fact be inhibition of the ventral giants.[74]

Ritzmann et al.[76,77] have demonstrated that activation of an individual dorsal giant axon can evoke activity in the leg motoneurons associated with running in order to avoid a predator, or, depending on the sensory input arising from the legs, the dorsal giants can initiate flight behavior by turning on the CPG for flight, driving the wing motoneurons instead of the limb motoneurons for walking or running. This is an example of the importance of sensory input for "gating" the activity arriving in a single giant axon to selectively activate one or the other CPG.

This work on giant axon activation of flight has led to some recent investigations of the flight system of the cockroach as a model for interneuronal activation of behavior. As in the case of the walking system, the flight musculature and its activity during bouts of flight have been described.[78] Some extremely large muscles which function as flight muscles are also active during rapid walking bouts.[78] Another interesting finding is that some of the flight muscles appear to be active continuously during flight rather than being active during only one phase of the flight pattern. The innervation of the flight musculature is also well known, allowing the precise description of motoneurons given above. It is clear that as a model system for studies of flight the locust currently offers a better model system. Nevertheless, the cockroach flight system is extremely useful for investigating the neural mechanisms responsible for the onset of flight activity.

Another extremely important aspect of the giant axons as an experimental model is that this system permits simultaneous microelectrode recordings from a single giant axon in the abdominal cord and from neuropilar processes of neurons within a thoracic ganglion. This arrangement allows the study of the synaptic interaction between giant axons and other neurons within the meso- and metathoracic ganglia. Ritzmann and co-workers[59,60] have recently completed an extensive investigation using this preparation, demonstrating synaptic interactions between giant fibers and other interganglionic interneurons. The DPG neurons receive excitatory input from the ventral giant axons and apparently are unaffected by dorsal giant stimulation. There is, however, a sidedness to the input, since ''Lambda'' responds to contralateral ventral giants 1 and 2 and ipsilateral 3, whereas ''Reverse-J'' and ''Cross'' are driven by ipsilateral 1 and 2. ''J'' neuron received input from ipsi- and contralateral ventral giant 2. The ''T'' cell, on the other hand, fires action potentials following activation of the ipsilateral dorsal giant axons 5 and 7 and most likely is unaffected by ventral giant input.

More recently, the effects of sensory input on at least the DPGs have been studied.[79] For a review of these data, as well as the role of sensory input on limb reflexes and on the control of leg movements, see Chapter 25 of this work.

V. GLIA

In addition to the neural elements of the CNS functioning as model systems for studies of basic physiological processes, the glia cells and their associated noncellular matrix, the neurolemma, surrounding the CNS of the cockroach have recently been studied thoroughly for their possible role in controlling the chemical and, more particularly, the ionic environment in and around the nerve cells and neuronal processes of the nervous system.[80-89] Lane[90] recently has written a review on the ultrastructure of the glial components associated with the CNS. Also presented in that review are some previously unpublished observations on the ultrastructure of unidentified synaptic connections in the neuropil of the thoracic ganglia. In addition, the glial elements play the primary role in the repair and regeneration of the sheath surrounding the CNS and the neural processes, respectively.

There are generally conceived to be two classes of glial cells surrounding the CNS: the perineural glial cells forming the outer sheath around the CNS and the neuroglia sending processes around the axons and into the neuropil of the CNS. The perineurium, generally assumed to be the blood-brain barrier,[85-87] consists of cells which are somewhat similar to epithelial cells in that there are extensive gap junctions between adjacent perineurium glial cells. Figure 9 illustrates the morphological relationships among the glial elements and neural elements of the CNS. For a thorough discussion of the role of the neural sheath and glial elements in control of the chemical milieu surrounding the neural elements and the repair function of glia, see Chapter 11 in this work and References 91 to 93.

In the peripheral nervous system, perineural glia cells are also found directly beneath the neurolemma, and subperineural glia is found surrounding individual motor axons. There may be as many as six or seven layers of glial sheath around individual motor axons, whereas there will be a glial sheath around pools of sensory axons rather than around single ones.

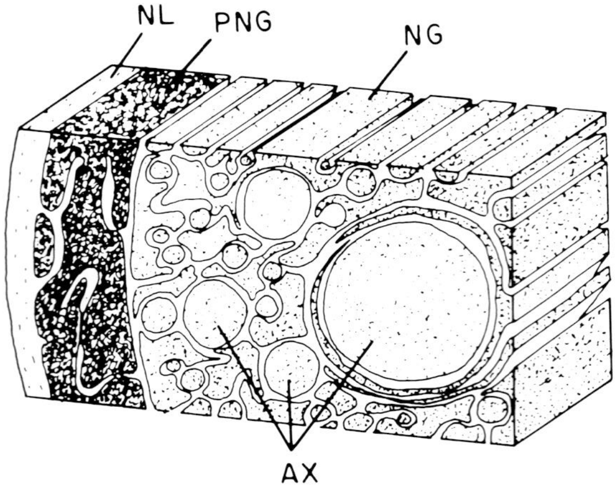

FIGURE 9. Three-dimensional schematic drawing of the nerve sheath surrounding the abdominal connectives. AX, axon; NL, neurolemma; PNG, perineuronal glia; NG, neuroglia. (From Treherne, J. E. and Schofield, J. E., *J. Exp. Biol.*, 95, 61, 1979. With permission.)

Figure 10A illustrates the morphological arrangement in a purely motor nerve containing axons innervating the extensor tibiae muscle of the cockroach leg. In addition, the glial membrane surrounding the motor axons envelops many regions of electron-dense material. In the normal glia there are occasional occurrences of microtubules, but these are usually located near an area where a nerve is splitting, sending a ramus to a group of muscle fibers.

Although the role of glia in repair and regeneration has been studied extensively in the CNS, studies on the role of glia in peripheral regeneration are lacking. In recent studies in our laboratory, we have shown that there is a remarkable transformation in the organization of the glial elements within 5 days of nerve damage.[95] The model for these studies is a small motor nerve innervating the muscle producing extension of the tibia (as seen in Figure 10B). The nerve contains four axonal profiles: the fast extensor tibiae (the largest); a somewhat smaller diameter axon, the slow extensor tibiae; and two small common inhibitors. The two excitors exit the thoracic ganglia via nerve root 3B, while the inhibitors exit via nerve 5, but ''cross over'' to 3B in an anastomosis near the end of the coxa.

The utility of this model system is that the two excitors or the two inhibitors can be damaged selectively and the effect of degenerating or regenerating fibers on intact fibers can be studied. One of our findings is that, following lesioning of the excitors, within 1 week there is hypertrophy in the cross-sectional diameter of the intact inhibitory axons. However, as mentioned above, the most remarkable occurrence is the transformation occurring in the subperineural glia. By day 5 following lesioning, the glia surrounding the degenerating excitors have invaded that axonal space and have begun to synthesize large numbers of microtubules. By 7 to 10 days after lesioning, the microtubules have begun to align axially along the nerve, and after 2 to 3 weeks the perineural glial has become an exquisite network of axially arranged microtubules (Figure 10C) which may serve in guiding the regenerating excitatory motor axons.[95]

VI. SUMMARY

With the development of the technical expertise to record simultaneously from various areas in the nervous system, the cockroach CNS has been and will continue to be a superb

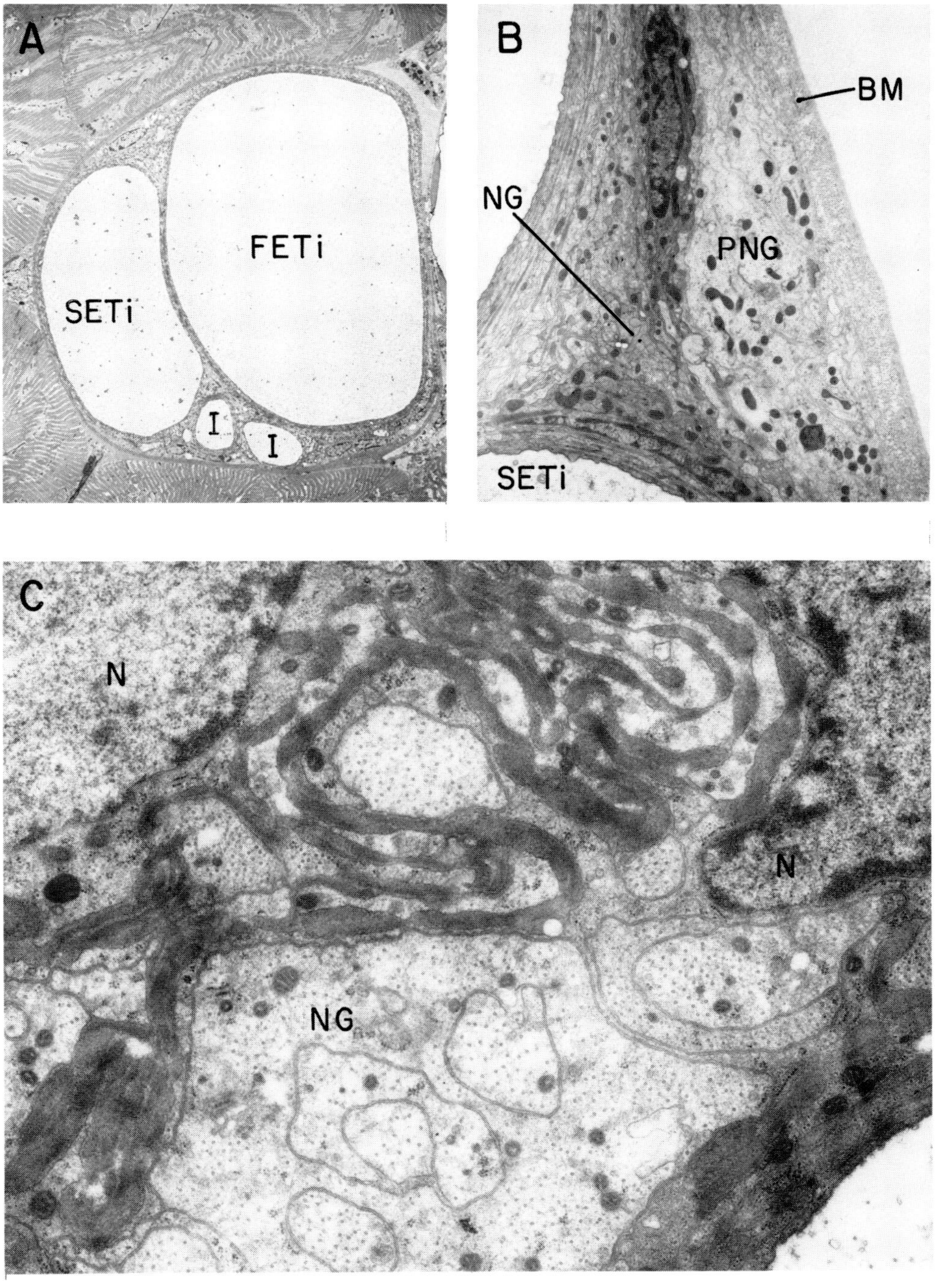

FIGURE 10. Electron microscopic illustration of peripheral motor nerve 3B11. (A) Cross section of the nerve, revealing the four motor units; FETi and SETi are the fast and slow extensor tibiae motor axons respectively; the two widespread common inhibitors = I. (Magnification × 1200.) (B) Cross section of the neuroglia surrounding SETi and FETi, revealing the basement membrane (BM), the perineural glia (PNG), lighter staining neural glial elements, the neural glia (NG), with denser staining cytoplasm and layered extracellular material circumferentially arranged around the axons. (Magnification × 7250.) (C) Cross section of transformed glia 14 d after nerve section, revealing glial processes with axially arranged microtubules; N = nucleus. (Magnification × 23,000.)

model for investigating neural events modulating behavior, neural mechanisms at the neuronal level, and regeneration of neural tissue. As a summary example, the case of the escape response of the cockroach offers just one instance of a complex neural response in which most of the neural mechanisms from sensory reception to motor activation are amenable to anatomical and physiological investigations. These studies will continue to serve as a basis for neurobiological mechanisms that span the animal kingdom.

REFERENCES

1. **Guthrie, D. M. and Tindall, A. R.,** *The Biology of the Cockroach,* Edward Arnold, London, 1968.
2. **Pipa, R. and Delcomyn, F.,** Nervous system, in *The American Cockroach,* Bell, W. J. and Adiyodi, K. G., Eds., Chapman and Hall, London, 1981, 175.
3. **Weevers, R. deG.,** The insect ganglia, in *Comprehensive Insect Physiology, Biochemistry and Pharmacology,* Vol. 5, Kerkut, G. A. and Gilbert, L. I., Eds., Pergamon Press, Oxford, 1985, 213.
4. **Krauthamer, V.,** Morphology of identified neurosecretory and nonneurosecretory cells in the cockroach pars intercerebralis, *J. Exp. Zool.,* 234, 221, 1985.
5. **Strausfeld, N. J.,** *Atlas of an Insect Brain,* Springer-Verlag, New York, 1976.
6. **Mobbs, P. G.,** Brain structure, in *Comprehensive Insect Physiology, Biochemistry and Pharmacology,* Vol. 5, Kerkut, G. A. and Gilbert, L. I., Eds., Pergamon Press, Oxford, 1985, 309.
7. **Weiss, M. J.,** Neuronal connections and the function of the corpora pedunculata in the brain of the American cockroach, *Periplaneta americana* (L.), *J. Morphol.,* 142, 21, 1974.
8. **Willey, R. B.,** The morphology of the stomodeal nervous system in *Periplaneta americana* (L.) and other Blattaria, *J. Morphol.,* 108, 219, 1962.
9. **Ernst, K.-D., Boeckh, J., and Boeckh, V.,** A neuroanatomical study on the organization of the central antennal pathway in insects. II. Deuterocerebral connections in *Locusta migratoria* and *Periplaneta americana, Cell Tissue Res.,* 176, 285, 1977.
10. **Bodenstein, D.,** Studies on nerve regeneration in *Periplaneta americana, J. Exp. Zool.,* 136, 89, 1957.
11. **Altman, J. S. and Kien, J.,** Functional organization of the subesophageal ganglion in arthropods, in *Arthropod Brain,* Gupta, A. P., Ed., John Wiley & Sons, New York, 1987, 265.
12. **Davis, N. T.,** Serial homologies of the motor neurons of the dorsal intersegmental muscles of the cockroach, *Periplaneta americana* (L.), *J. Morphol.,* 176, 197, 1983.
13. **Gupta, A. P., Ed.,** *Arthropod Brain,* John Wiley & Sons, New York, 1987.
14. **Dresden, D. A. and Nijenhuis, E. D.,** Fibre analysis of the nerves of the second thoracic leg in *Periplaneta americana, Proc. K. Ned. Akad. Wet.,* C61, 213, 1958.
15. **Nijenhuis, E. D. and Dresden, D.,** A micro-morphological study on the sensory supply of the mesothoracic leg of the American cockroach, *Periplaneta americana, Proc. K. Ned. Akad. Wet.,* C55, 300, 1955.
16. **Nijenhuis, E. D. and Dresden, D.,** On the topographical anatomy of the nervous system of the mesothoracic leg of the American cockroach *(Periplaneta americana), Proc. K. Ned. Akad. Wet.,* C58, 121, 1955.
17. **Pipa, R. L. and Cook, E. F.,** Studies on the hexapod nervous system. I. The peripheral distribution of the thoracic nerves of the adult cockroach, *Periplaneta americana, Ann. Entomol. Soc. Am.,* 52, 695, 1959.
18. **Pipa, R. L., Cook, E. F., and Richards, A. G.,** Studies on the hexapod nervous system. II. The histology of the thoracic ganglia of the adult cockroach, *Periplaneta americana (L.), J. Comp. Neurol.,* 113, 401, 1959.

18a. **Kaars, C. J.,** Innervation and control of tension in abdominal muscles of *Blaberus discoidalis* and *Gromphadorhina portentosa, J. Insect Physiol.,* 25, 209, 1983.

19. **Mizunami, M., Yamashita, S., and Tateda, H.,** Intracellular stainings of the large ocellar second order neurons in the cockroach, *J. Comp. Physiol.,* 149, 215, 1982.
20. **Mizunami, M. and Tateda, H.,** Classification of ocellar interneurones in the cockroach brain, *J. Exp. Biol.,* 125, 57, 1986.
21. **Gregory, G. E.,** Neuroanatomy of the mesothoracic ganglion of the cockroach, *Periplaneta americana* (L.). I. The roots of the peripheral nerves, *Philos. Trans. R. Soc. London,* 267, 421, 1974.
22. **Gregory, G. E.,** Silver staining of insect central nervous system by the Bodian protargal method, *Acta Zool. (Stockholm),* 51, 169, 1970.
23. **Krauthamer, V.,** Electrophysiology of identified neurosecretory and non-neurosecretory cells in the cockroach pars intercerebralis, *J. Exp. Zool.,* 234, 234, 1985.

24. **Chesler, M. and Fourtner, C. R.,** Mechanical properties of a slow muscle in the cockroach, *J. Neurobiol.,* 12, 391, 1981.
25. **Carr, C. E. and Fourtner, C. R.,** Pharmacological analysis of a monosynaptic reflex in the cockroach, *Periplaneta americana, J. Exp. Biol.,* 86, 259, 1980.
26. **Thomas, M. V. and Treherne, J. E.,** An electrophysiological analysis of extra-axonal sodium and potassium concentrations in the central nervous system of the cockroach *(Periplaneta americana* L.), *J. Exp. Biol.,* 63, 801, 1975.
27. **Treherne, J. E., Scofield, P. K., and Lane, N. J.,** Physiological and ultrastructural evidence for an extracellular anion matrix in the central nervous system of an insect *(Periplaneta americana), Brain Res.,* 247, 255, 1982.
28. **Cohen, M. J. and Jacklet, J. W.,** The functional organization of motor neurons in an insect ganglion, *Philos. Trans. R. Soc. London,* 252, 561, 1967.
29. **Young, D.,** The motor neurons of the mesothoracic ganglion of *Periplaneta americana, J. Insect Physiol.,* 15, 1175, 1969.
30. **Stretton, A. O. W. and Kravitz, E. A.,** Neuronal geometry: determination with a technique of intracellular dye injection, *Science,* 162, 132, 1968.
31. **Pitman, R. M., Tweedle, C. D., and Cohen, M. J.,** Branching of central neurons: intracellular cobalt injection for light and electron microscopy, *Science,* 176, 412, 1972.
32. **Tyrer, N. M. and Altman, J. S.,** Motor and sensory flight neurones in a locust demonstrated using cobalt chloride, *J. Comp. Neurol.,* 157, 117, 1974.
33. **Hoyle, G.,** Cellular mechanisms underlying behavior — neuroethology, *Adv. Insect Physiol.,* 7, 349, 1970.
34. **Pearson, K. G. and Bradley, A. B.,** Specific regeneration of excitatory motoneurons to leg muscles in the cockroach, *Brain Res.,* 47, 492, 1972.
35. **Ritzmann, R. E., Fourtner, C. R., and Pollack, A. J.,** Morphological and physiological identification of motor neurons innervating flight musculature in the cockroach, *Periplaneta americana, J. Exp. Zool.,* 225, 347, 1983.
36. **Fourtner, C. R. and Pearson, K. G.,** Morphological and physiological properties of motor neurons innervating insect leg muscles, in *Identified Neurons and Behavior of Arthropods,* Hoyle, G., Ed., Plenum Press, New York, 1977, 87.
37. **Pearson, K. G. and Fourtner, C. R.,** Identification of the somata of common inhibitory motoneurones in the metathoracic ganglion of the cockroach, *Can. J. Zool.,* 51, 859, 1973.
38. **Iles, J. F.,** Organization of motoneurones in the prothoracic ganglion of the cockroach, *Periplaneta americana* (L.), *Philos. Trans. R. Soc. London Ser. B,* 276, 205, 1976.
39. **Kandel, E. R. and Schwartz, J. H.,** *Principles of Neural Science,* Elsevier, New York, 1985.
40. **Delcomyn, F.,** Walking and running, in *Comprehensive Insect Physiology, Biochemistry and Pharmacology,* Vol. 5, Kerkut, G. A. and Gilbert, L. I., Eds., Pergamon Press, Oxford, 1985, 439.
41. **Wilson, D. M.,** Central nervous mechanism for the generation of rhythmic behavior in arthropods, *Symp. Soc. Exp. Biol.,* 20, 199, 1966.
42. **Delcomyn, F.,** Neural basis of rhythmic behavior in animals, *Science,* 210, 492, 1980.
43. **Fourtner, C. R.,** Central nervous control of cockroach walking, in *Neural Control of Locomotion,* Herman, R. M., Grillner, S., Stein, P. S. G., and Stuart, D. G., Eds., Plenum Press, New York, 1977, 401.
44. **Pearson, K. G.,** Interneurons in the ventral cord of insects, in *Identified Neurons and Behavior of Arthropods,* Hoyle, G., Ed., Plenum Press, New York, 1977, 329.
45. **Crossman, A. R., Kerkut, G. A., Pitman, R. M., and Walker, R. J.,** Electrically excitable nerve cell bodies in the central ganglia of two insect species, *Periplaneta americana* and *Schistocerca gregaria.* Investigation of cell geometry and morphology by intracellular dye injection, *Comp. Biochem. Physiol.,* 40A, 579, 1971.
46. **Crossman, A. R., Kerkut, G. A., and Walker, R. J.,** Electrophysiological studies on the axon pathways of specified nerve cells in the central ganglia of two insect species, *Periplaneta americana* and *Schistocerca gregaria, Comp. Biochem. Physiol.,* 43A, 393, 1972.
47. **Hoyle, G.,** A function for neurons (DUM) neurosecretory on skeletal muscle of insects, *J. Exp. Zool.,* 189, 401, 1974.
48. **Hoyle, G.,** The dorsal, unpaired median neurons of the locust metathoracic ganglion, *J. Neurobiol.,* 9, 43, 1978.
49. **Hoyle, G., Dagan, D., Moberly, B., and Colquhoun, W.,** Dorsal unpaired median insect neurons make neurosecretory endings on skeletal muscle, *J. Exp. Zool.,* 187, 159, 1974.
50. **Evans, P. D. and Meyers, C. M.,** Peptidergic and aminergic modulation of insect skeletal muscle, *J. Exp. Biol.,* 124, 143, 1986.
51. **Gregory, G. E.,** Neuroanatomy of the mesothoracic ganglion of the cockroach *Periplaneta americana.* II. Median neuron cell body groups, *Philos. Trans. R. Soc. London,* 306, 191, 1984.
52. **Pearson, K. G., Fourtner, C. R., and Wong, R. K.,** Nervous control of walking in the cockroach, in *Control of Posture and Locomotion,* Stein, R. B., Pearson, K. G., Smith, R. S., and Redford, J. B., Eds., Plenum Press, New York, 1973, 495.

53. **Pearson, K. G. and Fourtner, C. R.,** Nonspiking interneurons in walking system of the cockroach, *J. Neurophysiol.,* 38, 33, 1975.
54. **Watkins, B. L., Burrows, M., and Siegler, M. V. S.,** The structure of locust non-spiking interneurones in relation to the anatomy of a segmental ganglion, *J. Comp. Neurol.,* 240, 233, 1985.
55. **Burrows, M.,** The control of sets of motoneurones by local interneurones, *J. Physiol.,* 298, 213, 1980.
56. **Burrows, M. and Siegler, M. V. S.,** Graded synaptic transmission between local interneurons and motoneurones in the metathoracic ganglion of the locust, *J. Physiol.,* 285, 231, 1978.
57. **Roberts, A. and Bush, B. M. H.,** *Neurons without Impulses,* Cambridge University Press, London, 1981.
58. **Pearson, K. G.,** Nerve cells without action potentials, in *Simpler Networks and Behavior,* Fentress, J. C., Sinauer Associates, Sunderland, MA, 1976, 99.
59. **Ritzmann, R. E. and Pollack, A. J.,** Identification of thoracic interneurons that mediate giant interneuron-to-motor pathways in the cockroach, *J. Comp. Physiol. A,* 159, 639, 1986.
60. **Ritzmann, R. E., Pollack, A. J., and Westin, J.,** Integration of directional wind field information in thoracic interneurons of the cockroach, *Neurosci. Abstr.,* 13, 140, 1987.
61. **Stewart, W. W.,** Functional connections between cells as revealed by dye-coupling with a highly fluorescent napthalimide tracer, *Cell,* 14, 741, 1968.
62. **Miller, J. P. and Selverston, A. I.,** Rapid killing of single neurons by irradiation of intracellularly injected dye, *Science,* 206, 702, 1979.
63. **Fourtner, C. R. and Drewes, C. D.,** Excitation of the common inhibitory motor neuron: a possible role in the startle reflex of the cockroach, *Periplaneta americana, J. Neurobiol.,* 8, 477, 1977.
64. **Ritzmann, R. E.,** The cockroach escape response, in *Neural Mechanisms in Startle Behavior,* Eaton, R. C., Ed., Plenum Press, New York, 1984, 93.
65. **Roeder, K. D.,** Organization of the ascending giant fiber system in the cockroach *(Periplaneta americana), J. Exp. Zool.,* 108, 243, 1948.
66. **Parnas, I. and Dagan, D.,** Functional organizations of giant axons in the central nervous system of insects: new aspects, *Adv. Insect Physiol.,* 9, 95, 1971.
67. **Daley, D. L., Vardi, N., Appignani, B., and Camhi, J. M.,** Morphology of the giant interneurons and cercal nerve projections of the American cockroach, *J. Comp. Neurol.,* 196, 41, 1981.
68. **Westin, J., Langberg, J. L., and Camhi, J. M.,** Responses of giant interneurons of the cockroach, *Periplaneta americana* to wind puffs of different directions and velocities, *J. Comp. Physiol.,* 121, 307, 1977.
69. **Westin, J. and Ritzmann, R. E.,** The effect of single giant lesions on wind-evoked motor responses in the cockroach, *Periplaneta americana, J. Neurobiol.,* 13, 127, 1982.
70. **Ritzmann, R. E. and Camhi, J. M.,** Excitation of leg motor neurons by giant interneurons in the cockroach, *Periplaneta americana, J. Comp. Physiol.,* 125, 305, 1978.
71. **Ritzmann, R. E. and Pollack, A. J.,** Motor responses to paired stimulation of giant interneurons in the cockroach, *Periplaneta americana.* I. The dorsal interneurons, *J. Comp. Physiol.,* 143, 61, 1981.
71a. **Ritzmann, R. E.,** personal communication.
72. **Ritzmann, R. E.,** Motor responses to paired stimulation of giant interneurons in the cockroach, *Periplaneta americana.* II. The ventral interneurons, *J. Comp. Physiol.,* 143, 71, 1981.
73. **Daley, D. L. and Delcomyn, F.,** Modulation of excitability of cockroach giant interneurons during walking. I. Simultaneous excitation and inhibition, *J. Comp. Physiol.,* 138, 231, 1980.
74. **Daley, D. L. and Delcomyn, F.,** Modulation of excitability of cockroach giant interneurons during walking. II. Central and peripheral components, *J. Comp. Physiol.,* 138, 241, 1980.
75. **Delcomyn, F.,** Activity and structure of movement-signalling (corollary discharge) interneurons in a cockroach, *J. Comp. Physiol.,* 150, 185, 1983.
76. **Ritzmann, R. E., Tobias, M. L., and Fourtner, C. R.,** Flight activity initiated via giant interneurons of the cockroach: evidence for bifunctional trigger interneurons, *Science,* 210, 443, 1980.
77. **Ritzmann, R. E., Pollack, A. J., and Tobias, M. L.,** Flight activity mediated by intracellular stimulation of dorsal giant interneurons of the cockroach, *Periplaneta americana, J. Comp. Physiol.,* 147, 313, 1982.
78. **Fourtner, C. R. and Randall, J. B.,** Studies on cockroach flight: the role of continuous neural activation of non-flight muscles, *J. Exp. Zool.,* 221, 143, 1982.
79. **Murrain, M. P.,** Morphological and Physiological Analysis of Proprioceptive Inputs to DPG Interneurons in the Cockroach, Ph.D. dissertation, Case Western Reserve University, Cleveland, OH, 1987, 1.
80. **Ashhurst, D. E.,** The connective tissue of insects, *Annu. Rev. Entomol.,* 13, 45, 1968.
81. **Radojcic, T. and Pentreath, V. W.,** Invertebrate glia, *Prog. Neurobiol.,* 12, 115, 1979.
82. **Lane, N. J.** Invertebrate neuroglia — juctional structure and development, *J. Exp. Biol.,* 95, 7, 1981.
83. **Lane, N. J. and Treherne, J. E.,** Studies on perineurial junctional complexes and sites of uptake of microperoxidase and lanthanum in the cockroach central nervous system, *Tissue Cell,* 4, 427, 1972.
84. **Maddrell, S. H. P. and Treherne, J. E.,** The ultrastructure of the perineurium in two insect species, *Carausius morosus* and *Periplaneta americana, J. Cell Sci.,* 2, 119, 1967.

85. **Schofield, P. K. and Treherne, J. E.,** Localization of the blood-brain barrier of an insect: electrical model and analyses, *J. Exp. Biol.*, 109, 319, 1984.
86. **Schofield, P. K., Swales, L. S., and Treherne, J. E.,** Potentials associated with the blood-brain barrier of an insect: recordings from identified neuroglia, *J. Exp. Biol.*, 109, 307, 1984.
87. **Lane, N. J. and Treherne, J. E.,** Uptake of peroxidase by the cockroach central nervous system, *Tissue Cell*, 2, 413, 1970.
88. **Treherne, J. E. and Schofield, P. K.,** Mechanisms of ionic homeostasis in the central nervous system of an insect, *J. Exp. Biol.*, 95, 61, 1981.
89. **Treherne, J. E. and Pichon, Y.,** The insect blood-brain barrier, *Adv. Insect Physiol.*, 9, 257, 1972.
90. **Lane, N. J.,** Structure of components of the nervous system, in *Comprehensive Insect Physiology, Biochemistry and Pharmacology*, Vol. 5, Kerkut, G. A. and Gilbert, L. I., Eds., Pergamon Press, Oxford, 1985, 1.
91. **Treherne, J. E., Harrison, J. B., Treherne, J. M., and Lane, N. J.,** Glial repair in an insect central nervous system: effects of surgical lesioning, *J. Neurosci.*, 4, 2689, 1984.
92. **Smith, P. J. S., Leech, C. A., and Treherne, J. E.,** Glial repair in an insect central nervous system: effects of selective glial disruption, *J. Neurosci.*, 4, 2698, 1984.
93. **Meiri, H., Dormann, A., and Spira, M. E.,** Comparison of ultrastructural changes in proximal and distal segments of transected giant fibres in the cockroach, *Periplaneta americana*, *Brain Res.*, 263, 1, 1983.
94. **Treherne, J. E. and Schofield, J. E.,** Ionic homeostasis of the brain microenvironment in insects, *J. Exp. Biol.*, 95, 61, 1979.
95. **Fourtner, C. R. and Kaars, C. J.,** in preparation.

Chapter 5

INNERVATION AND ELECTROPHYSIOLOGY OF THE CORPUS ALLATUM

C. S. Thompson and S. S. Tobe

TABLE OF CONTENTS

I. INTRODUCTION

The corpora allata (CA) of insects are endocrine glands closely associated with and innervated by axons originating in the central nervous system (CNS). They are a major component of the retrocerebral system, but are also anatomically associated with the stomatogastric nervous system.[1] Indeed, these glands were incorrectly thought to be part of the nervous system when they were first described in the 19th century.[2] Numerous studies conducted since that time have established that the CA are composed of parenchymal cells which arise from the embryonic ectoderm and that they are the site of synthesis and release of juvenile hormone (JH).[3] JH plays a vital role in regulating the processes of metamorphosis and reproduction. The importance of JH in the regulation of these processes is reflected in the enormous literature dealing with JH and with the CA (see, for example, References 3 to 5).

An interesting feature of the brain-corpus cardiacum-corpus allatum system of insects is the superficial anatomical resemblance to the hypothalamic-pituitary system of vertebrates.[6] This analogy between the CA and the adenohypophysis suggests that the CA may be regulated at least in part by hormones produced by neurosecretory cells in the brain. This idea is supported by the observations that there are neurosecretory cells in the insect brain which have terminals in the CA and that neural connections between the brain and the CA are necessary for the normal functioning of these endocrine glands. However, there are important differences between the CA and the adenohypophysis. As far as is known, the CA produce only a single hormone rather than several different hormones, and JH is a lipid-soluble sesquiterpenoid. Thus, JH is not stored in membrane-bound vesicles as are the peptide hormones of the adenohypophysis. Rather, JH is released into the hemolymph as it is synthesized and appreciable quantities are not stored in the glands.[7,8] Thus, mechanisms acting to control JH biosynthesis may provide a model for mechanisms acting to control the rate of biosynthesis of the steroid hormones of vertebrates.[9]

The Pacific beetle cockroach, *Diploptera punctata* (Eschscholtz), has provided a useful model system in which to study factors controlling the biosynthesis and release of JH. In this species, the CA are relatively large glands (200 to 300 μm in adults)[10] and, when active, produce large amounts of JH in comparison to other species.[3] Furthermore, in adult females, JH is synthesized and released in a well-defined cycle associated with the vitellogenic growth and maturation of oocytes.[11] As reviewed in Chapter 21 of this work, several factors involved in the regulation of biosynthesis of JH by the CA of adult female *D. punctata* have been studied. Among these are neural connections with the brain, humoral factors from other sources, and the second messengers calcium and cyclic adenosine monophosphate (cAMP).

II. INNERVATION OF THE CORPORA ALLATA

The innervation of the CA is complex and has not been determined precisely for any species; i.e., the precise location and identity of neurons in the CNS remain to be described. Both conventional and neurosecretory neurons from many parts of the brain and various ganglia contribute axons to the nerves of the retrocerebral complex, and there is a great deal of species variability in the size, number, and location of these neurons as well as in the neuroanatomy of the retrocerebral complex itself. Whereas conventional staining techniques have revealed several groups of neurons which send axons to the CA in a variety of species, the functional significance of these neurons has been difficult to assess. Some axons may bypass the CA without giving off terminal branches, whereas some neurosecretory neurons may have terminal branches in the CA, but release a neurohormone which is not involved in the regulation of the CA. This is apparently the case in Lepidoptera in which cerebral neurosecretory neurons apparently containing the prothoracicotropic hormone have terminals in or on the CA.[12] The CA also may be regulated by neurosecretory neurons that release neurohormones at sites some distance from the CA.

In cockroaches, as in most other insect species, the CA appear to be innervated by axons from both the brain and the subesophageal ganglion. In *D. punctata,* the nickel chloride backfilling technique has been used to show that some neurons in both the medial and the lateral groups of neurosecretory cells in the protocerebrum send axons to the CA by way of the nervi corporis cardiaci I (NCC I) and the nervi corporis cardiaci II (NCC II), respectively. A few neurons in the tritocerebrum also send axons to the CA by way of the nervi corporis cardiaci III (NCC III), and about 20 neurons in the subesophageal ganglion send axons by way of the nervi corporis allati II (NCA II).[13] This is summarized in Figure 1. Similar groups of neurons have been shown to send axons to the CA in a variety of other species.[9] A comparison of brains backfilled from NCA I with brains backfilled from the postallatal nerves indicates that about 40 neurons in the contralateral pars intercerebralis (PI) — medial neurosecretory cells (MNSC) — and about 20 in the ipsilateral pars lateralis (PL) — lateral neurosecretory cells (LNSC) — actually terminate in each corpus allatum.[10,14] There is some evidence that neurons in these two groups are involved in regulation of the CA. Radiocautery of either group or of the axon tracts leading from them will stimulate JH biosynthesis by the CA of virgin females.[15] Neurons in the subesophageal ganglion may not be directly involved in the regulation of JH biosynthesis and release. In *Periplaneta americana* the function of the CA was not altered by severance of NCA II,[16] and in *D. punctata* this procedure did not prevent activation of the CA.[17] No information is available concerning the role of tritocerebral neurons in the regulation of CA activity.

More recently, intracellular recording and dye injection techniques have been employed to identify neurons with terminals in the CA of *D. punctata* and to study their physiology.[18] Neurosecretory neurons in the protocerebrum of cockroaches are located easily with microelectrodes because their somata are close to the surface of the brain and because they display long-duration somatic action potentials. Krauthamer[19,20] has shown that neurons in the PI of *P. americana* which generate somatic action potentials contain numerous large, electron-opaque neurosecretory granules, whereas those neurons which do not show somatic action potentials lack neurosecretory granules. Thompson et al.[18] injected the dye Lucifer yellow into electrically active somata in the protocerebrum of *D. punctata* to locate their terminal branches. Most of the neurons of the PI labeled in this manner showed extensive terminal arborizations in the contralateral corpus cardiacum (CC) (Figure 2B). In a few cases, axons of neurons passed through the contralateral CC and hypocerebral ganglion to the recurrent nerve and lacked any terminal branches, and a few neurons showed terminals in both the contralateral CC and CA. In no case were the terminals of the neurons originating in the PI restricted only to the contralateral CA. Some neurons of the PL had extensive terminal arborizations in the ipsilateral CA and, in some cases, the terminals appeared to be restricted to the CA (Figure 2A). It was also found that intrinsic neurons of the CC had terminal branches in the cap region of the CA as well as throughout the CC[18] (Figure 2C). Because of their close proximity to the CA, these neurons previously had been undetected by other methods of visualization.

One interesting feature of the protocerebral neurosecretory system of *D. punctata* is that small groups of neurons are coupled by low-resistance pathways. Thompson et al.[18] observed that injecting dye into one cell body labeled one or more other cell bodies about 50% of the time. Also, when electrical current was injected into one neuron, membrane responses could be recorded from other neurons nearby. This provides another parallel with the vertebrate hypothalamic-pituitary system, in which small groups of neurosecretory cells in the hypothalamus are coupled by low-resistance pathways.[22] Although the functional significance of this coupling is unknown, it has been proposed that such coupling may be a means of coordinating metabolic and electrical activity among neurons that secrete the same hormone.[22]

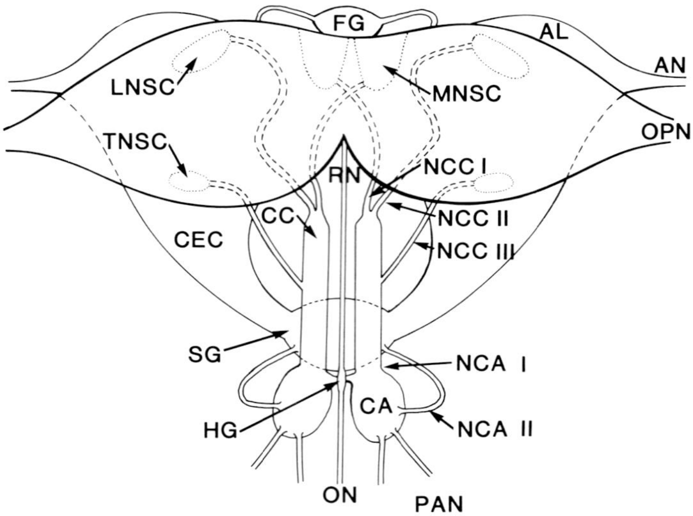

FIGURE 1. Diagramatic representation of brain-retrocerebral complex of *D. punctata* (dorsal view). AL: antennal lobe; AN: antennal nerve; CA: corpora allata; CC: corpora cardiaca; CEC: circumoesophageal connective; FG: frontal ganglion; HG: hypocerebral ganglion; LNSC: lateral neurosecretory cells located in the pars lateralis (PL); MNSC: medial neurosecretory cells located in the pars intercerebralis (PI); NCA: nervi corporis allati; NCC: nervi corporis cardiaci; ON: esophageal nerve; OPN: optic nerve; PAN: postallatal nerves; RN: recurrent nerve; SG: subesophageal ganglion; TNSC: tritocerebral neurosecretory cells.

III. CALCIUM AND CYCLIC AMP

A major advance in the study of JH production in insect species came with the development of an *in vitro* radiochemical assay sensitive enough to accurately measure the quantity of JH produced by a single pair of CA[7,23,24] over a short interval. This has permitted the precise determination of the rates of JH biosynthesis at different stages in the life cycle of various species, as well as the examination of factors which may be involved in the control of JH biosynthesis and release (see, for example, References 3 and 25). Recently, the role of the putative second messengers calcium and cAMP in these processes has been examined in *D. punctata*.

Kikukawa et al.[26] have shown that release of JH by the CA is almost totally inhibited in the absence of extracellular calcium ions and increases in a dose-dependent manner, with maximal release occurring between 3 and 5 m*M* calcium. Furthermore, when CA are incubated in medium containing either the calcium channel blocker lanthanum or the calcium ionophore A23187, JH release is strongly inhibited. Other agents known to alter calcium flux across the cell membrane are also able to modulate JH release *in vitro*. In calcium-free medium, rates of JH release fall to very low levels and only small quantities of JH are retained within the glands. However, stimulation of JH biosynthesis by the addition of the JH precursor farnesoic acid[23,24] in calcium-free medium results in substantial release of JH and in a significant accumulation of both JH and its immediate precursor, methyl farnesoate, within the glands. Thus, calcium ions play a role in the regulation of both the biosynthesis of JH and its release from the CA *in vitro*. Whether changes in intracellular calcium ion concentration regulate biosynthesis or release of JH *in vivo* remains to be determined. Provided that extracellular calcium is present *in vitro*, the CA of *D. punctata* store little JH (~10% of that released per hour) and the rate of release is precisely correlated with the rate of biosynthesis over the entire range of CA activity.[7,24]

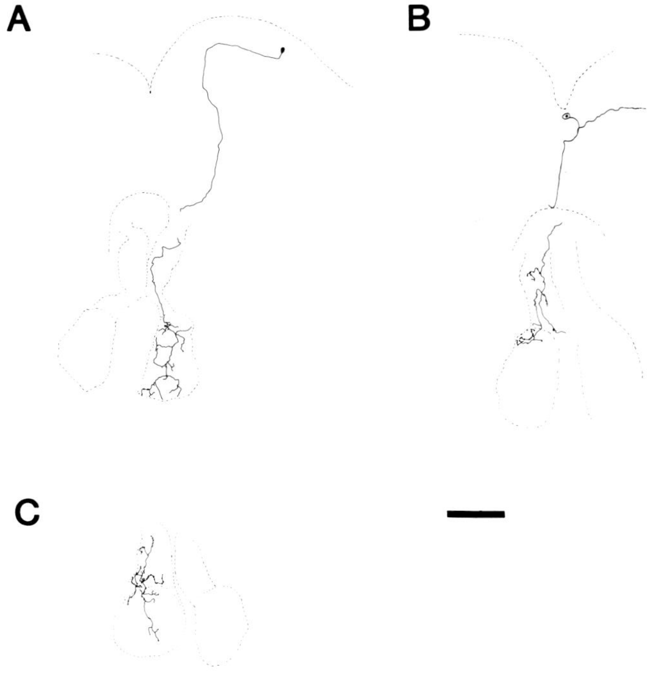

FIGURE 2. Neurons labeled by injecting Lucifer yellow into nerve terminal varicosities on the surface of the CA. (A) A lateral neurosecretory cell; (B) a medial neurosecretory cell; (C) an intrinsic neuron of the CC. Because these neurons were labeled by injecting terminal varicosities, the dendrites in the brain (A and B) were not labeled extensively. (Scale bar = 200 μm.)

The CA of *D. punctata* contain high levels of cAMP (up to 0.4 pmol per pair) and show a dose-dependent accumulation of cAMP following incubation in medium containing the adenylate cyclase activator forskolin and the phosphodiesterase inhibitor 3-isobutyl-1-methylxanthine (IBMX).[27] A number of observations indicate that cAMP may be involved in the regulation of JH biosynthesis and release by the CA. Incubation of CA of adult females with forskolin, IBMX, or 8-bromo-cAMP causes a rapid, dose-dependent inhibition of JH biosynthesis.[27] Incubation of CA in medium with an elevated potassium ion concentration also results in an increase in the cAMP content of the glands[28] and an inhibition of JH biosynthesis.[29] In addition, treatment of CA with aqueous extracts of the protocerebral lobes of the brain results in a dose-dependent elevation of cAMP content and an inhibition of JH biosynthesis.[30] These observations support the hypothesis that cAMP acts as a second messenger for an allatostatin that is released from neurosecretory cell terminals within the CA.

Additional experimental results indicate that calcium ions may play a role in the control of JH biosynthesis by cAMP. Incubation of CA in calcium-free medium causes a dramatic increase in cAMP content. On the other hand, elevated extracellular calcium levels substantially reduce the ability of brain extracts to inhibit JH biosynthesis *in vitro*, although such elevated levels of calcium do not block the ability of brain extracts to elevate the cAMP content of the glands.[30] These results suggest the existence of at least two separate regulatory pathways for JH biosynthesis, one dependent on and the other independent of calcium ions.

IV. ELECTROPHYSIOLOGY OF CORPORA ALLATA CELLS

The CA of adult female *D. punctata* are composed of several thousand irregularly shaped parenchymal cells. Most of these are up to 10 μm in length, although a few spindle-shaped cells are up to 20 μm in length. The parenchymal cells are coupled extensively by low-resistance intercellular pathways which permit the spread of electrical current and the fluorescent dye Lucifer yellow from cell to cell throughout the glands.[31] It is likely that such intercellular communication is mediated by gap junctions, since both transmission and freeze-fracture electron microscopy have revealed the presence of numerous gap junctions between CA cells.[31,32]

When electrical current is injected into a single, electrically isolated cell, a change in the membrane potential of the cell is generated as the current flows across the cell membrane to ground. The amplitude of this passive membrane response, or electrotonic potential, is the product of the current injected and the total membrane resistance (input resistance) of the cell (Ohm's law). When current is injected into a cell which is part of a network of electrically coupled cells, some current will flow through gap junctions into contiguous cells and across their cell membranes to ground, thus eliciting an electrotonic potential. The amplitude of this response will depend on both the gap junctional conductance and the conductance of the nonjunctional membrane in a manner which varies with the topology of the network of coupled cells.[33] Because of the small size of the CA cells of *D. punctata,* it has not been possible to insert two microelectrodes into the same cell without causing significant damage to the cell membrane. Thus, the input resistance of single CA cells and the coupling coefficient between cells (the ratio of the membrane response of a coupled cell to the membrane response to the injected cell during a current pulse) have not been determined accurately. It is possible to inject current into one cell and record membrane responses from other cells in the gland, however. In the CA of *D. punctata,* when current is injected into one cell and electrotonic potentials from other cells are recorded with a second microelectrode, there is little decrement in amplitude as the recording electrode is moved farther away from the current-passing electrode. The distance over which the electrotonic potential would show a significant decrease in amplitude is greater than the length of the gland itself.[31] Thus, it is likely that gap junctional conductance is large relative to nonjunctional conductance and that the coupling coefficient is close to unity. All of the cells of a gland appear to be coupled together, forming an electrical syncytium, since not a single instance of an uncoupled cell has been observed in thousands of penetrations.[44] However, the possibility that some cells in the center of a gland are uncoupled from the rest has not been eliminated. Electrical coupling was observed in every CA preparation from animals of the fourth stadium through day 13 of adulthood. The presence of such cell-to-cell coupling is a common feature of endocrine glands.[34] In many cases, however, electrical coupling is not as extensive as that found in the CA of *D. punctata.*[35]

When current pulses of differing polarity and amplitude are injected into one CA cell and electrotonic potentials from another cell are recorded, the resulting current/voltage (I/V) plot appears similar to that of many nonexcitable cell types (Figure 3). Current-voltage relationships are linear when negative current (hyperpolarizing) is injected, but show a pronounced decrease in slope with increasing amplitude of positive current (depolarizing) injection. This could be due to either an increase in the nonjunctional conductance of the cell membranes of the CA or a decrease in the gap junctional conductance, or both. The observation that the time constant of decay of the electrotonic potential (the time required for the electrotonic potential to decay to 1/e of its steady-state value) decreases with increasing amplitude of depolarizing current pulses (Figure 3) indicates that an increase in membrane conductance accounts for much of the decrease in slope. The time constant is the product of membrane resistance and capacitance and should not be influenced by gap junctional conductance, nor should membrane capacitance change. However, the possibility that a decrease in gap junctional conductance occurs during positive current injection cannot be excluded. In many tissues an elevation of intracellular calcium ion concentration has been

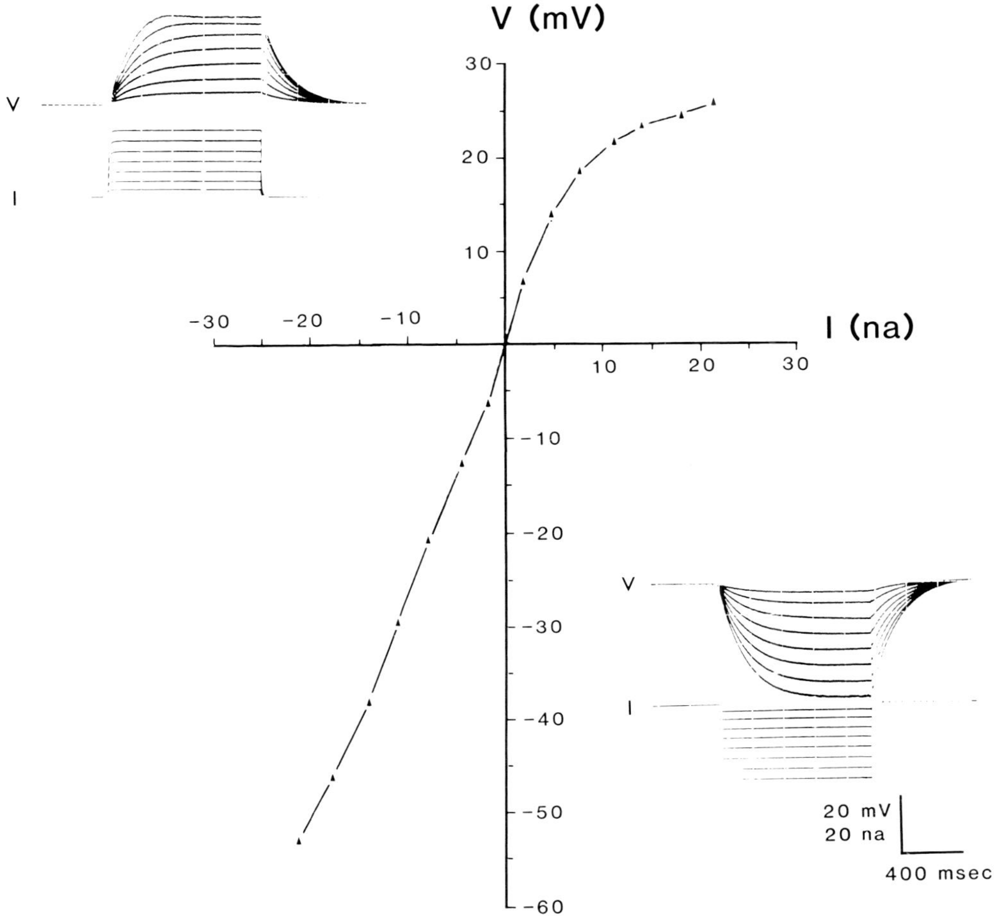

FIGURE 3. Typical current/voltage (I/V) relationships of the network of electrically coupled CA cells. I/V relationships are linear in the hyperpolarizing direction, but show a decrease in slope in the depolarizing direction. These data are from a corpus allatum of a day 11 mated female. Examples of the data used to construct such curves are shown in the upper left (depolarizing) and lower right (hyperpolarizing) quadrants; these were obtained from another preparation. Resting potential is set at 0 on the y axis.

shown to decrease gap junctional conductance, although very high levels of calcium typically are required to effect a significant change.[36] Thus, if calcium ions were to enter the CA cells during depolarizing current pulses (see below), a reduction in gap junctional conductance might occur. Also, although not a common phenomenon, several cases of rectifying gap junctions have been described.[36] If the gap junctions between CA cells were to pass negative current more easily than positive current, then a decrease in slope of the I/V curve might occur in the depolarizing direction.

In normal cockroach saline,[19] the "input resistance" (ratio of electrotonic potential to current injected) of the network of electrically coupled CA cells ranges from 1 to 3 MΩ.[31] Values of up to 10 MΩ have been observed following the addition of the potassium channel blockers tetraethylammonium (TEA) ions and 4-aminopyridine (4-AP) to the saline.[44] Under these conditions, positive current pulses may elicit active responses, or action potentials, from the cell membranes of the CA (Figure 4A).[37] In most other excitable cells, action potentials are attributable to the opening of voltage-sensitive cation channels followed by an influx of either sodium ions or calcium ions, or both, down their electrochemical gradients

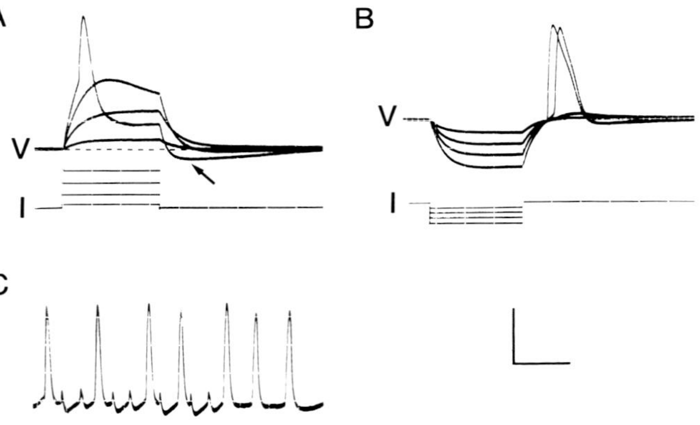

FIGURE 4. Examples of action potentials generated by CA cell membranes. (A) In the presence of TEA (40 m*M*) and 4-AP (200 μ*M*), depolarizing current pulses elicit action potentials above a threshold depolarization of about 25 mV. These action potentials are followed by a large decrease in the electronic potential and an after-hyperpolarization at the end of the current pulse (arrow). (B) In the presence of TEA, 4-AP, and barium ions (15 m*M*), action potentials may follow hyperpolarizing pulses. (C) In some preparations, spontaneous action potentials occur following perfusion with saline containing TEA, 4-AP, and barium ions. Scale bars: (A,B) vertical — upper trace, 20 mV; lower trace, 20 nA; horizontal — 1.0 s. (C) vertical — 10 mV; horizontal — 4.0 s.

into the cells.[38] In the CA cells of *D. punctata*, action potentials occur at a threshold depolarization of 15 to 30 mV and are followed by a large decrease in electrotonic potential and a pronounced after-hyperpolarization (Figure 4A). Several observations indicate that these action potentials are mediated by both sodium ions and calcium ions flowing through voltage-sensitive channels. The spikes are blocked by the sodium channel blocker tetrodotoxin and sodium-free saline. They are also blocked by the calcium channel blockers cobalt and lanthanum ions.[37] Barium ions, on the other hand, are known to pass through calcium channels easily and to block potassium channels.[38] In the presence of barium ions, the CA cells initially become hyperexcitable. Action potentials increase in amplitude and duration, and spontaneous spikes may occur (Figure 4B). Spikes also may occur following hyperpolarizing current pulses (Figure 4C). The decrease in electronic potential and the after-hyperpolarization which follows action potentials could be the result of the entry of calcium into the CA cells followed by the activation of a calcium-sensitive potassium conductance,[39] although this remains to be proved.

The above experimental findings are preliminary in nature, but they indicate that the CA cell membranes of *D. punctata* do contain a variety of ion channels and that some of them are voltage sensitive. Whether or not action potentials occur under normal circumstances *in vivo* is not known.

V. RESPONSES OF CORPORA ALLATA CELLS TO NERVE STIMULATION

Electrical stimulation of axons innervating the CA of *D. punctata* produces two distinct types of response. Stimulation of NCA I, containing axons of the medial and lateral protocerebral neurosecretory cells, elicits a hyperpolarization of the CA cells and an increase in the electrotonic potential (Figure 5A). These responses occur at a sharp threshold as stimulating voltage is increased from zero, and they have a relatively long latency. On the

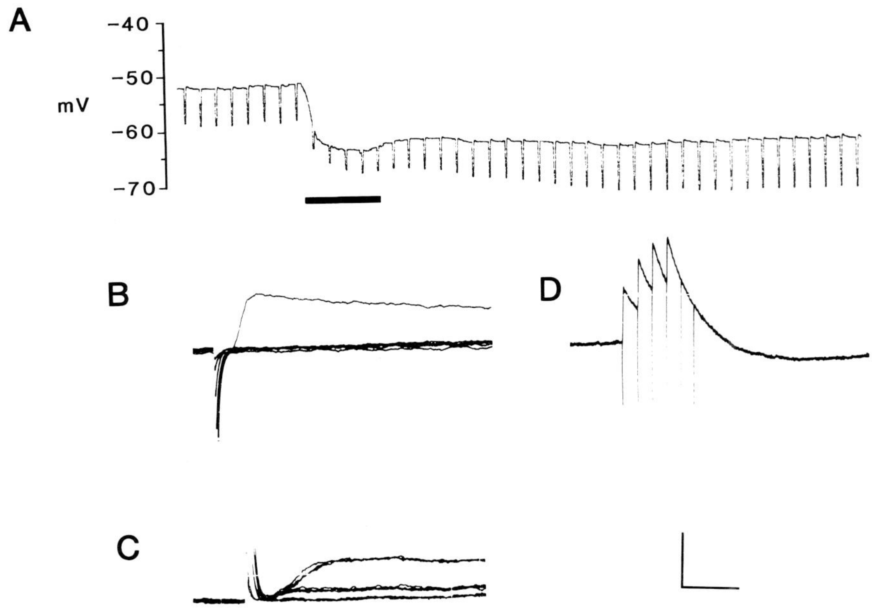

FIGURE 5. Responses of CA cell membranes to electrical stimulation of axons innervating the CA. (A) Current pulses (−5.0 nA, 1700-ms duration) were injected into one CA cell every 15 s, and electronic potentials (downward deflections) were recorded from another CA cell with a second microelectrode. Electrical stimulation of NCA I (0.5-ms pulses at 5 Hz; bar) elicited a hyperpolarization and a gradual increase in the size of the electrotonic potential. (B,C,D) stimulation of unidentified axons elicited apparent EPSPs in CA cells. (B) Six superimposed sweeps of the oscilloscope are shown. Stimulating voltage was increased slightly after each sweep. The final (largest) stimulus pulse elicited a short-latency depolarization of the CA cells. (C) Several sweeps of the oscilloscope are shown. Stimulus voltage was varied between each sweep. Two distinct amplitudes of EPSPs are apparent. (D) The all-or-none nature of the EPSPs is shown. Six stimulus pulses were delivered, but only the first four elicited an EPSP. Scale bars: (A) vertical — 10 mV; horizontal — 1.5 min; (B) 4 mV, 20 ms; (C) 4 mV, 10 ms; (D) 4 mV, 400 ms.

other hand, short-latency depolarizing responses also may be elicited by electrical stimulation (Figure 5B,C). These also occur at a sharp threshold and resemble conventional excitatory postsynaptic potentials (EPSPs).[40] The location of the axons eliciting the latter responses is uncertain. Higher stimulating voltages are typically required to elicit such responses, and fine axon branches within the CA may be activated by current spread from the nearby stimulating electrode.

These observations indicate that activity in neurons innervating the CA of *D. punctata* can produce both depolarizing and hyperpolarizing responses. The neurotransmitters involved and the ionic basis of these responses remain unknown, as does their role in the control of JH biosynthesis and release.

Changes in membrane potential of CA cells have been correlated with various conditions which alter JH release *in vitro*. High-potassium saline produces a marked inhibition of JH biosynthesis and release as well as an increase in the cAMP content of the glands, as reviewed above. High-potassium saline depolarizes CA cells and reduces the electrotonic potential (Figure 6B). On the other hand, aqueous extracts of the brain, which inhibit JH release and elevate cAMP content *in vitro*,[29] elicit a hyperpolarization of the CA cells and an increase in the electrotonic potential (Figure 6A). The application of forskolin and IBMX to the CA also causes a hyperpolarization of the CA cells and an increase in electrotonic potential, as well as an elevation of cAMP and an inhibition of JH release.[31]

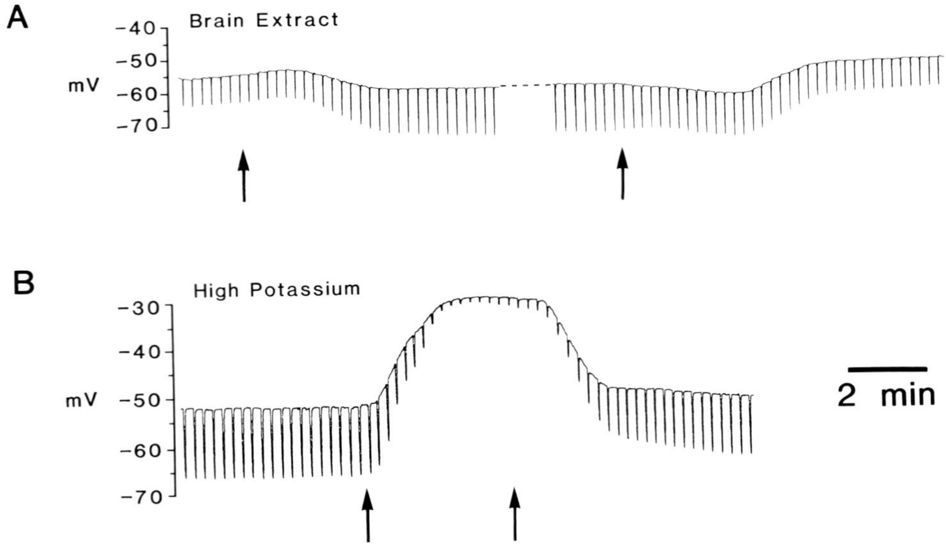

FIGURE 6. Responses of CA cells to brain extract and high-potassium saline. (A) Current pulses (−5.0 nA, 1700 ms) were injected into one CA cell and electrotonic potentials recorded from another cell with a second microelectrode. The preparation initially was perfused with normal saline. Brain extract (0.5 brain equivalents per 100 μl saline) entered the preparation dish at the first arrow. At the second arrow, perfusion with normal saline resumed. A total of 8.0 min was deleted from the continuous record. (B) Current pulses were injected into one CA cell, as in A. The preparation was perfused continually, first with normal saline (12 m*M* K^+), then with high-potassium saline (56 mM K^+) (first arrow), and, finally, with normal saline (second arrow).

Thus, it appears that JH biosynthesis and release bear no simple relationship to the membrane potential of the CA cells *in vitro*. *In vivo,* however, membrane potential may act to regulate calcium flux across CA cell membranes. It is likely that the level of intracellular calcium ions plays an important role in controlling JH biosynthesis and release as well as other cellular functions.

VI. SUMMARY AND CONCLUSIONS

Our understanding of the neural control of JH biosynthesis by the CA of *D. punctata* is far from complete; however, a general picture is beginning to emerge. Anatomical studies have revealed groups of neurons in three distinct locations within the nervous system which have terminal arborizations on or in the CA. Electrical stimulation of axon tracts near the CA indicates that at least some of these neurons make functional contact with CA cells: both long-term hyperpolarizing responses and brief depolarizing responses may be elicited by such stimulation. The ability to selectively stimulate these groups of neurons should aid in identifying which neurons make functional contact with CA cells and which neurotransmitters are involved. Also, measurements of CA cell membrane potentials and electrotonic potentials provide a relatively simple assay which can be used in attempting to purify neurotransmitters or neurohormones involved in control of the CA.

While the precise effects of electrical stimulation on JH biosynthesis remain unknown, there is some evidence for the involvement of Ca^{2+} and cAMP in the transduction of neural signals. The application of potassium channel blockers to CA cells reveals that they are not electrically unexcitable, as they initially appear to be, which is also the case with several other cell types.[41] This is consistent with the idea that CA cell membranes have voltage-sensitive calcium channels and that membrane potential is an important determinant of the level of ionized calcium in the cytoplasm. It also appears that CA cell membranes have receptors which are coupled to adenylate cyclase and that neurotransmitters or neurohormones act in the control of intracellular levels of cAMP.

The usefulness of the CA of cockroaches as potential biomedical models is somewhat limited by the fact that so much more is known about vertebrate endocrine glands than about insect CA. Also, CA cells are relatively small and have not yet been grown successfully in tissue culture. However, there appear to be many fundamental similarities between steroid-secreting endocrine cells of vertebrates and the CA cells of insects. The biosynthetic pathway for JH appears to be identical to that for sterol production up to the branchpoint of farnesyl pyrophosphate. Also, there is increasing evidence that intracellular signal molecules and control mechanisms have been highly conserved during the course of evolution. Thus, the CA may provide a system in which to develop alternate strategies to regulate sterol biosynthesis.

The parenchymal cells of the CA are intercoupled by gap junctions. This is a feature common to many endocrine glands, but coupling in the CA appears to be more extensive than in most vertebrate endocrine glands. It has recently been demonstrated that antibodies to gap junction proteins can alter the properties of gap junctions *in vivo* and produce developmental abnormalities in several invertebrate and vertebrate species.[42] The CA may provide a good model system in which to study the role of gap junction communication in the control of hormone biosynthesis by the nervous system, and other cellular processes as well. For example, when the CA of adult female cockroaches undergo a cycle of JH biosynthesis, there is an increase of about 50% in both cell size and cell number.[43] Thus, the nervous system is acting to control the size of the CA and, again, gap junction communication may be involved.

REFERENCES

1. **Willey, R. B.,** The morphology of the stomodeal nervous system in *Periplaneta americana* (L.) and other Blattaria, *J. Morphol.*, 13, 219, 1961.
2. **Cassier, P.,** The corpora allata of insects, *Int. Rev. Cytol.*, 57, 1, 1979.
3. **Tobe, S. S. and Stay, B.,** Structure and regulation of the corpus allatum, *Adv. Insect Physiol.*, 18, 305, 1985.
4. **Engelmann, F.,** *The Physiology of Insect Reproduction*, Pergamon Press, Oxford, 1970.
5. **de Kort, C. A. D. and Granger, N. A.,** Regulation of juvenile hormone titer, *Annu. Rev. Entomol.*, 26, 1, 1981.
6. **Wigglesworth, V. B.,** Historical perspectives, in *Comprehensive Insect Physiology, Biochemistry and Pharmacology*, Vol. 7, Kerkut, G. A. and Gilbert, L. I., Eds., Pergamon Press, Oxford, 1985, 1.
7. **Tobe, S. S. and Pratt, G. E.,** Dependence of juvenile hormone release from corpus allatum on intraglandular content, *Nature (London)*, 252, 474, 1974.
8. **Tobe, S. S. and Pratt, G. E.,** The influence of substrate concentrations on the rate of insect juvenile hormone biosynthesis by corpora allata of the desert locus *in vitro*, *Biochem. J.*, 114, 107, 1974.
9. **Feyereisen, R.,** Regulation of juvenile hormone titer: synthesis, in *Comprehensive Insect Physiology, Biochemistry and Pharmacology*, Vol. 7, Kerkut, G. A. and Gilbert, L. I., Eds., Pergamon Press, Oxford, 1985, 391.
10. **Lococo, D. J. and Tobe, S. S.,** Neuroanatomy of the retrocerebral complex, in particular the pars intercerebralis and partes laterales in the cockroach *Diploptera punctata* Eschscholtz (Dictyoptera: Blaberidae), *Int. J. Insect Morphol. Embryol.*, 13, 65, 1984.
11. **Tobe, S. S and Stay, B.,** Corpus allatum activity *in vitro* during the reproductive cycle of the viviparous cockroach *Diploptera punctata* (Eschscholtz), *Gen. Comp. Endocrinol.*, 31, 138, 1977.
12. **Carrow, G. M., Calabrese, R. L., and Williams, C. M.,** Spontaneous and evoked release of prothoracicotropin from multiple neurohemal organs of the tobacco hornworm, *Proc. Natl. Acad. Sci. U.S.A.*, 78, 5866, 1981.

13. **Lococo, D. J.**, Neuronal and Intercellular Communication in the Corpus Allatum of *Diploptera punctata*, Ph.D. thesis, University of Toronto, Toronto, Ontario, Canada, 1985.
14. **Lococo, D. J. and Tobe, S. S.**, Retrograde and orthograde axon transport by brief exposure to nickel chloride: methodology and parameters for success in the brain-retrocerebral complex of the cockroach *Diploptera punctata*, *J. Insect Physiol.*, 30, 635, 1984.
15. **Ruegg, R. P., Lococo, D. J., and Tobe, S. S.**, Control of corpus allatum activity in *Diploptera punctata:* roles of the pars intercerebralis and pars lateralis, *Experientia*, 39, 1329, 1983.
16. **Pipa, R. L.**, Neural influences on corpus allatum activity and egg maturation in starved virgin *Periplaneta americana*, *Physiol. Entomol.*, 7, 449, 1982.
17. **Stay, B. and Tobe, S. S.**, Control of juvenile hormone biosynthesis during the reproductive cycle of a viviparous cockroach. I. Activation and inhibition of corpora allata, *Gen. Comp. Endocrinol.*, 33, 531, 1977.
18. **Thompson, C. S., Lococo, D. J., and Tobe, S. S.**, Anatomy and physiology of neurons terminating in the corpora allata of the cockroach *Diploptera punctata*, *J. Comp. Neurol.*, 261, 120, 1987.
19. **Krauthamer, V.**, Electrophysiology of identified neurosecretory and non-neurosecretory cells in the cockroach pars intercerebralis, *J. Exp. Zool.*, 234, 207, 1985.
20. **Krauthamer, V.**, Morphology of identified neurosecretory and non-neurosecretory cells in the cockroach pars intercerebralis, *J. Exp. Zool.*, 234, 221, 1985.
21. **Scharrer, B.**, Neurosecretion. XIII. The ultrastructure of the corpus cardiacum of the insect *Leucophaea maderae*, *Z. Zellforsch.*, 60, 761, 1963.
22. **Renaud, L. P., Bourgne, C. W., Day, T. A., Ferguson, A. V., and Randle, J. C. R.**, Electrophysiology of mammalian hypothalamic supraoptic and paraventricular neurosecretory cells, in *The Electrophysiology of the Secretory Cell*, Poisner, A. M. and Trifaro, J. M., Eds., Elsevier, New York, 1985, 165.
23. **Pratt, G. E. and Tobe, S. S.**, Juvenile hormones radiobiosynthesized by corpora allata of adult female locusts *in vitro*, *Life Sci.*, 14, 575, 1974.
24. **Feyereisen, R., Tobe, S. S., and Friedel, T.**, Farnesoic acid stimulation of C_{16} juvenile hormone biosynthesis by corpora allata of adult female *Diploptera punctata*, *Insect Biochem.*, 11, 401, 1981.
25. **Kikukawa, S. and Tobe, S. S.**, Juvenile hormone biosynthesis in female larvae of *Diploptera punctata* and the effect of allatectomy on haemolymph ecdysteroid titre, *J. Insect Physiol.*, 32, 981, 1986.
26. **Kikukawa, S., Tobe, S. S., Solowiej, S., Rankin, S. M., and Stay, B.**, Calcium as a regulator of juvenile hormone biosynthesis and release in the cockroach *Diploptera punctata*, *Insect Biochem.*, 17, 179, 1987.
27. **Meller, V. H., Aucoin, R. R., Tobe, S. S., and Feyereisen, R.**, Evidence for an inhibitory role of cyclic AMP in the control of juvenile hormone biosynthesis by cockroach corpora allata, *Mol. Cell. Endocrinol.*, 43, 155, 1985.
28. **Aucoin, R. R.**, The Role of Cyclic AMP and Calcium in the Regulation of JH Biosynthesis in *Diploptera punctata*, Master's thesis, University of Toronto, Toronto, Ontario, Canada, 1986.
29. **Rankin, S. M., Stay, B., Aucoin, R. R., and Tobe, S. S.**, *In vitro* inhibition of juvenile hormone synthesis by corpora allata of the viviparous cockroach, *Diploptera punctata*, *J. Insect Physiol.*, 32, 151, 1986.
30. **Aucoin, R. R., Rankin, S. M., Stay, B., and Tobe, S. S.**, Calcium and cyclic AMP involvement in the regulation of juvenile hormone biosynthesis in *Diploptera punctata*, *Insect Biochem.*, 17, 965, 1987.
31. **Lococo, D. J., Thompson, C. S., and Tobe, S. S.**, Intercellular communication in an insect endocrine gland, *J. Exp. Biol.*, 121, 407, 1986.
32. **Johnson, G. D., Stay, B., and Rankin, S. M.**, Ultrastructure of corpora allata of known activity during the vitellogenic cycle in the cockroach *Diploptera punctata*, *Cell Tissue Res.*, 239, 317, 1985.
33. **Socolar, S. J.**, The coupling coefficient as an index of junctional conductance, *J. Membr. Biol.*, 34, 29, 1977.
34. **Meda, P., Perrelet, A., and Orci, C.**, Gap junctions and cell-to-cell coupling in endocrine glands, *Mod. Cell Biol.*, 3, 131, 1984.
35. **Petersen, O. H.**, *The Electrophysiology of Gland Cells*, Monogr. Physiol. Soc. No. 36, Academic Press, London, 1980, 253.
36. **Spray, D. C. and Bennett, M. V. L.**, Physiology and pharmacology of gap junctions, *Annu. Rev. Physiol.*, 47, 281, 1985.
37. **Thompson, C. S. and Tobe, S. S.**, Electrical properties of membranes of cells of the corpora allata of the cockroach *Diploptera punctata:* evidence for the presence of voltage-sensitive calcium channels, in *Insect Neurochemistry and Neurophysiology*, Borkovec, A. B. and Gelman, D. B., Eds., Humana Press, Clifton, NJ, 1986, 375.
38. **Hille, B.**, *Ionic Channels of Excitable Membranes*, Sinauer Associates, Sunderland, MA, 1984.
39. **Petersen, O. H. and Maruyama, Y.**, Calcium-activated potassium channels and their role in secretion, *Nature (London)*, 307, 693, 1984.
40. **Thompson, C. S., Lococo, D. J., and Tobe, S. S.**, Cell coupling and electrical activity in the corpora allata of the cockroach *Diploptera punctata*, *Neurosci. Abstr.*, 10, 480, 1984.

41. **O'Donnell, M. J.,** Potassium channel blockers unmask electrical excitability of insect follicles, *J. Exp. Zool.,* 245, 137, 1988.
42. **Marx, J. L.,** Symposium focuses on genes in development: gap junctions needed for development, *Science,* 239, 1493, 1988.
43. **Szibbo, C. M. and Tobe, S. S.,** Cellular and volumetric changes in relation to the activity cycle in the corpora allata of *Diploptera punctata, J. Insect Physiol.,* 27, 655, 1981.
44. **Thompson, C. S.,** unpublished data.

Chapter 6

NEUROANATOMICAL METHODS

Jonathan M. Blagburn and David B. Sattelle

TABLE OF CONTENTS

I. INTRODUCTION

With limited space available to cover an area to which several volumes have already been devoted, we will mainly focus on techniques with which we have personal experience and which we know are applicable to cockroaches. In view of the need for brevity, we apologize to those workers who would like to see more detailed methods included and refer them to more comprehensive works — in particular, two excellent volumes devoted to insect neuroanatomy, edited by Strausfeld and Miller[1] and by Strausfeld.[2]

In this chapter, we examine optical sectioning of unstained nervous tissue and general staining methods, together with more specific histochemical techniques for localizing cell types and cell components. Recent advances in immunocytochemistry offer the prospect of detailed mapping of cell constituents, especially following the successful application of monoclonal antibodies to particular peptides and even amino acids. Autoradiography and *in situ* hybridization techniques also offer complementary approaches for the localization of specific cell components. Approaches to neuronal morphology include the Golgi method, degeneration studies, and a wide range of intracellular marking techniques, most notably cobalt, horseradish peroxidase (HRP), and fluorescent dye injection. While advances in electron microscopy continue to add to our understanding of insect neuroanatomy, they are only discussed briefly in the present chapter.

The majority of the techniques described have been applied to vertebrate and invertebrate preparations; some, such as cobalt staining or fluorescent dye injection, were first attempted in insects or molluscs and subsequently acquired importance in the biomedical field. For studies which utilize intracellular microelectrodes to combine electrophysiological recording with dye injection, insect ganglia are particularly attractive because of the relatively small number of nerve cells, many of which are of large size and unique identity, organized with their cell bodies in a peripheral rind or cortex.

II. OPTICAL SECTIONING OF NERVOUS TISSUE WITH NOMARSKI OPTICS

This technique, although limited in its usefulness to tissues of favorable dimensions and degree of transparency, is the most rapid of the anatomical techniques since it is performed on living material. Optical sectioning with Nomarski optics is ideally suited to identifying cell bodies or other regions of neurons prior to impalement with intracellular microelectrodes[3-5] (Figure 1). A tissue which is amenable to being examined in this way is also likely to be a suitable candidate for studies utilizing voltage- or calcium-sensitive dyes and optical recording techniques.

Insect ganglia and those of many other invertebrates are organized so that the inner neuropil contains the axons and dendritic branches and their synaptic connections. The neuronal cell bodies are generally devoid of synaptic contacts and are concentrated in an outer layer, in contrast to the cell bodies of vertebrate neurons. This arrangement gives the advantage that the cell body forms a convenient and relatively easy site in which to place intracellular microelectrodes. However, it is the axons and dendrites which are the sites of functional interactions. The particular advantage of Nomarski optics is that the technique allows the experimenter to see structures which are inside the tissue; this is particularly useful when it is necessary to place microelectrodes in them. Vertebrate central nervous system (CNS) tissues are generally too large to be suitable for this technique, although there are notable exceptions in the cases of larval fish and amphibians. However, Nomarski optics can be useful in visualizing the neurons within brain slices.

It is necessary to use a microscope equipped with Nomarski (differential interference contrast) optics. Zeiss® Nomarski is recommended. A 40× water-immersion objective lens of approximately 0.6 to 0.7 numerical aperture (N.A.) gives a suitable magnification, allowing axons of 5 μm diameter to be visualized easily, but also giving a good overview of the tissue. In order to obtain good visibility within the tissue, it should be no thicker than 200 μm and should not contain pigment. It may be necessary to adjust the osmolarity of the saline slightly in order to improve the visibility of cells within the tissue. Under good conditions it is possible to visualize all neuronal cell bodies in a living ganglion from a first- or second-instar cockroach nymph and, by focusing through the ganglion, neurites, axons, and major dendritic branches are revealed.

If it is only necessary to examine the tissue, any type of microscope will suffice; however, if examination is to be coupled with intracellular recording or dye injection, then two important conditions should be fulfilled. First, the working distance of the objective should be at least 1 to 2 mm. This allows microelectrodes to be mounted at an angle of 30 to 45° such that they readily clear the walls of the experimental chamber. Suitable objectives include the Zeiss® Achromat 40×/0.75 N.A. water immersion. Second, if more than one cell is to be impaled with microelectrodes, the micromanipulators must be fixed relative to the stage. Most modern upright microscopes are designed to focus by moving the stage, an arrangement

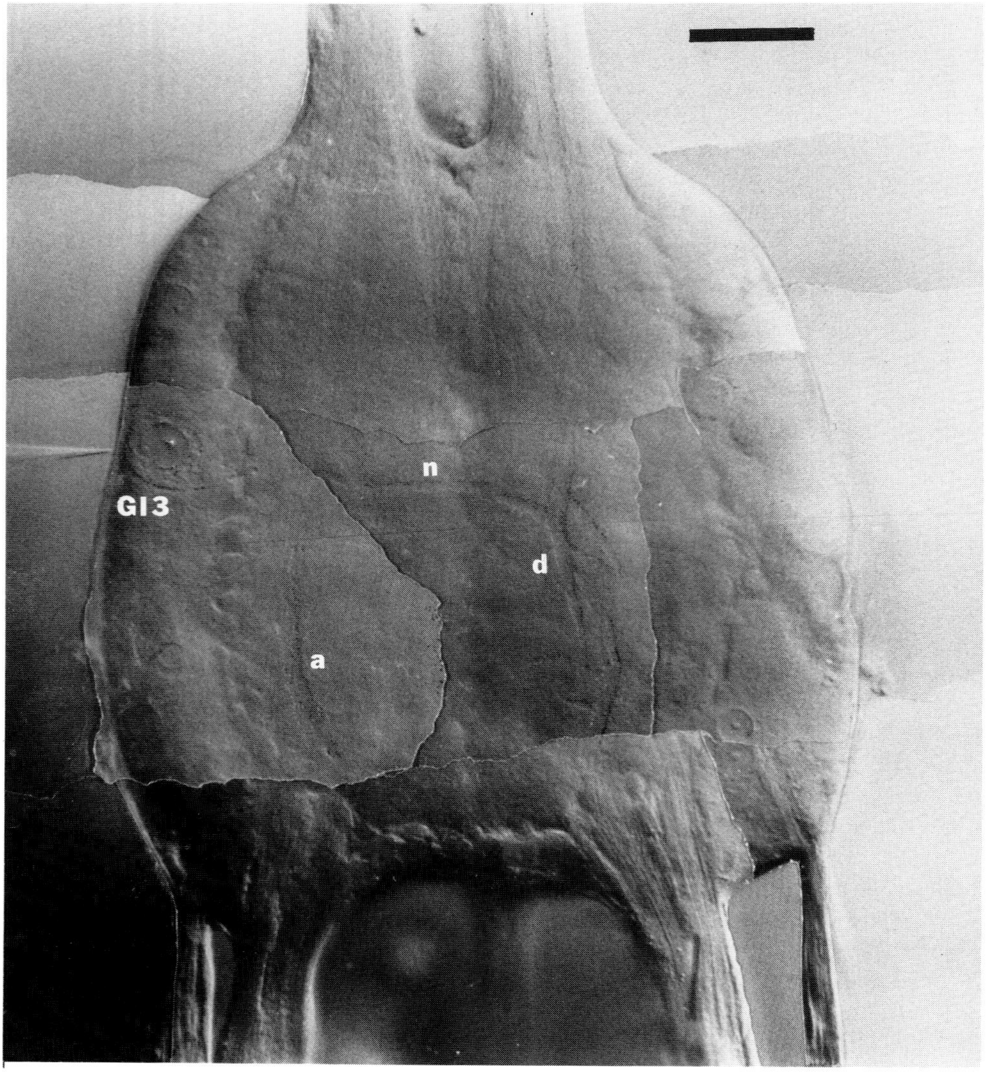

FIGURE 1. Montage of photographs of the terminal abdominal ganglion of first-instar *Periplaneta americana*, viewed with Nomarski optics. Some of the different parts of the picture were taken at slightly different planes of focus. The anterior of the ganglion gives rise to the interganglionic connectives (top), while the cercal nerves arise from the posterior. The tip of a microelectrode can been seen on the left, touching the cell body of giant interneuron 3, in which can be seen the nucleus and nucleolus. The cell body, located in the outer cortex of neuronal somata, sends a neurite (n) into the neuropil, where it crosses to the other side of the ganglion and forms dendritic branches. The primary dendrite (d) is visible. Large sensory axons (a) are also visible within the neuropil. (Scale bar: 50 μm.)

which precludes their use in multiple-electrode electrophysiological studies unless the manipulators holding the electrodes are mounted directly onto the stage. The Microtec® M2 manipulation microscope (Micro Instruments, Oxford, England) is designed specifically for such studies, while its counter-balanced head also allows the mounting of heavy camera equipment. Alternatively, an existing microscope can be modified in a workshop facility. An inverted microscope equipped with Nomarski optics may be used if the tissue is thin, but this type of microscope will not be suitable for visualizing cells on the upper surface of tissue more than about 100 μm thick.

III. GENERAL STAINING TECHNIQUES

A. METHYLENE BLUE

Intra vitam methylene blue staining can be used to stain nerve fibers in the insect nervous system, thereby providing an overview of the structures in the ganglionic neuropil.[6,7] Precise information about the shapes of individual nerve cells is best obtained using intracellular tracers (see Section IV.C).

Methylene blue (1 to 2.5 ml of a 0.08 to 1% solution of methylene blue in double-distilled water containing 0.75 to 2.25% NaCl or physiological saline) is injected into the hemocoel of the insect so that the body wall is distended. For peripheral elements, the insect is left at room temperature for 0.5 to 1 h before dissection (for central neurons, 1 to 5 h is sufficient). The best results are obtained with newly molted animals. After the staining period, the animal is rapidly dissected and flooded with fixative (12% ammonium molybdate). The tissues are transferred into fixative at 4°C for 7 to 16 h. The addition of two to four drops of 0.25% OsO_4 solution added to 50 ml ammonium molybdate improves fixation, possibly at the expense of staining density. After washing in distilled water, the tissue is dehydrated in three changes of absolute ethanol (the first change being no longer than 30 s because nonabsolute ethanol extracts the stain). After clearing in xylene the preparation is mounted in dammar-xylol, as Canada balsam, a more convenient mounting medium, also leaches out the stain.

B. ACRIDINE ORANGE (FOR LIVING CELLS)

Acridine orange (0.1 μg/ml in insect saline or culture medium) can be used to assess neuronal viability. Under UV illumination the nucleus appears orange and the cytoplasm green in living cells. This technique has been used to assess viability of dissociated cockroach neurons maintained in short-term culture[8] and is applicable to all types of dissociated neurons.

C. TOLUIDINE BLUE

Neuronal cell bodies can be localized in whole insect ganglia by staining with toluidine blue.[9,10] This technique serves as a useful preliminary to intracellular recording from identified somata and enables the morphology of single neurons (as delineated, for example, by cobalt staining [Section IV.C.1]) to be related to the overall distribution of cell bodies.

Freshly dissected ganglia are placed in staining solution at 50°C and left for 0.25 to 1 h (the optimal staining time should be determined for the particular preparation and staining conditions). The staining solution consists of 1 g toluidine blue, 6 g borax, and 1 g boric acid in 100 ml distilled water. The tissue is transferred into several changes (5 min each) of Bodian's fixative, which differentiates the stain. This should be continued under visual control until the cell bodies are seen clearly and the neuropil and nerves are white. The fixative consists of 5 ml formalin, 5 ml glacial acetic acid, and 90 ml 80% ethanol. The fixed tissue is dehydrated in two changes of 90% ethanol (5 min each) and two changes of 100% ethanol (10 min each), then cleared in methyl benzoate; the preparations are then mounted in neutral Canada balsam or a synthetic substitute.

D. ETHYL GALLATE

This staining technique is useful for revealing the tracts and commissures of insect ganglia.[11,12] The tissue is prepared as for electron microscopy (see Section V.A). After osmication, the tissue is washed in buffer and then stained for 2.5 h in ethyl trihydrobenzoate (ethyl gallate) before dehydrating and embedding in soft Araldite® and sectioning at 10 μm.

E. REDUCED SILVER IMPREGNATION TECHNIQUES

These somewhat complex methods provide a detailed picture of neuronal cell bodies, large fibers, fiber tracts, and neuropilar regions. A major disadvantage of these techniques is that they are capricious. The Bodian protargol technique is perhaps the simplest; a protocol is given in brief here. The Holmes and Ungewitter reduced silver techniques give similar results and have been dealt with in detail elsewhere.[13,14]

1. Bodian Protargol Technique

This technique has been described in detail by Gregory,[15,16] who has worked extensively with the nervous system of *Periplaneta americana.* A brief description of the technique is given here. The ganglia are fixed for 16 to 24 h at room temperature in "aged" alcoholic Bouin, which consists of 5 ml formaldehyde, 25 ml ethanol, 5 ml acetic acid, 5 ml ethyl acetate, 15 ml diethoxymethane, and 0.5 g picric acid, brought up to 100 ml with distilled water. The tissue is dehydrated, cleared, and embedded in paraffin wax; 20-μm sections are then cut, mounted on slides, and dewaxed.

The sections are impregnated for 18 to 24 h at 37°C in 2% protargol solution, adjusted to pH 8.4 with 10% ammonium hydroxide and containing 1 to 4 g copper; they are then washed briefly in distilled water. The stain is developed for 10 to 15 min in freshly made developer solution (containing 3 or 10% sodium sulfite in 1% hydroquinone, depending on the amount of copper used). After three 3-min washes in distilled water, all the previous stages are repeated. Intensification in 1% gold chloride solution for 15 min is followed by a rinse in distilled water, then reduction in 2% oxalic acid solution for 10 min. Finally, the sections are washed in water, then treated with 5% sodium thiosulfate solution for 10 to 15 min to remove residual silver. Three 3-min washes in distilled water are followed by dehydration, clearing, and mounting in the usual way.

IV. LOCALIZATION OF NEURONAL CONSTITUENTS

An important class of neuroanatomical methods reveal constituents of neurons. In most cases the molecule of interest is a neurotransmitter, neuromodulator, or neurohormone (or the synthetic or degradative enzyme of the above). However, it is also possible to localize the functional activity of neurons anatomically, using the deoxyglucose technique. In a few cases the substance of interest is spread throughout the neuron, and if the technique employed to localize it has high resolution, details of neuronal shape are obtained. More often, the localization technique must be coupled with other neuroanatomical methods if the anatomy of the neurons is to be studied.

A. *IN VIVO* OBSERVATIONS

Using dark-field or oblique illumination, many neurons containing neurosecretory material are opalescent in appearance — e.g., the giant dorsal bilateral cells in the third thoracic ganglion of the cockroach.[17]

B. HISTOCHEMICAL METHODS

1. Histofluorescence

a. Acridine Orange (for Neurosecretory Material)

Living nervous tissue can be stained in 0.1 mg/ml acridine orange in saline for 1 min.[18] On observation under UV light with a fluorescence microscope, neurosecretory material appears as red fluorescing granules on a background of pale green cytoplasm and darker green or orange nuclei. The fluorescence bleaches in about 5 min.

b. Falck-Hillarp Method for Biogenic Monoamines

The technique is used for detecting the catecholamines dopamine, noradrenaline, and adrenaline and the indolalkylamine 5-hydroxytryptamine (5-HT, serotonin). It has been used successfully on the brain of *P. americana.*[19,20] The following method is taken from Klemm.[21]

Tissues are normally freeze-dried, but small preparations such as the isolated nerve cord from embryos can be air dried at room temperature or in a vacuum. The frozen or dried specimens are placed in a 1-l glass vessel containing 5 g paraformaldehyde. The water content of the paraformaldehyde should be adjusted beforehand by heating it to 100°C for 1 h and then storing it for 1 week in a dessicator with a relative humidity of 70%. The vessel is sealed and heated to 80°C. Exposure to the resulting formaldehyde vapor for 1 to 1.5 h renders the monoamines fluorescent.

The tissues should be embedded immediately in paraffin wax or plastic in a vacuum.[22] Plastic allows thinner sections, but yields lower total fluorescence intensity and a higher background. Alternatively, tissues can be examined in whole-mounts. Under a fluorescence microscope, catecholamines exhibit a green or blue fluorescence (emission peak at 475 nm) which may appear yellow at high intensities. 5-HT produces yellow fluorescence (520 nm) and fades more rapidly.

c. Glyoxylic Acid Method for Biogenic Monoamines

This method has been developed for visualization of monoamines in whole-mounts of *Rhodnius* ganglia, but is equally applicable to the cockroach.[23] The ganglia are incubated in 5% glyoxylic acid in Ca-free, Mg-rich saline (pH 7) for 1 h at room temperature or for 8 to 12 h at 12°C. Excess liquid is removed from the tissues, which are stored over calcium carbonate for 3 to 5 d in the dark at room temperature. The dried ganglia are covered with a drop of mineral oil, then flash-heated on copper blocks at 85°C for 2 min. This should remove air from the tracheae; if it does not, the heating is repeated, although prolonged heating damages the tissue and increases background fluorescence. 5-HT fluorescence is yellow and fades rapidly; catecholamines fluoresce green.

In both the above techniques, preloading with monoamines or fluorogenic nonmetabolizable monoamine analogues facilitates the visualization of monoaminergic cells. Examples of nonmetabolizable analogues include α-methylnoradrenaline and 6-HT. The ganglia are incubated for 20 to 60 min at room temperature in 0.1 to 0.5 m*M* concentrations of the compounds.

2. Staining Techniques

a. In Vivo Neutral Red Staining for Biogenic Amines

This technique has been used in *P. americana* by Dymond and Evans[20] and by Adams and co-workers.[24] It has the advantage of allowing further physiological or dye-injection experiments after the amine-containing cells have been localized. Desheathed or intact ganglia are incubated at room temperature in 0.01 to 0.3 mg/ml neutral red in saline or tissue culture medium for 0.5 to 3 h. Alternatively, staining can be carried out at 4°C overnight. A destaining period (incubation in saline or culture medium for 4 to 6 h) reduces nonspecific background staining. Intracellular granules within the cell bodies of amine-containing cells stain red. Whole-mount preparations can be made by fixing in nonalcoholic fixative, followed by rapid dehydration and embedding.

b. Staining of Neurosecretory Cells

Numerous histochemical techniques exist for staining neurosecretory cells (NSC).[25] As an example, we will describe in detail only one of these methods — the paraformaldehyde-fuchsin stain for whole-mounts. This method may not stain all NSC; for a more rigorous study, other staining methods and wax sections should also be used.[25] The tissue is fixed in aqueous Bouin's fixative for 12 to 24 h, washed in lithium salt in 70% alcohol, and transferred to distilled water. A 1:1 mixture of 0.6% potassium permanganate and 0.6% sulfuric acid is used for 15 min to oxidize the material. After rinsing in water, the tissue is bleached in 2.5% sodium metabisulfite until completely white, then washed in distilled water for 5 to 10 min. Next, it is placed in 70% alcohol/paraldehyde-fuchsin (PF) solution (0.125 g PF crystals in 50 ml of 70% alcohol plus 1 ml acetic acid, adjusted with HCl to pH 1.6 to 1.7) for 30 min of staining. Differentiation of the stain in 0.5% HCl in absolute alcohol continues until no more superfluous stain is given off. Finally, the tissue is dehydrated, cleared, and mounted.

C. IMMUNOCYTOCHEMICAL METHODS

Immunocytochemical methods have several advantages over histochemical techniques: they are versatile, yet can be very specific, particularly with the recent advances in monoclonal antibodies whereby unequivocal identification and localization of cell components can be achieved.

1. Fixation

Fixation schedules vary according to the nature of the antigen. In some tissues, antigenicity is only preserved if no fixation is used; in these cases, frozen sections must be employed. Fixatives in general use on insect tissues include (1) 4% paraformaldehyde in buffer or saline for 2 to 12 h at 4 to 20°C,[17,26] with or without the inclusion of 1.4% lysine monohydrochloride and 0.2% sodium metaperiodate;[27] (2) 2.5% glutaraldehyde in 0.05 *M* phosphate buffer (pH 7.4) for 1 to 2 h (for substances conjugated with glutaraldehyde to a carrier protein);[28] (3) aqueous Bouin's (15:5:1 picric acid-formalin-glacial acetic acid) for 3 to 6 h; (4) 2% formaldehyde and 15% picric acid in 0.1 *M* phosphate buffer, pH 7.3; and (5) 15 ml picric acid, 5 ml 25% glutaraldehyde, and 0.1 ml glacial acetic acid.[29,30]

2. Tissue Preparation

As mentioned above, in some cases the tissue must be frozen and sections taken with a cryostat microtome. In the case of ganglia which are surrounded by a connective tissue sheath, this should be removed before fixation, using either dissecting instruments or 1 mg/ml collagenase. Other methods of rendering the tissue accessible to the large antibody molecules may be applied after fixation.

In some preparations the whole-mount method can be used. This is particularly suitable for small ganglia or thin sheets of tissue, which are pretreated for 1 to 2 h with a detergent such as 0.3% Triton® X-100 or saponin, made up in the same buffer as will be used later for antibody dilution. Primary antibody incubation times may have to be extended to 1 to 2 d at 4°C. The detergent is included in all later incubation steps in which large molecules are required to penetrate the tissue.

A common method of making the tissue accessible to the antibodies is to take frozen sections of fixed (or unfixed) tissue. The following method is taken from Buchner et al.[31] After fixation the tissue is washed for several hours in 25% sucrose in saline, both to remove the fixative and for cryoprotection. Tissue is mounted in 20% carboxymethylcellulose on a microtome chuck and then frozen in liquid nitrogen. Using a cryostat microtome, 10- to 12-μm sections are cut at −20°C and picked up on cold, gelatinized slides (or slides coated with chrome alum-gelatin). The sections are attached to the slide by briefly melting them and allowing them to air dry at −20°C for 15 min. After removal from the cold chamber, the slides are dried again for 30 min. The sections can be stained immediately or stored in saline at 4°C for several hours. They are incubated overnight at 4°C in the primary antibody.

Similarly, vibratome sections may be taken from pieces of fixed tissue and then incubated in the antisera as described above. Small pieces of tissue can be embedded in 5% agar. This technique has been found particularly useful in combined light and electron microscopy studies using pre-embedding staining, particularly in cases where larger (>1 mm) pieces of nervous tissue are available.

Alternatively, tissue can be fixed, dehydrated, and embedded in paraffin wax, then sectioned at 10 μm. After dewaxing in xylene the sections are rehydrated, then incubated in the antisera. Recently, more hydrophilic embedding media have been developed that eliminate the need for exposure to nonaqueous solvents, which may modify antigenicity. Tissue may be embedded in plastic or resin (e.g., Lowicryl® K4M, LR White, or LR Gold), sectioned at 0.5 to 4 μm, and the plastic sections exposed to antibody. The hydrophilic nature of these resins improves access of the antibodies to the tissue. Some plastics may be removed with solvents to increase access to the antibodies (e.g., Poly/REC, Polysciences, Inc., Warrington, PA).

3. Antibody Staining

a. Peroxidase-Antiperoxidase Method

The following method for peroxidase-antiperoxidase (PAP) staining of whole-mount ganglia is based on that for adult cockroaches and adult and embryonic grasshoppers.[17,32] The method for staining frozen or fixed sections is similar; however, detergent is omitted from the incubation medium and the primary antibody incubation time is reduced to approximately 12 h at 4°C. After overnight fixation at 4°C in 4% paraformaldehyde in 50 m*M* sodium phosphate buffer, pH 7.3, the tissues are washed, then agitated for 1 h at 4°C in a phosphate-buffered saline (PBS) consisting of 150 m*M* NaCl and 20 m*M* sodium phosphate with 0.2 to 0.3% Triton® X-100, pH 7.3 (PBS + T). The tissue is left in PBS + T containing 10% normal goat serum (NGS) or 1% bovine serum albumin (BSA) for 1 h at 4°C to reduce nonspecific staining.

The tissue is incubated in a suitable concentration (approximately 1:500 to 1:5000) of the primary antiserum (raised in an animal other than the goat), diluted in PBS + T with 1% NGS, for 12 to 72 h in the dark at 4°C. After a 1-h wash (six changes) in PBS + T, the tissue is incubated with the secondary unlabeled goat anti-rabbit antiserum, diluted 1:20 in PBS + T with 10% NGS, for 12 h at 4°C (or 1 h at room temperature). It is then washed for 1 h in PBS + T (several changes). Next, the tissue is incubated in a 1:40 dilution of rabbit antiperoxidase or PAP complex in PBS + T with 1% NGS for 12 h at 4°C (or 1 h at room temperature), then washed for 1 h with PBS and brought to room temperature. Finally, it is incubated for 1 h in the dark in 0.25 to 0.5 mg/ml 3,3′-diaminobenzidine (DAB) in PBS, then transferred to the same solution with the addition of 0.005 to 0.01% H_2O_2. The development of the reaction product is stopped by washing in PBS, and the tissues are dehydrated, cleared, and mounted for light microscopy. DAB should be handled with care using a fume hood and gloves. Utensils and work areas exposed to DAB should be washed with sodium hypochlorite (household bleach), which denatures this carcinogen. Old DAB solutions should be denatured prior to disposal.

b. Immunofluorescence

Alternatively, instead of incubating the tissue in unlabeled secondary antibody, a fluorochrome-conjugated secondary antibody is used. Commercial antibodies are labeled with either fluorescein isothiocyanate (FITC), which gives a green fluorescence under UV illumination, or tetramethyl rhodamine isothiocyanate (TRITC), which is red. Specific epifluorescence filter blocks can be obtained for each fluorochrome. Immunofluorescence gives more rapid results than peroxidase techniques, and the use of harmful chromogens is avoided; it is also suitable for double-labeling experiments. However, the method is less sensitive than the PAP technique and the fluorescence gradually fades, although fading is retarded by mounting the specimen in 5% *n*-propyl gallate in glycerol, pH 7.3.

c. ABC method

As an alternative to the PAP method, it is now possible to use a biotinylated secondary antibody followed by incubation with an avidin-biotinylated peroxidase complex (ABC method). The tissue is incubated with the reagents for 1 h at room temperature, and the DAB reaction product is developed as described above. If the tissue contains endogenous biotin, it can be preincubated with excess avidin, followed by excess biotin, before the antisera are applied.

d. Alkaline Phosphatase Labeling

In addition, it is now possible to use alkaline phosphatase as a marker enzyme, either in the form of a biotinylated enzyme-avidin complex or in a modification of the PAP method, using antiphosphatase antisera. This allows the production of red or blue reaction products, which can be distinguished from the brown or black DAB polymer, thus allowing double-labeling experiments to be carried out.[33]

4. Electron Microscopic Immunocytochemistry

A technique for immunocytochemical staining of proctolin in the cockroach has been described by Agricola et al.[30] The ganglia are fixed for 1 to 1.5 h in 4% formaldehyde in 0.1 *M* phosphate buffer, pH 7.3. For some antigens, 0.1 to 1% glutaraldehyde may be added to the fixative. Tissue can be fixed for electron microscopy in the mixture described by McLean and Nakane:[27] 4% paraformaldehyde, 1.4% lysine monohydrochloride, and 0.2% sodium metaperiodate. After fixation, ganglia are washed in buffer and embedded in 7% agar. Vibratome slices are cut at 100 μm and mounted on slides. The sections are first incubated in the antisera, then in PAP or ABC. Alternatively, the protein A-gold technique may be used.[29] Following incubation, sections are fixed with 1% OsO_4 and processed for electron microscopy.

Small ganglia or other pieces of tissue are sectioned this way less easily. Whole-mount staining techniques can be used for localizing cell-surface antigens in embryonic tissues,[34] but penetration of the antibodies may be a problem at later stages, even after ganglia are desheathed. Care is needed, as permeabilization of the membranes with detergents may have a detrimental effect on ultrastructure. Alternatively, after fixation the tissue can be cryoprotected with 30% sucrose in buffer, then frozen rapidly. The tissue is then allowed to thaw in buffer.

Postembedding staining techniques are effective for staining intracellular antigens. These methods have primarily been used on vertebrate tissues, although a postembedding technique has been described for the cockroach.[29] The use of more hydrophilic embedding media which can be polymerized at low temperatures (e.g., Lowicryl® K4M, LR Gold) has extended the use of this technique to more labile antigens. The main disadvantages of this approach include the need for scrupulous cleanliness when staining the ultrathin sections and the poor ultrastructural preservation caused by the lack of osmication. In the case of a few durable antigens, the tissue may be osmicated and the sections exposed to a strong oxidizing agent, a saturated solution of sodium metaperiodate, before incubation in the antisera.

D. AUTORADIOGRAPHIC METHODS

1. Localization of Functional Activity Using Deoxyglucose

This technique allows the state of physiological activity of nervous tissue to be localized. It relies on the fact that whereas glucose, on being taken up by neurons, enters the glycolytic pathway, deoxyglucose is taken up, but not metabolized. Glucose (and deoxyglucose) uptake is related to the electrical activity of a brain region, but also depends on the surface-to-volume ratio of cell processes in that region. Thus, small-diameter axons will be labeled more strongly than large axons with the same metabolic activity. Autoradiography limits the resolution of the technique to about 2 μm in the light microscope, but the activity of large numbers of neurons can be studied. The temporal resolution of the technique can be as short as 0.75 h for small insects. The method has been described for *Drosophila* by Buchner and Buchner,[35] and these authors have suggested modifications for other species.[36]

The deoxyglucose is introduced into the insect either by feeding or by injection into the hemocoel. In the latter case, 20 μCi/mg are used to give a concentration of 1 to 3 m*M* in the hemolymph. Higher concentrations may be necessary in the cockroach; prior starvation may increase uptake. The animal is then stimulated (e.g., visual or mechanosensory stimulation) for periods ranging from 0.75 to 8 h. For larger insects such as the cockroach, where the costs of high ^{3}H-deoxyglucose dosage are too great, longer periods of stimulus may be necessary. The ganglia are dissected out of the insect and frozen rapidly, then freeze-dried for several days. The dry tissue is fixed with osmium tetroxide vapor and embedded in Epon® under a slight vacuum. The blocks are sectioned at 2 to 3 μm using dry glass knives. The sections are stuck onto glass slides, which are then clamped against slides covered with dry autoradiography stripping film (see Reference 36 for details). After exposure for a suitable period (e.g., 10 d for 20 μCi), the slides are separated and the autoradiographs developed.

2. Localization of Specific Radiolabeled Probes

Detection of the distribution of specific radiolabeled ligands can be used to map neurotransmitter receptors and ion channels. Figure 2 shows the mapping of putative neuronal nicotinic receptors in the cockroach terminal ganglion.[37] Careful controls are needed in which binding is studied in the presence of a competing ligand. In this way, specific and nonspecific binding can be distinguished.

3. *In Situ* Hybridization Histochemistry

This new and powerful technique uses nucleotide probes (complementary DNA or RNA sequences) to detect specific nucleotide sequences (usually mRNAs) in slide-mounted cells or tissue sections.[38-40] Most probes are radiolabeled for detection by autoradiography, although nonradioactive tags are under development. Quantitative autoradiography allows the measurement of the mRNA levels under various physiological and pharmacological conditions. An example of the methods used is given.[41] Frozen sections of tissue are fixed in cold 95% ethanol for 1 h. Alternatively, wax-embedded tissue may be used. After rehydration (if necessary) the sections are equilibrated in hybridization buffer consisting of 450 m*M* NaCl, 45 m*M* sodium citrate (pH 7.2), 50% formamide, 0.02% polyvinyl pyrrolidone, 0.02% BSA, and 0.02% Ficoll®. Each section is incubated for 40 h with 30,000 cpm of (for example) tritium-labeled single-stranded DNA probe in hybridization buffer at 28°C. The sections are then washed with 300 m*M* NaCl and 30 m*M* sodium citrate for 4 h to remove nonhybridized probes. The slides are then coated with Kodak® Ilford L4 nuclear emulsion and exposed at 4°C for 12 weeks. For a detailed discussion of the methods for *in situ* hybridization, see Reference 42.

V. NEURONAL MORPHOLOGY

A. GOLGI TECHNIQUES

The Golgi stain allows the shapes of single neurons to be studied. However, this technique differs from those described later in that the experimenter has little or no control over which neurons are stained, although the region of the CNS in which the cells are located can be chosen by allowing only restricted access to staining solutions via holes made in the cuticle. Neurons apparently are impregnated at random, although some types of nerve cells may be refractive to the method. Some neurons may be only partly impregnated. The Golgi technique is not suited to quantitative studies unless the results can be corroborated using some other selective staining method. A variety of Golgi techniques for the insect nervous system have been described in detail by Strausfeld.[43]

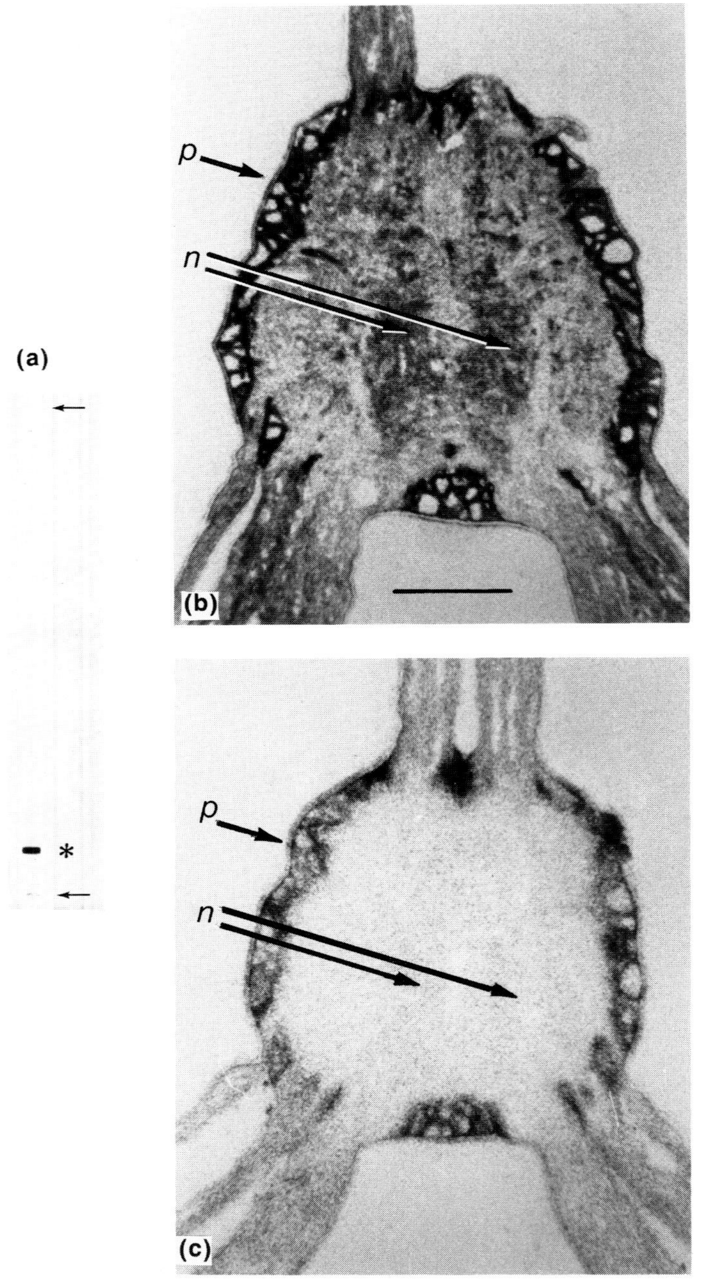

FIGURE 2. (a) SDS-polyacrylamide gel of 20 μg α-bungarotoxin. Top and bottom of the gel are indicated. The position of the α-bungarotoxin band is marked by an asterisk. (b,c) Distribution of ^{125}I-α-bungarotoxin binding in the terminal (sixth abdominal) ganglion of an adult male cockroach (*Periplaneta americana*). Horizontal frozen sections of the ganglia were prepared and autoradiographs were examined with bright-field illumination. (b) In this section through the central region of the ganglion, the silver grains are primarily located in two central patches in the neuropil (n) and in the periphery (p) of the ganglion. Some of the larger diameter neuronal cell bodies can also be seen in the periphery of the ganglion. (c) Pretreatment of sections with 1 m*M* D-tubocurarine chloride for 15 min removed all but the background binding of ^{125}I-α-bungarotoxin binding from the neuropil (n). Only a portion of the peripheral toxin binding (p) was removed. (Scale bar: 200 μm [b and c].) (From Sattelle, D. B., Harrow, I. D., Hue, B., Pelhate, M., Gepner, J. I., and Hall, L. M., *J. Exp. Biol.*, 107, 473, 1983. With permission.)

1. Buffered Golgi Rapid and Golgi-Colonnier Methods

These methods have proved reliable for a range of insects, including members of the Orthoptera. Fixation is in 2.5% glutaraldehyde in 0.1 *M* piperazine-*N,N'*-bis[2-ethanesulfonic acid] (PIPES) buffer at pH 7.0 to 7.4 for 3 to 12 h, followed by washing in several changes of PIPES buffer over 2 h and further washing in several changes of Holmes' buffered chromate, pH 5.4 to 5.6 (2.5% w/v sodium or potassium dichromate in 0.2 *M* boric acid buffer, adjusted to pH 7.4 with 0.05 *M* borax), over 2 h. For the Golgi rapid technique, the tissue is washed in Holmes' buffered dichromate-osmium (1 volume 2% osmium tetroxide + 20 volumes buffered chromate) for 12 to 72 h in the dark at room temperature. For the Golgi-Colonnier procedure, buffered dichromate-glutaraldehyde (1 volume 25% glutaraldehyde + 4 volumes buffered dichromate) is used. The tissue is rinsed several times in Holmes' buffered silver nitrate (0.75% w/v silver nitrate in boric acid/borax buffer) until the precipitate no longer appears; it is then left in buffered silver nitrate for 12 to 72 h. The chromation and silver impregnation stages may then be repeated if desired. Dehydration is followed by embedding in Araldite®, Epon®, or Spurr's resin.

2. Golgi for Electron Microscopy

This technique was developed by Ribi[44,45] for insects and works well on the optic lobes of flies and bees. Results on the cockroach CNS are not available. The tissue is fixed for up to 4 h at 4°C in 2% paraformaldehyde and 2.5% glutaraldehyde in Millonig's phosphate buffer (0.164 *M* monosodium phosphate, adjusted to pH 7.2 to 7.3 with 0.63 *M* NaOH, plus 0.0024% $MgCl_2$ and 1.3% sucrose). D-Glucose (1%) and $CaCl_2$ (0.009%) may be added to the fixative. Next, the fixed tissue is washed for 3 min in buffer; this is repeated twice. The tissue is then placed in a 1 to 2% OsO_4 solution (containing $CaCl_2$ and D-glucose) for 1 to 2 h at room temperature, followed by three 3-min buffer washes. The next step is to place it in 1% potassium chromate solution for 4 h in the dark at 4°C. The tissue is then washed briefly in distilled water, then in 0.5% silver nitrate solution until no precipitate remains, and, finally, is left in silver nitrate solution in the dark at 4°C for 30 min to 4 h, depending on the size of tissue and the density of impregnation desired.

The tissue is washed repeatedly in distilled water before dehydration in an ethanol series, clearing in propylene oxide, and embedding in Araldite® (see Section VI on electron microscopy). Thick sections (10 to 80 μm) can be cut on a rotary or sledge microtome if the block face is preheated to soften it (using an infrared heating bulb or soldering iron). Examination with light microscopy is followed by reembedding for electron microscopy.

B. DEGENERATION

Experimentally induced Wallerian degeneration is a useful tool for anatomical studies using light and electron microscopy.[46] It has been eclipsed to some extent by other techniques of marking neurons intracellularly, since it has the disadvantage that degenerating neurons cannot be recognized in whole-mounts of ganglia, but it still has a valuable role to play, particularly in studies where it is used along with other marking techniques or in cases where the neurons are inaccessible to other experimental procedures[47] (see also Chapter 11 in this work).

The rationale for using degeneration as a neuroanatomical tool lies in the changes which are produced in neuron processes 1 to 6 d after separation from the cell body. These changes involve first darkening and then shrinkage of the axoplasm and can be detected in sections from tissue prepared for light or electron microscopy following aldehyde-osmium fixation (see Section VI on electron microscopy techniques).

In *P. americana,* enucleated axons do not survive for long periods as they do in other orthopterous insects. However, regeneration of axons from the severed stumps is common after the first postoperative week. Care should be taken to control for possible spontaneous degenerative changes in neurons, along with changes due to aging.

C. INTRACELLULAR MARKERS

It has often been the case that new techniques for marking neurons have been developed first in invertebrate preparations and later are adapted for use in studies on vertebrates. This has been the case with Procion yellow and Lucifer yellow, as well as with cobalt and nickel ions, which were first used as intracellularly injected tracers in large, identified invertebrate neurons that were easily accessible to microelectrode penetration. Horseradish peroxidase, however, was first used in vertebrates as a retrograde axonal tracer and was later used for intracellular injection in vertebrates and invertebrates.

1. Cobalt and Nickel

a. Intracellular Microelectrode Injection

Glass micropipettes are filled with a solution of (1) 1 to 500 m*M* cobaltous chloride, cobalt nitrate, or cobalt acetate; (2) 3 to 6% hexamine cobalt chloride;[48] or (3) 250 m*M* nickel chloride. Nickel-filled electrodes are reported to have lower resistances compared to those filled with cobalt, and they are less prone to blockage.[49] These solutions can be combined with KCl or K acetate if lower resistance microelectrodes are necessary, although the presence of potassium ions will reduce the number of metal ions ejected by an iontophoretic current pulse.

These metal ions can be ejected from the microelectrode by means of pressure, but this necessitates a micropipette with a large tip, which may be precluded by the size of the neurons under investigation. A more commonly used technique is that of iontophoretic injection. Positive current should be used; to avoid blockage, the current is applied as 0.5-s pulses at a frequency of 1 to 2 Hz. The amplitude of the current and the injection time vary according to the size of the neurons and the desired density of the final fill. Large cockroach motoneurons impaled in the cell body have been injected with 50-nA pulses for 1 h to produce fills which are easily visible without silver intensification.[50] At the other extreme, embryonic cockroach neurons have been injected with 2-nA pulses for 3 min to give fills which are invisible until after intensification.[4]

It should be borne in mind that both Co^{2+} and Ni^{2+} are Ca^{2+} channel blockers and are toxic to neurons. Even with a good penetration the cell gradually will be poisoned as ions are injected, although many cells are able to sequester some of the ions in mitochondria. Serious damage to the neuron, particularly during impalement, often results in leakage of the metal ions out of the cell. In this regard, the input resistance of the neurons should be considered when selecting the amplitude of the injection current, since the higher the input resistance, the greater the damage sustained by the cell for a given amount of current.

It is most convenient if the amount of current passing out of the electrode is monitored during the injection period, since cobalt- and nickel-filled electrodes have a tendency to become blocked when passing large, positive currents. The current can be monitored using a current pump or a virtual-ground circuit. Blocked electrodes sometimes can be unblocked by temporary reversal of the current, but this strategy rarely is completely successful.

Both cobalt and nickel appear to spread through neurons at a rate of about 3 to 5 mm/h. The preparation may be left at room temperature for approximately 1 to 2 h in saline or tissue culture medium, or at 4°C for several hours, to allow for movement of the metal ions down long axons. Nickel ions are lost after 4 to 6 h.[49]

b. Introduction via Cut Axons

This technique has been described in detail by other authors.[51] It is not as specific as intracellular recording from a neuron followed by injection of dye, but has the advantage of requiring little in the way of apparatus. Basically, the cut ends of selected axons are placed in a reservoir of cobalt (or nickel) chloride solution, which must be isolated to prevent uptake into the rest of the nervous system. The cobalt ions migrate into the axons whether or not current is applied. The method can be applied to either *in vitro* or *in vivo* preparations.

In the first case the ganglia are dissected out of the animal, taking care to leave as long a length of the nerve of interest as possible and not to damage it by pulling. The ganglia are placed in a specially constructed dish containing two chambers connected by a shallow groove. This groove crosses a deeper gutter containing mineral oil, which provides a water-proof seal around the nerve. The ganglia are placed in one chamber, which contains saline or tissue culture medium, and the nerve of interest is draped across the shallow groove connecting the chambers so that its cut end enters the second chamber, containing the cobalt solution. This method also can be applied after sucrose- or oil-gap recording from giant interneurons in the cockroach nerve cord.[52] In that study, the cut ends of the axons were exposed to 2 to 5% cobalt chloride with the application of a 50-nA current for 8 h. This ensured the filling of the fine dendritic branches.

For *in vivo* staining, the nerve of interest is placed in an isolated pool of cobalt chloride solution, either within the body cavity or on the body surface. Such a reservoir is commonly made of purified white petroleum jelly, built up around the nerve using a syringe, and is filled with the cobalt solution. The nerve is then cut. Alternatively, the cut end of the nerve can be drawn into the end of a cobalt chloride-filled suction electrode.[53] Some sensory projections can be filled by cutting the sensory hairs and sealing a drop of cobalt chloride over them with wax[54] or by inserting them into a cobalt chloride-filled micropipette.[55] In cockroaches this is best achieved immediately after the animal has molted, since at this stage sensory hairs retain a lumen which connects to the sensory dendrite.[56]

Choice of cobalt concentration, filling time, and temperature depend on the preparation. The filling solution can be made up in insect Ringer solution or distilled water. For large axons up to 15% cobalt chloride can be used, while for small axons a 1% solution is satisfactory. Blagburn and Beadle[57] achieved satisfactory results filling giant interneurons and cercal sensory axons in the first-instar cockroach using 1% $CoCl_2$ with 0.13 mg/ml BSA (after Strausfeld and Obermayer[58]). The filling time depends on the temperature used and on the size of the neuron. Tyrer et al.[59] suggest that *in vitro* preparations are best left at room temperature for up to 8 h, while *in vivo* preparations are better at low temperatures (4°C) for periods of up to 48 h. In all cases, the preparations should be kept in a humid environment to prevent dessication.

c. *Injection into Ganglia*

A reservoir of extracellular cobalt ions can be produced inside insect ganglia by insertion of a broken-tipped microelectrode containing cobalt solution into the desired region. Cobalt is allowed to diffuse out of the micropipette for an empirically determined period; this is followed by a further diffusion period after the micropipette has been removed. Some neurons will take up cobalt ions and can be visualized after silver intensification. Bacon and Strausfeld[60] determined that the optimal micropipette tip size for Orthoptera was <1 μm, since larger tips result in an extensive extracellular Co^{2+} pool which later prevents the silver intensification of filled neurons. The micropipette, filled with 5 to 20% cobalt chloride, is held in place using a Plasticine® "micromanipulator" for a period of 30 min, followed by a 30-min diffusion period at room temperature.

d. *Preparation for Viewing*

The metal ions are precipitated using ammonium or sodium sulfide or rubeanic acid (dithiooxamide).[61] The tissue is first washed in saline to remove any excess of metal ions. Under a fume hood, ammonium or sodium sulfide is added to the saline to give a final concentration of about 1% (approximately one or two drops of concentrated solution usually suffice). After 5 to 10 min, the tissue is washed thoroughly with saline and transferred to fixative. Fresh ammonium sulfide should be used to avoid the formation of soluble polysulfide compounds. Alternatively, the tissue is flooded with a saturated ethanolic solution of rubeanic acid for 5 to 10 min. Rubeanic acid has the advantage of being nontoxic; however, Co- and Ni-rubeanic complexes may be soluble in acidic media or in ethanol and so should not be kept in either one for long periods. Exposure of small ganglia to ethanol may result in rapid shrinkage of the tissue.

In addition, rubeanic acid has the advantage of forming different-colored complexes with different metal ions (Co^{2+}, yellow; Ni^{2+}, blue; 50:50 Co:Ni, red). If sufficient amounts of the metal ions can be introduced into the neurons so as to avoid the need for intensification, then different neurons can be stained different colors using different combinations of metal ions.[62]

The preparation may be treated with one of several different fixatives. Alcoholic Bouin's (1 to 3 h) is perhaps the most effective. Glutaraldehyde-containing fixatives are not suitable if whole-mount intensification is to be used, since they render the ganglionic sheath less permeable to the intensification medium. The preparation is transferred to absolute ethanol and cleared in methyl salicylate, methyl benzoate, or cedarwood oil, then mounted in Canada balsam; alternatively, it can be viewed in the clearing medium.

e. Silver Intensification

Several modifications of silver intensification techniques have been described which allow the enhancement of cobalt (or nickel) deposits in whole ganglia or sections for light and electron microscopy.[59,63,64] The following method of whole-mount intensification for light microscopy is based on the modification of Bacon and Altman[65] and has been used to intensify cobalt- or nickel-filled neurons from orthopterous insects.

After fixation, the tissue is washed in distilled water, then incubated at 48°C (60°C in the original method) in developer stock solution for 1 h. The stock solution consists of 3 g gum arabic (acacia), 10 g sucrose, 0.43 g citric acid, and 0.17 g hydroquinone dissolved in 100 ml distilled water. Commercial powdered gum arabic may be contaminated with metal particles, which will cause immediate precipitation of silver. The tissue is transferred into a solution containing ten parts developer to one part 1% silver nitrate for a further 5 to 15 min at 48°C, until the neuron appears black and the rest of the tissue is yellow or golden brown. The reaction is stopped by washing in warm 30% ethanol, and the tissue is subsequently dehydrated, cleared, and mounted in Canada balsam. The cobalt fills may be photographed, but information from all focal planes of the specimen can best be conveyed in drawings made with a drawing tube or camera lucida (Figure 3).

Recently a new, rapid silver intensification technique has been described which can be carried out at room temperature.[66] This method uses Triton® X-100 as a stabilizer and therefore may not be suitable for electron microscopy. Following precipitation of cobalt sulfide, the tissue is fixed in alcoholic Bouin's for about 30 min for small preparations. Nonspecific silver deposits may be formed if aldehyde fixatives are used. After a 15-min wash in distilled water, the tissue is transferred to 2% sodium tungstate for 30 min, then into the developer solution in a petri dish at room temperature. The developer solution should be made freshly from three stock solutions, mixed in the proportions 8A:1B:1C;* to this is added 1 mg/ml silver nitrate. The intensification can be observed with a dissecting microscope until the cells are visible or the tissue begins to discolor (2 to 10 min). The tissue is then rinsed in distilled water, dehydrated, cleared, and mounted.

f. Electron Microscopy

Several techniques have been developed to intensify cobalt sulfide for electron microscopy. The earliest of these involved intensification of ultrathin sections on the grids or

* Solution A consists of 15 ml 1% Triton® X-100, 1.5 g sodium acetate, and 30 ml glacial acetic acid in 355 ml distilled water. Solution B is 5% sodium tungstate. Solution C is 0.25% ascorbic acid in distilled water, made up immediately before use.

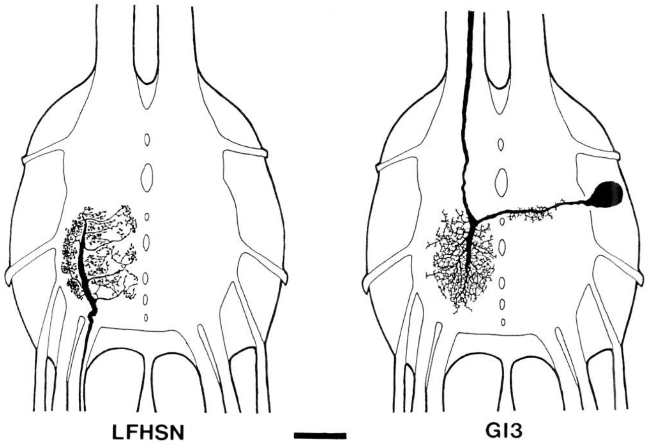

FIGURE 3. Drawings of identified neurons in the terminal ganglion of the first-instar cockroach. The cells were injected with cobalt chloride via intracellular electrodes, and the cobalt sulfide precipitate was subsequently silver-intensified. Left: the lateral filiform hair sensory neuron (LFHSN); right: giant interneuron 3 (GI3). (Scale bar: 50 μm.)

intensification of 1-μm plastic sections followed by resectioning.[59] Block intensification procedures are less time-consuming and give satisfactory results. We shall describe one of the methods here, but for more details and other variations of the method see Reference 64.

The cobalt is precipitated by placing the tissue for 2 min in four drops of ammonium sulfide in 10 ml 0.05 *M* cacodylate buffer containing 5% sucrose (pH 7.4), followed by washing in buffer for 2 min. The ganglia are fixed in glutaraldehyde in cacodylate buffer (see electron microscopy methods, Section VI). The tissue must be washed thoroughly with 5% sucrose solution to remove cacodylate, which may precipitate silver, and to remove glutaraldehyde, which may increase background staining. As alternative strategies to reduce ultrastructural damage from ammonium sulfide, cobalt precipitation and glutaraldehyde fixation can be combined or sodium sulfite in buffer can be used.

For block intensification, two solutions, A (0.05% hydroquinone, 0.7% citric acid, 2% gum arabic, 1% DMSO, and 5% sucrose in distilled water) and B (0.5% silver nitrate in distilled water), should be made. The tissues are then incubated for 1 h at 20°C in A, followed by 1 h at 37°C in 9.9 volumes of A + 0.1 volume of B, then 1 h at 37°C in 9.75 volumes of A + 0.25 volume of B, then 1.5 h at 37°C in 9.5 volumes of A + 0.5 volume B, and, finally, 1 h at 37°C in 9 volumes of A + 1 volume of B. Increasing the length of the last step to 2 to 3 h will increase the density of the intensification at the expense of increasing the size of the silver grains.

After four 15-min washes in 5% sucrose in the dark at 37°C, the tissue is transferred to buffer for 30 min (or overnight at 4°C) and then to 1% osmium tetroxide. The remaining steps are the same as for conventional electron microscopy. Lightly intensified neurons may not be visible inside osmicated blocks, so for preliminary light microscope reconstruction 1- to 20-μm sections should be taken, then reembedded after examination.

2. Fluorescent Dyes

The use of Procion dyes[67] (in particular, Procion yellow) has largely been superseded by the use of Lucifer yellow CH.[68] The latter is approximately 100 times brighter than Procion yellow and has a molecular weight of 457.3 (compared to 3000).

Lucifer yellow CH is divalent and negatively charged. In contrast with cobalt, the dye has a low toxicity and therefore can be used for longer-term electrophysiological recording.

One advantage of Lucifer yellow is that it can be used to demonstrate dye coupling between cells; this property can often be correlated with the presence of gap junctions and electrotonic coupling. The dye can also be used to kill whole cells or parts by UV illumination of the filled cell.[69] The most common method of filling cells is to inject the dye from an intracellular microelectrode, using negative current, although pressure can be used with large-tipped microelectrodes.

a. *Filling from Electrodes*

The tip of a fiber-fill micropipette is filled with a 3 to 5% solution of Lucifer yellow CH dissolved in 1% LiCl (precipitation occurs if KCl is used). The barrel of the electrode is filled with 3 *M* LiCl. The neuron is impaled with the electrode, and the required physiological experiments are carried out. To eject the dye from the electrode, a hyperpolarizing current (either constant or slow pulses) is applied for 3 to 10 min, depending on the size of the cell. The strength of the current may vary from −2 to −20 nA, again depending on the cell size. No diffusion time is necessary; in fact, delays between injection and fixation or viewing should be avoided.

Recently it was discovered that paraformaldehyde-fixed cells can be filled with Lucifer yellow from microelectrodes in the same manner as living cells.[70] This technique is particularly useful for filling very small cells which otherwise might be killed by microelectrode impalement, thus causing leakage of the dye. Since none of the physiological criteria of impalement are present, it is necessary to be able to visualize both cell and electrode tip when inserting the electrode.

b. *Filling through Cut Nerves*

This involves the same techniques as cobalt backfilling. Special attention should be paid to the tightness of the Vaseline® or grease seal around the Lucifer yellow pool, since the dye can penetrate the neural lamella. Filling takes place within minutes.

c. *Fixation and Mounting*

The neurons can be examined immediately after filling, but the dye fades rapidly. Permanent preparations are achieved by fixing with buffered aldehydes, 4% paraformaldehyde, or formaldehyde in 0.1 *M* phosphate buffer, producing less background fluorescence than glutaraldehyde. After fixing, the tissue is washed thoroughly in buffer.

For whole-mounts, the tissue is dehydrated in an alcohol series, cleared, and whole-mounted in methyl benzoate or methyl salicylate. Alternatively, the tissue is taken from the alcohol into acetone, then into Spurr's liquid embedding medium.[71] The latter is most convenient if the tissues are to be sectioned afterward. For sectioning, the Spurr's is polymerized at 70°C for 12 to 16 h. The block will become opaque. After cooling slowly, 10- to 12-μm sections are cut on a sliding microtome, collected on a microscope slide, and mounted in Gurr's Fluoromount. Sections can be counterstained with 1% methylene blue and azure blue.

d. *Examination*

Whole-mounts or sections are viewed using an epifluorescence microscope equipped with the following combinations of excitation and barrier filters: 450 to 490 nm plus 520 nm; 436 nm plus 490 nm; 350 to 460 nm plus 515 nm. A 270- to 300-nm excitation filter plus a 410- to 580-nm barrier filter will cut out Lucifer yellow fluorescence, leaving only tracheal autofluorescence.[71] Whole-mounts must be mounted on a depression slide or a specially constructed cavity slide. One disadvantage of Lucifer yellow is that bleaching may occur before a detailed drawing of the cell can be made. Serially focused photographs of the filled neuron are taken, allowing serial reconstruction from slides.

e. Lucifer Yellow Antibody

Whole-mounts or frozen sections of the tissue are incubated in antibodies raised against Lucifer yellow-albumin conjugates.[72] The antibodies are then labeled with a secondary antibody tagged with HRP, which is stained with DAB to give a permanent, brown reaction product.

f. Lucifer Yellow for Electron Microscopy

Photoexcitation of Lucifer yellow CH causes the oxidation of DAB, producing a flocculent, brown, electron-dense precipitate.[73] This technique was adapted successfully for use with vertebrate neurons, but to our knowledge has not been attempted in the cockroach. The Lucifer-DAB reaction could prove useful for double-labeling experiments, since it was reported that the precipitate could be distinguished from that produced by HRP. More recently, Sandell and Masland[74] have shown that a wide range of fluorescent molecules (e.g., fluorescein, rhodamine, 5,7-dihydroxytryptamine) can be photoconverted in the same way. The reaction can take place in unfixed or fixed tissue, although a good supply of oxygen to the tissue and strong irradiation are necessary.

The dye Procion Rubin EX-B (Polysciences Ltd.) has been used as an intracellular marker for electron microscopy and can be injected iontophoretically using hyperpolarizing current pulses.[75] To our knowledge this marker has not been used in insects.

g. Lucifer Yellow in Combination with Horseradish Peroxidase

This technique has not, to our knowledge, been tried in insects or vertebrates, but good results have been obtained in the leech nervous system.[76,77] One neuron is injected with 2% HRP (Type VI) in 0.2 *M* KCl containing 2% fast green FCF (for pressure injection), and the HRP is allowed to move through the cell for several hours. The second neuron is injected with 3% Lucifer yellow CH in 0.1 *M* LiCl using 5 to 10 nA pulsed negative current for 15 min at 1 Hz. The living tissue is incubated in a saturated solution of DAB in saline (pH 7); after 5 min, a few drops of 3% H_2O_2 are added to the solution, and the reaction is allowed to proceed under microscopic inspection for another 2 to 5 min. Following a rinse in saline and fixation for 1 h with 4% paraformaldehyde in 0.1 *M* phosphate buffer (pH 7.4), the ganglia are dehydrated in graded ethanol washes and mounted in methyl salicylate under a coverslip sealed with clear fingernail polish. The HRP- and Lucifer yellow-filled cells are examined simultaneously by balancing the intensity of the fluorescent epi-illumination and the transmitted light until both cells are clearly visible.

h. Lucifer Yellow and Other Fluorescent Dyes

Other fluorescent compounds have been used in conjunction with Lucifer yellow, e.g., SITS (4-acetamido-4′-isothiocyanato-stilbene-2,2′-disulfonic acid), which fluoresces blue and can be injected into cells using the same techniques as Lucifer yellow, although higher currents and longer injection times may be necessary.[78] Other such compounds include Texas red (Sulforhodamine 101)[79] and true blue. Strausfeld and Seyan[80] used Texas red in conjunction with Lucifer yellow and the nuclear stain bisbenzimide to stain neurons in the blowfly visual system. These authors reported that attempts to expel Texas red iontophoretically were unsuccessful.

One disadvantage of Lucifer yellow is that, because of its insolubility in K^+-containing solutions, the microelectrode must be filled with LiCl. In some preparations, intracellular Li^+ may be toxic. Also, there is a tendency for Lucifer yellow in the electrode tip to precipitate when exposed to external solutions containing a high concentration of K^+ (e.g., artificial seawater). For these reasons, the fluorescent dye 5,6-carboxyfluorescein recently has become popular as an intracellular marker.[80a] This compound is highly soluble in K^+-containing solutions, but is not as lipid soluble as fluorescein and so stays in the cell. It is strongly fluorescent under UV light (FITC filter blocks can be used).

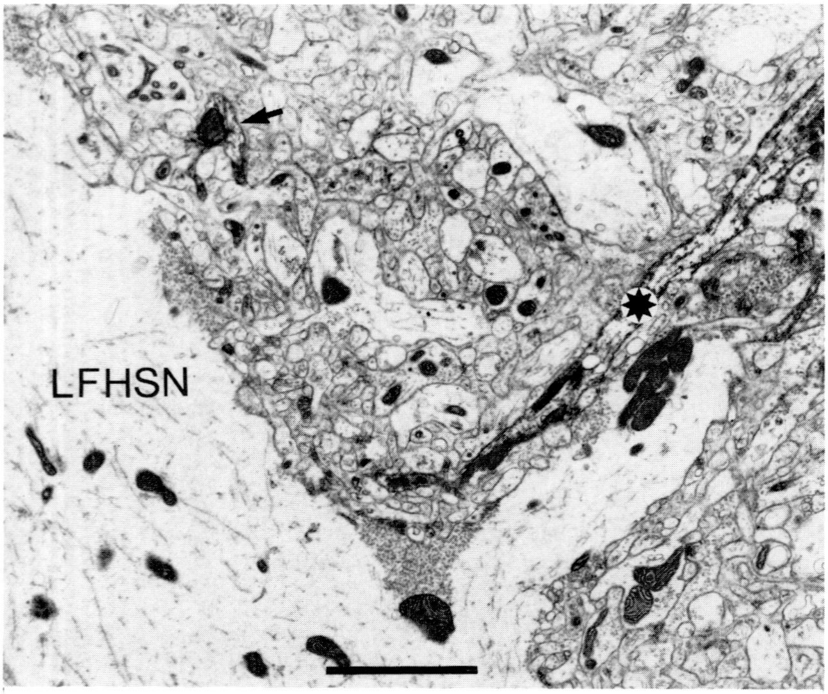

FIGURE 4. Electron micrograph of the terminal abdominal ganglion neuropil in the first-instar cockroach, sectioned horizontally. The large axon of the lateral filiform hair sensory neuron (LFHSN) is visible on the left of the picture. Giant interneuron 3 has been filled with horseradish peroxidase and an electron-opaque precipitate produced with diaminobenzidine. Dendritic branches of the interneuron (asterisk and arrow) form contacts with the LFHSN axon and receive synaptic input from it. (Scale bar: 2 μm.)

3. Horseradish Peroxidase and Other Enzymes

Intracellular labeling with HRP or other heme proteins such as cytochrome *c* and myoglobin has the advantage that the reaction product is visible using light and electron microscopy (Figure 4). The disadvantages are that the proteins do not migrate intracellularly as fast, or as far, as cobalt ions or the Lucifer yellow dye. Also, the best method for developing the reaction product involves the use of a probable carcinogen. Methods for introducing heme proteins into insect nerve cells, developing the reaction products, and processing the tissues for light and electron microscopy have been described by Nässel.[81]

a. Intracellular Injection

The tip of a glass micropipette is filled with a solution containing 2 to 10% HRP (type VI is used most frequently) with 0.1 to 0.3 *M* KCl in 0.2 *M* Tris or TES buffer, pH 7.4. The microelectrode is inserted into the neuron and, after physiological recording, square positive current pulses are applied, with similar regimes to those described for ejecting cobalt ions. We have found that it may be helpful to pass a few large-amplitude current pulses through the electrode before the cell is impaled. The preparation is left in saline or tissue culture medium for 10 min to 1 h to allow for movement of the protein through the cell. A good supply of oxygen appears to be necessary for transport of the HRP to take place. We rarely obtain good ultrastructural preservation of ganglia if they are left for long periods in saline after they have been desheathed; better results are obtained if the microelectrodes are inserted through the neural lamella which has first been softened by a brief (15 s to 1 min) exposure to 1 mg/ml collagenase or protease.

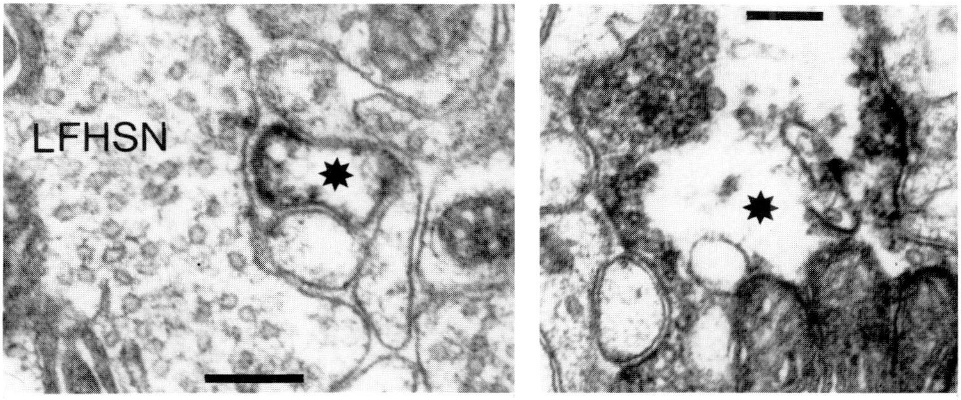

FIGURE 5. Electron micrographs of processes of giant interneuron 3 (GI 3) which has been injected with horseradish peroxidase (asterisks). Left: the axon of the lateral filiform hair sensory neuron (LFHSN) makes synaptic contacts with a GI 3 dendrite. Right: part of a larger branch of GI 3, forming output synapses with processes of unidentified neurons. The HRP-DAB reaction product does not obscure cytoplasmic details in GI 3. (Scale bars: 200 nm.)

The tissue is fixed for 2 h at 4°C in 2.5% glutaraldehyde in 0.1 *M* cacodylate or phosphate buffer (pH 7.4) containing 0.05 to 0.2 *M* sucrose. Paraformaldehyde irreversibly reduces the activity of the enzyme, while glutaraldehyde impairs it reversibly. The tissue therefore should be washed overnight in buffer. Alternatively, a 30-min wash in 0.1 *M* $NaBH_4$ in buffer will remove excess aldehydes if an overnight wash is precluded. The tissue is immersed in 0.1 *M* Tris buffer (pH 7.4) for 10 min, followed by treatment with 0.5% cobalt chloride (in Tris buffer) for 5 to 10 min to intensify the reaction product. After a 10-min wash in Tris buffer alone, the tissue is washed in cacodylate or phosphate buffer and then transferred to the incubation medium for 20 min to 1 h (until the neuron appears black or brown) at 30 to 37°C in the dark. The incubation medium consists of 10 ml 0.1 *M* phosphate or cacodylate buffer at pH 7.4, 4 mg ammonium chloride, 20 mg β-D-glucose, 20 to 30 units of glucose oxidase (type V), and 5 mg DAB.[82] After several washes in buffer, the tissue is osmicated and processed for electron microscopy as described in Section VI below.

In filled neurons which appear light brown or brown when examined with a light microscope, the granular DAB reaction product may be scattered throughout the cytoplasm of the filled cell, often preferentially associated with the cell membrane and microtubules, but allowing intracellular organelles to be distinguished (Figure 5). In heavier, black fills, however, the cytoplasmic organelles may be totally obscured by electron-dense deposits which do not penetrate the mitochondria.

VI. ELECTRON MICROSCOPY TECHNIQUES

A. TRANSMISSION ELECTRON MICROSCOPY

For the sake of brevity, techniques for tissue preparation and sectioning are described only briefly. Detailed descriptions of these techniques, as applied to the insect nervous system, have been given by Osborne,[83] and comprehensive reviews of insect neuronal and glial cell ultrastructure are available.[84] The following method has given good results in the embryonic, first-instar, and adult cockroach nervous systems. The tissues are fixed for 2 h at 4°C in 2.5% glutaraldehyde in 0.1 *M* phosphate or cacodylate buffer (pH 7.4) containing 0.2 *M* sucrose. Care must be taken with cacodylate since it contains arsenic. All preparative stages should be carried out under a fume hood. After three changes of cacodylate buffer containing sucrose (10 min per wash), the tissue is placed in 1% osmium tetroxide (in

0.1 *M* cacodylate buffer) for 1 to 2 h at 4°C. Following three 10-min washes in cacodylate buffer, the tissue is placed for 1 h in 1% uranyl acetate in distilled water, then washed in distilled water and dehydrated in an alcohol series (30 min each in 50, 70, and 90% ethanol, followed by two 30-min changes of absolute ethanol). The tissue is soaked in two changes of propylene oxide of about 15 min each and transferred to a 1:1 epoxy resin:propylene oxide mixture for 15 min. The tissue is then placed in fresh resin, left at room temperature for 24 h, and subsequently transferred to plastic embedding capsules. The resin is polymerized overnight at 60°C.

After removal from the embedding capsule, the block is mounted in a microtome chuck, trimmed with a razor blade, and sectioned using an ultramicrotome. Sections 1 to 4 μm thick can be used for orientation within the tissue; these are cut using a glass knife, placed on a microscope slide, and stained with toluidine blue. Thin sections (60 to 100 nm in thickness; interference colors, silver to gold) are cut using glass or diamond knives and collected on copper grids. Counterstaining of the sections is done using a saturated aqueous solution of uranyl acetate or a 30% solution in methanol for 2 h or 3 to 7 min, respectively, followed by 3 to 7 min of staining with Reynolds' lead citrate.[85] The sections are washed with 0.001% NaOH and distilled water, and after air drying they are examined with an electron microscope.

B. SCANNING ELECTRON MICROSCOPY

Scanning electron microscopy has limited use in insect neuroanatomical studies, although it is useful for visualizing arrays of cuticular sense organs.[86] Details of the preparative methods will not be given here. The technique has been used to visualize developing axons and their growth cones inside the embryonic grasshopper CNS.[87] This is achieved by preparing the tissue for transmission electron microscopy and embedding it in plastic, sectioning down through the block to the desired area, then etching away the plastic by immersion in a solution of 0.5% sodium methylate in 100% methanol for 2 d followed by a 1-d wash in 100% methanol. The block is critical-point dried and gold coated, as in normal scanning electron microscopy.

C. FREEZE FRACTURE

The freeze-fracture technique is well suited to investigations of the topography of cell membranes and intercellular junctions. Where possible, the results should be compared with those obtained from conventionally thin-sectioned material.

The following method has been used successfully on *Periplaneta* peripheral nerve.[88] Tissue is fixed for 20 to 30 min at room temperature in 2.5% glutaraldehyde in 0.1 *M* phosphate buffer, pH 7.2, containing 6% sucrose. After fixation, the tissue is washed in the same phosphate buffer, then placed in a cryoprotectant solution, 20% glycerol in 0.1 *M* phosphate buffer with 6% sucrose, for 15 to 20 min at room temperature. The tissue is supported with yeast paste and mounted on copper holders which are rapidly immersed in Freon® 22, cooled in liquid nitrogen, and remain there until used.

Fracturing of the tissue is performed in (for example) a Balzers freeze-fracturing device, in a vacuum of 2 μtorr (0.27 kPa) and at a temperature of −100°C. The fractured tissue is shadowed by evaporation from a tungsten-tantalum source followed by backing with carbon. After admission of air, the replicas are strengthened by coating with a drop of 2% collodion in amyl acetate while they are still on the cold table of the fracturing unit. The holders are removed from the Balzers device and the collodion allowed to dry before the replicas and tissue are removed from the holders. Tissue is cleaned from the replicas by immersing them in 2% sodium hypochlorite overnight followed by washing in distilled water. The replicas are then mounted on copper grids and the collodion removed by immersing the grids in undiluted amyl acetate for 30 min. The replicas are examined using a transmission electron microscope.

D. HIGH-VOLTAGE ELECTRON MICROSCOPY

The use of high-voltage electron microscopy (HVEM) can be advantageous in the study of insect nervous systems, particularly in the reconstruction of serial sections of neuropil and in the examination of cuticular sense organs.[89,90] The methods are similar to those used for conventional electron microscopy.[91] Sections for HVEM analysis are taken at 0.2 to 1 μm thickness, depending on the degree of resolution required, and are collected on grids covered with carbonized Formvar® film. Longer exposures to uranyl acetate and lead citrate are necessary to increase the contrast.

REFERENCES

1. **Strausfeld, N. J. and Miller, T. A., Eds.,** *Neuroanatomical Techniques. Insect Nervous System,* Springer-Verlag, New York, 1980.
2. **Strausfeld, N. J.,** *Functional Neuroanatomy,* Springer-Verlag, Berlin, 1983.
3. **Goodman, C. S. and Spitzer, N. C.,** Embryonic development of identified neurones: differentiation from neuroblast to neurone, *Nature (London),* 280, 208, 1979.
4. **Blagburn, J. M., Beadle, D. J., and Sattelle, D. B.,** Development of synapses between identified sensory neurones and giant interneurones in the cockroach, *Periplaneta americana, J. Embryol. Exp. Morphol.,* 86, 227, 1985.
5. **Blagburn, J. M. and Sattelle, D. B.,** Calcium conductance in an identified cholinergic synaptic terminal in the central nervous system of the cockroach, *J. Exp. Biol.,* 129, 347, 1987.
6. **Zawarzin, A. A.,** Histologische Studien über Insekten. 2. Das sensible Nervensystem der Aeschnalarven, *Z. Wiss. Zool. Abt. A,* 100, 245, 1912.
7. **Plotnikova, S. I. and Nevmyvaka, G. A.,** The methylene blue technique: classic and recent applications to the insect nervous system, in *Neuroanatomical Techniques. Insect Nervous System,* Strausfeld, N. J. and Miller, T. A., Eds., Springer-Verlag, New York, 1980, 1.
8. **Pinnock, R. D. and Sattelle, D. B.,** Dissociation and maintenance *in vitro* of neurones from adult cockroach *(Periplaneta americana)* and housefly *(Musca domestica)* neurones, *J. Neurosci. Methods,* 20, 195, 1987.
9. **Altman, J. S. and Bell, E. M.,** A rapid method for the demonstration of nerve cell bodies in invertebrate central nervous systems, *Brain Res.,* 63, 487, 1973.
10. **Altman, J. S.,** Toluidine blue as a rapid stain for nerve cell bodies in intact ganglia, in *Neuroanatomical Techniques. Insect Nervous System,* Strausfeld, N. J. and Miller, T. A., Eds., Springer-Verlag, New York, 1980, 21.
11. **Wigglesworth, V. B.,** The use of osmium in the fixation and staining of tissues, *Proc. R. Soc. Lond. Ser. B,* 147, 185, 1957.
12. **Watson, A. H. D. and Pflüger, H.-J.,** The distribution of GABA-like immunoreactivity in relation to ganglion structure in the abdominal nerve cord of the locust, *Cell Tissue Res.,* 249, 391, 1987.
13. **Blest, A. D.,** Reduced silver impregnations of the Ungewitter type, in *Functional Neuroanatomy,* Strausfeld, N. J., Ed., Springer-Verlag, Berlin, 1980, 119.
14. **Blest, A. D. and Davie, P. S.,** Reduced silver impregnations derived from the Holmes technique, in *Functional Neuroanatomy,* Strausfeld, N. J., Ed., Springer-Verlag, Berlin, 1980, 97.
15. **Gregory, G. E.,** Silver staining of insect central nervous systems by the Bodian protargol method, *Acta Zool. (Stockholm),* 51, 169, 1970.
16. **Gregory, G. E.,** The Bodian protargol technique, in *Neuroanatomical Techniques. Insect Nervous System,* Strausfeld, N. J. and Miller, T. A., Eds., Springer-Verlag, New York, 1980, 75.
17. **Bishop, C. A and O'Shea, M.,** Neuropeptide proctolin (H-Arg-Tyr-Leu-Pro-Thr-OH): immunocytochemical mapping of neurons in the central nervous system of the cockroach, *J. Comp. Neurol.,* 207, 223, 1982.
18. **Beattie, T. M.,** Vital staining of neurosecretory material with Acridine Orange in the insect, *Periplaneta americana, Experientia,* 27, 110, 1971.
19. **Frontali, N.,** Histochemical localization of catecholamines in the brain of normal and drug-treated cockroaches, *J. Insect Physiol.,* 14, 881, 1968.
20. **Dymond, G. R. and Evans, P. D.,** Biogenic amines in the nervous system of the cockroach, *Periplaneta americana:* association of octopamine with mushroom bodies and dorsal unpaired median (DUM) neurones, *Insect Biochem.,* 9, 535, 1979.
21. **Klemm, N.,** Histochemical demonstration of biogenic monoamines (Falck-Hillarp method) in the insect nervous system, in *Neuroanatomical Techniques. Insect Nervous System,* Strausfeld, N. J. and Miller, T. A., Eds., Springer-Verlag, New York, 1980, 51.

22. **Björklund, A., Falck, B., and Owman, C.,** Fluorescence microscopic and microspectrofluorometric techniques for the cellular localization and characterization of biogenic amines, in *The Thyroid and Biogenic Amines,* Rall, J. E. and Kopin, I. J., Eds., Elsevier/North-Holland, Amsterdam, 1972, 318.
23. **Flanagan, T. R. J.,** Monoaminergic innervation in a Hemipteran nervous system: a whole-mount histofluorescence study, in *Functional Neuroanatomy,* Strausfeld, N. J., Eds., Springer-Verlag, Berlin, 1983, 317.
24. **Adams, M. E., Bishop, C. A., and O'Shea, M.,** Strategies for the identification of amine- and peptide-containing neurons, in *Functional Neuroanatomy,* Strausfeld, N.J., Ed., Springer-Verlag, Berlin, 1983, 239.
25. **Panov, A. A.,** Demonstration of neurosecretory cells in the insect central nervous system, in *Neuroanatomical Techniques. Insect Nervous System,* Strausfeld, N. J. and Miller, T. A., Eds., Springer-Verlag, New York, 1980, 25.
26. **Sattelle, D. B., Ho, Y. W., Crawford, G. D., Salvaterra, P. M., and Mason, W. T.,** Immunocytochemical staining of central neurones in *Periplaneta americana* using monoclonal antibodies to choline acetyltransferase, *Tissue Cell,* 18, 51, 1986.
27. **McLean, J. D. and Nakane, P. K.,** Periodate-lysine-paraformaldehyde fixative: a new fixative for immunoelectron microscopy, *J. Histochem. Cytochem.,* 22, 1077, 1974.
28. **Watson, A. H. D. and Burrows, M.,** Immunocytochemical and pharmacological evidence for spiking local interneurons in the locust, *J. Neurosci.,* 7, 1741, 1987.
29. **Eckert, M. and Ude, J.,** Immunocytochemical techniques for the identification of peptidergic neurons, in *Functional Neuroanatomy,* Strausfeld, N. J., Ed., Springer-Verlag, Berlin, 1983, 267.
30. **Agricola, H., Eckert, M., Ude, J., Birkenbeil, H., and Penzlin, H.,** The distribution of a proctolin-like immunoreactive material in the terminal ganglion of the cockroach, *Periplaneta americana* (L.), *Cell Tissue Res.,* 239, 203, 1985.
31. **Buchner, E., Buchner, S., Crawford, G., Mason, W. T., Salvaterra, P. M., and Sattelle, D. B.,** Choline acetyltransferase-like immunoreactivity in the brain of *Drosophila melanogaster, Cell Tissue Res.,* 246, 57, 1986.
32. **Keshishian, H. and O'Shea, M.,** The distribution of a peptide neurotransmitter in the postembryonic grasshopper central nervous system, *J. Neurosci.,* 5, 992, 1985.
33. **Mason, D. Y. and Woolston, R.-E.,** Double immunoenzymatic labelling, in *Techniques in Immunocytochemistry,* Bullock, G. R. and Petrusz, P., Eds., Academic Press, London, 1982, 135.
34. **Bastiani, M. J., Harrelson, A. L., Snow, P. M., and Goodman, C. S.,** Expression of fasciclin I and II glycoproteins on subsets of axon pathways during neuronal development in the grasshopper, *Cell,* 48, 745, 1987.
35. **Buchner, E. and Buchner, S.,** Mapping stimulus-induced nervous activity in small brains by ^{3}H-2-deoxy-D-glucose, *Cell Tissue Res.,* 211, 51, 1980.
36. **Buchner, E. and Buchner, S.,** Anatomical localization of functional activity in flies using ^{3}H-2-deoxy-D-glucose, in *Functional Neuroanatomy,* Strausfeld, N. J., Ed., Springer-Verlag, Berlin, 1983, 225.
37. **Sattelle, D. B., Harrow, I. D., Hue, B., Pelhate, M., Gepner, J. I., and Hall, L. M.,** α-Bungarotoxin blocks excitatory synaptic transmission between between cercal sensory neurones and giant interneurone 2 of the cockroach, *Periplaneta americana, J. Exp. Biol.,* 107, 473, 1983.
38. **Singer, R. H., Lawrence, J. G., and Villenave, C.,** Optimization of *in-situ* hybridization using isotopic and non-isotopic detection methods, *Biotechniques,* 4, 230, 1986.
39. **Wuenschell, C. W., Fisher, R. S., Kaufman, D. L., and Tobin, A. J.,** *In situ* hybridization to localize mRNA encoding the neurotransmitter biosynthetic enzyme glutamate decarboxylase (GAD) in mouse cerebellum, *Proc. Natl. Acad. Sci. U.S.A.,* 83, 6193, 1986.
40. **Young, W. S., III,** *In-situ* hybridization, histochemistry and the study of the nervous system, *Trends Neurosci.,* November/December, 549, 1986.
41. **Roberts, J. L., Chen, C. C., Dionne, F. T., and Gee, C. E.,** Peptide hormone gene expression in heterogeneous tissues: the pro-opiomelanocortin system, in *Neurotransmitters in Action,* Bousfield, D., Ed., Elsevier, Amsterdam, 1985, 226.
42. **Higgins, G. A. and Wilson, M. C.,** *In situ* hybridization for mapping the neuroanatomical distribution of novel brain mRNAs, in *In Situ Hybridization in Neurobiology: Applications to Neurobiology,* Valentino, K. L., Eberwine, J. H., and Barchas, J. D., Eds., Oxford University Press, New York, 1987, 146.
43. **Strausfeld, N. J.,** The Golgi method: its application to the insect nervous system and the phenomenon of stochastic impregnation, in *Neuroanatomical Techniques. Insect Nervous System,* Strausfeld, N. J. and Miller, T. A., Eds., Springer-Verlag, New York, 1980, 131.
44. **Ribi, W. A.,** A Golgi-electron microscope method for insect nervous tissue, *Stain Technol.,* 51, 13, 1976.
45. **Ribi, W. A.,** Electron microscopy of Golgi-impregnated neurons, in *Functional Neuroanatomy,* Strausfeld, N. J., Ed., Springer-Verlag, Berlin, 1983, 1.

46. **Schürmann, F.-W.,** Experimental anterograde degeneration of nerve fibers: a tool for combined light- and electron-microscopic studies of the insect nervous system, in *Neuroanatomical Techniques. Insect Nervous System,* Strausfeld, N. J. and Miller, T. A., Eds., Springer-Verlag, New York, 1980, 263.
47. **Zill, S. N., Underwood, M. A., Rowley, J. C., III, and Moran, D. T.,** A somatotopic organization of groups of afferents in insect peripheral nerves, *Brain Res.,* 198, 253, 1980.
48. **Brogan, R. T. and Pitman, R. M.,** Axonal regeneration in an identified insect motor neuron, *J. Physiol. (London),* 319, 34P, 1981.
49. **Delcomyn, F.,** Intracellular staining with nickel chloride, in *Functional Neuroanatomy,* Strausfeld, N. J., Ed., Springer-Verlag, Berlin, 1983, 93.
50. **Pitman, R. M., Tweedle, C. D., and Cohen, M. J.,** Branching of central neurons: intracellular cobalt injection for light and electron microscopy, *Science,* 176, 412, 1972.
51. **Altman, J. S. and Tyrer, N. M.,** Filling selected neurons with cobalt through cut axons, in *Neuroanatomical Techniques. Insect Nervous System,* Strausfeld, N. J. and Miller, T. A., Eds., Springer-Verlag, New York, 1980, 373.
52. **Harrow, I. D., Hue, B., Pelhate, M., and Sattelle, D. B.,** Cockroach giant interneurones stained by cobalt-backfilling of dissected axons, *J. Exp. Biol.,* 84, 341, 1980.
53. **Mason, C. A. and Nishioka, R. S.,** The use of the cobalt chloride-ammonium sulfide precipitation technique for the delineation of invertebrate and vertebrate neurosecretory systems, in *Neurosecretion — The Final Neuroendocrine Pathway: Proceedings,* 6th Int. Symp. Neurosecretion, Knowles, F. and Vollrath, L., Eds., Springer-Verlag, New York, 1974, 48.
54. **Tyrer, N. M., Bacon, J. P., and Davies, C. A.,** Primary sensory projections from the wind-sensitive head hairs of the locust, *Schistocerca gregaria.* I. Distribution in the CNS, *Cell Tissue Res.,* 203, 79, 1979.
55. **Bacon, J. P. and Murphey, R. K.,** Receptive fields of cricket giant interneurones are related to their dendritic structure, *J. Physiol. (London),* 352, 601, 1984.
56. **Bacon, J. P.,** personal communication, 1987.
57. **Blagburn, J. M. and Beadle, D. J.,** Morphology of identified cercal afferents and giant interneurones in the hatchling cockroach *Periplaneta americana, J. Exp. Biol.,* 97, 421, 1982.
58. **Strausfeld, N. J. and Obermayer, M.,** Resolution of intraneuronal and transsynaptic migration of cobalt in the insect visual and central nervous systems, *J. Comp. Physiol.,* 110, 1, 1976.
59. **Tyrer, N. M., Shaw, M. K., and Altman, J. S.,** Intensification of cobalt-filled neurons in sections (light and electron microscopy), in *Neuroanatomical Techniques. Insect Nervous System,* Strausfeld, N. J. and Miller, T. A., Eds., Springer-Verlag, New York, 1980, 429.
60. **Bacon, J. P. and Strausfeld, N. J.,** Nonrandom resolution of neuron arrangements, in *Neuroanatomical Techniques. Insect Nervous System,* Strausfeld, N. J. and Miller, T. A., Eds., Springer-Verlag, New York, 1980, 357.
61. **Quicke, D. L. J. and Brace, R. C.,** Differential staining of cobalt- and nickel-filled neurons using rubeanic acid, *J. Microsc.,* 115, 161, 1979.
62. **Hackney, C. M. and Altman, J. S.,** Rubeanic acid and X-ray microanalysis for demonstrating metal ions in filled neurons, in *Functional Neuroanatomy,* Strausfeld, N. J., Ed., Springer-Verlag, Berlin, 1983, 132.
63. **Obermayer, M. and Strausfeld, N. J.,** Silver-staining cobalt sulfide deposits within neurons of intact ganglia, in *Neuroanatomical Techniques. Insect Nervous System,* Strausfeld, N. J. and Miller, T. A., Eds., Springer-Verlag, New York, 1980, 403.
64. **Bassemir, U. K. and Strausfeld, N. J.,** Block intensification and X-ray microanalysis of cobalt-filled neurons for electron microscopy, in *Functional Neuroanatomy,* Strausfeld, N. J., Ed., Springer-Verlag, Berlin, 1983, 19.
65. **Bacon, J. P. and Altman, J. S.,** A silver intensification method for cobalt-filled neurons in wholemount preparations, *Brain Res.,* 138, 359, 1977.
66. **Davis, N. T.,** Improved methods for cobalt filling and silver intensification of insect motor neurones, *Stain Technol.,* 57, 239, 1982.
67. **Stretton, A. O. W. and Kravitz, E. A.,** Intracellular dye injection: the selection of Procion Yellow and its application in preliminary studies of neuronal geometry in the lobster nervous system, in *Intracellular Staining in Neurobiology,* Kater, S. B. and Nicholson, C., Eds., Springer-Verlag, Berlin, 1973, 21.
68. **Stewart, W. W.,** Functional coupling between cells as revealed by dye-coupling with a highly fluorescent naphthalimide tracer, *Cell,* 14, 741, 1978.
69. **Miller, J. P. and Selverston, A. I.,** Rapid killing of single neurons by irradiation of intracellularly injected dyes, *Science,* 206, 702, 1979.
70. **Thomas, J. B., Bastiani, M. J., Bate, C. M., and Goodman, C. S.,** From grasshopper to *Drosophila:* a common plan for neuronal development, *Nature (London),* 310, 203, 1984.
71. **Strausfeld, N. J., Seyan, H. S., Wohlers, D., and Bacon, J. P.,** Lucifer Yellow histology, in *Functional Neuroanatomy,* Strausfeld, N. J., Ed., Springer-Verlag, Berlin, 1983, 132.

72. **Taghert, P. H., Bastiani, M., Ho, R. K., and Goodman, C. S.,** Guidance of pioneer growth cones: filopodial contacts and coupling revealed with antibody to Lucifer Yellow, *Dev. Biol.*, 94, 391, 1982.
73. **Maranto, A. R.,** Neuronal mapping: a photooxidation reaction makes Lucifer Yellow useful for electron microscopy, *Science*, 217, 953, 1982.
74. **Sandell, J. H. and Masland, R. H.,** Photocatalysed replacement of fluroescent markers with a diaminobenzidine reaction product, *Soc. Neurosci. Abstr.*, 13, 680, 1987.
75. **Mirolli, M., Cooke, I. M., Talbot, S. R., and Miller, M. W.,** Structure and localization of synaptic complexes in the cardiac ganglion of a portunid crab, *J. Neurocytol.*, 16, 115, 1987.
76. **DeReimer, S. A. and Macagno, E. R.,** Light microscopic analysis of contacts between pairs of identified neurons with combined use of horseradish peroxidase and Lucifer Yellow, *J. Neurosci.*, 1, 650, 1981.
77. **Macagno, E. R., Muller, K. J., Kristan, W. B., Jr., DeReimer, S., Stewart, R., and Granzow, B.,** Mapping of neuronal contacts with intracellular injection of horseradish peroxidase and Lucifer Yellow in combination, *Brain Res.*, 217, 143, 1981.
78. **Kettenmann, H. and Orkand, R. K.,** Intracellular SITS injection dye-uncouples mammalian oligodendrocytes in culture, *Neurosci. Lett.*, 39, 21, 1983.
79. **Pendleton, J. W. and Keshishian, H.,** Sulforhodamine 101 as a fluorescent probe for intracellular dye fills, diffusion backfills, and double labeling with Lucifer Yellow, *Soc. Neurosci. Abstr.*, 11, 440, 1985.
80. **Strausfeld, N. J. and Seyan, H. S.,** Resolution of complex neuronal arrangements in the blowfly visual system using triple fluorescence staining, *Cell Tissue Res.*, 247, 5, 1987.
80a. **Bablanian, G. M. and Page, C. H.,** Characterization of lobster abdominal postural interneurons and extensor and flexor motor neurons using the intracellular dye 5,6-carboxyfluorescein, *Soc. Neurosci. Abstr.*, 13, 680, 1987.
81. **Nässel, D. R.,** Horseradish peroxidase and other heme proteins as neuronal markers, in *Functional Neuroanatomy*, Strausfeld, N. J., Ed., Springer-Verlag, Berlin, 1983, 45.
82. **Watson, A. H. D. and Burrows, M.,** Input and output synapses on identified motor neurons of a locust revealed by the intracellular injection of horseradish peroxidase, *Cell Tissue Res.*, 215, 325, 1981.
83. **Osborne, M. P.,** Electron-microscopic methods for nervous tissues, in *Neuroanatomical Techniques. Insect Nervous System*, Strausfeld, N. J. and Miller, T. A., Eds., Springer-Verlag, New York, 1980, 205.
84. **Lane, N. J.,** Structure of components of the nervous system, in *Comprehensive Insect Biochemistry, Physiology and Pharmacology*, Vol. 5, Kerkut, G. A. and Gilbert, L. I., Eds., Pergamon Press, Oxford, 1985, 1.
85. **Reynolds, E. S.,** The use of lead citrate at high pH as an electron opaque stain in electron microscopy, *J. Cell Biol.*, 17, 208, 1963.
86. **Murphey, R. K., Bacon, J. P., Sakaguchi, D. S., and Johnson, S. E.,** Transplantation of cricket sensory neurons to ectopic locations: arborizations and synaptic connections, *J. Neurosci.*, 3, 659, 1983.
87. **Raper, J. A., Bastiani, M., and Goodman, C. S.,** Pathfinding by neuronal growth cones in grasshopper embryos. II. Selective fasiculation onto specific axonal pathways, *J. Neurosci.*, 3, 31, 1983.
88. **Blanco, R. E.,** personal communication, 1987.
89. **Chi, C. and Carlson, S. D.,** High voltage electron microscopy of the optic neuropile of the housefly, *Musca domestica*, *Cell Tissue Res.*, 167, 537, 1976.
90. **Moran, D. T. and Rowley, J. C., III,** High voltage and scanning electron microscopy, *J. Ultrastruct. Res.*, 50, 38, 1975.
91. **Chi, C.,** High voltage electron microscopy for insect neuroanatomy, in *Functional Neuroanatomy*, Strausfeld, N. J., Ed., Springer-Verlag, Berlin, 1983, 376.

Section III. Neuron Structure and Function

Chapter 7

COCKROACH AXONS AND CELL BODIES: ELECTRICAL ACTIVITY

Marcel Pelhate, Yves Pichon, and David J. Beadle

TABLE OF CONTENTS

I. INTRODUCTION

The existence of large interneurons in the abdominal nerve cord of *Periplaneta americana* justifies the choice of this insect for electrophysiological investigation. With external bipolar electrodes, the electrical activity originating from these axons is comparatively large (up to 3 mV) and, therefore, easy to record and to discriminate from the background activity. Their size is large enough to allow easy penetration with capillary microelectrodes and recording of full-sized resting and action potentials. Some cell bodies are also quite large and can be studied with the same techniques. Using this preparation, two techniques have been developed which enable very precise and detailed analysis of the biophysical and pharmacological properties of the neuronal membrane: the voltage-clamp technique for studying ionic conductances which underlie the electrical activity of the nerve cell and the patch-clamp technique to analyze activity of single transmembrane proteins, namely, the ionic channels.

In the following sections we shall summarize the results of three categories of experiments done with two neuronal preparations chosen for their ease of use and their accessibility to externally applied molecules: isolated giant axons and cultured neurons. The first section describes the ionic basis of electrical activity in axons and neurons; the second, the effects of channel blockers, channel modulators and insecticides; and the last, the response of the cultured neurons to externally applied neurotransmitters (acetylcholine [ACh], gamma-aminobutyric acid [GABA], and glutamate).

II. IONIC BASIS OF ELECTRICAL ACTIVITY IN AXONS AND CELL BODIES

The ionic basis of electrical activity was fully established on isolated axons of *P. americana* by Pichon,[1-3] using the voltage-clamp technique.

A. RESTING POTENTIAL IN AXONS

When the tip of a microelectrode is pushed through the membrane into the axoplasm of a giant axon of the cockroach, the potential moves abruptly to a new value which is about 70 mV more negative, demonstrating the existence of a resting potential difference between the two faces of the axonal membrane. This value is significantly smaller than E_K, the equilibrium potential for potassium ions, which lies at around −90 mV.[4]

The slope relating the resting potential to the logarithm of the external potassium concentrations is about 42 mV for a tenfold change. For potassium concentrations below 8 m*M* the curve tends to level off. The slope differs by 17 mV from the 59-mV slope predicted by the Nernst equation for potassium ions:

$$E_K = \frac{RT}{F} \log \frac{[K]o}{[K]i} \qquad (1)$$

where R is the gas constant, T is the absolute temperature, F is the Faraday, and [K]o and [K]i are the outside and inside potassium concentrations, respectively. This indicates that the axonal membrane does not behave as a perfect potassium electrode and that ions other than potassium participate in the generation of the resting potential.

B. RESTING POTENTIAL IN NERVE CELL BODIES

The resting potential (Er) recorded in nerve cell bodies is slightly lower than that recorded in axons under similar conditions. In *Periplaneta*[5] it lies at around −60 mV, vs. −46 mV in *Carausius*[6,7] and −50 mV in *Bombyx*.[8]

The relationship between Er and the external potassium concentration is similar to that

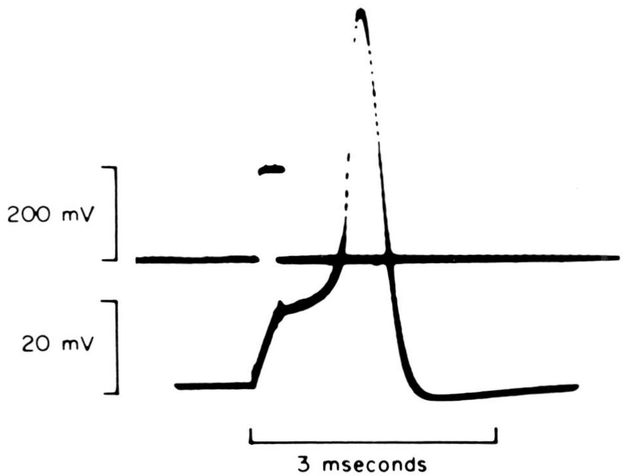

FIGURE 1. Membrane action potential recorded from an isolated giant axon of *Periplaneta americana* (lower tracing) induced by a short duration depolarizing pulse (upper tracing). (From Pichon, Y. and Boistel, J., *J. Exp. Biol.*, 47, 343, 1967. With permission.)

reported for the axon, the slope of the linear portion of the curve being about 42 mV in excitable nerve cell bodies of the cockroach.[5]

C. IONIC BASIS OF ELECTRICAL ACTIVITY IN AXONS AND NERVE CELL BODIES

The fundamental property of nerve cells is their excitability, i.e., their ability to respond to a depolarization with a graded and/or an all-or-none action potential. In both cases the effect of the stimulus is to modify the ionic permeability of the membrane. These modifications have been studied in isolated axons.

1. Graded Activity and Action Potentials in Axons and Nerve Cell Bodies

When the membrane of a cockroach axon is depolarized above a certain level, a hump appears on top of the initial phase of the catelectrotonic potential. This ''local response'', which is not actively propagated along the axon, increases with stimulus intensity and, above the critical depolarization level (which lies around −40 mV), gives rise to an all-or-none depolarization, the action potential. The action potential, an example of which is shown in Figure 1, is propagated along the axon. Its amplitude and time course are essentially similar to those recorded in other unmyelinated axons. In a series of experiments on isolated axons of *P. americana,* the mean amplitude of the spike (the large and fast portion of the action potential which follows the stimulus) was found to be approximately 95 mV, the membrane potential changing its polarity (overshooting) by about +35 mV.[9] The maximum rates of rise and fall of the action potential were 1370 and 640 V/s, respectively. The spike was followed by a slight posthyperpolarization (the positive after-potential), followed in turn by an even smaller and slower postdepolarization (the negative after-potential). Both after-potentials were significantly smaller in isolated axons than *in situ.*[9] Such a difference most probably reflects modifications in the driving force for potassium ions and in the geometry (absence of glial cells around the axons in the isolated axonal preparation).

Active electrical responses have also been recorded from nerve cell bodies. In some cases these action potentials arise spontaneously or following depolarization of the cell.[5,7,8,10,13] Most cell bodies in insect ganglia, however, are unexcitable under normal conditions; they can be purely passive or can respond to direct electrical stimulation by graded oscillatory responses. These cells can produce action potentials following various treatments such as axotomy, colchicine, injection of citrate, or superfusion with tetraethylammonium ions.[13,17]

2. Effects of Ions on Action Potentials in Axons and Nerve Cell Bodies

The action potential in cockroach axons is correlated with the presence of sodium ions in the external solution. The slope relating the overshoot of the action potential to the external sodium concentration is not significantly different from the theoretical 59-mV slope for a tenfold change predicted from the Nernst equation for sodium ions:

$$E_{Na} = \frac{RT}{F} \log \frac{[Na]o}{[Na]i} \qquad (2)$$

The action potential of nerve cell bodies was found to be sensitive to sodium in a few cases,[5,14] whereas calcium seemed to be responsible for action potential production in all other cases.[7,11,12,15,16]

The ionic channels which carry the inward current are different in axons and cell bodies. Axonal sodium channels were found to be permeable to sodium and lithium ions,[17a] but not to calcium ions. In a careful study of the effects of divalent cations on the maximum rate of rise of the action potential in giant axons of *Periplaneta,* Narahashi[18] found that calcium, magnesium, strontium, and barium ions did not contribute significantly to the action potential, but shifted the sodium inactivation curve (h_∞) toward lower membrane potentials.

Calcium channels from nerve cell bodies are blocked by magnesium ions[16] and are generally much less selective than sodium channels. Another important difference between sodium and calcium channels is the high sensitivity of the former to tetrodotoxin (TTX)[6,9,19] and saxitoxin (STX).[20]

3. Ionic Currents and Conductances in Axons

Voltage-clamp experiments have been carried out on isolated giant axons of *P. americana* using a double oil-gap technique.[21,22] Typical membrane currents generated under those conditions are illustrated in Figure 2. When the membrane potential is held at its resting level (Em = Eh = −60 mV), no current crosses the membrane. When the membrane potential is increased to 0 mV, a transient inward current begins; it reaches a peak value of about −5 mA/cm^2, then decreases and is followed by an outward current which reaches a plateau of about 6 mA/cm^2. When the membrane is reduced to its original holding potential value (Eh), the outward current returns to its resting value following an exponential time course.

Such current records can be obtained for various potential levels. The turning on and off of the inward current as well as the turning on of the delayed outward current become faster and faster as the membrane steps are made more positive. The inward current increases first, reaches a maximum for Em = −10 mV, and then decreases. The outward current increases continuously with step size. In all cases the outward tails of current which follow a return to the Eh level are exponential and proportional to the pulse size.

The current densities as well as the current-voltage characteristics are very similar[1,21] to those for the squid axon, described by Hodgkin et al.[23] As in other axons, the peak current is carried by sodium ions.[2,3,9] Since the sodium current is blocked selectively by TTX, it has been possible to separate the two ionic currents and study their kinetics.[3,9] These experiments have shown that the model proposed by Hodgkin and Huxley[24] to describe their experimental findings on the squid axon and to reconstruct most of the reactions of the nerve membrane can be used for the cockroach axon with only minor changes.

The cockroach nerve membrane can be described as being made of a capacitance (Cm) in parallel with three ionic channels: the sodium channel with a variable conductance to

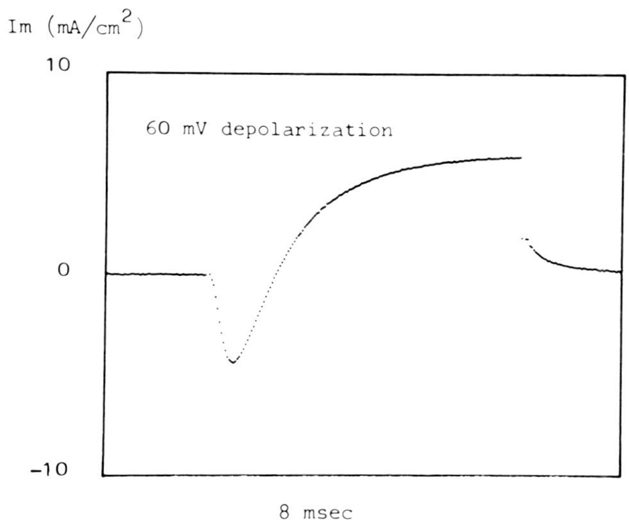

FIGURE 2. Ionic currents corresponding to a 5-ms depolarizing voltage pulse to 0 mV in a giant axon of *Periplaneta americana* under voltage-clamp conditions. Leak current subtracted. Holding potential (HP) = −60 mV. (Reprinted with permission from Pichon, Y. and Aschroft, F. M., in *Comprehensive Insect Physiology, Biochemistry and Pharmacology,* Vol. 5, Kerkut, G. A. and Gilbert, L. I., Eds., Copyright 1985, Pergamon Books Ltd.)

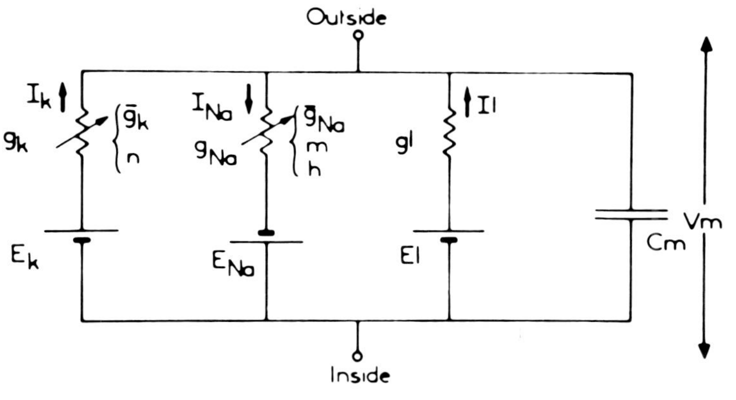

FIGURE 3. Electrical equivalent circuit of the axonal membrane based on the Hodgkin-Huxley model. (From Pichon, Y, in *Insect Neurobiology,* Treherne, J. E., Ed., Elsevier/North-Holland, Amsterdam, 1974, 73. With permission.)

sodium ions (gNa), which are more concentrated outside than inside (equilibrium potential E_{Na}); the potassium channel with a variable conductance to potassium ions (gK), which are more concentrated inside than outside (equilibrium potential E_K); and a leak channel exhibiting a constant and relatively low conductance (presumably from leaking chloride ions; see Figure 3).

According to the formulation of Hodgkin et al.,[23] the potassium and sodium variable conductances are a function of a maximum conductance ($\bar{g}K$ or $\bar{g}Na$), and their time course and voltage dependence can be described by a set of three time- and voltage-dependent parameters which can vary between 0 and 1. These parameters are potassium activation (n),

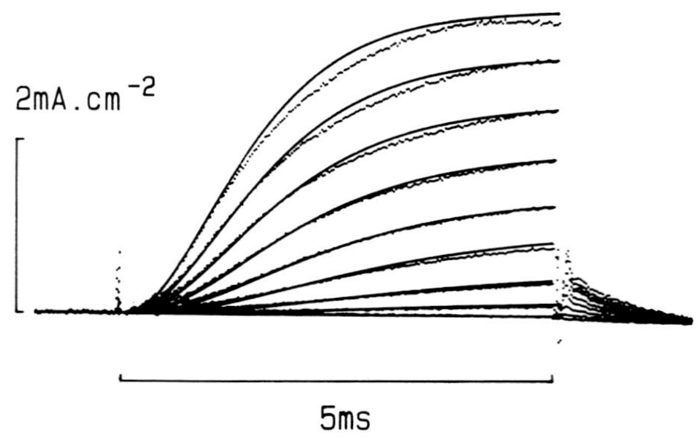

FIGURE 4. Superimposed tracings of the potassium currents corresponding to 5-ms step depolarizations of the axonal membrane of *Periplaneta americana* under voltage-clamp conditions. The membrane was brought from its holding value of −60 to −50 to +30 mV in 10-mV steps. Leak current substracted and Na^+ current blocked with 0.1 μ*M* TTX. (From Pichon, Y., Poussart, D., and Lees, G. V., in *Structure and Function in Excitable Cells,* Chang, D. C., Tasaki, I., Adelman, W. J., and Lenchtag, H. R., Eds., Plenum Press, New York, 1983, 211. With permission.)

sodium activation (m), and sodium inactivation (h). Each corresponds to the probability that a respective charge is located in a given region of the membrane. In cockroach axons, one potassium channel would open when three "n" charges simultaneously move to this region of the membrane. The potassium conductance would then be given by

$$gK = \bar{g}Kn^3 \tag{3}$$

Figure 4 illustrates the fit between potassium current traces recorded in the presence of 0.1 μ*M* TTX and theoretical curves.[4]

Steady-state values of n, m, and h for different potential values have been calculated.[9] From these data it has been possible to reconstruct, by calculation, the ionic currents corresponding to a step depolarization in a voltage-clamped cockroach axon and to reconstruct the action potential and the underlying conductance changes (Figure 5).

III. PHARMACOLOGICAL PROPERTIES OF COCKROACH AXONS

The majority of investigations on insect axons have used giant axons located in the connectives linking the abdominal ganglia in the central nervous system of *P. americana.*

The electrical properties of these axons and their modification by several drugs and insecticidal compounds were first examined by means of microelectrode recording techniques.[19,25,27] Later, a new recording method, the isolation of a single axon, was introduced by Pichon.[1] The ionic permeability changes underlying the observed electrical properties of these axons were characterized (as described in the previous section) together with the action of some pharmacological agents.[28]

The cockroach axonal model membrane preparation led to the discovery of an important new pharmacological tool in neurophysiology, 4-aminopyridine, a potent specific potassium conductance blocker.[29,32]

In the following section we shall describe the actions of specific sodium or potassium channel blockers, those of sea anemone and scorpion toxins, and the modifications induced by insecticide molecules.

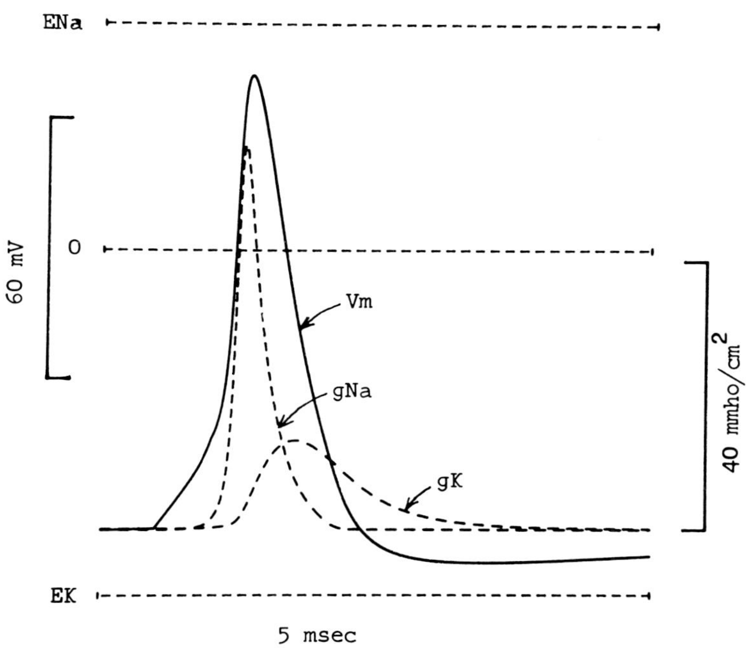

FIGURE 5. Computer reconstruction of a membrane action potential (continuous curve) and underlying conductance changes in a cockroach giant axon (interrupted curves). (Reprinted with permission from Pichon, Y. and Aschroft, F. M., in *Comprehensive Insect Physiology, Biochemistry and Pharmacology,* Vol. 5, Kerkut, G. A. and Gilbert, L. I., Eds., Copyright 1985, Pergamon Books Ltd.)

A. EFFECTS OF SODIUM AND POTASSIUM CHANNEL BLOCKERS

1. Sodium Channel Blockers

TTX, which blocks sodium channels in nerves from a wide range of animals, blocks action potentials in cockroach giant axons.[28,33] Voltage-clamp experiments on isolated axons have shown that TTX, at micromolar concentrations, selectively blocks the sodium current without affecting the delayed potassium current.[33]

STX, a naturally occurring compound obtained from the clam *Saxidomus giganteus,* the mussel *Mytilus californianus,* and from cultures of the dinoflagellate *Gonyaulax catenella,* is also a selective blocker of transient sodium in nerve membranes.[34] The structure of natural STX (nSTX) was established by X-ray crystallography, and the total synthesis of the toxin was reported by Tanino et al.[35] At 0.1 μM, synthetic STX (sSTX) applied externally to the isolated cockroach axon blocks the action potential in <3 min (Figure 6C). sSTX, like TTX, selectively blocks the sodium current of cockroach giant axons. STX is more potent than TTX, and its action is reversed more rapidly than that of TTX.[20] From the concentration dependence of sodium current suppression it is concluded that individual sodium channels are blocked by single molecules of STX which bind reversibly to part of the channel, with a dissociation constant of 3 n*M*.[20] The relative effectiveness of sSTX, nSTX, and natural TTX (nTTX) on cockroach axons has been determined; apparent K_d values are 3 n*M* (sSTX), 7 n*M* (nSTX), and 20 n*M* (nTTX). In the cockroach axonal preparation either STX or TTX can be used as the sodium channel blocker, although the former is preferred if reversibility is required (see Figure 2).

2. Potassium Channel Blockers

Until now, only a few chemicals and toxins were known to specifically block the voltage-dependent potassium conductance. In cockroach axonal preparations, tetraethylammonium

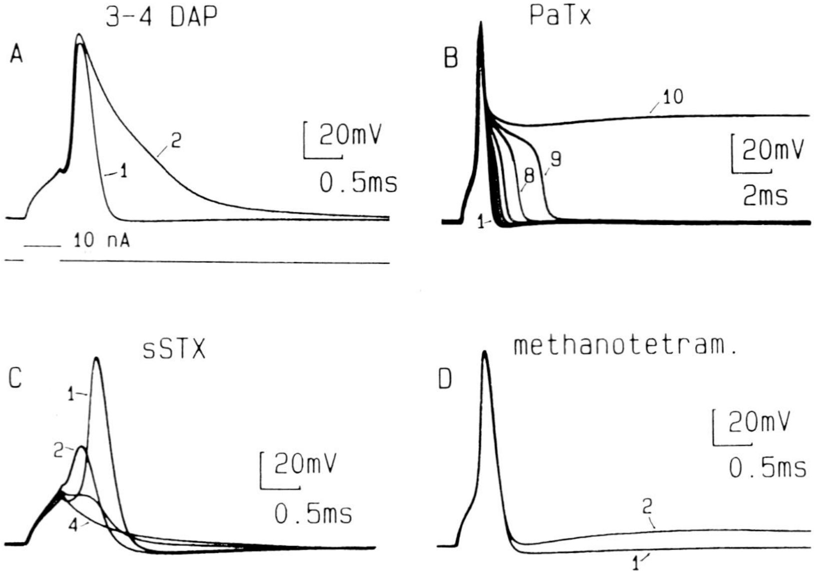

FIGURE 6. Membrane action potentials of isolated cockroach axons. (A) 1 = control; 2 = 5 min after application of 3,4-diaminopyridine (10 μ*M*). Lower tracing indicates the short duration (0.5-ms) depolarizing current pulse inducing the responses. (B) Superimposed recordings of the membrane action potentials during bath application of *Parasicyonis* toxin (1 μ*M*). An action potential was evoked every 10 s. 1 = control; 8, 9, and 10 = after 3 min 40 s, 3 min 50 s, and 4-min application, respectively. (C) Superimposed recordings of axonal responses during a 3-min application of synthetic STX (0.1 μ*M*), which progressively suppressed the action potential (1 to 4). (D) A control action potential (1) and an action potential followed by a postdepolarization (2) induced by an 8-min application of 2 μ*M* methanotetramethrin.

(TEA^+) causes marked prolongation of action potentials and suppresses the delayed potassium current when applied at high concentrations (>0.01 *M*).[9]

4-Aminopyridine, in micromolar concentrations, slows the repolarization phase of the action potential in the cockroach axon (Figure 6A) and, thus, increases excitability. These effects are the consequence of a specific decrease in the potassium conductance.[29,32] This specific property has been confirmed in similar preparations, and several isomers and derivatives of 4-aminopyridine have been found to be more or less potent and, similarly, more or less selective for the potassium channels.[36] Aminopyridines and their derivatives are widely used today as pharmacological tools. Moreover, particular interactions of potassium channel blockers, i.e., time, voltage, and frequency dependence of the blockade, have revealed novel structure-activity relationships.[37,41] In addition, these molecules have been used to differentiate several populations of potassium channels.[42] The contribution of the cockroach axonal preparation to the discovery of aminopyridine[42] is an illustration of the value of the insect system to neurobiological research.

At the present time, only one natural toxin from the scorpion *Centruroides noxius* has been described as a specific blocker of the potassium channels in the squid axon.[43] Several experiments with similar scorpion toxins[43a] did not reveal any potent natural toxin interacting specifically with potassium channels of the cockroach axon.

B. EFFECTS OF SEA ANEMONE AND SCORPION TOXINS

Among the animal toxins, sea anemone and scorpion toxins have been widely used to study ionic channels. To date, 4 sea anemone toxins and 12 scorpion toxins have been

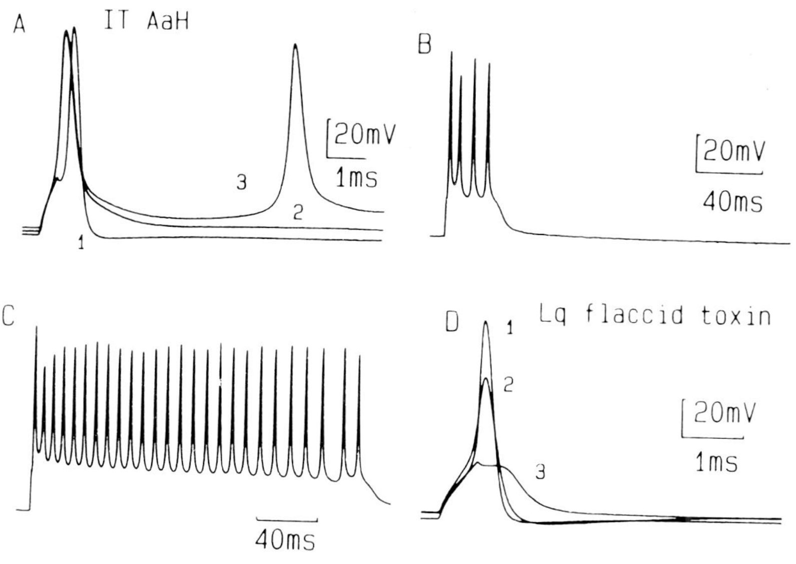

FIGURE 7. Membrane action potentials of isolated cockroach axons recorded in the presence of scorpion "insect" toxins. (A) 1 = control action potential; 2 and 3 = superimposed recordings of the responses after a 5-min application of 1.3 μM ITAaH (insect toxin from *Androctonus*). (B,C) After 7 and 12 min, respectively. (D) Three superimposed recordings (1 to 3) of the axonal responses during an application of 6 μM insect toxin II, a "flaccid" toxin from *Leiurus* venom.

studied for their membrane effects on the cockroach axon. The sea anemone toxins prolong the action potential considerably (Figure 6B), resulting in long "plateau" potentials,[44,46] through a delay in the turning off of the sodium current. Among the scorpion toxins, α-toxins such as toxin II from *Androctonus australis* (also called Mammal Toxin II [MTII AaH]) and crustacean toxin (CT AaH) cause "plateau" potentials, do not induce repetitive activity, and act in a manner similar to anemone toxins.[46] Another compound much more toxic to insects, called "insect toxin" (IT AaH) by Zlotkin et al.,[47,48] was found in *Androctonus* venom. This material, when applied externally to the isolated cockroach axon, depolarizes by 5 to 10 mV and induces a maintained repetitive activity of "short" action potentials (Figure 7A,B,C).[46] In voltage-clamp procedures IT AaH increases and prolongs the sodium conductance, mainly at negative membrane potential values, through a slowing of the sodium current turnoff.

Two groups of insect-selective toxins have been purified from each of the venoms of *Buthotus judaicus* and *Leiurus quinquestriatus*. The first toxins (Bj IT_1 and Lq IT_1) induce a fast, excitatory contraction paralysis of fly larvae and have been called "contractive" toxins. The second toxins (Bj IT_2 and Lq IT_2), causing a slow, depressant, flaccid paralysis of fly larvae, are called "flaccid" toxins.[49,50] The contractive toxins, when applied to the isolated cockroach axon, act like the insect toxin from *Androctonus,* inducing repetitive firing. The underlying phenomenon is also the induction of a slow, voltage-dependent sodium conductance which has its maximum around Em = −30 mV and decreases for larger depolarizations. Conversely, flaccid toxins depolarize the axonal membrane and block the action potential (Figure 7D) by inhibiting sodium conductance.[49,50]

It is not clear from these studies why toxins differing in origin (anemone or scorpion), amino acid sequence, and molecular weight have similar actions, while chemically similar toxins from the same venom show species specificity in insects and modify the sodium channels. Today, scorpion toxins are widely used as pharmacological tools to study populations of ionic channels. In the future, a better understanding of the "discrete" causes of specificity should help in inventing new strategies in pharmacology and insect control and in understanding the development of resistance to neurotoxic agents.

C. EFFECTS OF DDT AND PYRETHROID INSECTICIDES

Several years ago, the action of the organochlorine insecticide dichlorodiphenyltrichloroethane (DDT) on the cockroach axon was described.[9,27] Effects were concentration dependent, ranging from 1 to 10 μM, and had virtually no effect on the resting or action potentials, but postdepolarization increased in amplitude.[9] Larger amounts of DDT (up to 0.5 mM) increased both the amplitude and the duration of the postdepolarization.

The underlying changes in the ionic conductances were complex. A decrease in potassium conductance was noted. The peak current was delayed slightly, and the turning off of the sodium current during the clamp pulse was slowed down. The most obvious change, however, was the occurrence of a long exponential tail of sodium current after the pulse. Analysis of the voltage and time dependency of the characteristic tail of current suggested a close correlation between the development of the tail current and the turning on of the peak sodium conductance.[51] Although all sodium channels open normally (as predicted from the activation and inactivation processes), a small number of them may become insensitive to changes in membrane potential and, thus, remain open longer. The number of open channels was calculated by integrating the sodium conductance against time, and the proportion of modified channels was estimated as 20% of the total population. Assuming an overall exponential decay in conductance for these modified channels, an almost perfect fit to the experimental data was obtained.[51]

Allethrin, a synthetic derivative of pyrethrin, stimulates insect nerves by inducing repetitive discharges[52] and develops large negative after-potentials in cockroach axons.[53] *S*-Bioallethrin, kadethrin, biotetramethrin, cismethrin, biopermethrin, bioresmethrin, and methanotetramethrin, type I pyrethroids,[56] have been applied to the isolated axon.[54,55] (With regard to methanotetramethrin, see also Reference 58 in Chapter 18.) All induce depolarizing after-potentials of long duration (Figure 6D), slightly depolarize the axonal membrane, and induce repetitive firing in whole cockroach cercal sensory nerves.[56] Only cismethrin develops repetitive activity consisting of slow oscillations of the membrane potential.[54]

Among type II pyrethroids, deltamethrin, a very potent insecticide when applied topically, apparently has much smaller direct effects on the axonal membrane,[54] although the toxicity is well correlated with its ability to depolarize by a block in the open state of a very small population of sodium channels.[55]

Pyrethroids have been found to selectively increase the sodium conductance. The time and voltage dependency of this increase has been analyzed quantitatively for two molecules: *S*-bioallethrin[51] and methanotetramethrin,[57] both type I pyrethroids. For small concentrations (up to 5 μM), the most obvious effect is the development of low voltage-sensitive sodium conductance which turns off very slowly at the end of the depolarizing voltage pulses, producing large tails of inward current (Figure 8).[55,57] The time and voltage dependencies of this tail have been studied quantitatively during the onset of pyrethroid action. It was found that, during the first minutes, the voltage sensitivity of the tail current followed that of unmodified (fast) sodium conductance quite closely, although the activation curve usually was shifted to the right toward more positive potentials.[55] Later, the voltage sensitivity of the pyrethroid-induced current was found to decrease, as indicated by the decrease in the slope of the activation curve. The extent to which these modifications can account for the observed changes in nerve activity has been evaluated through computer reconstruction of nerve membrane behavior during the early or late phase of poisoning.[57]

The cockroach axon reveals itself as a good model for testing insecticide molecules. Because sodium channels play such a key role in nerve activity, they are and will remain

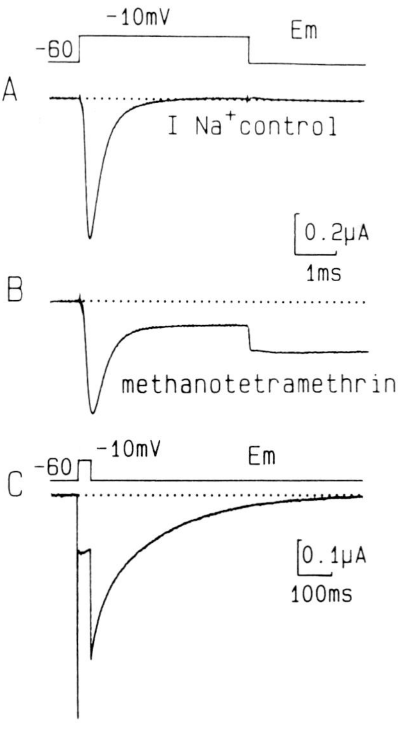

FIGURE 8. Sodium currents in a giant axon from a cockroach. (A) Control sodium current (lower tracing) after suppression of the potassium current in the presence of 3,4-diaminopyridine (0.2 m*M*). (B) Sodium current recorded after a 10-min application of methanotetramethrin (2 μ*M*). (C) Sodium current during a 40-ms duration pulse to − 10 mV and large "tail" sodium current turning off slowly at the end of the voltage pulse.

one of the best target sites for neurotoxic insecticides. Nevertheless, insecticides are also highly neurotoxic to mammals as well as to insects when applied directly to the nervous system, and it has been demonstrated[57a] that the pyrethroids share the same mechanism of action on voltage-dependent sodium channels in the frog nerve membrane (see Chapter 18 in this work). Much of the selectivity of insecticides would appear to be due to differences in penetration, pharmacokinetics, or metabolism.

IV. PHARMACOLOGICAL PROPERTIES OF COCKROACH NEURONS IN CULTURE

A. EFFECTS OF ACETYLCHOLINE AND RELATED MOLECULES

Cultured cockroach neurons possess α-bungarotoxin binding sites and are depolarized by pressure-applied ACh. The electrical events associated with the effect of ACh and carbamylcholine (CCh) have been analyzed with the patch-clamp technique.[59]

Pressure-injection of micromolar concentrations of the agonists in the vicinity of the neuron induces a depolarization. Under voltage-clamp conditions, in the "whole cell" configuration of the patch-clamp technique, the inward current which underlies this depolarization is associated with a large increase in "noise". The spectrum of these current fluctuations can be fitted with one (CCh, corner frequency = 30 Hz) or two (ACh, corner frequencies = 8 and 150 Hz) lorentzian curves.

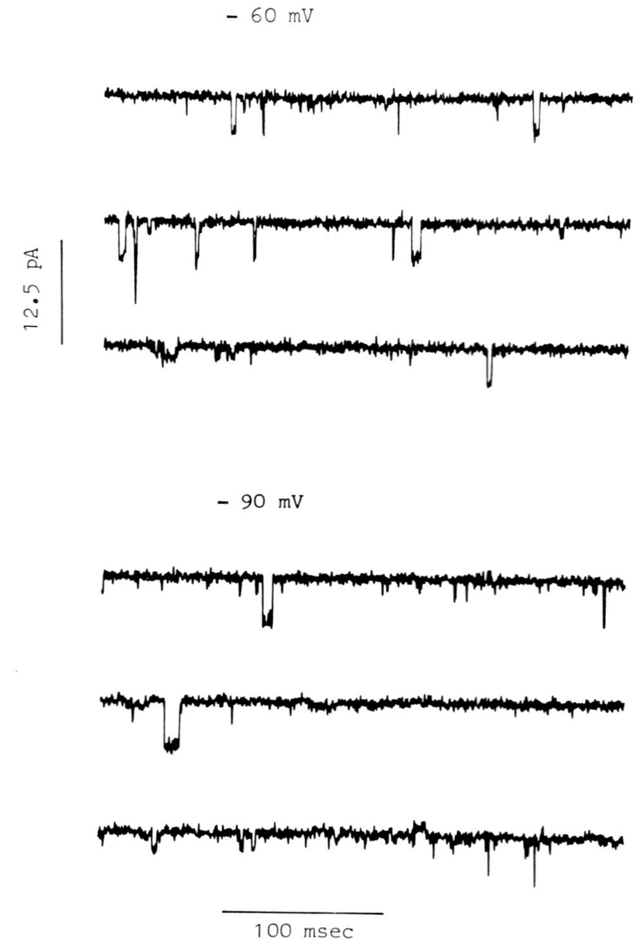

FIGURE 9. Single channel activity induced by 5 μ*M* carbamylcholine applied onto the membrane of a cultured cockroach neuron at two holding potential levels.[59]

Addition of micromolar concentrations of ACh or CCh to the solution contained in the patch pipette induces two categories of small unitary currents (Figure 9). The reversal potential of these unitary currents lies between −5 and −15 mV, with a conductance of ca. 18 to 48 pS. Their mean opening time is clearly longer in CCh than in ACh and, in the former case, can be fitted by two exponential functions. The time constant of these functions increases with membrane hyperpolarization.

B. EFFECTS OF GABA AND RELATED MOLECULES

About 60% of cultured neurons respond to externally applied GABA. The response, which is hyperpolarizing in the majority of the cells, corresponds to an increase in the chloride permeability of the cell membrane. It is blocked by 10 μ*M* picrotoxin, but is insensitive to 10 μ*M* bicuculline. In the whole-cell configuration the cells respond to the external application of GABA or muscimol with a small inward current on which is superimposed a large "noise" (Figure 10). Addition of 2 μ*M* flunitrazepam to the bathing solution increases the current as well as the current fluctuations (Figure 10D).

The power spectrum of these current fluctuations can be fitted with a single lorentzian

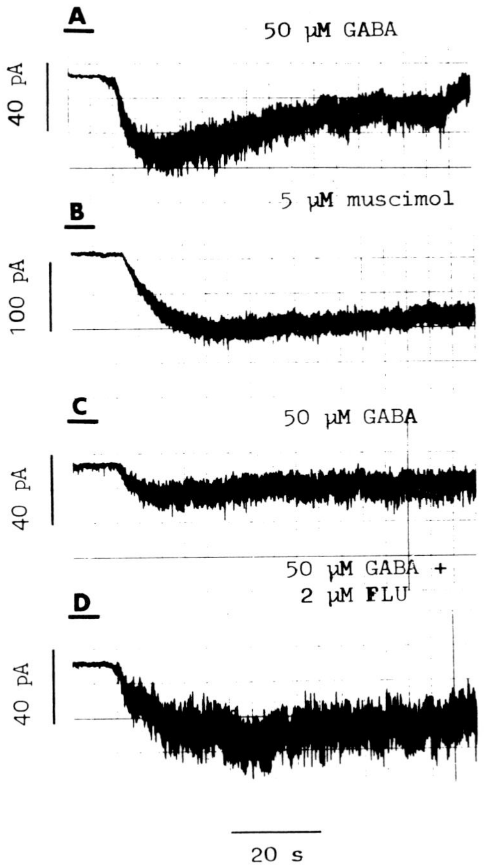

FIGURE 10. Examples of inward currents induced by prolonged application of GABA and muscimol onto insect neurons in the whole-cell configuration of the patch-clamp technique. (From Shimahara, T., Pichon, Y., Lees, G., Beadle, C. A., and Beadle, D. J., *J. Exp. Biol.*, 131, 231, 1987. With permission.)

function with corner frequencies of about 15 Hz (Figure 11). So far it has not been possible to record unitary currents which could be unequivocally related to the effects of GABA or its agonists.[60]

C. EFFECTS OF GLUTAMATE

Iontophoretic application of glutamate in the vicinity of the membrane of cultured cockroach neurons induces, in 30% of the cells tested, a depolarizing or a hyperpolarizing response corresponding to a decrease in membrane conductance. The equilibrium potential of the depolarizing response lies at around 0 mV, whereas that of the hyperpolarizing response is situated between −60 and −80 mV.[61]

L-Glutamate (1 to 10 μ*M*) in the solution contained in the patch pipette in the cell-attached configuration (patch-clamp technique) induces short inward unitary currents. In cells pretreated with 1 μ*M* concanavalin A (Con A) to reduce desensitization, the frequency of these unitary currents is very high and the events very short (mean open time ca. 0.2 ms; Figure 12). Single-event amplitude histograms are very broad, and there are indications that the corresponding channels may exhibit several conductance states.[61]

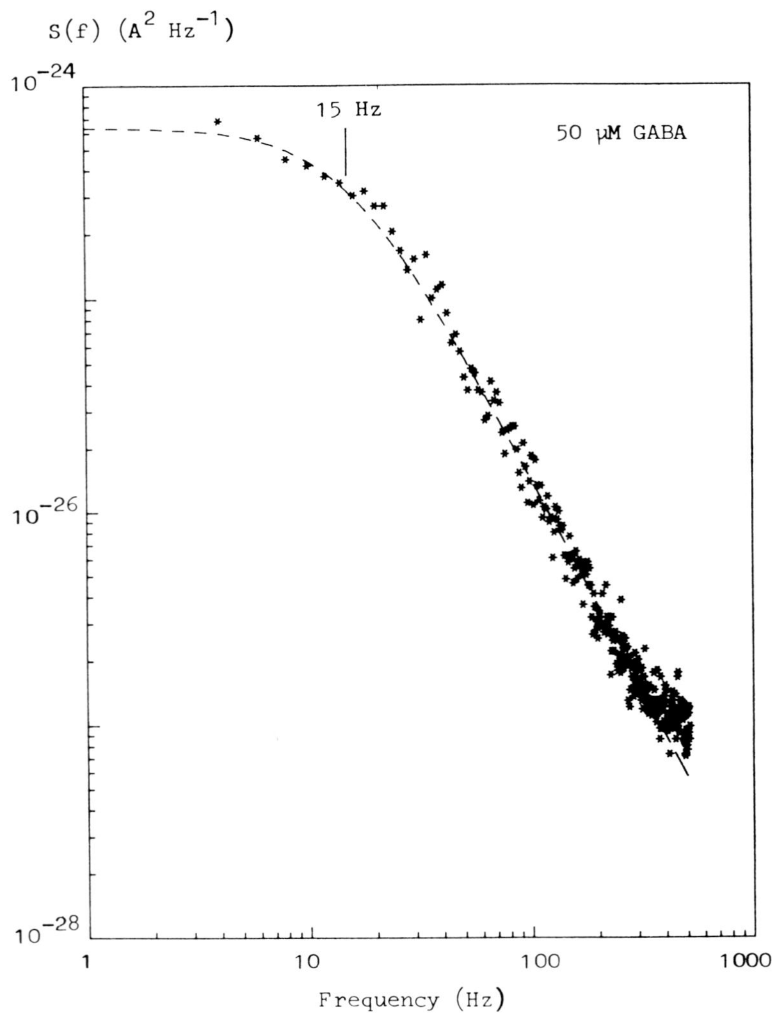

FIGURE 11. Power spectrum of the noise induced by application of 50 μM GABA onto cultured cockroach neurons: the experimental data points can be fitted with a single lorentzian component with a corner frequency of 15 Hz. (From Shimahara, T., Pichon, Y., Lees, G., Beadle, C. A., and Beadle, D. J., *J. Exp. Biol.*, 131, 231, 1987. With permission.)

V. CONCLUSION

Large axons from four phyla have been used extensively with the voltage clamp technique.

- Molluscs: squid axon[24]
- Arthropods: lobster,[62] crayfish,[63] and cockroach[21]
- Annelids: *Myxicola*[64]
- Vertebrates: amphibians[65] and mammals[66]

As stated by Bertil Hille,[67] "Apparently all axons use the two major channel types first described in the squid giant axon. The simplicity of the excitability mechanism of axons is in accord with the simplicity of their task: to propagate every impulse unconditionally."

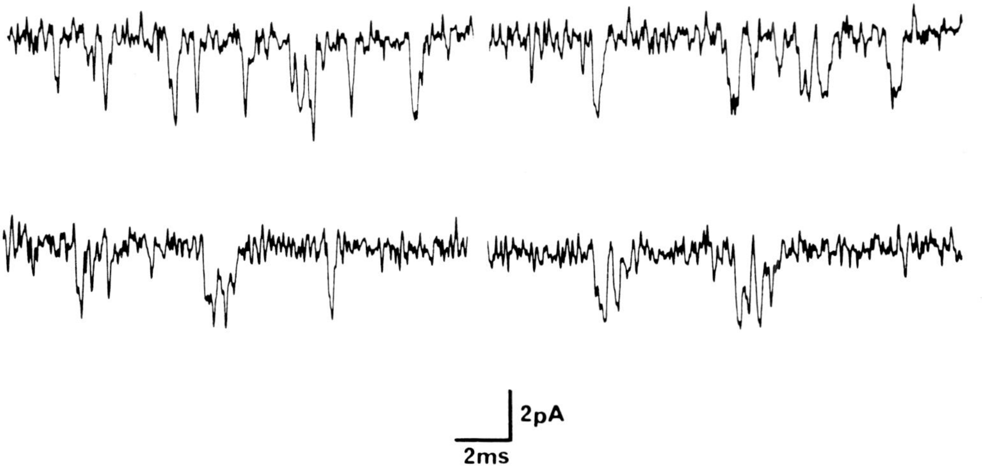

FIGURE 12. Single channel activity evoked by 0.1 m*M* glutamate in the presence of Con A in a cell-attached patch. (From Horseman, B. G., Seymour, C., Bermudez, I., and Beadle, D. J., *Neurosci. Lett.*, 85, 65, 1988. With permission.)

There are many advantages to using cockroaches as laboratory subjects. It takes little effort to rear large numbers, at low cost, in the laboratory; moreover, this is possible throughout the year. A wide range of neuroactive substances, such as aminopyridines, STX, TTX, anemone and scorpion venoms, pyrethroids, and other toxins, have similar effects in both cockroach and vertebrate preparations. The ease of making neural preparations coupled with the facts cited above justifies the use of cockroaches as biological models for vertebrates as well as for study in their own right. It is possible to study both axonal and synaptic responses in the neuron (see Chapter 8 in the present work). The electrophysiological actions of a compound on both voltage- and ligand-gated channels can be studied in such preparations. Newer data obtained with cell cultures of cockroach neurons (see also Chapter 12 in this work) supplement our methods of investigation and show that cockroach central nervous systems are choice animal models for biomedical studies.

REFERENCES

1. **Pichon, Y.,** Application de la technique du voltage imposé à l'étude de la fibre isolée d'insecte, *J. Physiol. (Paris)*, 9, 282, 1967.
2. **Pichon, Y.,** Nature des courants membranaires dans une fibre nerveuse d'insecte: l'axone de *Periplaneta americana, C.R. Soc. Biol.*, 162, 2233, 1968.
3. **Pichon, Y.,** Axonal conduction in insects, in *Insect Neurobiology*, Treherne, J. E., Ed., Elsevier/North-Holland, Amsterdam, 1974, 73.
4. **Pichon, Y., Poussart, D., and Lees, G. V.,** Membrane ionic currents, current noise and admittance in isolated cockroach axons, in *Structure and Function in Excitable Cells*, Chang, D. C., Tasaki, I., Adelman, W. J., and Lenchtag, H. R., Eds., Plenum Press, New York, 1983, 211.
5. **Jego, P., Callec, J. J., Pichon, Y., and Boistel, J.,** Etude électrophysiologique de corps cellulaires excitables du VIème ganglion abdominal de *Periplaneta americana;* aspects électriques et ioniques, *C.R. Soc. Biol.*, 164, 893, 1970.
6. **Treherne, J. E. and Maddrell, S. H. P.,** Membrane potentials in the central nervous system of the stick insect, *Carausius morosus, J. Exp. Biol.*, 46, 413, 1967.
7. **Orchard, I. and Finlayson, L. H.,** Electrical properties of identified neurosecretory cells in the stick insect, *Comp. Biochem. Physiol.*, 58, 87, 1977.

8. **Monticelli, G. and Depretts, G.,** Electrophysiological characteristics of *Bombyx mori* L. ventral nerve cord (effects of sodium and potassium on the membrane potential), *Experientia*, 35, 62, 1979.
9. **Pichon, Y.,** Aspects Électriques et Ioniques du Fonctionnement Nerveux chez les Insectes. Cas Particulier de la Chaîne Nerveuse Abdominale d'une Blatte, *Periplaneta americana* L., D.Sc. thesis, University of Rennes, Rennes, France, 1979.
10. **Callec, J. J. and Boistel., J.,** Etude de divers types d'activités électriques enregistrées par microélectrodes capillaires au niveau du dernier ganglion abdominal de *Periplaneta americana* L., *C.R. Soc. Biol.*, 160, 1943, 1966.
11. **Kerkut, G. A., Pitman, R. M., and Walker, R. J.,** Electrical activity in insect nerve cell bodies, *Life Sci.*, 7, 605, 1968.
12. **Orchard, I.,** Calcium-dependent action potentials in a peripheral neurosecretory cell of the stick insect, *J. Comp. Physiol.*, 112, 95, 1976.
13. **Jego, P., Callec, J. J., Pichon, Y., and Boistel, J.,** Etude électrophysiologique de corps cellulaires excitables du vième ganglion abdominal de *Periplaneta americana:* aspects électriques et ioniques, *C.R. Soc. Biol.* 164, 893, 1970.
14. **Goodman, C. and Heitler, W. J.,** Electrical properties of insect neurones with spiking and non-spiking somata: normal, axotomized, and colchicine treated neurones, *J. Exp. Biol.*, 83, 95, 1979.
15. **Pitman, R. M.,** The ionic dependence of action potentials induced by colchicine in an insect motorneurone cell body, *J. Physiol. (London)*, 247, 511, 1975.
16. **Pitman, R. M.,** Calcium-dependent action potentials in the cell body of an insect motorneurone, *J. Physiol. (London)*, 247, 62, 1975.
17. **Pitman, R. M.,** Intracellular citrate and externally applied tetraethylammonium ions produce calcium-dependent action potential in an insect motorneurone cell body, *J. Physiol. (London)*, 291, 327, 1979.

17a. **Pichon, Y. and Pelhate, M.,** unpublished data, 1974.

18. **Narahashi, T.,** Dependence of excitability of cockroach giant axons on external divalent cations, *Comp. Biochem. Physiol.*, 19, 759, 1966.
19. **Narahashi, T.,** The physiology of insect axons, in *The Physiology of the Insect Central Nervous System*, Treherne, J. E. and Beament, J. W. L., Eds., Academic Press, London, 1965, 1.
20. **Sattelle, D. B., Pelhate, M., and Hue, B.,** Pharmacological properties of axonal sodium channels in the cockroach *Periplaneta americana* L. I. Selective block by synthetic saxitoxin, *J. Exp. Biol.*, 83, 41, 1979.
21. **Pichon, Y. and Boistel, J.,** Current-voltage relations in the isolated giant axons of the cockroach under voltage-clamp conditions, *J. Exp. Biol.*, 47, 343, 1967.
22. **Pelhate, M. and Sattelle, D. B.,** Pharmacological properties of insect axons: a review, *J. Insect Physiol.*, 28, 889, 1982.
23. **Hodgkin, A. L., Huxley, A. F., and Katz, B.,** Measurement of current-voltage relations in the membrane of the giant axon of *Loligo*, *J. Physiol. (London)*, 116, 424, 1952.
24. **Hodgkin, A. L. and Huxley, A. F.,** Quantitative description of membrane current and its application to conduction and excitation in nerve, *J. Physiol. (London)*, 117, 500, 1952.
25. **Boistel, J. and Coraboeuf, E.,** Potentiel de membrane et potentials d'action de nerf d'insecte recueillis à l'aide de microélectrodes intracellulaires, *C.R. Acad. Sci.*, 238, 2116, 1954.
26. **Yamasaki, T. and Narahashi, T.,** Studies on the mechanism of action of insecticides. XIV. Intracellular microelectrode recordings of resting and action potentials from the insect axon and the effects of DDT on the action potential, *Botyu-Kagaku*, 22, 305, 1957.
27. **Narahashi, T.,** The properties of insect axons, *Adv. Insect Physiol.*, 1, 175, 1963.
28. **Pichon, Y.,** The pharmacology of the insect nervous system, in *The Physiology of Insecta*, Vol. 4, Rockstein, M., Ed., Academic Press, New York, 1974, 101.
29. **Pelhate, M., Hue, B., and Chanelet, J.,** Effets de la 4-aminopyridine sur le système nerveux d'un insecte: la Blatte (*Periplaneta americana* L.), *C.R. Soc. Biol.*, 166, 1598, 1972.
30. **Pelhate, M., Hue, B., and Chanelet, J.,** Modifications par la 4-aminopyridine des caractéristiques électriques de l'axone géant isolé d'un insecte, la Blatte (*Periplaneta americana* L.), *C.R. Soc. Biol.*, 168, 27, 1974.
31. **Pelhate, M., Hue, B., Pichon, Y., and Chanelet, J.,** Actions de la 4-aminopyrine sur la membrane de l'axone isolé d'insecte, *C.R. Acad. Sci.*, 278, 2807, 1974.
32. **Pelhate, M. and Pichon, Y.,** Selective inhibition of potassium current in the giant axon of the cockroach, *J. Physiol. (London)*, 242, 90, 1974.
33. **Pichon, Y.,** Effets de la tétrodotoxine (TTX) sur les caractéristiques de perméabilité membranaire de la fibre nerveuse isolée d'insecte, *C.R. Acad. Sci.*, 268, 1095, 1969.
34. **Narahashi, T., Haas, H. G., and Therrien, E. F.,** Saxitoxin and tetrodotoxin: comparison of nerve blocking mechanism, *Science*, 137, 1441, 1967.
35. **Tanino, H., Nakata, T., Kaneko, T., and Kishi, Y.,** A stereospecific total synthesis of d,l-saxitoxin, *J. Am. Chem. Soc.*, 99, 2818, 1977.

36. **Pichon, Y., Meves, H., and Pelhate, M.,** Effects of aminopyridines on ionic currents and ionic channel noise in unmyelinated axons, in *Aminopyridines and Similarly Acting Drugs — Effects on Nerves Muscles and Synapses,* Lechat, P., Thesleff, S., and Bowman, W. C., Eds, Pergamon Press, Oxford, 1982, 53.
37. **Yeh, J. Z., Oxford, G. S., Wu, C. H., and Narahashi, T.,** Interactions of aminopyridines with potassium channels of squid axon membranes, *Biophys. J.,* 16, 77, 1976.
38. **Yeh, J. Z., Oxford, G. S., Wu, C. H., and Narahashi, T.,** Dynamics of aminopyridine block of potassium channels in squid axon membrane, *J. Gen. Physiol.,* 68, 519, 1976.
39. **Meves, H. and Pichon, Y.,** The effect of internal and external 4-aminopyridine on the potassium currents in intracellularly perfused squid giant axons, *J. Physiol. (London),* 268, 511, 1977.
40. **Meves, H. and Pichon, Y.,** Modèle d'action de la 4-aminopyridine au niveau des "canaux" potassium de l'axone géant de Calmar (*Loligo forbesi* L.), *C.R. Acad. Sci.,* 284, 1325, 1977.
41. **Kirsch, G. E. and Narahashi, T.,** 3,4-Diaminopyridine. A potent new potassium channel blocker, *Biophys. J.,* 22, 507, 1978.
42. **Dubois, J. M.,** Evidence for the existence of three types of potassium channels in the frog ranvier node membrane, *J. Physiol. (London),* 318, 297, 1981.

42a. **Lechat, P., Thesleff, S., and Bowman, W. C., Eds.,** *Aminopyridines and Similarly Acting Drugs: Effects on Nerves, Muscles and Synapses,* Pergamon Press, Elmsford, New York, 1982.

43. **Carbone, E., Prestipino, G., Spadavecchia, L., Franciolini, F., and Possani, L. D.,** Blocking of the squid axon K^+ channel by noxius toxin: a toxin from the venom of the scorpion *Centruroides noxius, Pflügers Arch.,* 408, 423, 1987.

43a. **Pelhate, M.,** unpublished data, 1987.

44. **Pelhate, M., Hue, B., and Sattelle, D. B.,** Pharmacological properties of axonal sodium channels in the cockroach *Periplaneta americana* L. II. Slowing of sodium current turn-off by *Condylactis* toxin, *J. Exp. Biol.,* 83, 49, 1979.
45. **Pelhate, M., Laufer, J., Pichon, Y., and Zlotkin, E.,** Effects of several sea anemone and scorpion toxins on excitability and ionic currents in the giant axon of the cockroach, *J. Physiol. (Paris),* 79, 309, 1984.
46. **Pelhate, M. and Zlotkin, E.,** Actions of insect toxin and other toxins derived from the venom of the scorpion *Androctonus australis* on the isolated giant axons of the cockroach *(Periplaneta americana), J. Exp. Biol.,* 97, 67, 1982.
47. **Zlotkin, E., Miranda, F., Kupeyan, C., and Lissitzky, S.,** A new toxic protein in the venom of the scorpion *Androctonus australis* Hector, *Toxicon,* 9, 9, 1971.
48. **Zlotkin, E., Miranda, F., and Lissitzky, S.,** A toxic factor to crustacean in the venom of the scorpion *Androctonus australis* Hector, *Toxicon,* 10, 211, 1972.
49. **Lester, D., Lazarovici, P., Pelhate, M., and Zlotkin, E.,** Purification, characterization and action of two insect toxins from the venom of the scorpion *Buthotus judaicus, Biochim. Biophys. Acta,* 701, 370, 1982.
50. **Zlotkin, E., Kadouri, D., Gordon, D., Pelhate, M., Martin, M. F., and Rochat, H.,** An excitatory and a depressant insect toxin from scorpion venom both affect sodium conductance and possess a common binding site, *Arch. Biochem. Biophys.,* 240, 877, 1985.
51. **Pichon, Y., Guillet, J. Cl., Heilig, U., and Pelhate, M.,** Recent studies on the effects of DDT and pyrethroid insecticides on nervous activity in cockroaches, *Pestic. Sci.,* 16, 627, 1985.
52. **Narahashi, T.,** Effects of the insecticide allethrin on membrane potentials of cockroach giant axons, *J. Cell. Comp. Physiol.,* 59, 61, 1962.
53. **Narahashi, T.,** Nature of the negative after-potential increased by the insecticide allethrin in cockroach giant axons, *J. Cell. Comp. Physiol.,* 59, 67, 1962.
54. **Laufer, J., Roche, M., Pelhate, M., Elliott, M., Janes, N. F., and Sattelle, D. B.,** Pyrethroid insecticides: actions on insect axons, *J. Insect Physiol.,* 30, 341, 1984.
55. **Laufer, J., Pelhate, M., and Sattelle, D. B.,** Actions of pyrethroid insecticides on insect axonal sodium channels, *Pestic. Sci.,* 16, 651, 1985.
56. **Gammon, D. W., Brown, M. A., and Casida, J. E.,** Two classes of pyrethroid action in the cockroach, *Pestic. Biochem. Physiol.,* 15, 181, 1981.
57. **Pichon, Y., Pelhate, M., and Heilig, U.,** Sodium channels: primary targets for insecticide action?, *Am. Chem. Soc. Proc.,* 356, 212, 1987.

57a. **Vijverberg, H. P. M. and de Weille, J. R.,** The interaction of pyrethroids with voltage-dependent Na channels, *Neurotoxicology,* 6, 23, 1985.

58. **Pichon, Y. and Aschroft, F. M.,** Nerve and muscle: electrical activity, in *Comprehensive Insect Physiology, Biochemistry and Pharmacology,* Vol. 5, Kerkut, G. A. and Gilbert, L. I., Eds, Pergamon Press, Oxford, 1985, 85.
59. **Beadle, C. A., Beadle, D. J., Pichon, Y., and Shimahara, T.,** Patch-clamp and noise analysis studies of cholinergic properties of cultured cockroach neurones, *J. Physiol. (London),* 371, 145P, 1985.
60. **Shimahara, T., Pichon, Y., Lees, G., Beadle, C. A., and Beadle, D. J.,** Gamma-aminobutyric acid receptors on cultured cockroach brain neurones, *J. Exp. Biol.,* 131, 231, 1987.

61. **Horseman, B. G., Seymour, C., Bermudez, I., and Beadle, D. J.,** The effects of L-glutamate on cultured insect neurones, *Neurosci. Lett.,* 85, 65, 1988.
62. **Julian, F. J., Moore, J. W., and Goldman, D. E.,** Membrane potentials of the lobster giant axon obtained by use of the sucrose-gap technique, *J. Gen. Physiol.,* 45, 1195, 1962.
63. **Shrager, P.,** Ionic conductance changes in voltage clamped crayfish axons at low pH, *J. Gen. Physiol.,* 64, 666, 1974.
64. **Goldman, L. and Schauf, C. L.,** Quantitative description of sodium and potassium currents and computed action potentials in *Myxicola* giant axons, *J. Gen. Physiol.,* 61, 361, 1973.
65. **Dodge, F. A. and Frankenhaeuser, B.,** Membrane currents in isolated frog nerve fibre under voltage clamp conditions, *J. Physiol. (London),* 143, 76, 1958.
66. **Chiu, S. Y., Ritchie, J. M., Rogart, R. B., and Stagg, D.,** A quantitative description of membrane currents in rabbit myelinated nerve, *J. Physiol. (London),* 292, 149, 1979.
67. **Hille, B.,** *Ionic Channels of Excitable Membranes,* Sinauer Associates, Sunderland, MA, 1984.

Chapter 8

ELECTROPHYSIOLOGY AND PHARMACOLOGY OF SYNAPTIC TRANSMISSION IN THE CENTRAL NERVOUS SYSTEM OF THE COCKROACH

Bernard Hue and Jean-Jacques Callec

TABLE OF CONTENTS

I. INTRODUCTION

In experimental neuropharmacology, studies are based on the use of various nerve preparations. The objective of such research is the development of more refined methods of recording that minimize the use of complicated nervous structures. Obviously, giant cells offer a breadth of possibility for investigation at the single-cell level, but they present specific anatomy and physiology. In this respect, Arthropoda, the largest phylum in the animal kingdom, including Insecta, are well known for their contribution to electrophysiology.

The attractiveness of the giant fiber's relatively accessible structure and suitable size has focused research on the study of synaptic transmission in the terminal abdominal ganglion (A6) of the American cockroach, *Periplaneta americana*. This preparation has provided us with the earliest and the most numerous studies on insect synaptic function and is particularly useful for studies including the effects of diverse neuroactive agents on the synaptic membrane (e.g., neurotransmitter receptors and ion channels).

In this chapter we present evidence that this simple insect nervous preparation is suitable for neuropharmacological applications to biomedical research.

II. GENERAL PHYSIOLOGY AND PHARMACOLOGY

The central nervous system (CNS) of insects, confined to the nerve cord, consists of a pair of ganglia fused in the midline and linked together by paired interganglionic connectives. Each part is surrounded by the nerve sheath comprising an outer layer of connective tissue, the neural lamella. In ganglia, below the neural lamella, the perineurium is formed from a single inner layer of glial cells. The cell bodies of unipolar moto- and interneurons form more or less well-defined groups in the outer glial layer. Synaptic areas lie within the neuropil, which is usually restricted to the core of the ganglion.[1] The physiological importance of the nerve sheath has been demonstrated, especially in the cockroach nerve cord, where it is regarded as a specialized type of blood-brain barrier which restricts the penetration of extraneously applied substances within the ganglion.[2-4]

A. ANATOMY OF GIANT INTERNEURONS

Detailed neuroanatomical studies performed many years ago revealed the structural organization of the intraganglionic processes of giant interneurons (GI) of *P. americana*. Most of these investigations were related to the cercal afferent-GI system of the A6 ganglion. Cobalt chloride impregnation of fibers has been done either at the pre- or postsynaptic level, resulting in staining both cercal afferents and GI. Filling of GI was achieved with microelectrodes inserted into the GI cell body[5] or by retrograde migration via an isolated giant axon.[6,7] The morphologies of the three largest GI (referring to the diameter of the axon) of the ventral group were grossly similar. The cell body of each GI lies contralaterally to its axon, on the lateral edge of the ganglion. A neurite projects from each cell body and crosses to the opposite side of the ganglion. Numerous fine processes branch from each neurite in the contralateral half of the ganglion and near the midline (Figure 1). Each GI has many primary dendrites from which fine branches are seen to ramify through the neuropil constituting the dendritic tree. In contrast, the morphology of GI of the dorsal group is less homogeneous.[5,7] The simultaneous staining of afferent fibers issued from cercal nerves X and XI (nomenclature of Roeder et al.[8]) and ventral GI 1, 2, and 3 provided evidence that terminals of presynaptic fibers were distributed in the ipsilateral part of the A6 ganglion containing the dendritic tree. An overlap between the projections of cercal nerves X and XI and the dendritic area of GI was observed (Figure 1).

The localization of central synapses in this ganglion resulted not only from neuroanatomical studies, but especially from electrophysiological investigations developed by Callec.[9-11]

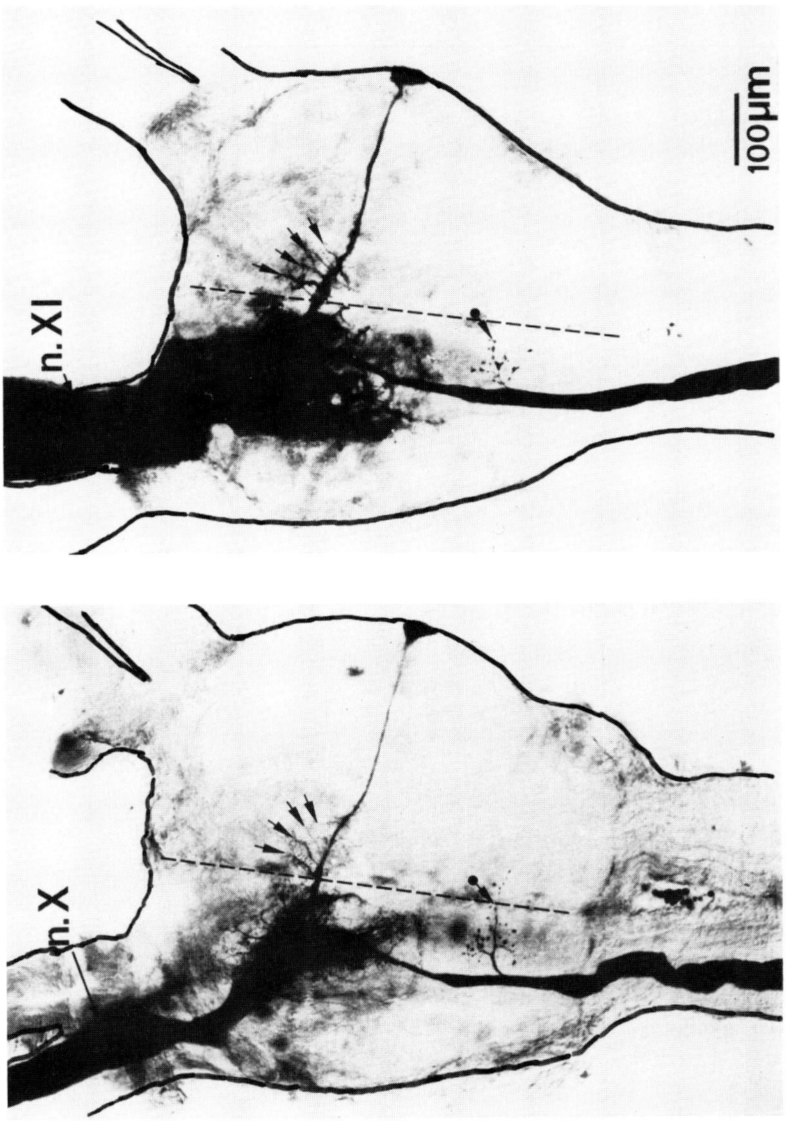

FIGURE 1. Photomicrograph of cobalt sulfide staining of cercal nerves (n.X and n.XI) and giant interneuron 2 in the A6 ganglion of the cockroach *Periplaneta americana*.[7] Note that the general pattern of tracts and distribution of nerve endings correspond with the dendritic tree of the GI. Dendritic processes of the GI neurite in the contralateral half of the ganglion are indicated by arrows. Arrows with dark circles indicate collaterals of the GI. In some cases they pass through the midline of the ganglion, indicated by the dashed line.

The main excitatory and inhibitory pathways of the cercal afferent-GI synaptic system have been identified accurately, promoting this preparation as a suitable model for detailed pharmacological investigations.[12]

B. ELECTROPHYSIOLOGICAL METHODS

Investigation of the electrophysiology of synaptic transmisssion in the CNS of the cockroach was pioneered by Roeder et al.[13] Electrophysiology studies of GI in the A6 ganglion started with the use of microelectrodes[14] and later advanced with the implementation of a more suitable technique, the single-fiber oil-gap method.[9]

1. Single-Fiber Oil-Gap Method

This technique necessitates the microdissection of a small length of a giant axon between the sixth and fifth abdominal ganglia of the nerve cord.

a. Principle

The principle of this method is based on the measurement of potential between two points via external electrodes separated by a high-resistance medium. The potential measured (V) by these electrodes will deviate from the true membrane potential (V_m) by a short-circuiting factor, such that

$$V = \frac{R_e}{R_e + R_i} V_m$$

where R_e = longitudinal resistance of the external medium and R_i = longitudinal resistance of the internal medium.

For this method (Figure 2), paraffin oil is employed to increase Re; the short-circuiting factor, $\frac{R_e}{R_e + R_i}$, is then close to unity.

Moreover, V sampled by the electrodes fades if the axon is dissected too far from synaptic sites. The space constant of GI is 1300 μm.[15] Recording of synaptic events in such favorable conditions exhibits high amplitude and is devoid of noise due to the low resistance of the recording electrode (<1 MΩ).

b. Dissection

The experiments normally were performed on adult male cockroaches. A portion of the nerve cord, including the A6 ganglion, the cercal nerves, and the cerci, was removed from the insect and placed on a glass slide in a drop of saline. If necessary, the nerve sheath was removed from the dorsal part of the A6 ganglion. One of the connectives linking the fifth and sixth ganglia was desheathed, and one of the GI of the ventral group was isolated under a binocular microscope. It appeared that selection during dissection of a particular GI was reasonably accurate. All other axons in the connective were cut, and the preparation was transferred to a Perspex experimental chamber. Finally, the intact connective was severed, leaving only one isolated axon linking the fifth and sixth abdominal ganglia bathing in paraffin oil.

A simplification of the oil-gap technique may be performed by recording all or a part of the electrical activity on the connective linking the fifth and sixth ganglia. It is referred to as the mannitol-gap technique since the resistance of the external medium is elevated with sucrose.[16] It is technically less demanding and readily enables the construction of dose-response curves.

In most of the experiments reported here, the A6 ganglion was permanently superfused by normal saline containing 208 m*M* NaCl, 3.1 m*M* KCl, and 5.4 m*M* $CaCl_2$. The pH value

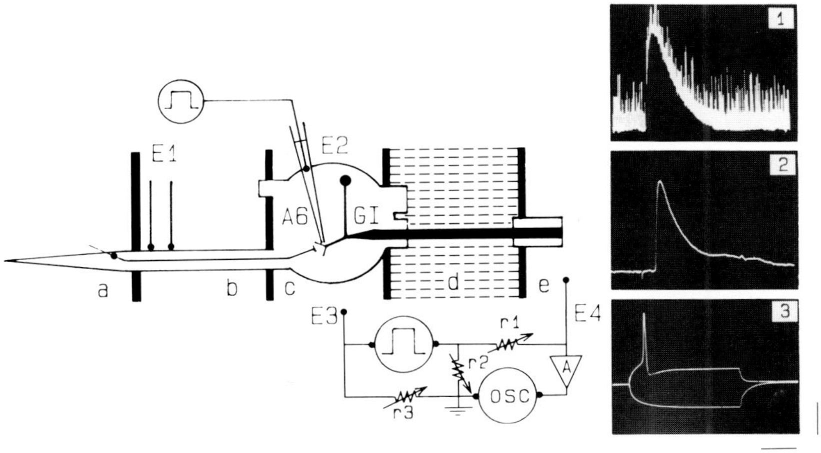

FIGURE 2. Schematic representation of the experimental recording apparatus used in the single-fiber oil-gap method.[93] Left: the experimental chamber and recording circuit; a = cercus compartment, b = cercal nerve compartment with electrical stimulating electrodes (E_1), c = A6 ganglion compartment irrigated with saline (E_3, reference electrode), d = gap compartment filled with paraffin oil in which the isolated giant axon is floating, e = recording electrode (E_4) compartment; iontophoretic electrode (E_2) is impaled within the neuropil of the A6 ganglion. A wheatstone bridge included in the recording circuit allows application of depolarizing or hyperpolarizing currents through the membrane; r1, r2, and r3 are variable resistances used to balance the wheatstone bridge. A = amplifier, osc = oscilloscope. Right: samples of recordings obtained with the single-fiber oil-gap method; 1 = ACh iontophoretic-induced potential superimposed by numerous unitary EPSPs, 2 = composite EPSP, 3 = action potential and hyperpolarizing potential produced by passing DC currents through the GI membrane. Horizontal scale: 1, 2 s; 2, 10 ms; 3, 10 ms. Vertical scale: 1, 2 mV; 2, 2 mV; 3, 40 mV.

was adjusted to 7.2 with either sodium bicarbonate buffer or HEPES. In some recent experiments 26 m*M* sucrose was added to the saline.[7]

2. Experimental Procedure

As we focus on pharmacological aspects, some details concerning the experimental procedure must be specified.

Because access to molecules within the neuropil most often is restricted by the nerve sheath, most of the pharmacological tests have been done on a desheathed A6 ganglion. Several modes of application of drugs have been used. First, the drugs can be solubilized in saline and applied externally by superfusion. Second, a large-tipped glass micropipette may be used for local topical application. Furthermore, iontophoretic injection has been employed for polar compounds such as acetylcholine (ACh). This method, developed by Callec[9] and later refined,[17] was particularly useful in demonstrating the putative postsynaptic effect of drugs.

C. ELECTROPHYSIOLOGY OF CERCAL AFFERENT-GIANT INTERNEURON SYNAPSES

The pharmacological investigations on synaptic mechanisms mentioned in this chapter will be devoted to bioelectrical activity and especially to postsynaptic potentials.

The electrophysiology of synapses has been studied particularly for GI 1, 2, and 3, which receive sensory information from afferent fibers reaching the A6 ganglion through cercal nerves X and XI. Cercal nerve X contains two groups of fibers; one emerges from a

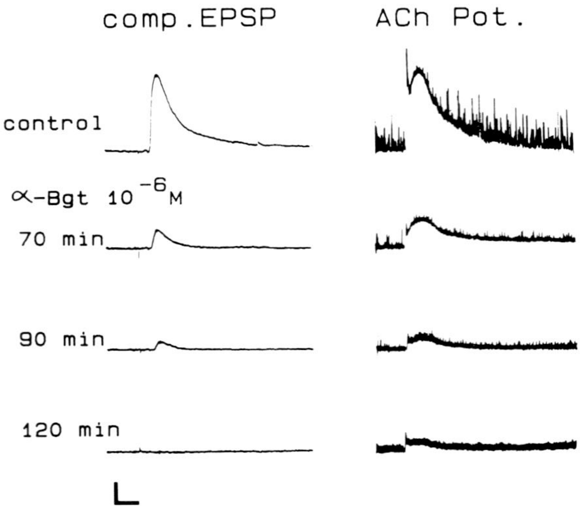

FIGURE 3. Effects of 1 μ*M* α-bungarotoxin on composite (comp.) EPSP and ACh iontophoretic-induced potential (ACh Pot.) in the A6 ganglion of the cockroach. A parallel decrease in amplitude of both phenomena is observed. Vertical scale: 1 mV; horizontal scale: 10-ms Comp. EPSP and 0.5-s ACh Pot.

chordotonal organ located at the base of each cercus. Electrical stimulation delivered on cercal nerve X triggers an inhibitory postsynaptic potential (IPSP) in GI. The other group is formed with axons derived from mechanoreceptors located on the ipsilateral half of the supra-anal plate and a few campaniform sensilla on the lateral extended portion of the same plate. The activation of this latter group of fibers evokes an excitatory postsynaptic potential (EPSP) in GI.[18]

Cercal nerve XI contains sensory fibers which originate from mechanoreceptors of the cerci and which project mainly in the ipsilateral half of the A6 ganglion. The mechanical or electrical activation of a mechanoreceptor or the electrical stimulation of presynaptic fibers give rise, respectively, to a monosynaptic unitary or composite EPSP in GI. The synaptic delay was estimated to be approximately 0.68 ms;[9] nevertheless, delayed polysynaptic depolarizations have been observed in response to increased stimulation.

The ionic mechanisms underlying the IPSP and EPSP were investigated by classical methods, i.e., estimation of the reversal potential under current-clamp conditions and modification of ionic composition of the saline. It appeared that the IPSP is mainly driven by chloride currents,[7,9] whereas the EPSP, whose reversal potential cannot be evaluated easily, is generated by sodium and potassium currents.[9]

D. PHARMACOLOGY OF RECEPTORS

1. Cholinergic Receptors

Extensive data have considerably strengthened the conclusion that ACh is an excitatory transmitter in the cockroach CNS.[4,9,19-24] It appeared that ACh effects were similar in part to those induced by nicotine and muscarine, suggesting the hypothesis of the existence of at least three distinct putative cholinergic receptors: nicotinic, muscarinic, and mixed. In cockroach GI, this pharmacological relationship has been characterized using radioligand binding, electrophysiology, and histoautoradiography.

As in other nervous tissues, the nicotinic probe α-bungarotoxin (α-Bgt) was used extensively in the characterization of the nicotinic synaptic component of insects (Figure 3). Recently reviewed[25] data concerning the biochemistry of insects indicate that the ^{125}I-α-Bgt binding component from extracts prepared from *Drosophila* heads closely resembles a nicotinic ACh receptor. Properties similar to those described for *Drosophila* have been reported in other insects, particularly in the abdominal nerve cord of *P. americana*. Table 1, derived from data of Lummis and Sattelle,[26] summarizes the pharmacology of α-Bgt binding sites of various animal tissue extracts.

Quinuclidinyl benzilate (QNB), a muscarinic ligand,[34] has been used in studies on CNS extracts of the locust[32] and of *P. americana*.[26] A saturable specific component of ^{3}H-QNB binding is present in these insects. Quite recently, a specific binding of *N*-methyl-^{3}H-scopolamine was found in the cockroach nerve cord.[35] Pharmacological properties of this component have led us to the conclusion that there is a muscarinic receptor in the insect CNS. Furthermore, a mixed "nicotinic-muscarinic" putative receptor has been characterized from the nerve cord of the cockroach *Gromphadorhina portentosa*.[36]

The American cockroach CNS preparation is of particular interest because it readily enables direct comparison between ligand binding data and electrophysiological experiments on GI. It was shown that α-Bgt (10 n*M*), when bath-applied to the A6 ganglion, fully and irreversibly blocked unitary as well as composite EPSPs without any modification of the resting potential and membrane resistance.[37,38] Of particular significance was the finding that the active concentration of α-Bgt used in physiological studies was close to its K_d determined in binding experiments.[39] In contrast, QNB was found ineffective on synaptic transmission, even at concentrations as high as 1 μ*M*.

Studies performed on neuronal tissue of head and thorax of *Drosophila*[40] had shown that the ^{125}I-α-Bgt binding is localized in synaptic areas of the neuropil. Parallel studies carried out in the cockroach revealed important binding on either half of the A6 ganglion,[38] corresponding to the bilateral localization of synaptic areas between cercal nerve endings and GI dendritic trees.

The good agreement between biochemical, electrophysiological, and autoradiographic results enhances the conclusion that the ^{125}I-α-Bgt binding component represents nicotinic receptors involved in postsynaptic functions in cercal afferent-GI synapses. There is no electrophysiological evidence for other (muscarinic) receptors.

2. GABA Receptors

The presence of γ-aminobutyric acid (GABA)[41,42] and its synthesizing enzyme, glutamic acid decarboxylase (GAD),[43] in relatively large amounts in the nerve cord of the cockroach, coupled with successful electrophysiological tests,[9,10] suggested to us that GABA was an inhibitory neurotransmitter in the insect CNS.

In order to characterize the pharmacological properties of the GABA receptor, biochemical investigations have been developed recently. They revealed specific properties quite different from those described in the vertebrate CNS. For example, it was shown that the separate classes of GABA receptors (i.e., $GABA_A$ and $GABA_B$) were not found in the insect CNS (Table 2). Furthermore, the recent investigation of binding of the benzodiazepine ligand ^{3}H-flunitrazepam and that of ^{35}S-*t*-butylbicyclophosphorothionate, a ligand which binds to the barbiturate/picrotoxin receptor site of the vertebrate GABA receptor complex, confirmed that insect receptors possess pharmacological properties distinct from mammalian GABA receptor subtypes.[35]

Electrophysiological data have revealed that GABA receptors are involved in inhibitory pathways between cercal nerve fibers and GI. When GABA was either bath-applied (3 to 10 m*M*) or iontophoretically injected into the A6 ganglion, a decrease in membrane resistance and hyperpolarization of the postsynaptic membrane of GI occurred. These effects were reversibly antagonized with picrotoxin.[9,10] Bath-applied high concentrations of GABA were required to induce an effect. It was later determined that this was due to an active uptake mechanism localized principally in glial cells.[45] The ionic mechanisms underlying effects

TABLE 1
Pharmacological Profiles of Putative Nicotinic Cholinergic Receptors Determined by Inhibition of Radiolabeled α-Bungarotoxin Binding to Tissue Extracts

	K_i(M) (binding)							V_{50}(M) (electrophysiology)
Ligand	**Chick optic lobe**[27]	**Rat diaphragm**[28]	***Aplysia californica* (CNS ganglia)**[29]	***Limulus polyphemus* (brain)**[30]	***Drosophila melanogaster* (heads)**[31]	***Locusta migratoria* (head ganglia)**[32]	***Periplaneta americana* (nerve cords)**[26]	***Periplaneta americana* (D_f)**[33]
α-Bungarotoxin	—	—	—	—	500 p*M*	700 p*M*	850 p*M*	64 n*M*
Nicotine	120 n*M*	—	—	7 μ*M*	800 n*M*	40 n*M*	3.5 μ*M*	2.3μ*M*[a]
d-Tubocurarine	250 n*M*	200 n*M*	2 μ*M*	2 μ*M*	2 μ*M*	30 μ*M*	850 n*M*	80 μ*M*
Gallamine	3.8 μ*M*	—	2 μ*M*	—	50 μ*M*	—	14 μ*M*	1.5 m*M*
Acetylcholine	37 μ*M*	470 n*M*	200 μ*M*	—	20 μ*M*	400 μ*M*	18 μ*M*[b]	11 μ*M*[a]
Atropine	85 μ*M*	—	30 μ*M*	300 μ*M*	50 μ*M*	100 μ*M*	20 μ*M*	100 μ*M*
Decamethonium	27 μ*M*	2.1 μ*M*	200 μ*M*	20 μ*M*	—	800 μ*M*	130 μ*M*	2.8 m*M*
Hexamethonium	250 μ*M*	120 μ*M*	300 μ*M*	—	—	800 μ*M*	140 μ*M*	800 μ*M*
Succinylcholine	—	—	—	—	—	—	280 μ*M*	2.8 m*M*
Pancuronium	—	110 n*M*	—	—	—	—	210 μ*M*	150 μ*M*
Mecamylamine	—	—	—	>1 m*M*	—	1 m*M*	25 μ*M*	—
Quinuclidinyl benzilate	—	—	—	—	—	—	140 μ*M*	160 μ*M*

Note: V_{50} was the concentration that produced 50% inhibition of the depolarizing response to iontophoretic application of acetylcholine onto the cell body membrane of the fast coxal depressor motoneuron (D_f).

[a] Concentration that produced 50% maximum depolarization of D_f.[26]
[b] IC_{50} used in estimation of K_i was determined from preincubation experiments.

Reprinted with permission from *Comp. Biochem. Physiol.*, Vol. 80C, Lummis, S. C. R. and Sattelle, D. B., *N*-[propionyl-^{3}H]-propionylated α-bungarotoxin and L-[benzilic-4-4′-^{3}H]-quinuclidinyl benzylate to CNS extracts of the cockroach *Periplaneta americana*, Copyright 1985.

TABLE 2
IC_{50} Values (Concentrations which Inhibit 50% of Specific ^{3}H-GABA Binding) for Various Ligands in the Cockroach CNS Preparation Compared with Values for a Vertebrate (Rat Brain) Preparation

	IC_{50}(M)		
		Vertebrate CNS	
Ligand	**Insect CNS**	**$GABA_A$**	**$GABA_B$**
GABA	130 ± 50 n*M*	120 ± 15 n*M*	80 ± 10 n*M*
Muscimol	710 ± 50 n*M*	15 ± 3 n*M*	5.4 ± 1.1 μ*M*
Thiomuscimol	30 ± 3 μ*M*	*120 ± 10 n*M*	n.t.
THIP	390 ± 60 μ*M*	*130 ± 5 n*M*	i.a.
3-APS	390 ± 20 μ*M*	110 ± 15 n*M*	10 ± 1 μ*M*
Isoguvacine	450 ± 90 μ*M*	190 ± 90 n*M*	i.a.
Bicuculline methiodide	i.a.	27 ± 3.6 μ*M*	i.a.
Baclofen	i.a.	i.a.	130 ± 50 n*M*
P4S	i.a.	*34 ± 2 n*M*	i.a.
Nipecotic acid	i.a.	i.a.	i.a.
Picrotoxin	i.a.	i.a.	i.a.
β-Alanine	i.a.	n.t.	n.t.

Note: Insect IC_{50} values (mean ± SEM) were determined with data from three or more experiments with each analogue using at least five different concentrations. Vertebrate data are taken from Reference 95, with the exception of data marked "*", which are from Reference 44. For vertebrate data: i.a. (inactive), IC_{50} > 100 μ*M*; n.t., not tested. For insect data: i.a., IC_{50} > 1 μ*M*. ^{3}H-GABA concentrations: insect CNS = 30 n*M*; vertebrate CNS — $GABA_A$ = 5 n*M*, $GABA_B$ = 10 n*M*. THIP = 4,5,6,7-tetrahydroisoxazolo[5,4-c]pyridin-3-ol; P4S = piperidine-4-sulfonic acid; 3-APS = 3-aminopropanesulfonic acid.

From Lummis, S. C. R. and Sattelle, D. B., *Neurosci. Lett.*, 9, 287, 1986. With permission.

of GABA and IPSP were mainly chloride dependent[7] (Figure 4). Autoradiographic analysis of ^{3}H-GABA binding sites in Na^+-free saline revealed a neuropilar localization of GABA receptor sites, which is in agreement with that of synaptic areas of the A6 ganglion.[26] Pharmacological tests also showed that bicuculline methiodide, commonly used as a GABA-receptor ligand in vertebrates, did not antagonize the IPSP or GABA effects.[7] Finally, the GABA receptor of the insect CNS seems to be formed from a complex molecule linked to the chloride ion channel. The GABA-sensitive site should be associated with benziodiazepine and barbiturate receptor sites linked to the chloride ionophore. Although this configuration was close to that described in vertebrates, the insect GABA receptor complex differs slightly in pharmacological properties. Furthermore, it has been suggested that GABA modulates ACh release at the cercal afferent-GI synapse via presynaptic inhibition.[46,47]

3. Other Receptors

The widespread distribution of the biogenic amine octopamine in the insect CNS does not necessarily account for a putative role in central neurotransmission, even if it is now well established that octopamine regulates neuromuscular transmission in the locust.[48] In the A6 ganglion of the cockroach, neutral red staining has been particularly successful in demonstrating the presence of octopamine in large dorsal unpaired median (DUM) cells.[49]

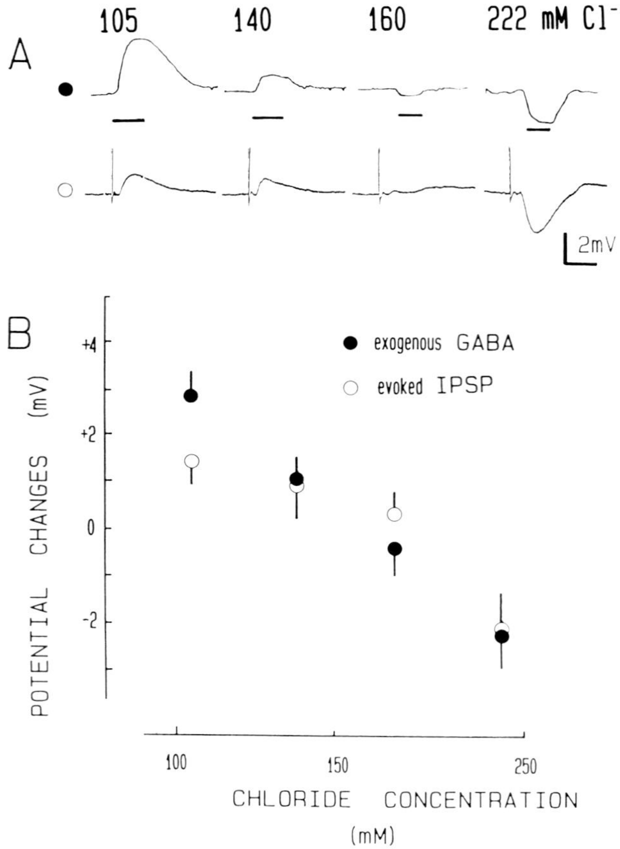

FIGURE 4. Effect of changing external chloride concentration on the evoked IPSP and the bath-applied GABA potential in a GI in the A6 ganglion of the cockroach *Periplaneta americana*. Decreasing chloride ion concentration reverses the two potentials at about the same value. Each point on the semilogarithmic plot is the mean ± SE of three to six experiments. Temperature, 20°C; GABA concentration, 20 m*M*. Horizontal scale: exogenous GABA, 2 min; evoked IPSP, 10 ms.

However, the lack of electrophysiological data in this field does not permit definitive conclusions on a putative role for octopamine in the regulation of cercal afferent-GI synaptic transmission.

III. SYNAPTIC ACTION OF NEUROACTIVE AGENTS

In this third section of this chapter we shall mainly be concerned with events at the cercal afferent-GI synapse of the cockroach, but will also briefly consider other pharmacological interactions which influence the insect CNS.

A. 4-AMINOPYRIDINE

It is now well known that some aminopyridines (APs), such as 4-aminopyridine (4-AP), are selective blockers of the potassium channel at the axonal membrane. This was first demonstrated at the GI axonal membrane of the cockroach by Pelhate et al.[50] Moreover, APs were known to facilitate synaptic transmission at central synapses[51] as well as at

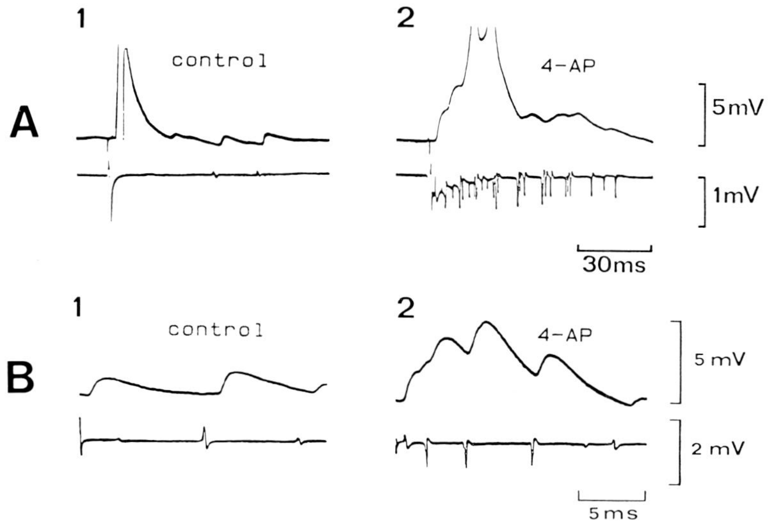

FIGURE 5. Effects of 50 μ*M* 4-AP on the composite EPSP (A) and unitary EPSP (B) recorded in GI no. 2. Note in A2 and B2 that EPSPs are enhanced and reinforced by action potentials spontaneously evoked in presynaptic afferents (cercal nerve XI) after 10 min of action. For each panel, the upper trace represents the postsynaptic event and the lower trace the presynaptic control derived from ipsilateral cercal nerve XI by external electrodes. (From Hue, B., D.Sc. thesis, University of Angers, Angers, France, 1983. With permission.)

peripheral synapses.[52] At the cercal afferent-GI synapse of the cockroach, APs (1 to 10 μ*M*) — particularly 4-AP, the most potent isomer — increased the postsynaptic background activity, i.e., the frequency and amplitude of unitary EPSPs and IPSPs. At the presynaptic level, 4-AP triggered action potentials in cercal nerve XI. These action potentials appeared singly or repetitively, especially after subthreshold electrical stimulation of cercal nerve XI (Figure 5). Monosynaptic correlation with unitary EPSPs was observed. The synaptic depressant effect of Mg^{2+} was antagonized by 4-AP. All of these events were observed without changes in the resting potential, the postsynaptic membrane resistance, the sensitivity of postsynaptic cholinoreceptors, or the equilibrium potentials of ions involved in postsynaptic events. It was suggested that APs facilitate central synaptic transmission by causing an increase in transmitter release.[51] Higher concentrations of AP (100 μ*M* to 10 m*M*) always induced depolarization of postsynaptic membranes, even after blockade of muscarinic and nicotinic receptors by atropine and *d*-tubocurarine, respectively.[53] 2- and 3-AP were approximately ten times less potent than 4-AP and produced a more reversible depolarization. These postsynaptic depolarizing effects of APs were not induced by activation of postsynaptic cholinoreceptors.

B. AMANTADINE AND PIRIBEDIL

We shall report here the synaptic actions of two agents (amantadine and piribedil) whose properties in the pharmacology of the vertebrate CNS have resulted in therapies for Parkinson's disease. Both molecules have been tested in synaptic transmission of *P. americana* in order to elucidate a putative role for cholinergic receptors.

The well-known antiviral agent amantadine[54] has also been used as an anti-Parkinson's drug.[55,56] The latter use was correlated with its pharmacological properties in the field of catecholamine metabolism, uptake, synthesis, or release.[57] However, binding experiments on membrane preparation from the electric organ of *Torpedo ocellata* showed that amantadine

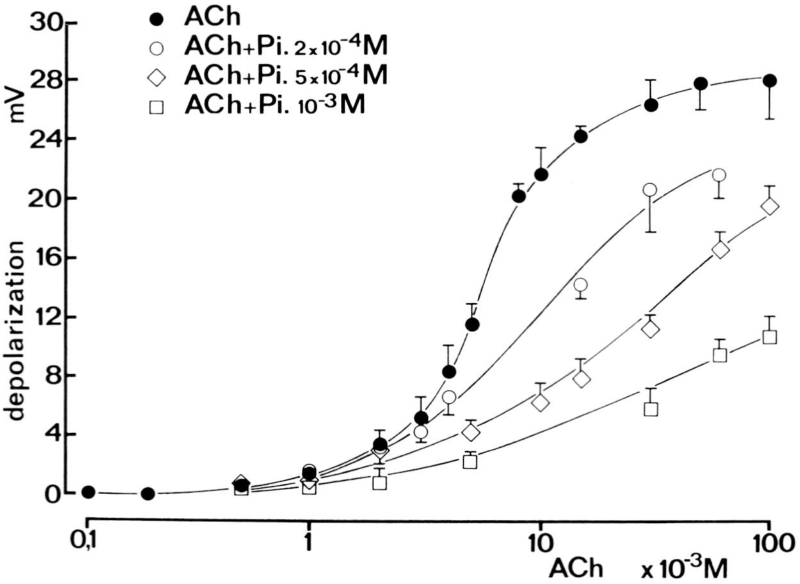

FIGURE 6. Semilogarithmic dose-response curves of the depolarizing action of ACh before and after pretreatment with three doses of piribedil (Pi.).[61] Each point is the mean ± SE of ten experiments.

interacted with the ionophore of the ACh-receptor complex,[58] and pharmacological investigations performed on the heads of flies demonstrated that binding of ^{3}H-H_{12}-histrionicotoxin was inhibited by amantadine.[59]

On the cercal afferent-GI synapses of the cockroach, amantadine (20 μ*M* to 1 m*M*) reduced the amplitude of both unitary and composite EPSPs, resulting in a blockade of synaptic transmission in a dose-dependent manner. A Hill coefficient of 0.94 calculated from dose-response curves revealed no cooperativity (i.e., an interaction of order greater than one) in the binding of amantadine to its site of action. Moreover, a shift to the right in the ACh dose-response curve was obtained after treatment of the A6 ganglion with amantadine (50 μ*M*). No protection against α-Bgt (50 μ*M*) blocking action was observed, indicating separate sites of binding for each molecule. Finally, tests on ACh extrasynaptic receptors in the cell body of the fast coxal depressor motoneuron (D_f) of the cockroach led to the conclusion that interaction of amantadine with the ACh-receptor complex was strongly dependent on membrane potential. An action of amantadine at the open ACh receptor-ion channel complex has been suggested.[60]

A parallel approach was carried out to investigate the putative interaction of piribedil (1-[2-pyrimidyl]4-piperonyl piperazine) with the ACh receptor of the cockroach CNS.[61] This molecule, known to possess a dopaminergic agonist action,[62] also has been promoted as an anti-Parkinson's drug.[63-65] However, a single agonist dopaminergic action of piribedil could not yield the observed pharmacological effects induced by this molecule.[66] The cockroach CNS has been employed to demonstrate the synaptic effect of this drug. Piribedil (10 μ*M* to 1 m*M*), which did not alter nervous conduction in pre- and postsynaptic fibers, depressed unitary and composite EPSPs in a dose-dependent relationship. No fluctuations in resting membrane potential or postsynaptic membrane resistance were noted. The piribedil dose-dependent shift to the right of the ACh semilogarithmic dose-response curve revealed a noncompetitive interaction of piribedil with postsynaptic ACh receptors (Figure 6).

C. TAURINE

Taurine has been proposed as a putative neurotransmitter in the vertebrate CNS, where

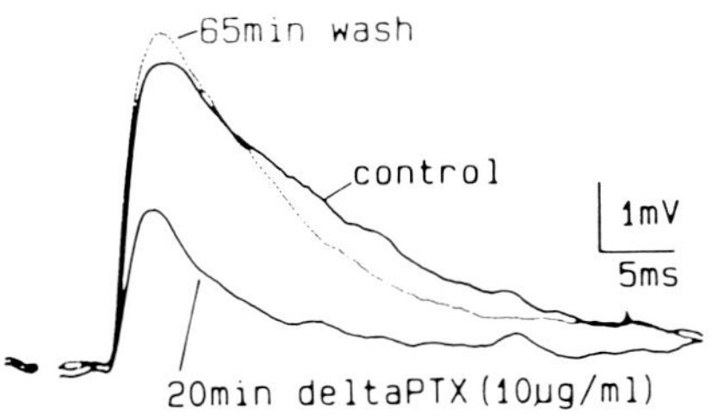

FIGURE 7. Effect of δ-philanthotoxin on the amplitude of subthreshold composite EPSP.[74] At a concentration of 10 μg/ml, the toxin causes a slowly reversible blockage of 50%.

pharmacological and electrophysiological investigations have shown that this amino acid mimics the action of GABA. Therefore, taurine has been considered a partial agonist of GABA. In insects, where GABA is also a possible inhibitory transmitter, taurine has been tested using mannitol- and oil-gap methods on the cercal afferent-GI synapse of the cockroach, where it depresses synaptic transmission.[67-69] On the postsynaptic membrane, taurine has hyperpolarizing effects brought about by an increase in chloride conductance.

On the presynaptic cercal afferents, taurine produces a depolarization antagonized by picrotoxin. Furthermore, inhibition of the taurine synaptic blockade by theophylline suggests competition between these molecules in the excitation-secretion coupling.

D. INVERTEBRATE AND VERTEBRATE TOXINS

The effects of toxins and venoms extracted from various animal species on synaptic transmission have been tested in the cockroach. We shall report two examples in order to illustrate both groups.

Evidence suggests that most of the solitary wasps (Hymenoptera Aculeata) sting their prey (i.e., insects, spiders) into or in the direction of the CNS.[70,71] The question then arises whether or not the venom injected into the ganglia can interfere with central synaptic transmission and/or axonal conduction. Morphological changes induced by the venom of *Philanthus triangulum* have been detected in the CNS of worker honeybees.[72]

The electrophysiological demonstration of central synaptic action of the venom and of the most active toxin, δ-philanthotoxin (δ-PTX), was performed on the cockroach A6 ganglion.[73,74] The venom of *P. triangulum,* at a concentration lower than the wasp might inject into the ganglion, induced a slight depolarization of the postsynaptic membrane of a GI test preparation. This was accompanied by a decrease in membrane resistance and evoked a composite EPSP. We now have increasing evidence that the main site of action of δ-PTX is at the central synaptic level. It appears that δ-PTX (20 μg/ml) does not affect the excitability of the cockroach giant axon significantly, but causes a slowly reversible block of synaptic transmission from cercal nerve XI to GI without any change in the resting membrane potential (Figure 7). Furthermore, investigations on the putative interaction with the ACh-receptor ionophore complex of GI have suggested that δ-PTX could interfere with the ion channel linked to the receptor protein. The ion channel blocking action of δ-PTX might be facilitated by an appropriate agonist, causing massive opening of channels in the postsynaptic membrane. In fact, it has been established that venoms of Hymenoptera often contain considerable amounts of agonists such as ACh, histamine, serotonin, and kinins.[70,71] When measured under δ-PTX treatment, the decay time of iontophoretic-induced ACh potential was not modified. This suggests that, as was first thought,[75] *Philanthus* toxin does not inhibit acetylcholinesterase in the insect CNS.

Small peptides extracted from the venom of the scoliid wasp *Megascolia flavifrons* have been studied recently using cockroach GI.[76] As a result of this study, five different active fractions have been recognized. Three fractions contained identified substances: histamine, Thr[6] bradykinin, and Thr[6] bradykinin-Lys-Ala (megascoliakinin). Two fractions were active, but have not been identified chemically; both of them, as well as bradykinin, blocked the cercal afferent-GI synapses.

A novel probe, κ-bungarotoxin (κ-Bgt), isolated from the venom of *Bungarus multicinctus* has been used on the cockroach CNS recently.[77] While α-toxins bind to pharmacologically nicotinic sites in autonomic ganglia and the CNS, this binding usually fails to inhibit the action of ACh at neuronal nicotinic receptors. For example, α-Bgt fails to block nicotinic transmission in rat and chick sympathetic ganglia and in the chick ciliary ganglion.[78] Nevertheless, there are CNS nicotinic receptors that are sensitive to this particular toxin, notably the nicotinic receptors of the fish and amphibian tectum[79] and those of certain invertebrates, including insects.[38] κ-Bgt blocked nicotinic synaptic transmission without affecting electrical excitability in the chick ciliary ganglion.[80] κ-Bgt, therefore, appeared to be a specific probe for postsynaptic neuronal nicotinic receptors in a wide variety of neuronal preparations.

In the cockroach CNS, the toxin (κ-Bgt) was applied by superfusion to the A6 ganglion. It was noted that κ-Bgt (0.1 to 1 μM) reduced and then blocked EPSPs and the iontophoretically induced ACh potential. The mean time required to achieve a full block was 180 min for 0.1 μM κ-Bgt. The fact that the ACh potential was fully blocked indicated an unequivocally postsynaptic action on nicotinic GI postsynaptic receptors. These results clearly demonstrate, for the first time, that insect neuronal nicotinic receptors are recognized by this novel probe. It is now conceivable that, unlike the situation in vertebrates, both κ- and α-Bgt bind to the same receptor molecule in invertebrates. In this context, there is evidence that insect receptors may have a different subunit composition than their vertebrate counterparts.[81,82] These latest results provide evidence for evolutionary changes in the recognition site of neuronal receptor protein.

E. INSECTICIDES

Most insecticides currently in use act on nervous tissue. Consequently, central synaptic effects involved in their neurotoxicity have been of considerable interest. In this section we address the neurophysiological effects of selected insecticides on cockroach GI synapses.

1. Isothiocyanate Compounds

The synthetic insecticide 2-isothiocyanotoethyltrimethylammonium iodide was characterized as a potent inhibitor of choline acetyltransferase purified from fly heads.[83] A direct putative postsynaptic effect on nicotinic receptors of the cockroach had been investigated in parallel experiments using biochemistry[84] and electrophysiology.[85] It was demonstrated that 2-isothiocyanotoethyltrimethylammonium iodide induced half-maximal blockage of ^{125}I-α-Bgt binding to insect extracts at concentrations similar to those required to produce half-maximal blockage of the composite EPSP recorded at the cercal afferent-GI synapses. The conclusion drawn was that the synaptic effects of this molecule could be explained by its binding to the α-Bgt-sensitive ACh receptors, as it inhibits toxin binding at physiologically relevant concentrations.

2. Nicotine

Still in use, nicotine is one of the oldest known insecticides. It is well known that nicotine acts at the nicotinic receptor level in the central and peripheral nervous systems. Although direct interaction between nicotine and ACh receptors by biochemical methods had been demonstrated,[86,87] the electrophysiological effects remained unknown. The electrophysiological proof was provided by experiments carried out on cercal afferent-GI synapses,[84] in which the concentration of nicotine needed to inhibit 50% of the composite EPSP was estimated to be 200 nM.

3. Nereistoxin

Nereistoxin (NTX) was initially isolated from a marine worm, *Lumbriconereis heteropoda*;[88] subsequently, synthetic NTX was prepared.[89] NTX is a potent neurotoxin in insects. At low concentrations (20 n*M* to 1 μ*M*), NTX produced partial blockage of cholinergic synaptic transmission in the A6 ganglion of the cockroach. Only partial recovery was observed. Further analysis performed in voltage-clamp experiments on the cockroach D_f indicated a potent voltage-dependent blocking action of NTX. Thus, it was concluded that NTX acted preferentially at the ionic channel linked to the ACh receptor.[90]

4. Pyrethroids

The main effect of pyrethroids on axonal membranes is to stabilize the sodium channel in the open state[91] (see also Chapter 18 of this work). Although this action may principally explain pyrethroid neurotoxicity, other properties were observed, especially at the cholinergic synaptic level. For example, using biochemical methods, Abbassy et al.[92] presented evidence of interaction of pyrethroids with the nicotinic ACh receptor complex of *Torpedo*. The effects of two pyrethroids (deltamethrin and tralomethrin) on the ACh receptor of GI have been investigated recently.[93] Both compounds, at 1 μ*M* concentrations, produced identical effects on synaptic transmission. Thus, in $<$10 to 20 min there was first observed a drastic increase in the amplitude of composite EPSP eliciting action potentials in the GI. In most cases, this was accompanied by a weak depolarization which eventually reached a plateau (up to 5 mV). This early facilitating phase was rapidly followed by a decrease of the EPSP leading to an irreversible blockade of the synapses. Furthermore, these pyrethroids induced a parallel decrease of the iontophoretic ACh-induced potential (Figure 8). At the presynaptic level, it was noted that deltamethrin strongly depolarized the cercal nerve endings.[94] Thus, two synaptic sites were suggested to account for the observed action of these molecules. A presynaptic site would be associated with a transient increase in synaptic efficiency and a long-term decrease in ACh release. The slowly developed synaptic interaction with post-synaptic ACh receptor could be ascribed to an irreversible block of the synapse.

IV. CONCLUSION

The cercal afferent-GI synapse of the cockroach, monitored with the latest electrophysiological techniques, provides a particularly suitable neurophysiological preparation for the routine study of the effects of numerous neuroactive agents, such as putative neurotransmitters, fractions of venoms and toxins, insecticides, and drugs. Cockroaches are readily available and, furthermore, do not present any ethical problems.

Finally, even though during the last 15 years enormous progress has been made in the area of GI anatomy, synaptic connections, and pharmacology of neurotransmitters, we need to devote more energy to increasing our knowledge of the physiology of the insect CNS.

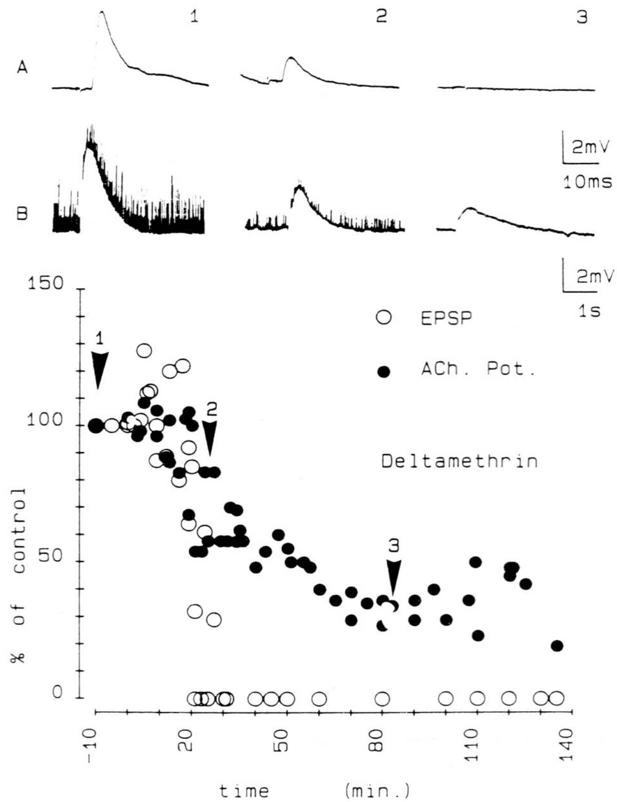

FIGURE 8. Time course of action of deltamethrin (1 μ*M*) on the composite EPSP and the depolarizing potential resulting from iontophoretic application of ACh (ACh. Pot.).[93] Normalized values (referred to control) were plotted against time. Control value was that obtained 10 min before the beginning of pyrethroid application. Test solution was applied during a space-time of 90 min from t = 0. Recording of EPSP (A) and ACh. Pot. (B) occurred at three periods (arrows) of the experiment. Open circles: EPSP amplitude; filled circles: ACh. Pot. amplitude.

REFERENCES

1. **Wigglesworth, V. B.,** The nutrition of the central nervous system of the cockroach *Periplaneta americana.* The mobilization of reserves, *J. Exp. Biol.,* 37, 500, 1960.
2. **Smith, D. S. and Treherne, J. E.,** Electron microscopical localization of acetylcholinesterase activity in the central nervous system of an insect *(Periplaneta americana), J. Cell Biol.,* 26, 445, 1965.
3. **O'Brien, R. D.,** *Insecticides, Action and Metabolism,* Academic Press, New York, 1967.
4. **Pitman, R. M.,** Transmitter substances in insects: a review, *Comp. Gen. Pharmacol.,* 2, 347, 1971.
5. **Daley, D. L., Vardi, N., Appignani, B., and Camhi, J. M.,** Morphology of the giant interneurons and cercal nerve projections of the American cockroach, *J. Comp. Neurol.,* 196, 41, 1981.
6. **Harrow, I. D., Hue, B., Pelhate, M., and Sattelle, D. B.,** Cockroach giant interneurones stained by cobalt-backfilling of dissected axons, *J. Exp. Biol.,* 84, 341, 1980.
7. **Hue, B.,** Electrophysiologie et Pharmacologie de la Transmission Synaptique dans le Système Nerveux Central de la Blatte, *Periplaneta americana,* D.Sc. thesis, University of Angers, Angers, France, 1983.
8. **Roeder, K. D., Tozian, L., and Weiant, E. A.,** Endogenous nerve activity and behaviour in the mantis and cockroach, *J. Insect Physiol.,* 15, 45, 1960.

9. **Callec, J. J.,** Etude de la Transmission Synaptique dans le Système Nerveux Central d'un Insecte *(Periplaneta americana).* Aspects Électrophysiologiques et Pharmacologiques, D.Sc. thesis, University of Rennes, Rennes, France, 1972.
10. **Callec, J. J.,** Synaptic transmission in the central nervous system of insects, in *Insect Neurobiology,* Treherne, J. E., Ed., Elsevier, New York, 1974, 119.
11. **Callec, J. J.,** Synaptic transmission in the central nervous system, in *Comprehensive Insect Physiology, Biochemistry and Pharmacology,* Vol. 5, Kerkut, G. A. and Gilbert, L. I., Eds., Pergamon Press, Oxford, 1985, 139.
12. **Callec, J. J. and Callec, M.,** Les poisons du système nerveux, *Pour la Science,* 92, 90, 1985.
13. **Roeder, K. D., Kennedy, N. K., and Samson, E. A.,** Synaptic conduction to giant fibers of the cockroach and the action of anticholinesterases, *J. Neurophysiol.,* 10, 1, 1947.
14. **Callec, J. J. and Boistel, J.,** Etude de divers types d'activités électriques enregistrées par microélectrodes capillaires au niveau du dernier ganglion abdominal de la Blatte, *Periplaneta americana* L., *C.R. Soc. Biol.,* 160, 1943, 1966.
15. **Boistel, J.,** Quelques caractéristiques électriques de la membrane de la fibre nerveuse au repos *(Periplaneta americana), C.R. Soc. Biol.,* 153, 1009, 1959.
16. **Callec, J. J. and Sattelle, D. B.,** A simple technique for monitoring the synaptic actions of pharmacological agents, *J. Exp. Biol.,* 59, 725, 1973.
17. **Callec, J. J., David, J. A., and Sattelle, D. B.,** Ionophoretic application of acetylcholine onto the dendrites of an identified giant interneurone (G.I. 1) in the cockroach *Periplaneta americana, J. Insect Physiol.,* 28, 1003, 1982.
18. **Bernard, J., Gobin, B., and Callec, J. J.,** A chordotonal organ inhibiting giant interneurones in the sixth abdominal ganglion of the cockroach, *J. Comp. Physiol.,* 153, 377, 1983.
19. **Farley, R. D. and Milburn, N. S.,** Structure and function of the giant fibre system in the cockroach, *Periplaneta americana, J. Insect Physiol.,* 15, 457, 1969.
20. **Colhoun, E. H.,** Distribution of choline acetylase in insect conductive tissue, *Nature (London),* 182, 1378, 1958.
21. **Colhoun, E. H.,** Acetylcholine in *Periplaneta americana.* I. Acetylcholine levels in nervous tissue, *J. Insect Physiol.,* 2, 108, 1958.
22. **Colhoun, E. H.,** Acetylcholine in *Periplaneta americana.* II. Acetylcholine and nervous activity, *J. Insect Physiol.,* 2, 117, 1958.
23. **Sattelle, D. B., McClay, A. S., Dowson, R. J., and Callec, J. J.,** The pharmacology of an insect ganglion: actions of carbamylcholine and acetylcholine, *J. Exp. Biol.,* 64, 13, 1976.
24. **Dagan, D. and Sarne, Y.,** Evidence for cholinergic nature of cockroach giant fibers. Use of specific degeneration, *J. Comp. Physiol.,* 126, 157, 1978.
25. **Sattelle, D. B.,** Acetylcholine receptors, in *Comprehensive Insect Physiology, Biochemistry and Pharmacology,* Vol. 11, Kerkut, G. A. and Gilbert, L. I., Eds., Pergamon Press, Oxford, 1985, 395.
26. **Lummis, S. C. R. and Sattelle, D. B.,** Binding of *N*-[propionyl-^{3}H]-propionylated α-bungarotoxin and L-[benzilic-4,4′-^{3}H]-quinuclidinyl benzilate to CNS extracts of the cockroach *Periplaneta americana, Comp. Biochem. Physiol.,* 80C, 75, 1985.
27. **Wang, G. K., Molinaro, S., and Schmidt, J.,** Ligand responses of α-bungarotoxin binding sites from skeletal muscle and optic lobe of the chick, *J. Biol. Chem.,* 253, 8507, 1978.
28. **Colquhoun, D. and Rang, H. P.,** Effects of inhibitors on the binding of iodinated α-bungarotoxin to acetylcholine receptors in rat muscle, *Mol. Pharmacol.,* 12, 519, 1976.
29. **Shain, W., Greene, L. A., Carpenter, D. O., Sykowski, A. J., and Vogel, Z.,** *Aplysia* acetylcholine receptors: blockade by a binding of α-bungarotoxin, *Brain Res.,* 72, 225, 1974.
30. **Thomas, W. E., Brady, R. N., and Townsel, J. G.,** A characterization of α-bungarotoxin binding in the brain of the horseshoe crab *Limulus polyphemus, Arch. Biochem. Biophys.,* 187, 53, 1978.
31. **Dudai, Y.,** Properties of an α-bungarotoxin-binding cholinergic nicotinic receptor from *Drosophila melanogaster, Biochim. Biophys. Acta,* 539, 505, 1978.
32. **Breer, H.,** Properties of putative nicotinic and muscarinic cholinergic receptors in the central nervous system of *Locusta migratoria, Neurochem. Int.,* 3, 43, 1981.
33. **David, J. A. and Sattelle, D. B.,** Actions of cholinergic pharmacological agents on the cell body membrane of the fast coxal depressor motoneurone of the cockroach *(Periplaneta americana), J. Exp. Biol.,* 108, 119, 1984.
34. **Yamamura, H. I. and Snyder, S. H.,** Muscarinic cholinergic binding in rat brain, *Proc. Natl. Acad. Sci. U.S.A.,* 71, 1725, 1985.
35. **Lummis, S. C. R. and Sattelle, D. B.,** Binding sites for [^{3}H]GABA, [^{3}H]flunitrazepam and [^{35}H]STBPS in insect CNS, *Neurosci. Lett.,* 9, 287, 1986.
36. **Aziz, S. A. and Eldefrawi, M. E.,** Cholinergic receptors of the central nervous system of insects, *Pestic. Biochem. Physiol.,* 3, 168, 1973.

37. **Sattelle, D. B., David, J. A., Harrow, I. D., and Hue, B.,** Actions of α-bungarotoxin on identified central neurones, in *Receptors for Neurotransmitters, Hormones and Pheromones in Insects,* Sattelle, D. B., Hall, L. M., and Hildebrand, J. G., Eds., Elsevier/North-Holland, Amsterdam, 1980, 125.
38. **Sattelle, D. B., Harrow, I. D., Hue, B., Pelhate, M., Gepner, J. I., and Hall, L. M.,** α-Bungarotoxin blocks excitatory synaptic transmission between cercal sensory neurones and giant interneurone 2 of the cockroach *Periplaneta americana, J. Exp. Biol.,* 107, 473, 1983.
39. **Sattelle, D. B.,** Acetylcholine receptors of insects, *Adv. Insect Physiol.,* 15, 215, 1980.
40. **Schmidt-Nielsen, B. K., Gepner, J. I., Teng, N. N. H., and Hall, L. M.,** Characterization of an α-bungarotoxin binding component from *Drosophila melanogaster, J. Neurochem.,* 29, 1013, 1977.
41. **Ray, J. W.,** The free amino acid pool of the cockroach *(Periplaneta americana)* central nervous system and the effect of insecticides, *J. Insect Physiol.,* 10, 587, 1964.
42. **Ray, J. W.,** The free amino acid pool of cockroach *(Periplaneta americana)* central nervous system, in *The Physiology of the Insect Central Nervous System,* Treherne, J. E. and Beament, J. W. L., Eds., Academic Press, London, 1965, 31.
43. **Oliver, G. W. O., Taberner, P. V., Rick, J. T., and Kerkut, G. A.,** Changes in GABA level, GAD and ChE activity in CNS of an insect during learning, *Comp. Biochem. Physiol.,* 38B, 529, 1971.
44. **Krosgaard-Larsen, P. and Falch, E.,** GABA agonists: development and interactions with the GABA receptor complex, *Mol. Cell. Biochem.,* 31, 105, 1981.
45. **Hue, B., Gabriel, A., and Le Patezour, A.,** Autoradiographic localization of [^{3}H]-GABA accumulation in the sixth abdominal ganglion of the cockroach, *Periplaneta americana* L., *J. Insect Physiol.,* 28, 753, 1982.
46. **Hue, B. and Callec, J. J.,** Presynaptic inhibition in the cercal afferent-giant interneurone synapses of the cockroach, *Periplaneta americana* L., *J. Insect Physiol.,* 29, 741, 1983.
47. **Blagburn, J. M. and Sattelle, D. B.,** Presynaptic depolarization mediates presynaptic inhibition at a synapse between an identified mechanosensory neurone and giant interneurone in the first instar cockroach, *Periplaneta americana, J. Exp. Biol.,* 127, 135, 1987.
48. **Evans, P. D.,** Octopamine, in *Comprehensive Insect Physiology, Biochemistry and Pharmacology,* Vol. 11, Kerkut, G. A. and Gilbert, L. E., Eds., Pergamon Press, Oxford, 1985, 499.
49. **Dymond, G. R. and Evans, P. D.,** Biogenic amines in the nervous system of the cockroach, *Periplaneta americana:* association of octopamine with mushroom bodies and dorsal unpaired median (DUM) neurones, *Insect Biochem.,* 9, 535, 1979.
50. **Pelhate, M., Hue, B., Pichon, Y., and Chanelet, J.,** Action de la 4-aminopyridine sur la membrane de l'axone isolé d'insecte, *C.R. Acad. Sci.,* 278, 2807, 1974.
51. **Hue, B., Pelhate, M., Callec, J. J., and Chanelet, J.,** Synaptic transmission in the sixth ganglion of the cockroach: action of 4-aminopyridine, *J. Exp. Biol.,* 65, 517, 1976.
52. **Molgo, J., Lemeignan, M., and Lechat, P.,** Effects of 4-aminopyridine at the frog neuromuscular junction, *J. Pharmacol. Exp. Ther.,* 203, 653, 1977.
53. **Hue, B., Pelhate, M., Callec J. J., and Chanelet, J.,** Postsynaptic effects of 4-aminopyridine in the sixth abdominal ganglion of the cockroach, *Eur. J. Pharmacol.,* 49, 327, 1978.
54. **Jackson, G. G., Muldoon, R. L., and Akers, L. W.,** Serological evidence for prevention of influenza infection in volunteers by an anti-influenza drug, adamantanamine hydrochloride, *Antimicrob. Agents Chemother.,* 3, 703, 1963.
55. **Schwab, R. S., England, A. C., Poskancer, D. C., and Young, R. R.,** Amantadine in the treatment of Parkinson's disease, *J. Am. Med. Assoc.,* 208, 1168, 1969.
56. **Grelak, R. P., Clark, R., Stump, J. M., and Vernier, V. G.,** Amantadinedopamine interaction: possible mode of action in Parkinsonism, *Science,* 169, 203, 1970.
57. **Brown, F. and Redfern, P. H.,** Studies on the mechanism of action of amantadine, *Br. J. Pharmacol.,* 58, 561, 1976.
58. **Warnick, J. E., Maleque, M. A., Bakry, N., Eldefrawi, A. T., and Albuquerque, E. X.,** Structure activity relationships of amantadine. I. Interaction of the *N*-alkyl analogues with the ionic channels of the nicotinic acetylcholine receptor and electrically excitable membrane, *Mol. Pharmacol.,* 22, 82, 1982.
59. **Eldefrawi, A. T., Shaker, N., and Eldefrawi, M. E.,** Binding of acetylcholine receptor/channel probes to housefly head membranes, *Neuropharmacol. Insects,* 88, 137, 1982.
60. **Artola, A., Callec, J. J., Hue, B., David, J. A., and Sattelle, D. B.,** Actions of amantadine at synaptic and extrasynaptic cholinergic receptors in the central nervous system of the cockroach, *Periplaneta americana, J. Insect Physiol.,* 30, 185, 1984.
61. **Hue, B., Pelhate, M., and Chanelet, J.,** Effets du piribédil sur la transmission synaptique dans le système nerveux central d'un insecte, la Blatte (*Periplaneta americana* L.), *J. Pharmacol.,* 4, 455, 1981.
62. **Corrodi, H., Farnebo, L. O., Fuxe, K., Hamberger, B., and Understedt, U.,** ET 495 and brain catecholamine mechanisms: evidence for stimulation of dopamine receptors, *Eur. J. Pharmacol.,* 20, 195, 1972.

63. **Truelle, J. L., Chanelet, J., Bastard, J., Six, P., and Emile, J.,** Enregistrement polygraphique du tremblement et des phénomènes d'hypertonie dans la maladie de Parkinson. Application à l'action de différentes drogues, *Nouv. Presse Med.,* 6, 2987, 1977.
64. **Chase, T. N., Woods, A. C., and Glaubiger, G. A.,** Parkinson disease treated with a suspected dopamine receptor agonist, *Arch. Neurol.,* 30, 383, 1974.
65. **Corsini, G. U., Del Zompo, M., Spissu, A., Mangoni, A., and Gessa, G. L.,** Parkinsonism by haloperidol and piribedil, *Psychopharmacology,* 59, 139, 1978.
66. **Chanelet, J. and Lonchampt, P.,** Action de l'acétylcholine sur la moelle épinière du chat, *J. Physiol. (Paris),* 63, 186A, 1971.
67. **Hue, B., Pelhate, M., and Chanelet, J.,** Sensitivity of postsynaptic neurons of the insect central nervous system to externally applied taurine, in *Taurine and Neurological Disorders,* Barbeau, A. and Huxtable, R. J., Eds., Raven Press, New York, 1978, 225.
68. **Hue, B., Pelhate, M., and Chanelet, J.,** Pre- and postsynaptic effects of taurine and GABA in the cockroach central nervous system, *J. Can. Sci. Neurol.,* 6, 243, 1979.
69. **Hue, B., Pelhate, M., and Callec, J. J.,** GABA and taurine sensitivity of cockroach giant interneurone synapses: indirect evidence for the existence of a GABA uptake mechanism, *J. Insect Physiol.,* 27, 357, 1981.
70. **Piek, T.,** Insect venoms and toxins, in *Comprehensive Insect Physiology, Biochemistry and Pharmacology,* Vol. 11, Kerkut, G. A. and Gilbert, L. I., Eds., Pergamon Press, Oxford, 1985, 595.
71. **Piek, T.,** *Venoms of the Hymenoptera,* Academic Press, New York, 1986.
72. **Rathmayer, W.,** Das Paralysierungsproblem beim Bienenwolf *Philanthus triangulum* F. (Hym. Sphec.), *Z. Vgl. Physiol.,* 45, 413, 1962.
73. **Piek, T., May, T. E., and Spanjer, W.,** Paralysis of locomotion in insects by the venom of the digger wasp *Philanthus triangulum* F., in *Insect Neurobiology and Pesticide Action,* Society for Chemical Industry, London, 1980, 219.
74. **Piek, T., Mantel, P., and Van Ginkel, C. J. W.,** Megascoliakinin, a bradykinin-like compound in the venom of *Megascolia flavifrons* (Hymenoptora: Scoliidae), *Comp. Biochem. Physiol.,* 78C, 473, 1984.
75. **Piek, T., Spanjer, W., Veldsema-Currie, R. D., Van Groen, T., De Haan, N., and Mantel, P.,** Effect of the venom of the digger wasp, *Philanthus triangulum* F., on the sixth abdominal ganglion of the cockroach, *Comp. Biochem. Physiol.,* 71C, 159, 1982.
76. **Piek, T., Hue, B., Mony, L., Nakajima, T., Pelhate, M., and Yasuhara, T.,** Block of synaptic transmission in insect CNS by toxins from the venom of the wasp *Megascolia flavifrons* (Fab.), *Comp. Biochem. Physiol.,* 87C, 287, 1987.
77. **Chiappinelli, V. A., Hue, B., Mony, L., and Sattelle, D. B.,** κ-Bungarotoxin blocks nicotinic transmission at an identified invertebrate central synapse, *J. Exp. Biol.,* 141, 61, 1989.
78. **Chiappinelli, V. A.,** Actions of snake venom toxins of neuronal nicotinic receptors and other neuronal receptors, *Pharmacol. Ther.,* 30, 1, 1986.
79. **Freeman, J. A., Schmidt, J. T., and Oswald, R. E.,** Effect of bungarotoxin on retinotectal synaptic transmission in the goldfish and the toad, *Neuroscience,* 5, 929, 1980.
80. **Dryer, S. E. and Chiappinelli, V. A.,** Kappa-bungarotoxin: an intracellular study demonstrating blockade of neuronal nicotinic receptors by a snake neurotoxin, *Brain Res.,* 289, 317, 1983.
81. **Sattelle, D. B. and Breer, H.,** Purification by affinity-chromatography of a nicotinic acetylcholine receptor from the CNS of the cockroach *Periplaneta americana, Comp. Biochem. Physiol.,* 82C, 349, 1985.
82. **Breer, H., Kleene, R., and Behnke, D.,** Isolation of a putative nicotinic acetylcholine receptor from the central nervous system of *Locusta migratoria, Neurosci. Lett.,* 46, 323, 1984.
83. **Baillie, A. C., Corbett, J. R., Dowsett, J. R., Sattelle, D. B., and Callec, J. J.,** Inhibitors of choline acetyltransferase as potential insecticides, *Pestic. Sci.,* 6, 645, 1975.
84. **Gepner, J. I., Hall, L. M., and Sattelle, D. B.,** Insect acetylcholine receptors as a site of insecticide action, *Nature (London),* 276, 188, 1978.
85. **Sattelle, D. B. and Callec, J. J.,** Actions of isothiocyanates on the central nervous system of *Periplaneta americana, Pestic. Sci.,* 8, 735, 1977.
86. **Eldefrawi, M. E., Abbassy, M. A., and Eldefrawi, A. T.,** Effects of environmental toxicants on nicotinic acetylcholine receptors: action of pyrethroids, *Cellular and Molecular Neurotoxicology,* Narahashi, T., Ed., Raven Press, New York, 1984, 177.
87. **Baillie, A. C. and Wright, K.,** Biochemical pharmacology, in *Comprehensive Insect Physiology, Biochemistry and Pharmacology,* Vol. 11, Kerkut G. A. and Gilbert, L. I., Eds., Pergamon Press, Oxford, 1985, 323.
88. **Nitta, S.,** Uber Nereistoxin, einen giftigen Bestandteil von *Lumbriconereis heteropoda* Marenz (Eunicidae), *Yakagaku Zasshi,* 54, 648, 1934.
89. **Okaichi, T. and Hashimoto, Y.,** The structure of nereistoxin, *Agric. Biol. Chem.,* 26, 224, 1962.
90. **Sattelle, D. B., Harrow, I. D., David, J. A., Pelhate, M., Callec, J. J., Gepner, J. I., and Hall, L. M.,** Nereistoxin: actions on a CNS acetylcholine receptor/ion channel in the cockroach *Periplaneta americana, J. Exp. Biol.,* 118, 37, 1985.

91. **Lund, A. E. and Narahashi, T.,** Kinetics of sodium channel modification as the basis for the variation in the nerve membrane effects of pyrethroids and DDT analogs, *Pestic. Biochem. Physiol.*, 20, 203, 1983.
92. **Abbassy, M. A., Eldefrawi, M. E., and Eldefrawi, A. T.,** Allethrin interactions with the nicotinic acetylcholine receptor channel, *Life Sci.*, 31, 1547, 1983.
93. **Hue, B. and Mony, L.,** Actions of deltamethrin and tralomethrin on cholinergic synaptic transmission in the central nervous system of the cockroach *(Periplaneta americana)*, *Comp. Biochem. Physiol.*, 86, 349, 1987.
94. **Hue, B. and Mony, L.,** Effets pré et postsynaptiques de la deltamethrin dans le système nerveux central de la Blatte, *Periplaneta americana*, *C.R. Groupe Fr. Pestic.*, 16, 35, 1987.
95. **Bowery, N. J., Hill, D. R., and Hudson, A. L.,** Characteristics of $GABA_B$ receptor binding sites on rat whole brain synaptic membranes, *Br. J. Pharmacol.*, 78, 191, 1983.

Chapter 9

NEUROMUSCULAR TRANSMISSION IN THE COCKROACH

Hiroshi Washio

TABLE OF CONTENTS

I. INTRODUCTION

Nerve cells are characterized by two specialized properties which distinguish them from all other cells. The first is the propagation of impulses along the axons of nerve cells, depending upon the ionic fluxes across the neuronal membrane. The second is the transmission of nerve impulses to other nerve and muscle cells. Therefore, to understand the neural control of insect behavior, research on synaptic and neuromuscular transmission as well as on the ion channels mediating neural ion fluxes may be considered essential, one of the central themes of insect neurobiology.

In this chapter on neuromuscular transmission in the cockroach, the reader will be referred to recent results obtained from cockroach preparations as well as from locust preparations. Many refined studies on neuromuscular transmission have been made in locust muscles. Recently, the patch-clamp technique[1,2] has been used to examine electrical changes associated with the opening and closing of a single ion channel activated by L-glutamate in locust muscle fibers.[3,4] These works undoubtedly have contributed to the great progress in the understanding of insect neuromuscular transmission.

With our work on insect preparations we recognize two research strategies: the comparative and the general. For example, neuromuscular systems of insects as well as crustaceans have an advantage for work on inhibitory mechanisms because of the existence of peripheral inhibitory innervation. The substantial evidence that γ-aminobutyric acid (GABA) is an inhibitory neurotransmitter was obtained in studies on arthropod neuromuscular preparations. In this context, you are referred to Hoyle's description[5] of the relationship between comparative and general studies: "Comparative and general studies should always proceed together, being continually and reciprocally related."

The muscle and muscular activity, including neuromuscular transmission of cockroaches, have been reviewed previously by Delcomyn.[6] An excellent account of the pharmacology of neurotransmission and neuromodulation of insect muscle has been offered by Piek.[7] A more detailed treatment of the major problems of neurotransmitters in insect muscles has been provided by Usherwood.[8]

II. TRANSMISSION AT EXCITATORY NEUROMUSCULAR JUNCTIONS

Much work on excitatory neuromuscular junctions has been done with locusts.[9-12] Although not yet confirmed, it is generally agreed that L-glutamate is the transmitter at excitatory neuromuscular junctions in insect skeletal muscles.[8,13] In the cockroach *Periplaneta americana,* bath application of L-glutamate was found to cause muscle contraction.[14] The threshold concentration was 340 n*M*, and higher concentrations brought about greater and more sustained contractions.

Iontophoretic application of L-glutamate by Beránek and Miller[9] and by Usherwood and Machili[10] was made to record depolarizations of the membrane from locust leg muscles. Usherwood and Machili[10] also worked on the metathoracic retractor unguis muscle of the giant South American cockroach, *Blaberus giganteus,* and the results were identical depolarizations. Transient depolarizations (glutamate potentials) to brief application of L-glutamate were recorded only at the synaptic sites. Beránek and Miller[9] found that the equilibrium potential for L-glutamate coincided with the equilibrium potential for miniature excitatory postsynaptic potentials (mEPSPs) and lay between −10 and −25 mV (Figure 1). The pharmacological properties of excitatory neuromuscular junctions in the locust and cockroach have been studied by Usherwood and Machili.[10] L-Glutamate was the most active excitatory substance tested, while D-glutamate was inactive. Among the other amino acids tested, L-aspartate, *N*-methyl-D,L-aspartate, L-asparagine, and L-glutamine were about 1000 times less

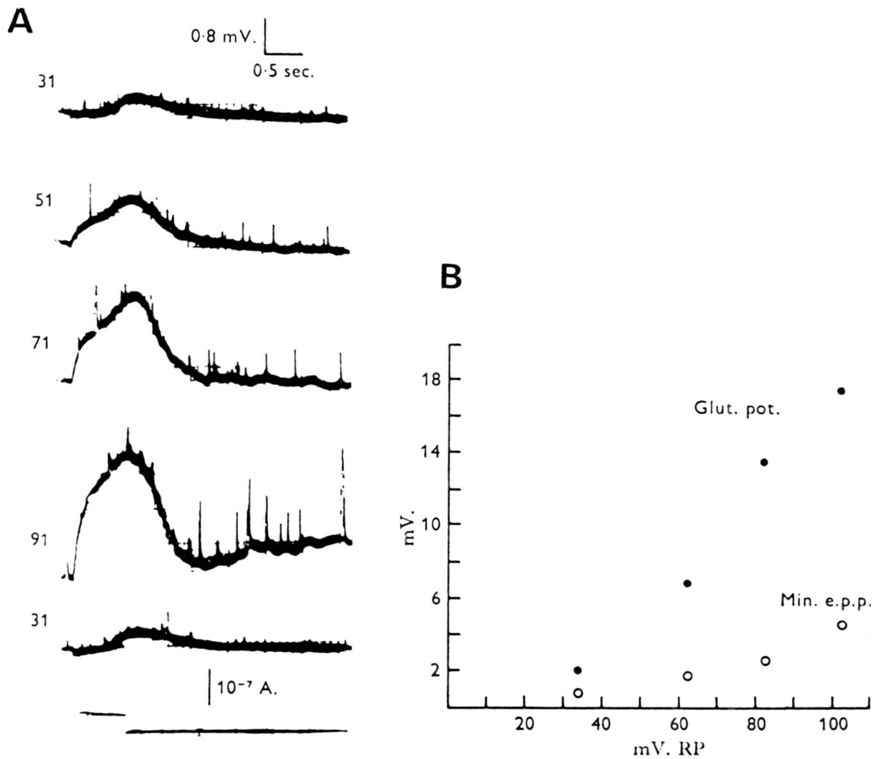

FIGURE 1. (A) Glutamate potentials and miniature EJPs recorded simultaneously by an intracellular electrode, while the resting membrane potential is displaced to various values which are indicated over the corresponding recording in mV. (B) Diagrammatic representation of the experiment from A. Ordinate, amplitude of glutamate potentials (Glut. Pot.), and miniature EJPs (Min. e.p.p.) in mV; abscissa, corresponding resting potentials. The minus signs of membrane potential are omitted. (From Beránek, R. and Miller, P. L., *J. Exp. Biol.*, 49, 83, 1968. With permission.)

potent. Four acidic amino acids (glycine, alanine, aspartate, and L-glutamate) were identified in the perfusate from isolated retractor unguis muscles, but only L-glutamate increased in concentration during nerve stimulation.[10]

At the larval neuromuscular junction of the fruit fly *Drosophila melanogaster,* Jan and Jan[15] also found that iontophoretically applied L-glutamate produced muscle depolarization when L-glutamate was applied close to the synaptic junctions. Other putative transmitter substances and chemicals were tested and found ineffective in mimicking the neurally evoked excitatory junction potential (EJP); they were D-glutamate, L-glutamine, L-aspartate, L-asparagine, glycine, octopamine, serotonin, GABA, acetylcholine (ACh), and substance P. The reversal potentials of the neurally evoked EJP and L-glutamate were virtually identical (-1 and -0.2 mV, respectively). Thus, it was concluded that L-glutamate is the excitatory transmitter at *Drosophila* larval neuromuscular junctions.[15]

The ionic basis underlying the neurally evoked EPSP and the depolarization evoked by L-glutamate at locust extensor tibiae muscles was later studied by Anwyl[11,16] using the voltage-clamp technique. He showed that the reversal potentials of the excitatory postsynaptic current (EPSC) and the glutamate current were $+3$ and $+4$ mV, respectively, and that the neural stimulation and L-glutamate caused an increase in permeability to Na and K, with an approximately equal permeability increase to both of these ions, but not to Cl. This conclusion was derived from the finding that the reversal potential of the synaptic current was approximately midway between the equilibrium potentials for Na and K when the concentration of Na, K, and Cl ions in the perfusate was altered. Additionally, a very small permeability increase to Ca was found when the external concentration of Na was lowered.

Analysis of the membrane noise which resulted from the activation of glutamate receptors at the locust nerve-muscle junction was made by Anderson et al.[17,18] Their work showed that glutamate channels had a conductance of about 125 pS and a mean open time of about 2 ms at a temperature of 21°C. The duration of a channel opening was found to decrease with hyperpolarization of the membrane, which was in marked contrast to the end-plate channels opened by ACh.[17,18] Direct measurement of the currents passing through glutamate-activated single channels in the extrajunctional membrane of locust muscle was first made by Patlak et al.[3] and Cull-Candy et al.[4] using the patch-clamp technique[1,2] (Figure 2). The conductance of the single extrajunctional channels was between 130 and 150 pS, which was similar to the conductance of junctional channels obtained in the membrane noise analysis.[18] The study on the kinetics of the channel's opening and closing revealed that the open times of the channels are distributed roughly exponentially with a mean time constant of about 2 ms, which was again similar to the mean open time measured by noise analysis and by the decay of miniature excitatory junction currents (mEJCs).[16,17]

The presence of glutamate receptors on the nonsynaptic (extrajunctional) membrane was shown in the extensor tibiae muscle of the locust *Schistocerca gregaria*.[19,20] Iontophoretic application of L-glutamate to the extrajunctional membrane was found to produce a biphasic response — depolarization (D-response) followed by hyperpolarization (H-response). Cull-Candy[19,20] showed that glutamate and ibotenate, the glutamate analogue, are able to distinguish two subpopulations of extrajunctional glutamate receptors; glutamate produces a two-component response, but ibotenate produces only an H-response. Study of the equilibrium potential (ca. −60 mV) for the H-response indicates that an increase in chloride conductance is responsible. An increase in the permeability to sodium and potassium is responsible for the extrajunctional D-response. So far, there have been no reports regarding the presence of extrajunctional glutamate receptors in the cockroach muscle fiber.

The long-term application of transmitter molecules to postsynaptic receptors has been known to reduce the amplitude of junctional responses. This process has been called desensitization by Katz and Thesleff,[21] who worked on the frog motor end plate. Desensitization of glutamate receptor populations at excitatory neuromuscular junctions in the extensor tibiae muscle of *S. gregaria* has been studied by Daoud and Usherwood.[22] Their work using a double-barreled iontophoretic electrode indicates that the time course of onset of and recovery from desensitization is very rapid and that the rate of onset of desensitization increases with the concentration of L-glutamate. Later, however, Anis et al.[23] showed that the kinetics of recovery from desensitization are agonist dependent, which is different from the desensitization of ACh receptors on vertebrate muscle.[24] Application of L-glutamate (1 m*M*) has been found to completely abolish the extrajunctional H- and D-receptor responses on locust leg muscle fibers.[19,20] On the other hand, concanavalin A (Con A), the lectin from the seed of the jack bean (*Canavalin ensiformis*), has been shown to specifically inhibit the desensitization of junctional and extrajunctional D-receptors, but not of extrajunctional H-receptors.[25,26] The effect of Con A has been found to persist up to 7 h after washing the muscle with Con A-free saline. It is conceivable that Con A may bind directly to junctional and D-receptor molecules, thereby inhibiting desensitization transitions in these molecules.[25,26]

It is generally accepted that ACh is not involved in neuromuscular transmission in insects. Therefore, it seems reasonable to conclude that curare (*d*-tubocurarine) does not affect the insect neuromuscular junctions. In fact, 1 m*M* curare failed to alter the indirect response of leg muscles in *P. americana*.[27] Also, curare had no effect on the spontaneous miniature end-plate potentials (mEPPs) in the locust and cockroach muscle fibers, even at concentrations as high as 10 m*M*.[28] However, Yamamoto and Washio[29] recently reported that curare

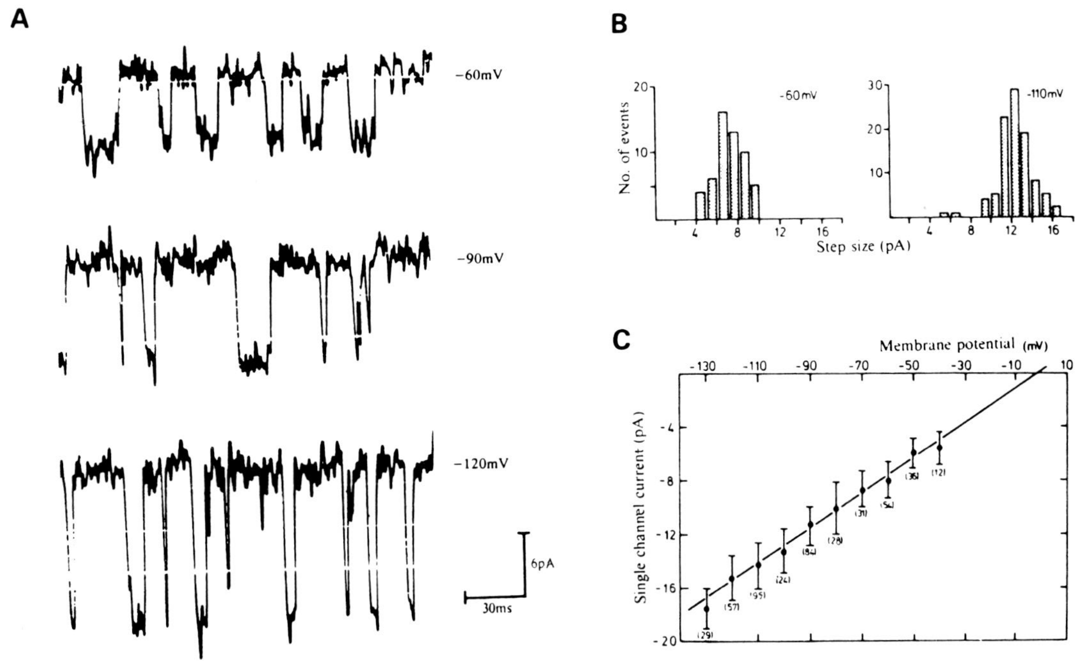

FIGURE 2. (A) The currents through single glutamate-activated channels in the metathoracic extensor tibiae muscle of a 6-d denervated *L. migratoria*. The muscle cell was voltage clamped, the membrane potentials being held at the value shown next to each trace. A patch electrode was filled with physiological saline containing 100 μM L-glutamate. All three records were from the same patch. The temperature was 21°C. The cell had been treated previously for 15 min with 1 μM Con A in saline, then washed with normal saline for more than 30 min. (B) Histograms of measured step sizes of the single-channel currents at two membrane potentials, −60 and −110 mV. (C) The mean of the measured step sizes and the standard deviation plotted as a function of the holding potential. Numbers in brackets are the number of individual events analyzed at each potential. The solid line is the least-square fit to the mean value at each holding potential. The slope of this line was 133 pS. The zero-current intercept was −2.5 mV. The recording conditions were the same as described in A. (Reprinted by permission from *Nature*, Vol. 278, pp. 643—645, ©1979, Macmillan Magazines Limited.)

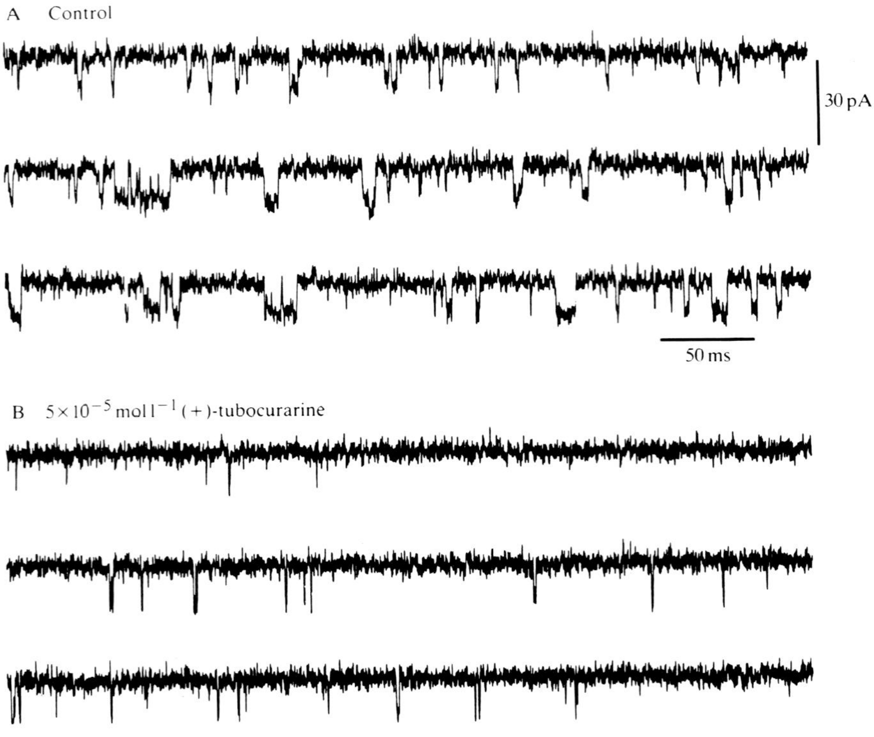

FIGURE 3. Single-channel recordings from an adult locust extensor tibiae muscle fiber showing channels gated by 100 μ*M* sodium L-glutamate in the absence (A) and in the presence (B) of (+)-tubocurarine (TC). A and B are parts of a continuous recording from separate recording sites on a muscle fiber obtained using megohm seals, with the muscle fiber voltage clamped at −100 mV. The data were filtered at 3 kHz. Comparison of B with A shows the absence of long open events when TC is present in the patch electrode along with sodium L-glutamate. (From Kerry, C. J., Ramsey, R. L., Sansom, M. S. P., Usherwood, P. N. R., and Washio, H., *J. Exp. Biol.*, 127, 121, 1987. With permission.)

blocked transmission at the neuromuscular junction of ventral longitudinal muscle fibers of the larval mealworm, *Tenebrio molitor*, by antagonizing the transmitter at the postsynaptic sites in a voltage-dependent manner. The noncompetitive antagonism of curare against glutamate-induced depolarization has been taken as evidence that the drug blocks ionic channels mediated by the transmitter rather than acting on the receptor itself. Also, Cull-Candy and Miledi[30] have observed that curare and gallamine, both cholinergic antagonists, have an increasingly pronounced effect on the excitatory junction current (EJC) and the glutamate-induced current at hyperpolarized membrane potentials in the extensor tibiae muscle fiber of the *S. gregaria* metathoracic leg. Thus, the result suggests that cholinergic antagonists may be able to block amino acid-activated channels by acting after receptor activation. More recently, the effect of curare on the extrajunctional ionic channel gated by glutamate D-receptors[18] in voltage-clamped locust muscle fibers has been examined using the patch-clamp technique,[30] as shown in Figure 3. (+)-Tubocurarine (5 to 500 μ*M*) caused a concentration-dependent decrease in the mean open time, as well as in the probability of the channel being in the open state, for channels opened by 100 μ*M* glutamate. The results support the view that the cholinergic antagonist curare blocks glutamate receptor channels on locust muscle membrane when they are open, in a manner similar to the action of local anesthetics.[31]

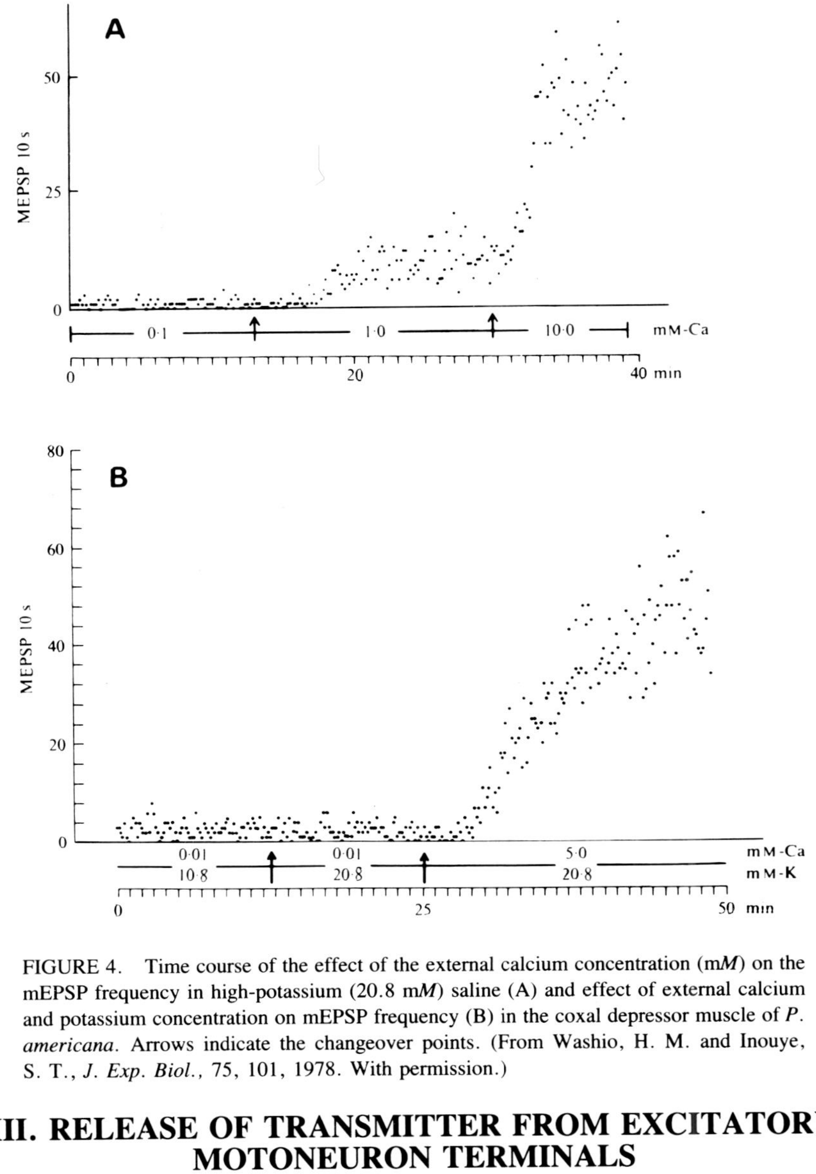

FIGURE 4. Time course of the effect of the external calcium concentration (m*M*) on the mEPSP frequency in high-potassium (20.8 m*M*) saline (A) and effect of external calcium and potassium concentration on mEPSP frequency (B) in the coxal depressor muscle of *P. americana*. Arrows indicate the changeover points. (From Washio, H. M. and Inouye, S. T., *J. Exp. Biol.*, 75, 101, 1978. With permission.)

III. RELEASE OF TRANSMITTER FROM EXCITATORY MOTONEURON TERMINALS

The release of transmitter from nerve terminals on the insect muscle is quantal in nature. The excitatory mEPPs in insect muscle fibers were first recorded intracellularly from the extensor tibiae in *B. giganteus* by Usherwood.[32] The frequency of the spontaneous discharges is significantly increased by calcium and decreased by magnesium.[32] The interaction of calcium with potassium was studied in *P. americana*.[33] The effect of raising the potassium concentration in saline containing 10 μM $CaCl_2$ was negligible. When calcium was added to the saline, the acceleration of the frequency was rapid and substantial. Thus, the increase in the rate of spontaneous transmitter release upon depolarization is highly dependent on the extracellular calcium concentration in the cockroach muscle (Figure 4), as it is in vertebrate neuromuscular junctions.[34-38]

Calcium ions are essential for the release of transmitter quanta at the neuromuscular junction,[39] and this action of calcium is antagonized competitively by magnesium.[40-42] The effect of magnesium as a competitive antagonist of the spontaneous release of transmitter was studied in cockroach muscle.[43] In this work, in the absence of magnesium the slope of double logarithmic plots of miniature EPSP (mEPSP) frequency vs. calcium concentration varied from 0.5 to 2.4 over ten preparations. Although the slope was much less than the value obtained in the *Drosophila* larval neuromuscular junction,[44] in the presence of 2 m*M* Mg^{2+} the slope remained the same, but the frequency of mEPSPs was depressed. The results are taken as evidence that Ca^{2+} and Mg^{2+} compete for a common site.[40,41] The equilibrium dissociation constant for Mg^{2+} was obtained from the concentration of extracellular Ca ions that produces a frequency of mEPSPs in the presence of 2 m*M* Mg^{2+} identical to that in the absence of Mg.[43] The value for Mg^{2+} was 4.0 m*M*. In comparison, the value for Co^{2+} ranged from 0.40 to 0.65 m*M*, implying that Co^{2+} is much more potent than Mg^{2+} in suppressing spontaneous transmitter release (Figure 5).

On the other hand, at relatively low extracellular calcium concentrations the depressed mEPSP frequency produced by Co^{2+} was followed by an increase in frequency.[43] The delayed accelerating action of Co^{2+} may result from the entry of Co^{2+} into the nerve terminal via voltage-sensitive channels and subsequent activation of the release process.[45-47]

In frog neuromuscular junctions the spontaneous release of transmitter was shown to occur at random intervals, indicating no interaction between consecutive potentials. By counting the number of mEPPs per unit time interval, Gage and Hubbard[49] showed that spontaneous transmitter release followed a Poisson distribution under a variety of experimental conditions at the mammalian neuromuscular junction. Recently, a more sophisticated statistical analysis[50] has been applied to the mode of transmitter release at vertebrate[51,52] and invertebrate[53] neuromuscular junctions and at the mammalian ganglionic synapse.[54]

A great deal of statistical analysis has been done on the release of transmitter at excitatory neuromuscular junctions in the insect muscle.[55-59] In locust neuromuscular junctions[55,56,58] it has been shown that spontaneous transmitter release is rarely a random process, bursts of mEPSPs being recorded frequently. Hodgkiss and Usherwood[58] analyzed spontaneous and evoked transmitter release from terminals of excitatory motor axons on locust muscle fibers using intra- and extracellular recording combined with the Ca-electrode technique.[60-62] This enabled them to activate the release of transmitter from single nerve terminals on locust muscle fibers innervated multiterminally. Intracellular recording of spontaneous mEPSPs at active spots on these muscle fibers showed nonrandom occurrence with frequent bursts of mEPSPs. Later, Washio and Inouye[59] analyzed the time course of mEPSPs detected at the cockroach neuromuscular junctions, using focal extracellular recording by the method of Cox and Lewis.[50] The time intervals between mEPSPs showed that the variance-to-mean curve for the extracellular data sets lay significantly above the line predicted for a Poisson process (Figure 6B). In contrast with the curves for the extracellular data, those for the intracellular data sets in which the frequency of the mEPSPs did not change with time were found to lie around the line for a Poisson prediction (Figure 6A). In comparison with the intensity function, the departure from the random process appeared to be mainly due to a large positive correlation of intervals.[59] Shuffling procedures, which randomize the order of intervals, made the series almost random. This was clearly shown in variance-to-mean curves (Figure 6C) for shuffled series which lay around the straight line for a Poisson prediction. These results suggest that observed transmitter release may be the pooled output of a finite number of transmitter release sites,[48,50,51,63] each of which releases quanta at insect neuromuscular junctions in a manner more clustered in nature than Poisson.

IV. TRANSMISSION AT INHIBITORY NEUROMUSCULAR JUNCTIONS

Evidence that GABA is an inhibitory transmitter at the metathoracic extensor tibiae muscles of the locust *S. gregaria* and the grasshopper *Romalea microptera* was presented

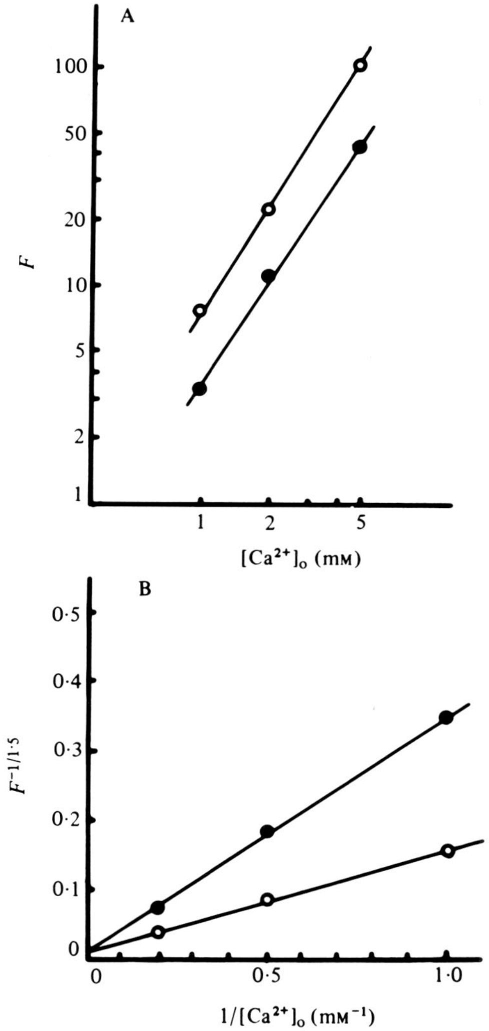

FIGURE 5. The competitive interaction of calcium and cobalt in spontaneous transmitter release in elevated-K^+ (20.8 m*M*) saline in the coxal depressor muscle of *P. americana*. (A) Double logarithmic plots of mEPSP frequency per 10 S (F, in ordinate) against the external calcium concentration in the absence (○) and presence (●) of 0.5 m*M* Co ions. (B) Modified Lineweaver-Burk plot of data shown in A, in the absence (○) and presence (●) of 0.5 m*M* Co ions. (From Washio, H., *J. Exp. Biol.*, 98, 353, 1982. With permission.)

by Usherwood and Grundfest.[64] They showed that the postsynaptic effect of stimulating the inhibitory neuron resulted from an increased permeability to chloride ions and that picrotoxin blocked both the inhibitory postsynaptic potential (IPSP) and the effect of GABA, but did not affect the EPSP.

The iontophoretic application of GABA to locust coxal adductor muscle fibers revealed that the sensitive sites for GABA corresponded to the inhibitory synapses and that the reversal potential for GABA was close to the resting potential, which was similar to the reversal potential for the IPSP.[65] Thus, the similarity of action of GABA to that of the inhibitory transmitter has been shown in the locust muscle.

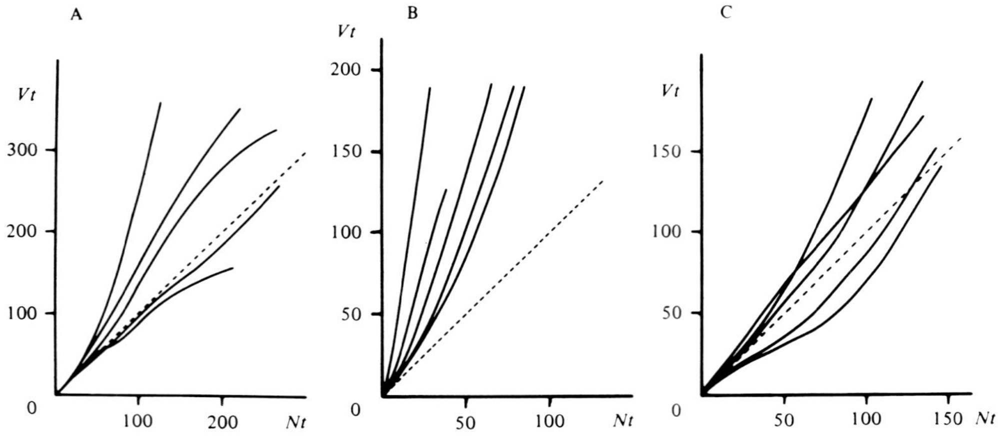

FIGURE 6. The variance-to-mean curves calculated from the data recorded intracellularly from five cells. (A) Recorded extracellularly at five active spots before (B) and after (C), applying a shuffling procedure. The dashed lines show the relationship expected for a Poisson process. The abscissa is the mean number, *Nt*, of the events occurring in any duration of time *t*. The ordinate is the variance, *Vt*, of the event in that duration. (From Washio, H. M. and Inouye, S. T., *J. Exp. Biol.*, 87, 195, 1980. With permission.)

In *P. americana,* Kerkut et al.[14] showed that bath application of GABA at the threshold concentration (5 μg/ml) inhibited the leg's response to electrical stimulation by the thoracic ganglion. Iontophoretic application of GABA was found to decrease both neurally evoked contractions and contraction caused by bath application of glutamate in leg muscles of *P. americana*. The amplitude and frequency of the mEPSPs were also decreased by iontophoretic application of GABA.[66]

Iontophoretic application of GABA to the extensor tibiae muscle of *S. gregaria* was performed by Cull-Candy and Miledi[67] using a voltage-clamp technique. In their work, the presence of both junctional and extrajunctional GABA receptors was determined. The equilibrium potential for GABA action was close to −60 mV, similar to the chloride equilibrium potential for these muscle fibers (Figure 7C,D). Interestingly, extrajunctional GABA receptors have even been demonstrated on a fiber that lacks inhibitory innervation.[67] The pharmacological property of the extrajunctional GABA receptors is identical to the junctional GABA receptors, as picrotoxin and Cl-deficient medium abolish the extrajunctional GABA responses.[67]

Analysis of the membrane current noise resulting from the activation of GABA receptors showed that for junctional and extrajunctional GABA receptors the mean channel lifetime was 3 to 4 ms at −80 mV, longer than that of glutamate receptors.[67] The single-channel conductance was about 22 pS at −80 mV and 21°C, much smaller than that of the excitatory glutamate-activated channels in locust muscle fibers. Inhibitory synaptic currents in voltage-clamped locust muscle fibers were also recorded in the extensor tibiae muscle where the excitatory synapses were desensitized with glutamate to abolish EJCs (Figure 7A,B). The time constant of decay of the nerve-impulse-evoked inhibitory junction currents (IJCs) was found to be 7.7 ms at −80 mV, roughly three times longer than that of the EJC at −80 mV.[68] The relationship between amplitude and membrane potential was nonlinear at hyperpolarization (Figure 7B). The nonlinear relationship was also observed between the amplitude of GABA-induced membrane current and the membrane potential (Figure 7D). Interestingly, a direct recording of miniature IJCs (mIJCs) was made in some fibers. The time constant of decay of mIJCs, measured from spectral analysis of these currents, was 6.6 ms at −75 mV, roughly equivalent to that of the IJCs.

In *P. americana* muscles, Smyth et al.[69] recorded the miniature IPSPs (mIPSPs) from

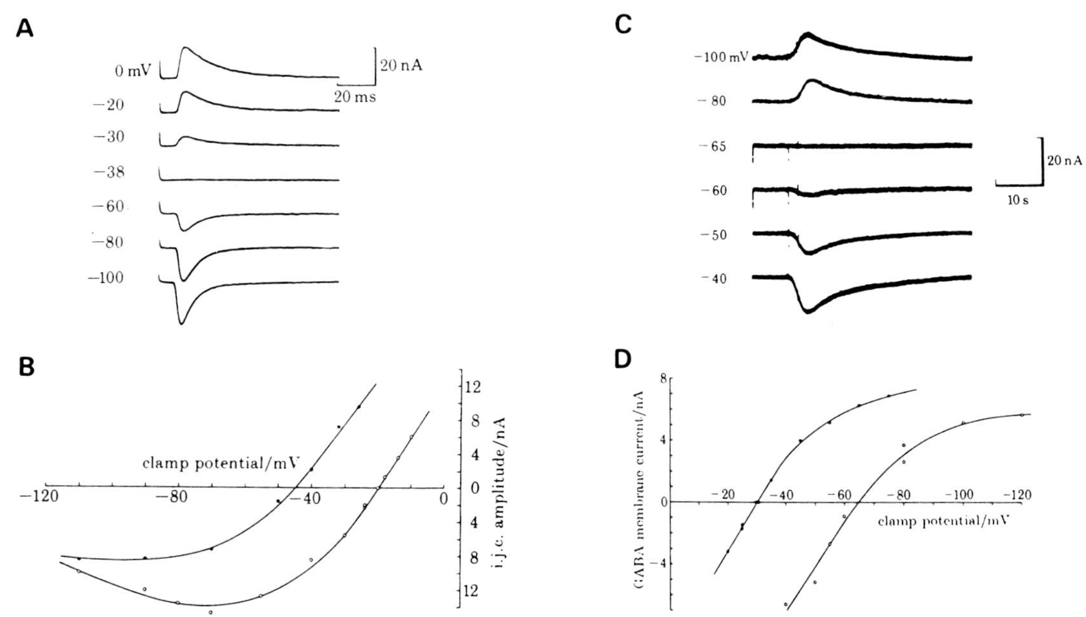

FIGURE 7. (A) Inhibitory junctional currents (IJCs) at various holding potentials in the presence of 1 m*M* glutamate in the extensor tibiae muscle of *S. gregaria* at 22°C. An upward direction of the current is outward. Decay time (and rise time) is decreased upon hyperpolarization. (B) Dependence of IJC amplitude on membrane potential in two fibers. In one of the fibers (open circle) the equilibrium potential has been shifted from its normal level to −20 mV by holding the cell at less negative potentials. (C) GABA-induced membrane currents over a range of clamp-holding potentials. An upward direction of the current is inward. The equilibrium potential for GABA action occurs close to −65 mV. (D) Relation between amplitude of GABA-induced membrane currents and membrane potential in two muscle fibers. Peak conductance change induced by GABA becomes progressively smaller as the membrane potential is hyperpolarized. (A and B from Cull-Candy, S. G., *Proc. R. Soc. London Ser. B,* 221, 375, 1984; C and D from Cull-Candy, S. G. and Miledi, R., *Proc. R. Soc. London Ser. B*, 211, 527, 1981. With permission.) With permission.)

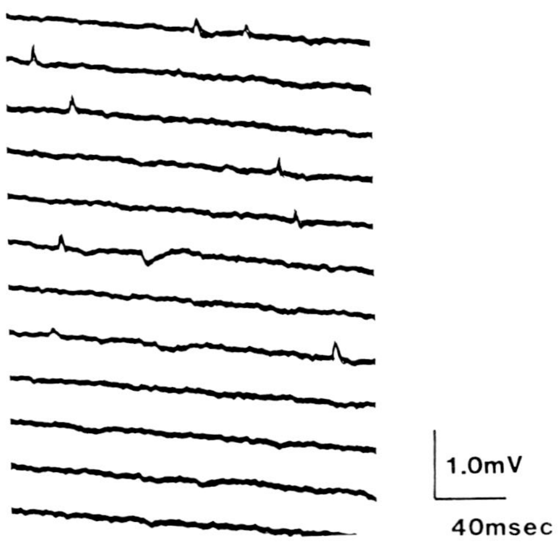

FIGURE 8. Miniature excitatory and inhibitory junction potentials recorded from the posterior coxal levator (182 D) muscle of *P. americana*.[71a] Membrane potential, −42 mV.

metathoracic femoral muscles. The results suggest that the release of the inhibitory transmitter is quantal in nature. They reported the occurrence of small spontaneous hyperpolarizations and a decrease in amplitude in the saline containing low Ca and high Mg concentrations. Since the inhibitory equilibrium potential is generally close to the resting potential, it is difficult to record the mIPSP from the insect muscle fiber innervated by the inhibitory motoneuron. Later, Atwood et al.[70] showed the recording location and the electrical response in the metathoracic extensor tibiae muscle of *P. americana*. They recorded the hyperpolarizing IPSP from the group of fibers at the distal end of the muscle which were innervated by fast, slow, and inhibitory axons. However, they could not record the mIPSP in this muscle fiber.

Recently, Washio[71] recorded mIPSPs from posterior coxal levator (182C,D)* and coxal depressor (177d,e) muscles in the metathoracic legs of *P. americana* using K-citrate electrodes (Figure 8). Frequent depolarizing and occasional hyperpolarizing potential changes were recorded at the level of a resting potential (−45 mV). The hyperpolarizing spontaneous potentials had a relatively low frequency (<1.0 per second) and a slow time course. When the muscle was bathed in 100 μM glutamate saline the depolarizing spontaneous potentials gradually disappeared. On the other hand, hyperpolarizing spontaneous potentials were abolished in Cl-free saline and a saline containing 1 μM picrotoxin. Increasing extracellular K and/or Ca concentrations increased the frequency of the hyperpolarizing spontaneous potentials. The effect of high K concentration may be due to an increase in intracellular Ca concentration of the inhibitory nerve terminal, similar to the excitatory nerve terminal described above. Injection of hyperpolarizing DC current through a recording K-citrate electrode using a conventional bridge circuit converted the hyperpolarizing potential to a depolarizing potential. The reversal potential of the miniature potentials was about 10 mV more negative than the resting potential when recorded with a K-citrate electrode. All results indicate that the hyperpolarizing spontaneous potentials observed in these cockroach muscle fibers are due to the spontaneous release of transmitter from inhibitory nerve terminals and that the transmitter substance significantly increases the Cl permeability of the postsynaptic membrane.

It is of interest to note that GABA acts presynaptically on the excitatory nerve terminals

* Notation of Carbonell.[72]

of the phallic neuromuscular system of *P. americana.*[73] Application of GABA causes the reduction of the amplitude of EPSPs and the absence of change in the amplitude histogram of mEPPs, and it suggests that the phallic muscles are innervated by inhibitory axons which act presynaptically to reduce the release of the excitatory transmitter.[73]

V. DEGENERATION AND REGENERATION OF NEUROMUSCULAR TRANSMISSION

Systematic studies on the effect of denervation on the neuromuscular transmission of the extensor tibiae muscle of the locust *S. gregaria* have been done by Usherwood.[74,75] In his experiments the effects of independently cutting the fast and slow axons which innervate the extensor tibiae were studied. The results suggest that the motor nerve innervating the locust muscle exerts a "trophic" influence on this muscle. Impulse transmission failed within about 2 weeks after nerve section at 20°C. It is important to note that the site of the breakdown was in the region of the nerve-muscle junction, since the isolated peripheral portions of the severed motor nerves were still able to conduct impulses and the muscle was electrically excitable after transmission had failed.[75] In the metathoracic extensor tibiae muscle of the locust, the fibers were permanently silent electrically after the failure of transmission. Thus, a resumption of miniature activity during the later postoperative days was not observed. However, in the metathoracic coxal muscles of *P. americana,* locomotor activity resumed and the miniature EJPs (mEJPs) and evoked electrical response reappeared within about 40 d after the leg nerves were injured.[76]

Cockroach nerves and muscles are more readily induced to regenerate than those of the locust. *P. americana* was found to possess a great ability to regenerate nerves, even in the adult animal[77] (see also Chapter 11 in this work). The time course of degeneration and regeneration was studied carefully in the coxal depressor muscle of *P. americana.*[78] This time course depended largely on the temperature. Spontaneous subthreshold activity ceased in about 2 d at 26°C (16 d at 12°C) in the coxal depressor muscle. Also, the onset of the failure was partly dependent on the length of the transected distal stump, since the normal stage (the time during which neuromuscular transmission in the denervated muscle remains normal after nerve section) was shorter in the mesothoracic than in the metathoracic coxal muscle at the same temperature.[78] In the coxal depressor muscle of the metathoracic leg the earliest reappearance of mEPSPs was observed 13 d after the nerve crush at 26°C, and the resumption of the miniature potentials generally was accompanied by a response to nerve stimulation (Figure 9). The result was in contrast to the finding on the frog end plate, at which no response to nerve stimulation was observed while mEPPs were resumed.[79] The Schwann cell, a type of glial cell, established close contact with the muscle fiber at this stage, suggesting that the spontaneous release of ACh from the Schwann cell caused the mEPPs.[80,81] On the other hand, in the denervated coxal muscle of *P. americana,* three cellular layers — the glial cells (which surround the nerve terminal on one side and separate it from the hemolymph), the nerve terminal, and the muscle fiber[82] — reappeared. However, the nerve terminals were fewer in number per section of the regenerating preparation, and the glial cells seemed to take up much more space in the synaptic cleft within 18 d of nerve crush (Figure 10).[83] At this stage the mEPSPs were observed accompanied by a response to nerve stimulation. Thus, the resumed spontaneous release of transmitter in the cockroach muscle may be derived from the regenerative nerve terminals, not from the glial cells which once engulfed the nerve terminal.

Insect muscle fibers are innervated multiterminally, and some of them are polyneural. The effect of denervation on the dual nervous control at the peripheral level was examined using red fibers of the metathoracic retractor unguis (RU) muscle of *S. gregaria.*[85] These fibers are supplied by both fast excitatory and inhibitory motor axons. For some parts of

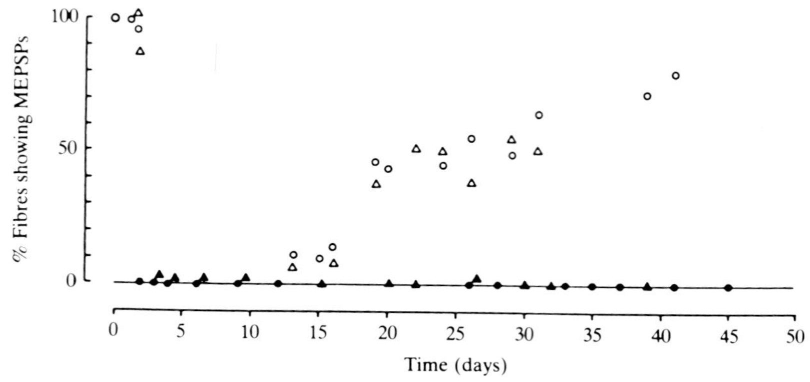

FIGURE 9. Time course of changes in neuromuscular transmission after axons were crushed at 26°C. Abscissa, time in days after operation; ordinate, percentage of fibers impaled which showed mEPSPs from one muscle. At least 25 muscle fibers were tested. Data are from the metathoracic (circles) and mesothoracic (triangles) coxal muscles of *P. americana*. Open symbols show fibers with transmission functioning; filled symbols indicate fibers in which no response to nerve stimulation was obtained. (From Washio, H., *J. Exp. Biol.*, 120, 143, 1986. With permission.)

the study the RU muscle was denervated by sectioning metathoracic nerve 5, while for others it was done by sectioning excitatory and inhibitory axons selectively, keeping operated animals at 30°C.[85] Excitatory motor nerve terminal degeneration was complete 2 to 3 d after nerve section. Interestingly, however, IPSPs were recorded from the denervated red fibers of RU muscles up to 18 d after nerve section. These results suggested that the inhibitory inputs to the RU muscle were intact despite the disappearance of the excitatory innervation. The difference between disappearance time courses of the excitatory and inhibitory innervation was probably caused by more rapid degeneration of the large excitatory axons compared with the small inhibitory ones.[84,85] In general, the inhibitory axons are much smaller in diameter than the excitatory axons in insect muscle. However, in coxal depressor muscles 177d and e of *P. americana*, which are innervated by one slow motoneuron and three common inhibitory motoneurons,[86] both excitatory and inhibitory transmission failed almost simultaneously at 27°C within about 24 h after chronic denervation of both excitatory and inhibitory axons by severing nerve 5.[71a] mEJPs and miniature IJPs (mIJPs) disappeared at almost the same time. The cross-sectional figure of nerve 5r1 in *P. americana,* shown by Pearson and Iles,[86] indicates that the largest is a fast axon, the second is a slow one, and the three smaller ones are all inhibitory. If the degeneration of the motoneuron is dependent on its diameter, the degeneration of the inhibitory axons innervating the coxal muscle should be delayed in comparison with the excitatory axon. Further studies are needed to evaluate the effect of denervation of the inhibitory and excitatory axons in the coxal depressor muscle in relation to the size of axons which innervate the muscle.

In the chronically denervated muscles of the locust, sensitivity of the muscle membrane to L-glutamate was prominently enhanced and the receptor sites of L-glutamate spread along the fibers.[12,87] Usherwood[87] has shown that a high density of glutamate receptors is found at the extrajunctional membrane, but there is no increase in the glutamate sensitivity of the synaptic membrane in denervated muscles. The changes in glutamate sensitivity were first observed about 5 d after nerve section and were most noticeable 32 d after nerve section at 30°C.[87] Later, Cull-Candy[12] worked in detail on the glutamate sensitivity of the extrajunctional membrane of denervated locust muscles and found that after denervation the entire extrajunctional sensitivity to L-glutamate was increased about 100 times above control levels. Distribution of extrajunctional receptor sensitivity following denervation revealed that the increased sensitivity resulted from an increase in D-receptors, not in H-receptors.

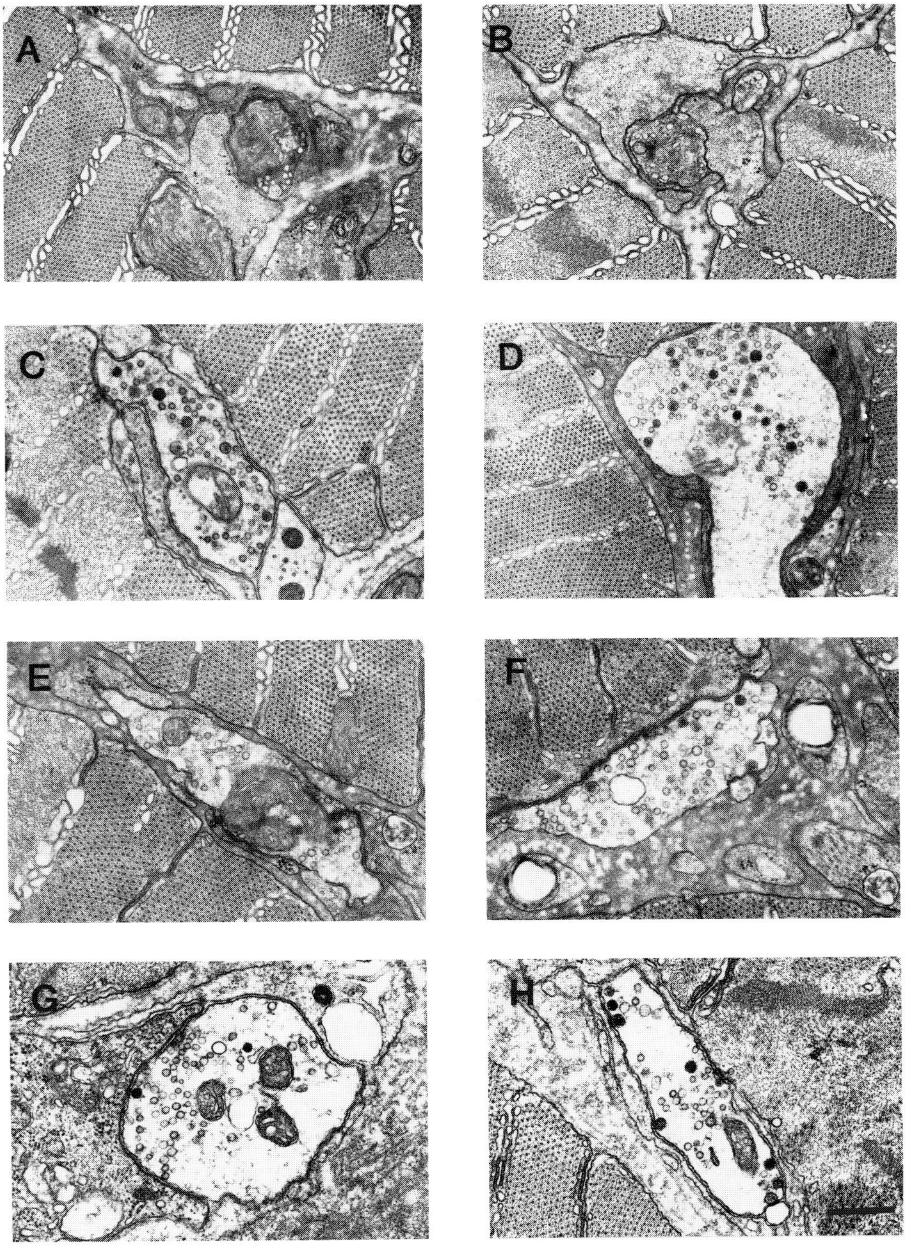

FIGURE 10. Cross section of degenerating and regenerating nerve terminal following nerve crush in the coxal depressor muscle of *P. americana*; (A,B) 4 d, (C,D) 18 d, (E,F) 24 d, (G,H) 36 d after nerve crush. The glial cell is in close contact with the muscle in A and B. A three-cellular arrangement is clearly seen in C and D. Also, many large dense-core vesicles are visible in C and D. Accumulation of synaptic vesicles at the presynaptic membrane is seen in G and H. The bar in H indicates 0.5 μm for all parts of this figure. (Reprinted with permission from *Comp. Biochem. Physiol.*, Vol. 86A, Washio, H. and Nihonmatsu, I., Structural changes in the denervated neuromuscular junction in coxal muscles of the cockroach *Periplaneta americana,* ©1987, Pergamon Journals Ltd.)

Recently, the effect of denervation on the mechanical responses to various concentrations of L-glutamate in the RU muscle of *P. americana* has been examined and compared with contractions induced by high-potassium saline.[88] By 16 to 20 d after denervation, the tension ratios of the maximum glutamate-to-potassium contractions were significantly higher than that of the innervated muscles. Thus, it has been suggested that the sensitivity of the muscle membrane to L-glutamate in cockroach muscle fibers is increased in the denervated muscle in accordance with the results obtained from previous work in the locust muscle fibers.[12,87]

VI. CONCLUDING SUMMARY

L-Glutamate and GABA fulfill most of the criteria for the identification of excitatory and inhibitory transmitters, respectively, in both the insect and the crustacean neuromuscular junction. Also, L-glutamate and GABA are regarded as the major excitatory and inhibitory transmitters, respectively, in many regions of the mammalian brain. It has now been about 30 years since L-glutamate was found to depolarize and excite individual neurons in the mammalian central nervous system (CNS).[89] At present, glutamate receptors have been classified into at least three types of receptors based on agonist specificity in the mammalian CNS. These are *N*-methyl-D-aspartate (NMDA), kainate, and quisqualate. However, there has been no clear evidence to show the involvement of these glutamate receptors in insect muscles.

On the other hand, GABA receptors on vertebrate neurons have been characterized as being of two distinct types, $GABA_A$ and $GABA_B$.[90,91] $GABA_A$ receptors are linked to changes in chloride conductance and are antagonized by bicuculline, whereas $GABA_B$ receptors are inactive to bicuculline.[91] $GABA_B$ is sensitive to baclofen as an agonist, but $GABA_A$ is not.[91] Unfortunately, there has been little work to date on GABA receptors on insect muscle. Recently, GABA receptors on the locust muscle have been reported to lack sensitivity to bicuculline, unlike the vertebrate $GABA_A$ receptor,[92] suggesting that there are pharmacological differences in GABA receptors on insect muscle fibers and neurons, since bicuculline reduces GABA-induced chloride currents in an identified motoneuron of *P. americana*.[93] Thus, there is evidence for considerable heterogeneity of GABA receptors in both vertebrates and invertebrates. However, the recent findings[94] on the interaction between the GABA receptor-chloride channel complex in the CNS of mammals and the sites of action of pyrethroid insecticides have contributed to characterizing the properties of the GABA receptor site in insects.[95] A detailed pharmacological study on the insect GABA receptor-ionophore complex in the peripheral and central nervous systems may provide possibilities for the design of novel insecticides[93] because we have been confronted with the fact that the study of insect neuromuscular junctions has been impeded by the lack of an antagonist which can be used to block excitatory junctional transmission by a specific postsynaptic action.

ACKNOWLEDGMENTS

I thank Drs. F. Delcomyn and I. Huber for giving me the opportunity to write this chapter. Thanks are also due to Mrs. Matsumura for her preparation of the manuscript.

REFERENCES

1. **Neher, E. and Sakmann, B.,** Single-channel currents recorded from membrane of denervated frog muscle fibers, *Nature (London)*, 260, 799, 1976.
2. **Neher, D., Sackmann, B., and Steinback, J. H.,** The extracellular patch clamp: a method for resolving currents through individual open channels in biological membrane, *Pflugers Arch.*, 375, 219, 1978.

3. **Patlak, J. B., Gration, K. A. F., and Usherwood, P. N. R.,** Single glutamate-activated channels in locust muscle, *Nature (London),* 278, 643, 1979.
4. **Cull-Candy, S. G., Miledi, R., and Parker, I.,** Single glutamate-activated channels recorded from locust muscle fibres with perfused patch-clamp electrodes, *J. Physiol. (London),* 321, 195, 1980.
5. **Hoyle, G.,** *Muscles and Their Neural Control,* John Wiley & Sons, New York, 1983.
6. **Delcomyn, F.,** Muscle and muscular activity, in *The American Cockroach,* Bell, W. J. and Adiyodi, K. G., Eds., Chapman and Hall, London, 1981, chap. 11.
7. **Piek, T.,** Neurotransmission and neuromodulation of skeletal muscles, in *Comprehensive Insect Physiology, Biochemistry and Pharmacology,* Vol. 11, Kerkut, G. A. and Gilbert, L. I., Eds., Pergamon Press, Oxford, 1985, 55.
8. **Usherwood, P. N. R.,** Amino acids as neurotransmitters, in *Advances in Comparative Physiology and Biochemistry,* Vol. 7, Lowenstein, O., Ed., Academic Press, New York, 1978, 227.
9. **Beránek, R. and Miller, P. L.,** The action of iontophoretically applied glutamate on insect muscle fibres, *J. Exp. Biol.,* 49, 83, 1968.
10. **Usherwood, P. N. R. and Machili, P.,** Pharmacological properties of excitatory neuromuscular synapses in the locust, *J. Exp. Biol.,* 49, 341, 1968.
11. **Anwyl, R.,** Permeability of the post-synaptic membrane of an excitatory glutamate synapse to sodium and potassium, *J. Physiol. (London),* 273, 367, 1977.
12. **Cull-Candy, S. G.,** Glutamate sensitivity and distribution of receptors along normal and denervated locust muscle fibres, *J. Physiol. (London),* 276, 165, 1978.
13. **Usherwood, P. N. R. and Cull-Candy, S. G.,** Pharmacology of somatic nerve-muscle synapses, in *Insect Muscle,* Usherwood, P. N. R., Ed., Academic Press, London, 1975, chap. 4.
14. **Kerkut, G. A., Shapira, A., and Walker, R. J.,** The effect of acetylcholine, glutamic acid and GABA on the contractions of the perfused cockroach leg, *Comp. Biochem. Physiol.,* 16, 37, 1965.
15. **Jan, L. Y. and Jan, Y. N.,** L-Glutamate as an excitatory transmitter at the *Drosophila* larval neuromuscular junction, *J. Physiol. (London),* 262, 215, 1976.
16. **Anwyl, R.,** The effect of foreign cations, pH and pharmacological agents on the ionic permeability of a excitatory glutamate synapse, *J. Physiol. (London),* 273, 389, 1977.
17. **Anderson, C. R., Cull-Candy, S. G., and Miledi, R.,** Glutamate and quisqualate noise in voltage-clamped locust muscle fibres, *Nature (London),* 261, 151, 1976.
18. **Anderson, C. R., Cull-Candy, S. G., and Miledi, R.,** Glutamate current noise: postsynaptic channel kinetics investigated under voltage clamp, *J. Physiol. (London),* 282, 219, 1978.
19. **Cull-Candy, S. G. and Usherwood, P. N. R.,** Two populations of L-glutamate receptors on locust muscle fibres, *Nature (London),* 246, 62, 1973.
20. **Cull-Candy, S. G.,** Two types of extrajunctional L-glutamate receptors in locust muscle fibres, *J. Physiol. (London),* 255, 449, 1976.
21. **Katz, B. and Thesleff, S.,** A study of "desensitization" produced by acetylcholine at the motor end plate, *J. Physiol. (London),* 138, 63, 1957.
22. **Daoud, M. A. R. and Usherwood, P. N. R.,** Desensitization and potentiation during glutamate application to locust skeletal muscle, *Comp. Biochem. Physiol.,* 59C, 105, 1978.
23. **Anis, N. A., Clark, R. B., Gration, K. A. F., and Usherwood, P.** N.R., Influence of agonists on desensitization of glutamate receptors on locust muscle, *J. Physiol. (London),* 312, 345, 1981.
24. **Rang, H. P. and Ritter, J. M.,** On the mechanism of desensitization at cholinergic synapses, *Mol. Pharmacol.,* 6, 357, 1970.
25. **Mather, D. A. and Usherwood, P. N. R.,** Concanavalin A blocks desensitization of glutamate receptors on insect muscle fibres, *Nature (London),* 259, 409, 1976.
26. **Mather, D. A. and Usherwood, P. N. R.,** Effect of concanavalin A on junctional and extrajunctional L-glutamate receptors on locust skeletal muscle fibres, *Comp. Biochem. Physiol.,* 59C, 151, 1978.
27. **Roeder, K. D. and Weiant, E. A.,** The electrical and mechanical events of neuromuscular transmission in the cockroach, *Periplaneta americana* (L.), *J. Exp. Biol.,* 27, 1, 1950.
28. **Usherwood, P. N. R.,** Spontaneous miniature potentials from insect skeletal muscle fibres, *J. Physiol. (London),* 169, 149, 1963.
29. **Yamamoto, D. and Washio, H.,** Curare has a voltage-dependent blocking action on the glutamate synapse, *Nature (London),* 281, 372, 1979.
30. **Cull-Candy, S. G. and Miledi, R.,** Blocks of glutamate-activated synaptic channels by curare and gallamine, *Proc. R. Soc. London Ser. B,* 218, 111, 1983.
31. **Kerry, C. J., Ramsey, R. L., Sansom, M. S. P., Usherwood, P. N. R., and Washio, H.,** Single-channel studies of the action of (+)-tubocurarine on locust muscle glutamate receptor, *J. Exp. Biol.,* 127, 121, 1987.
32. **Usherwood, P. N. R.,** Spontaneous miniature potentials from insect muscle fibres, *Nature (London),* 191, 814, 1961.

33. **Washio, H. M. and Inouye, S. T.,** The effect of calcium and magnesium on the spontaneous release of transmitter at insect motor nerve terminals, *J. Exp. Biol.,* 75, 101, 1978.
34. **del Castillo, J. and Katz, B.,** Quantal components of the end-plate potential, *J. Physiol. (London),* 124, 560, 1954.
35. **del Castillo, J. and Katz, B.** Statistical factors involved in neuromuscular facilitation and depression, *J. Physiol. (London),* 124, 574, 1954.
36. **del Castillo, J. and Katz, B.,** Changes in end-plate activity produced by presynaptic polarization, *J. Physiol. (London),* 124, 586, 1954.
37. **Hubbard, J. I., Jones, S. F., and Landau, E. M.,** On the mechanism by which calcium and magnesium affect the spontaneous release of transmitter from mammalian motor nerve terminals, *J. Physiol. (London),* 194, 355, 1968.
38. **Cooke, J. D., Okamoto, K., and Quastel, D. M. J.,** The role of calcium in depolarization-secretion coupling at the motor nerve terminal, *J. Physiol. (London),* 228, 459, 1973.
39. **Katz, B.,** *The Release of Neural Transmitter Substances,* Charles C Thomas, Springfield, IL, 1969.
40. **Jenkinson, D. H.,** The nature of the antagonism between calcium and magnesium ions at the neuromuscular junction, *J. Physiol. (London),* 138, 434, 1957.
41. **Dodge, F. A. and Rahamimoff, R.,** Co-operative action of calcium ions in transmitter release at the neuromuscular junction, *J. Physiol. (London),* 193, 419, 1967.
42. **Hubbard, J. I., Jones, S. F., and Landau, E. M.,** On the mechanism by which calcium and magnesium affect the release of transmitter by nerve impulses, *J. Physiol. (London),* 196, 75, 1968.
43. **Washio, H.,** A dual effect of cobalt ions on the spontaneous release of transmitter at insect motor nerve terminals, *J. Exp. Biol.,* 98, 353, 1982.
44. **Jan, L. Y. and Jan, Y. N.,** Properties of the larval neuromuscular junction in *Drosophila melanogaster, J. Physiol. (London),* 262, 189, 1976.
45. **Kita, H. and Van der Kloot, W.,** Action of Co and Ni at the frog neuromuscular junction, *Nature New Biol.,* 245, 52, 1973.
46. **Alnaes, E. and Rahamimoff, R.,** On the role of mitochondria in transmitter release from motor nerve terminals, *J. Physiol. (London),* 248, 285, 1975.
47. **Kita, H. and Van der Kloot, W.,** Effect of the ionophore X-537A on acetylcholine release at the frog neuromuscular junction, *J. Physiol. (London),* 259, 177, 1976.
48. **Fatt, P. and Katz, B.,** Spontaneous subthreshold activity at motor nerve endings, *J. Physiol. (London),* 117, 109, 1952.
49. **Gage, P. W. and Hubbard, J. I.,** Evidence for Poisson distribution of miniature end-plate potentials and some implications, *Nature (London),* 208, 395, 1965.
50. **Cox, D. R. and Lewis, P. A. W.,** *The Statistical Analysis of Series of Events,* Methuen, London, 1966.
51. **Hubbard, J. I. and Jones, S. F.,** Spontaneous quantal transmitter release: a statistical analysis and some implications, *J. Physiol. (London),* 232, 1, 1973.
52. **Cohen, I., Kita, H., and Van der Kloot, W.,** The stochastic properties of spontaneous quantal release of transmitter at the frog neuromuscular junction, *J. Physiol. (London),* 236, 341, 1974.
53. **Cohen, I., Kita, H., and Van der Kloot, W.,** Stochastic properties of spontaneous transmitter release at the crayfish neuromuscular junction, *J. Physiol. (London),* 236, 363, 1974.
54. **Bornstein, J. C.,** Spontaneous multiquantal release at synapse in guinea-pig hypogastric ganglia: evidence that release can occur in bursts, *J. Physiol. (London),* 282, 375, 1978.
55. **Usherwood, P. N. R.,** Transmitter release from insect excitatory motor nerve terminals, *J. Physiol. (London),* 227, 527, 1972.
56. **Rees, D.,** The spontaneous release of transmitter from insect nerve terminals as predicted by the negative binomial theorem, *J. Physiol. (London),* 236, 129, 1974.
57. **Washio, H. and Inouye, S. T.,** The mode of spontaneous transmitter release at the insect neuromuscular junction, *Can. J. Physiol. Pharmacol.,* 53, 679, 1975.
58. **Hodgkiss, J. P. and Usherwood, P. N. R.,** Transmitter release from normal and degenerating locust motor nerve terminals, *J. Physiol. (London),* 285, 113, 1978.
59. **Washio, H. M. and Inouye, S. T.,** The statistical analysis of spontaneous transmitter release at individual junctions on cockroach muscle, *J. Exp. Biol.,* 87, 195, 1980.
60. **Katz, B. and Miledi, R.,** Propagation of electric activity in motor nerve terminals, *Proc. R. Soc. London Ser. B,* 161, 453, 1965.
61. **Katz, B. and Miledi, R.,** The measurement of synaptic delay and the time course of acetylcholine release at the neuromuscular junction, *Proc. R. Soc. London Ser. B,* 161, 483, 1965.
62. **Katz, B. and Miledi, R.,** The effect of calcium on acetylcholine release from motor nerve terminals, *Proc. R. Soc. London Ser. B,* 161, 496, 1965.
63. **Vere-Jones, D.,** Simple stochastic models for the release of quanta of transmitter from a nerve terminal, *Aust. J. Stat.,* 8, 53, 1966.

64. **Usherwood, P. N. R. and Grundfest, H.,** Peripheral inhibition in skeletal muscle of insects, *J. Neurophysiol.,* 28, 497, 1965.
65. **Usherwood, P. N. R.,** Action of iontophoretically applied gamma-aminobutyric acid on locust muscle fibres, *Comp. Biochem. Physiol.,* 44A, 663, 1973.
66. **Kerkut, G. A. and Walker, R. J.,** The effect of iontophoretic injection of L-glutamic acid and γ-amino-*N*-butyric acid on the miniature end-plate potentials and contractures of the coxal muscles of the cockroach *Periplaneta americana* L., *Comp. Biochem. Physiol.,* 20, 999, 1967.
67. **Cull-Candy, S. G. and Miledi, R.,** Junctional and extrajunctional membrane channels activated by GABA in locust muscle fibres, *Proc. R. Soc. London Ser. B,* 211, 527, 1981.
68. **Cull-Candy, S. G.,** Inhibitory synaptic currents in voltage-clamped locust muscle fibres desensitized to their excitatory transmitter, *Proc. R. Soc. London Ser. B,* 221, 375, 1984.
69. **Smyth, T., Jr., Greer, M. H., and Griffiths, J. G.,** Insect neuromuscular synapses, *Am. Zool.,* 13, 315, 1974.
70. **Atwood, H. L., Smyth, T., Jr., and Johnston, H. S.,** Neuromuscular synapsis in the cockroach extensor tibiae muscle, *J. Insect Physiol.,* 15, 529, 1969.
71. **Washio, H.,** Inhibitory miniature potentials in insect leg muscle fibers, *J. Physiol. Soc. Jpn.,* 49, 378, 1987.

71a. **Washio, H.,** unpublished data, 1987.

72. **Carbonell, C. S.,** The thoracic muscles of the cockroach, *Periplaneta americana, Smithson. Misc. Collect.,* 107, 1, 1947.
73. **Parnas, I. and Grossman, Y.,** Presynaptic inhibition in the phallic neuromuscular system of the cockroach *Periplaneta americana, J. Comp. Physiol.,* 82, 23, 1973.
74. **Usherwood, P. N. R.,** Response of insect muscle to denervation. I. Resting potential changes, *J. Insect Physiol.,* 9, 247, 1963.
75. **Usherwood, P. N. R.,** Response of insect muscle to denervation. II. Changes in neuromuscular transmission, *J. Insect Physiol.,* 9, 811, 1963.
76. **Jacklet, J. W. and Cohen, M. J.,** Nerve regeneration: correlation of electrical, histological and behavioral events, *Science,* 156, 1640, 1967.
77. **Bodenstein, D.,** Studies on nerve regeneration in *Periplaneta americana, J. Exp. Zool.,* 136, 89, 1957.
78. **Washio, H.,** Spontaneous miniature potentials in denervated coxal muscle fibres of the American cockroach, *J. Exp. Biol.,* 120, 143, 1986.
79. **Birks, R., Katz, B., and Miledi, R.,** Physiological and structural changes at the amphibian myoneural junctions in the course of nerve degeneration, *J. Physiol. (London),* 150, 145, 1960.
80. **Miledi, R. and Slater, C. R.,** Electrophysiology and electron microscopy of rat neuromuscular junction after nerve degeneration, *Proc. R. Soc. London Ser. B,* 169, 289, 1968.
81. **Miledi, R. and Slater, C. R.,** On the degeneration of rat neuromuscular junction after nerve section, *J. Physiol. (London),* 207, 507, 1970.
82. **Osborne, M. P.,** The ultrastructure of nerve-muscle synapses, in *Insect Muscle,* Usherwood, P. N. R., Ed., Academic Press, London, 1975, chap. 3.
83. **Washio, H. and Nihonmatsu, I.,** Structural changes in the denervated neuromuscular junction in coxal muscles of the cockroach *Periplaneta americana, Comp. Biochem. Physiol.,* 86A, 643, 1987.
84. **Rees, D. and Usherwood, P. N. R.,** Fine structure of normal and degenerating motor axons and nerve-muscle synapses in the locust, *Schistocerca gregaria, Comp. Biochem. Physiol.,* 43A, 83, 1972.
85. **Clark, R. B., Gration, K. A. F., and Usherwood, P. N. R.,** Relative 'trophic' influences of excitatory and inhibitory innervation of locust skeletal muscle fibres, *Nature (London),* 280, 679, 1979.
86. **Pearson, K. G. and Iles, J. F.,** Innervation of coxal depressor muscles in the cockroach, *Periplaneta americana, J. Exp. Biol.,* 54, 215, 1971.
87. **Usherwood, P. N. R.,** Glutamate sensitivity of denervated insect muscle fibres, *Nature (London),* 223, 411, 1969.
88. **Yamaguchi, T., Tsuru, A., and Washio, H.,** L-Glutamate and potassium induced contractures in denervated cockroach muscles, *Comp. Biochem. Physiol.,* 87C, 401, 1987.
89. **Curtis, D. R., Phillis, J. W., and Watkins, J. C.,** Chemical excitation of spinal neurones, *Nature (London),* 183, 611, 1959.
90. **Hill, D. R. and Bowery, N. G.,** ^{3}H-Baclofen and ^{3}H-GABA bind to bicuculline-insensitive GABA sites in rat brain, *Nature (London),* 290, 149, 1981.
91. **Simmonds, M. A.,** Multiple GABA receptors and associated regulatory sites, *Trends Neurosci.,* 6, 279, 1983.
92. **Scott, R. H. and Duce, I. R.,** Pharmacology of GABA receptors on skeletal muscle fibres of the locust *(Schistocerca gregaria), Comp. Biochem. Physiol.,* 86C, 305, 1987.
93. **David, J. A., Lummis, S. C. R., and Sattelle, D. B.,** Receptors for GABA and acetylcholine in the central nervous system of an insect, *Br. J. Pharmacol.,* 80, 432P, 1983.

94. **Lawrence, L. J. and Casida, J. E.,** Stereospecific action of pyrethroid insecticides on the γ-aminobutyric acid receptor-ionophore complex, *Science*, 221, 1399, 1983.
95. **Lummis, S. C. R. and Sattelle, D. B.,** Binding sites for 4-aminobutyric acid and benzodiazepines in the central nervous system of insects, *Pestic. Sci.*, 16, 695, 1985.

Chapter 10

ELECTRICAL AND MECHANICAL RECORDING METHODS

Moray Anderson, Karel Sláma, and Thomas Miller

TABLE OF CONTENTS

I. INTRODUCTION

In years past, the electrophysiological laboratory contained standard and predictable equipment; today, the versatility of instruments is much greater. The biggest changes have been the switch to digitalized designs and the advent of the microprocessor.

Previously, one needed a microelectrode puller and amplifier, an oscilloscope (most preferred the storage oscilloscope), and some form of recording device, which varied from frequency-modulated (FM) tape recorders to cameras. The Brush recorder became very popular because of its reliability. There were (and still are) any number of AC preamplifiers (preamps) available, and a number of designs for hand-crafted AC preamps are readily available to those who are inclined to "do it themselves".[1]

The choice of instrumentation for an electrophysiological laboratory will continue to change because of advances in microprocessors and associated software. Microelectrode pullers now come designed to make patch-clamp electrodes in addition to ordinary intracellular microelectrodes. Accordingly, microelectrode amplifiers can be purchased for ordinary recording with or without current injection provisions and can now be used for patch recording.

The choice for signal processing has become bewildering. The fainthearted might still purchase a storage oscilloscope and then decide how to make permanent records of traces. The more adventurous might buy a processor-based system (such as DATA-PAC from Run Technologies, 5768 W. Pico Blvd., Los Angeles, CA 90019) and add to it as advances are made in software. Tape recorders are still used, but the video cassette recorder (VCR) has now entered the electrophysiological laboratory. Also, a good pen recorder is still useful.

Electrophysiology, like electron microscopy, is not for the novice or occasional user. It is most efficiently used by those who can afford the time and attention necessary to perform tissue preparation and to manipulate signal processing information. This normally takes time to learn and must be accompanied by a continuing investment of time for retention as well as for keeping up with new developments.

II. EXTRACELLULAR RECORDING

A. WIRE ELECTRODES

When recording the nervous activity from any nerve in the cockroach, a great deal can often be learned from a recording using simple wire electrodes. The wire used varies according to the particular situation. For example, if the cuticle is to be penetrated, then it is best to use a tough stainless steel or tungsten wire, whereas if it is a delicate visceral nerve from which the recording is to be made, then it is better to use fine copper or silver wires.[1]

When using wire electrodes, the most common problem is the need to reduce the information arriving at the electrode. To keep out extraneous interference (information), it is necessary to shield the wire recording electrode by thorough insulation of all but the recording tip. This is best achieved by painting on a lacquer of some type, making the recording area as large or as small as desired.

The method of recording extracellular nerve activity using wire electrodes is well illustrated in the preparation where output from the tibial mechanoreceptive hairs on the cockroach leg can be recorded and their responses to mechanical distortion monitored. This simple yet extremely effective method uses stainless steel entomological pins as the recording electrodes. The pins are inserted into the tibia of the cockroach (Figure 1).

A large amount of nerve activity is often present initially, but after a short time this can be seen to decrease. By careful manipulation of the stimulating device, in this case a loudspeaker-driven probe, it is possible to single out the activity from one sensillum. Clear accommodation to the stimulating signal can often be illustrated (Figure 2).

FIGURE 1. A photograph of a device for stimulating and obtaining recordings via wire electrodes from an isolated cockroach leg. (1) Stainless steel recording electrodes; (2) retaining pins for attachment of the positive and indifferent recording wires; (3) magnets which allow for infinitely variable positioning of the leg when held by recording electrodes; (4) wooden stimulating rod driven by a small loudspeaker within the metal box; (5) control knob which alters degree of stimulation; (6) fine positioning control knob to line up stimulator with the leg spine.

When recording from nerves in the abdomen of the cockroach with wire electrodes, it is often essential to pick the loop of the nerve clear from the saline in order to obtain a clear recording. Desiccation problems can be overcome using various greases and light waxes, in a manner similar to the grease electrodes. (Grease electrodes are described below in Section II.C).

Recording from chemosensitive sensilla with wire electrodes is often possible. In this case, it is best to use sharpened tungsten wire. The fine tip is best produced by sharpening the tungsten wire electrolytically in hydrochloric acid, potassium hydroxide, or some other corrosive solution. The electrodes can then be placed into the sensillum close to the sensory cells, and clear recordings can subsequently be obtained (Figure 3).

B. SUCTION ELECTRODES

Suction electrodes, drawn-out glass or plastic tubes fitting onto nerves, have been used for a number of years by many research workers both to record nervous activity and to stimulate nerves. There are two ways in which the suction electrode can be used, either by drawing the cut end of the nerve into the electrode or by taking in a loop of the intact nerve; the latter is termed an "en passant" recording. In both cases, success of the method depends on the snugness of the fit of the nerve in the electrode tip, providing a high-resistance pathway from inside the electrode to the outside.

There are a number of detailed descriptions of producing good-fitting suction electrodes in the book by Miller,[1] including the best ways to achieve a good "suck" to hold the nerve in the electrode when recording or stimulation is taking place. Experience has shown that some of the very simple designs are, in fact, often the most suitable.

To produce the correct diameter of the glass suction electrode for the best fit, it is often

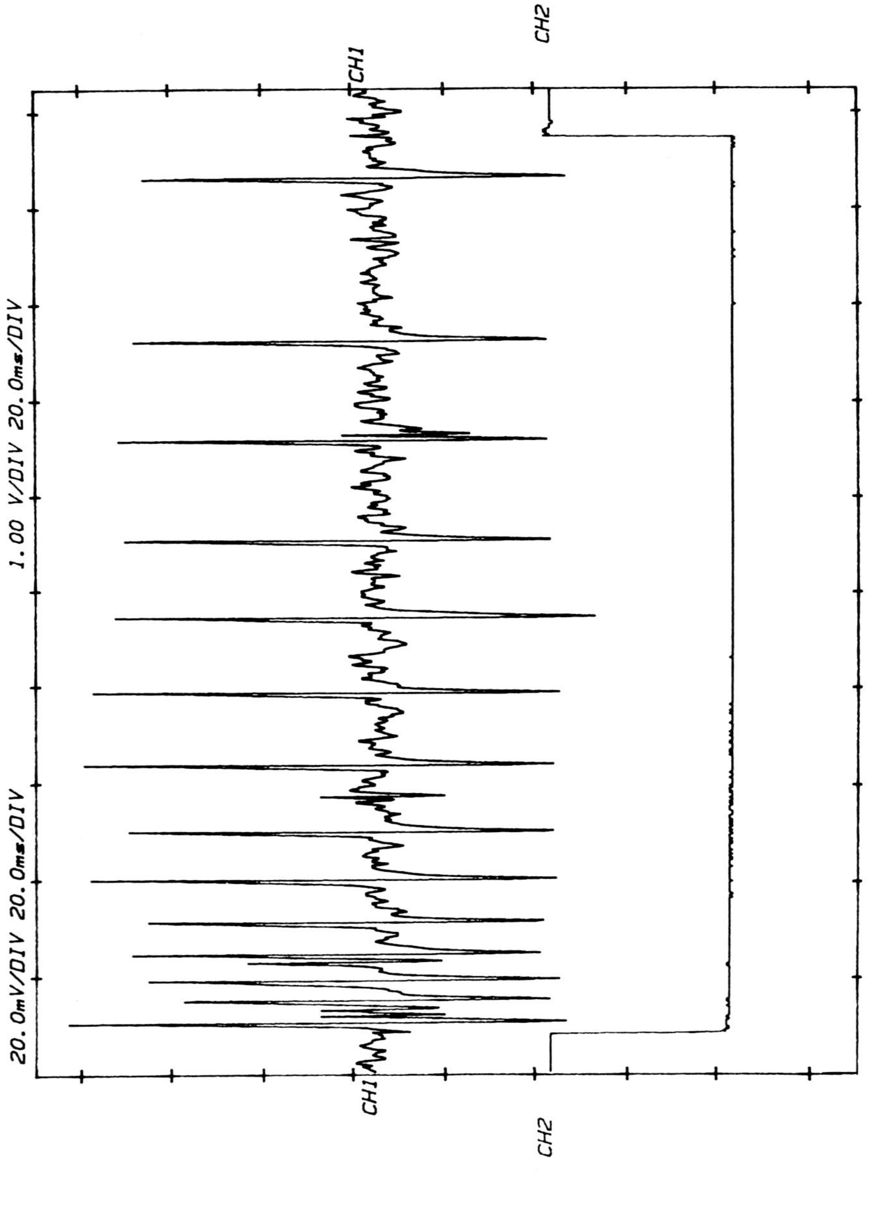

FIGURE 2. A trace taken from a wire electrode recording (see Figure 1) of mechanoreceptor output from a cockroach tibial spine. The signal shows clear accommodation to the stimulus. CH1, trace; CH2, stimulus applied.

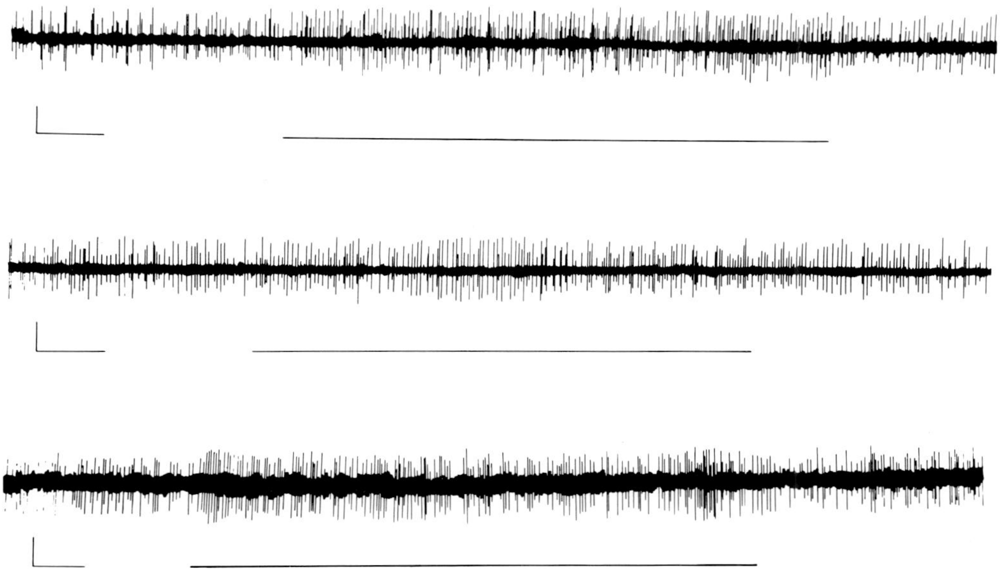

FIGURE 3. Traces from a chemoreceptive sensillum on the antenna of a cockroach, recorded with sharpened tungsten wire electrodes. Lines below traces refer to where odor stimulus has been applied to antenna. Bars: vertical, 0.1 mV; horizontal, 0.4 s.

useful to melt the tip with some sort of furnace (Figure 4). The device in the illustration allows constant monitoring of the glass electrode as it is being closed by the heat from the nichrome wire loop. Satisfactory electrodes can be made in this way.

Those who have used suction electrodes know that in practice a great deal of time can be spent obtaining measurements of nerve diameter, making the electrode fit precisely that size only to find that a small piece of fat body gets into the electrode tip just as the nerve is being drawn in. Thus, no space remains for the nerve. Considered another way, the fat body can be a good ''plug'' to wedge a nerve into a slightly large electrode.

To produce a suction tip, many workers now prefer to use some sort of plastic/polyethylene, drawn out in a small flame and cut cleanly with a scalpel. These tips have the great advantage of being unbreakable and are easy to clean in an ultrasonic cleaner should the tip become blocked.

The suction can be applied most simply with a disposable syringe. The simple design shown in Figure 5 works effectively and has the great advantage of being quick to fabricate and service.

Within the suction electrode the signal is transmitted to the recording device by means of an Ag/AgCl wire. Sealing around the electrical connections for suction-proof junctions is very important. The modern silicone rubber adhesives and sealants are perfect for this function.

C. GREASE ELECTRODES

The grease electrode can be regarded as being a hybrid of the wire and suction electrodes. In this case, a metal hook is used for recording or stimulating, with the nerve being drawn into a nonconducting environment such as mineral oil. The advantage of this method is clear because there is a great reduction in the desiccation of the nerve, which makes extended recording or stimulating times possible.

D. ELECTROMYOGRAMS

Extracellular recording of muscle activity has been accomplished in a number of

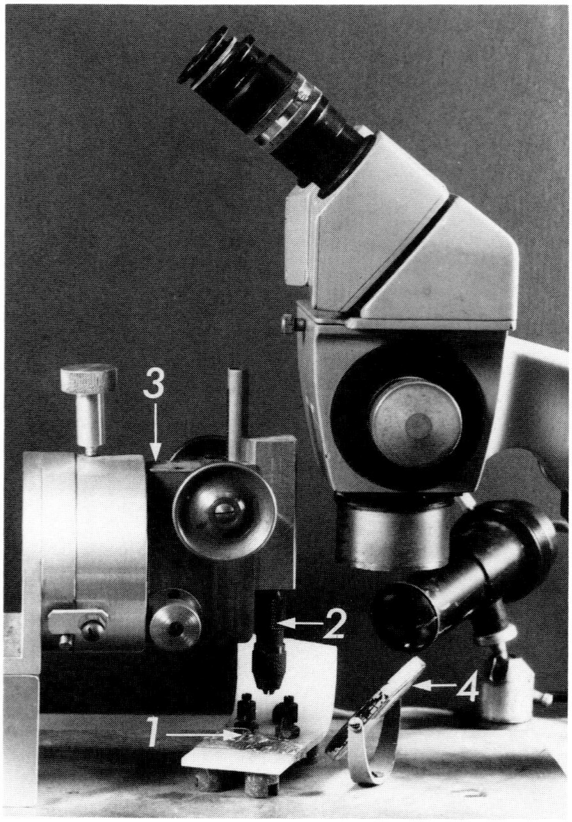

FIGURE 4. A photograph of a furnace and inspection microscope for producing glass suction electrodes. (1) Nichrome wire loop comprising the furnace; (2) drill chuck to hold glass of various diameters; (3) swivel holder with varying height adjuster to lower the glass into the furnace loop; (4) adjustable mirror which allows microscopic inspection of the glass as it is constricted by the furnace.

preparations using the cockroach. None of these are more elegant than the free-walking preparations of Gammon[2] and of Delcomyn and Usherwood.[3] These authors have used insulated wire (silver and copper) to record muscular activity while the animals are free-moving. Such techniques of relatively nonintrusive recording have distinct advantages when recordings are to be made of ''normal'' behavior before and altered behavior after the application of some test chemical.

The major difficulties with these techniques are those associated with the precise placement of the electrodes. It is relatively straightforward to insert the electrodes and to obtain a recording, but often it is not so easy to ascertain the exact muscle from which the recording has been obtained.

E. AMPLIFIERS FOR EXTRACELLULAR RECORDINGS

For electrophysiological experiments, the amplifiers can be divided into those required for intracellular recording and those needed for extracellular recording. Those for extracellular recording are much less demanding in their specifications. The important features are high-impedance differential inputs, low noise, and low drift. It is also desirable to have dual

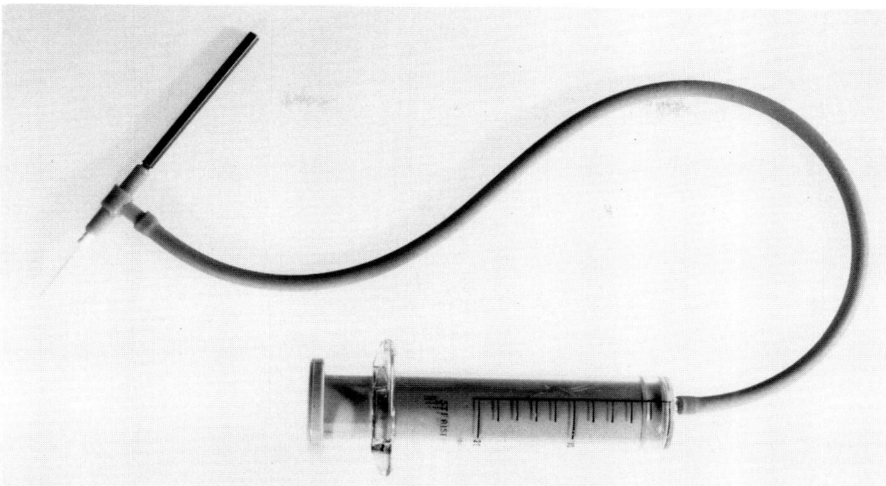

FIGURE 5. A simple suction electrode setup. The suction is produced by the syringe, and the plastic T-piece can accommodate the positive and negative electrodes. The joints are sealed with silicone sealer.

filters for high and low frequencies. There are a number of such amplifiers on the market, manufactured by Dagan (2855 Park Ave., Minneapolis, MN 55407), Grass Instruments (101 Old Colony Ave., Quincy, MA 02169), U.F.I. (545 Main, Suite C2, Morro Bay, CA 93442), and WPI (375 Quinnipiac Ave., New Haven, CT 06513), for example. The DAM 50 from WPI is shown in Figure 6.

III. INTRACELLULAR RECORDING

The first serious steps in the use of micropipettes for recording were taken by Gerard and co-workers[4,5] when they measured the membrane potentials of frog muscle. However, it was Hodgkin and Huxley[6] who first recorded from nerve in the giant squid axon using a large (0.1-mm) glass pipette as the recording electrode. Although these early attempts provided the impetus for intracellular recordings, they were limited to large cells. These large-diameter tips often gave unstable recordings. It was not until instrumentation was developed for the fabrication of fine electrodes that stable recordings for long time periods were obtainable, especially from small cells.

A. GLASS FOR MICROPIPETTES

Glass is an excellent electrical insulator, and when fabricated as a microelectrode it provides a cavity for confining various solutions as well as for isolating electrical signals. Once it penetrates a cell and seals the cell membrane against the glass, it provides a channel into the cell, allowing the stable measurement of internal electrical activity for considerable periods.

1. Tubing Design

In common use in neurophysiology is Corning® No. 7740 tubing, which contains a solid glass fiber 0.1 mm in diameter. Another type of tubing in common use is the Omega Dot tubing. Tubing with heavy Omega Dot or thick walls (high O.D./I.D. ratio) requires a greater heat output from a heating filament for drawing into micropipettes.

Capillary tubing is most often used for microinjection (microiontophoresis) or for cell penetration and intracellular recording. The tip opening can be enlarged by lowering the O.D./I.D. ratio. A standard O.D./I.D. ratio of 2 for Corning® glass is the norm for mi-

FIGURE 6. A photograph of the WPI DAM 50 differential amplifier. The input connector is an ordinary standard telephone cable connection which can be purchased anywhere in the U.S. Signals may be either AC or DC recorded, making the DAM 50 versatile. A test button on the front sounds a tone if the battery is good.

croinjection. Ratios too much higher give fragile tips. A short, stiff shank and very small-diameter tips aid in penetration. Thus, the O.D./I.D. ratio should be $\geqslant 2$ to give a tip stiff enough to be able to penetrate cells and tissue.

B. MICROPIPETTE PULLERS

The preparation of the micropipette is the most fundamental step in preparing for intracellular work. An authoritative discussion of micropipette technology, written by Brown and Flaming,[7] is available and should be referred to by anyone starting this work. The following sections on pulling and beveling microelectrodes were written after a thorough reading of the Brown and Flaming work referred to above and also include the junior author's own experiences.

When drawing micropipettes, it is important to control the temperature of the glass. Glass temperatures that are too high give long, flimsy tips, while cool glass temperatures result in large, broken tips. Control of both timing and filament temperature is critical to achieve reproducibility in drawing micropipettes.[8] Consistent results are usually achieved only through the use of automated commercial instruments.

To form the fine tip, it is necessary that the pull begins slowly under weak tension and then accelerates with a stronger pull. By quickly removing the tip from the heating element the tip is allowed to cool, thus preventing the formation of tips too long to be usable. It is also important to hold the glass tubing straight during the pull in order to produce straight tips. The Du Bois micropipette puller[9] was one of the earliest automated microforges which incorporated the above characteristics in a horizontal puller, and it became a model for the design of other pullers. It is an excellent design for making microneedles for microdissection.

Another horizontal puller, designed by Alexander and Nastuk,[10] incorporates aspects of the older pullers as well as new features. It uses an electromagnetic solenoid to supply the pulling force and pulls reasonably straight and fine micropipettes. However, very fine tips for small cell work are difficult to make because insufficient velocity is obtained in the second pull. This instrument is still available with little modification through Industrial Science Associates, Inc. (25-44 163rd St., Flushing, NY 11358).

The Brown-Flaming micropipette puller incorporated many of the older characteristics which were desirable in a horizontal puller and minimized the variables by utilizing a microprocessor to control heat, pull velocity, cooling time, and pull strength.[7] These workers incorporated the use of an air stream similar to the Chowdhury puller in their design.[11] In the Chowdhury puller, air is blown over the softened glass to shorten the cooling time during the strong pull, this action resulting in two identical ultrafine (0.05-μm) tips with short shanks.

The Brown-Flaming puller as supplied by Sutter Instruments (P.O. Box 3592, San Rafael, CA 94912) is presently a very sophisticated puller capable of pulling a complete range of micropipettes including micropipettes for patch clamping. Two models are available, one incorporating the air-jet principle and a less expensive unit without this feature. These pullers now produce tips of a given size with high reliability. All the advantages of short tips (lower electrical resistance, decreased noise levels, greater stiffness for beveling) are obtainable in combination with ultrafine tips.

C. BEVELING

Beveling has been used by many workers to modify the tip of a micropipette to improve penetration into cells and tissues without causing significant damage. Early beveling methods consisted of breaking the micropipette tip to achieve an angle different from 90°, employing a rotating abrasive stone disk,[12] applying diamond dust as the grinding abrasive on a spinning quartz rod,[13] utilizing fine (0.05-μm) alumina particles suspended in a salt solution rotating on a wobble-free glass plate,[14] and using embedded particles of alumina or variously sized diamond dust in a polyurethane matrix rotating at a slow speed.[7,15]

In each case, progress was made toward the beveling of finer and more consistent micropipette tips. The necessary requirements in each case were sufficient contact of the tip with the abrasive surface and sufficient pressure to shape the tip. Long tips, such as those made in an Alexander-Nastuk puller, were sufficiently flexible so that tips did not break, with tip sizes of 0.2 μm obtained routinely. The short, stiff fine tips of the Brown-Flaming puller were initially troublesome to bevel, but using more uniform adhesive coating procedures now allows satisfactory beveling of these very fine micropipettes.

The K.T. Brown Micro-Pipette Beveler available through Sutter Instruments can conveniently bevel tips from 0.06 μm to very large sizes.[16] It is possible to bevel tips of 2 to 3 μm in a few minutes. This instrument uses a wet grinding process, so it is possible to continuously monitor the electrical resistance of the micropipette as the beveling proceeds. A gradual, continuous decrease in electrical resistance indicates a smooth abrasive action on the tip, while sharp drops in resistance indicate chipping of the tip.

The usual range of tip angles on beveled electrodes seldom exceeds 20 to 35°. It should be realized that beveling enlarges the tip opening and, thus, lowers the electrical resistance of a micropipette. The method of choice is to bevel the micropipette the minimum amount required for improved penetration of cells. Increases in resistance indicate plugging of the tip by abrasive particles. Some improvement in beveling of fine tips can be achieved by filling micropipettes overnight prior to actual beveling.

Methods for beveling of tips without instrumentation are also available. To obtain reasonably sharp tips, Baldwin[17] used sheets of tough plastic film (Imperial Lapping Film) with an adhesive backing containing 0.3-μm embedded alumina (Thomas Scientific, P.O. Box 779, Philadelphia, PA 19105-0779) on a rotating device. This method was used by Werblin[18] to selectively bevel one side of a dual-channel micropipette and decrease the electrical coupling between the two channels. Tauchi and Kikuchi[19] used ceric oxide, a polishing compound, in an agar matrix placed on a phonograph turntable to achieve tips of very nice quality. A jet stream microbeveler described by Ogden et al.[20] has also been used to obtain useful beveled tips. A very simple slurry beveling method[21] has also been shown to achieve bevel tips that allow penetration into small cells. The abrasives used in these methods can be obtained from Buehler Ltd. (2120 Greenwood St., Evanston, IL 60204).

The quality of a beveled tip depends on the type of abrasive and how it is applied to the micropipette tip. Abrasive suspended in saline results in slow beveling of tips because the abrasive particles are not anchored and, hence, can move about freely in the medium. Since there is freedom of movement, there is also difficulty in controlling the bevel angle. When the abrasive is concentrated in a slurry or other soft medium to restrict abrasive-particle movement, beveling time is reduced. However, the increased density of the medium causes greater forces on the tip, and bevel angle is again difficult to control. Embedded abrasive particles further decrease the beveling time, and accurate tip angles can be obtained by presetting the bevel angle. All three methods have their value, but the embedded abrasive method has been shown to produce a somewhat sharper cutting edge more reliably.

The advantage of beveling is that the micropipette can more easily penetrate cells without significant damage. The sharp cutting edge should allow a clean cut into the membrane and subsequent better sealing of the membrane to the tip, with higher membrane potentials and preparation stability the result.

Beveling can also reduce the signal/noise ratio by reducing the electrical resistance of the micropipette. This reduction in electrical resistance can also reduce the resistance experienced during iontophoretic injection of compounds onto or into the cell. Beveling has also been found to be advantageous for penetration through tough overlying membranes and for improving performance of dual-channel micropipettes.

D. MICROMANIPULATORS

When positioning a micropipette, a mechanism free of vibration and with the ability to control the movement of the micropipette in three dimensions is necessary. Inadvertent movements caused by hand motion are then not important. Although coarse controls are required for two dimensions, fine control is required when advancing toward a cell or tissue. Many types of micromanipulators have been described.[1,22,23]

IV. RECORDING MOVEMENT

Problems concerned with measuring the locomotory movements of cockroaches were described earlier,[24] and techniques used for measuring locomotory movements of insects in general were also described previously.[1]

In this section, we concentrate our attention on methods of recording movements of a different kind. It has been shown that insects periodically "pump" their abdomens, resulting in sharp increases and decreases in hemocoel pressure.[25,26] The change in pressure is thought to be caused by contractions of intersegmental muscles which connect adjacent flexible abdominal segments. The action of the intersegmental muscles, in turn, causes minute movements of the abdomen. These pressure pulses are thought to participate in the regulation of certain autonomic functions, such as hemolymph circulation through appendages, movements of visceral organs, and tracheal ventilation. Therefore, the recording of movements is another way of recording suspected autonomic functions.

Long ago, Harker[27] noticed that decapitated adult cockroaches could remain alive for many days or weeks. More recently, it was found that houseflies could be tested for 24-h mortality after treatment with organophosphate insecticides regardless of whether their heads were on or off.[28] This suggests that the head or brain of insects does not support a vital function aside from eating and orientation and that homeostasis may be maintained for some considerable time in a vegetative state.[29]

The body of the decapitated cockroach or of an immobilized cockroach is a very convenient model system to monitor changes in hemocoel pressure and other autonomic functions because of the extensive physiological information already available.[30]

A. ELEMENTS OF THE STRAIN GAUGE TECHNIQUE

Very small mechanical changes can be measured by strain gauge elements (tensiometers). These are usually made from a thin wire, metallic foil, or semiconductor silicon material. In principle, a mechanical deformation of the strain gauge shape would cause a corresponding change in its resistance. If the gauge were a part of a Wheatstone resistance bridge, the deformation would cause deviations in the balance current of the bridge. The imbalance could then be amplified and recorded.

B. GAUGE FACTOR

The efficiency of transformation of mechanical deformity into resistance of the strain gauge has been expressed conventionally as a gauge factor. The gauge factor usually ranges from 2 in the wire and foil gauges to 120 or more in the silicon gauges. The latter type thus provides a 60-fold more sensitive measurement than the former at the same amplification. This is especially convenient for the construction of strain or movement transducers, using the commercially available, low-cost, operational amplifiers.

C. WHEATSTONE BRIDGE AND AMPLIFIER CIRCUITRY

A simple circuit for an inexpensive strain gauge transducer and amplifier is shown in Figure 7. The circuit is suitable for recording changes in hemocoel pressure of insects. The matched pair of strain gauges, T_1 and T_2, form half of a Wheatstone resistance bridge. They are connected in neighboring branches of the bridge so that the effects of small temperature changes acting on both sensors should be eliminated. When the gauges are placed on opposite

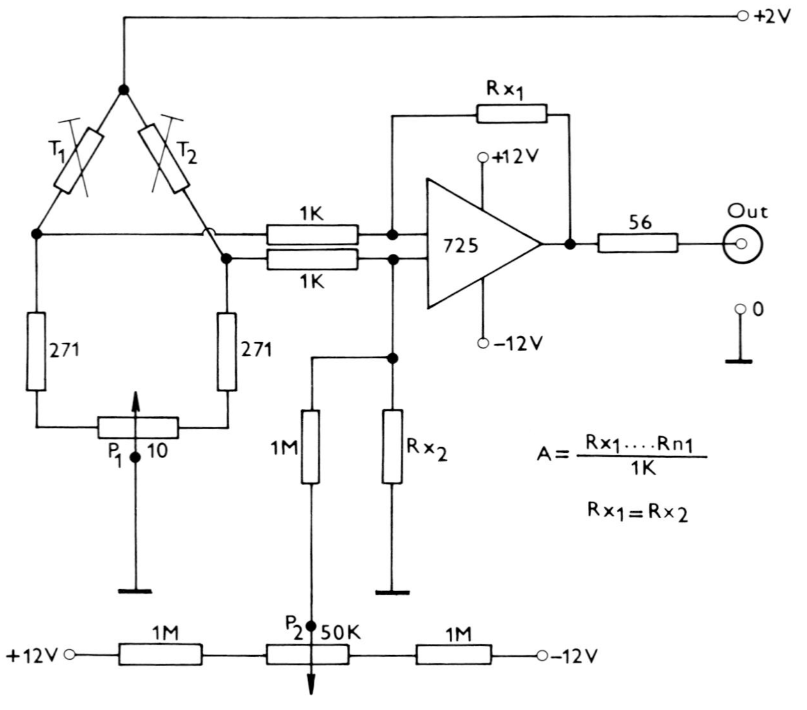

FIGURE 7. Circuit diagram for an ordinary Wheatstone bridge amplifier of the type used in force transducer. T_1 and T_2 are strain gauges. P = potentiometer; A = the formula for amplification. Resistor values are shown, Rx resistors are selected, and 725 is the operational amplifier designation for analog devices or the equivalent.

sides of the foil, they are both "active" and additive; one causes an increase and the other a decrease in the current flowing through the bridge. A still more efficient system can be created from four gauges forming the full resistance bridge, where the sensors on one side alternate with the sensors on the other side of the foil.

The scheme described in Figure 7 uses a DC bridge-balancing current for simplicity. However, the amplifier will amplify the electrical "noise" along with the signal, and the sensitivity of this arrangement is limited. Greater stability and sensitivity are achieved using 5 kHz AC for the bridge-balancing current. When filtered, decoded (or demodulated), and converted into a DC output, the signal contains very little electrical "noise".

D. SOURCES

Manufacturers of electronic instruments offer many types of foil or silicon strain gauges, including prefabricated foil strain gauges. They can be purchased, for instance, from Hottinger Baldwin Measurements (139 Newbury St., Framingham, MA 01701, or P. O. Box 4235, D-6100 Darmstadt 1, Federal Republic of Germany), Philips GmbH (D-3500 Kassel, Federal Republic of Germany), Kyowa Electronic Instruments (Tokyo, Japan), and other suppliers. The silicon strain gauges used in the transducers described here were purchased from Podnik Služeb (76000 Gottwaldov, Czechoslovakia).

E. CONSTRUCTION TIP

The attachment of a strain gauge to the measuring membrane or foil is a very important step in fabricating the gauge. Any special instructions given by the manufacturer should be followed. In principle, since we mainly use dynamic (not static) measurements, any cyanoacrylic or hard epoxy resin adhesive with low hysteresis can be used. Foils made from beryllium bronze or stainless steel are suitable.

F. AN EXAMPLE OF AN ISOTONIC MOVEMENT TRANSDUCER

A very convenient strain gauge transducer that is capable of sensing small movements of the cockroach body has been described by Sláma.[26] This anisometric or partly isotonic transducer is shown on the right side of Figure 8. It is built from a flat box of Plexiglas®, with an internal width of 10 mm. The most important component is a thin beryllium-bronze foil (**r**), which contains at its base a matched pair of stain gauges (**s**) forming half of a resistance bridge.

The transducer records movements with minimum loading and can reliably detect displacements as small as 10 nm. Its advantage is that it can be used for prolonged recordings (for days and weeks) on unrestrained insects, without the necessity of penetrating the integument. Occasional twisting and large sideward movements of the abdomen can be minimized using a long wire to connect the transducer to the subject insect.

In most cases, the transducer foil is gently polarized initially by taking up the slack in the transducer connection to the insect and putting a few micrograms of tension on the foil. In certain cases, especially when the integument is very soft, the transducer can be used in a ''pushing mode'' with the membrane polarized in the opposite direction by pushing the transducer pin against the surface of the body.

G. IMPEDANCE CONVERTERS

Impedance converters remain among the most versatile of measuring instruments available to the physiologist. Originally designed for measuring pulmonary activity of crash victims during the ambulance ride to a hospital, the converter can be used for a number of other applications, especially in measuring movements or force in the cockroach.

The UFI® Model 2991 impedance converter is designed to measure impedance changes. A 1% change in source resistance, capacitance, or inductance causes the output of the converter to vary by about 0.5 V. Impedance changes at the input modulate the amplitude of a 50-kilocycle oscillator. The oscillator output is then demodulated and filtered. The output voltage is either directly proportional to the input changes and shows positional information, or it is filtered to show faster changes and is insensitive to very slow changes in impedance.

Compared to electrical measurements, impedance conversion has few electrical noise problems and is remarkably free from interference. The device is very sensitive in recording movements in insects, but the output recorded depends on the geometry of recording electrodes and on the physical location of the electrodes.

The strain gauges described above may be connected directly to the input of the impedance converter. As long as the output signal is calibrated, the need for amplifiers and Wheatstone bridge circuitry is eliminated. For rapid setup, versatility, and convenience, it is hard to improve upon the impedance converter. In addition to its other advantages, it is also rugged enough to be student proof.

H. THE VERO TRANSDUCER

The need to measure very small forces produced by visceral muscles in insects led to the design and construction of a transducer based on an inductance feedback principle. The basic idea for an inductance transducer was converted by Vero into a very sensitive device with a linear output.[31] This device is now being developed by U.F.I. as Microgage (Model 1630).

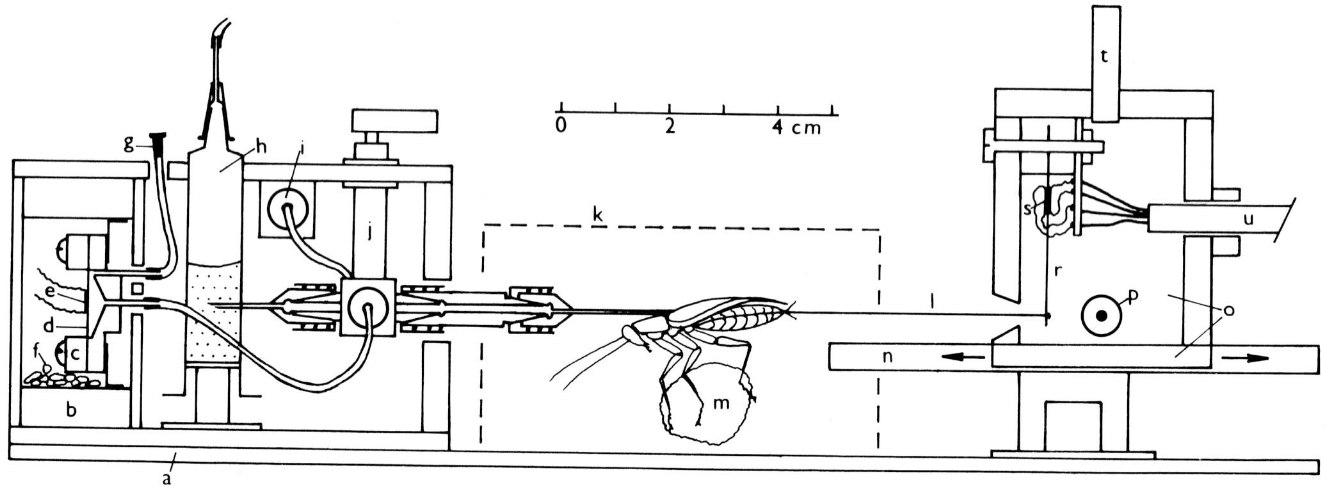

FIGURE 8. Diagram of pressure transducer (left) and force transducer (right). (a) Supporting plate; (b) Plexiglas® body of the hydraulic transducer; (c) stainless steel drum of the transducer (25 mm O.D., 11 mm I.D.); (d) beryllium-bronze transducer arm; (e) silicon strain gauges, type AP 120-3-12, size 0.1 × 3 mm, gauge factor 120 (Podnik Služeb, Gottwaldov, Czechoslovakia); (f) silicic acid crystals; (g) outlet for washing out air bubbles; (h) reservoir of hydraulic fluid connected to water manometer by Teflon® tubing; (i) 701 N Hamilton® syringe; (j) 1XP Hamilton® inert valve; (k) light and humidity cover; (l) hooked wire connecting a small triangle cemented to the abdomen with transducer membrane; (m) polystyrene ball; (n) adjustable arm to position transducer and tighten with knurled knob; (o) 2-mm-thick side wall of a transducer box; (p) screw compressing the side wall for tightening; (r) beryllium-bronze membrane, 0.09 × 5 × 30 mm; (s) matched pair of silicon strain guages (AP 120-6-12), 6 mm length, 120 ohm resistance (Podnik Služeb, Gottwaldov, Czechoslovakia); (t) metallic rod for micromanipulator attachment; (u) shielded outlet cable. See text for explanation (Section IV.F).

V. RECORDING PRESSURE

A. HYDRAULIC TRANSDUCER

Prolonged monitoring of changes in hemocoel pressure in cockroaches may be achieved with the hydraulic transducer shown on the left side of Figure 8. The hemocoel cavity is connected with the hydraulic fluid of the transducer through a stainless steel syringe needle of a size appropriate for the measured object. The fluid may be distilled water, Ringer's solution, or light mineral oil, and it must be free from air bubbles (evacuated). A viscous material such as lanolin, adhesive gum, nonirritating cement, etc. should be applied around the tip of the needle to prevent bleeding during insertion of the needle into the body.

An essential component of the pressure transducer is a Hamilton® Model 1XP miniature inert valve, which connects the hemocoel with either the transducer or a syringe. The function of the syringe is to flush the needle used to penetrate the insect with hydraulic fluid.

In addition, solutions to be tested for their effects on hemocoel pressure can be taken into the syringe needle prior to its insertion into the hemocoel (usually separated on both sides with 1 μl of oil). Following a period of equilibration and control measurements, the Hamilton valve may be turned, switching the syringe and the attached insect from the pressure transducer to a test solution. After injecting the solution into the body, the valve may then be returned immediately to the transducer setting with very little interruption of measurements.

By moving the valve to its second position, the pressure sensors are connected to a reservoir of hydraulic fluid. The level of fluid in the reservoir (stippled portion of **h** in Figure 8) is adjusted to about 5 mm above the orifice of the needle. This hydrostatic pressure is arbitrarily taken as a zero (5 mm above local barometric) pressure. Using a manometric tube filled with water and connected to the reservoir, we can apply various hydrostatic pressures to the reservoir, which can be used for calibration or for direct manometric pressure control during the recordings. More details concerning the construction of the hydraulic transducers can be found in the paper by Sláma.[25]

VI. EXAMPLES OF RECORDING

Adult cockroaches are normally very active, especially when handled, and it may be necessary to narcotize them by submersion in water for 5 to 10 min before attachment of the transducers. For decapitation, the neck is ligated and the head separated from the body immediately after removal from the water. For direct recording with the hydraulic transducer, the needle is inserted into the body cavity, usually through the meso- or metanotal region.

For indirect recording of hemocoel pulsations with the isotonic transducer, a small wire triangle was cemented to the tip of the abdomen, using a quick-setting cyanoacrylate glue, before the cockroaches recovered from water narcosis. The triangle was hooked by a wire attached to the strain gauge foil. Decapitated cockroaches maintained their cleaning responses and by reflex wiping could use the hind-leg spines to remove any object not properly cemented to the abdomen.

After being immobilized by decapitation the cockroaches were placed in a horizontal position, either dorsal or ventral side uppermost. They could be affixed to a substrate between strips of cardboard and held in place by attachment to the mesonotum or at the tips of the wings. Females bearing oothecae could be studied by recording oothecal movements directly. Specimens to be measured continuously for several days should be isolated from light, sound, or air movement since these cause behavioral responses.

Recording from intact (nondecapitated) *Blattella germanica* adult cockroaches is difficult because of their struggling. However, during periods of quiescence, hemocoel pressure changes as a regular pattern of repeated bursts (Figure 9, top trace). The pressure pulses

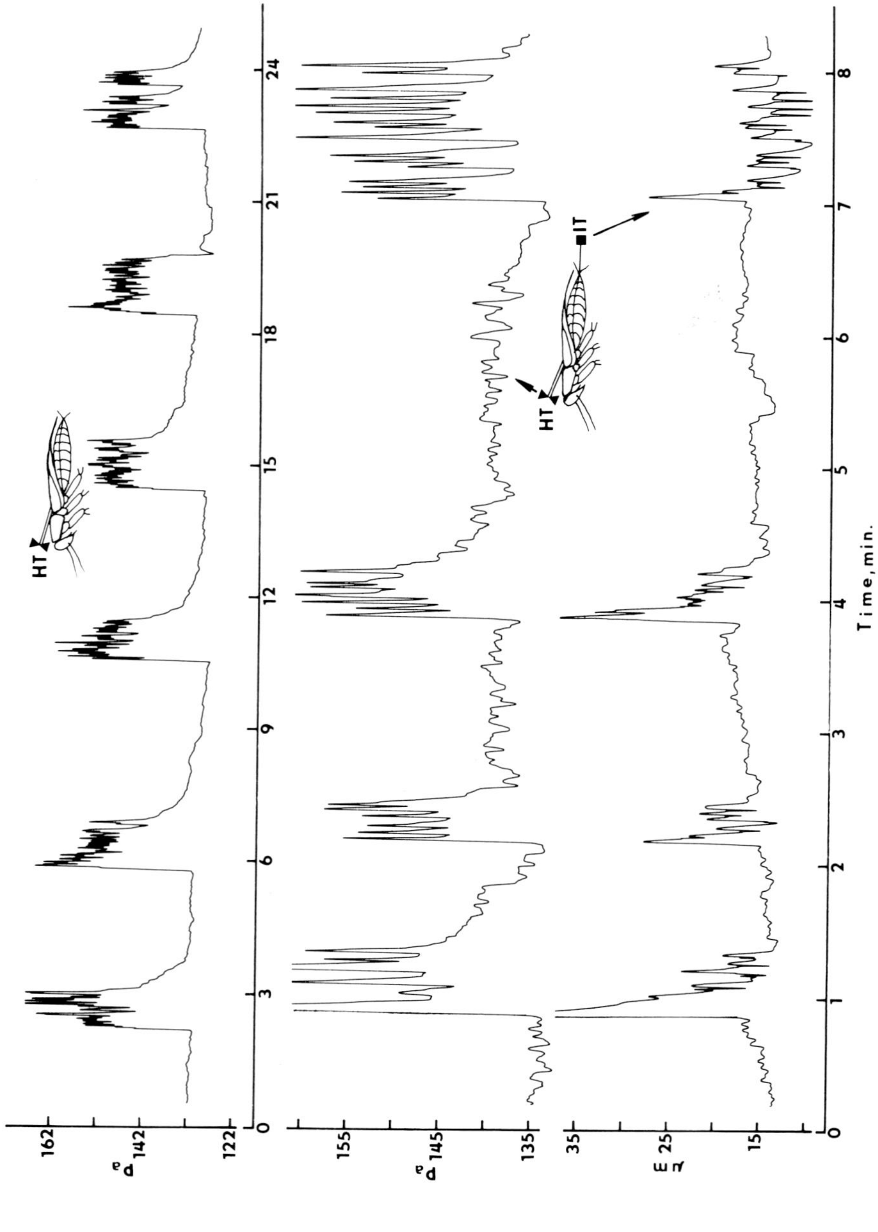

FIGURE 9. Recordings of pressure of the hemocoel (HT) and movement (IT) from adult cockroach. See text for explanation (Section VI).

are shown to be associated with movements of the abdomen in simultaneous recordings (Figure 9 — middle trace, pressure; bottom trace, isotonic tension).

A difference in amplitude can be seen when comparing the pressure and movement records in Figure 9. This is thought to be due to spiracular activity during respiratory pumping. A 1-μm movement is roughly correlated with 4 Pa of pressure.

Nauphoeta cinerea is a convenient species with which to work. Unlike *Periplaneta americana,* it has no pungent odor when cultured. Pressure recordings from decapitated adults are similar to those from intact insects, i.e., a series of repeated pulse bursts (Figure 10, top trace). Movement recordings from *Nauphoeta* show that the bursts of abdominal pumping accompanying pressure bursts are sometimes interspersed with gross movements of the abdomen (Figure 10, middle and bottom traces, respectively).

Recordings from decapitated *P. americana* are very similar to those from the other insects. They show a series of repeated pulses that continue for as long as they are recorded (Figure 11, top trace). Simultaneous recording of pressure and movement also shows the coordination between these activities (Figure 11, bottom and middle traces, respectively). Although the amplitudes of the pressure pulses remain similar throughout the burst, the movement trace shows low-amplitude pulses first, then full-amplitude pulses, followed by low-amplitude pulses at the end of the burst.

The different amplitudes of movement in the middle trace (isotonic force record) of Figure 11 are thought to be due to the spiracles being either open or closed during the pressure pulses. If the spiracles are open during increasing pressure, air will be vented from the tracheal system much like from a bellows, with a corresponding reduction in the overt abdominal displacement.

In general, females produce pulsations more often or with larger amplitude than do males. This may be due to higher metabolic rates in the former. Any change in metabolic rate is immediately reflected in changes in the number and frequency of pressure pulses, such as would occur after a meal.

VII. SUMMARY

The cercal nerve/giant interneuron synapse in the sixth abdominal ganglion of the American cockroach has been used by zoologists as a model cholinergic synapse. The "giant" axons in this preparation produce enough electrical activity that they may be used *in situ* in the ventral nerve cord with an appropriately miniaturized nerve chamber to study the propagation of the nerve impulse.

The cockroach antenna has provided a model olfactory organ for the study of odor perception. Other unique tissue preparations from the cockroach, including salivary glands, heart, ventral diaphragm, and central nervous system, provide convenient model systems to study pharmacological responses to specific drugs.

Aside from necessary miniaturization in some cases, the techniques described here are identical to those used for studying any vertebrate; indeed, many are borrowed or modified directly from vertebrate studies and are therefore interchangeable. Force or movement transducers are possibly the most unique adaptation of all physiological techniques to study insect tissues, and these have been described here in great detail.

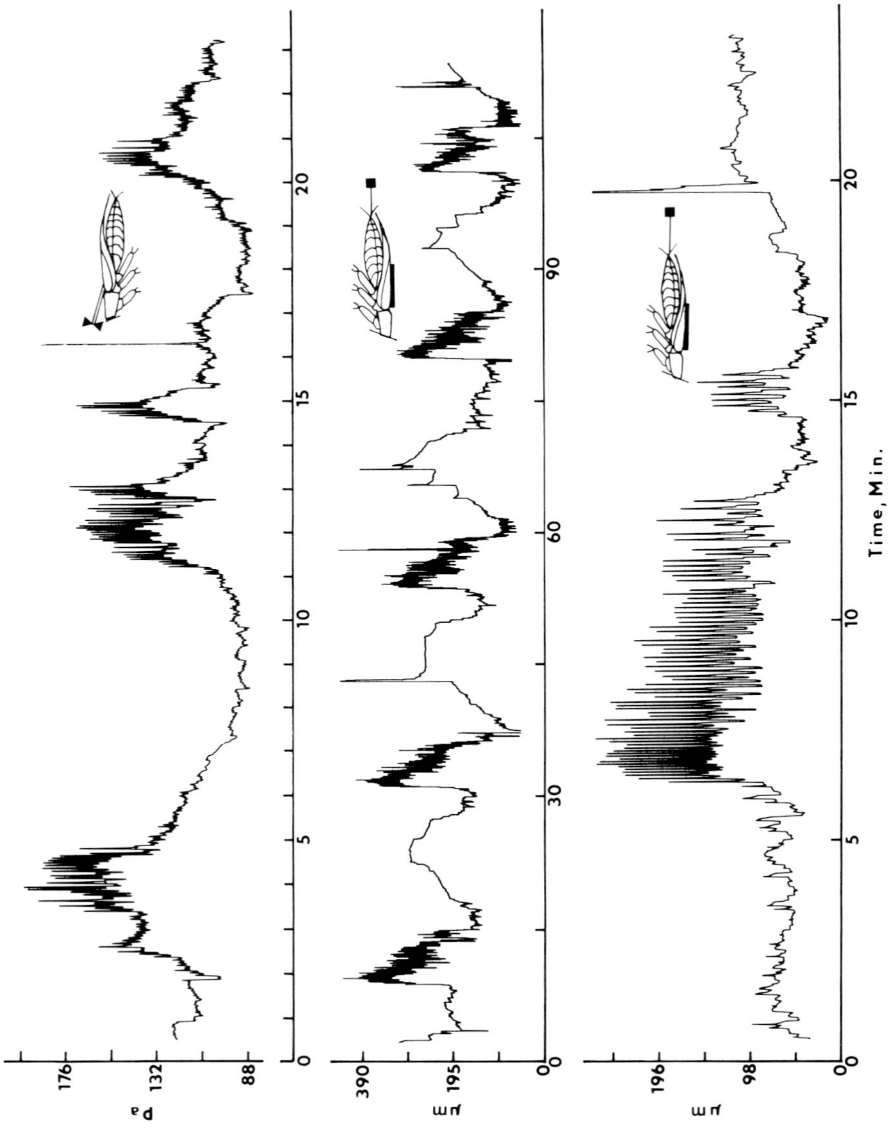

FIGURE 10. Recordings of internal pressure (top trace) and movements (middle and bottom traces) from decapitated cockroaches. See text for explanation (Section VI).

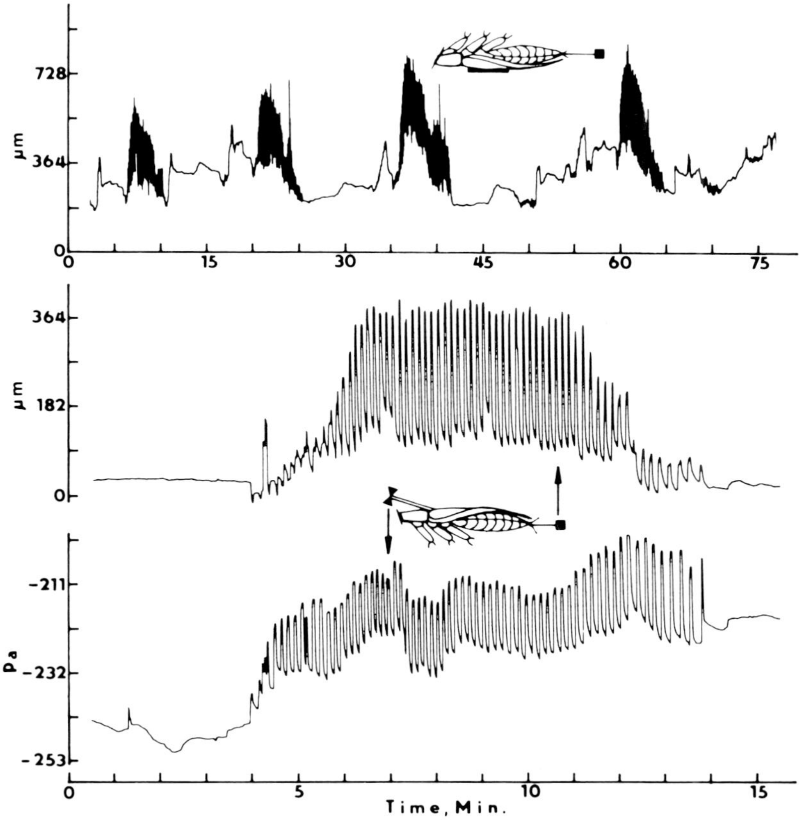

FIGURE 11. This is the same type of recording as Figure 10. See text for explanation (Section VI).

REFERENCES

1. **Miller, T. A.,** *Insect Neurophysiological Techniques,* Springer-Verlag, New York, 1979.
2. **Gammon, D.,** Nervous effects of toxins on an intact insect: a method, *Pestic. Biochem. Physiol.,* 7, 1, 1977.
3. **Delcomyn, F. and Usherwood, P. N. R.,** Motor activity during walking in the cockroach, *Periplaneta americana, J. Exp. Biol.,* 59, 629, 1973.
4. **Graham, J. and Gerard, R. W.,** Membrane potentials and excitation of impaled single muscle fibers, *J. Cell. Comp. Physiol.,* 28, 99, 1946.
5. **Ling, G. and Gerard, R. W.,** The normal membrane potential of frog sartorious fibers, *J. Cell. Comp. Physiol.,* 34, 383, 1949.
6. **Hodgkin, A. L. and Huxley, A. F.,** Resting and action potentials in single nerve fibers, *J. Physiol. (London),* 104, 176, 1945.
7. **Brown, K. T. and Flaming, D. G.,** *Advanced Micropipette Technology for Cell Physiology,* John Wiley & Sons, New York, 1986.
8. **Flaming, D. G. and Brown, K. T.** Micropipette design: form of the heating filament and effects of filament width on tip length and diameter, *J. Neurosci. Methods,* 6, 91, 1982.
9. **Du Bois, D.,** A machine for pulling glass micropipettes and needles, *Science,* 73, 344, 1931.
10. **Alexander, J. T. and Nastuk, W. L.,** An instrument for the production of microelectrodes used in electrophysiological studies, *Rev. Sci. Instrum.,* 24, 528, 1953.
11. **Chowdhury, T. K.,** Fabrication of extremely fine glass micropipette electrodes, *J. Sci. Instrum.,* 2, 1087, 1969.

12. **Barrett, J. N. and Graubaud, K.,** Fluorescent staining of cat motoneurons *in vivo* with beveled micropipettes, *Brain Res.,* 18, 565, 1970.
13. **Barrett, J. and Whitlock, D. G.,** Technique for beveling glass microelectrodes, in *Intracellular Staining in Neurobiology,* Kater, S. B. and Nicholson, C., Eds., Springer-Verlag, New York, 1973.
14. **Kripke, B. R. and Ogden, T. E.,** A technique for beveling fine micropipettes, *Electroencephelogr. Clin. Neurophysiol.,* 36, 323, 1974.
15. **Brown, K. T. and Flaming, D. G.,** Technique for precision beveling of relatively large micropipettes, *J. Neurosci. Methods,* 1, 25, 1979.
16. **Brown, K. T. and Flaming, D. G.,** Instrumentation and technique for beveling fine micropipette electrodes, *Brain Res.,* 86, 172, 1975.
17. **Baldwin, D. J.,** Dry beveling of micropipette electrode, *J. Neurosci. Methods,* 2, 153, 1980.
18. **Werblin, F. S.,** Regenerative hyperpolarization in rods, *J. Physiol. (London),* 244, 53, 1975.
19. **Tauchi, M. and Kikuchi, R.,** A simple method for beveling micropipettes for intracellular recording and current injection, *Pfluegers Arch.,* 368, 153, 1977.
20. **Ogden, T. E., Citron, M. C., and Pierantoni, R.,** The jet stream microbeveler: an inexpensive way to bevel ultrafine glass micropipettes, *Science,* 201, 469, 1978.
21. **Lederer, W. J., Spindler, A. J., and Eisner, D. A.,** Thick slurry bevelling, *Pfluegers Arch.,* 381, 287, 1979.
22. **Ellis, G. W.,** Piezoelectric micromanipulators, *Science,* 138, 84, 1962.
23. **Corey, D. P. and Hudspeth, A. J.,** Mechanical stimulation and micromanipulation with piezoelectric bimorph elements, *J. Neurosci. Methods,* 3, 183, 1980.
24. **Cornwell, P. B.,** *The Cockroach,* Vol. 1, Hutchinson, London, 1968, 391.
25. **Sláma, K.,** Insect haemolymph pressure and its determination, *Acta Entomol. Bohemoslov.,* 74, 362, 1976.
26. **Sláma, K.,** Recording of haemolymph pressure pulsations from the insect body surface, *J. Comp. Physiol.,* 154B, 635, 1984.
27. **Harker, J. E.,** Control of diurnal rhythms of activity in *Periplaneta americana* (L.), *Nature (London),* 175, 733, 1955.
28. **Mengle, D. C. and Casida, J. E.,** Biochemical factors in the acquired resistance of house flies to organophosphate insecticides, *J. Agric. Food Chem.,* 8, 431, 1960.
29. **Sláma, K. and Miller, T. A.,** Insecticide poisoning: disruption of a possible autonomic function in pupae of *Tenebrio molitor, Pestic. Biochem. Physiol.,* 29, 25, 1987.
30. **Bell, W. J. and Adiyodi, K. G., Eds.,** *The American Cockroach,* Chapman and Hall, New York, 1981.
31. **Vero, M. and Miller, T.,** Sensitive tension and force transducer, *Med. Biol. Eng. Comput.,* 17, 662, 1979.

Chapter 11

CELLULAR RESPONSES TO NERVE INJURY IN THE COCKROACH NERVOUS SYSTEM

Robert M. Pitman

TABLE OF CONTENTS

I. INTRODUCTION

The limited ability of the human nervous system to achieve functional recovery following trauma presents a major clinical problem. Strictly speaking, for full recovery to occur, individual neurons should be able to re-form the connections that have been interrupted by a lesion. This, of course, is impossible if neurons have been damaged irreversibly, since neurons in the adult nervous system are no longer capable of cell division to replace any cell loss. However, even following a lesion that interrupts axonal pathways without causing any significant neuron destruction, recovery is normally far from complete; the reasons for this are apparently different in the human central and peripheral nervous systems. Although damaged mammalian peripheral nerves apparently possess considerable potential for axonal growth, functional recovery is often very poor after peripheral nerve trauma because the regenerating fibers show only a weak preference for their original target and, as a consequence, many inappropriate synapses may be formed.[1] Restricted regeneration in the central nervous system (CNS), on the other hand, appears to result largely from failure of central neurons to extend over long distances within the CNS. Although a number of hypotheses have been proposed to account for abortive regenerative growth within the CNS,[2] the causes have not been established. Failure of axons in the adult mammalian CNS to regenerate over long distances is not, however, a result of an inherent inability of central neurons to produce axonal elongation; axons of central neurons can extend over considerable distances when directed into peripheral nervous tissue grafts.[3-5]

In order to understand the mechanisms by which regenerative growth may occur, many research workers have directed their attention to animal preparations in which the nervous system has a well-developed capacity for regenerative recovery. Through studies on such preparations the mechanisms leading to restoration of appropriate connections may be elucidated. This knowledge may then be applied to determine the causes underlying abortive regeneration in man. Many invertebrates present a number of important advantages as experimental models in which to study regeneration. Their nervous systems are relatively simple by comparison with those of even the simpler vertebrates, even though they operate on most of the same mechanistic principles. Since the nervous systems of many invertebrates contain relatively few large neurons, it has become routine to adopt the "identified neuron" approach. In the context of regeneration, this approach allows a range of methods to be used to investigate responses to damage of a neuron in which the functional and morphological characteristics have been clearly defined. Thus, in such preparations, unlike most vertebrate systems, the need to work with populations of similar but nonidentical neurons is eliminated. Among the invertebrates, insects present a number of practical advantages for studying nerve regeneration. For example, they are extremely robust and can be easily manipulated experimentally. In addition, the insect nervous system is bilaterally symmetrical, allowing a neuron on one side to serve as a control for its contralateral homologue on the other in a variety of different types of experiments.

In this review, attention will be focused on the cockroach as an experimental model for studying cellular consequences of nerve damage. Although a neuron is a functional entity, axonal, dendritic, and somatic responses are discussed separately since each may be considered a different functional compartment within the neuron itself. Nerve lesions may operate through different mechanisms to generate appropriate responses in each compartment of the cell. It should be remembered, however, that different regions of a neuron interact with each other, so that the nature of a response in one compartment doubtless will have far-reaching consequences for other parts of that neuron. Since normal neural function can be maintained only if the chemical environment of the nervous system is controlled closely, changes in satellite cells during repair are also considered since these cells form the blood-brain barrier.

II. REGENERATIVE GROWTH OF AXONS

Although specific reinnervation of denervated targets does not occur in the mammalian nervous system, many lower vertebrates, such as amphibians, fish, and birds, are capable of extensive regeneration. Much work has been done on the re-formation of retinotectal connections after lesions to the optic pathways of lower vertebrates. Such studies have shown that following regeneration the topographical projection of retinal axons onto the optic tectum map is restored. It was originally proposed that this restoration could occur because retinal axons and their tectal target neurons carry complementary surface labels which confer upon them a specific chemoaffinity for each other.[6] However, several different lines of evidence, including experiments in which size disparities are introduced between the retina and tectum, now indicate that the topographical projection results from competition between retinal axonal terminals which tend to array themselves across the tectum such that they maintain their neighbor relationships.[7]

During regeneration of the neuromuscular system in lower vertebrates, as in the CNS, competition between neurons may be important in enabling muscles to become reinnervated by the ''correct'' motoneurons.[8-11] A ''foreign'' nerve can form neuromuscular junctions if it is directed into a urodele limb muscle after removal of the normal innervation. However, as the original nerve reinnervates the muscle, ''foreign'' synapses show a reduction in quantal content, indicating that release of neurotransmitter from presynaptic terminals of ''foreign'' motor axons has been suppressed.[12] However, the mechanisms responsible for determining the pattern of neuromuscular connections that are reestablished during regeneration in lower vertebrates are still unclear. While it is possible that there is an intrinsic tendency for muscles to be reinnervated by the ''correct'' motoneurons, the connections that form during regeneration may be controlled by other factors, such as whether or not a motoneuron is supporting an overexpanded peripheral field. According to the latter proposal, a motoneuron with an enlarged peripheral innervation field would be unable to compete successfully with a motoneuron having a reduced field.[13]

Evidence has been available for some time which indicates that damaged cockroach motor axons are able to regenerate successfully, producing a high degree of functional recovery in the majority of animals.[14-17] Return of function in the coxal depressor muscles is achieved by specific reinnervation with the appropriate ''fast'' motoneuron (D_f) and probably also by the original ''slow'' motoneuron (D_s).[18] The specificity with which neuromuscular junctions are formed during regeneration, however, is not absolute, since coxal depressor muscles of a metathoracic leg transplanted into the mesothoracic segment become innervated specifically by mesothoracic homologues of the motoneurons which would normally innervate them.[19] Coxal depressor motoneurons are also able to selectively innervate contralateral homologues of their normal target muscles if nerve 5 (which contains the axons of these neurons) is cut and directed across the midline.[20,21]

Although the studies described above indicate that the normal innervation pattern of coxal muscles may be reestablished eventually, they do not address the question of the manner in which this is achieved. Denburg et al.[22] studied reinnervation of coxal depressor muscles by backfilling regenerating nerves with cobalt at a range of time intervals after lesions were created at different locations on nerve 5 of the metathoracic leg (Figure 1). Using this method, they found little evidence of axonal growth within 12 d following nerve lesioning. Once axonal elongation had started, the growth rate appeared to be approximately 0.9 mm/d. These workers also presented evidence that regenerating axons of coxal depressor motoneurons frequently grew along nerve 5 some distance distal to the branch (nerve 5 ramus 1 [N5r1]) into which they normally would travel to their target muscles. In addition, some unidentified neurons whose axons normally would not enter N5r1 initially did so during regeneration. The number of motoneurons with ''misdirected'' axons fell off with increasing

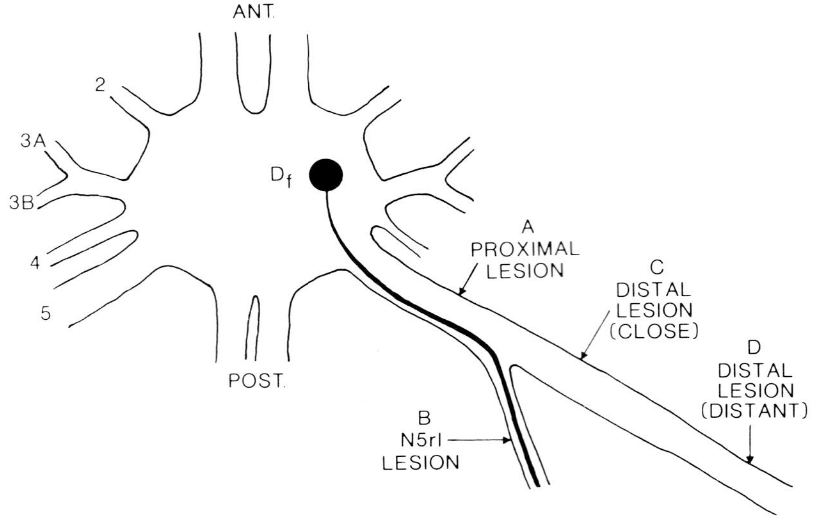

FIGURE 1. Diagram showing the different positions at which nerve 5 of the cockroach metathoracic ganglion may be interrupted to gain information about the responses of coxal motoneurons to damage. The paths taken by regenerating axons of coxal motoneurons have been determined by cobalt backfilling through nerve 5 cut at each of the positions A to D. Cobalt staining was performed at different times after a nerve crush at position A.[22] Lesions also have been performed at positions A to D to study dendritic sprouting in the coxal depressor motoneuron D_f. (See text for details.) In normal, unoperated preparations, coxal depressor motoneurons would be axotomized by lesions at position A or B, while motoneurons to distal limb muscles would be axotomized by lesions at A, C, or D. A lesion at B would axotomize the coxal depressor motoneurons while sparing the axons of distal limb motoneurons. Lesions at C or D would spare the axons of coxal depressor motoneurons, but axotomize distal limb motoneurons. Abbreviations: D_f = "fast" coxal depressor motoneuron; N5rl = nerve 5 ramus 1 (the first nerve branch leaving the main trunk of nerve 5); ANT. = anterior connectives; POST. = posterior connectives. Numbers 2 to 5 indicate the numbering of peripheral nerves leaving the cockroach metathoracic ganglion.

time after the nerve lesion. Further evidence that regenerating cockroach leg motoneurons do not regenerate unerringly along the appropriate trajectory to reestablish connections with the correct target muscle has been provided by electrophysiological experiments.[23] In the initial stages of regeneration, both the D_f and D_s motoneurons innervated incorrect as well as correct coxal depressor muscles in the majority of preparations. In many cases, coxal depressor muscles also received synaptic contacts from unidentified "foreign" motoneurons that could elicit small excitatory junction potentials in these muscles. The results also indicated that the incidence of inappropriate connections made by coxal depressor and unidentified "foreign" motoneurons declined with increasing recovery time after nerve lesions. These observations, therefore, like those based on cobalt backfilling,[22] are consistent with the conclusion that regenerating motoneuron axons that take the wrong path or make incorrect synaptic connections may be eliminated selectively.

More recently, reinnervation of coxal depressor muscle 178[24] has been studied by determining which motoneuron cell bodies become stained after injecting horseradish peroxidase-conjugated wheat germ agglutinin (WGA-HRP) into the muscle.[25] Muscle 178 was selected since it normally receives innervation only from motoneuron D_f. Following nerve crushing, evidence for muscle reinnervation was obtained within 5 d, suggesting that the apparent 12-d lag before the onset of axonal outgrowth reported previously[22] was attributable to the relative insensitivity of the cobalt backfilling method. Another interesting observation made in this study was that after nerve 5 was crushed bilaterally, injection of WGA-HRP into muscle 178 on one side resulted in bilateral staining of motoneuron cell bodies in the metathoracic ganglion. Furthermore, bilateral cell body staining could also be seen if WGA-HRP was injected into a nondenervated muscle 178 contralateral to a crushed nerve 5. It was concluded that these observations could be accounted for most readily if motoneurons sent out processes bilaterally during regeneration. There was a progressive decrease in the number of inappropriate contralateral cell bodies that stained as the time interval between nerve crush and intramuscular injection of WGA-HRP was increased. The transience of the contralateral branches was attributed to the presence of the normal innervation on these muscles, with which the "foreign" branches presumably must compete. Perhaps relevant in this context is the observation that if the distal leg segments are removed, motoneurons which normally innervate muscles of distal limb segments via nerve 3B may send axonal branches into coxal depressor muscle 178. Such branches may develop even if nerve 5 is not crushed, so that the normal nerve supply to the muscle remains intact. Following distal limb amputation combined with nerve 5 crush, branches of "foreign" distal limb motoneurons in muscle 178 may persist indefinitely. Conversely, inappropriate contacts made by the D_s motoneuron upon this muscle are eliminated,[26] presumably because distal limb motoneurons have been deprived of their normal target muscles while motoneuron D_s has not.

The foregoing work indicates that reinnervation of cockroach muscles apparently is not achieved through absolute specificity between individual motoneurons and muscles. Instead, the "overextension hypothesis"[13] may account for many of the observations. Although a motoneuron may have some preference for a particular muscle, it may form and maintain neuromuscular synapses more avidly as the number of connections is reduced. Conversely, the ability to maintain synapses may fall as the total synaptic load increases. A motoneuron with a reduced target load would have a competitive advantage over a motoneuron with an enlarged target. Thus, if regenerating motoneurons grew out of their ganglion bilaterally, as proposed above,[25] they would have to form and maintain extra processes in contralateral limb muscles in addition to terminals in their correct (ipsilateral) target muscle. Muscles contralateral to a lesion are already innervated by the appropriate motoneurons, which would be at a competitive advantage because their undamaged axons would not have extended into "foreign" muscles. Damaged motoneurons which regenerate bilaterally will be overextended and, therefore, will tend to form stable connections only with denervated muscle, while their connections with muscles which already have an intact motor supply will be transient. On the other hand, a motoneuron deprived of its normal target would exhibit an increased tendency to branch into foreign muscles and form persistent connections. Thus, when distal segments of a limb are amputated, motoneurons that normally innervate distal limb muscles will have an increased tendency to form persistent contacts on "foreign" proximal muscles. Although the above results have been discussed in the context of overextension and interneuronal competition, it should be remembered that the observations depend upon neuronal uptake of intramuscularly injected WGA-HRP and that the mechanisms and conditions required for WGA-HRP uptake into nerves in this system are unknown. Therefore, although following injection of WGA-HRP into a particular muscle any motoneuron soma that is stained presumably has axonal branches in that muscle, those branches may not be functional. Thus, branches of different motoneurons that coexist anatomically in a muscle may not be competing functionally.

To directly observe axonal regeneration of individual identified motoneurons, the motoneuron D_f has been studied by intracellular injection of hexamminecobaltic chloride into the nerve cell body at different intervals after nerve lesioning.[27] It was found that at intervals of 1 to 2 d after crushing nerve 5 the crushed end of the axon was frequently indistinct, apparently because cobalt had leaked out into the interstitium. In those cases which allowed clear visualization, no sprouting was visible from the proximal axonal stump. The first sign of any outgrowth was seen on the third postoperative day and took the form of delicate varicose processes emanating from the crushed region of the axon. Thus, the delay between crushing nerve 5 and the onset of axonal outgrowth observed in this study was slightly longer than that reported by Denburg.[25] In the subsequent days a profusion of processes developed from the proximal axonal stump. Except in the immediate region of the lesion, fibers grew axially. In most preparations many processes could be seen growing distally, although a large number of fibers also extended toward the ganglion (Figures 2B and C). Therefore, in the initial stages of regenerative growth, there seems to be little proximodistal preference for growth along nerve 5 for the regenerating axonal processes of motoneuron D_f. By the sixth or seventh postoperative day, fine processes from the axon of motoneuron D_f could be seen entering N5r1, the branch into which this axon normally travels, although, as indicated by the observations of Denburg et al.,[22] many fibers normally grew beyond this point (Figure 2C). At postoperative periods of several weeks, processes frequently could be seen extending along nerve 5 more than 1 mm distal to N5r1. At progressively longer times after nerve crush, processes entering the appropriate nerve branch could be seen to increase in diameter, while "misdirected" processes remained thin and varicose (Figure 2D). The regenerated axon of motoneuron D_f never regained its normal appearance, even at intervals as long as 17 months after crushing nerve 5 (the longest postoperative period tested). Instead of sending a single axon into each subbranch of N5r1, many processes continued to travel in parallel in a single nerve branch. Observations on axonal outgrowth from unidentified motoneurons with axons in nerve 5 indicated that, although these neurons were able to send processes into N5r1, they were less likely to do so than motoneuron D_f (Figure 3). These observations directly confirm the conclusions of Denburg et al.[22] based on backfilling. However, in experiments where nerve 5 received a single crush lesion, contrary to the observations of Denburg,[26] axonal processes were never seen crossing the midline to exit through contralateral peripheral nerves. At this time the explanation for the difference between these two studies is unclear.

The above observations made by intracellular injection of motoneuron D_f suggest that during regeneration the axon produces a profusion of processes which do not grow in a highly directed manner. Processes that take the correct route and presumably form synaptic connections with the target muscle persist and become enlarged, while misdirected processes either remain thin or are lost. The apparent preference of motoneuron D_f for N5r1, therefore, does not appear to result from an ability of its regenerating neurites to grow unerringly along the appropriate path, but rather from being able to sprout profusely soon after damage and then consolidate appropriately directed processes. The fact that a relatively small proportion of regenerating processes from motoneurons to distal leg muscles enter N5r1, while many from motoneuron D_f do so, suggests that regenerating fibers may show some preference in the route they take. This effect, however, is most probably a result of the relative positions of the axons of different motoneurons in the main trunk of nerve 5. The axon of D_f runs on the margin of nerve 5 from which N5r1 originates. Its regenerating processes therefore will have easier access to this small nerve branch than will neurites growing from axons that are normally on the opposite side of nerve 5 from the exit point of N5r1. Thus, the relatively small proportion of "foreign" neurites that grow into N5r1 probably can be ascribed to geometric factors.

III. DENDRITIC CHANGES FOLLOWING LESIONS

In the vertebrate nervous system there is considerable evidence to indicate that dendrites may exhibit significant morphological plasticity both during development and in adulthood. Pyramidal cells of the mammalian cerebral cortex possess spines along their dendrites, each of which represents a synaptic specialization. Any change in the density of these dendritic

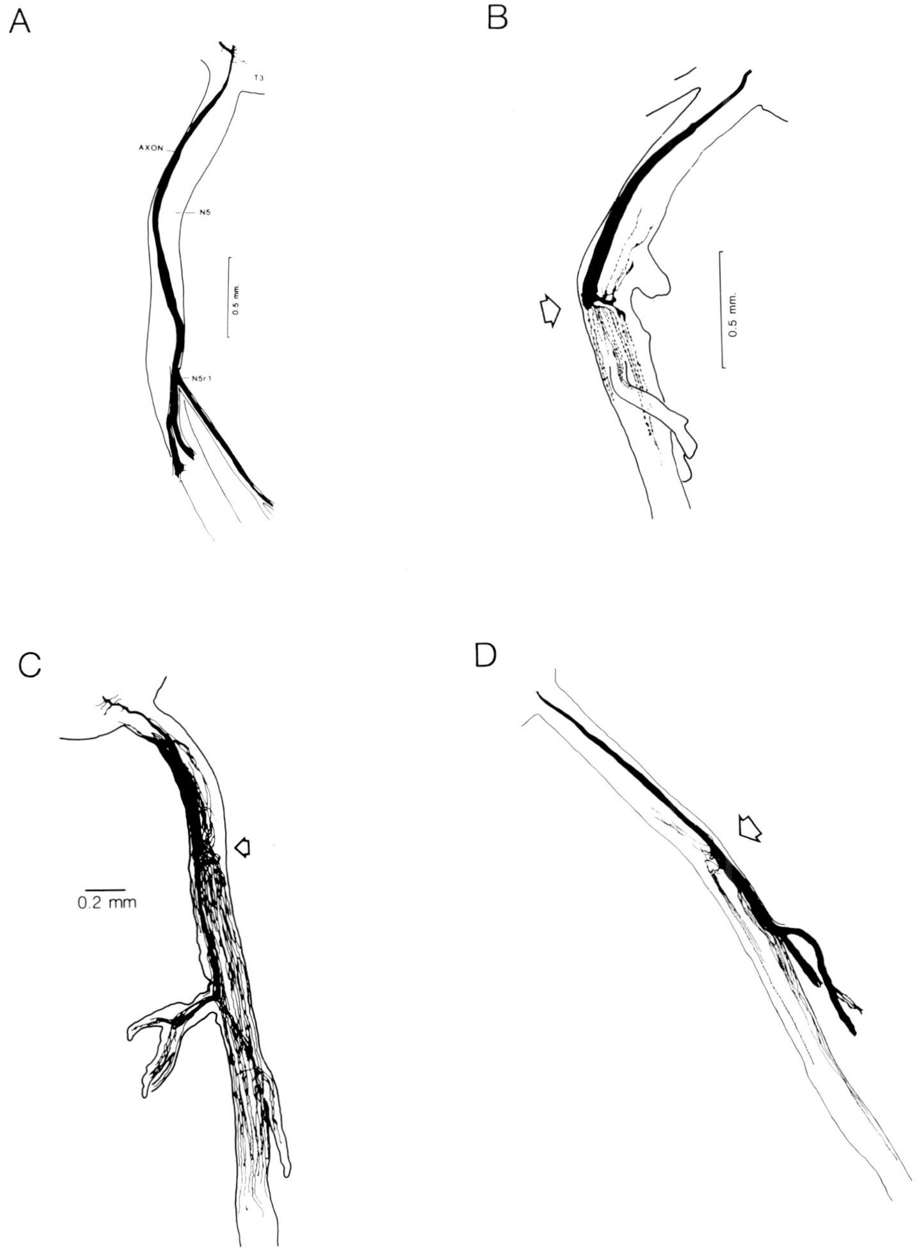

FIGURE 2. Camera lucida drawings showing axonal outgrowth from motoneuron D_f after crushing metathoracic nerve 5 (N5), demonstrated by intracellular injection of hexamminecobaltic chloride. (A) Normal appearance of the axon in an uncrushed preparation. The axon travels along N5 and enters N5r1, in which it divides to enter the individual secondary branches of this nerve. (B) Axonal sprouting observed 7 d after crushing N5 (at arrow). Fine processes travel proximally and distally in the nerve; some fibers enter N5r1, while others extend beyond it. (C) Axonal outgrowth 14 d after nerve crush (at arrow); numerous fibers enter N5r1, while misdirected fibers can be seen growing beyond the branch point of N5r1. Processes also extend proximally as far as the ganglion (the margin of which is shown at the top of the figure). (D) Axon of D_f 103 d after crushing N5 (at arrow). Many fibers enter each branch of nerve N5r1. Those which have done so have a larger diameter than misdirected fibers. Calibrations for panels A and D are the same.

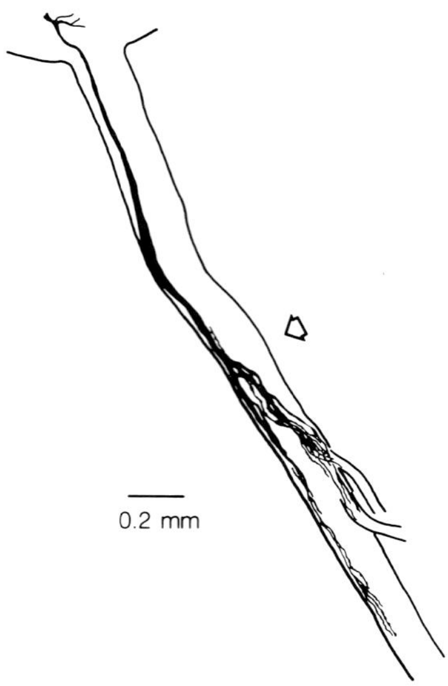

FIGURE 3. Camera lucida drawing of axonal sprouting in an unidentified distal limb muscle motoneuron 42 d after nerve 5 was crushed (at arrow). The neuron was stained by intracellular injection of hexamminecobaltic chloride. Some processes can be seen near the junction of N5r1 with the main trunk of nerve 5; other nerve fibers extend beyond the branch point.

spines is therefore a measure of the number of presynaptic fibers that converge upon the pyramidal cell dendritic tree. At birth, dendritic spines are absent from pyramidal cells of the cerebral cortex in kittens and rats. The number of spines normally increases in postnatal life. This increase is, however, under environmental influence. There is a reduction in the number of dendritic spines on apical dendrites of pyramidal cells in the visual cortex of mice reared in darkness from birth until 22 to 25 d.[28,29] On the other hand, pyramidal cells in the visual and auditory cerebral cortices of young rats exposed to a diverse battery of sensory stimuli on a regular daily basis exhibited an increase in the number of spines relative to neurons of unstimulated controls. Sensory experience had a more marked effect upon basilar and oblique dendrites than upon apical dendrites.[30]

In addition to changes observed during postnatal development, the dendrites of vertebrate neurons may be affected morphologically by lesions suffered in adulthood. Axotomy is normally associated with dendritic retraction. Cerf and Chacko[31] showed that 8 to 11 d after bullfrog motoneurons were axotomized both the number and diameter of their dendrites were reduced. Similar observations have been made about mammalian motoneurons,[32,33] which show a similar dendritic response after their terminals have been treated with botulinum toxin (which produces a similar response in the neuron cell body to that of axotomy; see Section IV below). Reexpansion of the dendritic arbor was dependent upon reinnervation of muscle. If contact of the regenerating nerve with denervated muscle was delayed, dendritic spread did not occur until the time of reinnervation. This change in the extent of the dendritic arbor is accompanied by suppression of synaptic transmission to the injured neuron[34-37] and insertion of glial cells between the injured neuron and its presynaptic boutons. Synapses on the soma are affected to a greater extent than those on dendrites.[33,38]

Deafferentation in adult mammals, like axotomy, is normally associated with a reduction in the extent of dendritic arborizations. For example, lesions at different levels in the olfactory pathway produce significant alterations in the morphology of postsynaptic neurons. Observations on the rabbit accessory olfactory bulb, using Bodian's protargol staining method, have shown that dendrites of mitral and tufted cells undergo atrophy within 1 month of when they are deafferented by destruction of the olfactory mucosa. This atrophy consists of a reduction in number, length, and diameter of dendrites.[39] Interruption of fibers from the olfactory bulb to the prepiriform cortex of the rat causes a reduction in higher order dendritic branches of pyramidal cells in the denervated cortex.[40] Dendritic changes also have been reported following deafferentation of other regions of the mammalian CNS.[41]

Dendritic morphology of at least some neurons in immature insects, like those of mammals, is influenced by the connections they receive. For example, dendrites of cricket interneurons deprived of their presynaptic nerve fibers at nymphal stages remain smaller than normal.[42-45] The effects of partial deafferentation are very localized, since dendritic regions that are not deafferented are indistinguishable from those of control neurons.[42-44] At least one type of cricket interneuron, however, is capable of exhibiting dendritic growth in response to deafferentation.[46] If the distal segments of one prothoracic leg are amputated soon after hatching from the egg, the CNS will be unilaterally deprived of acoustic input, as the ears (which normally do not begin to develop until the third nymphal instar) are located on the tibial segment of the prothoracic legs. A pair of identified acoustic interneurons (interneurons 1) normally have a dendritic arbor restricted to one side of the prothoracic ganglion. This arbor receives input exclusively from the ipsilateral ear. After removal of this ear, the interneuron 1 deprived of its synaptic input projects dendritic branches contralaterally, where they selectively form functional connections with acoustic afferent neurons.[44]

Early experiments on neurons of adult insects, however, indicated that their dendrites apparently showed less plasticity than did dendrites of mammalian neurons. Although there was evidence that axotomy could be followed by changes in efficacy of synaptic input,[47] studies on the morphology of dendrites indicated that in adult animals they remained stable following either axotomy or deafferentation.[48] In fact, on the basis of such experiments, it had been suggested that neurons of adult insects may have become specialized for stability to enable the generation of stereotypical behaviors characteristic of many insects.[49] Cause to doubt these early observations first appeared when it was found that extensive dendritic sprouting could be induced in adult insect neurons by lesions involving a number of nerve trunks leaving a ganglion.[50] Under these circumstances it was observed that the motoneuron D_f could undergo extensive dendritic sprouting (Figure 4). Supernumerary dendrites frequently originated at or near the ends of branches of the normal dendritic arbor. These new branches could extend well outside the normal territory of this neuron, traveling into connectives or into segmental nerves which under normal circumstances would never contain either axonal or dendritic branches of this neuron. Near their origins, dendritic sprouts frequently had extremely convoluted trajectories, although at greater distances from their origin their course frequently straightened, suggesting that they were perhaps following preexisting pathways within the tissue. Although it was originally presumed that damage to a neuron would probably be the most effective stimulus of dendritic sprouting, it soon became apparent that some lesions which did not directly damage the axon of the motoneuron could serve as an equally effective stimulus.

Even though it is important to understand the factors which might trigger growth of supernumerary dendrites, the lesions used to induce dendritic sprouting in early experiments were sufficiently extensive to make a detailed examination virtually impossible. Fortunately, more recent work has indicated that extensive sprouting can take place after far more restricted lesions than those used originally. Although dendritic sprouting normally does not occur in motoneuron D_f after single cut or crush lesions to peripheral nerves through which the axon of this motoneuron runs, extensive sprouting can be observed after ligation or repetitive crushing of these peripheral nerves. It appears that development of supernumerary dendrites critically depends upon delaying the process of repair. If, for example, a single nerve trunk was simply crushed or cut, regeneration could occur relatively rapidly. In such preparations,

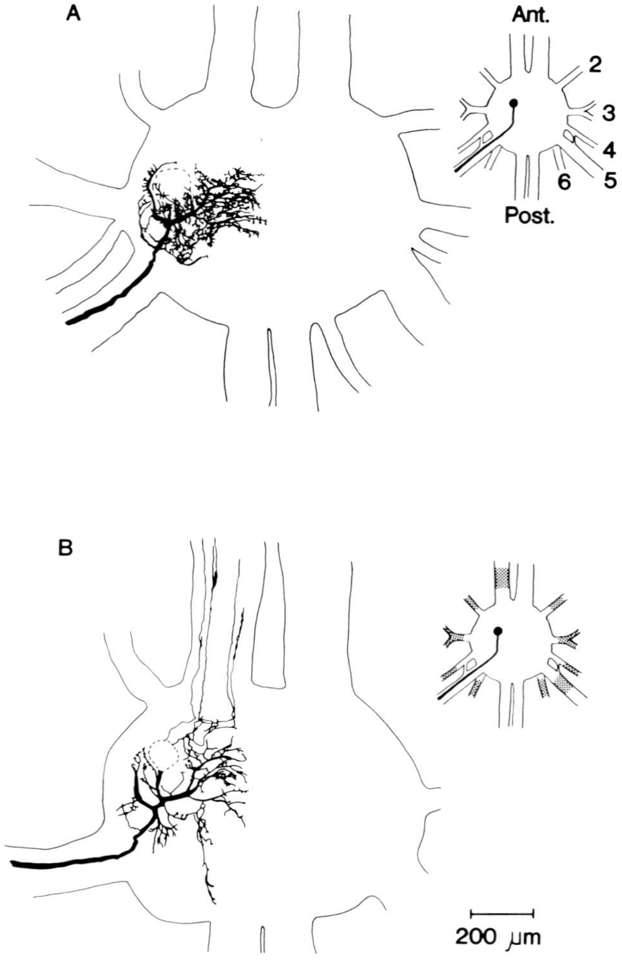

FIGURE 4. Camera lucida drawings of motoneuron D_f stained by intracellular cobalt injection, showing dendritic sprouting after lesions were made on peripheral nerves. The motoneuron soma is indicated by a dotted line to avoid obscuring dendritic branches. (A) Neuron from a preparation in which the metathoracic ganglion had been exposed and the tracheal supply over the ganglion disrupted 35 d before staining. The dendritic arbor is indistinguishable from that of D_f in unlesioned preparations. (B) Neuron stained 21 d after cutting the nerve trunks, indicated by stippling in the inset. The axon of the stained neuron was not severed. Some dendrites of the motoneuron extend into the anterior connectives, while others extend within the ganglion beyond the limits of the normal arbor. Abbreviations in inset: Ant. = anterior connective; Post. = posterior connective. Peripheral nerves numbered according to standard nomenclature. (From Pitman, R. M. and Rand, K. A., *J. Exp. Biol.*, 96, 125, 1982. With permission.)

sprouting only occurred in a small minority of neurons studied, and even then its extent was very limited. On the other hand, if the same nerve trunk was ligated so that motor axons were physically prevented from regaining contact with their peripheral targets and sensory neurons (which in insects have peripherally located cell bodies) were kept from making connections with their central targets, extensive dendritic sprouting occurred in the majority

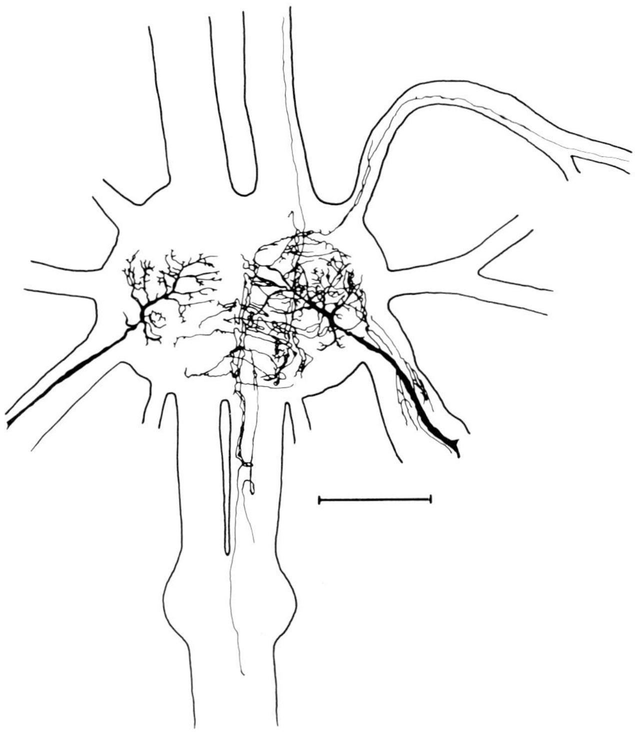

FIGURE 5. Camera lucida drawing of a pair of motoneurons D_f, stained by intracellular cobalt injection 44 d after a ligature was placed on the right nerve 5 (at position A in Figure 1). The dendrites of the motoneuron ipsilateral to the ligation show extensive intraganglionic sprouting and extend into the connectives and nerve 2 ipsilateral to the lesion. The dendrites of the motoneuron contralateral to the lesion show no sprouting. The somata of the two neurons have been omitted to provide an unobscured view of the dendritic branches. (Calibration: 0.5 mm.)

of neurons studied. Figure 5 shows a typical example of dendritic sprouting in motoneuron D_f following ligation of nerve 5 proximal to the exit point of ramus 1 containing the axons of the cell.

Using this model to routinely induce dendritic sprouting in adult neurons following relatively restricted lesions, efforts have been made to determine whether there was any one stimulus that could account for dendritic sprouting in this motoneuron. Therefore, the factors most likely to trigger dendritic sprouting have been listed to help establish which may be consistent with the experimental observations:

1. Direct damage to the neuron itself
2. Interruption of contact with target muscles
3. Partial deafferentation of the motoneuron
4. Release of "factors" from damaged muscle
5. Chemical signals resulting from glial cell proliferation around the site of the nerve lesion
6. General trauma resulting from damage associated with operations (e.g., disruption of the tracheal supply to the tissues)

To distinguish between these different alternatives, lesioning was performed as illustrated in Figure 1. Thus, confirmation was obtained for earlier observations that extensive dendritic

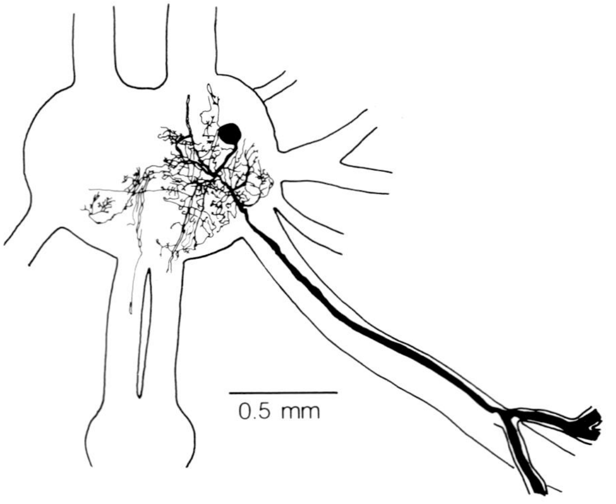

FIGURE 6. Camera lucida drawing of motoneuron D_f stained by intracellular cobalt injection 45 d after a ligature was placed on the right-hand nerve 5 distal to N5r1 (at position C in Figure 1). Although this lesion did not axotomize the neuron, its dendrites have sprouted ipsilaterally and contralaterally within the metathoracic ganglion, and one process extends into the contralateral posterior connective.

sprouting could occur in the absence of direct surgical damage to the neuron itself.[50] A ligature placed on nerve 5 distal to the nerve branch (N5r1) through which the axon of motoneuron D_f exits (lesion C, Figure 1) was able to induce dendritic sprouting (Figure 6). Sprouting observed under these circumstances could not be attributed to trigger candidates 1, 2, or 4. Neither general trauma (option 6) nor any other systemic effect appeared to be primarily responsible for sprouting. No supernumerary dendrites were formed following sham operations in which the metathoracic ganglion and nerve 5 were exposed and the tracheal supply disturbed while all nerves were left intact (Figure 4A). If sprouting were triggered by a systemic mechanism, sprouting normally would occur bilaterally. However, dendritic sprouting only was found ipsilateral to unilateral lesions (see Figure 5). Furthermore, it is also hard to accept the proposal that partial deafferentation constitutes the only stimulus for sprouting (option 3). Ligation of the minor nerve branch, into which the axon of D_f travels (N5r1) and which contains few sensory axons, also can lead to dendritic sprouting. It therefore appears unlikely that partial deafferentation is the trigger for growth of supernumerary dendritic branches in these experiments. To gain further evidence in this respect, motoneuron 3 (a basalar motoneuron of the hindwing, identical to cell 3 in the map of Cohen and Jacklet[51]) also has been studied. The axon of this neuron travels in nerve 4, which has a purely motor function.[52] Therefore, if dendritic sprouting were observed in this neuron after ligation of nerve 4 it could not be attributed to partial deafferentation. In fact, ligation of nerve 4 constituted a very strong stimulus for dendritic sprouting in cell 3. Therefore, in the above series of experiments, the only single mechanism among the proposed candidates that could account for dendritic sprouting under all the experimental conditions would be a signal related to a glial injury reaction (option 5). An alternative proposal, however, is that sprouting may be initiated by more than one type of trigger mechanism. Lesions that affect neurons in different ways therefore may induce sprouting via differing mechanisms. Thus, axotomy may induce sprouting by one mechanism or a group of them, while partial deafferentation may operate through others.

Another example of dendritic sprouting in an adult insect neuron has been reported for an identified giant interneuron of the cricket *Acheta domesticus*. Dendritic sprouting has been observed several days after axotomy performed close to the terminal abdominal ganglion in which the cell body of this interneuron is located.[53] There is apparently a reciprocal relationship between the extent of dendritic and of axonal sprouting after this type of lesioning. Thus, neurons which exhibit extensive axonal sprouting generate little dendritic sprouting, while cells in which many supernumerary dendrites are present have few axonal sprouts. The authors concluded from their observations that axotomized neurons produce new membrane at a constant rate and that whether this is used to produce axonal or dendritic sprouts may be determined by the extent of retrograde degeneration of the axon after a lesion has been made. When axonal "die back" is marked, most new membrane will be used to produce new dendritic branches. At the present time it appears that dendritic sprouting may be regulated by different factors in giant interneurons of the cricket and in cockroach motoneurons. Dendritic sprouting only occurs in the majority of cricket giant interneurons when axotomy is performed very close (200 to 300 μm) to the cell body. Sprouting is never observed after an interganglionic connective has been cut at a distance >2.0 mm from the terminal ganglion, even if axonal growth is abortive. Conversely, in cockroach motoneuron D_f, dendritic sprouting is only seen after sustained lesioning (repeated nerve crushes or ligations). These, however, can be very effective even when made at distances >2.0 mm from the neuron soma.

IV. SOMATIC RESPONSES TO INJURY

When the axon of either a vertebrate or invertebrate neuron is damaged, there frequently follows a constellation of changes in the soma which may be termed the "axon reaction" or "cell body reaction". Changes may include a major redistribution of RNA within the cell body cytoplasm, an increase in nucleolar and nuclear volume, movement of the nucleus to the periphery of the cell body, and swelling of the cell body.[54-56] The relative timing of each of these somatic changes may vary between different types of neuron. Although the functional significance of most of the above changes is still not entirely clear, RNA changes have received a great deal of attention, as the cell body is the major site of protein synthesis within a neuron. Alterations in RNA handling may reflect changes in protein synthesis which may be significant for regeneration.

The exact nature of the change in distribution of RNA seen histologically varies considerably from one type of neuron to another. The cell body cytoplasm of many vertebrate neurons normally contains aggregations of basophilic "Nissl substance" which give it a granular appearance. This Nissl substance consists of ribosomes attached to cisternae of rough endoplasmic reticulum. Classically, the response of vertebrate neurons to axotomy is known as "chromatolysis" and consists of a dispersion of Nissl substance, causing the cell body cytoplasm to lose its granular appearance.[57] Ultrastructurally, these changes are seen as disorganization of rough endoplasmic reticulum and an increase in the number of free polyribosomes. In most neurons the response develops within 2 to 3 d of axonal injury, but is dependent upon the site and exact nature of the axonal lesion. At this time there is an increase in incorporation of nucleosides into RNA,[58-62] while RNA breakdown is also enhanced.[61] After axonal injury most neurons show an overall increase in protein synthesis. This, however, does not represent a global stimulation of synthetic activity, as synthesis of some individual proteins may undergo a decline. Among those proteins that show reduced synthesis are enzymes involved in neurotransmitter metabolism. In cholinergic neurons, acetylcholinesterase[63-65] and choline acetyltransferase[66] production may fall, while in aminergic neurons there may be a fall in dopamine β-hydroxylase, tyrosine hydroxylase, and monoamine oxidase synthesis.[67-69] On the other hand, synthesis of protein components of the cytoskeleton increases significantly.[70-72]

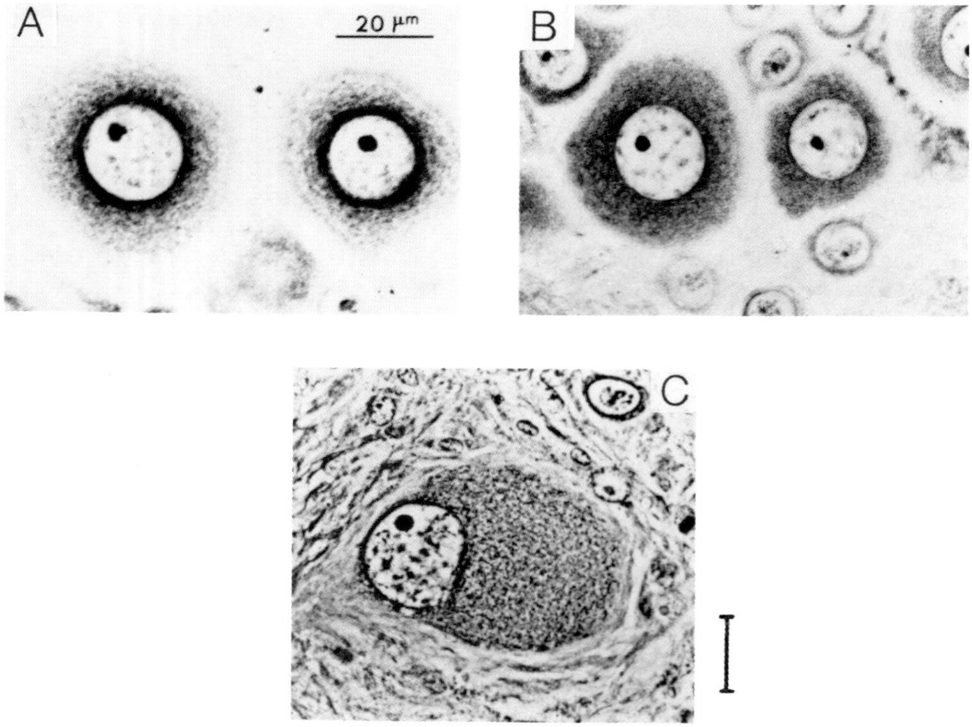

FIGURE 7. Effect of axotomy on the histological appearance of cockroach motoneurons. Sections were stained with pyronine-malachite green to show RNA. (A) Cells which had been axotomized 2 d before fixation, showing a densely staining shell of material around the nucleus. (B) Bilaterally matched control cells, the axons of which remained intact up to the time of fixation. These cells lack any densely staining perinuclear shell. (C) Cell body of a neuron 24 d after axotomy. By this stage, there is no longer a perinuclear shell of dense staining and the nucleus has taken up an eccentric position. (Calibration: panels A and B are at the same magnification; scale bar in C = 20 μm.) (A and B from Cohen, M. J., in *Chemistry of Learning,* Corning, W. C. and Ratner, S. C., Eds., Plenum Press, New York, 1967, 407; C from Jacklet, J. W. and Cohen, M. J., *Science,* 156, 1640, Copyright 1967 by the AAAS. With permission.)

The response of insect nerve cell bodies to axotomy differs in some respects from the classical vertebrate pattern. Although it has been reported that insect neurons normally lack Nissl granules,[73-76] Byers[77] was able to observe Nissl granules in the cell bodies of a pair of identified neurons in the metathoracic ganglion of the cockroach *Diploptera punctata* using both light and electron microscopy. These granules are small by comparison with those observed in many types of vertebrate neuron. Byers[77] considered it likely that the use of relatively thick paraffin sections had caused Nissl granules to be overlooked. In such sections the cell body cytoplasm of intact neurons appears uniformly stained, with the exception of a very thin shell of higher density around the nucleus.[76] Within 12 h of axotomy the density and thickness of the perinuclear shell increase markedly, reaching a maximum after 2 to 3 d (Figure 7). After about 2 weeks the distribution of staining material in the cell body has returned to normal. The densely staining perinuclear shell is apparently RNA, since it is absent from sections exposed to ribonuclease before staining.[76] In addition to a change in distribution of basophilic material, at relatively short intervals (about 2 to 5 d) after axonal damage, cell bodies appear swollen and Golgi bodies decrease in size. At longer intervals (about 3 to 5 weeks), the cell body undergoes further changes in histological appearance. The perinuclear basophilia declines, the nuclear and nucleolar volumes increase, the nucleus moves toward the axon hillock region of the cell (Figure 7C), lysosomes become more numerous, and the Golgi bodies, which initially had decreased in size, now enlarge.[76,77]

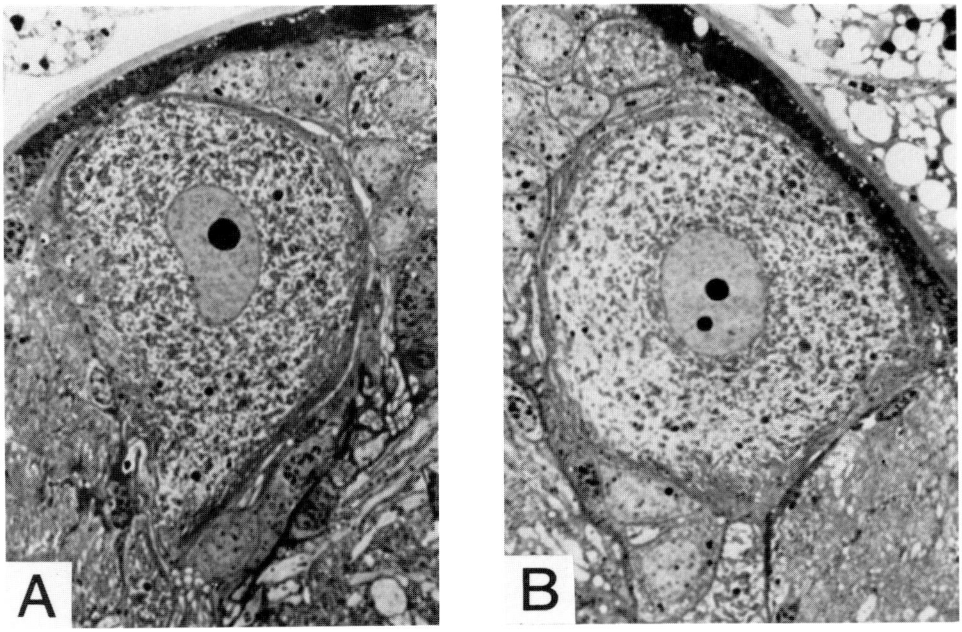

FIGURE 8. Distribution of Nissl granules in cockroach motoneurons. (A) Normal cell, showing even distribution of Nissl granules in the cell body cytoplasm. (B) Soma of a neuron that was axotomized 3 d before fixation. Nissl granules are more dense near the nucleus and less dense near the cell membrane. (Calibration: 25 μm.) (From Byers, M. R., *Tissue Cell*, 2, 255, 1970. With permission.)

There are some differences in ultrastructural changes reported by different laboratories. Young et al.[79] found that although there was a generalized increase in the amount of endoplasmic reticulum in the cell body cytoplasm, accompanied by a dilatation of the individual endoplasmic reticulum cisternae, there was no preferential increase close to the nucleus as might have been expected from the changes in basophilia observed in light microscopy. Byers,[77] however, was able to observe an increase in the size and number of Nissl granules (rough endoplasmic reticulum) close to the nucleus and a corresponding decline at the periphery of the cell body cytoplasm (Figure 8). Autoradiographic data from the same laboratory indicate that the perinuclear basophilia and the change in distribution of Nissl granules are produced by a movement of available RNA rather than from accumulation of newly synthesized RNA around the nucleus. These conclusions, however, are not consistent with preliminary data reported by Cohen,[80] which indicated that the perinuclear shell of RNA seen after axotomy resulted from *de novo* synthesis.

Denburg and Hood[81] have used autoradiography to determine whether insect neurons undergo an increase in protein synthesis after axotomy. They found an increase in grain density over the cell body of axotomized coxal depressor motoneurons (D_f) compared to their contralateral unoperated partners. This difference in grain density first became statistically significant 12 d after axotomy and was still present at 16 d. It is surprising, however, that these authors were unable to detect any similar difference in grain density between the paired somata of a neuron (cell 27) adjacent to D_f. This should have been axotomized at the same time as D_f since, according to the ganglionic map of Cohen and Jacklet,[51] its axon travels in the same peripheral nerve as that of D_f. However, Denburg and Hood[81] state that the axon of this neuron runs in a different peripheral nerve trunk.

The significance of the above changes in the appearance of the cell body has attracted some attention and speculation. At the time when much of the work was done on the "axon reaction" in insect neurons, it was believed that there was a delay of several weeks after an axonal lesioning before the onset of any axonal outgrowth.[14,15,17,22] On this basis, Jacklet and Cohen[17] suggested that the perinuclear shell of RNA had to disperse before any significant synthesis of new proteins could occur. At this stage axonal sprouting could begin. The observations of Denburg and Hood[81] confirm that the first detectable increase in protein synthesis occurs after the perinuclear accumulation of RNA has dispersed. On the other hand, the results on axonal sprouting of the motoneuron D_f, which have been discussed previously in this review, indicate that axonal sprouting begins within 3 d of axotomy, well before the dispersion of RNA or the first detectable signs of an increase in protein synthesis. Significant sprouting takes place within 1 week of injury, while at least some axonal processes may regain physical contact with their target muscle within 2 weeks. However, although extensive axonal outgrowth occurs relatively soon after axonal injury, most new processes are extremely fine at this stage. It is therefore likely that the increase in protein synthesis, which first becomes detectable 12 d after injury,[81] may be primarily associated with an increase in the diameter of regenerated fibers and the elaboration of new neuromuscular junctions rather than with supplying materials for the initial phase of axonal extension.

The nature of the signal or signals that trigger(s) the cell body reaction has been the subject of a great deal of interest and speculation (see References 82 and 83 for reviews). Although it was originally thought that axonal injury may produce a single signal that triggers the cell body reaction, more recently it has become clear that more than one signal may be involved. Bisby[83] has proposed that two main classes of signals are involved: (1) signals resulting from disconnection of the neuron from its target and (2) signals resulting from a fall in the total volume of axoplasm when distal parts of the axon are separated from the cell body. The first signal is assumed to be associated with changes in some trophic influences normally released from target structures. The best model for the operation of such a signal is provided by nerve growth factor (NGF) that in sympathetic neurons is able to prevent at least some, but not all, of the changes associated with axotomy.[84-87] It is possible that the amount of such trophic material available to a neuron is dependent upon the activity of the neuron and its target. The cell body reaction appears in motoneurons either following blockade of impulses with tetrodotoxin[88] or after treatment with botulinum toxin, which prevents acetylcholine release from motoneuron terminals.[69] Such results might be expected if the cell body reaction is normally prevented by trophic substances which are released from target structures, then internalized into neuronal terminals during recycling of synaptic vesicle membrane. If this mechanism were involved, suppression of transmitter release might reduce membrane recycling and sufficiently diminish uptake of trophic substances into nerve terminals to allow the cell body reaction to occur.

Bisby[83] considered the signal generated by a reduction in axoplasmic volume to most likely involve (1) reduction in the level of trophic materials derived from supporting cells, which normally would be transferred to the axon along its entire length; (2) influx of materials at the injury site; or (3) premature return of materials by retrograde axonal transport. Although there is little evidence for or against the first mechanism, it is known that injured axons may show an increase in the uptake of externally applied marker molecules.[90,91] However, uptake of actual signal molecules by this method has not been demonstrated. The observation that colchicine can stimulate the cell body reaction in neurons of the chick ciliary ganglion[92] or in cockroach motoneurons[93] without blocking axonal action potential conduction implies that under these conditions there is no generalized breakdown of membrane permeability. Entry of extracellular signal molecules into the axon would have to involve specific uptake processes. The effects of colchicine are more readily explained in terms of disruption of axonal transport.

Since cockroach motoneurons can exhibit dendritic sprouting in the absence of direct injury,[50] experiments have been performed to determine whether such sprouting is accom-

TABLE 1
Grain Densities in Control and Experimental Neurons Measured from Autoradiograms Obtained after Injection of ^{3}H-Uridine

Neuron	$\frac{\text{Grain density in perinuclear cytoplasm}}{\text{Grain density in peripheral cytoplasm}} \times 100$
1. Control	8.7% ± 2.3 (n = 16)
2. Axotomized	30.8% ± 6.1 (n = 11)
3. Lesioned, but not axotomized	19.0% ± 4.4 (n = 11)

Note: Relative densities of grains in perinuclear and peripheral regions of soma cytoplasm were determined in motoneurons from (1) control animals; (2) animals which had received neural lesions that caused axotomy of the neurons; and (3) animals which had received lesions that did not axotomize the cells, but constituted a strong stimulus for dendritic sprouting.[50] ^{3}H-Uridine was injected 2 weeks before lesioning, and tissue was prepared for autoradiography 5 d after lesioning. Values expressed as means ± standard error; n = number of neurons contributing to each group.

panied by a cell body reaction. If this reaction is normally part of a response to meet an increase in the demand for axoplasmic and membrane components, dendritic sprouting might be expected to be accompanied by somatic changes similar to those observed after axotomy. The lesioning protocol employed was to section one anterior connective and all the segmental nerves exiting the metathoracic ganglion, with the exception of one nerve 5. This type of extensive lesion was used since it had previously been shown to constitute a powerful stimulus for dendritic sprouting.[50] To assess the cell body reaction in insect motoneurons, the distribution of RNA in the cell body cytoplasm of the metathoracic motoneuron D_f was measured autoradiographically. Animals were injected with ^{3}H-uridine 2 weeks before lesioning, since Byers[77] reported that if the labeled uridine was injected at this time the distribution of grains in autoradiograms reflected the perinuclear shell of basophilia seen after axotomy. Measurement of grain distribution constitutes a good index of the cell body reaction. Animals were sacrificed and tissues prepared for autoradiography 5 d after lesioning, at which time the axon reaction produced by axotomy was maximal. Autoradiogram grain densities were determined for the zone of cytoplasm immediately surrounding the nucleus (perinuclear) and for that just below the cell body membrane (peripheral). The grain density of the perinuclear zone was then expressed as a percentage above that in the peripheral cytoplasm (Table 1). Both axotomized neurons and those from ganglia that had received neural lesions which spared their axons showed a statistically significant increase in the relative grain density in the perinuclear cytoplasm relative to control neurons (p <0.001 and 0.05, respectively). On the other hand, there was no statistically significant difference between the two experimental groups (p >0.1). These experiments demonstrate that the cell body reaction can be stimulated in an identified insect motoneuron either by axotomy or by lesions which induce dendritic sprouting without axotomy. It therefore may be concluded that the cell body response studied here, like the cell body reaction observed in some other preparations, may be triggered by stimuli other than axonal damage. Although the exact nature of the stimulus is still uncertain, the observations are consistent with a trigger related to an increased biosynthetic demand.

V. ROLE OF SATELLITE CELLS IN REPAIR

There is evidence to indicate that, in at least some preparations, nonneuronal cells such as glia play a crucial role in determining the consequences of damage to the CNS. In mammals, for example, glia may have a significant influence on axonal growth. First, it has been proposed that glia may form a scar in the damaged tissue through which regenerating axons are unable to grow.[94] Second, normal glia in the CNS may provide a poor environment for axonal growth, since mammalian central neurons grow well when directed into peripheral nerve tissue, but exhibit only limited growth within the CNS.[3-5] The converse is also true. Growth of peripheral nerves may be inhibited when they are directed into CNS glial grafts.[95] In addition to influencing the extent of axonal growth, glial cells also may provide a substrate that assists in neuronal guidance. In fish, which have a good capacity for central regeneration, transplanted optic nerves and Mauthner neurons grow along stereotypical routes. It is proposed that they are guided along pathways provided by ependyma and glia.[96,97] After many types of neural lesions it is also important that the blood-brain barrier be restored so that the CNS is protected from materials in the blood circulation. To understand the overall process of CNS repair it is also essential to know the functions and origins of the different cell types involved.

The insect CNS serves as a good model for studying the function of supporting cells during recovery from neural damage, since these cells can undergo proliferation at lesion sites.[98] In addition, the insect CNS normally has a well-developed blood-brain barrier which can be disrupted by lesions, but which can subsequently be repaired. The insect CNS is surrounded by a nerve sheath which consists of two components, the outer acellular neural lamella and an inner cellular layer termed the perineurium. The blood-brain barrier is formed by perineurial glial cells which are coupled together by occluding junctional complexes.[99] The insect blood-brain barrier prevents access of water-soluble cations.[100,101] If undamaged nerve cords are bathed in high-potassium solutions, axonal action potentials can persist without any reduction in amplitude.

Ultrastructural observations have also demonstrated that the blood-brain barrier excludes lanthanum from undamaged nerve cords.[99,102] However, following sectioning of interganglionic connectives the blood-brain barrier is disrupted, enabling potassium and lanthanum to penetrate the CNS. The ability of the blood-brain barrier to exclude potassium and lanthanum was not restored until 4 to 6 months after the creation of such lesions.[103] During this time a number of ultrastructural changes were visible in the nerve cord. Among the most prominent of these changes were the appearance in the damaged region of large amounts of extracellular matrix material and of cells with prominent electron-dense granules in their cytoplasms. The granule cells, which resemble hemocytes, were particularly abundant in the superficial layers of the nerve cord, but also could be found more deeply. Some of these cells appear to be involved in phagocytosis. In order to further understand the role of different cell types in neural repair, glial cells have been disrupted selectively using ethidium bromide.[104] This drug selectively binds to RNA and DNA. When applied to interganglionic connectives it produces both nuclear and cytoplasmic damage in glia, but has relatively little effect upon the structure or function of axons. The cell bodies of these axons are located in ganglia and not connectives. Following selective glial disruption with ethidium bromide, as after surgical damage to connectives, granule cells appear. Initially, these cells seem to be involved in phagocytosis of cellular debris in the damaged region, but later they appear to become directly involved in repair, especially in the perineurial layers of the nerve sheath. However, responses to glial disruption and to surgical damage differ in several significant ways. Ethidium bromide treatment is followed by neither the increase in the amount of extracellular matrix material nor the extensive glial proliferation seen after surgical damage. Another important difference is that restoration of the ionic selectivity of the blood-brain barrier begins much more rapidly after glial disruption. Lanthanum is excluded as early as 4 d after ethidium bromide treatment, while potassium ion exclusion is reduced within 1 week, but takes about 4 weeks to return to normal. During the period in which the properties of the blood-brain barrier return to normal there are parallel changes in the ultrastructure of the perineurium. Initially, granule cells are present beneath the neural lamella, where they appear to contribute to the perineurium. However, the number of granule cells gradually falls, so that by 1 month after treatment with ethidium bromide these cells are absent and perineurial cells have gained cytological characteristics similar to those of cells in undamaged connectives. It is unclear, however, whether granule cells are gradually replaced by glia or are themselves transformed into glia during repair.

Whichever of the above possibilities turns out to be correct, experimental evidence indicates that granule cells do appear to be intimately involved in the process of repair. Perineurial repair is delayed if the antimitotic drug bleomycin is injected into the hemolymph before glial disruption with ethidium bromide. Under these conditions, cellular debris persists longer in damaged connectives, while the perineurium remains loosely packed and contains large extracellular spaces. Bleomycin also causes a concomitant marked reduction in the number of granule cells in damaged regions.[105] Restoration of the perineurium is also disrupted if inert microspheres are injected into the hemolymph before ethidium bromide treatment. Although the microspheres are rapidly removed from circulation by uptake into circulating hemocytes, they cause marked abnormalities in perineurial ultrastructure. A particularly notable feature is that granule cells are completely absent from the CNS during the repair following injection of microspheres into the hemolymph. On the other hand, other ultrastructural features of the perineurium are similar to those seen in preparations treated with bleomycin. Debris persists longer and cells are arranged loosely, with large extracellular spaces.[106] Since microspheres are unable to gain access to the CNS, they must exert their effects outside the nervous system. Because injecting microspheres into the hemolymph prevents the appearance of granule cells in the damaged perineurium, these cells presumably must originate from outside the CNS, where they would be accessible to the injected microspheres. When damage occurs they must then migrate into the CNS. It is likely that granule cells are derived from hemocytes, since the two cell types have similar cytological characteristics and since uptake of microspheres into hemocytes is associated with failure of granule cell recruitment into the CNS during repair. It is interesting that, despite the ultrastructural abnormalities of the repairing perineurium seen after injection of microspheres into the hemolymph, restoration of the blood-brain barrier follows a normal time course. This suggests that, although granule cells may normally play an important part in contributing to perineurial repair, other types of cells are capable of generating an adequate functional barrier. During repair of the insect blood-brain barrier, therefore, it appears that cells both endogenous and exogenous to the CNS may contribute to complete structural and functional restoration.

VI. CONCLUSION

From the material presented above, several conclusions can be made about the responses of cockroach neurons to nerve damage. The axons of insect motoneurons generate a profusion of new processes within 3 to 4 d of peripheral nerve injury. Initial growth apparently shows little directional preference so that, at relatively short postoperative periods, muscles may receive branches from appropriate and inappropriate motoneurons. As longer times elapse after lesioning, the number of motoneurons that branch into "foreign" muscles declines. Elimination of these contacts made by "foreign" motoneurons may be the result of competition with motoneurons that normally innervate the muscle.

The dendritic arbor of adult cockroach motoneurons is hardly altered by lesions such as nerve crushes that allow subsequent rapid recovery. Dramatic dendritic sprouting can occur, however, after sustained lesions such as ligation of peripheral nerves and under some circumstances that do not damage the cell directly. The effects of different types of experimental lesions suggest that there may be several alternative mechanisms by which dendritic sprouting can be triggered. However, the possibility that sprouting results from a stimulus released by glial cells in damaged peripheral nerves has not been eliminated.

Cockroach motoneuron cell bodies exhibit a histochemical response to axotomy which consists of a redistribution of RNA in the perinuclear cytoplasm. Although the stimulus for this reaction has not been identified, it is unlikely to result from loss of axoplasm following axotomy. The cell body reaction can be observed after treatment with colchicine, which does not break down the axonal membrane. The cell body reaction in cockroach neurons can be evoked not only by axotomy, but also by lesions which spare the axon yet act as powerful stimuli for dendritic sprouting. It is possible, therefore, that the axon reaction in this preparation reflects a response to a change in the biosynthetic needs of the neuron.

The insect nervous system normally has an efficient blood-brain barrier which is restored after neural lesioning. There is now evidence to indicate that cells both endogenous and exogenous to the CNS participate in repair. Glial cells provide the former category, while hemocytes appear to migrate into the CNS, where they may be involved both in the phagocytosis of cellular debris and in the actual repair of the blood-brain barrier.

ACKNOWLEDGMENTS

I thank Dr. K. A. Rand for her critical reading of a previous draft of this review.

REFERENCES

1. **Purves, D.,** Long term regulation in the vertebrate peripheral nervous system, *Int. Rev. Physiol.,* 10, 125, 1976.
2. **Kiernan, J. A.,** Hypotheses concerned with axonal regeneration in the mammalian nervous system, *Biol. Rev.,* 54, 155, 1979.
3. **Richardson, P. M., McGuinness, U. M., and Aguayo, A. J.,** Axons from CNS neurons regenerate into PNS grafts, *Nature (London),* 284, 264, 1980.
4. **David, S. and Aguayo, A. J.,** Axonal elongation into peripheral nervous system "bridges" after central nervous system injury in adult rats, *Science,* 214, 931, 1981.
5. **Aguayo, A. J., David, S., and Bray, G. M.,** Influence of the glial environments on the elongation of axons after injury, *J. Exp. Biol.,* 95, 231, 1981.
6. **Sperry, R. W.,** Chemoaffinity in the orderly growth of nerve fiber patterns and connections, *Proc. Natl. Acad. Sci. U.S.A.,* 50, 703, 1963.
7. **Easter, S. S., Purves, D., Rakic, P., and Spitzer, N. C.,** The changing view of neural specificity, *Science,* 230, 507, 1985.
8. **Cass, D. T. and Mark, R. F.,** Re-innervation of axolotl limbs. I. Motor nerves, *Proc. R. Soc. London Ser. B,* 190, 45, 1975.
9. **Fangbonner, R. F. and Vanable, J. W.,** Formation and regression of inappropriate nerve sprouts during trochlear nerve regeneration in *Xenopus laevis, J. Comp. Neurol.,* 157, 391, 1974.
10. **Bennett, M. R. and Raftos, J.,** The formation and regression of synapses during the re-innervation of axolotl striated muscles, *J. Physiol. (London),* 265, 261, 1977.
11. **Genat, B. R. and Mark, R. F.,** Electrophysiological experiments on the mechanism and accuracy of neuromuscular specificity in the axolotl, *Philos. Trans. R. Soc. London Ser. B,* 278, 335, 1977.
12. **Dennis, M. J. and Yip, J. W.,** Formation and elimination of foreign synapses on adult salamander muscle, *J. Physiol. (London),* 274, 299, 1978.
13. **Landmesser, L. T.,** The generation of neuromuscular specificity, *Annu. Rev. Neurosci.,* 3, 279, 1980.
14. **Bodenstein, D.,** Studies on nerve regeneration in *Periplaneta americana, J. Exp. Zool.,* 136, 89, 1957.
15. **Guthrie, D. M.,** Regenerative growth in insect nerve axons, *J. Insect Physiol.,* 8, 79, 1962.
16. **Guthrie, D. M.,** The regeneration of motor axons in an insect, *J. Insect Physiol.,* 13, 1593, 1967.
17. **Jacklet, J. W. and Cohen, M. J.,** Nerve regeneration: correlation of electrical, histological and behavioral events, *Science,* 156, 1640, 1967.
18. **Pearson, K. G. and Bradley, A. B.,** Specific regeneration of excitatory motoneurons to leg muscles in the cockroach, *Brain Res.,* 47, 492, 1972.

19. **Young, D.,** Specific re-inervation of limbs transplanted between segments in the cockroach, *Periplaneta americana, J. Exp. Biol.*, 57, 305, 1972.
20. **Fourtner, C. R., Holtzmann, T. W., and Drewes, C. D.,** Specific contralateral afferent and efferent regeneration in the cockroach, *Am. Zool.*, 17, 962, 1977.
21. **Fourtner, C. R., Drewes, C. D., and Holtzmann, T. W.,** Specificity of afferent and efferent regeneration in the cockroach: establishment of a reflex pathway between contralaterally homologous target cells, *J. Neurophysiol.*, 41, 885, 1978.
22. **Denburg, J. L., Seecof, R. L., and Horridge, G. A.,** The path and rate of growth of regenerating motor neurons in the cockroach, *Brain Res.*, 125, 213, 1977.
23. **Whitington, P. M.,** The specificity of innervation of regenerating motor neurons in the cockroach, *J. Comp. Neurol.*, 186, 465, 1979.
24. **Carbonell, C. S.,** The thoracic muscles of the cockroach *Periplaneta americana* (L.), *Smithson. Misc. Collect.*, 107, 1, 1947.
25. **Denburg, J. L.,** Elimination of inappropriate axonal branches of regenerating cockroach motor neurons as detected by retrograde transport of horseradish peroxidase conjugated wheat germ agglutinin, *Brain Res.*, 248, 1, 1982.
26. **Denburg, J. L.,** Plasticity in the cockroach neuromuscular system, *Dev. Biol.*, 306, 1985.
27. **Brogan, R. T. and Pitman, R. M.,** Axonal regeneration in an identified insect motoneurone, *J. Physiol. (London)*, 319, 34P, 1981.
28. **Valverde, F.,** Apical dendritic spines of the visual cortex and light deprivation in the mouse, *Exp. Brain Res.*, 3, 337, 1967.
29. **Ruiz-Marcos, A. and Valverde, F.,** The temporal evolution of the distribution of dendritic spines in the visual cortex of normal and dark raised mice, *Exp. Brain Res.*, 8, 284, 1969.
30. **Schapiro, S. and Vukovitch, K. R.,** Early experience effects upon cortical dendrites: a proposed model for development, *Science*, 167, 292, 1970.
31. **Cerf, J. A. and Chacko, L. W.,** Retrograde reaction in motoneuron dendrites following ventral root section in the frog, *J. Comp. Neurol.*, 109, 205, 1958.
32. **Sumner, B. E. H. and Watson, W. E.,** Retraction and expansion of the dendritic tree of motor neurones of adult rats induced *in vivo, Nature (London)*, 233, 273, 1971.
33. **Sumner, B. E. H. and Sutherland, F. I.,** Quantitative electron microscopy on the injured hypoglossal nucleus in the rat, *J. Neurocytol.*, 2, 315, 1973.
34. **Eccles, J. C., Libet, B., and Young, R. R.,** The behaviour of chromatolysed motoneurones studied by intracellular recording, *J. Physiol. (London)*, 143, 11, 1958.
35. **McIntyre, A. K., Bradley, K., and Brock, L. G.,** Responses of motoneurones undergoing chromatolysis, *J. Gen. Physiol.*, 42, 931, 1959.
36. **Shapovalov, A. I. and Grantyn, A. A.,** Nadsegmentarnye sinapticheskie vliianiia na khromatolizirovannye motoneurony, *Biofizika*, 13, 260, 1968.
37. **Kuno, M. and Llinas, R.,** Alterations of synaptic action in chromatolysed motoneurones of the cat, *J. Physiol. (London)*, 210, 823, 1970.
38. **Blinzinger, K. and Kreutzberg, G.,** Displacement of synaptic terminals from regenerating motoneurons by microglial cells, *Z. Zellforsch. Mikrosk. Anat.*, 85, 145, 1968.
39. **Matthews, M. R. and Powell, T. P. S.,** Some observations on transneuronal cell degeneration in the olfactory bulb of the rabbit, *J. Anat.*, 96, 89, 1962.
40. **Jones, W. H. and Thomas, D. B.,** Changes in the dendritic organization of neurons in the cerebral cortex following deafferentation, *J. Anat.*, 96, 375, 1962.
41. **Cowan, W. M.,** Anterograde and retrograde transneuronal degeneration in the central and peripheral nervous system, in *Contemporary Research Methods in Neuroanatomy*, Nauta, W. J. H. and Ebbsson, S. O. E., Eds., Springer-Verlag, New York, 1970, 217.
42. **Murphey, R. K., Mendenhall, B., Palka, J., and Edwards, J. S.,** Deafferentation slows the growth of specific dendrites of identified giant interneurons, *J. Comp. Neurol.*, 159, 407, 1975.
43. **Murphey, R. K., Matsumoto, S. G., and Mendenhall, B.,** Recovery from deafferentation by cricket interneurons after reinnervation by their peripheral field, *J. Comp. Neurol.*, 169, 335, 1976.
44. **Murphey, R. K. and Levine, R. B.,** Mechanisms responsible for changes observed in response properties of partially deafferented insect interneurons, *J. Neurophysiol.*, 43, 367, 1980.
45. **Shankland, M., Bentley, D., and Goodman, C. S.,** Afferent innervation shapes the dendritic branching pattern of the medial giant interneuron in grasshopper embryos raised in culture, *Dev. Biol.*, 92, 507, 1982.
46. **Hoy, R. R., Nolen, T. G., and Casaday, G. C.,** Dendritic sprouting and compensatory synaptogenesis in an identified interneuron following auditory deprivation in a cricket, *Proc. Natl. Acad. Sci. U.S.A.*, 82, 7772, 1985.
47. **Horridge, G. A. and Burrows, M.,** Synapses upon motoneurons of locusts during retrograde degeneration, *Philos. Trans. R. Soc. London Ser. B*, 269, 95, 1974.

48. **Tweedle, C. D., Pitman, R. M., and Cohen, M. J.,** Dendritic stability of insect central neurons subjected to axotomy and deafferentation, *Brain Res.*, 60, 471, 1973.
49. **Cohen, M. J.,** Cellular events in the evolution of behavior, in *Simpler Networks and Behavior,* Fentress J. C., Ed., Sinauer Associates, Sunderland, MA, 1976, 39.
50. **Pitman, R. M. and Rand, K. A.,** Neural lesions can cause dendritic sprouting of an undamaged adult insect motoneurone, *J. Exp. Biol.*, 96, 125, 1982.
51. **Cohen, M. J. and Jacklet, J. W.,** The functional organization of motor neurons in an insect ganglion, *Philos. Trans. R. Soc. London Ser. B,* 252, 561, 1967.
52. **Dresden, D. and Nijenhuis, E. D.,** Fibre analysis of the nerve of the second thoracic leg in *Periplaneta americana, K. Ned. Akad. Versl. Gewone Vergud. Afd. Natuurk.*, 61C, 213, 1958.
53. **Roederer, E. and Cohen, M. J.,** Regeneration of an identified central neuron in the cricket. I. Control of sprouting from soma, dendrites and axon, *J. Neurosci.*, 3, 1835, 1983.
54. **Lieberman, A. R.,** The axon reaction: a review of the principal features of perikaryal responses to axonal injury, *Int. Rev. Neurobiol.*, 14, 49, 1971.
55. **Lieberman, A. R.,** Some factors affecting retrograde neuronal responses to axonal lesions, in *Essays on the Nervous System,* Bellairs, R. and Gray, E. G., Eds., Clarendon, Oxford, 1974, 71.
56. **Grafstein, B.,** The nerve cell body response to axotomy, *Exp. Neurol.*, 48, 32, 1975.
57. **Nissl, F.,** Uber die Veranderungen der Ganglienzellen am Facialiskern des Kaninchens nach Ausreissung der Nerven, *Allgem. Z. Psychiat. Ihre Grenzgeb.*, 48, 197, 1892.
58. **Brattgard, S.-O., Hyden, H., and Sjostrand, J.,** Incorporation of orotic acid-^{14}C and lysine-^{14}C in regenerating single nerve cells, *Nature (London),* 182, 801, 1958.
59. **Porter, K. R. and Bowers, M. B.,** A study of chromatolysis in motor neurons of the frog *Rana pipiens, J. Cell Biol.*, 19, 56A, 1963.
60. **Watson, W. E.,** An autoradiographic study of the incorporation of nucleic-acid precursors by neurones and glia during nerve regeneration, *J. Physiol. (London),* 180, 741, 1965.
61. **Watson, W. E.,** Observations on the nucleolar and total cell body nucleic acid of injured nerve cells, *J. Physiol. (London),* 196, 655, 1968.
62. **Haddad, A., Iucif, S., and Cruz, A. R.,** Synthesis of RNA in neurons of the hypoglossal nerve nucleus after section of the axon in mice, *J. Neurochem.*, 16, 865, 1969.
63. **Flumerfelt, B. A. and Lewis, P. R.,** Cholinesterase activity in the hypoglossal nucleus of the rat and the changes produced by axotomy: a light and electron microscopic study, *J. Anat.*, 119, 309, 1975.
64. **Watson, W. E.,** Quantitative observations upon acetylcholine hydrolase activity of nerve cells after axotomy, *J. Neurochem.*, 13, 1549, 1966.
65. **Sinicropi, D. V., Michels, K., and McIlwain, D. L.,** Acetylcholinesterase distribution in axotomized frog motoneurones, *J. Neurochem.*, 38, 1099, 1982.
66. **Frizell, M. and Sjostrand, J.,** Transport of proteins, glycoproteins, and cholinergic enzymes in regenerating hypoglossal neurons, *J. Neurochem.*, 22, 845, 1974.
67. **Cheah, T. B. and Geffen, L. B.,** Effects of axonal injury on norepinephrine, tyrosine hydroxylase and monoamine oxidase levels in sympathetic ganglia, *J. Neurobiol.*, 4, 443, 1973.
68. **Kopin, I. J. and Silberstein, S. D.,** Axons of sympathetic neurons: transport of enzymes *in vivo* and properties of axonal sprouts *in vitro, Pharmacol. Rev.*, 24, 245, 1972.
69. **Reis, D. J. and Ross, R. A.,** Dynamic changes in brain dopamine-beta-hydroxylase activity during anterograde and retrograde reactions to injury of central noradrenergic neurons, *Brain Res.*, 57, 307, 1973.
70. **Heacock, A. M. and Agranoff, B. W.,** Enhanced labeling of a retinal protein during regeneration of optic nerve in goldfish, *Proc. Natl. Acad. Sci. U.S.A.*, 73, 828, 1976.
71. **Lasek, R. J. and Hoffman, P. N.,** The neuronal cytoskeleton, axonal transport and axonal growth, in *Cell Motility, Book C, Microtubules and Related Proteins,* Pollard, T. and Rosenbaum, J., Eds., Cold Spring Harbor Laboratory, Cold Spring Harbor, NY, 1976, 1021.
72. **Hoffman, P. N. and Lasek, R. J.,** Axonal transport of the cytoskeleton in regenerating motor neurons: constancy and change, *Brain Res.*, 202, 317, 1980.
73. **Hess, A.,** Fine structure of nerve cells and fibers, neuroglia, and sheaths of the ganglion chain in the cockroach *(Periplaneta americana), J. Biophys. Biochem. Cytol.*, 4, 731, 1958.
74. **Wigglesworth, V. B.,** Axon structure and the dictysomes (Golgi bodies) in the neurones of the cockroach *(Periplaneta americana), Q. J. Microsc. Sci.*, 101, 381, 1960.
75. **Ashhurst, D. E.,** The cytology and histochemistry of the neurones of *(Periplaneta americana), Q. J. Microsc. Sci.*, 102, 399, 1961.
76. **Cohen, M. J. and Jacklet, J. W.,** Neurons of insects: RNA changes during injury and regeneration, *Science,* 148, 1227, 1965.
77. **Byers, M. R.,** Chromatolysis in a pair of identifiable metathoracic neurons in the cockroach *Diploptera punctata, Tissue Cell,* 2, 255, 1970.
78. **Cohen, M. J.,** Some cellular correlates of behavior controlled by an insect central ganglion, in *Chemistry of Learning,* Corning, W. C. and Ratner, S. C., Eds., Plenum Press, New York, 1967, 407.

79. **Young, D., Ashhurst, D. E., and Cohen, M. J.,** The injury response of the neurones of *Periplaneta americana, Tissue Cell,* 2, 387, 1970.
80. **Cohen, M. J.,** Correlations between structure, function and RNA metabolism in central neurons of insects, in *Invertebrate Nervous Systems. Their Significance for Mammalian Neurophysiology,* Wiersma, C. A. G., Ed., University of Chicago Press, Chicago, 1967, 65.
81. **Denburg, J. L. and Hood, N. A.,** Protein synthesis in regenerating motor neurons in the cockroach, *Brain Res.,* 125, 227, 1977.
82. **Cragg, B. G.,** What is the signal for chromatolysis?, *Brain Res.,* 23, 1, 1970.
83. **Bisby, M. A.,** Retrograde transport and regeneration studies, in *Axoplasmic Transport,* Iqbal, Z., Ed., CRC Press, Boca Raton, FL, 1986, 249.
84. **Banks, B. E. C. and Walter, S. J.,** The effects of postganglionic axotomy and nerve growth factor on the superior cervical ganglia of developing mice, *J. Neurocytol.,* 6, 287, 1977.
85. **Hamberger, V., Brunsobechtold, J. K., and Yip, J. W.,** Neuronal death in the spinal ganglia of the chick embryo and its reduction by nerve growth factor, *J. Neurosci.,* 1, 60, 1981.
86. **Hendry, I. A.,** The response of adrenergic neurons to axotomy and nerve growth factor, *Brain Res.,* 94, 87, 1975.
87. **Nja, A. and Purves, D.,** Effects of nerve growth factor and its antiserum on synapses in the superior cervical ganglion of the guinea pig, *J. Physiol. (London),* 227, 53, 1978.
88. **Czeh, G., Galego, R., Kudo, N., and Kuno, M.,** Evidence for the maintenance of motoneuron properties by muscle activity, *J. Physiol. (London),* 281, 239, 1978.
89. **Watson, W. E.,** The response of motor neurons to intramuscular injection of botulinum toxin, *J. Physiol. (London),* 202, 611, 1969.
90. **Kristensson, K. and Olsson, Y.,** Retrograde transport of horseradish peroxidase in transected axon. III. Entry into injured axons and subsequent localization in perikaryon, *Brain Res.,* 126, 154, 1977.
91. **Anderson, P. N., Mitchell, J., and Mayor, D.,** On the mechanism of the uptake of horseradish peroxidase into the retrograde transport system of ligated postganglionic sympathetic nerves *in vitro, J. Anat.,* 133, 371, 1981.
92. **Pilar, G. and Landmesser, L.,** Axotomy mimicked by localized colchicine application, *Science,* 177, 1116, 1972.
93. **Pitman, R. M., Tweedle, C. D., and Cohen, M. J.,** Electrical responses of insect central neruons: augmentation by nerve section and colchicine, *Science,* 178, 507, 1972.
94. **Windle, W. F.,** Regeneration of axons in the vertebrate central nervous system, *Physiol. Rev.,* 36, 426, 1956.
95. **Aguayo, A. J., Bray, G. M., Perkins, C. S., and Duncan, I. D.,** Axon-sheath cell interactions in peripheral and central nervous system transplants, *Soc. Neurosci. Symp.,* 4, 361, 1979.
96. **Katz, M. J. and Lasek, R. J.,** Substrate pathways which guide growing axons in *Xenopus* embryos, *J. Comp. Neurol.,* 183, 817, 1979.
97. **Katz, M. J. and Lasek, R. J.,** Substrate pathways demonstrated by transplanted Mauthner axons, *J. Comp. Neurol.,* 195, 627, 1981.
98. **Pitman, R. M.,** unpublished data, 1985.
99. **Lane, N. J. and Treherne, J. E.,** Studies on perineurial junctional complexes and sites of uptake of microperoxidase and lanthanum in the cockroach central nervous system, *Tissue Cell,* 4, 427, 1972.
100. **Treherne, J. E. and Schofield, P. K.,** Mechanisms of ionic homeostasis in the central nervous system of an insect, *J. Exp. Biol.,* 95, 61, 1981.
101. **Schofield, P. K. and Treherne, J. E.,** Localization of the blood-brain barrier of an insect: electrical model and analysis, *J. Exp. Biol.,* 109, 319, 1984.
102. **Lane, N. J.,** Invertebrate neuroglia: junctional structure and development, *J. Exp. Biol.,* 95, 7, 1981.
103. **Treherne, J. E., Harrison, J. B., Treherne, J. M., and Lane, N. J.,** Glial repair in an insect central nervous system: effects of surgical lesioning, *J. Neurosci.,* 4, 2689, 1984.
104. **Smith, P. J. S., Leech, C. A., and Treherne, J. E.,** Glial repair in an insect central nervous system: effects of selective glial disruption, *J. Neurosci.,* 4, 2698, 1984.
105. **Treherne, J. E., Howes, E. A., Leech, C. A., and Smith, P. J. S.,** The effects of an anti-mitotic drug, bleomycin, on glial repair in an insect central nervous system, *Cell Tissue Res.,* 243, 375, 1985.
106. **Smith, P. J. S., Howes, E. A., Leech, C. A., and Treherne, J. E.,** Haemocyte involvement in the repair of the insect central nervous system after selective glial disruption, *Cell Tissue Res.,* 243, 367, 1986.

Chapter 12

PERIPLANETA AMERICANA DISSOCIATED PRIMARY CULTURES

Isabel Bermudez and David J. Beadle

TABLE OF CONTENTS

I. INTRODUCTION

Cockroach embryos are ideal sources of both nerve and muscle tissue for the production of dispersed monolayer cultures. Techniques are now available for producing neuronal-, muscle-, and glial-enriched cultures from embryonic cockroach tissues, and these can be used for developmental, biochemical, and pharmacological studies. The cultures are ideal for biochemical investigations, as there is no possibility of biochemical processes from other cell types complicating the experimental results. The monolayer nature of the cultures and the lack of any barriers to the application of drugs also make them particularly suitable for electrophysiological studies. The cells are readily impaled with high-impedance glass microelectrodes, and drugs can be applied onto specific sites while the whole system is under continual observation. The clean plasma membrane and the architecture of the cells also make them suitable for the patch-clamp technique in both the cell-attached mode for single-channel analysis and in the whole-cell voltage-clamp mode for the analysis of macroscopic currents. It is also possible to grow mixed cultures such as nerve and muscle in which the individual cells interact in a normal physiological manner. These cultures provide ideal models for investigating the interactions of different cell types and the factors that govern them, since the investigator has almost total control over the environment in which the cells are growing.

II. METHODOLOGY

A. SOURCE OF TISSUE

Embryos of the cockroach *Periplaneta americana* are the source of all cells in our culture systems. The American cockroach is a particularly convenient insect to use for culturing because the eggs are laid in an ootheca that may contain up to 16 embryos. This facilitates sterilizing the eggs and permits the harvesting of relatively large quantities of developmentally homogeneous embryonic tissue. The female carries the ootheca around for about 24 h, so the oothecae can either be removed directly from her or collected every day to ensure that the eggs can be dated to within 24 h. The oothecae are stored at 29°C in plastic petri dishes in a moist atmosphere until required. Under these conditions, unused oothecae hatch after 30 to 31 d.

B. CULTURE TECHNIQUES

All operations are performed under aseptic conditions inside a laminar-flow cabinet. The walls and work surface of the cabinet are washed with absolute alcohol at the beginning and the end of culture work; instruments are sterilized by autoclaving or flaming. Culture media and all other solutions used are sterilized by filtration through 0.22-μm multipore filter units (Millipore, Bedford, MA), applying positive pressure from disposable syringes.

Cells are grown on Falcon® tight-seal petri dishes (50 × 12 mm) using a modification of the hanging column technique described by Shields et al.[1] Dissociated cultures will also grow on glass coated with polylysine (1 μg/ml) or protamine (0.1 mg/ml). The culture dishes are prepared by coating three pieces of broken glass microscope slide with Vaseline® and placing them in the dish as illustrated in Figure 1. The dissociated cells are plated between the three glass fragments, and then a sterile glass coverslip (40 × 20 mm) is lowered onto the glass fragments and gently pressed down until the medium is firmly positioned between the coverslip and the bottom of the dish (Figure 1). It is important to keep the culture drop in place, as excessive movement tends to dislodge the cells from the surface of the dish, thereby reducing cell density and growth. All cultures are grown in humidified air at 29°C until required for experimentation.

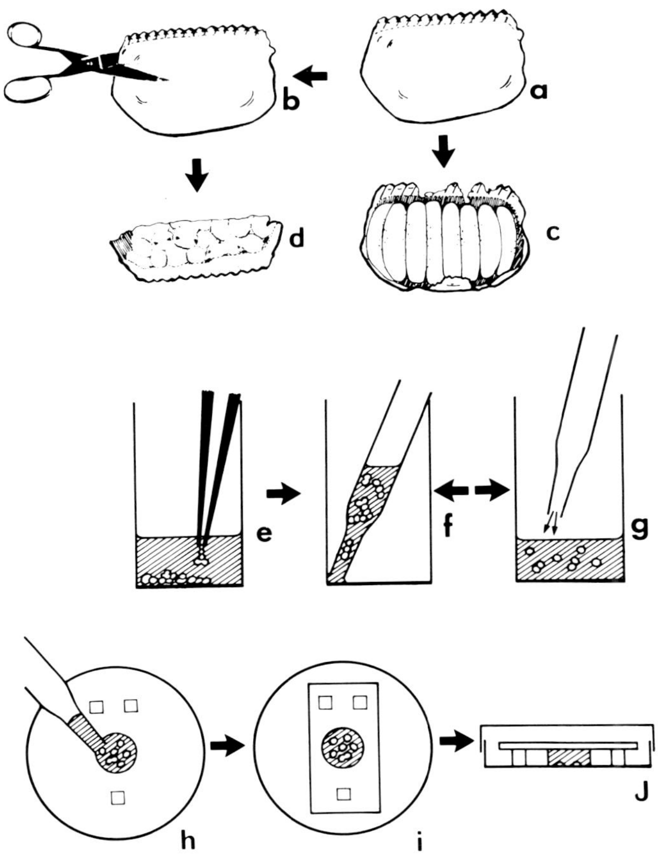

FIGURE 1. Preparation of dissociated neuronal cultures. (a) *P. americana* ootheca, 23 to 26 d old. (b) Embryos consistently arrayed within oothecae (as shown in c) are decapitated using sterile scissors. (d) Each ootheca contains 12 to 16 heads from which cerebral ganglia are isolated using watchmakers' no. 5 forceps. (e to g) Intact cerebral ganglia are pooled and suspended in 5 + 4 medium prior to trituration by Pasteur pipette. (h) Four or three drops of the cell suspension are seeded onto a 50-mm Falcon® petri dish. (i and j) A glass coverslip is stuck down onto glass fragments to support the culture medium in the form of a hanging column. (From Beadle, D. J. and Lees, G., in *Neuropharmacology and Neurobiology*, Ellis Horwood, Chichester, England, 1986, 423. With permission.)

C. CULTURE MEDIA

Cockroach ganglia and muscle somites are dissociated in the 5 + 4 medium of Levi-Montalcini et al.[2] This consists of five parts Schneider's revised *Drosophila* medium and four parts Eagle's basal medium in Hepes buffer, containing 100 IU/ml penicillin, 100 μg/ml streptomycin (pH 6.8 to 7, 370 to 390 mosmol). Disaggregation of tissue in serum-containing medium should be avoided as it causes frothing of the cell suspension and loss of substrate adherence.

All cells are plated into the culture dishes in the 5 + 4 medium. After 1 h, during which time they attach to the floor of the culture dish, muscle cultures and glial cultures are transferred to their specific growth media. Neuronal cultures are usually grown for 4 to 6 d in the 5 + 4 medium before transfer to the neuronal culture medium.

The neuronal medium consists of equal parts of Leibovitz L-15 medium and Yunker's modified Grace's medium supplemented with 100 IU/ml penicillin and 100 μg/ml streptomycin. This medium appears to allow complete neuronal differentiation.

Muscle cultures and nerve-muscle cocultures are grown in a medium consisting of equal parts of Leibovitz L-15 medium and Yunker's modified Grace's medium, 10% horse serum, 10 μg/ml β-ecdysone, 100 IU/ml penicillin, and 100 μg/ml streptomycin. This medium not only permits full differentiation of muscle cells, but also inhibits the growth of fibroblast-like cells.

The growth medium of glial cultures consists of equal parts of Leibovitz L-15 medium and Grace's medium, supplemented with 10% fetal calf serum, 100 IU/ml penicillin, and 100 μg/ml streptomycin.

The medium for all cultures is changed every 7 or 14 d. More frequent replenishment of nutrients is deleterious to the differentiation of all cells, presumably due to the removal of trophic substances. The coverslip is removed gently under sterile conditions and the spent medium aspirated with a Pasteur pipette. Four drops of the fresh medium are added and a new coverslip placed in position.

D. NEURONAL CULTURES

For dissociated neuronal cultures the cerebral ganglia of 23- to 26-day-old embryos are used. The whole nerve cord can be dissected and used to produce cultures, but this is very time-consuming if large numbers of cultures need to be produced for biochemical studies. The technique illustrated in Figure 1 is recommended. The ootheca is sterilized in alcohol and then cut with sharp sterile scissors approximately one third of the distance from the dorsal edge, producing a fragment that contains 12 to 16 sacs, each containing one brain. The sacs are removed with sterile forceps and stored in Schneider's revised *Drosophila* medium at room temperature until the required number has been obtained. These can be dissected rapidly to provide large numbers of brains that can be dissociated to produce neuronal cultures.

To dissociate the brains a sterile Pasteur pipette is wetted thoroughly by sucking the 5 + 4 culture medium up and down to prevent the nerve tissue from sticking to the glass, and this is used to transfer the brains to a bijou bottle containing one drop of 5 + 4 medium (0.025 ml) for every two brains. A second Pasteur pipette is then wetted, and the medium containing the tissue is drawn into it and expelled vigorously against the bottom of the bottle. This is repeated four or five times. A third Pasteur pipette is then flame polished by rotating its tip in a Bunsen flame to reduce the aperture by about 50%, cooled, wetted, and the dissociation procedure repeated five or six times. Finally, a fourth pipette is used to place three drops of the suspension in the center of a Falcon® petri dish prepared as described in Section II.B. After the coverslip has been lowered onto the culture drop, the culture is left for at least 1 h to allow the neurons to attach to the bottom of the dish. After this time they are inverted and stored in large containers as discussed above. The inversion of the dishes prevents fine debris from settling on the growing surface as this inhibits neurite growth. Cultures prepared in this way can be maintained for at least 6 weeks. The culture medium is changed once a week, the first change being into the neuronal culture medium. The dissociation process and the initial growth of the cells in the 5 + 4 medium destroy the vast majority of the glial cells in the ganglion, while the neurons survive (Figure 2a). Cultures prepared in this manner are comprised of at least 90% neurons.[3] Such preparations have many advantages compared to vertebrate preparations, which are often contaminated by other types of cell.

E. GLIAL CELL CULTURES

For dissociated glial cell cultures the cerebral ganglia of 18- to 20-day-old embryos are used. The ganglia are dissected and stored in 5 + 4 medium as described for neuronal

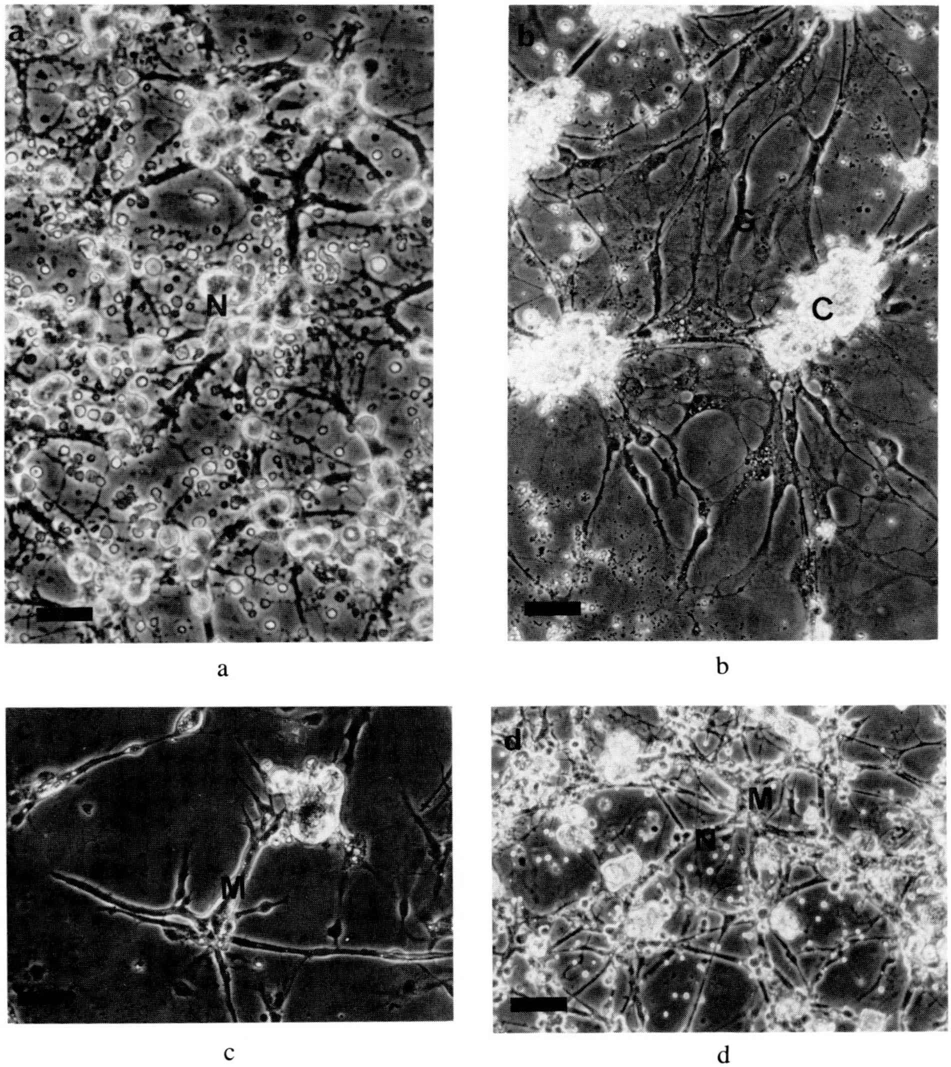

FIGURE 2. Phase-contrast micrographs of *P. americana* dissociated cultures. (a) Neuronal culture prepared from 23-day-old embryonic brains after 26 d *in vitro*. Small clumps or single spherical somata (N) are connected by bundles of neuronal processes. No other type of cell is present. (Scale bar: 100 μm.) (b) Glial cell cultures after 7 d of growth in culture, showing the glial cells (G) flattened out on the floor of the culture dish. Clumps of cells (C) are apparent and these contain neuronal cell bodies. (Scale bar: 20 μm.) (c) Muscle culture after 17 d *in vitro* showing the morphology of the cultured muscle cells (M). (Scale bar: 30 μm.) (d) A nerve-muscle coculture after 14 d of growth in culture showing neurons (N) and muscle cells (M) connected via neuronal processes (P). (Scale bar: 20 μm.)

cultures. The ganglia are then treated in 5 + 4 medium containing collagenase/dispase at 2 mg/ml (Boehringer-Mannheim Corp.) for 1 h at room temperature prior to dissociation by gentle trituration with a Pasteur pipette. The enzyme treatment allows the ganglia to be dissociated with much less force than that used during the preparation of neuronal cultures, and this results in the survival of the glial cells (Figure 2b). The dissociated cells are plated onto Falcon® petri dishes and allowed to attach to the substratum for 1 h. After this time the medium is replaced with the glial culture medium.[4] The glial cells that develop in this preparation do not cross-react with any of the monoclonal antibodies that have been used as markers for vertebrate glia.[4a] Thus, the relationship between insect and vertebrate glial cells needs to be elucidated in more detail.

F. MUSCLE CULTURES

For dissociated muscle cultures, thoracic and abdominal muscle somites of 10- to 11-day-old embryos are used. It is also possible to prepare cultures from the legs of 11-day-old embryos, but the dissection is quite time-consuming and is only suitable for the production of a small number of cultures containing $\leq$10 μg of protein for electrophysiological studies.

When 10- to 11-day-old oothecae are available, they are removed from the incubator and surface sterilized in absolute ethanol for 5 to 10 min. They are then allowed to dry, and the outer case is removed with sterile forceps under a dissecting microscope in a laminar-flow cabinet. The embryos, in their sacs, are transferred to a petri dish containing sterile calcium- and magnesium-free phosphate-buffered saline (dissecting saline). This saline is used to facilitate dissociation of the tissue.[5] The egg sacs are removed carefully with sterile fine forceps to reveal the embryonic cockroaches, and these must then be oriented with their dorsal midline uppermost.

The embryos are opened along their midline, exposing the yolk sac, which is cleaned away carefully. Below the yolk sac are the thoracic and abdominal muscle somites running along both sides of the nerve cord. The muscle somites are excised and then incubated for 15 min at 37°C in 2 to 3 ml of dissecting saline containing 0.05% trypsin (Figure 3). The tissue is then washed with 3 to 5 ml of Leibovitz L-15 medium supplemented with 20% fetal calf serum to arrest trypsinization and then with 3 ml of 5 + 4 medium to remove the serum.

The trypsinized tissue is then resuspended in 5 + 4 medium at a rate of one drop of medium per dissected embryo, and the cells are dissociated by repeated trituration with a fire-polished Pasteur pipette until no large tissue fragments are visible (usually ten times). The cell suspension is then plated onto culture dishes at a rate of three drops per dish and kept at room temperature for 50 to 60 min, when the medium is removed gently and replaced with three to four drops of muscle medium. This procedure removes unattached cells and debris. Cultures prepared in this manner are comprised of approximately 80% myogenic cells (Figure 2c).

The plating density is approximately 6 to 7 $\times$ 10^4 cells per drop of suspension, and each embryo yields ca. 6 $\times$ 10^4 cells. At lower densities, cell viability and differentiation are impaired. Higher densities can be used, although the visualization of individual cells becomes more difficult.

G. NERVE-MUSCLE CULTURES

These cultures are obtained by plating myoblastic cells onto culture dishes which have previously been seeded with cerebral neurons of 23-day-old *P. americana* embryos as described in Section II.D. The neuronal cultures are grown in 5 + 4 medium for 4 to 5 d and then seeded with three drops of freshly prepared muscle cell suspension. The 5 + 4 medium is removed gently 1 h after seeding and replaced with muscle medium. After 14 d of growth *in vitro*, the neurons and muscle cells can be seen connected to each other, forming a complex cellular network (Figure 2d). It is also possible to obtain nerve-muscle cultures by dissociating whole 11-day-old cockroach embryos, although this type of culture will also contain a number of other cell types.

III. APPLICATIONS

A. DEVELOPMENTAL STUDIES

Dissociated cultures prepared from embryonic *P. americana* myoblasts and/or neurons are particularly useful for developmental studies, where problems such as complex cellular

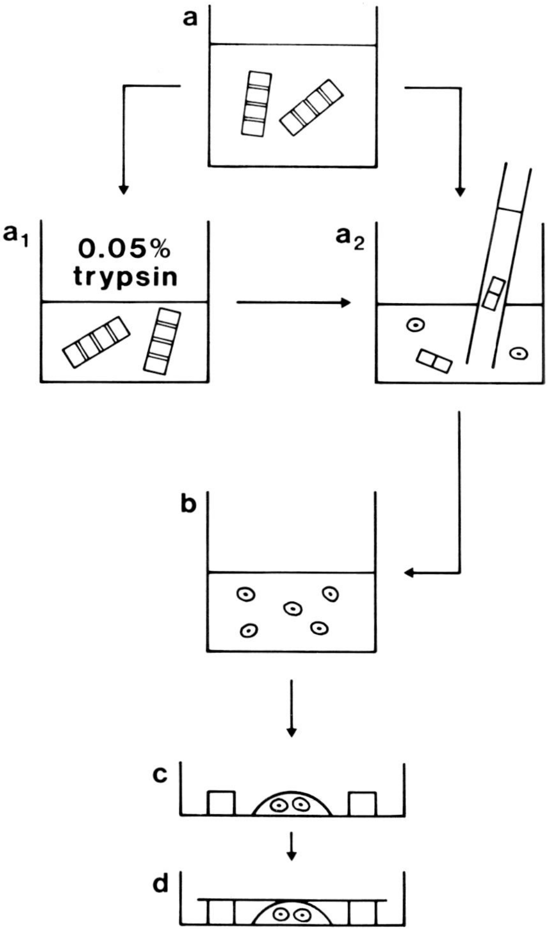

FIGURE 3. Preparation of dissociated muscle cultures. (a) Thoracic and abdominal muscle somites from 11-day-old *P. americana* embryos are suspended in a calcium/magnesium-free phosphate-buffered saline. The tissue can be dissociated by trypsin treatment (a_1) followed by trituration (a_2) or by trituration only (a_2). (b) The resulting cell suspension is dispensed onto culture dishes (three drops per culture dish) and kept at room temperature for 1 h. (c) After 1 h the medium is removed and replaced by muscle medium. (d) A coverslip is lowered onto the glass support to keep the culture drop firmly in place.

networks and accessibility can make *in vivo* experiments difficult to perform and can complicate the interpretation of experimental data. The cultured cells can be followed closely throughout development using biochemical, electrophysiological, or morphological techniques. In addition, the conditions of growth (culture media, cell density, etc.) can be easily altered and their effects on differentiation readily assessed.

Muscle cultures have been used successfully to investigate the differentiation of rounded, mononucleated myoblasts into contractile, multinucleated muscle fibers.[6] The sequence of events accompanying this process is similar to that characterizing the differentiation of vertebrate muscle and consists of one or two cycles of cell division followed by elongation, fusion, and synthesis and assembly of the contractile apparatus of the muscle cells.[6] In nerve-muscle cocultures, muscle cells are connected to neurons via neuromuscular synapses.

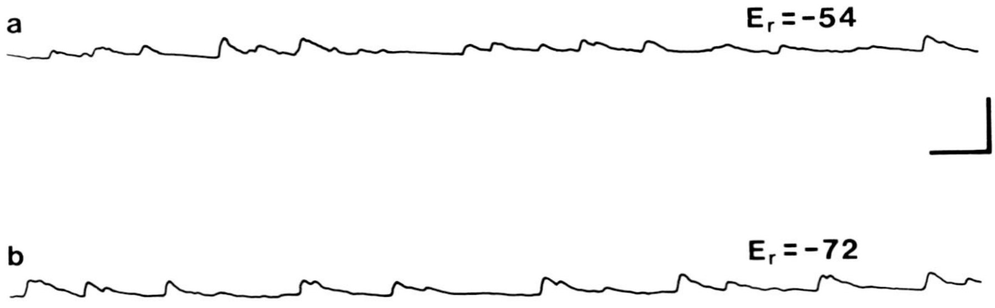

FIGURE 4. Intracellular current-clamp recordings showing spontaneous excitatory synaptic potentials. (a) These synaptic potentials were recorded from a muscle cell present in a 16-day-old nerve-muscle coculture. (Calibration: 20 mV, 100 ms.) (b) Synaptic potentials recorded from a neuron after 26 d of growth in a neuronal culture. (Calibration as in a.)

The *in vitro* synapses are morphologically similar to those seen in insect muscles *in vivo*[6] and are physiologically functional (Figure 4a). We have not studied in detail the effect of coculturing on the properties of both muscle cells and neurons. Preliminary studies have shown that coculturing improves the survival of muscle cells and increases the number of both L-glutamate-sensitive neurons and γ-aminobutyric acid (GABA) transport sites on neurons.

The development of neurons has been investigated morphologically[7] and electrophysiologically.[8] The cultured neurons possess a typically neuronal fine structure.[7] The cells have a rounded soma with a prominent nucleus and nucleolus, as well as small clumps of heterochromatin that are not associated with the nuclear envelope. The cytoplasm contains large numbers of ribosomes and polysomes, mitochondria, lysosomes, and Golgi bodies. The cells produce long, narrow processes that contain microtubules, mitochondria, and elements of smooth endoplasmic reticulum, and they terminate in regions that contain synaptic vesicles. Synaptic profiles are usually encountered in neuronal cultures. The presynaptic terminals contain large numbers of spherical, clear vesicles, 40 to 50 nm in diameter, and presynaptic dense bars associated with clusters of vesicles at points of contact with the postsynaptic membrane thickenings.

Conventional intracellular recording studies have shown that the development of the active membrane properties of cultured *P. americana* neurons is similar to that of insect neurons during normal embryogenesis.[9] During the first 12 d in culture the cells cannot be stimulated to produce action potentials. The characteristic membrane response of these young cells to the injection of depolarizing current pulses is the activation of outward, voltage-dependent potassium channels. However, after this time, increasing numbers of cells can be stimulated (by appropriate current pulses) to produce sodium action potentials. Spontaneous activity can only be seen after about 3 weeks in culture. At this time excitatory, spontaneous synaptic potentials are frequently encountered (Figure 4b).

B. UPTAKE STUDIES

High-affinity uptake systems may play a role in transmitter inactivation (e.g., amino acid neurotransmitters) and/or the synthesis of neurotransmitter (e.g., acetylcholine). We therefore have taken advantage of the cellular homogeneity of our culture systems to characterize the uptake of choline[10] and amino acid neurotransmitters[4,11] in cockroach neurons, myotubes, and glial cells. Cultured neurons show both high- and low-affinity uptake of ^{3}H-choline.[10] The high-affinity transport mechanism is sodium dependent (Figure 5a,b), temperature sensitive, and inhibited by hemicholinium-3, but relatively insensitive to metabolic inhibitors. This uptake system is associated with a considerable degree of acetylcholine synthesis, indicating that the cultures contain a high proportion of cholinergic neurons.

Our studies on amino acid neurotransmitter uptake systems have advanced our knowledge

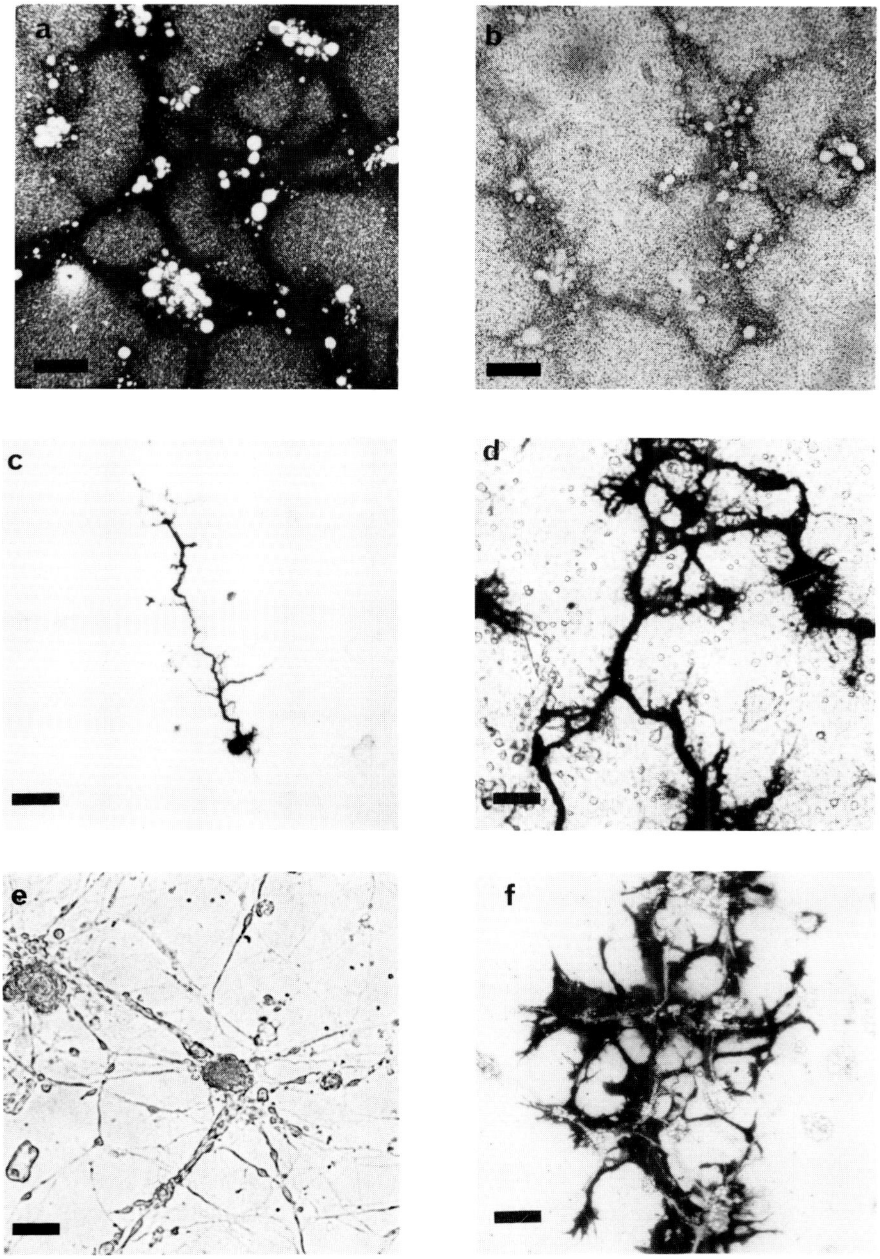

FIGURE 5. Autoradiographs of dissociated *P. americana* cultures. (a) Neuronal culture after incubation in a sodium phosphate-buffered saline containing 0.5 μM ^{3}H-choline. Label is confined to the neuronal processes of a large population of neurons comprising approximately 90% of the cultured neurons. As shown in (b), ^{3}H-choline labeling decreases markedly in the absence of sodium ions. (Scale bars for a and b: 50 μm.) (c) Neuronal culture (14 d old) after incubation in a sodium phosphate-buffered saline containing 1 μM ^{3}H-GABA. (Scale: 20 μm.) (d) Neuronal culture (18 d old) after incubation in a sodium-containing saline supplemented with 1 μM ^{3}H-glutamate. (Scale bar: 30 μm.) (e) Muscle culture (18 d old) after incubation in saline containing sodium ions and 1 μM ^{3}H-GABA. These cells have not accumulated the amino acid. (Scale bar: 15 μm.) (f) Glial cells after incubation in ^{3}H-GABA; all glial cells have accumulated the amino acid. (Scale bar: 20 μm.)

in this area significantly, since it has been possible for the first time to analyze the glial, neuronal, and muscle uptake systems of insects separately. The most significant findings indicate that a high-affinity, sodium-dependent GABA uptake system resides in a small subpopulation of the cultured neurons and that muscle cells do not seem to be endowed with such a system[11] (Figure 5c,e). Both neurons and muscle cells show sodium-dependent, saturable uptake of L-glutamate, but their properties differ markedly in terms of kinetic parameters. Thus, for example, the neuronal uptake is of high affinity (K_m = 1.39 μM; Figure 5d), whereas the muscular transport system is of low affinity (K_m = 29.2 μM). GABA and L-glutamate are also transported into glial cells via sodium-dependent transport mechanisms[4] (Figure 5f). The structural requirements of the glial and neuronal transport systems are similar, suggesting that there may not be marked differences in the mechanisms subserving amino acid transport in these two cell types.

C. PHARMACOLOGICAL STUDIES

The majority of studies of neurotransmitters using these culture systems have concentrated on acetylcholine, the major excitatory transmitter in the insect central nervous system;[12] GABA, the transmitter at insect inhibitory central nervous synapses[3] and neuromuscular junctions;[13] and L-glutamate, the transmitter at the insect excitatory neuromuscular junction.[13]

The majority of the neurons in these cultures are depolarized by acetylcholine and nicotine when these agonists are applied to the cells by pressure ejection at concentrations as low as 0.1 to 5 μM (Figure 6a). The pharmacological properties of the receptors mediating these responses are nicotinic, with α-bungarotoxin and tubocurare being the most potent antagonists and atropine and quinuclidinyl benzilate being less so.[3] The neurons in these cultures are suitable for patch clamping, and studies have revealed unitary currents, evoked by acetylcholine and carbamylcholine, that have two conductance states. The neurons are inhibited by low concentrations of GABA, and the responses are antagonized by picrotoxin, but not by bicuculline (Figure 6b). The GABA responses are potentiated by the benzodiazepine flunitrazepam, suggesting that they resemble the vertebrate $GABA_A$ receptor.[14] A small number of neurons are also sensitive to L-glutamate[15] (Figure 6c).

L-Glutamate (100 μM) depolarizes all cultured myotubes when applied by pressure ejection, the depolarization being accompanied by a marked decrease in membrane input resistance (Figure 6d). L-Glutamate potentials are readily desensitized in the absence of concanavalin A and reverse at about −3 mV. Qualitatively similar responses are obtained to pressure-applied 1 mM quisqualate. The muscle cells are also hyperpolarized by GABA with a pharmacology similar to that observed with the neurons (Figure 6e).

IV. CONCLUDING REMARKS

P. americana glial cells, myotubes, and neurons can now be successfully grown and maintained in culture for several weeks. These cultured cells differentiate and display morphological and physiological characteristics that are essentially similar to those of their *in vivo* counterparts. Although at present it is not possible to determine the completeness of the *in vitro* differentiation of these cells, the presence of physiologically active synapses and neuromuscular junctions in neuronal and nerve-muscle cultures, respectively, indicates that an advanced level of development is reached under our culture conditions. Therefore, these cultures should provide a powerful tool with which to examine the properties of insect glial cells, myotubes, and neurons. Since they also possess neurotransmitter receptor and uptake mechanisms that are essentially similar to those of their vertebrate counterparts, they provide ideal models for comparing the pharmacology of the nervous systems of these two important animal groups.

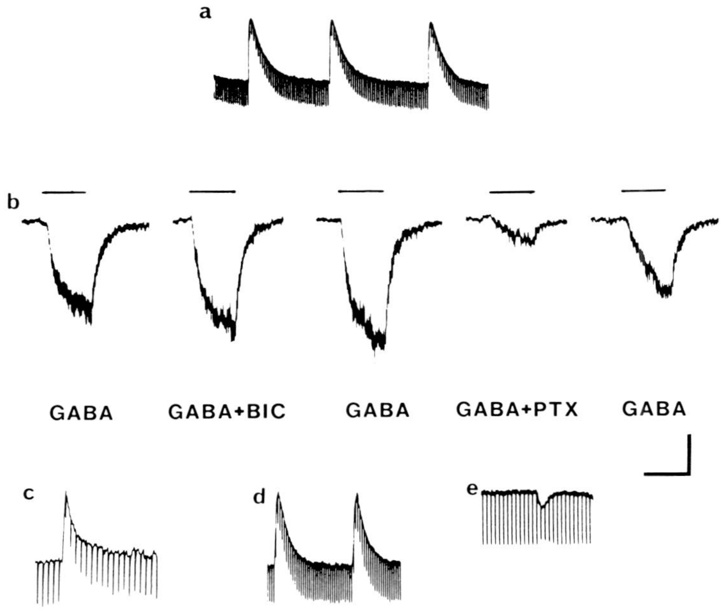

FIGURE 6. Response types evoked by application of putative neurotransmitters onto 14- to 20-day-old cultured muscle cells and neurons. (a) Typical neuronal response to 0.1 m*M* ACh pressure ejected onto the neuronal somata for 1 s. The resting potential (E_R) of the cell was −54 mV. (Calibration: 20 mV, 10 s.) (b) GABA (1 m*M*) applied onto the somata of whole-cell voltage-clamped neurons evokes an inward current which reverses at the equilibrium potential predicted by the Nernst equation for chloride ions. Bicuculline (10 μ*M*) has no effect on this current, whereas 10 μ*M* picrotoxin decreases it markedly. (Clamping potential: −50 mV; calibration: 50 pA, 10 s.) Bars indicate pulse duration. (c) Response of a whole-cell voltage-clamped neuron to pressure-applied 1 m*M* L-glutamate. E_R = −60 mV, calibration = 20 mV, 5 s. (d) Current-clamp record of a typical L-glutamate response in a muscle cell. The amino acid (0.1 m*M*) was applied onto the cell by pressure ejection. Clamping potential = −50 mV; calibration as in a. (e) Typical response to 1 m*M* GABA applied by pressure ejection onto a muscle cell. E_R = −56 mV; calibration = 10 mV, 10 s.

REFERENCES

1. **Shields, G., Dubendorfer, A., and Sang, J. H.,** Differentiation *in vitro* of larval cell types from early embryonic cells of *Drosophila melanogaster, J. Embryol. Exp. Morphol.,* 33, 159, 1975.
2. **Levi-Montalcini, R., Chen, J. S., Seshan, K. R., and Aloe, L.,** An *in vitro* approach to the insect nervous system, in *Developmental Neurobiology of Arthropods,* Young, D., Ed., Cambridge University Press, London, 1973, 5.
3. **Beadle, D. J. and Lees, G.,** Insect neuronal cultures — a new tool in insect neuropharmacology, in *Neuropharmacology and Pesticide Action,* Ford, M. G., Lunt, G. G., Reay, R. C., and Usherwood, P. N. R., Eds., Ellis Horwood, Chichester, England, 1986, 423.
4. **Beadle, C. A., Bermudez, I., and Beadle, D. J.,** Amino acid uptake by neurones and glial cells from embryonic cockroach brain growing *in vitro, J. Insect Physiol.,* 33, 761, 1987.

4a. **Bermudez, I. and Beadle, D. J.,** unpublished observation, 1987.

5. **Moscona, A., Trowell, O. A., and Willmer, E. N.,** Chemical dissociation, in *Cells and Tissues in Culture,* Vol. 1, Willmer, E. N., Ed., Academic Press, New York, 1965, 52.
6. **Bermudez, I., Lees, G., Botham, R. P., and Beadle, D. J.,** Myogenesis and neuromuscular junction formation in cultures of *Periplaneta americana* myoblasts and neurones, *Dev. Biol.,* 116, 467, 1986.

7. **Beadle, D. J., Hicks, D., and Middleton, C.,** Fine structure of neurones from embryonic *Periplaneta americana* growing in long term culture, *J. Neurocytol.,* 11, 611, 1982.
8. **Lees, G., Beadle, D. J., Botham, R. P., and Kelly, J. S.,** Excitable properties of insect neurones from the central nervous system of embryonic cockroaches, *J. Insect Physiol.,* 31, 135, 1985.
9. **Goodman, C. S. and Spitzer, N. C.,** The development of electrical properties of identified neurones in grasshopper embryos, *J. Physiol.,* 313, 369, 1981.
10. **Bermudez, I., Lees, G., Middleton, C., Botham, R. P., and Beadle, D. J.,** Choline uptake by cultured neurones from the central nervous system of embryonic cockroaches, *Insect Biochem.,* 15, 427, 1985.
11. **Bermudez, I., Botham, R. P., and Beadle, D. J.,** High- and low-affinity uptake of amino acid transmitters in cultured neurones and muscle cells of the cockroach *Periplaneta americana, Insect Biochem.,* 18, 249, 1988.
12. **Sattelle, D. B.,** Acetylcholine receptors in insects, *Adv. Insect Physiol.,* 15, 215, 1980.
13. **Piek, T.,** Neurotransmission and neuromodulation of skeletal muscle, in *Comprehensive Insect Physiology, Biochemistry and Pharmacology* , Vol. 11, Kerkut, G. A. and Gilbert, L. I., Eds., Pergamon Press, Oxford, 1985, 55.
14. **Shimahara, T., Pichon, Y., Lees, G., Beadle, C. A., and Beadle, D. J.,** Gamma-aminobutyric acid receptors on cultured cockroach brain neurones, *J. Exp. Biol.,* 131, 231, 1987.
15. **Horseman, B. G., Seymour, C., Bermudez, I., and Beadle, D. J.,** The effects of L-glutamate on cultured insect neurones, *Neurosci. Lett.,* 85, 65, 1988.

Index

INDEX

VOLUMES I AND II

A

B

C

D

E

F

G

H

I

J

K

L

M

N

O

P

Q

R

S

T

U

V

W

X

Cockroaches as Models for Neurobiology: Applications in Biomedical Research

Volume II

Editors

Ivan Huber, Ph.D.
Professor
Department of Biological and Allied Health Sciences
Fairleigh Dickinson University
Florham Park — Madison Campus
Madison, New Jersey

Edward P. Masler, Ph.D.
Research Physiologist
Insect Reproduction Laboratory
Agricultural Research Service
U.S. Department of Agriculture
Beltsville, Maryland

B. R. Rao, Ph.D.
Professor
Department of Biological Sciences
East Stroudsburg University
East Stroudsburg, Pennsylvania

CRC

CRC Press, Inc.
Boca Raton, Florida

Library of Congress Cataloging-in-Publication Data

Cockroaches as models for neurobiology: applications in biomedical research/editors, Ivan Huber, Edward P. Masler, B. R. Rao.
p. cm.
Includes bibliographical references.
ISBN 0-8493-4838-2 (v. 1) — ISBN 0-8493-4839-0 (v. 2)
1. Neurobiology—Research—Methodology. 2. Plectoptera (Cockroach)—Physiology. I. Huber, Ivan, 1931- . II. Masler, Edward P., 1948- . III. Rao, B. R., 1936- .
QP357.C63 1990
595.7′22′0724—dc20

89-22149
CIP

Direct all inquiries to CRC Press, Inc., 2000 Corporate Blvd., N.W., Boca Raton, Florida 33431.

International Standard Book Number 0-8493-4838-2
International Standard Book Number 0-8493-4839-0

Library of Congress Card Number 89-22149
Printed in the United States

THE EDITORS

Ivan Huber, Ph.D., is Professor of Biology at Fairleigh Dickinson University, Madison, New Jersey.

Dr. Huber received his A.B. degree in Zoology from Cornell University, Ithaca, New York in 1954 and his Doctorate in Entomology from the University of Kansas, Lawrence in 1968.

Dr. Huber is a member of the Entomological Society of America, The New York Entomological Society, the Society for the Study of Evolution, the Society of Systematic Zoology, and the honorary society Sigma Xi. He has been the recipient of research grants from the Merck Institute for Therapeutic Research.

Dr. Huber has written on allomones in chigger mites and the ecological genetics of flour beetles, but most of his work has been on cockroaches. He has published articles on the taxonomy, population biology, and oviposition behavior of this group. He currently is studying learning and memory in cockroaches.

Edward P. Masler, Ph.D., is a Research Physiologist in the Insect Reproduction Laboratory, Agricultural Research Service, U.S. Department of Agriculture, Beltsville, Maryland. He received his A.B. in Biology from St. Anselm College, Manchester, New Hampshire (1970) and an M.S. in Genetics from the University of New Hampshire, Durham (1973). He was awarded a Ph.D. in Biology from the University of Notre Dame, South Bend, Indiana (1978), where he became interested in the physiology of insect development and reproduction. He was awarded a postdoctoral fellowship to the Roche Institute of Molecular Biology, where he spent 2 years in the Department of Biochemistry studying vitellogenic protein structure and modification in the developing oocyte. A postdoctoral appointment to the Department of Entomology, Cornell University, Ithaca, New York followed, where he investigated neuropeptides responsible for ecdysteroid and vitellogenin production. Dr. Masler has been a Research Physiologist with the U.S. Department of Agriculture since 1982 and currently is interested in the biochemistry and molecular biology of development and reproduction in selected dipteran and lepidopteran species. Special emphasis is placed on the identities and roles of neuropeptides in metamorphosis and ovarian maturation. Memberships include the International Brain Research Organization, the International Society of Invertebrate Reproduction and Development, The Society for Developmental Biology, the Society for Neuroscience, and the Society of Sigma Xi.

Balakrishna R. Rao, Ph.D., is Professor of Biology at East Stroudsburg University, East Stroudsburg, Pennsylvania. He received his B.S. in Agriculture from Banaras Hindu University, Varanasi, India in 1957, his M.S. in Agriculture from Karnatak University, Dharwad, Karnataka, India in 1959, and a Ph.D. in Entomology from The Ohio State University, Columbus in 1964. His doctoral work in insect physiology culminated in his dissertation, entitled "Trypsin Activity Associated with the Reproductive Development in the Female Tampa Cockroach, *Nauphoeta cinerea* Oliv.". Dr. Rao had been a Postdoctoral Fellow at The Johns Hopkins University, the University of Connecticut, and The Marine Biological Laboratories, Woods Hole, Massachusetts prior to becoming Associate Professor of Biology at East Stroudsburg University. He is a member of the Entomological Society of Pennsylvania, the Entomological Society of America, Sigma Xi, the the Commonwealth of Pennsylvania University Biologists. Dr. Rao's current interests include reproductive physiology and digestive enzymes in cockroaches as well as mosquito reproductive physiology.

CONTRIBUTORS

Volume I

David W. Alsop, Ph.D.
Department of Biology
Queens College, CUNY
Flushing, New York

Moray Anderson, Ph.D.
Department of Zoology and Comparative Physiology
University of Birmingham
Birmingham, England

David J. Beadle, Ph.D.
School of Biological and Molecular Sciences
Oxford Polytechnic
Headington, Oxford
England

William J. Bell, Ph.D.
Department of Entomology
University of Kansas
Lawrence, Kansas

Isabel Bermudez, Ph.D.
School of Biological and Molecular Sciences
Oxford Polytechnic
Headington, Oxford
England

Jonathan M. Blagburn, Ph.D.
Institute of Neurobiology
University of Puerto Rico
San Juan, Puerto Rico

Jean-Jacques Callec, Ph.D.
Laboratory of Animal Physiology
University of Rennes I
Rennes, France

Yesu T. Das, Ph.D.
Innovative Scientific Services, Inc.
Piscataway, New Jersey

Charles R. Fourtner, Ph.D.
Department of Biological Sciences
State University of New York
Buffalo, New York

Ayodhya P. Gupta, Ph.D.
Department of Entomology
Rutgers University
New Brunswick, New Jersey

Ivan Huber, Ph.D.
Department of Biological and Allied Health Sciences
Fairleigh Dickinson University
Florham Park — Madison Campus
Madison, New Jersey

Bernard Hue, Ph.D.
Laboratory of Physiology
URA CNRS 611
University of Angers
Angers, France

Charles J. Kaars, Ph.D.
Department of Biological Sciences
State University of New York
Buffalo, New York

E. P. Masler, Ph.D.
Agricultural Research Service
U.S. Department of Agriculture
Beltsville, Maryland

Thomas A. Miller, Ph.D.
Department of Entomology
University of California
Riverside, California

Marcel Pelhate, Ph.D.
Laboratory of Physiology
URA CNRS 611
University of Angers
Angers, France

Yves Pichon, Ph.D.
Laboratory of Cellular Neurobiology
Department of Biophysics
C.N.R.S.
Gif-sur-Yvette, France

Robert M. Pitman, Ph.D.
Department of Biology and
Preclinical Medicine
Gatty Marine Laboratory
St. Andrews, Fife
Scotland

B. R. Rao, Ph.D.
Department of Biological Sciences
East Stroudsburg University
East Stroudsburg, Pennsylvania

David B. Sattelle, Ph.D.
Unit of Insect Neurophysiology
and Pharmacology
Department of Zoology
University of Cambridge
Cambridge, England

Karel Sláma, Ph.D.
Insect Chemical Ecology Unit
Institute of Organic Chemistry
Czechoslovak Academy of Sciences
Prague, Czechoslovakia

C. S. Thompson, Ph.D.
Department of Zoology
University of Toronto
Toronto, Ontario
Canada

S. S. Tobe, Ph.D.
Department of Zoology
University of Toronto
Toronto, Ontario
Canada

Hiroshi Washio, Ph.D.
Laboratory of Neurophysiology
Mitsubishi-Kasei Institute
of Life Sciences
Machida, Tokyo
Japan

CONTRIBUTORS

Volume II

Michael E. Adams, Ph.D.
Division of Toxicology and Physiology
Department of Entomology
University of California
Riverside, California

Cynthia A. Bishop, Ph.D.
Department of Psychology
Stanford University
Stanford, California

Benjamin J. Cook, Ph.D.
Veterinary Toxicology and Entomology Research Laboratory
Agricultural Research Service
U.S. Department of Agriculture
College Station, Texas

Roger G. H. Downer, Ph.D.
Department of Biology
University of Waterloo
Waterloo, Ontario
Canada

Franz Engelmann, Ph.D.
Department of Biology
University of California
Los Angeles, California

Derek W. Gammon, Ph.D.
Agricultural Research Division
American Cyanamid Co.
Princeton, New Jersey

Ivan Huber, Ph.D.
Department of Biological and Allied Health Sciences
Fairleigh Dickinson University
Florham Park — Madison Campus
Madison, New Jersey

Larry L. Keeley, Ph.D.
Laboratories for Invertebrate Neuroendocrine Research
Department of Entomology
Texas Agricultural Experiment Station
Texas A&M University
College Station, Texas

Manfred J. Kern, Ph.D.
Pflanzenschutzforschung-Biologie
Hoechst Aktiengesellschaft
Frankfurt am Main, Federal Republic of Germany

Michael K. Leung, Ph.D.
Multidisciplinary Center for the Study of Aging and Chemistry/Physics Program
SUNY/College at Old Westbury
Old Westbury, New York

Michael I. Mote, Ph.D.
Department of Biology
Temple University
Philadelphia, Pennsylvania

Michael O'Shea, Ph.D.
Laboratory of Neurobiology
University of Geneva
Geneva, Switzerland

Terry L. Page, Ph.D.
Department of Biology
Vanderbilt University
Nashville, Tennessee

Susan M. Rankin, Ph.D.
Department of Entomology
Texas A&M University
College Station, Texas

Coby Schal, Ph.D.
Department of Entomology
Cook College
Rutgers University
New Brunswick, New Jersey

Berta Scharrer, Ph.D.
Department of Anatomy and Structural Biology
Albert Einstein College of Medicine
Bronx, New York

Günter Seelinger, Ph.D.
Institute for Zoology
University of Regensburg
Regensburg, Federal Republic of Germany

Alan F. Smith, Ph.D.
Department of Entomology
Cook College
Rutgers University
New Brunswick, New Jersey

Rajindar S. Sohal, Ph.D.
Department of Biological Sciences
Southern Methodist University
Dallas, Texas

George B. Stefano, Ph.D.
Multidisciplinary Center for the Study of Aging and Biological Sciences Program
SUNY/College at Old Westbury
Old Westbury, New York

Renée M. Wagner, Ph.D.
Veterinary Toxicology and Entomology Research Laboratory
Agricultural Research Service
U.S. Department of Agriculture
College Station, Texas

Jane L. Witten, Ph.D.
Laboratory of Neurobiology
University of Geneva
Geneva, Switzerland

Stephen Zawistowski, Ph.D.
Division of Social Sciences
St. John's University
Staten Island, New York

Sasha N. Zill, Ph.D.
Department of Anatomy
Marshall University School of Medicine
Huntington, West Virginia

TABLE OF CONTENTS

Volume I

TABLE OF CONTENTS

Volume II

VI. Sense Organs, Plasticity, and Behavior

Section IV. Neurohormones and Neurotransmitters

Chapter 13

PEPTIDES AS CHEMICAL SIGNALS: HORMONES TO TRANSMITTERS

Michael E. Adams

TABLE OF CONTENTS

I. INTRODUCTION

A. NEUROPEPTIDES: A MAJOR CLASS OF CHEMICAL MESSENGER

One of the most significant trends in neuroscience over the past 15 years has been the steady increase in the number of novel chemical messengers discovered in neurons. The number of neural substances involved in synaptic transmission, modulation, or endocrine functions is thought to exceed 100, an estimate that many investigators regard as conservative. This remarkable increase in putative chemical messengers can be attributed to the virtual explosion of newly discovered neuropeptides acting as "nonconventional" signals. The designation "nonconventional" contrasts these agents with the "conventional" transmitters such as acetylcholine, γ-aminobutyric acid (GABA), or glutamate, which elicit responses in the millisecond time frame, usually through activation of ion channels. As will be pointed out in this chapter, peptides differ from conventional transmitters by their ability to diffuse considerable distances from sites of release and by their interaction with diverse second messenger systems. In other words, whereas the conventional transmitters may be regarded primarily as synaptic signals with well-defined and predictable functions, peptides function in a variety of contexts as transmitters, modulators, or hormones.

The proliferation of reports on the widespread occurrence of peptides has stimulated new flexibility in our conceptual thinking about the rules governing signal transmission. In a rather short period of time, the primary association of peptides with neurosecretion has been expanded to include a wide range of neuroeffector actions, including corelease and joint action with conventional transmitters at synaptic junctions. Moreover, it is also likely that the involvement of peptides in diverse aspects of nervous system and gut function may be more the rule than the exception. The activation of ion channels and second messengers by the release of peptides and conventional transmitters provides for a complexity and subtlety in target regulation by single neurons that was scarcely imagined previously.

Our understanding of peptide chemistry and distribution in the nervous system is progressing rapidly due to ever more refined techniques for their isolation, sequencing, and immunocytochemical localization. Nevertheless, our knowledge of precise functional roles for neuropeptides remains limited by comparison, due to the complexities of execution and interpretation of experiments on their cellular roles. In this regard, the use of invertebrate animals as paradigms for studying the functions of peptides at the cellular level offers distinct advantages over more complex vertebrate preparations.

B. INSECT MODELS FOR STUDIES OF NEUROPEPTIDES

The involvement of neuropeptides in the regulation of developmental, metabolic, and behavioral processes in diverse phylogenetic groups is increasingly apparent. Their widespread distribution in animal groups ranging from coelenterates to man implies that peptidergic regulation may be of general importance and argues that comparative studies are likely to reveal general rules governing such regulation.[1] As this chapter will illustrate, the chemistry and cellular actions of neuropeptides are directly comparable in vertebrates and invertebrates. For example, sequence homologies between neuropeptides occurring in the two groups are high. Peptidergic neurons show common patterns of structure, cytology, electrical properties, and release mechanisms. The responses of target cells and tissues to peptidergic neurons involve modulation of second messenger systems, including cyclic nucleotides, phosphoinositides, and calcium, that appear to be indistinguishable between the two groups. Thus, the use of invertebrate animals as cellular models for studying neuropeptide function offers opportunities to develop principles governing synthesis, release, physiological actions, and degradation in relatively simple paradigms that can then be tested in the more complex systems of higher animals. Finally, the increased cost and time necessary to utilize vertebrate animals in experimental neuroscience mandates consideration of alternative animal models where possible.

Several reviews of the insect neuropeptide literature prior to 1985 are recommended to the interested reader.[2-7] This review will concentrate on recent studies of peptidergic systems in the cockroach (*Periplaneta americana, Leucophaea maderae*), locust (*Schistocerca gregaria, Locusta migratoria*), and lepidopterous insects (*Manduca sexta, Bombyx mori, Heliothis virescens*), giving particular emphasis to their functional analyses using identified neurons. The use of recent experimental techniques, particularly dye filling for morphological studies, antibodies and high-performance liquid chromatography (HPLC) for biochemical studies, and intracellular recording for physiological studies, will be emphasized, since they are of particular utility in the development of new concepts governing the types and functions of peptides and peptidergic neurons. Where examples in these insects are not available, knowledge gained from studies in other invertebrates will be mentioned in order to complete a thought or illustrate a general cellular principle.

C. IDENTIFIED PEPTIDERGIC NEURONS

The concept of the "uniquely identifiable neuron" has afforded powerful insights into mechanisms governing the invertebrate nervous system. Such cells can be unambiguously recognized in each individual animal by their unique combination of anatomical, biochemical, and physiological attributes. Upon repeated scrutiny at each of these levels, precise information on connections, transmitter profile, and target responses is gained.

For example, the neuropeptide proctolin is associated with several types of functionally distinct neuronal types. It was first demonstrated in extracts of individually identified neurosecretory cells[8] innervating the heart and perisympathetic organs, both of which serve as neurohemal sites. This suggested a possible endocrine role for proctolin. However, the heart is sensitive to low concentrations of proctolin, suggesting as well a close range neuroeffector function for the Lateral White (LW) neurons and proctolin (see Section III.B.2.a). Subsequently, proctolin was detected in motoneurons innervating skeletal[9] and visceral[10,11] muscle, where it is implicated as a cotransmitter in the regulation of tonic and rhythmic tension. In summary, studies of proctolin in identified neurons suggest its role in diverse functions ranging from purely endocrine to neuroeffector (see also Chapter 14 in this work).

Cellular analyses promise to define diverse roles for the adipokinetic hormones (AKH) and related peptides, which are circulatory neurohormones that regulate energy mobilization.[12] They are released by neurosecretory cells in the corpus cardiacum (CC), an endocrine gland analagous to the pituitary of vertebrates. A wider range of physiological roles for

AKH-like peptides is indicated by immunohistochemical evidence of their presence in neuronal cell bodies in the brain and ventral ganglia, as well as in neuromuscular terminals in the cockroach foregut.[13] These observations are consistent with the myotropic actions of AKH-like peptides on the heart[14] and gut.[15]

The foregoing examples illustrate the insights which can be gained by the precise cellular analysis of peptide localization, release, and target actions. Studies on proctolin and AKH systems illustrate that peptides can act as hormones or transmitters, depending on cellular architecture and the proximity of the target receptors to points of secretion.

The following sections summarize the identified insect neuropeptides and, in some instances, their relationship to vertebrate peptides. Specific neuropeptides and their associations with identified peptidergic neurons are described, together with references to methods for microelectrode recording and dye filling, immunohistochemical staining, single cell dissection, and bioassay where possible.

II. CHEMISTRY AND OCCURRENCE OF INSECT NEUROPEPTIDES

Some 22 biologically active neuropeptides have been purified from insect tissues and sequenced (Table 1). Of these, 15 were identified from cockroaches, 4 from locusts, and 3 from moths. Although most of these peptides have not been assigned *in vivo* physiological functions, they are known to regulate muscle tension and rhythmicity, water balance, energy metabolism, and development. The following sections summarize the chemistry and biological actions of insect peptides, many of which are related to vertebrate peptides and have astonishingly analogous functions.

A. MYOTROPIC PEPTIDES

1. Proctolin

Proctolin was originally characterized as a "gut factor", located principally in the fore- and hindgut of *P. americana*.[16] Its eventual structure elucidation from extracts of 125,000 whole cockroaches[17] and its proposal as a *bona fide* gut neurotransmitter[18] essentially initiated the era of neuropeptide research in insect physiology. Shortly thereafter, the occurrence of proctolin was documented in six orders of insects,[19] and subsequent studies have demonstrated either authentic proctolin or proctolin-like immunoreactivity (PLI) in the Crustacea, Annelida, and Mammalia. Direct or modulatory effects of proctolin are documented in all types of insect muscle, including visceral,[11,18,20,21] cardiac,[22] and skeletal.[9,23-25] The impressive potency of proctolin as a cardioaccelerator is illustrated by *in vivo* infusion of 1 to 10 pmol into an adult cockroach (Figure 1). At this dose, a sustained increase in the amplitude and frequency of heartbeat can be observed for 15 to 20 min.

The association of proctolin with neurons began with its detection in identified neurosecretory cells[8] using single cell dissection and bioassay. Subsequent development of antisera against proctolin[26,27] provided pivotal insights into its distribution in the central nervous system (CNS) and was the key step in the discovery of proctolinergic motoneurons.[9,28] Considerable evidence has been gathered since then in support of Brown's original contention that proctolin is a transmitter substance in insect muscle[11,21,23,25,29] (see Chapter 14 of this work). Although the physiological significance of proctolin in the insect CNS remains unclear, the occurrence of PLI in presynaptic terminals in the sixth abdominal ganglion[30] is consistent with a transmitter or modulator role. Central actions for proctolin have been demonstrated in the lobster, where it modulates central pattern generation in the stomatogastric system.[31-34] The presence of PLI in the rat brain[35] raises provocative yet unanswered questions about its physiological significance in higher animals.

TABLE 1
Sequenced Insect Neuropeptides

Neuropeptides	Sequence	Source	Physiological actions	Ref.
Myotropic peptides				
Proctolin	R Y L P T	CNS	Elevates rhythmic and tonic tension in muscle	18, 113
Leucopyrokinin	p-E T S F T P R L-NH_2	Head	Causes hindgut contraction	15
Leucokinins				
I	D P A F N S W G-NH_2	Head	Causes hindgut contraction	43
II	D P G F S S W G-NH_2	Head	Causes hindgut contraction	43
III	D N G F N S W G-NH_2	Head	Causes hindgut contraction	44
IV	D A S F H S W G-NH_2	Head	Causes hindgut contraction	44
V	G S G F S S W G-NH_2	Head	Causes hindgut contraction	45
VI	p-E S S F H S W G-NH_2	Head	Causes hindgut contraction	45
VII	D P A F S S W G-NH_2	Head	Causes hindgut contraction	46
VIII	G A S F Y S W G-NH_2	Head	Causes hindgut contraction	46
Leucomyosuppressin	p-E D V D H V F L R F-NH_2	Head	Causes hindgut contraction	49
Leucosulfakinin	E N F E D Y G H M R F-NH_2 (Y with SO_3H)	Head	Causes hindgut contraction	47
Leucosulfakinin II	p-E S D D Y G H M R F-NH_2 (Y with SO_3H)	Head	Causes hindgut contraction	48
FMRFamide	F M R F-NH_2	CNS	Modulates neuromuscular transmission	36, 41, 42
Enkephalin[a]	Y G G F M	Vertebrate CNS	Inhibits neuronal firing in vertebrate CNS	143
Metabolic peptides				
AKH I	p-E L N F T P N W G T-NH_2	Corpus cardiacum	Lipid mobilization	12, 50
AKH IIS	p-E L N F S A G W-NH_2	Corpus cardiacum	Lipid mobilization	51
AKH IIR	p-E L N F S T G W-NH_2	Corpus cardiacum	Lipid mobilization	51
AKH-M/H	p-E L T F T S S W G-NH_2	Corpus cardiacum	Lipid mobilization	56, 57
Neurohormone D (CC1; MI)	p-E V N F S P N W-NH_2	Corpus cardiacum	Glycogen mobilization, cardioacceleration	52—55
CC2 (MII)	p-E L T F T P N W-NH_2	Corpus cardiacum	Glycogen mobilization, cardioacceleration	53—55

TABLE 1 (continued)
Sequenced Insect Neuropeptides

Neuropeptides	Sequence	Source	Physiological actions	Ref.
Glucagon[a]	S Q G T F T S D Y S K	Pancreas	Glycogen mobilization	53
AVP-like DH	C L I T N C P R G–NH_2	CNS	Diuretic hormone	59
	C L I T N C P R G–NH_2			
Arginine vasopressin (AVP)[a]	C Y F Q N C P R G–NH_2	Vertebrate CNS	Antidiuretic hormone	145
Developmental peptides				
Eclosion hormone	N P A I A T G Y D P M E I	Corpus cardiacum	Triggers eclosion and ecdysis behavior	67, 68
	C I E N C A N C K K M L G			
	A W F E G P L C A E S C I			
	K F K G K L I P E C E D F			
	A S I A P F L N K L–OH			
Prothoracicotropic Hormone				
A chains				
	5 10			
Human insulin[a]	G I V E Q C C T S I C	Pancreas, CNS	Promotes glycogen synthesis	144
4K-PTTH-II	G I V D E C C L R P C	CNS		
	15 20		Stimulates ecdysone secretion secretion	64, 65
	S L Y E L E N Y C N–OH			
	S V D V L L S Y C–OH			
B chains				
Human insulin[a]	5 10	See above	See above	
4K-PTTH-II	F V N Q H L C G S			
	p–E Q P Q A V H T Y C G R			
	20 25 30			
	H L V E A L Y L V C G E			
	H L A R T L A D L C W E			
	35 40			
	R G L F F Y T P K T–OH			
	A G V D–OH			

TABLE 1 (continued)
Sequenced Insect Neuropeptides

Note: The amino acid symbols are as follows:

Amino acid	Three-letter symbol	One-letter symbol
Alanine	Ala	A
Arginine	Arg	R
Asparagine	Asn	N
Aspartic Acid	Asp	D
Cysteine	Cys	C
Glutamine	Gln	Q
Glutamic acid	Glu	E
Glycine	Gly	G
Histidine	His	H
Isoleucine	Ile	I
Leucine	Leu	L
Lysine	Lys	K
Methionine	Met	M
Phenylalanine	Phe	F
Proline	Pro	P
Serine	Ser	S
Threonine	Thr	T
Tryptophan	Trp	W
Tyrosine	Tyr	Y
Valine	Val	V

[a] Neuropeptides found in vertebrates which show sequence homology to insect neuropeptides.

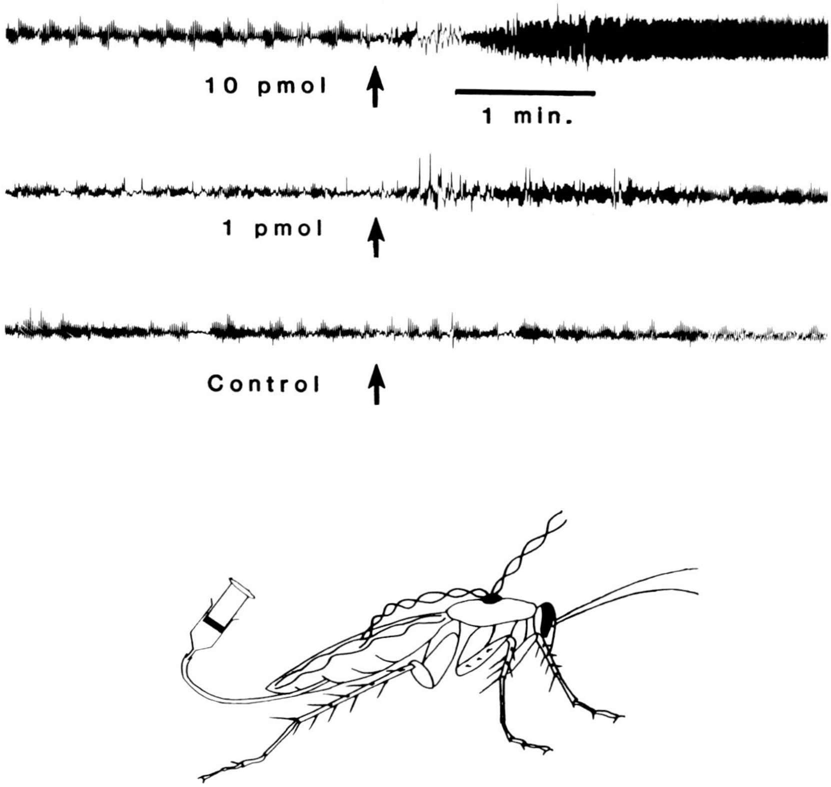

FIGURE 1. *In vivo* modulation of the cockroach heart by the neuropeptide proctolin. Copper electrodes insulated to the tip are inserted through the dorsal cuticle and positioned close to, but not touching, the heart. Movement of the myocardium is transduced and amplified using an impedence converter. A cannula implanted between two abdominal sclerites in close proximity to the myocardium provides the entry point for proctolin infusion. Infusion of 10 μl of physiological saline (lower trace) elicits no response. Injection of the same volume containing 1 and 10 pmol of proctolin (middle and upper traces, respectively) causes increased amplitude and frequency of heartbeat lasting 10 to 15 min.

2. FMRFamide

Originally identified as a cardioaccelerator from molluscan heart,[36] FMRFamide (Phe–Met–Arg–Phe–NH_2) belongs to a large ''RF'' peptide family (cholecystokinin, gastrin, pancreatic polypeptide), the distribution of which appears to extend from coelenterates to higher chordates. In insects, FMRFamide-like immunoreactivity is present in central neurons and neurohemal structures of the locust[37] and moth,[38] but not in the terminals of skeletal motoneurons. However, FMRFamide modulates synaptic transmission in the locust extensor tibiae muscle, acting presynaptically to increase conventional transmitter release and postsynaptically to cause membrane depolarization, elevated membrane resistance, and increased twitch amplitude and rate of relaxation.[39,40] The combined actions of FMRFamide are therefore indirect, amplifying the effects of normal motoneuron activity. The accumulated evidence thus suggests a neuromodulatory role for FMRFamide in insect neuromuscular transmission.

The presence of authentic FMRFamide in insects recently was demonstrated by its extraction and sequencing from whole *Drosophila*. Simultaneously, genes encoding FMRFamide were identified,[41,42] paving the way for molecular genetic studies of its developmental and physiological significance.

3. Cephalomyotropins

In a *tour de force* of peptide isolation and sequencing, 12 novel myotropic peptides affecting the cockroach hindgut recently were identified from head extracts of the cockroach *Leucophaea maderae* by Holman et al. (Table 1; see Chapter 15 of this work). Four classes of these peptides are recognized: leucopyrokinin,[15] leucokinins[43-46] (I to VIII), leucosulfakinins[47,48] (I and II), and leucomyosuppressin.[49] All cephalomyotropins are amidated at the C-terminus, and many possess a blocked glutamic acid residue at the N-terminus. Apparent sequence homologies relate the cephalomyotropins to AKH-like peptides and to FMRFamide-related peptides. In particular, leucopyrokinin and the leucokinins are closely related to AKH-like peptides identified in orthopterous and lepidopterous insects (Table 1). Leucomyosuppressin, leucosulfakinins, and FMRFamide share common C-terminal LRFamide or MRFamide sequences. These similarities lend support to the notion that many peptide sequences in insects may be expressed by a small number of genes.

Low concentrations of the cephalomyotropins stimulate contractions in the cockroach hindgut; an exception is leucomyosuppressin, which has an inhibitory action. Cellular localization and physiological actions of the cephalomyotropins have not yet been reported.

B. METABOLIC PEPTIDES

1. Adipokinetic Hormones (AKH)

AKH-like peptides (Table 1) occur widely in the class Insecta. Structures with related sequences occur in tissues of locusts,[50,51] cockroaches,[52-55] and lepidopterous insects.[56,57] The primary function of AKH in locusts is the mobilization of triglyceride from fat body lipid to fuel flight. This is complemented by enhancement of diglyceride penetration into flight muscle.[58] Likewise, cockroach AKH-like peptides periplanetin CC1 (MI, Neurohormone D) and CC2 (MII) mobilize energy stores by stimulating the production of the insect sugar trehalose from fat body glycogen stores. Significant amino acid sequence similarity between CC2 and glucagon (Table 1) represents remarkable biochemical and functional analogy between distant animal phyla.

AKH peptides therefore may regulate the release of stored energy, its transport through the circulatory system, and its efficient utilization by muscle simultaneously. Simultaneous regulation of diverse target tissues — fat body, heart, and flight muscle — illustrates high-level orchestration of complex physiology and behavior by neuropeptide hormones (see also Chapter 19 of this work).

2. Diuretic Hormones (DH)

Peptides controlling water elimination in insects have been described most extensively in the locust. Several peptides are implicated in the stimulation of urine secretion by Malpighian tubules, the functional analogues of the vertebrate kidney. The first DH to be sequenced was "arginine vasopressin-like diuretic hormone" (AVP-like DH; Table 1). Its isolation from subesophageal and thoracic ganglia of the locust *Locusta migratoria* was facilitated by the use of an antiserum recognizing the vertebrate neuropeptide arginine vasopressin.[59] AVP-like DH is an antiparallel homodimer with sequence homology to arginine vasopressin and to arginine vasotocin, which is the ancestral form found in lower vertebrates. AVP-like DH and two other diuretic peptides isolated from locust CC[60,61] stimulate both cyclic AMP and urine secretion in Malpighian tubules. One of the unidentified DH peptides is recognized by antisera specific for the vertebrate peptide adrenocorticotropic hormone (ACTH).[60]

C. DEVELOPMENTAL PEPTIDES

Peptide hormones have been known for decades to control critical stages of insect development. Two key developmental hormones have been isolated and identified in recent

years: the prothoracicotropic hormone (PTTH) and eclosion hormone (EH). Although both of these hormones have been identified in lepidopterous insects, it is likely that similar hormones are present in cockroaches.

1. Prothoracicotropic Hormone (PTTH)

PTTH, or "brain hormone", was the first *bona fide* neuroendocrine substance to be described.[62,63] It is synthesized by neurosecretory cells in the brain and released from the storage lobe of the CC into the circulatory system. PTTH stimulates the prothoracic glands to secrete the steroid molting hormone ecdysone. In the moth *B. mori,* peptides with PTTH-like activity occur in two sizes: 22 kDa and 4 kDa. Similar forms have been described in other lepidopterous insects. The 4-kDa form of *B. mori* can be resolved further into three molecular species: 4K-PTTH-I, II, and III.[64] One of these, 4K-PTTH-II, has recently been sequenced in its entirety.[65] Its sequence is closely related to insulin (Table 1), a finding that may account for the reports over the years of insulin-like peptides detected with immuno-histochemical methods in insects.

2. Eclosion Hormone (EH)

EH is secreted by brain neurosecretory cells into the bloodstream,[66] from which it acts on the CNS to trigger the complex motor program responsible for cuticular shedding at metamorphosis.[66] The EH of the sphinx moth, *M. sexta,* recently was isolated and sequenced.[67,68] It consists of 62 amino acids and has a molecular mass of 6813 Da.

The question of whether EH controls ecdysis in nonlepidopterous insects has not been clearly resolved. However, Truman et al.[69] reported that extracts from the nervous systems of five orders of insects tested positive on an *M. sexta* EH assay, suggesting that EH or a peptide similar to it may indeed be important in the ecdysis of all insects.

3. Allatotropins and Allatostatins

Considerable evidence demonstrates the regulation of juvenile hormone (JH) secretion by neuropeptides. Although peptide sequences are not yet published, several factors have been characterized which promote (allatotropic[70,71]) or inhibit (allatostatic[72,73]) JH secretion. Because the timing of JH release is important both in the determination of form in developing insects and in the reproductive cycle of adult females, its regulation is critical in developmental and reproductive processes.

Allatotropin I of *Locusta migratoria* is apparently a small (700 to 2,000 Da), heat-stable peptide located primarily in the brain and CC.[71] This contrasts with the considerably larger (40,000 Da) allatotropin characterized in the moth *M. sexta.*[70] Allatostatic peptides of *M. sexta* are reported to reach the corpus allatum (CA) both through the bloodstream and by direct delivery via axon projections from the brain.[72] In the viviparous cockroach *Diploptera punctata,* an allatostatin occurring at high concentration in the brain appears to regulate specifically the synthetic step prior to methylation and epoxidation.[73]

4. Bursicon

Bursicon, a peptide originally associated with the cuticular tanning process in blowflies,[74] has been associated with sclerotization, plasticization, cuticle deposition, and programmed cell death (see review of Reynolds[75]). Bursicon in cockroaches[76-78] and moths[75,79] has received considerable attention. In the cockroach *P. americana* it is distributed throughout the nervous system and occurs as well in extracts of fore- and hindgut.[80]

Bursicon shows an impressive level of cross-reactivity among insect groups, indicating that its structure and/or the structure of its receptor may be conserved during evolution. The molecular weight of bursicon has been estimated to be between 20 and 60 kDa.[75] The purification and structural elucidation of bursicon has been hampered by stability problems,

and its ultimate identification may depend on a combination of partial sequence information and cDNA probing to isolate the gene.

D. VERTEBRATE PEPTIDES IN INSECTS

A considerable body of data, mostly from immunohistochemical studies, indicates that neuropeptides initially identified in the vertebrate nervous system or gut are present in at least some related form in the Insecta. The foregoing discussion of insect peptides cited four instances of apparent sequence homology between identified insect neuropeptides and vertebrate peptides: leucosulfakinins (gastrin/CCK), AVP-like DH (arginine vasopressin), periplanetin CC2 (glucagon), and PTTH (insulin). These examples are few compared to the number of reports demonstrating immunohistochemical staining of insect cells with vertebrate peptide antisera. Such evidence indicates that the following peptides or related forms may be chemical messengers in insects: oxytocin, vasopressin, neurophysin, somatostatin, ACTH, substance P, hypothalamic growth hormone-releasing factor (GRF), glucagon, insulin, gastrin/cholecystokinin, vasoactive intestinal polypeptide, pancreatic polypeptide, secretin, and endogenous opioids. Because this literature is too extensive to cover in any detail here, the reader is referred elsewhere for further information.[4,81] Studies on the presence of opioid-like peptides and corresponding receptors in the cockroach nervous system are reviewed in this work by Stefano et al. (Chapter 16).

III. PEPTIDERGIC SIGNALING IN INSECTS: CELLULAR ELEMENTS AND PHYSIOLOGICAL TARGETS

A. NEUROSECRETORY CELLS

The novel idea that the nervous system serves an endocrine function was first suggested by Kopeć,[62,63] who described a secretion from the insect brain promoting pupation in moths. Subsequent recognition that magnocellular neurosecretory cells in teleost fish secrete vasopressin and oxytocin[82] led to the development of neurosecretion as a concept, at that time a rather radical one. During the intervening decades, neurosecretion has evolved considerably to include a wide array of secretory strategies affecting both long- and short-range targets.

Classical neurosecretory cells (NSCs) engage in endocrine-like secretion of peptides into the circulatory system. They are distinguished from typical neurons by synthesis and release of peptides on a large scale, one sufficient to achieve physiologically relevant concentrations in the bloodstream. Hence, neurosecretory somata display a highly developed synthetic machinery which is typical of endocrine cells. At the same time, they show morphological and electrical properties typical of neurons, thus constituting a class of hybrid cell with both endocrine and neural characteristics.

The vast majority of NSCs in insects are known from classical histological stains that react with cytoplasmic polypeptides thought to be primarily carrier proteins associated with bioactive peptides. These methods and their uses in insects have been dealt with cogently by Rowell[83] and Raabe[3,4] and will not be reviewed here. Suffice it to say that, at the light microscope level, NSCs stained by acidic dyes are "Type A", basic dyes, "Type B", and other dyes, "Type C". Whereas such histological stains depict NSCs as separate types, ultrastructural studies reveal that all contain cytoplasmic osmiophilic secretory granules ranging from 50 to 300 nm. Attempts to associate larger neurosecretory granules (150 to 300 nm) with peptides and smaller granules (<150 nm) with amines[4] have not proven successful.

Type A, B, and C NSCs have been mapped in the brain and ventral nerve cord of the cockroach,[84] locust,[85] and cricket.[86] These studies reveal a remarkable constancy and similarity of NSC type and position,[83] implying that they may be uniquely identifiable and, thus, amenable to studies at anatomical, physiological, and biochemical levels. Indeed, many

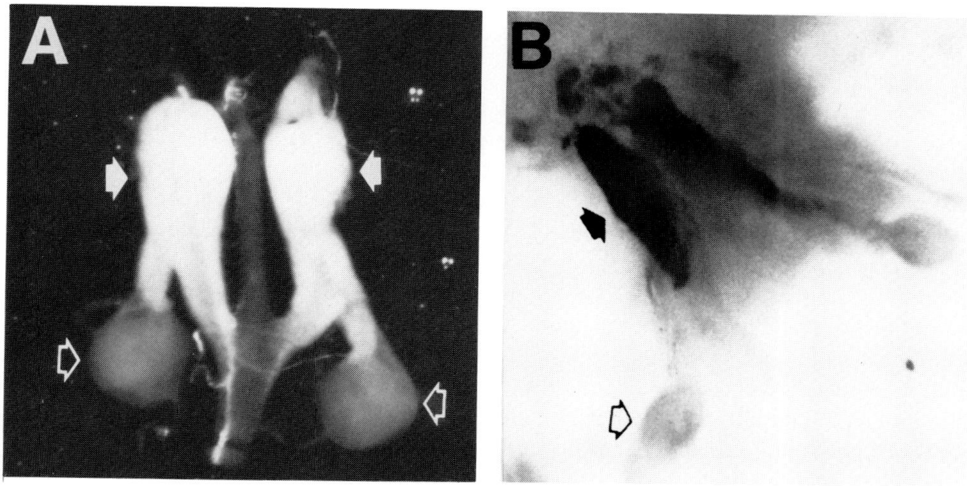

FIGURE 2. The retrocerebral complex of *Periplaneta americana*, a major neurohemal organ for liberation of hormonal peptides into the hemolymph, consists of paired corpora cardiaca (CC) and corpora allata (CA). (A) A freshly dissected gland pair, showing the elongated, whitish CC (solid arrows) and, attached to their distal ends, the bulbous CA (open arrows). Peptidergic neurons in the brain project to the CC/CA complex for release of their contents into the bloodstream. The CC also contains intrinsic glandular cells which release a variety of neuropeptides. (B) A similar view of the glands following staining with an antiserum which recognizes periplanetin CC1 and CC2. Note the heavy reaction in the CC (solid arrow). No staining of the CA (open arrow) is apparent. (Figure 2A courtesy of Dr. S. J. Kramer, Zoecon Research Institute, Sandoz Crop Protection, Palo Alto, CA.)

NSCs have been uniquely identified in recent years as intracellular recording, dye infusion, and immunohistochemical staining techniques have improved.

Although designation as "neurosecretory" emphasizes a primarily endocrine function for these cells, it should be emphasized that they arborize within the CNS, where they appear on ultrastructural grounds to make synaptic junctions with postsynaptic elements. Such apparent synaptic junctions, characterized by neurosecretory granules in presynaptic terminals, were termed "synaptoid" structures by Bargmann et al.[87] Whether the physiological properties of synaptoids differ from those of typical synapses cannot be said with certainty at the present time. Nevertheless, these observations strongly suggest that NSCs are capable of both neurosecretion as well as direct delivery of peptides, depending on the architecture of their secretory specializations and their spatial proximity to target receptors. Likewise, neuropeptides themselves play multiple roles: as neurohormones when released into the blood and as neurotransmitters when released at synaptic junctions. The occurrence of more than one type of specialization within a single identified NSC has not to this author's knowledge been described, but one anticipates that such will emerge as analyses of uniquely identified NSCs reach higher levels of sophistication.

B. NEUROHEMAL ORGANS

Large quantities of neurosecretory material in the cockroach are released from the retrocerebral complex (Figure 2A), a pituitary-like gland located immediately behind the brain. It is composed of paired CC and CA, which have both neural and glandular portions. The neural components of the CC and CA are release sites for NSCs of the brain[88,89] and subesophageal ganglion.[90] Glandular cells intrinsic to the CC and CA account for a substantial portion of the secreted material from the retrocerebral complex (see also Chapters 5 and 21 in this work).

The major secretory products of the CC in the cockroach are the AKH-like peptides MI and MII[54] (periplanetin CC1 and CC2[53]). These peptides are present at levels of 100 and 50

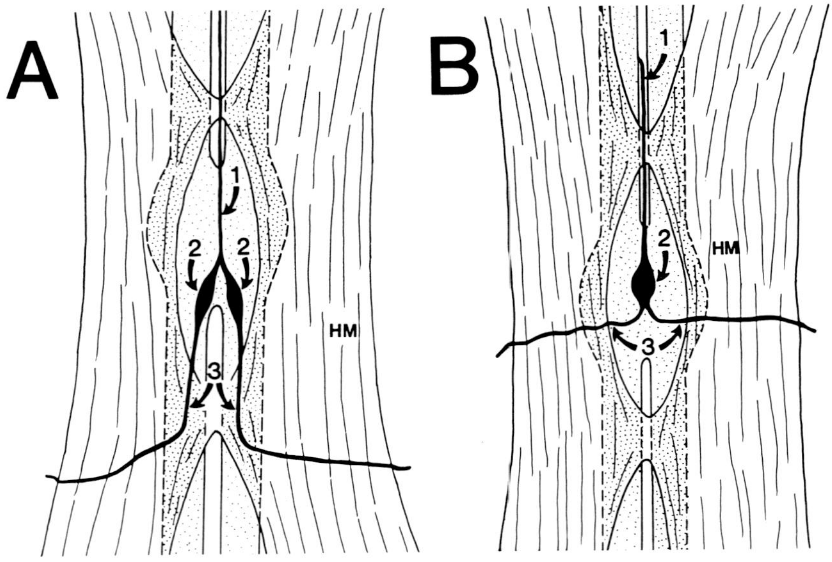

FIGURE 3. Perisympathetic organs on the dorsal surface of each segment ganglion provide sites of neurosecretory release. Two examples of their varied configurations are depicted here on the dorsal surfaces of the second (A) and third (B) unfused abdominal ganglia of the cockroach *Periplaneta americana*. Each drawing outlines the ganglion (dashed line) and overlaying hyperneural muscle (HM) to which the transverse nerves (arrows, 3) are attached. (A) The perisympathetic organs (arrows, 2) occur as paired swellings of the transverse nerves (arrows, 3) at their junction with the median nerve (arrows, 1). (B) In the next posterior segment, a single perisympathetic organ (arrow, 2) appears as a swelling of the median nerve (arrow, 1) at its junction with the transverse nerve (arrow, 3).

pmol per gland pair, respectively. Immunohistochemical detection of MI and MII demonstrates their presence in the glandular lobes of both locust and cockroach CC[13,91] (Figure 2B). Synthesis of this material in the glandular lobe of the CC has been demonstrated by incorporation of ^{3}H-tryptophan into the fully processed AKH-like peptides, which are released upon depolarization with high potassium.[54,92] The CC of the locust, enriched in AKH I (500 pmol per gland pair) and AKH II, are a favorable preparation for labeling and identifying precursor proteins which are processed to the mature bioactive peptides.[92] Despite evidence that the glandular lobes account for synthesis and release of AKH peptides, it should be mentioned that at least one report suggests the presence of AKH in the neural lobes of the locust CC.[93] Brain neurosecretory cells, therefore, may contribute a portion of the total store in the glands.

Neurosecretory cells of the ventral nerve cord project to two obvious neurohemal structures: the perisympathetic organs and the cardiac nervous system. The most conspicuous perisympathetic organs occur as swellings of the transverse and median nerves at their junction on the dorsal surface of each segmental ganglion (Figure 3). Ultrastructural evidence from a variety of studies shows that these are regions specialized for release of neurosecretory substances directly into the hemolymph.[3,4] Neurosecretory axons may reach the perisympathetic organs via two routes: (1) via the segmental nerve (Figure 4D) and link nerve to the transverse nerve and (2) directly via the median nerve.[94]

The cardiac neurosecretory system is composed of paired lateral cardiac nerve cords and segmental vessels (Figure 4A). These appear to serve as end organs for peptidergic elements of the segmental ganglia.[95] Within each body segment, paired segmental nerves projecting from the ventral nerve cord make connections with the lateral cardiac nerve cord. Each segmental nerve contains two types of axon.[96,97] Extracellular potentials of large

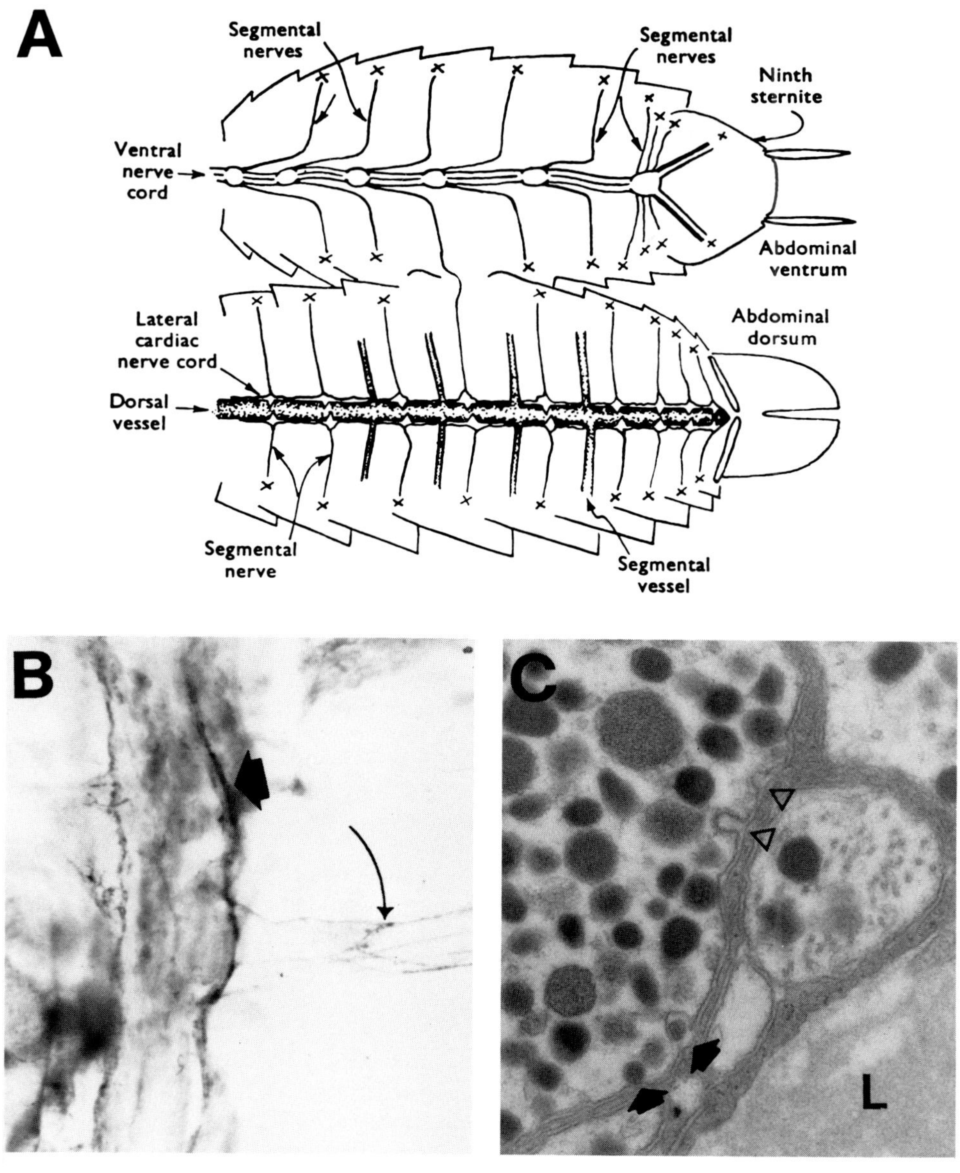

FIGURE 4. The cardiac system in *P. americana* serves as a neurohemal organ for elaboration of neuropeptides into the bloodstream. (A) Segmental nerves originating from the CNS extend into the heart and bifurcate before joining the lateral cardiac nerve cords. ''X'' marks indicate cut nerves. Ostial valves located at these bifurcations are entry points for hemolymph, which is pumped anteriorly to the head as well as laterally via paired segmental vessels. (B) Immunohistochemical staining with an antiserum against periplanetin CC1/CC2 reveals immunoreactivity in the lateral cardiac nerve (arrow) and in nerves located in the walls of segmental vessels (curved arrows). (Magnification × 500.) (C) Ultrastructure of axon profiles in a segmental vessel wall shows electron-dense neurosecretory granules and possible sites of release (solid arrows) and membrane reuptake (open arrows) situated close to the vessel lumen (L). (Magnification × 40,000). (Figure 4A from Miller, T. and Usherwood, P. N. R., *J. Exp. Biol.*, 54, 329, 1971. With permission.)

amplitude travel at 0.5 to 0.7 m/s and correspond to "ordinary" motor axons which make typical neuromuscular junctions on the myocardium. In addition to such ordinary axons, varicose neurosecretory axons containing electron-dense granules conduct lower amplitude spikes at 0.2 to 0.3 m/s and contribute to excitatory electrogenesis in the myocardium.[97] Finally, extrinsic neurosecretory cells intrinsic to the lateral cardiac nerve cords have been observed,[98,99] but their physiological properties remain as yet undescribed.

Extending laterally from the myocardium are the paired segmental vessels (Figure 4A). The walls of the vessels are noncontractile, but they contain bundles of neurosecretory axons which may release their contents into the lumen (Figure 4C). Immunohistochemical staining of the cardiac system using anti-periplanetin antisera[13] reveals (an) immunoreactive substance(s) in the lateral cardiac nerve cords and in neurosecretory axons of the segmental vessel walls (Figure 4B). No data are available on the release of immunoreactive material from the lateral cardiac nerves or segmental vessels, but its presence suggests that the cardiac system may be a major neurohemal organ in the cockroach.

1. Brain Neurosecretory Cells

The majority of NSCs occurring in the cockroach brain are located in the medial and lateral regions of the protocerebrum (Figure 5); smaller numbers are observed in the deuto- and tritocerebrum and optic lobes. A rough count of the NSCs in the cockroach brain is estimated to be on the order of 100 to 200. This number was based on several studies which employed either classical neurosecretory stains or retrograde dye infusion from release sites in the retrocerebral complex.[83,88,89] While most of these cells are distributed in the medial and lateral pars intercerebralis of the protocerebrum, additional cells have been located in the tritocerebrum by cobalt backfilling via the nervi corporis cardiaci I (NCCI).

Individual median NSCs of cockroach and cricket brain have been identified anatomically and physiologically using intracellular microelectrode techniques.[100,101] Two types of cells have been observed in the cricket brain: those projecting axons through the NCCI to the retrocerebral complex and beyond, and local cells whose projections do not leave the brain.[101] Little is known regarding the biochemical content of these cells or their functional significance. However, it may be expected by analogy with homologous structures in other insects that these cells release PTTH, EH, DH, bursicon, and a variety of peptides first discovered in the vertebrate nervous system or gut.

The first biochemical identification of brain NSCs involved localization of PTTH.[102] Some 62 years after the demonstration by Kopeć[62] of PTTH-like activity in moth brain, the peptide was located in a single pair of lateral protocerebral NSCs by single cell dissection and bioassay.[102] Later, cells considered to be identical to these PTTH neurons were dye-injected and shown to project to release sites in the CA.[103]

A rather comprehensive identification of EH-containing neurons in *M. sexta* brain was provided by Copenhaver and Truman.[66] Hormonal activity was demonstrated by single cell dissection, bioassay, and staining with EH antisera. Intracellular staining of EH neurons *in situ* revealed axons projecting to the retrocerebral complex; spiking activity in these neurons led to the appearance of EH bioactivity in the bathing medium. Thus identified anatomically and biochemically, these neurons were assigned unequivocal physiological functions.

Neurosecretory somata and endings occurring in the stomodeal ganglia and associated musculature have been observed with classical stains and electron microscopy, but individual neurons have yet to be identified. Immunohistochemical staining using an antiserum recognizing both periplanetin CC1 and CC2 (MI and MII) revealed immunoreactive neuronal somata of the brain, esophageal nerve, and ingluvial ganglion.[13] Cell bodies in the brain occur in the medial and lateral regions of the protocerebrum and at the bases of the optic lobes. Immunoreactive innervation in the paired proventricular ganglia and on anterior foregut musculature (Figure 6) suggests a widespread regulatory role for these peptides in the stomodeal nervous system and musculature.

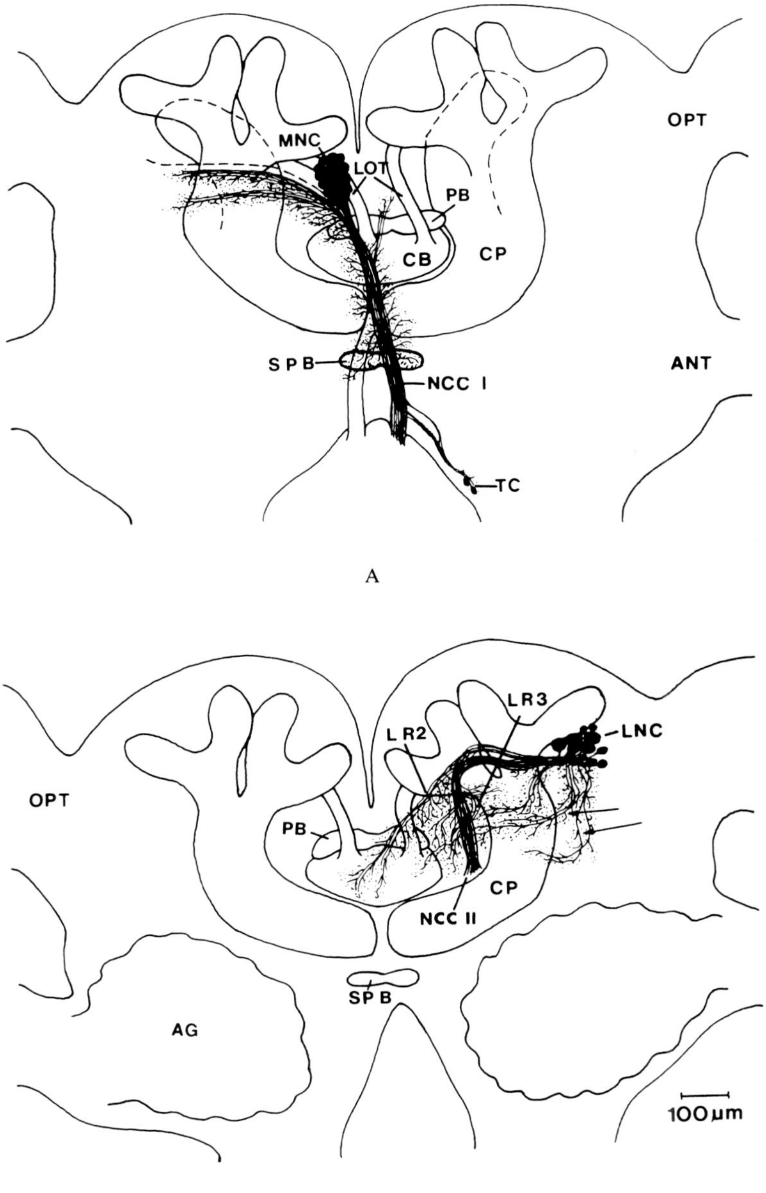

FIGURE 5. Drawings of cerebral NSCs in the cockroach *P. americana,* revealed by retrograde cobalt infusion through the nervi corpora cardiaca I (NCCI) and II (NCCII). (A) The median neurosecretory cells (MNC) occur as a cluster between the paired corpora pedunculata (CP) and project axons to the retrocerebral complex via NCCI. (B) Lateral neurosecretory cells (LNC), positioned anterior to the lateral calyx of the corpora pedunculata, project axons to the retrocerebral complex via the NCCII. Additional abbreviations: LOT, lateral ocellar tract; PB, protocerebral bridge; CB, central body; SPB, subpeduncular body; TC, tritocerebral cell group LR2; LR3: see Reference 88; AG, antennal glomeruli; OPT, optic lobe. (From Koontz, M. and Edwards, J. S., *J. Morphol.,* 165, 285, 1980. With permission.)

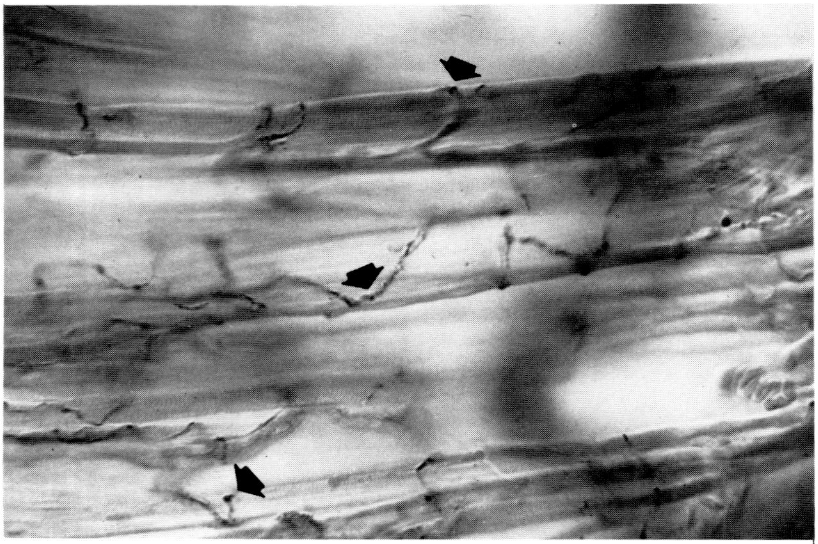

FIGURE 6. An antiserum specific for periplanetin CC1/CC2 stains presynaptic terminals (arrows) innervating the foregut musculature of *P. americana*. Immunoreactivity of both the retrocerebral complex and nerve terminals innervating muscle suggests that AKH-like peptides may act either hormonally or as synaptic transmitter substances.

2. Neurosecretory Cells of the Ventral Ganglia

a. Lateral White (LW) Neurons

These NSCs were recognized in paraldehyde-fuchsin stains as lateral Type A cells in the ventral ganglia of several orthopterous insects.[84-86] LW somata are easily visualized in live preparations as whitish, bilaterally paired somata in the first three unfused abdominal ganglia (Figure 7A). The somata of these neurons are further distinguished from others in the ganglion by their vacuolated appearance in stained preparations[86,104] (Figure 7B). The vacuoles may serve as a reservoir for precursor material.[104] The abundance of neurosecretory granules 150 to 300 nm in diameter suggests active synthesis of peptides by these cells.

Upon dissection of individual LW somata and extraction in acid, a biologically active peptide indistinguishable from proctolin was detected using HPLC fractionation and bioassay[8] (see also Chapter 4 of this work). Subsequent morphological, immunological, and biochemical analyses of LW neuron somata suggest that they contain peptides in addition to proctolin. Extraction and bioassay of LW somata show them to contain bursicon bioactivity.[80] This information correlates with central and peripheral projection patterns which are quite similar to bursicon-containing neurons in *M. sexta*.[105] Furthermore, the LW neurons show positive staining using antisera recognizing periplanetin CC1-like immunoreactivity.[13]

A typical cockroach LW neuron, reconstructed from intracellular cobalt infusion followed by silver intensification, is shown in Figure 7D. The cell projects to the medial area of the neuropil, where it arborizes and makes apparent connections with fiber tracts passing through the ganglion. Fine projections also are observed to extend outside of the neuropil, several hundred microns into the connectives; the precise functions for these extraneuropilar projections are unknown. Intracellular recordings from LW neurons indicate that they receive synchronous presynaptic input from segmental interneurons passing though each segmental ganglion.[105a] Possibly the LW neurons are recruited as a functional unit.

Each LW neuron in the cockroach sends collateral axons via multiple segmental nerves to peripheral targets (Figure 7C). Two end organs for the LW neurons are known at the present time. Each axon projects through the segmental nerve to its junction with the link

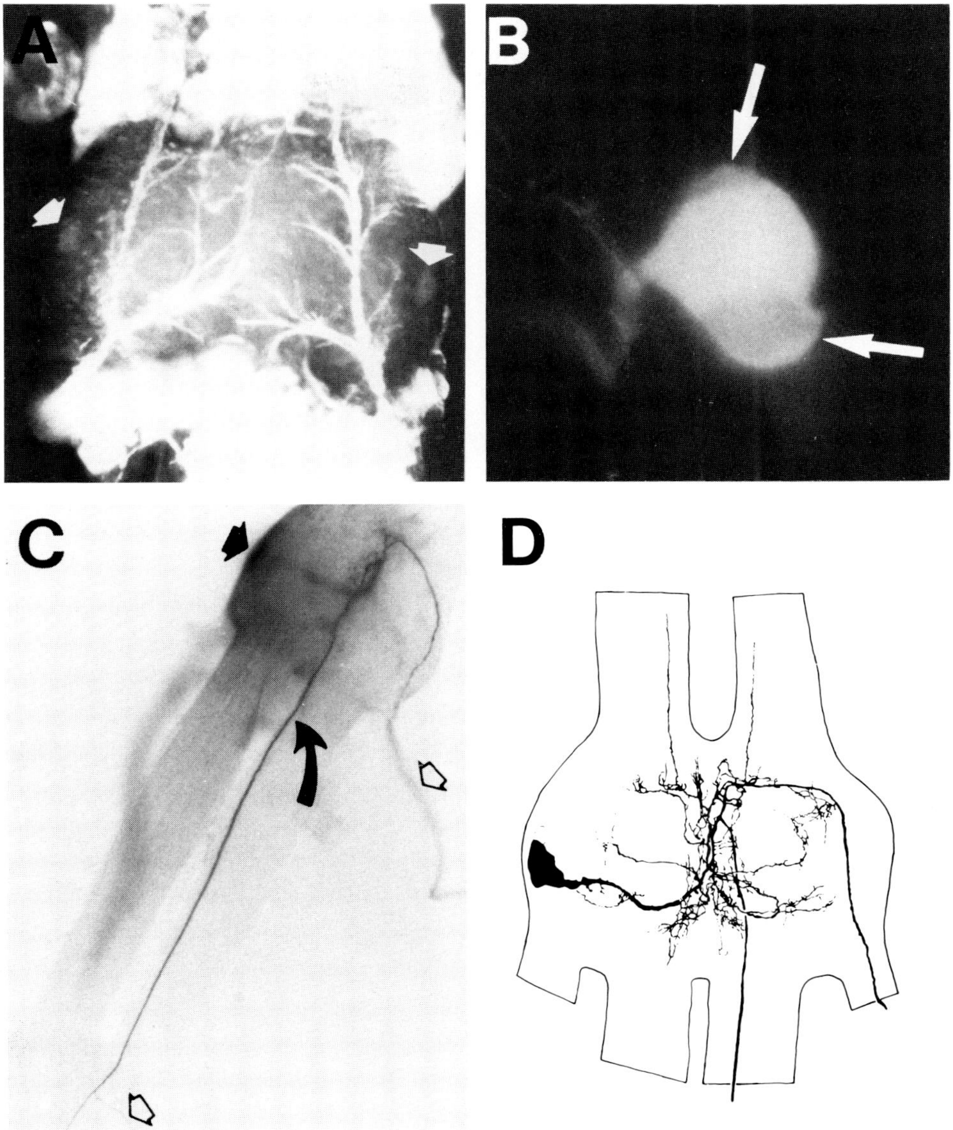

FIGURE 7. The Lateral White (LW) neurons of *P. americana* are bilaterally paired, segmentally repeated neurosecretory cells which occur in the unfused abdominal ganglia. (A) The whitish somata of the LW (arrows) neurons are visible in freshly dissected abdominal ganglia, permitting microelectrode penetration, dye filling, and single-cell dissection for biochemical analysis. (B) An LW soma which has been injected with the fluorescent dye Lucifer yellow. Arrows point to lightly stained vacuoles associated with the soma which may be involved in storage or processing of precursor proteins. (C) Injection of Lucifer yellow into a single LW neuron reveals a cell body (arrow) and a contralateral axon (curved arrow). Branches (open arrows) from the descending axon project out of the CNS in each ganglion. (D) Following injection of intracellular dye, the anatomy of an LW neuron is revealed in a camera lucida drawing. The cell has extensive arborizations both within and outside of the neuropil.

nerve, where it bifurcates. One branch enters the transverse nerve and makes endings in perisympathetic organs; the other branch extends to the lateral cardiac nerve, where it again bifurcates, entering this nerve both anteriorly and posteriorly. Spiking activity in the LW neurons leads to an increase in the force and frequency of heartbeat contractions,[105b] an effect which is consistent with the release of the cardioactive peptides proctolin and periplanetin.

b. AVP-Immunoreactive Neurons

A single pair of distinctive Type A NSCs are reported[84-86] to occur on the ventral surface of the subesophageal ganglion. These cells were described by Delphin[85] as being "the most distinctive neurosecretory cells in the entire nervous system". Recent immunohistochemical studies demonstrate that these two neurons account for the total vasopressin-like immunoreactivity in the nervous system.[106] Anti-AVP-like immunoreactivity has been detected in projections of these neurons which extend to all ganglia of the nervous system, including the brain and optic lobes. Neurohemal endings are observed at the bases of nerve roots in each segmental ganglion, indicating that the AVP-like DH described by Proux and colleagues[59] may be elaborated in all segments by this single pair of NSCs. A functional connection between activity in these cells and secretion of the diuretic peptide remains to be demonstrated.

c. Dorsal Unpaired Median (DUM) Neurons

This unique class of NSCs, first identified in locust ventral ganglia to be octopaminergic[105] (see also Chapter 17 of this work), is present in a variety of insects. Secretory activity by one of these cells, DUMETi (dorsal unpaired median [DUM] neuron innervating the extensor tibiae muscle), modulates skeletal neuromuscular transmission in the locust.[108] Homologous, segmentally repeated clusters of DUM neurons also occur in cockroach segmental ganglia (Figure 8) and appear to be octopaminergic.[109]

Evidence from immunohistochemical staining indicates that some DUM neurons in cockroach abdominal ganglia are peptidergic.[13] Figure 8B depicts two periplanetin-immunoreactive DUM neurons in the fourth unfused abdominal ganglion. This result suggests the possible colocalization of octopamine and AKH-like peptides in these DUM neurons. The physiological significance of this apparent colocalization is not known.

C. PEPTIDERGIC MOTONEURONS

The discovery of neurosecretion established peptides as chemical signals produced by and released from the nervous system. However, the emphasis on endocrine secretion tended to reinforce the notion that neuropeptides were strictly hormonal in nature. In fact, the possibility that peptides could act as synaptic transmitters in insects was suggested for many years by ultrastructural evidence depicting electron-dense granules at what appeared to be presynaptic specializations contacting gland and muscle cells.[110,111] Unfortunately, interpretation of such morphological evidence was hampered by the virtual absence of chemical or physiological information regarding the contents of these so-called neurosecretory granules until the latter 1970s. There was uncertainty and even reluctance to accept the notion that peptidergic endings viewed ultrastructurally could belong to typical motoneurons responsible for normal contractile events in muscle targets. Instead, terminals containing electron-dense granules were postulated to be "neurosecretomotor" or "synaptoid" structures to distinguish them from typical synaptic structures.

The introduction of antisera specific for proctolin led to its unequivocal localization in identified motoneurons of the cockroach[9] and locust.[28,112,113] Subsequently, proctolin-immunoreactive nerve terminals were documented on both skeletal and visceral musculature, including the coxal depressor muscle, intersegmental body wall muscle, ventral diaphragm,[114] proctodeum,[10] and oviducts.[11] Ultrastructural immunocytochemistry also revealed proctolin-like immunoreactivity in presynaptic terminals within the sixth abdominal ganglion of *Periplaneta,*[30] indicating that proctolin may be a CNS neurotransmitter as well.

The total number of neurons exhibiting proctolin-like immunoreactivity in the cockroach nervous system amounts to 2% or less of the total neuronal pool in the ventral nerve cord.[113] This small proportion is in line with the other neuronal populations which contain octopamine, serotonin, and AKH-like neuropeptides and may indicate that a large number of neurotransmitter substances remain to be identified.

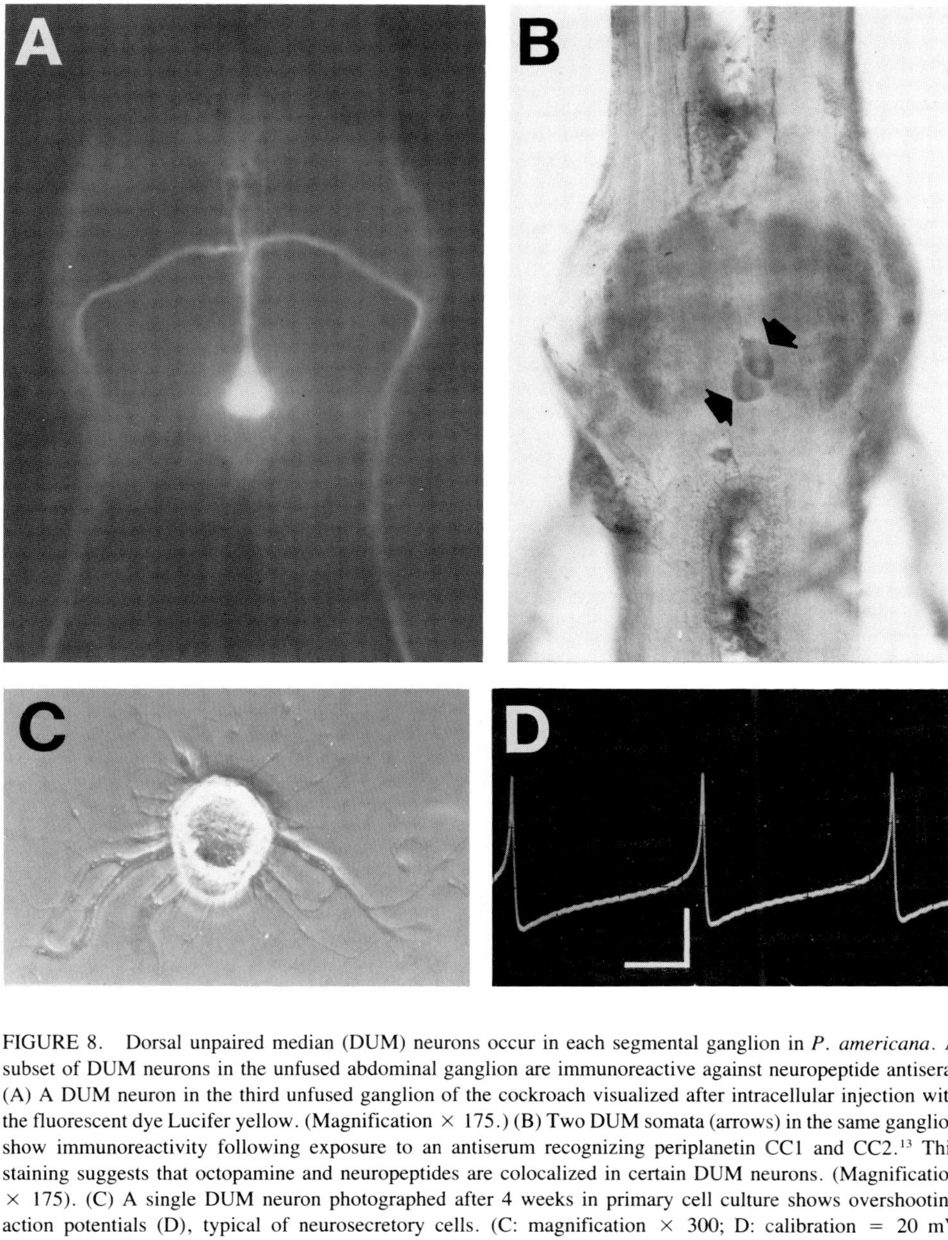

FIGURE 8. Dorsal unpaired median (DUM) neurons occur in each segmental ganglion in *P. americana*. A subset of DUM neurons in the unfused abdominal ganglion are immunoreactive against neuropeptide antisera. (A) A DUM neuron in the third unfused ganglion of the cockroach visualized after intracellular injection with the fluorescent dye Lucifer yellow. (Magnification × 175.) (B) Two DUM somata (arrows) in the same ganglion show immunoreactivity following exposure to an antiserum recognizing periplanetin CC1 and CC2.[13] This staining suggests that octopamine and neuropeptides are colocalized in certain DUM neurons. (Magnification × 175). (C) A single DUM neuron photographed after 4 weeks in primary cell culture shows overshooting action potentials (D), typical of neurosecretory cells. (C: magnification × 300; D: calibration = 20 mV [vertical], 20 ms [horizontal]).

Periplanetin-like immunoreactivity is also observed in presynaptic terminals on muscle. Antisera recognizing both periplanetin CC1 and CC2 (MI and MII) stain presynaptic terminals on foregut musculature[13] (Figure 6), suggesting that peptides other than proctolin may function as synaptic transmitters.

1. D_s Motoneuron

This motoneuron, "the slow depressor of the coxa" (or D_s), was the first proctolinergic motoneuron to be identified. Conclusive evidence for peptidergic transmission was presented in a series of papers on the role of proctolin at the D_s motoneuron-coxal depressor neuro-

muscular junction of the cockroach *P. americana*[9,23,112,113] (see also Chapter 14 of this work). Several lines of evidence support the role of proctolin as a cotransmitter in this motor unit: (1) the D_s motoneuron is proctolin immunoreactive and contains an extractable substance that is biochemically indistinguishable from proctolin,[9] (2) proctolin is released upon nerve stimulation or potassium depolarization in a calcium-dependent manner,[23] and (3) proctolin elicits contractile events which resemble those obtained by selective activation of the motoneuron.[23] Interestingly, the D_s motoneuron and presumably other motoneurons as well do not react with classical neurosecretory stains and show none of the classical glandular attributes typical of NSCs. This would seem to indicate that peptidergic motoneurons, the targets of which are close-range synaptic targets, synthesize far less peptide than NSCs.

2. SETi Motoneuron

Recent reports that the SETi (slow extensor of the extensor tibiae) motoneuron of the locust contains proctolin[28,112] help to explain a number of previous observations on the contractile responses in the extensor tibiae muscle (ETi).[115,116] For example, rhythmic activity in a myogenic bundle within the extensor tibiae muscle is accelerated by extremely low levels of proctolin[8,115] or by SETi motoneuron activation.[112] Stimulation of this motoneuron also induces several types of tonic tension which may not be explained adequately by the release of conventional transmitter alone. For example, smooth tonic tension occurs at stimulation frequencies as low as 3 Hz, and a strong facilitation of the mechanical response is measured when low-frequency SETi motoneuron activity is repeated following a 1- to 2-min pause. Based on previous findings,[23] these results are consistent with the cotransmitter actions of proctolin in the extensor tibiae muscle.

3. Visceral Motoneurons

Proctolin activates many types of visceral muscle in the cockroach, including the proctodeum,[18] hyperneural muscle,[21] myocardium,[22] antennal heart,[117] and oviduct.[117a] In one of the earliest accounts of proctolin immunohistochemistry, Eckert and colleagues[10] showed immunoreactive axons and presynaptic terminals associated with the proctodeum. Additional instances of PLI were demonstrated in synaptic endings on the hyperneural muscle[114] and oviduct.[11,29] The identification of individual proctolinergic visceral motoneurons remains to be accomplished.

D. PEPTIDERGIC INTERNEURONS

Little is known about peptidergic interneurons in insects. However, in what appears to be the sole reference to them, Keshishian and O'Shea[118] demonstrated PLI in interneurons of the postembryonic locust CNS. Confirmation of authentic proctolin content has not been achieved. Central integrative roles for neuropeptides can only be speculated on at this point, but this report provides an impetus for pursuing the challenge of identifying these interneurons physiologically.

E. PERIPHERAL PEPTIDERGIC CELLS

NSCs situated along peripheral nerve roots are known to occur in a variety of insects.[119] These cells have not yet been uniquely identified using intracellular dyes, nor have the substances (presumably peptides) contained within these cells been eludicated structurally. The most detailed studies of peripheral NSCs have been conducted by Orchard[5] using the stick insect *Carausius*. The cells are multipolar, sending what appear to be efferent processes along the surfaces of nerve fibers. Ultrastructural investigations provide morphological evidence for release sites along these processes, indicating that these cells elaborate their contents directly into the circulatory system.

F. PEPTIDE NEURONS IN CULTURE

A few attempts to culture NSCs from the cockroach nervous system have been made. Unidentified medial NSCs explanted and grown in long-term primary cultures demonstrated an impressive ability to extend processes, form connections, and exhibit spontaneous electrical activity for long periods *in vitro*.[120] More recently, Beadle and Lees[121] have achieved impressive successes in maintaining neurons from the embryonic cockroach nervous system in long-term cultures. These exhibit apparently normal excitatory activity and show considerable promise for single-channel analysis of ion channels and receptors. Whether these cultures contain peptidergic neurons is not clear, but it may be possible to incorporate immunohistochemical staining with the electrophysiological work which has already been demonstrated (see also Chapter 12 of this work).

In vitro culturing of individual neurosecretory neurons in insects has been accomplished in a pilot study.[121a] Figure 7C shows a DUM cell after 3 weeks *in vitro*. Recordings from this cell showed normal resting potentials and characteristic action potentials (Figure 8D) upon current injection. Similar results were obtained with LW neurons. It appears feasible to study the connections formed by individually identified peptidergic neurons in culture.

IV. RELEASE OF NEUROPEPTIDES

A *bona fide* chemical messenger is expected to be secreted into the bathing medium following cellular activity. Demonstrations of peptidergic release from stimulated NSC groups or tracts are reported to be calcium dependent,[5] supporting the notion that peptides are released by a vesicular mechanism. However, the common thread among release studies performed prior to the 1980s was the uncertain nature of both the cellular elements and the factors that they released. This situation has changed with the recent structure elucidation of numerous neuropeptide factors and the identification of neurons which synthesize and release them. The following summarizes several reports which document the release of proctolin, AKH-like peptides, and EH.

A. CORPUS CARDIACUM (CC)

The CC of the cockroach and locust are particularly rich in neuropeptides, making them excellent experimental systems for studies of peptide release. AKH is released from the locust CC following any of the following three treatments: electrical stimulation, potassium depolarization, or incubation of the glands in saline containing octopamine.[122] These experiments demonstrate not only that AKH is released, but that octopamine may act as a neurotransmitter to cause release under natural conditions. This hypothesis is reinforced by the localization of octopamine in terminals innervating the glandular lobe of the CC.[122]

O'Shea and colleagues[54] demonstrated the release of MI and MII (periplanetin CC1, CC2) from potassium-depolarized CC of the cockroach. The released peptides were identified by HPLC and mass spectroscopy.[55] Similarly, AKH I and AKH II are released upon potassium depolarization of locust CC.[92] In addition to characterizing the release process, Hekimi and O'Shea[92] have exploited the locust CC as a model system for investigating precursor proteins involved in the biosynthesis of the AKH peptides. Through incorporation of radioactive amino acids into the glands, AKH precursor proteins have been identified as well as novel, fully processed peptides that are coprocessed with the AKHs.[123]

B. RELEASE FROM IDENTIFIED PEPTIDERGIC NEURONS

Proctolin is released from the identified motoneuron D_s upon electrical stimulation of the neuron or by potassium depolarization of its terminals on the coxal depressor muscle.[23] The method used for potassium depolarization of the coxal depressor muscle is shown in Figure 9. Individual coxal depressor muscles are excised and placed in a perfusion cartridge.

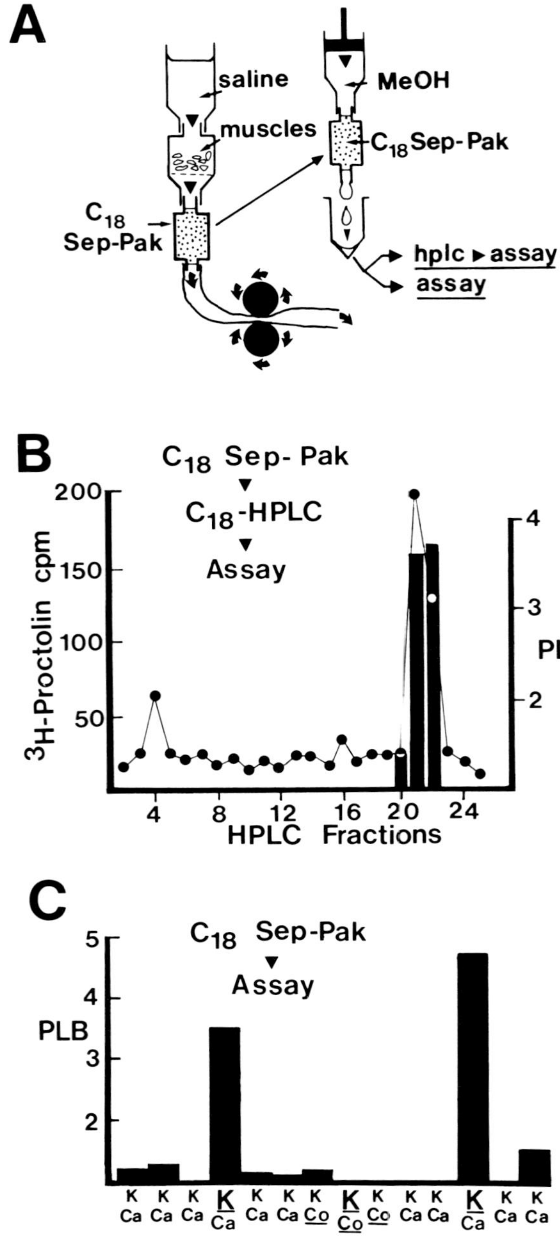

FIGURE 9. Characterization of released proctolin from coxal depressor muscles of the cockroach *P. americana*. (A) Diagram of the apparatus used to depolarize nerve terminals with high-potassium saline. Released peptides are trapped downstream in a C_{18} Sep-Pak cartridge and subsequently eluted with 80% methanol. The eluant was either purified by reversed-phase HPLC (B) or bioassayed without purification. (B) HPLC of Sep-Pak eluant from a 100 m*M* potassium perfusion of 15 coxal depressor muscles. Proctolin-like bioactivity (PLB) and the tritiated proctolin standard elute together in fractions 20 to 22. (C) Calcium dependence of proctolin release is demonstrated by replacement of calcium with 5 m*M* cobalt. Inhibition of release by cobalt saline is reversed following reintroduction of calcium. Each black bar represents the eluted material from a different Sep-Pak in a series of 10-min exposures to 20 ml of saline. Normal physiological saline is represented by a small K and Ca; altered salines are underlined (elevated-potassium saline is represented by a large K and elevated cobalt by Co). (From Adams, M. E. and O'Shea, M., *Science*, 221, 286, Copyright 1983 by the AAAS. With permission.)

Saline solutions of varying ionic compositions are drawn past the muscles and through a C_{18} reversed-phase Sep-Pak cartridge (Waters Associates, Milford, MA) by a peristaltic pump. The Sep-Pak cartridge traps peptides released into the bathing medium following exposure of the tissue to high-potassium saline (Figure 9A). For each saline condition, the cartridge is replaced and the trapped peptides are subsequently eluted with methanol.

Depolarization of the coxal depressor muscles with high-potassium saline causes release of proctolin-like bioactivity (PLB; Figure 9C). The released bioactivity displays identical elution characteristics upon purification by reversed-phase HPLC against a tritiated synthetic proctolin standard (Figure 9B). Replacement of calcium by cobalt in the bathing medium reversibly blocks proctolin release (Figure 9C). These results show that proctolin is released from the coxal depressor muscle by a calcium-dependent process.

Subsequent demonstrations of calcium-dependent proctolin release from proctodeum and oviduct nerve terminals[29,124] provide further support for proctolin being both a skeletal and a visceral neurotransmitter in insects.

V. TRANSDUCTION OF NEUROPEPTIDE SIGNALS

Reports on the receptor transduction of neuropeptide signals in insects are only beginning to emerge. Unlike conventional neurotransmitters which activate postsynaptic ion channel proteins, peptides most often have been shown to modulate intracellular levels of cyclic nucleotides or phosphoinositides. Nevertheless, recent evidence argues that peptides may also modulate membrane ion channels — in particular, calcium.

A. CYCLIC NUCLEOTIDES

Proctolin may modulate cyclic nucleotides in certain muscle types, but the available evidence is somewhat ambiguous. No direct measurement supporting cyclic nucleotide elevation is available, but agents which elevate cyclic AMP levels in the *P. americana* proctodeum potentiate both neurally evoked and proctolin-induced contraction.[125] Somewhat paradoxically, the same agents applied to the *Leucophaea maderae* proctodeum apparently cause inhibitory actions.[125] Evidence from other laboratories argues against a role for cyclic nucleotides in proctolin regulation of insect muscle. Worden and O'Shea[126] found no change in cyclic nucleotide levels of the locust extensor tibiae muscle following exposure to proctolin. Likewise, proctolin produces no measurable effect on cyclic nucleotides in locust oviduct muscle[127] or in the proctodeum, myocardium, or coxal depressor muscles of *P. americana*.[121a]

In contrast to the above, the actions of certain metabolic neuropeptides are clearly associated with elevated levels of cyclic AMP in target tissues. Lipid mobilization from locust fat body by AKH involves simultaneous elevation of cyclic AMP and inhibition of protein synthesis.[128,129] Similarly, all three diuretic hormones described in the locust — AVP-like DH,[59] an ACTH-like peptide,[60] and CC diuretic hormones[61] — stimulate cyclic AMP elevation in Malpighian tubules.

Elevation of cyclic GMP is reported to be associated with the actions of EH. EH promotes eclosion and programmed cell death in larval muscles in *M. sexta* by elevating cyclic GMP levels.[130,131] More recently, Morton and Truman[132,133] have correlated the action of EH on the CNS of *M. sexta* with cyclic GMP-mediated phosphorylation of specific proteins.

B. PHOSPHOINOSITIDE METABOLISM

Preliminary evidence suggests that myoactive peptides stimulate phosphoinositide metabolism in certain muscles. For example, proctolin-induced tension in the locust extensor tibiae muscle is correlated with the elevation of inositol trisphosphate (IP_3).[126] Likewise, cardioacceleration caused by the cardioactive peptides of *M. sexta* is associated with elevation of IP_3.[134] Treatment with lithium, an inhibitor of inositol recycling, antagonizes the effect of these peptides.

C. ION CHANNELS

Proctolin appears to increase calcium permeability of insect muscle, and this may involve direct effects of the peptide on calcium channels. Thus, the effects of proctolin on proctodeal,[135] hyperneural,[21] and oviduct[122] muscles are abolished or greatly reduced in the absence of extracellular calcium ions. Other observations that proctolin increases membrane resistance in coxal depressor[23] and hyperneural muscles[21] suggest inhibition of potassium channels by the peptide. In both instances, the observed resistance increase was calcium dependent. These results are interpreted in terms of a calcium-dependent inhibition of potassium conductance via phosphorylation of intracellular proteins.[21] Such phosphorylation could involve second messengers such as cyclic nucleotides or phosphoinositides. Alternatively, elevated intracellular calcium, either from the extracellular medium or from intracellular stores, may promote protein phosphorylation directly.

VI. DEGRADATION OF RELEASED NEUROPEPTIDES

Precise physiological degradation pathways for insect neuropeptides have not been demonstrated convincingly. The termination of a peptide signal may be simple diffusion away from the receptor area and eventual proteolysis. However, several studies have examined the possibility that specific enzymes located in target tissues may play a significant role in the inactivation of proctolin and periplanetin CC2.

A. PROCTOLIN

Proctolin in cockroach hemolymph is rapidly hydrolyzed by soluble proteases.[136-138] When high levels of proctolin (0.13 to 31 μm) were injected into individual cockroaches, proteolysis occurred primarily at N-terminal arginine.[136] A subsequent study examined the fate of physiological concentrations of proctolin (3.5 n*M*) when incubated in hemolymph or various tissue homogenates.[137] These results showed that cleavage occurred at both the Arg–Tyr and Tyr–Leu bonds, with the Tyr–Leu cleavage predominating. In tissues which contained proctolinergic innervation, the ratio of Arg–Tyr/Tyr–Leu–Pro–Thr fragments ranged from 3.1 (coxal depressor muscle 177d) to 13 (proctodeum), while tissues receiving little or no innervation exhibited ratios of 1.3 (coxal depressors 178, 179) to 1.8 (midgut). These results indicate that proctolin-innervated tissues may contain (a) specific enzyme(s) which is/are designed to degrade the peptide by cleavage at the Tyr–Leu bond. The proteolytic activity of soluble enzyme(s) present in the proctodeum homogenate was studied in some detail.[137] Using conventional Michaelis-Menten kinetics, the K_m values for the enzyme(s) Tyr–Leu cleavage were determined to be in the 50 to 150 n*M* range.

Subsequently, Isaac[139] pointed out that such (a) soluble enzymes(s) is/are not likely to constitute a specific synaptic proctolinase, since the released peptide would not be expected to come into contact with the cytosolic fraction. He characterized a membrane-bound aminopeptidase from the locust CNS which has an apparent K_m of 23 μ*M* and displays a high affinity for proctolin (0.3 μ*M*). Further work is needed to demonstrate the involvement of this enzyme in the degradation of synaptically released proctolin.

B. PERIPLANETIN CC2 (MII)

The *in vivo* and *in vitro* degradation of [4-^{3}H-Phe]CC2 was studied in the cockroach *P. americana*.[140] Unlike proctolin, this neuropeptide is blocked at both termini, has a much more hydrophobic character, and presumably functions as a hormone. Degradation was examined in three different tissues: hemolymph, fat body homogenate (a presumed target tissue[53]), and Malpighian tubule homogenates (a presumed site of degradation[141]). When incubated at 6 n*M* in tissue homogenates, CC2 degradation proceeded according to a $t_{1/2}$ of ~1 h. The major degradation products resulted from cleavage at the Pro–Asn and Phe–Tyr

bonds. Tritiated phenylalanine was also found to be a major degradation product. The rate of *in vivo* degradation was estimated to be 1 to 2 h.

VII. CONCLUSIONS AND PERSPECTIVE

Research into the occurrence and physiological significance of neuropeptides has uncovered remarkable new avenues of information transfer in the neuroendocrine system. Peptides issuing from the nervous system as hormones can orchestrate multiple target tissues participating in behavioral, developmental, and metabolic events. At the same time, peptides released at or around synaptic contacts (for example, within the nervous system or at neuromuscular junctions) modulate the strengths of connections, adding depth and versatility to existing neural networks. Peptides often defy our tendency to classify them as transmitters or hormones, for they can be both.

The recent surge of interest in peptides reflects an awareness of their importance, but is also a consequence of improved methods for identifying not only the peptides themselves, but the cells which synthesize and release them. Brian Brown's commitment to the identification of proctolin led to the widespread availability of the first insect neuropeptide.[17-19] This, in turn, permitted the generation of proctolin-specific antisera for labeling cells, the measurement of its occurrence in tissues, and the development of model preparations for analysis of its physiological significance. These advances allowed identification of proctolinergic neurons and physiological experiments on the cellular interactions involving the peptide.

At this moment, the identification of peptidergic neurons has been accomplished only to a limited extent. Even peptides known for many years, such as proctolin and AKH, have only been associated with a handful of identified neurons. Our knowledge of peptide signal transduction and degradation mechanisms is at an even earlier stage. In view of the importance attached to peptidergic mechanisms, these are fertile areas for investigation.

REFERENCES

1. **Scharrer, B.,** Neurosecretion: beginnings and new directions in neuropeptide research, *Annu. Rev. Neurosci.,* 10, 1, 1987.
2. **Tobe, S. S. and Stay, B.,** Neurosecretions and hormones, in *The American Cockroach,* Bell, W. J. and Adiyodi, K. G., Eds., Chapman and Hall, New York, 1982, chap. 12.
3. **Raabe, M.,** *Insect Neurohormones,* Plenum Press, New York, 1982.
4. **Raabe, M.,** The neurosecretory-neurohaemal system of insects; anatomical, structural and physiological data, *Adv. Insect Physiol.,* 17, 205, 1983.
5. **Orchard, I.,** Neurosecretion: morphology and physiology, in *Endocrinology of Insects,* Downer, R. G. H. and Laufer, H., Eds., Alan R. Liss, New York, 1983, 13.
6. **Truman, J. W. and Taghert, P. H.,** Neuropeptides in insects, in *Brain Peptides,* Krieger, D. T., Brownstein, M. J., and Martin, J. B., Eds., John Wiley & Sons, New York, 1983, 166.
7. **O'Shea, M. and Schaffer, M.,** Neuropeptide function: the invertebrate contribution, *Annu. Rev. Neurosci.,* 8, 171, 1985.
8. **O'Shea, M. and Adams, M. E.,** Pentapeptide (proctolin) associated with an identified neuron, *Science,* 213, 567, 1981.
9. **O'Shea, M. and Bishop, C. S.,** Neuropeptide proctolin associated with an identified skeletal motoneuron, *J. Neurosci.,* 2, 1242, 1982.
10. **Eckert, M., Agricola, H., and Penzlin, H.,** Immunocytochemical identification of proctolin-like immunoreactivity in the terminal ganglion and hindgut of the cockroach, *Periplaneta americana* (L.), *Cell Tissue Res.,* 217, 633, 1981.
11. **Lange, A. B., Orchard, I., and Adams, M. E.,** Peptidergic innervation of insect reproductive tissue: the association of proctolin with oviduct visceral musculature, *J. Comp. Neurol.,* 254, 279, 1986.

12. **Goldsworthy, G. J.,** The endocrine control of flight metabolism in locusts, *Adv. Insect Physiol.*, 17, 149, 1983.
13. **Adams, M. E., Ray, M. F., Ho, R. K., Kramer, S. J., and Marbach, P.,** Immunohistochemical localization and physiological actions of two related insect neuropeptides (CC1 and CC2), *Soc. Neurosci. Abstr.*, 11, 942, 1985.
14. **Baumann, E. and Gersch, M.,** Purification and identification of neurohormone D, a heart accelerating peptide from the corpora cardiaca of the cockroach, *Periplaneta americana, Insect Biochem.*, 12, 7, 1982.
15. **Holman, G. M., Cook, B. J., and Nachman, R. J.,** Primary structure and synthesis of a blocked myotropic neuropeptide isolated from the cockroach, *Leucophaea maderae, Comp. Biochem. Physiol.*, 85C, 219, 1986.
16. **Brown, B. E.,** Neuromuscular transmitter substance in insect visceral muscle, *Science*, 155, 595, 1967.
17. **Starratt, A. N. and Brown, B. E.,** Structure of the pentapeptide proctolin, a proposed neurotransmitter in insects, *Life Sci.*, 17, 1253, 1975.
18. **Brown, B. E. and Starratt, A. N.,** Proctolin: a peptide transmitter candidate in insects, *Life Sci.*, 17, 1241, 1975.
19. **Brown, B. E.,** Occurrence of proctolin in six orders of insects, *J. Insect Physiol.*, 23, 861, 1977.
20. **Cook, B. J. and Holman, G. M.,** The action of proctolin and L-glutamic acid on the visceral muscles of the hindgut of the cockroach *Leucophaea maderae, Comp. Biochem. Physiol.*, 64C, 21, 1979.
21. **Hertel, W. and Penzlin, H.,** Electrophysiological studies of the effect of the neuropeptide proctolin on the hyperneural muscle of *Periplaneta americana, J. Insect Physiol.*, 32, 239, 1986.
22. **Miller, T.,** Nervous versus neurohormonal control of insect heartbeat, *Am. Zool.*, 19, 77, 1979.
23. **Adams, M. E. and O'Shea, M.,** Peptide cotransmitter at a neuromuscular junction, *Science*, 221, 286, 1983.
24. **Bishop, C. A., Wine, J. J., and O'Shea, M.,** Neuropeptide proctolin in postural motoneurons of the crayfish, *J. Neurosci.*, 4, 2001, 1984.
25. **Bishop, C. A., Wine, J. J., Nagy, F., and O'Shea, M.,** Physiological consequences of a peptide cotransmitter in a crayfish nerve-muscle preparation, *J. Neurosci.*, 7, 1769, 1987.
26. **Bishop, C. A., O'Shea, M., and Miller, R. J.,** Neuropeptide proctolin (H–Arg–Tyr–Leu–Pro–Thr–OH): immunological detection and neuronal localization in the insect central nervous system, *Proc. Natl. Acad. Sci. U.S.A.*, 78, 5899, 1981.
27. **Bishop, C. A. and O'Shea, M.,** Neuropeptide proctolin (H–Arg–Tyr–Leu–Pro–Thr–OH): immunocytochemical mapping of neurons in the central nervous system of the cockroach, *J. Comp. Neurol.*, 207, 223, 1982.
28. **Worden, M. K., Witten, J. L., and O'Shea, M.,** Proctolin is co-transmitter for the SETi motoneurone, *Soc. Neurosci. Abstr.*, 11, 327, 1985.
29. **Orchard, I. and Lange, A. B.,** Cockroach oviducts: the presence and release of octopamine and proctolin, *J. Insect Physiol.*, 33, 265, 1987.
30. **Agricola, H. M., Ude, J., Birkenbeil, H., and Penzlin, H.,** The distribution of proctolin-like immunoreactive material in the terminal ganglion of the cockroach, *Periplaneta americana, Cell Tissue Res.*, 239, 203, 1985.
31. **Marder, E., Hooper, S. L., and Siwicki, K. K.,** Modulatory action and distribution of the neuropeptide proctolin in the crustacean stomatogastric nervous system, *J. Comp. Neurol.*, 243, 454, 1986.
32. **Hooper, S. L. and Marder, E.,** Modulation of the lobster pyloric rhythm by the peptide proctolin, *J. Neurosci.*, 7, 2097, 1987.
33. **Heinzel, H. G. and Selverston, A. I.,** Gastric mill activity in the lobster. III. Effects of proctolin on the isolated central pattern generator, *J. Neurophysiol.*, 59, 566, 1988.
34. **Heinzel, H. G.,** Gastric mill activity in the lobster. II. Proctolin and octopamine initiate and modulate chewing, *J. Neurophysiol.*, 59, 551, 1988.
35. **Bernstein, H. G., Eckert, M., Penzlin, H., and Dorn, A.,** Proctolin-related material in the mouse brain as revealed by immunohistochemistry, *Neurosci. Lett.*, 45, 229, 1984.
36. **Price, D. A. and Greenberg, M. J.,** Structure of a molluscan cardioexcitatory neuropeptide, *Science*, 197, 670, 1977.
37. **Myers, C. M. and Evans, P. D.,** An FMRF-amide antiserum differentiates between populations of antigens in the central nervous system of the locust, *Schistocerca gregaria, Cell Tissue Res.*, 242, 109, 1985.
38. **Carroll, L. S., Carrow, G. M., and Calabrese, R. L.,** Localization and release of FMRFamide-like immunoreactivity in the cerebral neuroendocrine system of *Manduca sexta, J. Exp. Biol.*, 126, 1, 1986.
39. **Evans, P. D. and Myers, C. M.,** Peptidergic and aminergic modulation of insect skeletal muscle, *J. Exp. Biol.*, 126, 403, 1986.
40. **Walther, C. and Schiebe, M.,** FMRF-NH_2-like factor from neurohaemal organ modulates neuromuscular transmission in the locust, *Neurosci. Lett.*, 77, 209, 1987.
41. **Nambu, J. R., Murphy-Erdosh, C., Andrews, P. C., Feistner, G. J., and Scheller, R. H.,** Isolation and characterization of a *Drosophila* neuropeptide gene, *Neuron*, 1, 55, 1988.

42. **Schneider, L. E. and Taghert, P. H.,** Isolation and characterization of a *Drosophila* gene that encodes multiple neuropeptides related to Phe–Met–Arg–Phe–NH_2 (FMRFamide), *Proc. Natl. Acad. Sci. U.S.A.*, 85, 1993, 1988.
43. **Holman, G. M., Cook, B. J., and Nachman, R. J.,** Isolation, primary structure and synthesis of two neuropeptides from *Leucophaea maderae:* members of a new family of cephalomyotropins, *Comp. Biochem. Physiol.*, 84C, 205, 1986.
44. **Holman, G. M., Cook, B. J., and Nachman, R. J.,** Primary structure and synthesis of two additional neuropeptides from *Leucophaea maderae:* members of a new family of cephalomyotropins, *Comp. Biochem. Physiol.*, 84C, 271, 1986.
45. **Holman, G. M., Cook, B. J., and Nachman, R. J.,** Isolation, primary structure and synthesis of leucokinins V and VI: myotropic peptides of *Leucophaea maderae, Comp. Biochem. Physiol.*, 88C, 27, 1987.
46. **Holman, G. M., Cook, B. J., and Nachman, R. J.,** Isolation, primary structure and synthesis of leucokinins VII and VIII: the final members of this new family of cephalomyotropic peptides isolated from head extracts of *Leucophaea maderae, Comp. Biochem. Physiol.*, 88C, 31, 1987.
47. **Nachman, R. J., Holman, G. M., Cook, B. J., Haddon, W., and Ling, N.,** Leucosulfakinin II, a blocked sulfated insect neuropeptide with homology to cholecystokinin and gastrin, *Biochem. Biophys. Res. Commun.*, 140, 357, 1986.
48. **Nachman, R. J., Holman, G. M., Haddon, W. F., and Ling, N.,** Leucosulfakinin, a sulfated insect neuropeptide with homology to gastrin and cholecystokinin, *Science*, 234, 71, 1986.
49. **Holman, G. M., Cook, B. J., and Nachman, R. J.,** Isolation, primary structure and synthesis of leucomyosuppressin, an insect neuropeptide that inhibits spontaneous contraction of the cockroach hindgut, *Comp. Biochem. Physiol.*, 85C, 329, 1986.
50. **Stone, J. V., Mordue, W., Batley, K. E., and Morris, H. R.,** Structure of locust adipokinetic hormone that regulates lipid utilisation during flight, *Nature (London)*, 263, 207, 1976.
51. **Siegert, K., Morgan, P., and Mordue, W.,** Primary structures of locust adipokinetic hormones. II, *Biol. Chem. Hoppe-Seyler*, 366, 723, 1985.
52. **Baumann, E. and Penzlin, H.,** Sequence analysis of neurohormone D, a neuropeptide of an insect, *Periplaneta americana, Biomed. Biochim. Acta*, 43, 13, 1984.
53. **Scarborough, R. M., Jamieson, G. C., Kalish, F., Kramer, S. J., McEnroe, G. A., Miller, C. A., and Schooley, D. A.,** Isolation and primary structure of two peptides with cardioacceleratory and hyperglycemic activity from the corpora cardiaca of *Periplaneta americana, Proc. Natl. Acad. Sci. U.S.A.*, 81, 5575, 1984.
54. **O'Shea, M., Witten, J., and Schaffer, M.,** Isolation and characterization of two myoactive neuropeptides: further evidence of an invertebrate peptide family, *J. Neurosci.*, 4, 521, 1984.
55. **Witten, J. L., Schaffer, M. H., O'Shea, M., Carter Cook, J., Hemling, M. E., and Rinehart, K. L., Jr.,** Structures of two cockroach neuropeptides assigned by fast atom bombardment mass spectrometry, *Biochem. Biophys. Res. Commun.*, 124, 350, 1984.
56. **Ziegler, R., Eckart, K., Schwarz, H., and Keller, R.,** Amino acid sequence of *Manduca sexta* adipokinetic hormone elucidated by fast atom bombardment (FAB)/tandem mass spectrometry, *Biochem. Biophys. Res. Commun.*, 133, 337, 1985.
57. **Jaffe, H., Raina, A. K., Riley, C. T., Fraser, B. A., Holman, G. M., Wagner, R. M., Ridgway, R. L., and Hayes, D. K.,** Isolation and primary structure of a peptide from the corpora cardiaca of *Heliothis zea* with adipokinetic activity, *Biochem. Biophys. Res. Commun.*, 135, 622, 1986.
58. **Robinson, N. L. and Goldsworthy, G. J.,** A possible site of action of adipokinetic hormone on the flight muscle of locusts, *J. Insect Physiol.*, 23, 153, 1977.
59. **Proux, J. P., Miller, C. A., Li, J. P., Carney, R. L., Girardie, A., Delaage, M., and Schooley, D. A.,** Identification of an arginine vasopressin-like diuretic hormone from *Locusta migratoria, Biochem. Biophys. Res. Commun.*, 149, 180, 1987.
60. **Rafaeli, A., Moshitzky, P., and Applebaum, S. W.,** Diuretic action and immunological cross-reactivity of corticotropin and locust diuretic hormone, *Gen. Comp. Endocrinol.*, 67, 1, 1987.
61. **Morgan, P. J., Siegert, K. J., and Mordue, W.,** Preliminary characterization of locust diuretic peptide (DP-1) and another corpus cardiacum peptide (LCCP), *Insect Biochem.*, 17, 383, 1987.
62. **Kopeć, S.,** Experiments on metamorphosis on insects, *Bull. Acad. Sci. Cracovie Classe Sci. Math. Nat. Ser. B*, 1917, 57, 1917.
63. **Kopeć, S.,** Studies on the necessity of the brain for the inception of insect metamorphosis, *Biol. Bull.*, 42, 323, 1922.
64. **Nagasawa, H., Kataoka, H., Isogai, A., Tamura, S., Suzuki, A., Ishizaki, H., Mizoguchi, A., Fujiwara, Y., and Suzuki, A.,** Amino-terminal amino acid sequence of the silkworm prothoracicotropic hormone: homology with insulin, *Science*, 226, 1344, 1984.
65. **Nagasawa, H., Kataoka, H., Isogai, A., Tamura, S., Suzuki, A., Mizoguchi, A., Fujiwara, Y., Suzuki, A., Takahashi, S. Y., and Ishizaki, H.,** Amino acid sequence of a prothoracicotropic hormone of the silkworm *Bombyx mori, Proc. Natl. Acad. Sci. U.S.A.*, 83, 5840, 1986.

66. **Copenhaver, P. F. and Truman, J. W.,** Identification of the cerebral neurosecretory cells that contain eclosion hormone in the moth, *Manduca sexta, J. Neurosci.,* 6, 1738, 1986.
67. **Kataoka, H., Troetschler, R. G., Kramer, S. J., Cesarin, B. J., and Schooley, D. A.,** Isolation and primary structure of the eclosion hormone of the tobacco hornworm, *Manduca sexta, Biochem. Biophys. Res. Commun.,* 146, 746, 1987.
68. **Marti, T., Takio, K., Walsh, K. A., Terzi, G., and Truman, J. W.,** Microanalysis of the amino acid sequence of the eclosion hormone from the tobacco hornworm, *Manduca sexta, FEBS Lett.,* 219, 415, 1987.
69. **Truman, J. W., Taghert, P. H., Copenhaver, P. F., Tublitz, N. J., and Schwartz, L. M.,** Eclosion hormone may control all ecdyses in insects, *Nature (London),* 291, 70, 1981.
70. **Granger, N. A., Mitchell, L. J., Janzen, W. P., and Bollenbacher, W. E.,** Activation of *Manduca sexta* corpora allata *in vitro* by a cerebral neuropeptide, *Mol. Cell. Endocrinol.,* 37, 349, 1984.
71. **Gadot, M., Rafaeli, A., and Applebaum, S. W.,** Partial purification and characterization of locust allatotropin, *Arch. Insect Biochem. Physiol.,* 4, 213, 1987.
72. **Bhaskaran, G., Jones, G., and Jones, D.,** Neuroendocrine regulation of corpus allatum activity in *Manduca sexta:* sequential neurohormonal and nervous inhibition in the last-instar larva, *Proc. Natl. Acad. Sci. U.S.A.,* 77, 4407, 1980.
73. **Rankin, S. M. and Stay, B.,** Distribution of allatostatin in the adult cockroach, *Diploptera punctata* and effects on corpora allata *in vitro, J. Insect Physiol.,* 33, 551, 1987.
74. **Fraenkel, G. and Hsaio, C.,** Bursicon, a hormone which mediates tanning of the cuticle in the adult fly and other insects, *J. Insect Physiol.,* 11, 513, 1965.
75. **Reynolds, S.,** Bursicon, in *Endocrinology of Insects,* Downer, R. G. H. and Laufer, H., Eds., Alan R. Liss, New York, 1983, 235.
76. **Mills, R. R., Mathur, R. B., and Guerra, A. A.,** Studies on the hormonal control of tanning in the American cockroach. I. Release of an activation factor from the terminal abdominal ganglion, *J. Insect Physiol.,* 11, 1047, 1965.
77. **Mills, R. R. and Lake, C. R.,** Hormonal control of tanning in the American cockroach. IV. Preliminary purification of the hormone, *J. Insect Physiol.,* 12, 1395, 1966.
78. **Srivastava, B. B. L. and Hopkins, T. L.,** Bursicon release and activity in haemolymph during metamorphosis of the cockroach, *Leucophaea maderae, J. Insect Physiol.,* 21, 1985, 1975.
79. **Taghert, P. H. and Truman, J. W.,** The distribution and molecular characteristics of the tanning hormone, bursicon in the tobacco hornworm, *Manduca sexta, J. Exp. Biol.,* 98, 373, 1982.
80. **Adams, M. E. and Phelps, M. N.,** Co-localization of bursicon bioactivity and proctolin in identified neurons, *Soc. Neurosci. Abstr.,* 9, 313, 1983.
81. **Scharrer, B.,** Insects as models for neuroendocrine research, *Annu. Rev. Entomol.,* 32, 1, 1987.
82. **Scharrer, E.,** Die Lichtempfindlichkeit blinder Elritzen (Untersuchungen über das Zwischenhirn der Fische), *Z. Vgl. Physiol.,* 7, 1, 1928.
83. **Rowell, H. F.,** The cells of the insect neurosecretory system: constancy, variability, and the concept of the unique identifiable neuron, *Adv. Insect Physiol.,* 12, 63, 1976.
84. **Füller, H. B.,** Morphologische und experimentelle Untersuchungen über die neurosekretorischen Verhältnisse im zentral Nervensystem von Blattiden und Culiciden, *Zool. Jahrb. Physiol.,* 69, 438, 1960.
85. **Delphin, F.,** The histology and possible functions of neurosecretory cells in the ventral ganglia of *Schistocerca gregaria* Forskål, *Trans. R. Entomol. Soc. London,* 117, 167, 1965.
86. **Gaude, H.,** Histologische Untersuchungen zur Struktur und Funktion des neurosekretorischen Systems der Hausgrille *Acheta domesticus* L., *Zool. Anz.,* 194, 151, 1975.
87. **Bargmann, W., Lindner, E., and Andres, K. H.,** Über Synapsen an endokrinen Epithelzellen und die Definition sekretorischer Neurone. Untersuchungen am Zwischenlappen der Katzenhypophyse, *Z. Zellforsch. Mikrosk. Anat.,* 77, 282, 1967.
88. **Pipa, R. L.,** Locations and central projections of neurons associated with the retrocerebral neuroendocrine complex of the cockroach, *Periplaneta americana, Cell Tissue Res.,* 193, 443, 1978.
89. **Koontz, M. and Edwards, J. S.,** The projection of neuroendocrine fibers (NCCI and II) in the brains of three orthopteroid insects, *J. Morphol.,* 165, 285, 1980.
90. **Pipa, R. L. and Novak, F. J.,** Pathways and fine structure of neurons forming the *nervi corporis allati II* of the cockroach *Periplaneta americana* (L.), *Cell Tissue Res.,* 201, 227, 1979.
91. **Schooneveld, H., Romberg-Privee, H. M., and Veenstra, J. A.,** Immunocytochemical differentiation between adipokinetic hormone (AKH)-like peptides in neurons and glandular cells in the corpus cardiacum of *Locusta migratoria* and *Periplaneta americana* with C-terminal and N-terminal specific antisera to AKH, *Cell Tissue Res.,* 243, 9, 1986.
92. **Hekimi, S. and O'Shea, M.,** Identification and purification of two precursors of the insect neuropeptide adipokinetic hormone, *J. Neurosci.,* 7, 2773, 1987.
93. **Highnam, K. C. and Goldsworthy, G. J.,** Regenerated corpora cardiaca and hyperglycemic factor in *Locusta migratoria, Gen. Comp. Endocrinol.,* 18, 83, 1972.

94. **Ali, Z. I. and Pipa, R.,** The abdominal perisympathetic neurohemal organs of the cockroach *Periplaneta americana:* innervation revealed by cobalt chloride diffusion, *Gen. Comp. Endocrinol.*, 36, 396, 1978.
95. **Johnson, B.,** Fine structure of the lateral cardiac nerves of the cockroach *Periplaneta americana* (L.), *J. Insect Physiol.*, 12, 645, 1966.
96. **Miller, T. and Usherwood, P. N. R.,** Studies of cardio-regulation in the cockroach, *Periplaneta americana, J. Exp. Biol.*, 54, 329, 1971.
97. **Miller, T. and Rees, D.,** Excitatory transmission in insect neuromuscular systems, *Am. Zool.*, 13, 299, 1973.
98. **Miller, T.,** Role of cardiac neurons in the cockroach heartbeat, *J. Insect Physiol.*, 14, 1265, 1968.
99. **Miller, T. and Thomson, W. W.,** Ultrastructure of cockroach cardiac innervation, *J. Insect Physiol.*, 14, 1099, 1968.
100. **Krauthammer, V.,** Electrophysiology of identified neurosecretory and nonneurosecretory cells in the cockroach pars intercerebralis, *J. Exp. Zool.*, 234, 207, 1985.
101. **Zaretsky, M. and Loher, W.,** Anatomy and electrophysiology of individual neurosecretory cells of an insect brain, *J. Comp. Neurol.*, 216, 253, 1983.
102. **Agui, N., Granger, N. A., Gilbert, L. I., and Bollenbacher, W. E.,** Cellular localization of the insect prothoracicotropic hormone: *in vitro* assay of a single neurosecretory cell, *Proc. Natl. Acad. Sci. U.S.A.*, 76, 5694, 1979.
103. **Carrow, G. M., Calabrese, R. L., and Williams, C. M.,** Architecture and physiology of insect cerebral neurosecretory cells, *J. Neurosci.*, 4, 1034, 1984.
104. **Adams, M. E. and O'Shea, M.,** Vacuolation of an identified peptidergic (proctolin-containing) neuron, *Brain Res.*, 230, 439, 1981.
105. **Taghert, P. H. and Truman, J. W.,** Identification of the bursicon-containing neurones in abdominal ganglia of the tobacco hornworm, *Manduca sexta, J. Exp. Biol.*, 98, 385, 1982.
105a. **Adams, M. E. and O'Shea, M.,** unpublished data, 1979.
105b. **Adams, M. E.,** unpublished data, 1983.
106. **Remy, C. and Girardie, J.,** Anatomical organization of two vasopressin-neurophysin-like neurosecretory cells throughout the central nervous system of the migratory locust, *Gen. Comp. Endocrinol.*, 40, 27, 1980.
107. **Evans, P. D. and O'Shea, M.,** The identification of an octopaminergic neuron and the modulation of a myogenic rhythm in the locust, *J. Exp. Biol.*, 73, 235, 1978.
108. **O'Shea, M. and Evans, P. D.,** Potentiation of neuromuscular transmission by an octopaminergic neuron in the locust, *J. Exp. Biol.*, 79, 169, 1979.
109. **Dymond, G. R. and Evans, P. D.,** Biogenic amines in the nervous system of the cockroach, *Periplaneta americana:* association of octopamine with mushroom bodies and dorsal unpaired median (DUM) neurons, *Insect Biochem.*, 9, 535, 1979.
110. **Miller, T. A.,** Insect visceral muscle, in *Insect Muscle*, Usherwood, P. N. R., Ed., Academic Press, New York, 1975, 545.
111. **Osborne, M. P., Finlayson, L. H., and Rice, M. J.,** Neurosecretory endings associated with striated muscles in three insects (*Schistocerca, Carausius* and *Phormia*) and a frog *(Rana), Z. Zellforsch. Mikrosk. Anat.*, 116, 391, 1971.
112. **O'Shea, M.,** Are skeletal motoneurons in arthropods peptidergic?, in *Model Neural Networks and Behavior*, Selverston, A. I., Ed., Plenum Press, New York, 1985, 401.
113. **O'Shea, M. and Adams, M. E.,** Proctolin: from "gut factor" to model neuropeptide, *Adv. Insect Physiol.*, 19, 1, 1986.
114. **Witten, J. L. and O'Shea, M.,** Peptidergic innervation of insect skeletal muscle: immunochemical observations, *J. Comp. Neurol.*, 242, 93, 1985.
115. **Piek, T. and Mantel, P.,** Myogenic contractions in locust muscle induced by proctolin and by wasp, *Philanthus triangulum* venom, *J. Insect Physiol.*, 23, 321, 1977.
116. **Hoyle, G.,** Distributions of nerve and muscle fibre types in locust jumping muscle, *J. Exp. Biol.*, 73, 205, 1978.
117. **Hertel, W., Pass, G., and Penzlin, H.,** Electrophysiological investigation of the antennal heart of *Periplaneta americana* and its reactions to proctolin, *J. Insect Physiol.*, 31, 563, 1985.
117a. **Adams, M. E.,** unpublished data, 1984.
118. **Keshishian, H. and O'Shea, M.,** The distribution of a peptide neurotransmitter in the postembryonic grasshopper central nervous system, *J. Neurosci.*, 5, 992, 1985.
119. **Finlayson, L. H.,** Distribution and function of neurosecretory cells, in *Insect Neurobiology and Pesticide Action (Neurotox '79)*, Society of Chemical Industry, London, 1980, 297.
120. **Seshan, K. R., Provine, R. R., and Levi-Montalcini, R.,** Structural and electrophysiological properties of nymphal and adult insect medial neurosecretory cells: an *in vitro* analysis, *Brain Res.*, 78, 359, 1974.
121. **Beadle, D. J. and Lees, G.,** Insect neuronal cultures, a new tool in insect neuropharmacology, in *Neuropharmacology and Neurobiology*, Ford, M. G., Usherwood, P. N. R., Reay, R. C., and Lunt, G. G., Eds., Ellis Horwood, London, 1986, 423.

121a. **Adams, M. E.,** unpublished data, 1984.

122. **Orchard, I. and Loughton, B. G.,** Is octopamine a transmitter mediating hormone release in insects?, *J. Neurobiol.,* 12, 143, 1981.

123. **Hekimi, S. and O'Shea, M.,** Neuropeptide biosynthesis: model system in an insect, *Soc. Neurosci. Abstr.,* 13, 1256, 1987.

124. **Keshishian, H. and O'Shea, M.,** The acquisition and expression of a peptidergic phenotype in the grasshopper embryo, *J. Neurosci.,* 5, 1005, 1985.

125. **Jennings, K. R., Steele, R. W., and Starratt, A. N.,** Cyclic AMP actions on proctolin- and neurally-induced contractions of the cockroach hindgut, *Comp. Biochem. Physiol.,* 74C, 69, 1983.

126. **Worden, M. K. and O'Shea, M.,** Evidence for stimulation of muscle phosphatidylinositol metabolism by an identified skeletal motoneuron, *Soc. Neurosci. Abstr.,* 12, 948, 1986.

127. **Lange, A. B., Orchard, I., and Lam, W.,** Mode of action of proctolin on locust visceral muscle, *Arch. Insect Biochem. Physiol.,* 5, 285, 1987.

128. **Orchard, I., Carlisle, J. A., Loughton, B. G., Gole, J. W. D., and Downer, R. G. H.,** *In vitro* studies on the effects of octopamine on locust fat body, *Gen. Comp. Endocrinol.,* 48, 7, 1982.

129. **Asher, C., Moshitzky, P., Ramachandran, J., and Applebaum, S. W.,** The effects of synthetic locust adipokinetic hormone on dispersed locust fat body cell preparations: c-AMP induction, lipid mobilization and inhibition of protein synthesis, *Gen. Comp. Endocrinol.,* 55, 167, 1984.

130. **Truman, J. W., Mumby, S. M., and Welch, S. K.,** Involvement of cyclic GMP in the release of stereotyped behavior patterns in moths by a peptide hormone, *J. Exp. Biol.,* 84, 201, 1979.

131. **Schwartz, L. M. and Truman, J. W.,** Cyclic GMP may serve as a second messenger in peptide-induced muscle degeneration in an insect, *Proc. Natl. Acad. Sci. U.S.A.,* 81, 6718, 1984.

132. **Morton, D. B. and Truman, J. W.,** Steroid regulation of the peptide-mediated increase in cyclic GMP in the nervous system of the hawkmoth, *Manduca sexta, J. Comp. Physiol. A,* 157, 423, 1985.

133. **Morton, D. B. and Truman, J. W.,** The EGPs: the eclosion hormone and cyclic GMP-regulated phosphoproteins. I. Appearance and partial characterization in the CNS of *Manduca sexta, J. Neurosci.,* 8, 1326, 1988.

134. **Tublitz, N. J. and Trombley, P. Q.,** Peptide action on insect cardiac muscle is mediated by inositol trisphosphate (IP_3), *Soc. Neurosci. Abstr.,* 13, 285, 1987.

135. **Cook, B. J. and Holman, G. M.,** The role of proctolin and glutamate in the excitation-contraction coupling of insect visceral muscle, *Comp. Biochem. Physiol.,* 80C, 65, 1985.

136. **Starratt, A. N. and Steele, R. W.,** *In vivo* inactivation of the insect neuropeptide proctolin in *Periplaneta americana, Insect Biochem.,* 14, 97, 1984.

137. **Quistad, G. B., Adams, M. E., Scarborough, R. M., Carney, R. L., and Schooley, D. A.,** Metabolism of proctolin, a pentapeptide neurotransmitter in insects, *Life Sci.,* 34, 569, 1984.

138. **Steele, R. W. and Starratt, A. N.,** *In vitro* inactivation of the insect neuropeptide proctolin in hemolymph from *Periplaneta americana, Insect Biochem.,* 15, 511, 1985.

139. **Isaac, R. E.,** Proctolin degradation by membrane peptidases from nervous tissues of the desert locust *(Schistocerca gregaria), Biochem. J.,* 245, 365, 1987.

140. **Skinner, W. S., Quistad, G. B., Adams, M. E., and Schooley, D. A.,** Metabolic degradation of the peptide periplanetin CC-2 in the cockroach, *Periplaneta americana, Insect Biochem.,* 17, 433, 1987.

141. **Baumann, E. and Gersch, M.,** Untersuchungen zur Stabilität des Neurohormons D, *Zool. Jahrb. Physiol.,* 77, 153, 1973.

142. **Orchard, I. and Lange, A. B.,** The release of octopamine and proctolin from an insect visceral muscle: effects of high potassium saline and neural stimulation, *Brain Res.,* 413, 251, 1987.

143. **Goodman, R. R., Fricker, L. D., and Snyder, S. H.,** Enkephalins, in *Brain Peptides,* Krieger, D. T., Brownstein, J. J., and Martin, J. B., Eds., John Wiley & Sons, New York, 1983, 827.

144. **Hendricks, S. A., Roth, J., Rishi, S., and Becker, K. L.,** Insulin in the nervous system, in *Brain Peptides,* Krieger, D. T., Brownstein, M. J., and Martin, J. B., Eds., John Wiley & Sons, New York, 1983, 903.

145. **Zimmerman, E. A.,** Oxytocin, vasopressin, and neurophysins, in *Brain Peptides,* Krieger, D. T., Brownstein, M. J. and Martin, J. B., Eds., John Wiley & Sons, New York, 1983, 597.

Chapter 14

PROCTOLIN IN THE COCKROACH: PROVIDING MODEL SYSTEMS FOR STUDYING NEUROPEPTIDE TRANSMISSION

Cynthia A. Bishop, Jane L. Witten, and Michael O'Shea

TABLE OF CONTENTS

I. INTRODUCTION

It is now apparent that a general strategy used by the nervous system is to release peptides which often act in concert with conventional neurotransmitters to modify the response of the target cells. The emergence of peptides as a class of neurotransmitters has raised numerous challenging questions concerning chemical signaling in the nervous system. Often the functional significance of a neuropeptide is not known because targets are not defined or because technical difficulties prevent the physiological study of these peptidergic neurons. Little is known about molecular mechanisms underlying peptide function. Since neuropeptides often occur with conventional neurotransmitters in neurons, questions concerning mechanisms of release of multiple transmitters and the coordination of postsynaptic action of each transmitter are raised. In summary, we have little understanding of the need for the increased complexity of the nervous system that has been revealed by the discovery of peptide neurotransmitters.

Synaptic preparations in invertebrates offer excellent experimental models for studies of peptidergic transmission. Their relative simplicity, as well as the size and accessibility of neurons in invertebrate systems, allows identification of individual neurons for anatomical, biochemical, and physiological study. In this chapter we shall describe the study of one peptidergic system in the cockroach *Periplaneta americana* which clearly demonstrates these convenient features. We shall summarize the techniques used to identify neurons in the cockroach which contain the peptide proctolin and shall catalogue putative proctolinergic neurons in the central nervous system (CNS) and a number of visceral and skeletal muscles which receive innervation from proctolin-immunoreactive neurons. We shall show conclusively that proctolin serves as cotransmitter at one skeletal neuromuscular junction and shall describe the peptide's function. We shall also describe how two other preparations have emerged subsequent to these cockroach studies which may provide important model systems in which some of the salient questions associated with multiple neurotransmitters (one or more of which may be a peptide) can now be examined.

II. PROCTOLIN: A COCKROACH GUT NEUROPEPTIDE

Proctolin was first discovered in the American cockroach by Brown and was postulated to be a neurotransmitter in the hindgut, where it occurs in high concentrations.[1-3] Starratt and Brown[4] determined the amino acid sequence of proctolin (a pentapeptide) purified from an extract of about 125,000 cockroaches (Figure 1). Proctolin is present in at least five other insect orders and is active at low concentrations in a variety of neuronal and muscular preparations from insects,[6-10] crustaceans,[11-13] and the chelicerate *Limulus*,[14,15] as well as in rats and mice.[16,17] Such widespread bioactivity strongly suggested that proctolin was more than a "gut peptide", and this was confirmed when the distribution of the peptide was analyzed further. This analysis was initiated in the cockroach. Interestingly, the first neurons in which proctolin was localized were not those innervating the gut, but putative neurosecretory neurons (Lateral White [LW] neurons) found in the anterior abdominal ganglia of the CNS of the cockroach.

III. THE LATERAL WHITE NEURON: THE FIRST DEMONSTRATION OF AN IDENTIFIED PROCTOLINERGIC NEURON

The LW cells were first noted for their size (ca. 60 μm) and opacity in the abdominal ganglia of the cricket. Similar bilateral cells were subsequently found in the first three abdominal ganglia of the cockroach (Figure 2).[18] These cells also stained with neutral red, suggesting the presence of a monoamine. Together the features of these cells suggested a

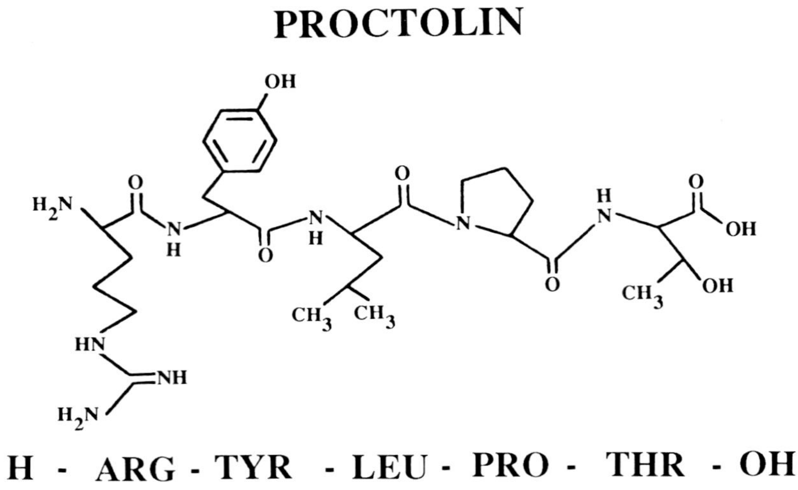

FIGURE 1. Chemical structure of the neuropeptide proctolin.[5]

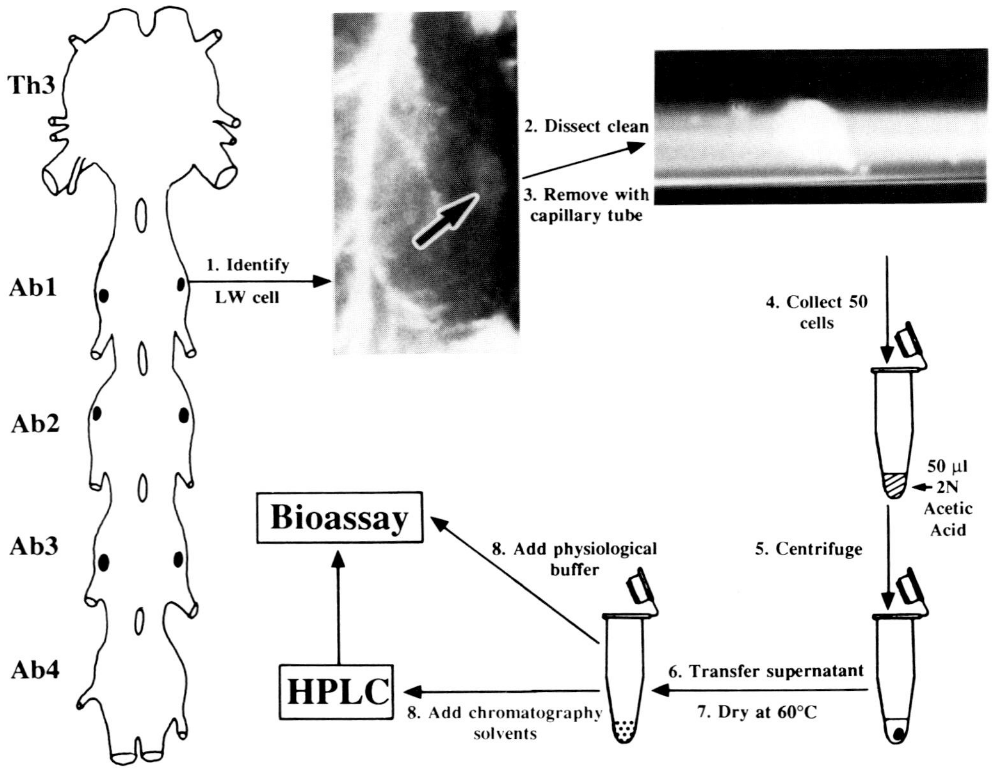

FIGURE 2. Schematic (left) showing the portion of the cockroach CNS containing the large (ca. 60 μm) proctolinergic Lateral White neurons.[18] These cells were easily identified by their size and opacity and by their position along the lateral edges of the first three abdominal ganglia. These identified neurons were subsequently dissected free and sucked into a capillary tube. Cells were then pooled, processed, and analyzed for the presence of proctolin using HPLC and a sensitive, quantifiable bioassay; the bioassay procedure is described in more detail in Figure 3. Th = thoracic ganglion; Ab = abdominal ganglion.

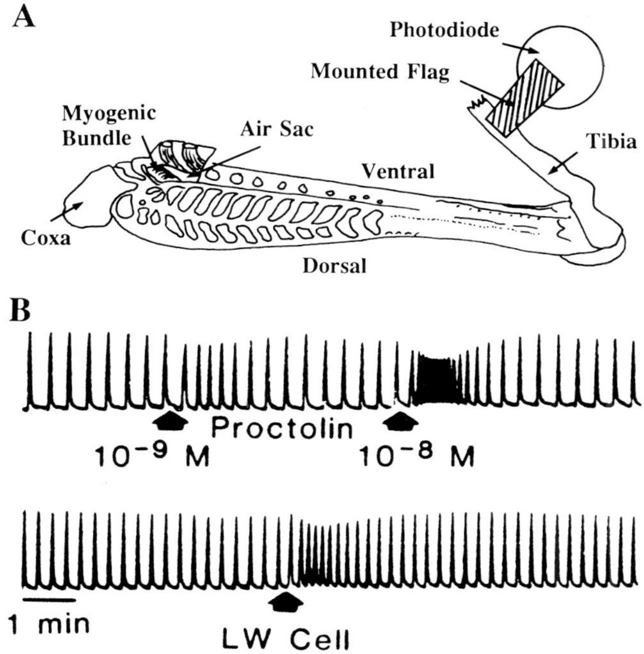

FIGURE 3. The locust extensor-tibia preparation, a sensitive and quantifiable bioassay for proctolin.[18,19] The bioassay depends upon proctolin's ability to modify the myogenic oscillations of a small subset of muscle fibers within the main extensor muscle of the locust hindleg. Proctolin in femtomolar concentrations increases the frequency of contractions in the muscle bundle in a log-linear manner. Only 1 μl of a sample is required for this bioassay. (A) Illustration of bioassay preparation. A flag is mounted on the tibia with wax and positioned in front of a photodiode to detect the oscillatory movements of the tibia. A small window in the exoskeleton is made to expose the myogenic bundle of fibers controlling these movements. (B) Oscillatory movements of the tibia, as detected by the photodiode. Upper trace: A 1-μl aliquot of 1 n*M* proctolin increases the frequency of this oscillation. The frequency is further increased by an aliquot of 10 n*M* proctolin. Lower trace: A 1-μl aliquot of an LW cell extract containing $^{1}/_{50}$ of one cell produces a proctolin-like increase in the myogenic rhythm.

neurosecretory function and a possible location for the peptide proctolin. (A more extensive discussion of these neurons is given in this work by Adams in Chapter 13).

The large size and unambiguous identification of these neurons made biochemical analysis of cell peptide content possible.[18] Individual LW cell bodies were dissected free of the cockroach CNS, sucked into a capillary tube, and collected for analysis by reverse-phase high-pressure liquid chromatography (HPLC; Figure 2). This analysis was greatly enhanced by the development of a sensitive, low-volume, quantifiable bioassay that could be used specifically to detect low levels of proctolin from cell extracts before and after HPLC purification (Figure 3).[18] Each LW cell body was shown to contain about 0.1 pmol proctolin.[18] These results confirmed that proctolin was indeed located within neurons in the cockroach and that the peptide was associated with neurons other than those which innervate the gut. Therefore, a more general view of proctolin distribution in the cockroach CNS was sought. Proctolin antisera were produced and immunochemical techniques developed for this purpose.

IV. IMMUNOCHEMICAL TECHNIQUES INDICATE WIDESPREAD DISTRIBUTION OF PROCTOLINERGIC NEURONS IN THE COCKROACH CNS

Immunochemical techniques are convenient primary probes for transmitter presence. Once a specific antiserum has been obtained, the distribution of proctolin (or antigens similar to proctolin) in tissue are quantified using radioimmunoassay (RIA) and then localized to neurons using immunohistochemical techniques.

Proctolin antisera were produced by immunizing 20 female New Zealand White rabbits subcutaneously with conjugates of synthetic proctolin (Sigma Chemical Company, St. Louis, MO) and bovine serum albumin emulsified with Freund's adjuvant. Serum titer, sensitivity, and specificity were checked by RIA.[19] Several antisera were produced which were highly specific for proctolin. The most sensitive one, capable of detecting 50 fmol of proctolin, was used in the RIA to quantify the distribution of proctolin-like immunoreactivity (PLI) in the CNS of the cockroach[19] (see Figure 4). PLI is unevenly distributed throughout the CNS of the cockroach: the highest concentration occurs in the terminal or genital ganglion and the lowest concentration in the cerebral ganglia or brain.

Somewhat surprisingly, a second antiserum with a much lower titer and sensitivity in the RIA provided superior staining results in immunohistochemistry.[19] This antiserum was used in a whole-mount immunohistochemical procedure to map the distribution of proctolin-immunoreactive neurons in the cockroach CNS.[20] The map revealed the presence of neurons immunoreactive to the proctolin antiserum throughout the entire CNS (Figure 4; Table 1). This distribution correlated well with the CNS distribution of proctolin-like immunoreactivity determined by RIA. For example, the terminal ganglion contains the highest number of immunoreactive cell bodies (about 185) and the highest levels of PLI (about 140 pmol per ganglion). The brain, which contains the fewest immunoreactive cells (about 45, primarily located in the tritocerebrum), has the lowest levels of PLI (about 10 pmol). Other ganglia contain intermediate numbers (60 to 80 cells) and PLI ranging from 20 to 40 pmol per ganglion.

Immunoreactive cell bodies were found in the dorsal, ventral, and lateral regions of the ganglia (Figure 4; Table 1). Immunoreactive processes were detected in all interganglionic connectives and in many ganglionic nerve roots, including those which innervate the proctodeum (the posterior portion of the hindgut). Dense ramifications of immunoreactive processes and varicosities were detected in portions of all ganglia.

The immunochemical map revealed that about 5% of the neurons in the cockroach CNS are possibly proctolinergic. Of these cells, 10% have efferent processes, suggesting that they might be motoneurons. Proctolin may also be associated with sensory functions, since proctolin-immunoreactive processes were detected in the second nerve of the thoracic ganglion, which is thought to be purely sensory.[20] The large number and wide distribution of cells with PLI support the idea that proctolin probably has diverse central and peripheral functions.

Among the group of cells that stain with the antiserum, large and accessible neurons were sought that could be recognized and isolated for biochemical and physiological analysis. An example of such a cell is the giant dorsal bilateral cell (GDB; Figure 4B) which is present in each of the thoracic ganglia. In addition to being an excellent candidate for biochemical analysis, the GDB neuron appeared to be identical in size to and copositional with the identified slow motoneuron D_s, which innervates the coxal depressor muscles.[21,22] The possibility that the GDB neuron was D_s was additionally exciting for two reasons. First, it suggested that the proctolin-immunoreactive neuron was an identified neuron whose target was known. Second, it indicated that a peptide might be a transmitter at an insect skeletal neuromuscular junction. This was a novel concept because, as in vertebrates, transmitters

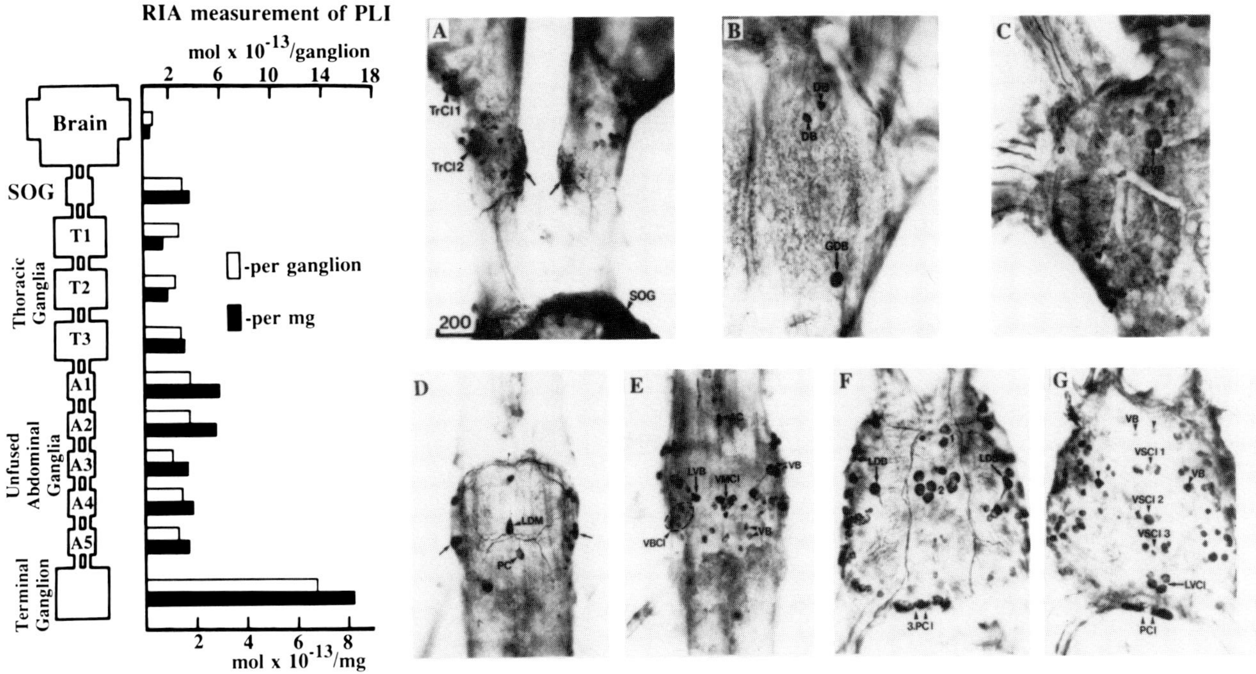

FIGURE 4. Distribution of proctolin-line immunoreactivity (PLI) in the cockroach CNS, as determined by RIA (left histogram) and immunohistochemistry (A to G).[19,20] Individual ganglia were pooled for RIA. Cell bodies and processes immunoreactive to the proctolin antiserum were visualized in whole-mount, after treatment with colchicine and desheathing of the ganglia, using the peroxidase-antiperoxidase technique. (Refer to articles by Bishop et al.[19] and Bishop and O'Shea[20] for a complete description of these procedures.) (A) Anterior view of the cockroach brain and circumesophageal connectives. Unmarked arrows indicate tritocerebral regions of the brain, which contain dense immunoreactive fibers and varicosities (SOG = subesophageal ganglion; TrCl 1 and 2 = tritocerebral cell body clusters). (B) Dorsal view, right half of the first thoracic ganglion (DB = dorsal bilateral cell body; GDB = giant dorsal bilateral cell body). (C) Ventral view, right half of the first thoracic ganglion (GVB = giant ventral bilateral cell body). (D) Dorsal view of the first unfused abdominal ganglion (LDM = large dorsal medial cell body; PC = posterior crotch cell body). (E) Ventral view of the first unfused abdominal ganglion (AC = anterior crotch cell body; LVB = large ventral bilateral cell body; VB = ventral bilateral cell body; VBCl = ventral bilateral cluster, VMCl = ventral medial cluster). (F) Dorsal view of the terminal ganglion (LDB = large dorsal bilateral cell body; PCl = posterior cluster). (G) Ventral view of the terminal ganglion (LVCl = large ventral cluster; PCl = posterior cluster; VB = ventral bilateral cell body; VSCl = ventral segmental cluster).

TABLE 1
Summary of Proctolin-Like Immunoreactive Neurons in the Cockroach *Periplaneta americana*[20]

	Staining intensity (S) and variability (V)				
Cells and clusters[a]	**Strong S, low V**	**Medium S, more V**	**Light S, high V**	**Number[b]**	**Diameter (μm)**
Cerebral ganglion (brain)					
Protocerebrum					
Right pair		X		2	20—30
Left pair		X		2	20—30
Medial pair		X		2	20—30
Tritocerebrum					
Dorsal bilateral cluster	X			20	30
Ventral bilateral cluster	X			16	20
Posterior bilateral pair 1—2			X	2 ea.	20
Subesophageal ganglion					
Dorsal					
Medial pair		X		2	30
Large bilateral pair	X			2	35
Small bilateral pair 1		X		2 ea.	20—25
Ventral					
Large anterior medial cluster	X			12	30—40
Large posterior medial cluster		X		3	40
Bilateral cluster 1	X			8	25
Bilateral cluster 2			X	12	20
Bilateral cluster 3			X	10	20
Bilateral cluster 4			X	8	20
Bilateral cluster 5		X		4	25
Bilateral cluster 6	X			8	25
Bilateral cluster 7		X		12	25
Bilateral pair 1—6		X		2 ea.	20—25
Thoracic ganglion 1					
Dorsal					
Giant posterior bilateral pair	X			2	60
Small posterior bilateral pair			X	2	15
Anterior bilateral pair 1	X			2	25
Anterior bilateral pair 2	X			2	25
Anterior bilateral pair 3		X		2	20
Anterior bilateral pair 4	X			2	10
Ventral					
Giant anterior bilateral pair		X		2	60
Bilateral cluster 1—4			X	8—16 ea.	10—30
Thoracic ganglion 2					
Dorsal					
Giant posterior bilateral pair	X			2	60
Small posterior bilateral pair			X	2	15
Small anterior bilateral pair 1—3			X	2 ea.	10
Medial pair		X		2	20
Ventral					
Giant anterior bilateral pair		X		2	60
Anterior bilateral pair			X	23	10—25
Posterior bilateral cluster			X	14	10—25
Thoracic ganglion 3					
Dorsal					
Giant posterior bilateral pair	X			2	60
Posterior bilateral pair 1		X		2	25
Posterior bilateral pair 2		X		2	20
Posterior bilateral pair 1		X		2	15

TABLE 1 (continued)
Summary of Proctolin-Like Immunoreactive Neurons in the Cockroach *Periplaneta americana*[20]

	Staining intensity (S) and variability (V)				
Cells and clusters[a]	Strong S, low V	Medium S, more V	Light S, high V	Number[b]	Diameter (μm)
Posterior medial pair 1—3	X			2 ea.	10—20
Small bilateral pair			X	2	10
Large anterior pair			X	2	45
Small anterior pair 1—2			X	2 ea.	10—15
Ventral					
Giant anterior bilateral pair		X		2	60
Anterior bilateral cluster			X	29	10—25
Posterior bilateral cluster			X	21	10—25
Abdominal ganglia 1—5[c]					
Lateral edge					
Anterior bilateral cluster	X			10	10—50
Medial bilateral cluster		X		8	10—30
Posterior bilateral cluster			X	10	10—40
Dorsal					
Anterior pair		X		2	35—45
Posterior pair		X		2	35—45
Unpaired medial cells		X		6	10—50
Ventral					
Medial cluster		X		8	25—30
Anterior bilateral cluster		X		6	20—40
Posterior bilateral cluster			X	20	15—30
Terminal ganglion					
Dorsal					
Large medial cluster 1		X		3	40—50
Large medial cluster 2	X			7	40—50
Large medial cluster 3	X			14	40—50
Small medial cluster 1—2			X	2—3 ea.	10—20
Small medial cluster 3			X	13	50
Large bilateral pair	X			4	20—30
Bilateral cluster 1	X			4	20—30
Bilateral cluster 2			X	8	20—30
Bilateral cluster 3			X	4	20—30
Bilateral cluster 4		X		14	15—25
Bilateral cluster 5			X	11	15—25
Bilateral pair 1—5		X		2 ea.	25—45
Ventral					
Medial cluster 1—4	X			4—7 ea.	15—50
Bilateral pair 1—9		X		2 ea.	25—50
Bilateral cluster 1—4			X	10—13 ea.	20—45

[a] Individual cells or cell clusters are identified in terms of size and location in the ganglia.
[b] Numbers reflect what is most commonly observed in six preparations.
[c] Abdominal ganglia 1—5 each contain the cells described here. A few individual differences occur among these ganglia, which are noted elsewhere.[20]

were thought to be known in insect skeletal motoneurons. Indeed, there was good evidence that the excitatory neurotransmitter for motoneurons is L-glutamic acid[23,24] or aspartic acid.[25] The inhibitory motoneurons use γ-aminobutyric acid (GABA),[26] and the modulatory dorsal unpaired motoneurons are octopaminergic.[27] We identified the immunoreactive cell GDB as the identified motoneuron D_s using immunohistochemical staining of the terminals of the neuron on the muscle[28] as described in the next section.

V. IMMUNOCHEMICAL STUDIES REVEAL PROCTOLIN-LIKE IMMUNOREACTIVITY IN SEVERAL SKELETAL MUSCLES AND MULTIPLE GUT REGIONS

A. ASSOCIATION OF PROCTOLIN IMMUNOREACTIVITY WITH THE COXAL DEPRESSOR MUSCLE

Immunoreactive staining in the CNS suggests that proctolin immunoreactivity is associated with at least one skeletal motoneuron, the slow depressor motoneuron which innervates the coxal depressor muscles that control leg extension during walking and stance.[29] There are two coxal depressor slow muscles (capable of generating slow, graded contractions), numbered 177D and 177E, and two fast muscles (capable of generating a rapid-twitch contraction), numbered 178 and 179. Muscles 177D and E are further divided into subgroups 177d, d′, e, and e′.[21] Both subgroups of the two slow muscles are innervated by the slow depressor motoneuron D_s. Inhibitory motoneurons D_1, D_2, and D_3 also innervate 177d and 177e, while 177d′ and 177e′ also receive innervation from the fast depressor motoneuron D_f.[21] In the metathoracic segment, all five coxal depressor motoneurons extend to these muscles via the first branch of the fifth ganglionic nerve. This system lent itself to determining whether terminals of the D_s (as well as other identified motoneurons in the system) were proctolin immunoreactive.

Only one axon stained in the thoracic ganglionic fifth root branch through which all five coxal depressor motoneurons extend. Proctolin immunoreactivity, as measured by RIA, was present at a significantly higher level in slow muscles 177D and E (95 ± 12 fmol/mg dry weight), which are innervated by D_s, than in fast muscle 178 (18 ± 31 fmol) or 179 (6.5 ± 6 fmol), which are not innervated by D_s (Figure 5). The amount of proctolin immunoreactivity in the slow muscles was reduced to that found in the fast muscles by denervation (Figure 5). Correspondingly, nerve processes with PLI were associated with all portions of slow muscles 177D and E, but were not found on fast muscle 178 or 179 (Figure 5). Together these results indicate that proctolin immunoreactivity is associated only with the axons and terminals of the slow depressor motoneuron and only with the slow muscles of the coxal depressor muscle system (Table 2).

B. ASSOCIATION OF PROCTOLIN IMMUNOREACTIVITY WITH OTHER SLOW SKELETAL MUSCLES

The association of proctolinergic motoneurons with three other skeletal muscles was examined with immunohistochemistry.[28] Among these were two slow muscles (abdominal intersegmentals and hyperneural) and a fast muscle (wing muscle 177C). Both slow muscles had proctolin-immunoreactive processes associated with them, but the fast muscle did not (Table 2). These results, combined with the coxal depressor muscle results, imply an association of proctolin immunoreactivity with a subset of the slow motoneuronal population in the cockroach. It is unlikely, however, that all slow motoneurons in the cockroach are proctolinergic, since the total number of proctolin-immunoreactive efferent neurons from the CNS is less than the total likely slow motoneurons in the animal.[28]

C. ASSOCIATION OF PROCTOLIN IMMUNOREACTIVITY WITH THE GUT

Similar immunochemical studies were performed in the three gut regions (fore-, mid-, and hindgut) to determine the possible distribution of proctolin innervation there. Proctolin was found to be unevenly distributed in the three gut regions. RIA results showed the highest level of proctolin immunoreactivity to be in the proctodeum (rectum) portion of the hindgut (5.7 ± 0.8 pmol/mg dry weight of muscle). Substantial amounts of proctolin were also found in the foregut (2.0 ± 0.4 pmol/mg dry weight). However, there was virtually no proctolin immunoreactivity (0.001 ± 0.002 pmol/mg) in the midgut (Figure 6). Whole-

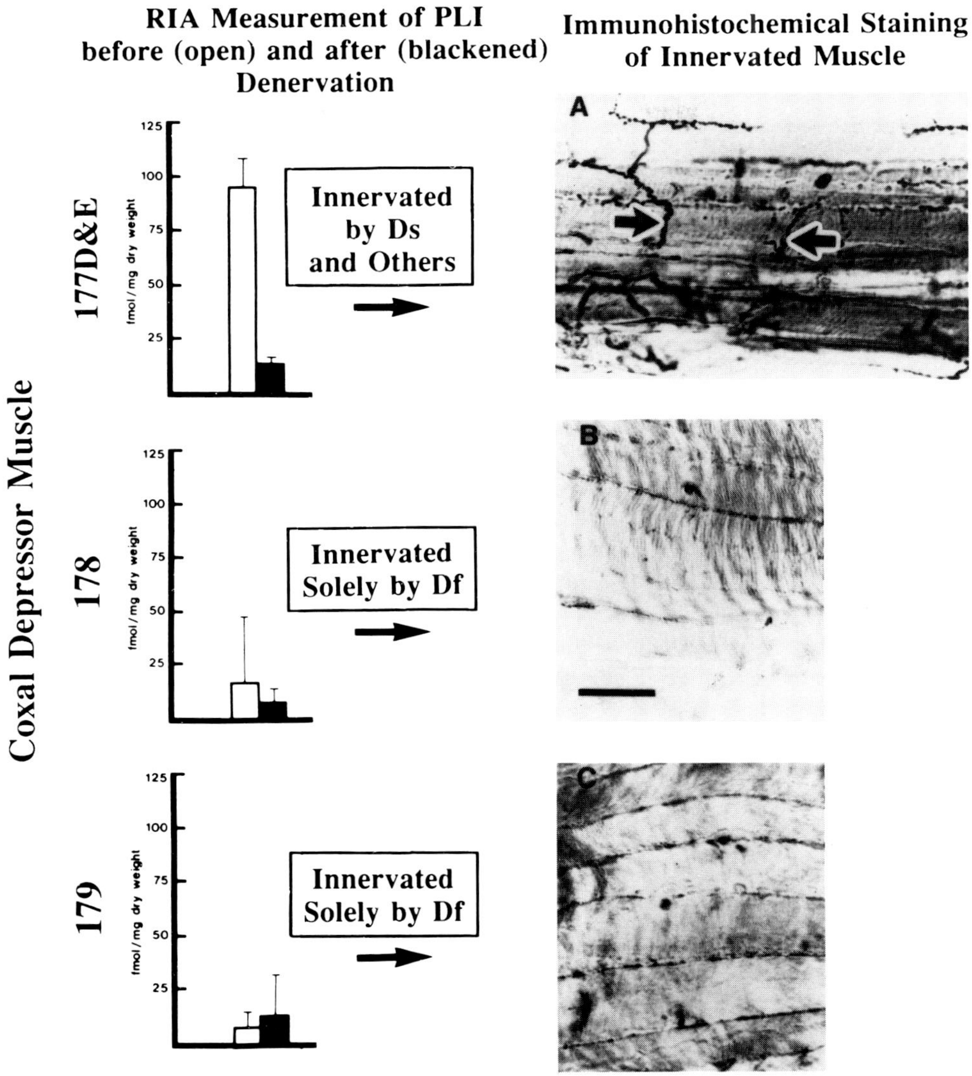

FIGURE 5. Distribution of proctolin-like immunoreactivity (PLI) in the metathoracic coxal depressor muscles.[28] Similar results were obtained in the prothoracic and mesothoracic coxal depressor muscles. Proctolin immunoreactivity, measured by radioimmunassay (graphs labeled RIA) or visualized with immunohistochemistry using whole-mount peroxidase-antiperoxidase procedure (micrographs A to C), was associated with the slow muscles 177D and E, both innervated by the slow depressor motoneuron D_s. Fast muscles 178 and 179, which are innervated by the fast motoneuron D_f, had no proctolin-immunoreactive innervation. (Arrows indicate main immunoreactive axons; scale bar = 50 μm; n = 3 for each RIA measurement.)

mount immunohistochemistry reflected a similar distribution of proctolin immunoreactivity in the gut regions and demonstrated that immunoreactivity in these muscles was associated with innervating axons. The highest density of immunoreactive nerve processes was found in the proctodeum, then the foregut, and, finally, the hindgut anterior to the proctodeum (intestine). Very sparse immunoreactive staining was associated with the midgut, and only near the hindgut junction (Figure 6),[30] suggesting innervation by hindgut proctolin-immunoreactive neurons. The presence of immunoreactive neurons in the terminal ganglion of the CNS copositional with hindgut motoneurons,[31] of immunoreactive axons in the proctodeal nerve, and of axons and terminals innervating the hindgut supports Brown's hypothesis that

TABLE 2
Summary of the Association of Proctolin-Like Immunoreactivity with Skeletal Muscles[28]

Muscle	Muscle type	Innervated with immunoreactive neuron(s)
Coxal depressor muscle		
177D	Slow	Yes
177E	Slow	Yes
178	Fast	No
179	Fast	No
Abdominal intersegmentals	Slow	Yes
Hyperneural muscle	Slow	Yes
Wing muscle 177C	Fast	No

proctolin is a transmitter for hindgut motoneurons.[3] Interestingly, the foregut is not innervated by neurons of the CNS, but by neurons located in the peripheral stomatogastric nervous system.[32] Staining of the stomatogastric nervous system reveals dense immunoreactive cell bodies and processes in all five unpaired ganglia of the system innervating the foregut and immunoreactive processes in the corpora cardiaca and allata, which are neurosecretory portions of the stomatogastric nervous system located in the head (Figure 6).

VI. PROCTOLIN IS A COTRANSMITTER FOR THE IDENTIFIED SLOW SKELETAL MUSCLE D_s

The immunohistochemical results described above strongly suggest that the slow depressor motoneuron D_s contains proctolin. To prove this, an analysis similar to that used for the LW cell (Figure 2) was undertaken.

To make certain that the D_s cell body could be identified in the living ganglion so it could be extracted from the ganglion for biochemical analysis, a cell body corresponding in size and position to the immunoreactive D_s cell (Figure 7A) was impaled with a microelectrode in a living ganglion and filled with the fluorescent dye Lucifer yellow (Figure 7B). The preparation was then processed for whole-mount immunohistochemistry and was shown to be the same cell as the proctolin-immunoreactive D_s neuron (Figure 7C).[33] D_s cell bodies were then collected and analyzed for proctolin on the basis of their bioactivity in the locust extensor-tibia bioassay before and after HPLC. The bioactivity of the D_s cell extract was identical to that of proctolin (Figure 8A and B), and bioactivity was found to coelute with ^{3}H-proctolin in reverse-phase liquid chromatography (Figure 8C). The D_s cell body contained about 0.04 pmol of proctolin. These results provided convincing support to the immunochemical data indicating proctolin's presence in the neuron.[33]

Could this preparation be used to demonstrate a specific physiological role of neurally localized proctolin? Evidence from physiological experiments on the D_s motoneuron indicated that proctolin functions as a cotransmitter at this insect neuromuscular junction.[34] Stimulation of the D_s motoneuron at low frequency (Figure 9A) resulted in a biphasic response in the coxal depressor muscle 177d. The first part of the response, which was characterized by transient tension, correlated with transient depolarization of the membrane (or the excitatory junctional potential), which was the response to the conventional neurotransmitter of the motoneuron (Figure 9B). The second part of the response was a delayed, gradual, and persistent tension not associated with a change in membrane potential (Figure 7B). That the latter response is due to release of proctolin from the motoneuron is supported by the fact that bath-applied proctolin produces a similar effect (Figure 9C) and that nerve stimulation (either electrically or by increasing the potassium in the saline) increases the level of proctolin in the saline bath.[34]

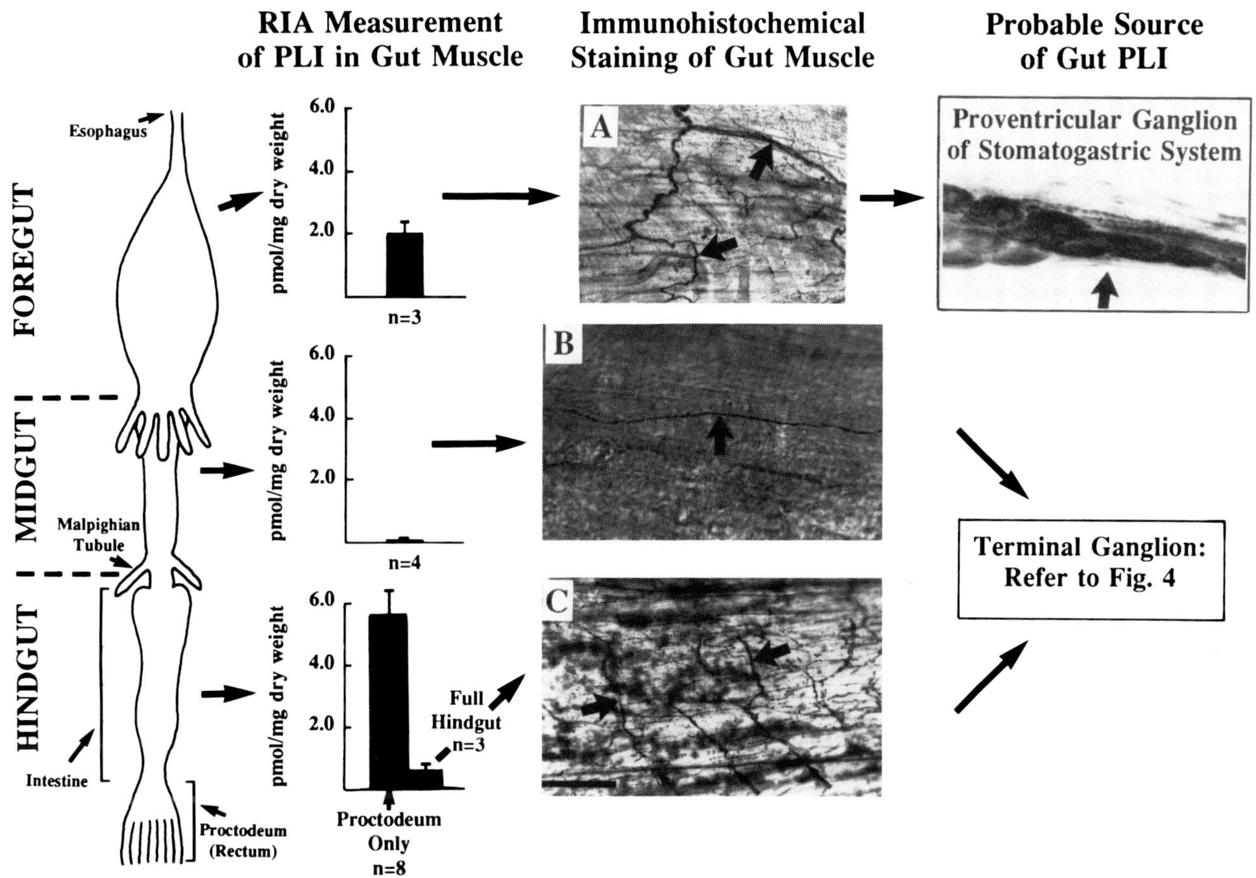

FIGURE 6. Distribution of proctolin-like immunoreactivity (PLI) in the gut.[30] Highest PLI, as detected by RIA, was recorded in the proctodeum of the hindgut. Substantial PLI was also found in the rest of the hind- and foregut, but virtually none was detected in the midgut. Immunocytochemistry indicated proctolin innervation was likewise greatest in the hindgut and virtually absent from the midgut. Arrows in micrographs indicate main immunoreactive axon(s). Scale bar = 200 μm for fore- (A) and hindgut (B), 100 μm for midgut (C), and 60 μm for proventricular ganglion.

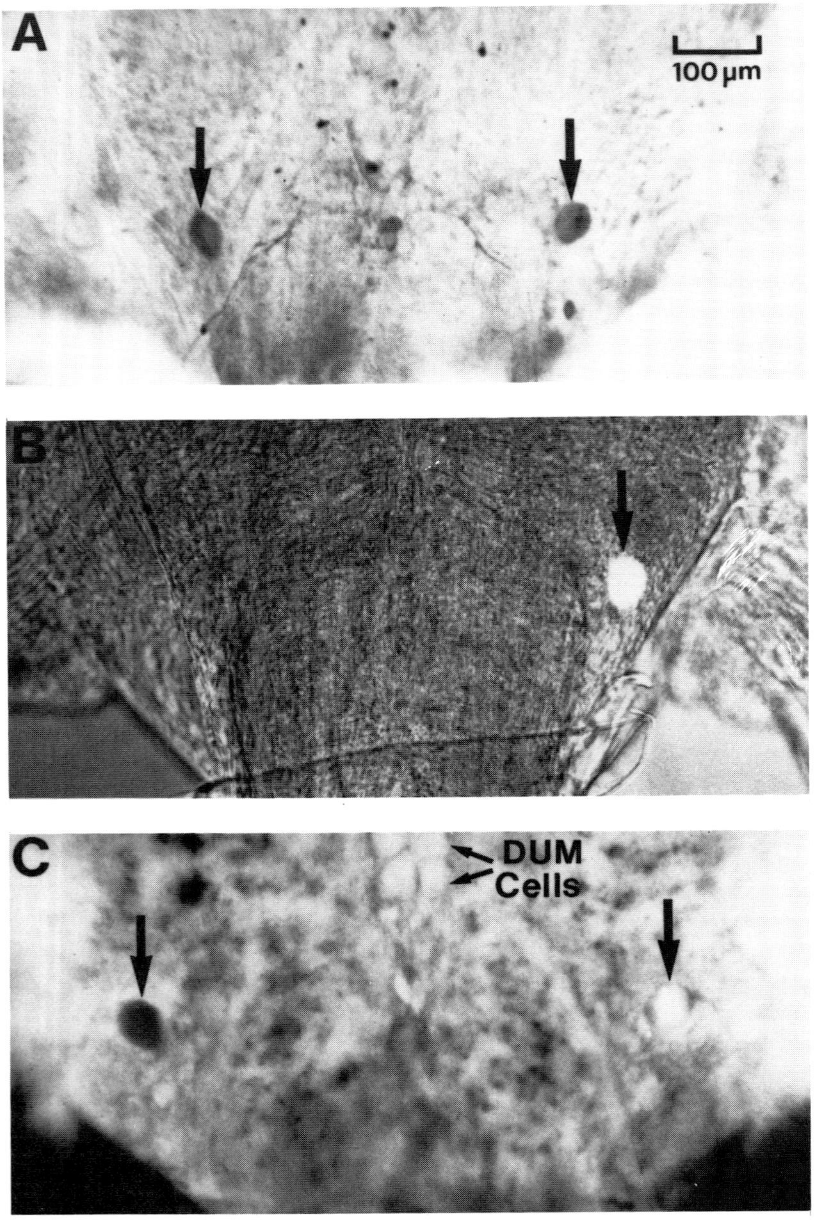

FIGURE 7. Double staining of the slow depressor motoneuron D_s to verify its identification in the living ganglion:[33] a cell body similar in size and position to the densely proctolin-immunoreactive bilateral D_s cells (arrows in A) was impaled with a Lucifer yellow-filled microelectrode (4% Lucifer yellow in 1 *M* lithium chloride) and filled with the dye by passing hyperpolarizing current pulses (500 ms duration, 1 to 5 nA) at 1 Hz for about 20 min. The dye-filled cell was then visualized with epifluorescence (arrow in B). The same ganglion was fixed and stained for proctolin immunoreactivity using the peroxidase-antiperoxidase procedure (C). The bilateral, uninjected homologue of the Lucifer yellow-injected cell was clearly proctolin immunoreactive (left arrow in C). The injected cell (right arrow in C) shows a fluorescent interior and an immunoreactive rim. The dorsal unpaired medial (DUM) cells (shown in C) serve as one of the landmarks in the ganglion for locating the D_s neurons.

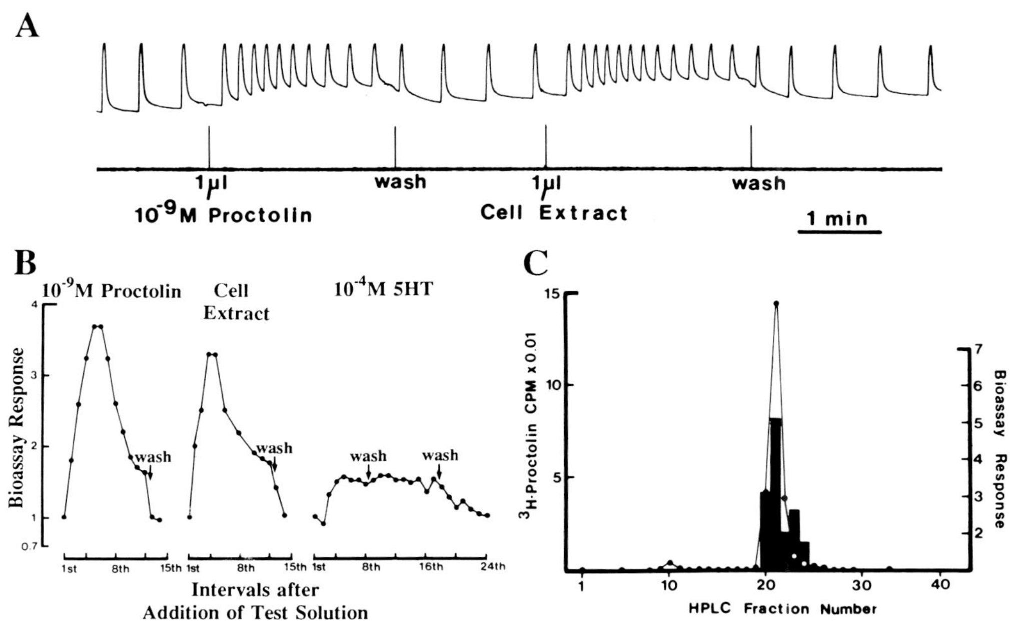

FIGURE 8. Confirming evidence for the presence of proctolin in the slow depressor motoneuron D_s.[33] D_s cell bodies were identified as described in Figure 7 and then isolated. An extract of 30 cells was prepared as described in Figure 2 and then applied to the proctolin bioassay (refer to Figure 3) before and after HPLC. (A) Qualitative comparison of proctolin and the cell extract on the bioassay, showing considerable similarity. In this example, 1 µl of cell extract contained the acid-soluble content of 0.3 cells. (B) Quantitative comparison of bioassay response to proctolin, serotonin (5-HT), and D_s cell extract. The bioassay was quantified by dividing the length of the interval between oscillations immediately before adding the test solution by the length between oscillations during various intervals after the addition. The responses to proctolin and the D_s cell extract were identical, while the response to serotonin, which also increases the frequency of the bioassay's myogenic rhythm, was quite different. (C) Proctolin-like bioactivity in the cell extract (solid bars), as measured by the bioassay, was also found to coelute with ^{3}H-proctolin (circles) in reverse-phase HPLC.

Thus, the cockroach depressor nerve-muscle preparation provided the first demonstration of a specific physiological role of neurally localized proctolin.

VII. SUMMARY AND PERSPECTIVE

The establishment of cotransmitter function in the slow depressor nerve-muscle preparation of the cockroach has led to the discovery of a similar phenomenon in nerve-muscle preparations of two other arthropods. In both cases, the presence of slow proctolinergic motoneurons was first suggested by immunohistochemical staining and subsequently supported by HPLC data. In each case, release of the peptide was also demonstrated in response to neural stimulation. In the first of the two preparations, proctolin released from the locust slow extensor tibia motoneuron, which innervates the extensor tibia muscle (used for the proctolin bioassay described in Figure 3), was found to increase the myogenic rhythm and produce a persistent tension in the muscle.[35,36] In the second case, proctolin from three of five excitatory tonic flexor motoneurons innervating the abdominal tonic flexor muscles of the crayfish (which help control abdominal posture) was found to modulate the tension generated by the released conventional neurotransmitter.[37]

Thus, the original cockroach system involving the peptide proctolin has led to the generation of two other model preparations in which some of the currently unsolved questions regarding peptide neurotransmission and multiple transmitter release may be examined. The

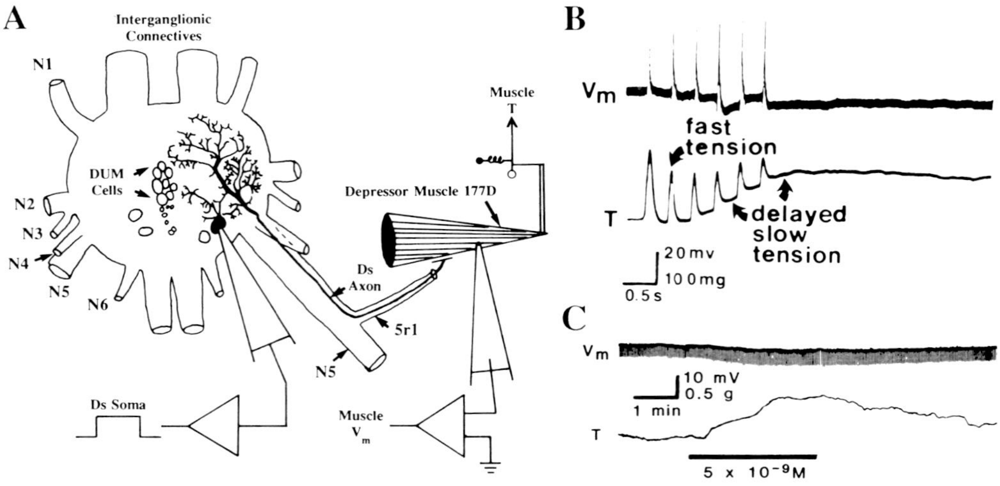

FIGURE 9. Physiological action of proctolinergic motoneuron D_s: evidence for the peptide's cotransmitter function.[34] (A) The D_s cell was stimulated by a microelectrode in its soma. Membrane potential (V_m) and tension (T) of the coxal depressor muscle 177D were measured. (B) Stimulated by a brief burst, D_s produced 1:1 excitatory junction potentials (EJPs; upper trace) and 1:1 transient tensions (lower trace), consistent with the action of the excitatory neurotransmitter L-glutamate. However, the stimulation also induced a delayed, persistent, and slow tension which was not accompanied by a change in membrane potential. (C) The slow tension response in B could be imitated by applying proctolin exogenously: bath application of 50 m*M* proctolin caused a tonic contracture (lower trace) not associated with muscle depolarization (upper trace).

possibility that proctolin may be present in mammals,[16,17,38-40] as well as the evidence that mammalian motoneurons may also release dual transmitters[41] and peptide transmitters,[42] increases the importance of studies of these preparations.

The studies of the distribution of the peptide proctolin which have just been described were initiated in part because proctolin was, at the time, one of the few sequenced invertebrate peptides available in sufficient amounts to generate antisera for immunochemical studies. Recently, however, more insect neuropeptides, including the three adipokinetic hormones, AKH I, AKH II S, and AKH II L,[43] the two related myotropic peptides, MI and MII,[36,44] and the gastrin and cholecystokinin-like myotropic peptide leucosulfakinin,[45] have been sequenced and have become available. Similar studies with these may likewise yield useful preparations for the study of the extremely important, unsolved questions concerning peptide neurotransmission. (For a general discussion of peptides as chemical signals, see Chapter 13 of this work.)

REFERENCES

1. **Brown, B. E.,** Proctolin: a peptide transmitter candidate in insects, *Life Sci.,* 17, 1241, 1975.
2. **Brown, B. E. and Nagai, T.,** Insect visceral muscle: neural relations of the proctodeal muscles of the cockroach, *J. Insect Physiol.,* 15, 1767, 1969.
3. **Brown, B. E.,** Neurotransmitter substance in insect visceral muscle, *Science,* 155, 595, 1967.
4. **Starratt, A. N. and Brown, B. E.,** Structure of the pentapeptide proctolin, a proposed neurotransmitter in insects, *Life Sci.,* 17, 1253, 1975.
5. **Miller, T.,** The properties and pharmacology of proctolin, in *Invertebrate Endocrinology: Endocrinology of Insects,* Vol. 1, Downer, R. G. H. and Laufer, H., Eds., Alan R. Liss, New York, 1983, 101.
6. **Piek, T. and Mantel, P.,** Myogenic contractions in locust muscle induced by proctolin and by wasp, *Philanthus triangulum,* venom, *J. Insect Physiol.,* 23, 321, 1977.

7. **May, T. E., Brown, B. E., and Clements, A. N.,** Experimental studies upon a bundle of tonic fibres in the locust extensor tibialis muscle, *J. Insect Physiol.,* 25, 169, 1979.
8. **Miller, T.,** Nervous versus neurohormonal control of insect heartbeat, *Am. Zool.,* 17, 77, 1979.
9. **Cook, B. J. and Holman, G. M.,** Activation of potassium depolarized visceral muscles by proctolin and caffeine in the cockroach *Leucophaea maderae, Comp. Biochem. Physiol.,* 67C, 115, 1980.
10. **Walker, R. J., James, V. A., Roberts, C. J., and Kerkut, G. A.,** Neurotransmitter receptors in invertebrates, in *Receptors for Neurotransmitters, Hormones, and Pheromones in Insects,* Sattelle, D. B., Hall, T. M., and Hildebrand, J. G., Eds., Elsevier, Amsterdam, 1980, 41.
11. **Lingle, C. J.,** The Effects of Acetylcholine, Glutamate, and Biogenic Amines on Muscle and Neuromuscular Transmission in the Stomatogastric System of the Spiny Lobster, *Panulirus interruptus,* Ph.D. thesis, University of Oregon, Eugene, 1979.
12. **Schwarz, T. L., Harris-Warrick, R. M., Glusman, S., and Kravitz, E. A.,** A peptide action in a lobster neuromuscular preparation, *J. Neurobiol.,* 11, 623, 1980.
13. **Miller, M. W. and Sullivan, R. E.,** Some effects of proctolin on the cardiac ganglion of the Maine lobster, *Homarus americanus* (Milne Edwards), *J. Neurobiol.,* 12, 629, 1981.
14. **Benson, J. A., Sullivan, R. E., Watson, W. H., III, and Augustine, G. J., Jr.,** The neuropeptide proctolin acts directly on *Limulus* cardiac muscle to increase the amplitude of contraction, *Brain Res.,* 213, 449, 1981.
15. **Watson, W. H., III, Augustine, G. J., Benson, J. A., and Sullivan, R. E.,** Proctolin and an endogenous proctolin-like peptide enhance the contractility of the *Limulus* heart, *J. Exp. Biol.,* 103, 55, 1983.
16. **Penzlin, H., Agricola, H., Eckert, M., and Kusch, T.,** Distribution of proctolin in the sixth abdominal ganglion of *Periplaneta americana* (L.) and the effect of proctolin on the ileum of mammals, *Adv. Physiol. Sci.,* 22, 525, 1981.
17. **Schulz, H., Schwarzberg, H., and Penzlin, H.,** The insect neuropeptide proctolin can affect the CNS and the smooth muscle of mammals, *Acta Biol. Med. Ger.,* 40, K1, 1981.
18. **O'Shea, M. and Adams, M. E.,** Pentapeptide (proctolin) associated with an identified neuron, *Science,* 213, 567, 1981.
19. **Bishop, C. A., O'Shea, M., and Miller, R. J.,** Neuropeptide proctolin (H–Arg–Tyr–Leu–Pro–Thr–OH): immunological detection and neuronal localization in the insect central nervous system, *Proc. Natl. Acad. Sci. U.S.A.,* 78, 5899, 1981.
20. **Bishop, C. A. and O'Shea, M.,** Neuropeptide proctolin (H–Arg–Tyr–Leu–Pro–Thr–OH): immunocytochemical mapping of neurons in the central nervous system of the cockroach, *J. Comp. Neurol.,* 207, 223, 1982.
21. **Pearson, K. G. and Iles, J. F.,** Innervation of coxal depressor muscle in cockroach, *Periplaneta americana, J. Exp. Biol.,* 54, 215, 1971.
22. **Pearson, K. G. and Fourtner, C. R.,** Non-spiking interneurons in walking system of the cockroach, *J. Neurophysiol.,* 38, 33, 1974.
23. **Usherwood, P. N. R.,** Amino acids as transmitters, *Adv. Comp. Physiol. Biochem.,* 7, 227, 1978.
24. **Usherwood, P. N. R. and Cull-Candy, S. G.,** Pharmacology of somatic nerve muscle synapses, in *Insect Muscle,* Usherwood, P. N. R., Ed., Academic Press, London, 1975, 207.
25. **Irving, S. N. and Miller, T. A.,** Aspartate and glutamate as possible transmitters at the "slow" and "fast" neuromuscular junctions of the body wall muscles of *Musca* larvae, *J. Comp. Physiol.,* 135, 299, 1980.
26. **Emson, P. C., Burrows, M., and Fonnum, F.,** Levels of glutamate decarboxylase, choline acetyltransferase and acetylcholinesterase in identified motoneurons of the locust, *J. Neurobiol.,* 5, 33, 1974.
27. **Evans, P. D. and O'Shea, M.,** An octopaminergic neuron modulates neuromuscular transmission in the locust, *Nature (London),* 270, 257, 1977.
28. **Witten, J. and O'Shea, M.,** Peptidergic innervation of insect skeletal muscle: immunochemical observations, *J. Comp. Neurol.,* 242, 93, 1985.
29. **Pearson, K. G., Fourtner, C. R., and Wong, R. K.,** Nervous control of walking in the cockroach, in *Control of Posture and Locomotion,* Stein, R. B., Pearson, K. G., Smith, R. S., and Redford, J. B., Eds., Plenum Press, New York, 1973, 495.
30. **Witten, J. L.,** The Identification of Peptidergic Neuromuscular Systems in Insects, Ph.D. thesis, University of Chicago, Chicago, IL, 1984.
31. **Eckert, M., Agricola, H., and Penzlin, H.,** Immunocytochemical identification of proctolin-like immunoreactivity in the terminal ganglion and hindgut of the cockroach *Periplaneta americana* (L.), *Cell Tissue Res.,* 217, 633, 1981.
32. **Willey, R. B.,** The morphology of the stomatodeal nervous system in *Periplaneta americana* (L.) and other blattaria, *J. Morphol.,* 108, 219, 1961.
33. **O'Shea, M. and Bishop, C. A.,** Neuropeptide proctolin associated with an identified skeletal motoneuron, *J. Neurosci.,* 2, 1242, 1982.
34. **Adams, M. E. and O'Shea, M.,** Peptide co-transmitter at a neuromuscular junction, *Science,* 221, 286, 1983.

35. **Worden, M. K., Witten, J. L., and O'Shea, M.,** Proctolin is a co-transmitter for SETi motoneuron, *Soc. Neurosci. Abstr.*, 11, 327, 1985.
36. **O'Shea, M., Adams, M. E., Bishop, C., Witten, J., and Worden, M. K.,** Model peptidergic systems at the insect neuromuscular junction, *Peptides*, 6, 417, 1985.
37. **Bishop, C. A., Wine, J. J., Nagy, F., and O'Shea, M. R.,** Physiological consequences of a peptide cotransmitter in a crayfish nerve-muscle preparation, *J. Neurosci.*, 7, 1769, 1987.
38. **Holets, V. R., Hökfelt, T., Ude, J., Eckert, M., and Hansen, S.,** Co-existence of proctolin with TRH and 5-HT in the rat CNS, *Soc. Neurosci. Abstr.*, 10, 692, 1984.
39. **Bernstein, H. G., Eckert, M., Penzlin, H., and Dorn, A.,** Proctolin related material in the mouse brain as revealed by immunohistochemistry, *Neurosci. Lett.*, 45, 229, 1984.
40. **Bernstein, H. G., Eckert, M., Penzlin, H., Vieweg, U., Röse, I., and Dorn, A.,** Proctolin immunoreactive neurons in the human brain stem, *Acta Histochem.*, 80, 111, 1986.
41. **Chan-Palay, V., Engel, A. G., Wu, J.-Y., and Palay, S. L.,** Coexistence in human and primate neuromuscular junctions of enzymes synthesizing acetylcholine, catecholamine, taurine and γ-aminobutyric acid, *Proc. Natl. Acad. Sci. U.S.A.*, 79, 7027, 1982.
42. **Gold, M. R.,** The action of vasoactive intestinal peptide at the frog's neuromuscular junction, in *Coexistence of Neuroactive Substances in Neurons*, Chan-Palay, V. and Palay, S. L., Eds., John Wiley & Sons, New York, 1984, 161.
43. **Mordue, W.,** Strategies for the isolation of insect peptides, *Peptides*, 6, 407, 1985.
44. **Witten, J., Schaffer, M. H., O'Shea, M., Cook, J. C., Hemling, M. E., and Rinehart, K. L., Jr.,** Structures of two cockroach neuropeptides assigned by fast atom bombardment mass spectrometry, *Biochem. Biophys. Res. Commun.*, 124, 350, 1984.
45. **Nachman, R. J., Holman, G. M., Haddon, W. F., and Ling, N.,** Leucosulfakinin, a sulfated insect neuropeptide with homology to gastrin and cholecystokinin, *Science*, 234, 71, 1986.

Chapter 15

ISOLATION AND CHEMICAL CHARACTERIZATION OF COCKROACH NEUROPEPTIDES: THE MYOTROPIC AND HYPERGLYCEMIC PEPTIDES

Benjamin J. Cook and Renée M. Wagner

TABLE OF CONTENTS

INTRODUCTION

To exist, every living organism must possess the functions of assimilation, autonomy, and reproduction. Without these vital activities an organism would quickly lose its individuality. Clearly, a constant and delicate equilibrium is required to ensure an organism's maintenance. In insects, as in most animals, this task is accomplished by the synergistic action of the nervous and endocrine systems. The cells of these two systems are uniquely fashioned to transmit information from either the internal organic centers of command or the environment to target organs and tissues within the organism.

Neurons are remarkably adapted to conduct electrical signals over their long cellular extensions, or axons. In fact, many millimeters or centimeters may be traversed before an interruption occurs in the pathway at a junction between cells. In some instances these junctions or synapses are "tight" enough to permit the electrical signal to jump the gap, but the cleft between most cells does not permit this, and a chemical messenger is released into the synaptic gap. These messengers, in turn, open ion channels or initiate a cascading response of secondary messengers in the effector cells by interaction with receptor proteins at the surface membrane.

The second channel for cellular coordination in insects is composed of endocrine cells and organs. Hormones released from these sources are distributed to target cells by the circulatory system. Here again, specific receptor proteins at the surface of the target cells mediate the responses.

A third avenue of cellular regulation consists of a specialized group of neurons within the central nervous system (CNS), the neurosecretory cells. Although these cells are able to conduct electrical signals in the same manner as ordinary neurons, they generally release their secretory products into the circulatory system in a manner similar to endocrine cells. The release occurs at specialized end structures called neurohemal organs. These organs are formed by a profuse branching of terminal axons, the swollen ends of which are separated from hemolymph only by an acellular basement membrane.[1] The corpus cardiacum (CC) is generally the largest neurohemal organ found in insects, and it serves in part as a storage site for neurosecretions synthesized in the brain. This neuroendocrine complex is analogous to the X-organ sinus gland of crustaceans and the hypothalamo-hypophyseal system of vertebrates. Indeed, the nearly universal occurrence of peptide-producing neurosecretory cells in the animal kingdom[1a] provides occasion for the biomedical community to seriously consider cockroaches as a model for the study of the basic aspects of neurosecretion (see also Chapter 13 of this work). Although these insects are small, their neurosecretory system is not as complex as those found in vertebrates, at least in terms of the total number of cells involved. Moreover, experiments with such animals are not encumbered by many of the regulatory considerations that must be addressed for vertebrate experimental animals.

Since the discovery of substance P by von Euler and Gaddum,[1b] several additional vertebrate peptide hormones that regulate the motility of the alimentary tract have been identified. Two important facts have emerged from this research:

1. The isolation and purification of these peptides is often accomplished by the use of some type of visceral organ assay.
2. Although such peptides as the enkephalins, the vasoactive intestinal peptide cholecystokinin, and gastrin have activity on the digestive tract, they are also found in cells of the mammalian brain.[1c]

These observations gave some of the first hints that peptides were multifunctional regulators, and it was established from subsequent research that substance P is a "brain-gut" peptide that serves as a sensory neurotransmitter in addition to stimulating the gut. Research on cockroach neuropeptides seems to have uncovered similarly multifunctional patterns.

In 1953, Cameron[2] discovered that extracts of the CC from the American cockroach, *Periplaneta americana,* could stimulate the visceral muscles of that insect. In a subsequent study of the same neuroendocrine organs in *P. americana,* Brown[3] demonstrated that three peptides are responsible for most of this myotropic activity. One of them is active on both the hindgut and heart, and it was designated P_1. The other two peptides, P_2 and P_3, are specific activators of the hindgut and heart, respectively. Shortly afterward, Brown[4] reported the presence of an unidentified substance in extracts of the hindgut which causes a slow-type graded contraction of the longitudinal muscles of that organ. He also noted a high specific activity for this factor in extracts of the proctodeal nerve. This substance is not inactivated by chymotrypsin. In this respect, it was considered distinct from the peptides found in the CC.

In an effort to substantiate Brown's findings in another cockroach species, Holman and Cook[5] found three materials in hindgut tissue extracts of the Madeira cockroach, *Leucophaea maderae,* that stimulated the muscular activity of the isolated hindgut. Two of the compounds were identified as L-glutamic acid and L-aspartic acid. Application of either of these amino acids to an isolated hindgut causes a single slow contraction that is indistinguishable from the response produced by electrical stimulation of the nerves innervating the hindgut. An application of the third material results in a prolonged and complex stimulation of the hindgut which is similar in character to the responses reported by both Davey[6] and Brown[3] for CC extracts. A further study of this third material by Holman and Cook[7] showed that its biological activity is destroyed by pronase, but not by chymotrypsin. Moreover, this peptide is found in extracts of hindguts, terminal ganglia, proctodeal nerves, and heads of *L. maderae*. Finally, the persistent physiological effects of the peptide on the hindgut and the presence of neurosecretion in the proctodeal nerve suggested that the peptide might be a type of neurohormone.

When Brown and Starratt[8-11] isolated and identified proctolin (Arg–Tyr–Leu–Pro–Thr) from the cockroach *P. americana*, they suggested that this peptide and the hindgut-stimulating neurohormone (HSN) described by Holman and Cook[7] were identical. This suggestion was based on a number of similarities in the chemical and biological properties of the two peptides. However, this proposed identity did not account for some of the reported differences between the two materials. Although the "gut factor", or proctolin, was reported to be present in extracts of the fore- and hindgut by Brown and Starratt,[4,10] it is absent from extracts of the head and pharmacologically and chemically distinct from factors P_1 and P_2 found in the CC. HSN, by comparison, is found in both head and hindgut extracts and can be synthesized *in vitro* by the brain for storage and release from the CC.[12,13] In addition, the response of the hindgut to HSN showed a close resemblance to those for factors P_1 and P_2. These differences posed two important and interrelated questions:

1. Is proctolin indeed absent from extracts of the head as suggested?
2. Are there other myotropic peptides in the head that can account for the activation of the hindgut?

An effective approach to answering these questions was provided by the development of reverse-phase high-performance liquid chromatography (RP-HPLC). Proctolin was detected (by bioassay and by absorbance at a specific retention time) in hindgut extracts of *L. maderae* at a concentration of 3.3 ng per organ, but no evidence of this peptide could be found in head extracts from the same insect after HPLC analysis.[14] Nevertheless, a hindgut-stimulating fraction with the same retention time as proctolin and an activity equivalent to 1.5 ng per head was found. When this fraction was derivatized with phenylisothiocyanate (PITC), a substance with a slightly shorter retention time than PITC-proctolin was obtained. Recently, a reverse-phase system was developed that provided a clear separation of five

active fractions from head extracts of *L. maderae*.[15] This initial finding eventually resulted in the purification, amino acid analysis, and primary sequence determination of 11 myotropic peptides[16-22] and of one that suppresses contractile activity[23] (Table 1).

The capacity of living cells for continuous chemical change consists of two unique processes: (1) those physiochemical transformations that release energy to support life and (2) those molecular rearrangements and reconstructions that produce organic substances. If these processes are to proceed in an orderly fashion, they must be subject to regulation. The primary agents for this kind of regulation are hormones.

One of the first hormones recognized in insects as a regulator of metabolism was the hyperglycemic factor of the American cockroach. Steele[24] showed that extracts of the CC from this species could elevate hemolymph trehalose levels at the expense of fat body glycogen. This depletion of glycogen is caused by the activation of the enzyme glycogen phosphorylase through adenylate cyclase.[25] Early attempts to isolate and purify this factor established the fact that two separate fractions from extracts of the CC have hyperglycemic activity.[3,26] However, as a consequence of the poor resolution of the separation techniques available at the time, the chemical structures of these factors could not be determined. Once the method of RP-HPLC was introduced, the necessary resolution for isolation[27] and identification of these two peptides was obtained, and two research groups[28,29] accomplished this task almost simultaneously in 1984. Both peptides have hyperglycemic and heart-acceleratory effects and are remarkably similar in structure to locust adipokinetic hormones I and II. In fact, these cockroach peptides exhibit adipokinetic action when injected into the locust *Schistocerca nitens*.[28]

The structures of the two hyperglycemic hormones in *P. americana* have recently been confirmed by a third group using enzymatic and gas-phase sequencing techniques.[30] Moreover, Siegert et al.[31] have further demonstrated that both hormones can activate fat body glycogen phosphorylase. Another hyperglycemic hormone that possesses cardiotropic properties has been isolated and characterized from the cockroach *Blaberus discoidalis*.[32] This hormone, unlike the octapeptides found in *P. americana*, is a decapeptide. Nevertheless, the sequence for the first eight amino acids is almost identical to hyperglycemic hormone I found in *P. americana*; the only difference between the two is the presence of glycine instead of asparagine at position 7 in the *Blaberus* peptide. A hyperglycemic peptide of identical structure has also been identified in the cockroach *Nauphoeta cinerea*[33] (Table 1).

II. THE BIOLOGICAL ASSAYS

The search for endogenous chemical mediators in living systems ultimately depends on the detection of some critical physiological or biochemical change in the organism that results from exposure to a tissue extract. Although there is no assurance that a single chemical in the initial extract is responsible for a specific physiological response, the probability of it increases with each step of the purification. Moreover, additional confidence is gained in favor of this assumption if one selects a simple isolated preparation for the biological assay. Such a precaution minimizes the number of available sites for chemical-tissue interactions. As it happens, many preparations of insect visceral muscle provide this opportunity, and once Cameron[2] had discovered that the products of the neuroendocrine system of *P. americana* stimulate these muscles, attempts to isolate and characterize the chemicals that evoke contractile events were soon undertaken. Davey[6] was the first to describe the effects of CC homogenates on the spontaneous contractile activity of the hindgut. In this preparation, the whole hindgut is suspended in an organ bath by fine threads between a fixed point and a lever. Such an arrangement provides an isotonic recording of longitudinal muscle movement in the hindgut, and the myographic records shows changes in the frequency, amplitude, and tonus of these contractions. From this simple and straightforward preparation the necessary

TABLE 1
Chemical and Biological Properties of Myotropic and Hyperglycemic Peptides from the Cockroach

Peptide[a]	Activities[b]	Structure	Threshold of synthetic product	Ref.
Leucokinins (*L. maderae*)	MT			
LK I		Asp-Pro-Ala-Phe-Asn-Ser-Trp-Gly-NH_2	200 p*M*	16
LK II		Asp-Pro-Gly-Phe-Ser-Ser-Trp-Gly-NH_2	160 p*M*	16
LK III		Asp-Gln-Gly-Phe-Asn-Ser-Trp-Gly-NH_2	72 p*M*	17
LK IV		Asp-Ala-Ser-Phe-His-Ser-Trp-Gly-NH_2	140 p*M*	17
LK V		Gly-Ser-Gly-Phe-Ser-Ser-Trp-Gly-NH_2	52 p*M*	21
LK VI		pGlu-Ser-Ser-Phe-His-Ser-Trp-Gly-NH_2	56 p*M*	21
LK VII		Asp-Pro-Ala-Phe-Ser-Ser-Trp-Gly-NH_2	120 p*M*	22
LK VIII		Gly-Ala-Asp-Phe-Tyr-Ser-Trp-Gly-NH_2	29 p*M*	22
Leucopyrokinin (LPK) (*L. maderae*)	MT	pGlu-Thr-Ser-Phe-Thr-Pro-Arg-Leu-NH_2	600 p*M*	18
Leucosulfakinins (*L. maderae*)	MT			
LSK I		Glu-Gln-Phe-Glu-Asp-Tyr-(SO_3H)-Gly-His-Met-Arg-Phe-NH_2	220 p*M*	19
LSK II		pGlu-Ser-Asp-Asp-Tyr-(SO_3H)-Gly-His-Met-Arg-Phe-NH_2	45 p*M*	20
Leucomyosuppressin (LMS) (*L. maderae*)	MS	pGlu-Asp-Val-Asp-His-Val-Phe-Leu-Arg-Phe-NH_2	78 p*M*	23
Proctolin (*P. americana*)	MT, CA	Arg-Tyr-Leu-Pro-Thr	1 n*M*	8, 10, 11
Hyperglycemic hormones (*P. americana*)	MT, CA, HT, HG		5—10 n*M*	27, 28, 29, 81
Neurohormone D (MI, CCI, HGH I, HTH I)		pGlu-Val-Asn-Phe-Ser-Pro-Asn-Trp-NH_2		
Myoactive factor II (MII) (CCII, HGH II, HTH II)		pGlu-Leu-Thr-Phe-Thr-Pro-Asn-Trp-NH_2		
Hypertrehalosemic hormone (HTH) (*B. discoidalis, N. cinerea*)	HT	pGlu-Val-Asn-Phe-Ser-Pro-Gly-Trp-Gly-Thr-NH_2	0.05 pmol	32, 33, 77

[a] Abbreviations: CC, corpora cardiaca factor; HGH, hyperglycemic hormone; HTH, hypertrehalosemic hormone; M, myoactive factor; HG, hyperglycemic.
[b] Abbreviations: MT, myotropic; CA, cardioacceleratory; MS, myosuppressive; HG, hyperglycemic.

physiological and pharmacological methods were developed to isolate and identify most of the myotropic neuropeptides found in cockroaches.

A. BIOASSAYS FOR THE MYOTROPINS

This section describes in detail the methods developed for the detection and analysis of myotropic neuropeptides found in *L. maderae*; the procedures may be readily adapted to other cockroach species. Although there is an emphasis on the hindgut preparation, the methods for the isolation of the foregut and oviduct are also included because the same physiological apparatus may be used to record the responses of these organs, with only slight modifications. The preparation of the cockroach heart and the measurement of heartbeat by impedance conversion are also described.

1. Procedures for Organ Isolation and Preparation

The cockroaches used for the biological assay were taken from stock colonies maintained at 27°C and a relative humidity of 50%. The insects were fed dry dog food and water *ad libitum*. All isolated organs were perfused with saline solution that had the following composition: Na, 154 m*M*; K, 2.7 m*M*; Ca, 1.8 m*M*; Cl, 160 m*M*; and glucose, 22 m*M*. The pH was adjusted to 6.8 with sodium hydroxide.

a. Hindgut

The hindgut of *L. maderae* is divided into four distinct parts[34] (Figure 1): the ileum (not shown), a short, narrow section of the gut extending from the evagination of the Malpighian tubules to the abrupt expansion of the intestinal wall, which marks the anterior limit of the colon; the colon, which is separated into anterior and posterior limbs by a median sphincter; and the rectum, which is readily identified by the six symmetrically arranged longitudinal muscles.

Six groups of muscles that can dilate the rectum arise from the posterior terminations of the superior bundles of the longitudinal strap muscles (Figure 1). The dorsal pair of rectal suspending muscles (RSD) are inserted on the terminal abdominal tergite just above the rectum. The lateral suspending muscles of the rectum (RSL) are attached to the same tergite, but at a more lateral position, and the ventral pair of rectal suspending muscles (RSV) are attached to the lateral borders of a ventral genital sclerite in the male.

A second group of suspensory muscles is associated with the anal aperture (AS; Figure 1). A single dorsal depressor is attached to the supra-anal plate, and one pair each of extensors and depressors are attached to the perianal plates on either side of the anal aperture.

The hindgut of the Madeira cockroach is innervated bilaterally from a pair of dorsally directed branches of the cercal nerves. Shortly after these proctodeal nerves (proc. nerves) reach the posterior lateral surface of the rectum, they bifurcate and give rise to anterior (A) and posterior (P) branches (Figure 1). The posterior branch innervates the dorsal and lateral dilator muscles of the rectum, the suspensory muscles of the anal aperture, and the circular muscles of the posterior rectum. The anterior branch innervates the circular and longitudinal muscles of the anterior rectum. At the constriction between the rectum and posterior colon (rectal valve), these branches bifurcate to give rise to two nerves on either side that innervate the surface of the colon as they proceed toward the anterior limit of the organ.

Muscle preparation — Adult cockroaches of either sex were decapitated and the legs and wings were removed. While carefully squeezing the lateral borders of the abdomen, a dorsal incision was made at the midline, just anterior to the last abdominal sclerite, and continued throughout the abdomen (Figure 2A). The abdomen was then severed from the thorax, and the cuticle was pinned open to expose the viscera (Figure 2B). The point at which the Malpighian tubules attach to the hindgut was exposed by pulling up on the foregut in an anterior direction with a pair of forceps (Figure 2C). The hindgut was separated from

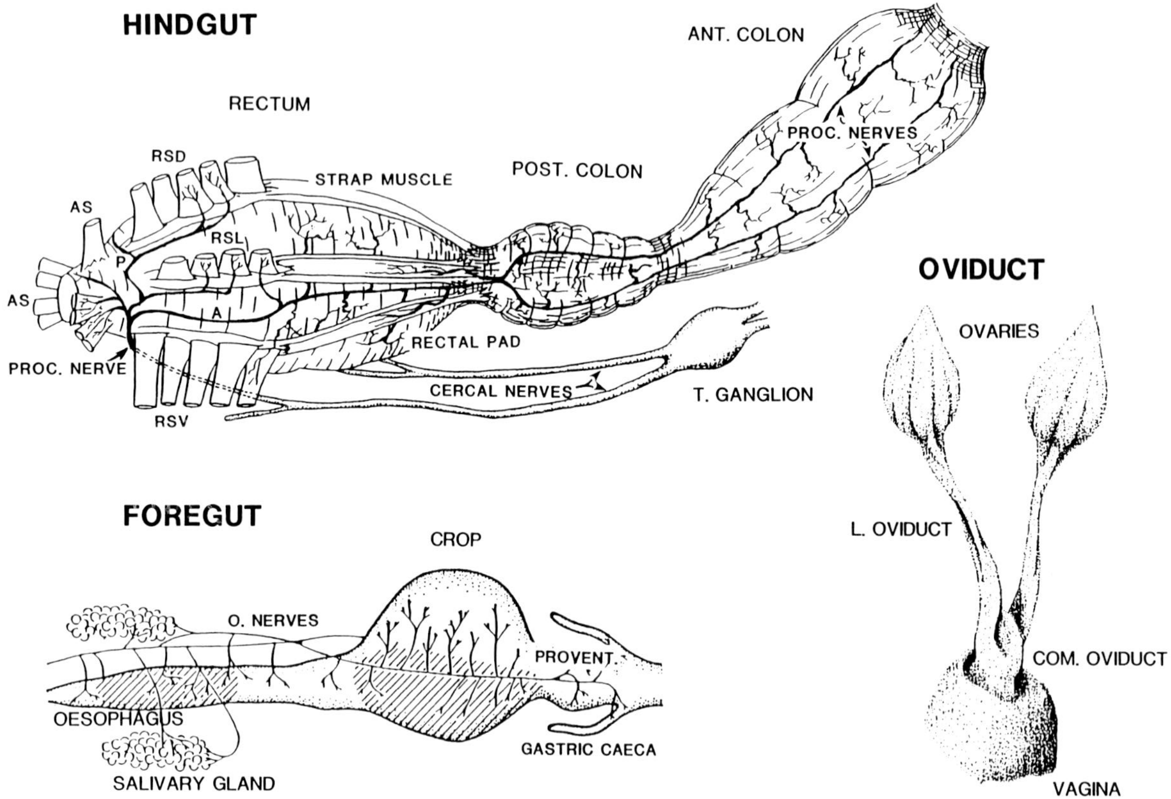

FIGURE 1. **Hindgut:** dorsolateral view of the basic subdivisions and innervation pattern of the hindgut of *Leucophaea maderae*. The rectum is suspended from the cuticular exoskeleton by dorsal, lateral, and ventral groups of muscle (RSD, RSL, and RSV, respectively) on either side of the rectum. Four groups of muscles (AS) suspend the anal aperture. Rectal pads are covered with a layer of circular muscle. The proctodeum is innervated bilaterally from proctodeal nerves (PROC. NERVES) that branch off the cercal nerves. The proctodeal nerves have anterior (A) and posterior (P) branches on the surface of the posterior rectum. Terminal (T.) ganglion. **Foregut:** dorsolateral view of the basic subdivisions and innervation pattern of the foregut of *L. maderae*. O. NERVES = esophageal nerves; PROVENT. = proventriculus. Hatched areas indicate probable pacemaker regions. **Oviduct:** an extended view of the oviduct of *L. maderae* as it appears in the muscle chamber. Lateral (L.) oviducts; common (COM.) oviduct.

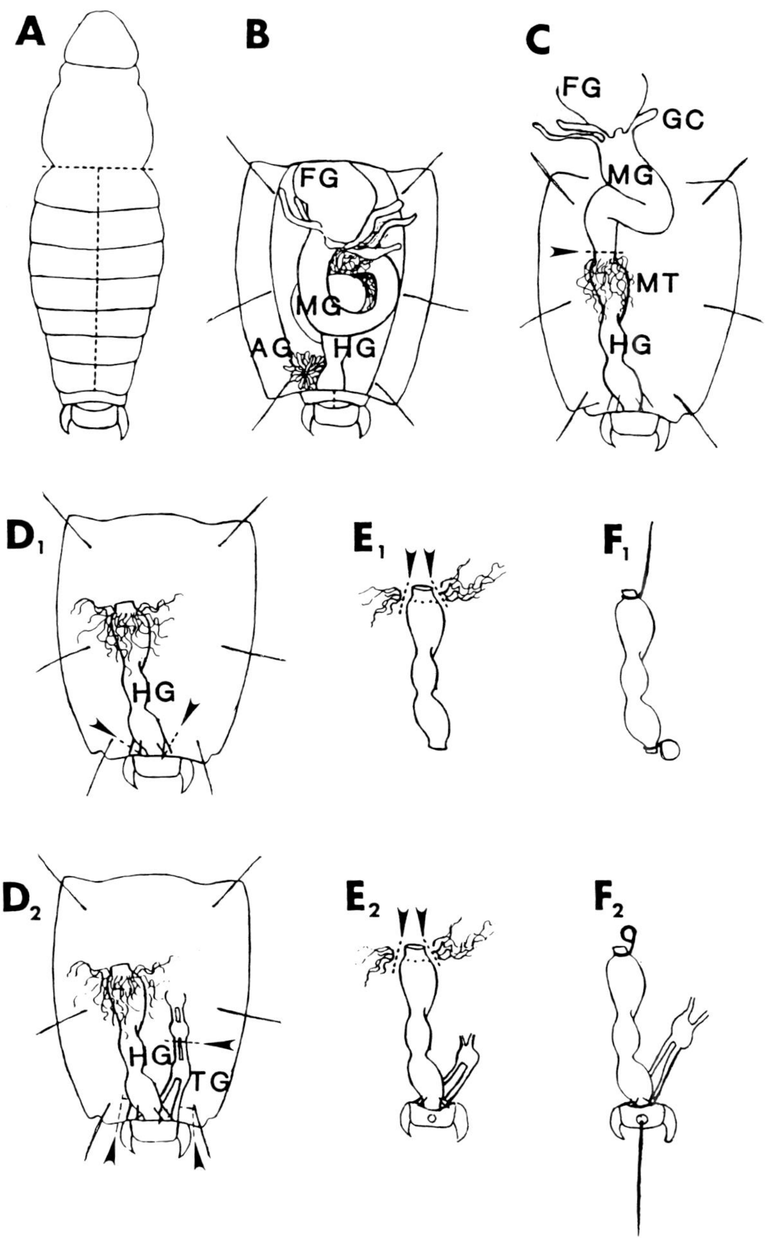

FIGURE 2. A series of line drawings that illustrate (dorsal aspect) the successive steps in the preparation of the isolated hindgut (A,B,C) for spontaneous myographic recording (D_1,E_1,F_1) and neurally evoked responses (D_2,E_2,F_2). Detailed procedures are described in the text. Abbreviations: foregut (FG), midgut (MG), accessory gland (AG), hindgut (HG), Malpighian tubules (MT), gastric caeca (GC), terminal ganglion (TG).

the midgut just anterior to this evagination. Once the tissues (fat body, large tracheae, etc.) that adhere to the hindgut were cut and/or pulled away from its surface, the posterior end of the organ could be carefully released from its attachments to the terminal tergites by cutting the dorsal and lateral suspending muscles of the rectum (RSD and RSL; Figures 1 and $2D_1$). The hindgut was then pulled completely free from the abdomen by cutting the

suspending muscles that remain along the ventral surface of the rectum (RSV) together with the muscles (AS) and the thin cuticular membrane around the anal aperture. The isolated hindgut was then transferred to a paraffin-filled petri dish (Figure $2E_1$), where a thread was tied about the end of the midgut just above the evagination of the Malpighian tubules (Figure $2F_1$). Another thread was tied around the hindgut close to the anal aperture. Both threads were used to suspend the organ in a muscle chamber.

Nerve-muscle preparation — Male cockroaches were generally used for innervated hindgut preparations because of the smaller amounts of fat body present and the more discrete nature of the reproductive tract. Once the hindgut was severed just anterior to the Malpighian tubules, the accessory gland was removed and the organs of copulation were separated from their attachments to the ventral genital sclerite by pulling the posterior portion of these organs in an anterior direction. At this point, the last two abdominal ganglia could be freed from their tracheal (Figure $2D_2$) and peripheral neural attachments in the ventral portion of the abdomen. The cercal nerves containing the proctodeal nerves were carefully separated from their connections to the lateral borders of the ventral genital sclerite. In the process of releasing the posterior hindgut from the abdomen, the intersegmental membranes between the eighth and tenth abdominal tergites and the seventh and eighth abdominal sternites were severed, along with the minor attachments between the ventral genital sclerite and the underlying abdominal sternites 6 and 7. Here one must be cautious with the proctodeal nerve because it runs just beneath the group of suspending muscles that attach the ventral genital sclerite to the rectum (Figure 1, RSV). The isolated hindgut and adhering terminal sclerites were transferred to a wax-filled petri dish (Figure $2E_2$). Threads were tied around the anterior and just above the Malpighian tubules. A pinhole was punched through the tenth abdominal tergite (Figure $2F_2$) and the ninth sternite, and a thread was fed through both sclerites and secured by a loop.

b. Foregut

When the dorsal surface of the cockroach was opened (both the abdomen and thorax), the salivary glands and tracheae were removed from the surface of the foregut (Figure 1). The foregut was then removed from the insect by cutting the alimentary canal between the proventriculus and the gastric ceca. One thread was then tied about the anterior esophagus and another between the crop and proventriculus. The proventricular thread was attached to a metal hook that could be lowered into a muscle chamber by means of a micromanipulator. The esophageal thread was looped over a balsa-wood lever and fastened down in soft dental wax. The beam of the transducer was sufficiently counterweighted to produce a tension of approximately 180 mg on the suspended organ. The amount of food in the foregut did not seem to have any effect on the level of spontaneous activity recorded.

c. Oviduct

Adult female cockroaches of varying age and reproductive state were decapitated and the legs and wings removed. A dorsal incision was then made, as shown in Figures 2A and B, and the hindgut and the Malpighian tubules were removed to expose the lateral oviducts and ovaries beneath. Once the apical suspending ligaments and the lateral tracheal attachments of the ovaries were severed, it was possible to loosen the lateral oviduct from the side wall of the vagina by gently drawing the ovaries in a slightly anterior direction while cutting the remaining attachments. After both oviducts were freed in this manner, a dorsal patch of the vagina was cut out around the junction with the common oviduct (Figure 1). This released the preparation from the insect, and a thread was tied about the junction of the vagina with the common oviduct. Another thread was fastened about both ovaries just above the lateral oviducts. The first thread was attached to a metal hook that could be lowered into a saline-filled muscle chamber by means of a micromanipulator. The second was fixed

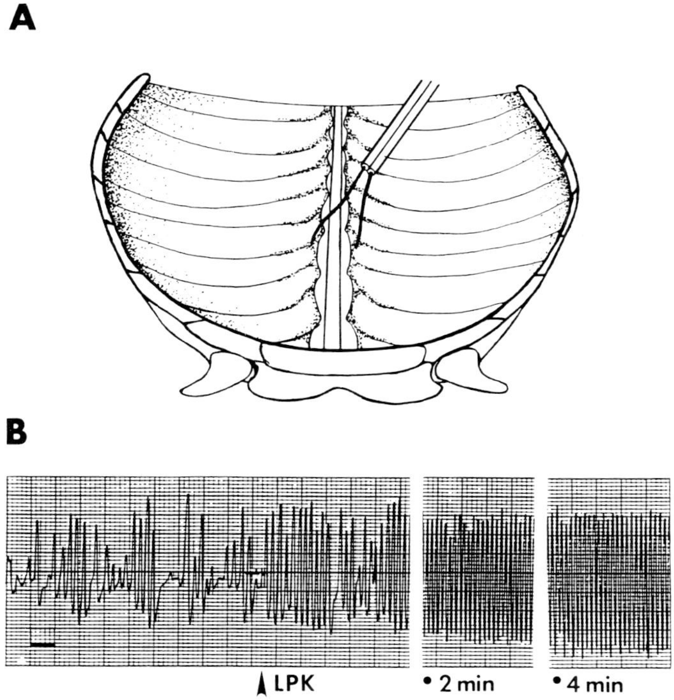

FIGURE 3. (A) Drawing of the abdominal dorsum of the cockroach *L. maderae* showing the heart and the position of the electrodes for recording the heartbeat; (B) response of the cockroach myocardium to the neuropeptide leucopyrokinin (LPK). An appropriate amount of the peptide was added (arrow) to give a final concentration of 1 n*M* in the 100-μl bath that covered the heart. Activity of the heart was monitored for 30 s every minute after addition of the peptide.[34a]

with wax to a balsa-wood lever that activated the muscle transducer as described in Section II.A.2. The balsa-wood beam was counterweighted with 40 to 85 mg to produce sufficient tension on the suspended organ.

d. Heart

Adult cockroaches of either sex were decapitated, and the abdomens were removed as close to the metathorax as possible. The ventral cuticular surface of the abdomen was then carefully trimmed away so that the lateral spiracular structures remained attached to the dorsal sclerites. This precaution ensured a sufficient supply of air to the heart. At this point, the dorsum was pinned out in a wax-filled petri dish and perfused with 100 μl of saline solution. The heartbeat was recorded by positioning two 36-gauge silver wires on either side of the heart tube (Figure 3A). Leads from these wires were attached to a UFI Model 2991 impedance converter (UFI, Morro Bay, CA) and then to a chart recorder to obtain myocardiograms (Figure 3B).

2. Physiological Apparatus

A 5-ml muscle chamber (Figure 4) was fabricated from a 6-cm length of Pyrex® glass tubing with an inside diameter of 1 cm. This tube was closed at one end with a rubber stopper, and a length of glass tubing was inserted through the stopper to permit the rapid release of solutions from the chamber. The chamber was also continuously aerated through a 27-gauge hypodermic needle inserted through the rubber stopper.

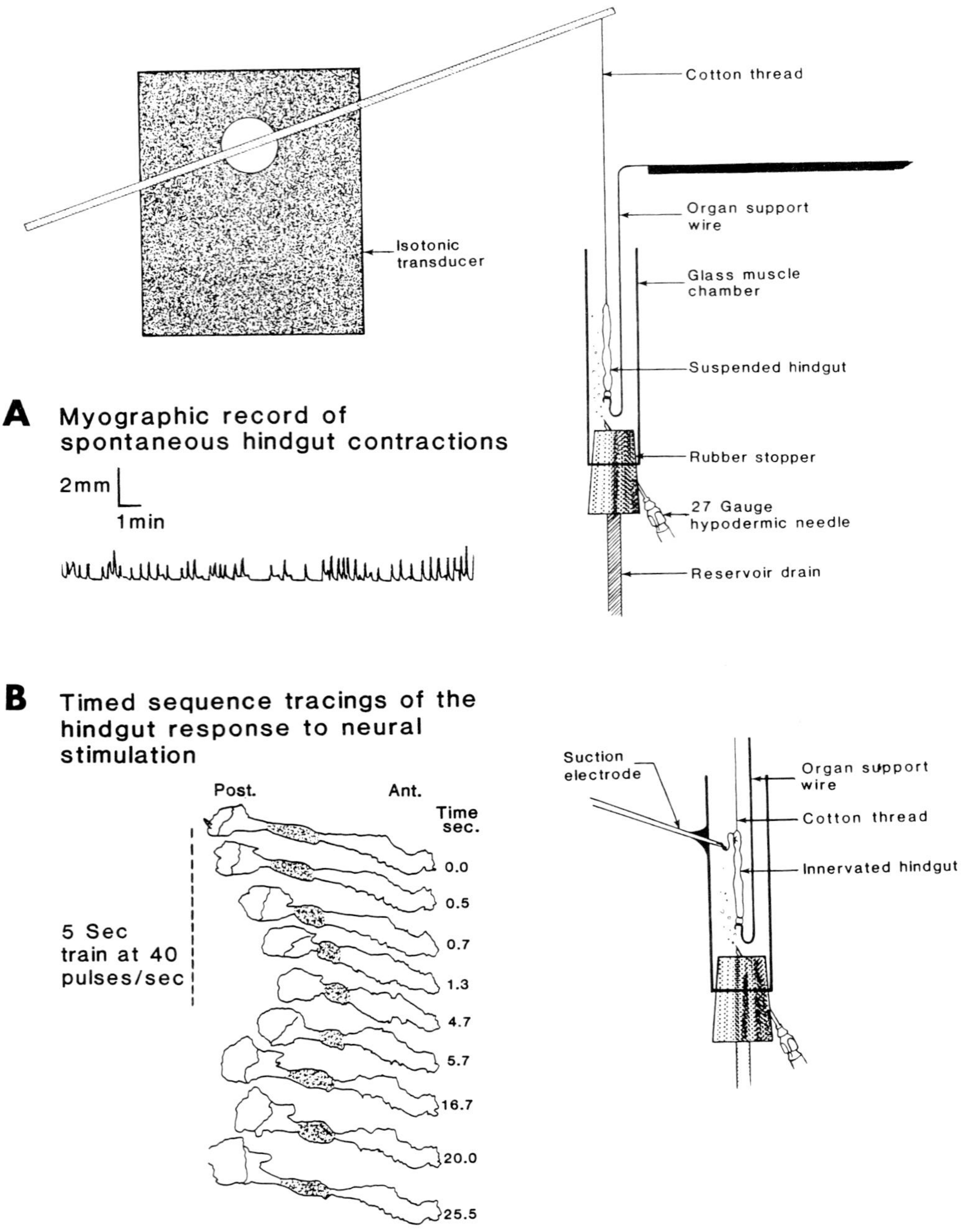

FIGURE 4. Diagram of the muscle chamber assemblies used to detect spontaneous (A) and neurally evoked (B) contractions, together with recorded examples of the two types of responses shown on the left. The stippled areas in the timed sequence tracings (B) represent changes in the posterior colon. The glass chamber was 60 × 11 mm I.D. The suction electrode (1.5-mm tubing) was sealed in a side port with dental wax. Suction was provided by a 5-ml syringe.

An L-shaped Jackson support stand was used to hold both the transducer and the muscle chamber in proper relationship to each other. The stainless steel support hook for suspending hindguts in the bath was fastened to a Prior micromanipulator (Stoelting Co., Wood Dale, IL), which in turn was placed on a platform jack support. This arrangement allowed for rapid three-dimensional adjustments of the hindgut in the bath.

Isolated hindguts were suspended in the bath between the stainless steel support hook and a metal beam by means of cotton threads (Figure 4A). This beam was inserted into a metal hub that served as a fulcrum to register change in torque on a Metripak® isotonic muscle transducer (model 793341-X4042, Gould Electronics, Cleveland, OH). The beam was then sufficiently counterweighted to produce a tension of approximately 180 mg on the hindgut. Myographs were recorded by connecting the transducer to a pen recorder.

Innervated hindguts were hung upside down in a muscle chamber with an elliptical hole in the side wall to allow for positioning of a stimulating electrode (Figure 4B). A loop in the thread around the anterior colon was attached to the hook of the stainless steel support wire, and the thread from the terminal sclerites was looped over the transducer lever. Once the suction electrode was in place in the side wall of the chamber, it was secured with dental wax. The posterior rectum could then be positioned in front of the electrode so that it was possible to leave the nerve rather loose in the bath after the terminal ganglion was drawn into the tip. The electrode was connected to a Grass® S48 stimulator (Grass Instruments, Quincy, MA) through an isolation unit, and the circuit was completed by placing an indifferent lead on the metal hook in the chamber.

3. Myographic Recordings and Threshold Detection

Although the movements of insect visceral organs are generally rhythmical, they are often complex. In fact, a cinematographic analysis of hindgut motion in *L. maderae* has revealed the following basic types of movement:[34]

1. Compression (shortening of the viscera)
2. Segmentation (an annular constriction of the gut with progression)
3. Peristalsis (an annular constriction passing along the tube in a posterior direction)
4. Reverse peristalsis (an annular constriction that passes along the gut in an anterior direction).

Moreover, these defined modes of activity appear to be completely myogenic in nature, since all categories are readily observed in hindguts 30 min after treatment with tetrodotoxin (1 μg/ml).[34] Compression is clearly the dominant form of activity in most preparations, followed by segmentation. Fortunately, the suspension of the isolated hindgut in the muscle chamber provides an effective means of detecting any longitudinal shortening of the viscera. The general phasic character of these spontaneous contractions is shown in Figure 4A.

When the proctodeal nerve is stimulated through the terminal ganglion, the innervated hindgut shows a sensitivity to the frequency of the stimulating pulse.[35] The maximum amplitude of contraction occurs at a frequency of 40 pulses per second. The recorded mechanical response to this type of stimulation has several characteristics. The vigorous contractions of longitudinal muscles commence about 500 ms after the initiation of electrical pulses, and peak tension is often maintained for 1 to 2 s after the passage of current. After a brief relaxation, the hindgut undergoes a series of three or four oscillatory (rebound) contractions of diminishing magnitude.

Generally, when the frequency is less than ten pulses per second, a single phasic response is obtained, but if the frequency of stimulation exceeds 10 Hz, the initial contraction is followed by a series of two or three rebound contractions. Variations in voltage of the stimulating current between 1.5 and 10 V produce contractions of approximately the same amplitude, but hindgut preparations often fail to respond if the duration of the pulse is <0.5 ms.

Figure 4B shows the time course of a hindgut response to stimulation, reconstructed from time-lapse photographs. The rectum and the posterior colon show a noticeable shortening at 0.5 s, but the anterior colon is slower to respond, with a detectable shortening at

0.7 s. Such a cinematographic analysis confirms the high density of motor nerve terminals in the rectal sphincter.

The sensitivity of the hindgut to pharmacological agents is detected by one or more of the following changes in the character of spontaneous contractile activity: (1) an increase or decrease in the amplitude of phasic contractions, (2) any change in the frequency of these events, or (3) any sustained change in the baseline tonus. Moreover, a substance is not considered active unless the amplitude of the response is 10% above the background of unstimulated activity for 2 min.

Residues from biologically active HPLC fractions are taken up in cockroach saline, and an aliquot is added to the bioassay chamber. A positive response is indicated by an increase in frequency or amplitude of contractions of the gut within 1 min following application. The threshold activity is defined as the minimum number of head equivalents necessary to produce a positive response. (One head-equivalent is equal to the amount of material found in the head of one cockroach.) This value also represents one activity unit.

4. Pharmacological Analysis

When a series of agonists are found that interact with a certain physiological preparation in a graded manner, some initial measure of comparison is needed to define a mode of action. The dose-response relationship has an obvious value toward this end. It can give a measure of individual affinities of the agonists for their receptors, and it can detect differences in the intrinsic activity of agonists, i.e., the maximum response that each agonist can elicit. Although dose-response curves do not directly measure pharmacological changes at the molecular level, they are crucial in the development of any hypothesis on the mode of action of agonist. These primary measurements define the nature of peptide action, and all subsequent studies must be consistent with them.

Although the effects of proctolin on insect visceral muscle have been explored for more than 10 years, only a few studies[36,37] have employed the dose-response relationship to analyze the questions of structure-function interaction and mode of action. Myographs of the hindgut revealed two characteristic patterns of activity: (1) simple phasic contractions of brief duration and (2) sustained tonic contractions, which are recognized as any extended change in the baseline of activity. Both patterns need to be considered in any attempt to construct a dose-response curve. This is especially true with proctolin and other peptides that cause changes in both the phasic and tonic contractions of the hindgut.

A typical profile of these changes that occur with increasing amounts of the neuropeptide leucokinin II is shown in Figure 5A.[37a] Dose-response curves can be constructed from five or more experiments such as this one by measuring the mean contraction height for spontaneous phasic contractions (Figure $5B_1$). Subsequently, responses of the hindgut to increasing amounts of the peptide are measured as illustrated in Figure $5B_{2,3}$. The measurements may then be converted to a percentage of the maximum response obtained for the concentration series in each replicate and plotted (Figure 5C) as a function of concentration.

B. ASSAYS OF THE HYPERGLYCEMIC NEUROPEPTIDES

The hyperglycemic/hypertrehalosemic hormones of cockroaches were discovered and isolated on the basis of one or more of the following activities: increase in hemolymph carbohydrate concentration, mobilization of lipids from hemolymph to fat body, activation of glycogen phosphorylase in the fat body, increase in carbohydrate concentration in the fat body, and cardioacceleratory activity (Figure 3) or other muscle contraction. The preparation of visceral muscle and subsequent myotropic bioassays have been described previously in this section.

Before discussing the biochemical assays listed above, it is necessary to distinguish between the terms ''hyperglycemic'' and ''hypertrehalosemic''. Although most investigators

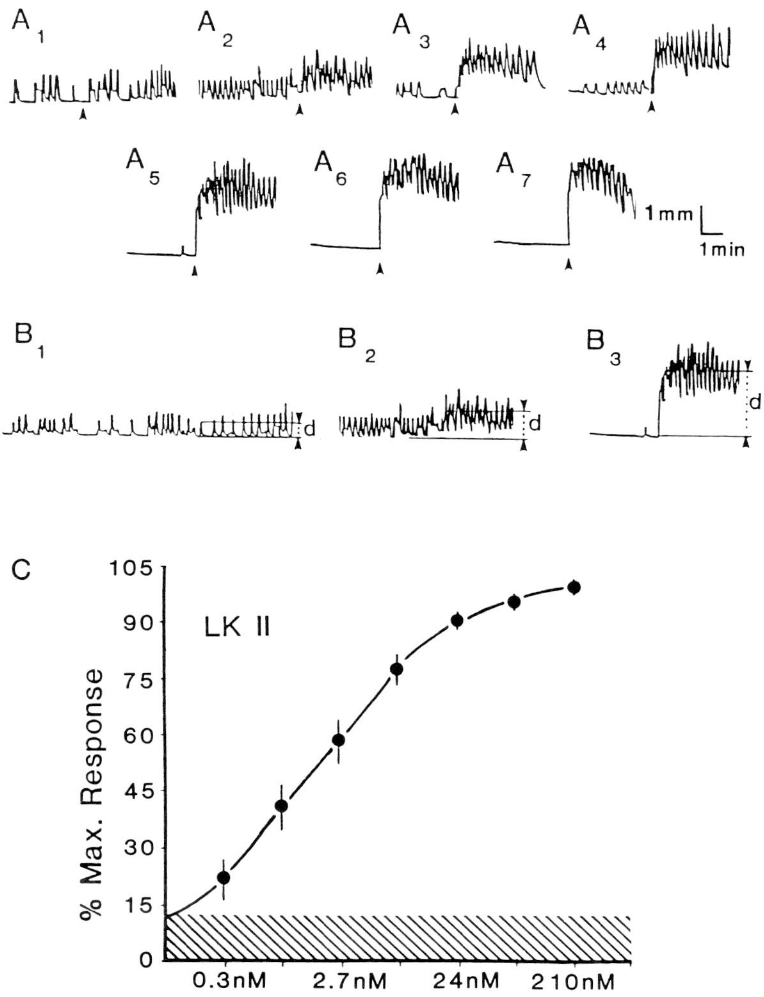

FIGURE 5. Hindgut myographic response profile to increasing amounts of the neuropeptide leucokinin II and the methods used to construct a dose-response curve. (A) Arrows indicate time from application of neuropeptide. (A_1) Response of hindgut to 0.3 n*M* LK II; (A_2) 0.9 n*M*; (A_3) 2.7 n*M*; (A_4) 8 n*M*; (A_5) 24 n*M*; (A_6) 72 n*M*; (A_7) 210 n*M*. (B_1) Spontaneous phasic contractions of the hindgut; d = mean amplitude of these contractions. (B_2) Response of hindgut to 0.9 n*M* LK II; d = mean amplitude of the peak response. (B_3) Response of hindgut to 24 n*M* LK II. (C) Dose-response curve for leucokinin II; hatched area represents the mean amplitude of spontaneous phasic contractions.[37a]

assume that trehalose accounts for almost all of the measurable carbohydrate in insect hemolymph, no attempt will be made in this chapter to assign appropriate nomenclature; instead, the terminology of the original investigators will be presented. In general, glycogen from the fat body is converted to trehalose, which is the major circulating form of glucose in the hemolymph.[38] The hyperglycemic factor is injected into the cockroach in saline, and after a given length of time an aliquot of hemolymph is removed for analysis. After deproteinization, the carbohydrate content of the hemolymph is determined and compared to an aliquot of hemolymph taken prior to injection.[24] Usually, a general colorimetric assay for total carbohydrate content is used. This assay is based on the reaction of carbohydrates with

anthrone in a sulfuric acid medium and is measured spectrometrically at 620 or 585 nm.[39,40] This test is not specific for trehalose, as most carbohydrates react with anthrone, but with differing color yields. In order to compensate for this discrepancy, Steele,[24,25] for example, measured carbohydrate content before and after acid hydrolysis of hemolymph samples. The assumption is that the difference between these two measurements represents conversion of the trehalose content of the sample to glucose. Glucose may also be measured before and after hydrolysis by a coupled glucose oxidase assay[41-43] which specifically oxidizes only glucose.

Other investigators, such as Downer,[44] have used an *in vitro* assay of carbohydrate content in fat body preparations. Hayes and Keeley[45] also prepared fat body fragments in Ringer's saline from cockroaches that had been decapitated 24 h earlier. The production of carbohydrate is measured in sections of the same fat body fragments before and after incubation with hyperglycemic hormone. Trehalose and glucose from the fat body are separated by thin-layer chromatography on cellulose or silica gel layers. The anthrone colorimetric assay is again used for quantitation of each carbohydrate. It is assumed that glucose and trehalose are the only carbohydrates present and that trehalose accounts for most of the carbohydrate measured by the anthrone assay.

Another assay involving carbohydrate metabolism and regulation is the activation of the enzyme glycogen phosphorylase.[46-50] Upon conversion of the inactive *b* form of the enzyme to the active *a* form, glycogen is converted to glucose-1-phosphate units. This compound is then converted to glucose-6-phosphate units and finally to trehalose, sequentially, by the activities of phosphoglucomutase, trehalose-6-phosphate synthetase, and trehalose-6-phosphatase, respectively.[51-53] However, activation of glycogen phosphorylase does not always result in the production of trehalose. As there is no known hormonal control between the production of glucose-6-phosphate and the final conversion to trehalose, and because hyperglycemic hormones appear to affect only the phosphorylase enzyme in the pathway to trehalose synthesis, measurement of glycogen phosphorylase activity in the fat body before and after injection of the animal with a hyperglycemic factor provides a measure of the hormonal regulation of conversion of glycogen to trehalose. Fat body preparations are homogenized and centrifuged, and the precipitate is incubated with ^{14}C-uridine diphosphate glucose and glycogen. The products may then be separated by paper chromatography and the radioactive products identified.[54] Alternatively, inorganic phosphate released during conversion of trehalose-6-phosphate to trehalose may be quantitated as a measure of overall conversion of glycogen to trehalose,[55] but only in animals in which trehalose levels increase with increasing phosphorylase activity.

Finally, many of these hyperglycemic/hypertrehalosemic hormones have also been tested in locusts for adipokinetic (lipid-mobilizing) activity. While injection of most of these factors stimulates uptake of glycerides from hemolymph into the fat body, none have exhibited hyperlipemic activity in the cockroach species studied so far.[56-58]

III. CHEMISTRY OF THE COCKROACH NEUROPEPTIDES

A. EXTRACTION AND ISOLATION OF THE NEUROPEPTIDES

The first objective in any scheme for the isolation and purification of peptide hormones is to obtain solubilization of the active agent. Once this is achieved, the peptide of interest may be extracted from cells and tissues, but the arduous task of separating it from the other components in the extraction solvent still remains. Here the principles of differential solubility often provide the initial steps for separation. In fact, such vertebrate hormones as insulin and glucagon have been successfully purified by the extraction of aqueous polypeptide solutions with various organic solvents and/or neutral salts under controlled conditions of pH and temperature.[59] However, many of the brain-gut peptides, such as gastrin and cho-

lecystokinin, could not be isolated by these methods alone.[59] The recent improvements in the techniques of countercurrent distribution, ion-exchange chromatography, partition and adsorption chromatography, molecular sieving, and various types of electrophoresis were necessary. Even with these additional techniques, the structures of very few insect peptides could be determined for two principal reasons:[60]

1. Most active peptides from insects occur in picomole quantities. Thus, large numbers (from 10,000 to several hundred thousand) of individuals and/or tissues are required in the initial extracts.
2. Ultramicroanalytical techniques to determine the amino acid composition and sequence of subnanomole amounts of pure peptides were not sufficiently refined until 1982.

The rapid development of HPLC technology since 1975 and the introduction of gas-phase protein sequencing and fast atom bombardment mass spectrometry (FABMS) in the early 1980s, however, have greatly diminished these difficulties, so it is now possible to obtain a complete structural determination with only 100 to 200 pmol of pure material. Indeed, these advances have already accelerated progress in the understanding of the primary structure of many insect neuropeptides, as this chapter documents. Nevertheless, there is no single solution to the problem of extraction of these materials because each peptide has its own unique properties. For example, the natriuretic factors found in the head of the mosquito *Aedes aegypti*[61] were easily extracted in plain Ringer's saline, while the cytochromogenic[62] and hypertrehalosemic[32,45] hormones of *B. discoidalis* were best solubilized from tissues in a mixture of 90% acetone and 10% 0.1 *N* HCl with 0.01% thiodiglycol. In the latter instance, the acetone in the solvent system precipitated most large proteins while the acid allowed solubility of the hormones in the acetone. Although the primary structure of proctolin was obtained from pure material extracted from *P. americana* with 7% perchloric acid,[63] a system of methanol, water, and acetic acid (90:9:1) appears to be more efficient.[14] In fact, Stone and Mordue[60] remark that methanolic extracts of the glandular lobes of the locust CC are efficient for separation of the adipokinetic hormones from many proteins and other constituents. However, cold 15% trifluoroacetic acid (TFA) was more effective than methanol-water-acetic acid for the extraction of the natriuretic peptides from the mosquito.[64]

In summary, the best plan for final purification of any peptide emerges only after considerable trial and error with fractionation efforts on extracts from small amounts of tissue. Nevertheless, the basic plan of work generally follows the outline shown in Figure 6. The remainder of this section compares principal methods that have been used in a number of instances to isolate and purify cockroach neuropeptides.

1. Solubilization, Partition, and Solid-Phase Extraction

The fact that neurohormones are often concentrated in specific insect tissues may be of great utility in their isolation. For example, the structure of the locust adipokinetic hormone was elucidated from material obtained solely from CC.[65] This was also true for the recently determined primary structure of the two hyperglycemic hormones from *P. americana*.[28,29] Certainly, such circumstances eliminate the introduction of many interfering substances into the initial extracts. However, neuropeptides with a broader distribution or a lower cellular titer usually require extraction from specific body regions[15] or whole insects.[10]

Such was the case with proctolin,[10] the first insect neuropeptide to be characterized. Although the hindgut is a reasonably rich source for this peptide, the amounts needed for structure determination in the mid-1970s could not be provided by this tissue alone. Thus, whole-body extracts were prepared. As already mentioned, proctolin was extracted from cockroach tissues by homogenization in cold 7% perchloric acid (approximately 1 ml of solvent per insect). The thick suspension that resulted was filtered through cheesecloth and

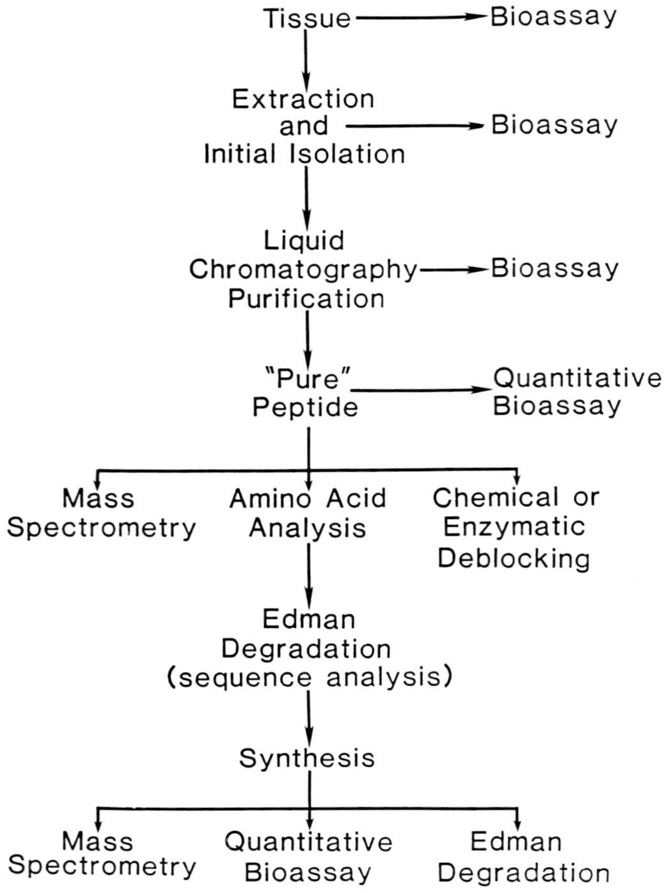

FIGURE 6. General protocol for the isolation and chemical characterization of insect neuropeptides.

filter paper under reduced pressure in the cold. This step, in more recent adaptations, has been replaced by centrifugation at 10,000 × g for 5 to 30 min, depending on the ratio of solvent to tissue. Steele and Starratt[66] found that proctolin and other myotropically active substances from cockroach tissue could be extracted with about equal facility by each of the following solvent mixtures: (1) TFA (20%, v/v), (2) methanol-water (90:10, v/v), (3) methanol-water-acetic acid (90:9:1, v/v), and (4) acetonitrile-water-acetic acid (60:10:10, v/v). The efficiencies of these systems were judged by measurements of [^{3}H-Tyr^2]-proctolin after HPLC. Extracts initially spiked with radioactive peptide gave recovery values of 78.5%.

Although extracting solvents such as methanol, ethanol, and acetone precipitate many extraneous proteins, whole-body extracts present the additional problem of interfering lipids. In the isolation of proctolin, Brown and Starratt[10] eliminated these by the use of pyridine, an elution solvent in ion-exchange procedures. However, a simple mixing and partitioning of the aqueous phase with a sequence of organic solvents such as ethyl acetate, hexane, or diethyl ether may also be quite effective.[15,62,67] The residual organic solvents in the aqueous phase may be removed by rotary vacuum evaporation, and the water that remains may be removed by lyophilization.

At this point, the implementation of a solid-phase system of chromatography is generally useful. Although alumina adsorption chromatography has been used in the case of proctolin, the recent development of small plastic cartridges that contain silica or various reverse-phase

adsorbents offers almost universal application for the early steps of purification of any peptide. These miniature, prepacked, low-resolution chromatographic columns easily attach to standard luer syringes and provide many of the following important functions: (1) concentration of bioactive peptides from dilute extraction solutions, (2) removal of extraneous proteins and other organic substances, (3) removal of inorganic salts from samples collected in physiological saline solution, and (4) removal of fine cellular particulates which block HPLC columns. Once the lyophilized samples are dissolved in water they may be easily applied to these cartidges, and bioactive peptides may then be eluted from the cartridge with appropriate solvents.

2. High-Performance Liquid Chromatographic Purification

The separation of peptides by HPLC has been especially successful in reverse-phase systems. In such systems a hydrophobic stationary phase interacts with peptides dissolved in a polar solvent (the mobile phase). It is the hydrophobic properties of the individual peptides that determine the specific pattern of differential distribution between the two phases of the system. The final result is reflected in a specific retention time for each peptide eluting from the column.

The work of O'Shea et al.[27] represents one of the first successful applications of HPLC technology to the purification and characterization of insect neuropeptides. In this instance, two myotropic factors from the CC were separated on a reverse-phase C_{18} column according to the conditions described in Figure 7. These two factors (myoactive factors I and II [MI and MII]) were not restricted to the CC, but were also present in the foregut and the CNS. Biological activity was detected by monitoring movements of the principal muscle (extensor tibialis) of the locust hindleg. This muscle performs two functions: it extends the tibia when motoneurons are stimulated, and it possesses a myogenic heart-like rhythm in the absence of any neural input. Successful assays could be performed with as little as 0.002 of a single CC. Both MI and MII induce contraction of the extensor muscle (upward movement of the lower trace in Figure 7) and an increase in the myogenic beating frequency of the muscle. The estimated threshold sensitivity for the muscle bioassay is about 5 to 10 n*M* for both peptides. Large-scale purifications were achieved in two steps after extraction. Dried extracts were dissolved in ammonium acetate (pH 4.5) and chromatographed using a linear gradient of acetonitrile. The two peaks of optical density that correspond to the biological activity were rechromatographed separately using unbuffered solvent.

Scarborough et al.[28] reported the isolation and primary structure of two cardioaccelerator peptides from *P. americana* that had hyperglycemic activity. These two octapeptides were identical in structure to the peptides originally reported by O'Shea et al.[27] Although the peptides were separated by HPLC, the conditions were different. A 218 TP column (Vydac, Hesperia, CA) and gradient elution of 0.1% TFA and acetonitrile were used. Biological activity eluted from the column in two peaks, one at 17 min (corpora cardiaca factor I [CCI]) and the other at 34 min (corpora cardiaca factor II [CCII]). These fractions were rechro matographed separately on a C_{18} support (Aquapore RP-300, Brownlee Laboratories, Santa Clara, CA) with 9% *n*-propanol in 0.1% TFA for CCI and 12% *n*-propanol with the same amount of TFA for CCII. A third group[31] recently confirmed the primary structure of these two cockroach peptides by enzymatic and gas-phase sequencing techniques, but used different procedures to isolate and purify them. Methanolic extracts of the CC were separated on a molecular size exclusion column (TSK column, 30 cm, Type G 2000 SW, Toya Soda, Tokyo) using 0.1% TFA as the solvent. Biologically active fractions were pooled and chromatographed on a reverse-phase column (Aquapore RP-300, 25 cm) using a 0.1% TFA and acetonitrile gradient. Two major peaks were eluted from this column, at 13.7 and 21.3 min.

In summary, reverse-phase columns have provided adequate separation of the cardi-

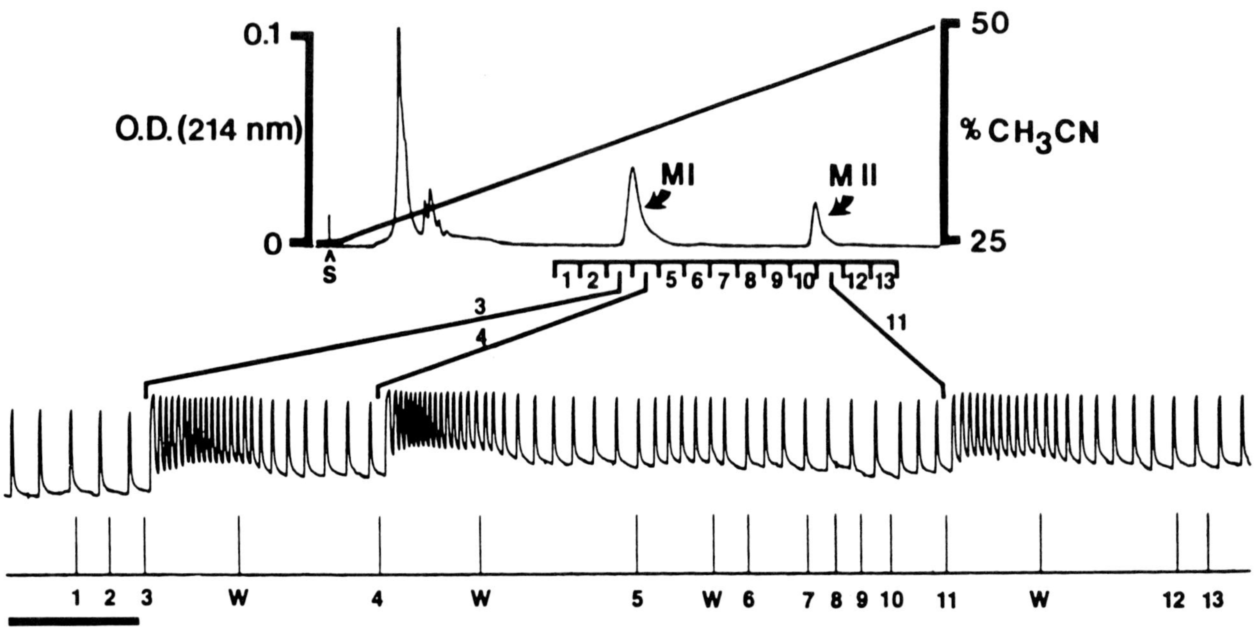

FIGURE 7. Identification of two myoactive factors (MI and MII) in an HPLC fractionation of an extract of ten corpora cardiaca. The upper part of the figure shows the chromatographic record; S indicates the start (O.D. = optical density). Eluting buffer was 1 m*M* ammonium acetate (pH 4.5) and a linear gradient (gradient indicated) of between 25 and 50% acetonitrile. Fractions numbered 1 to 13 (1-ml fractions) were collected and bioassayed (lower part of figure). The numbers under the lowest trace indicate when that fraction was applied. The W indicates saline wash. The monitor of muscle movement (contraction is an upward deflection) showed that fractions 3, 4, and 11 were most active and caused an increase in beating frequency and a tonic muscle contraction. Time scale (horizontal bar) is 5 min. (From O'Shea, M., Witten, J., and Schaffer, M., *J. Neurosci.*, 4, 521, 1984. With permission from the Society for Neuroscience.)

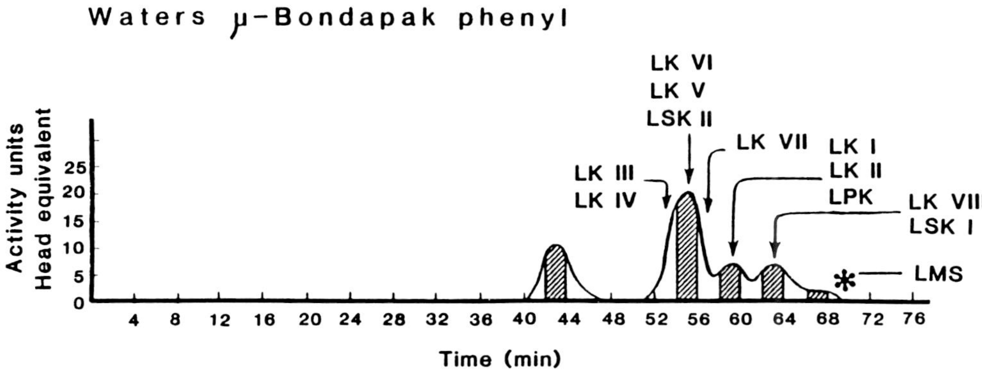

FIGURE 8. Comparative activities and temporal distribution of neuropeptides purified from head extracts of *Leucophaea maderae* by μ-Bondapak® phenyl HPLC (4.6 × 25 cm; Waters). A linear gradient from 0.1% TFA to 25% acetonitrile (containing 0.1% TFA) over 1 h was used to elute these components after 8 min at initial conditions. Hatched areas indicate fractions with peak hindgut contractile activity. (One head-equivalent is equal to the amount of peptide found in the head of one cockroach.) Neuropeptides = leucokinins (LK I to VIII), leucosulfakinins (LSK I and LSK II), leucopyrokinin (LPK). (*For LMS, measured activity is inhibitory.) The peak eluting at 40 to 44 min contains proctolin and another unidentified active component. The presence of proctolin in these head extracts is consistent with the identification of proctolin in cockroach brain and subesophageal ganglia by radioimmunoassay.[69] (Redrawn with permission from Reference 15.)

otropins or hyperglycemic hormones from the American cockroach. Additionally, a wider range of the reverse-phase resolution now available in liquid chromatographic systems has been illustrated in the isolation of a series of myotropic peptides from the cockroach *L. maderae* by Holman and colleagues.[15-22] Initially, five peaks of biological activity were separated by μ-Bondapak® phenyl chromatography from head extracts of the cockroach (Figure 8). The first peak eluted between 42 and 44 min. The other four peaks of activity had longer retention times, the last eluting between 66 and 68 min. The initial solvent was 0.1% TFA, and a linear gradient of acetonitrile (containing 0.1% TFA) was used to initially separate these components. The active fractions obtained from the μ-Bondapak® phenyl column were further purified by a series of three HPLC columns having different separation characteristics. The columns and operating conditions described by Holman et al.[16] are summarized in sequence as follows:

1. Microsorb C-1 (Rainin, Woburn, MA) using a TFA (0.1% in water) and acetonitrile gradient (Figure 9)
2. Techsphere 3μ C_{18} (Phenomenex, Rancho Palos Verdes, CA) using gradient elution with the same solvents as for the C_1 system (Figure 10)
3. 125 Protein-Pac Column (Waters, Milford, MA) using an acetonitrile/0.01% TFA gradient (Figure 11)

This sequence of chromatographic separations was necessary to resolve the initial five peaks of activity from the μ-Bondapak® phenyl column into these 12 active peptides (one of which has an inhibitory action [66 to 68 min; see asterisk in Figure 8]) and an additional component (42 to 44 min; Figure 8) which has not yet been identified.

B. CHARACTERIZATION AND STRUCTURE DETERMINATION

After separation into discrete peaks of activity, it is necessary to define the structural characteristics which are essential for this activity. The characterization and structural determination of the aforementioned myotropic and hyperglycemic peptides (Table 1) were not as straightforward as anticipated because of the amounts of material available for analysis

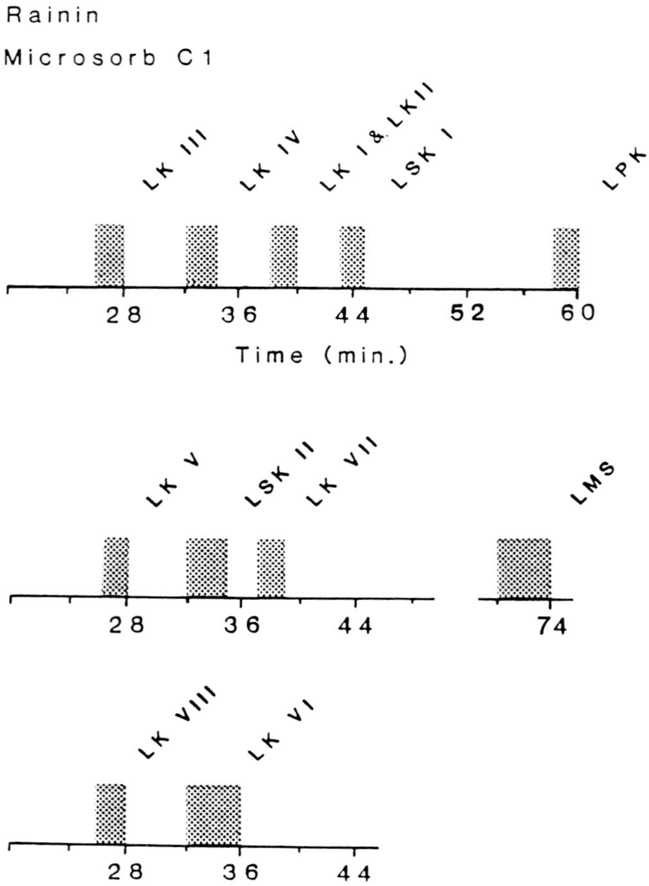

FIGURE 9. Comparative retention times for 11 cephalomyotropins and leucomyosuppressin (LMS) found in head extracts of *Leucophaea maderae* after HPLC on Microsorb C-1 (4.6 mm × 25 cm; Rainin). Peptides were eluted using a gradient from 0.1% TFA to 12.5% acetonitrile (containing 0.1% TFA) over 1 h after 8 min at initial conditions. Hatched areas correspond to elution of active fractions. Three profiles are shown to indicate overlap of peaks. Leucokinins: LK I to VIII; leucosulfakinins: LSK I and LSK II; leucopyrokinin: LPK.

and because of the presence of carboxamide, pyroglutamate, and sulfated tyrosine moieties. Comprehensive strategies were therefore needed to identify these peptides. Different combinations of HPLC, amino acid analyses, enzymatic digestions, mass spectrometry, and chemical syntheses were necessary in order to obtain complete structural details (see Figure 6).

1. Hydrolysis and Amino Acid Analysis

a. Peptide Verification

As evidenced by the preceding figures of chromatographic separation of myotropic peptides from *L. maderae*, the presence of a single peak does not imply a single component. Therefore, other means are needed to determine the purity and uniqueness of a sample of a given chromatographic retention time. In order to determine the proper methods to be used in structural analysis, it is imperative to prove that the sample of interest is, in fact, a peptide. To this end, the use of peptidase digestion in conjunction with a quantitative assay provides

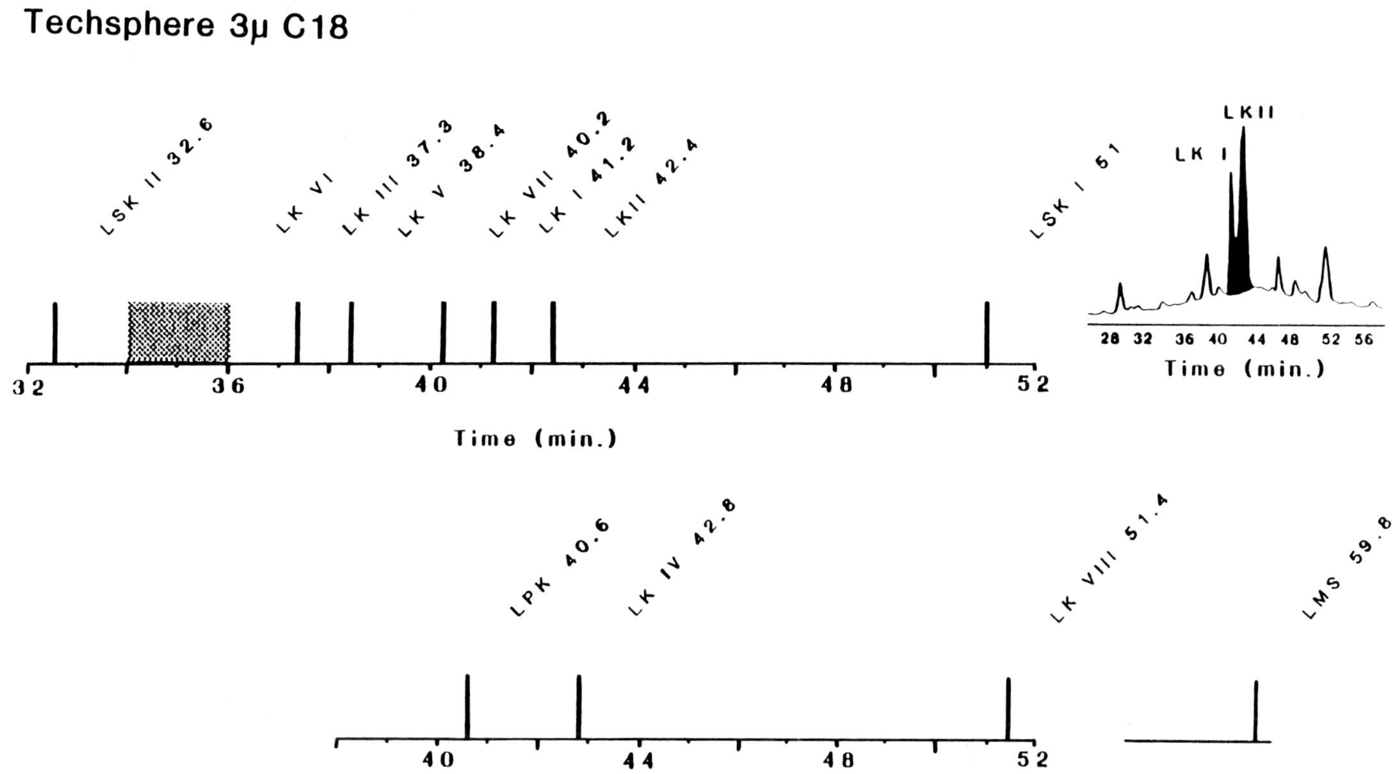

FIGURE 10. Comparative retention times for 11 cephalomyotropins and leucomyosuppressin (LMS) found in head extracts of *Leucophaea maderae* after HPLC on Techsphere 3μ C_{18} (4.6 mm × 15 cm; Phenomenex). Peptides were eluted using a gradient from 12.5% acetonitrile/0.1% TFA to 25% acetonitrile/0.1% TFA over 40 min after 8 min at initial conditions. Bars corresponding to fractions with biological activity are shown in two tracings to indicate overlap. Leucokinins: LK I to VIII; leucosulfakinins: LSK I and LSK II; leucopyrokinin: LPK. Peak absorbance values of inset are 30.6 and 41.4% of 0.5 AUFS (absorbance units full scale) for LK I and LK II, respectively. (Chromatogram inset excerpted with permission from Reference 16.)

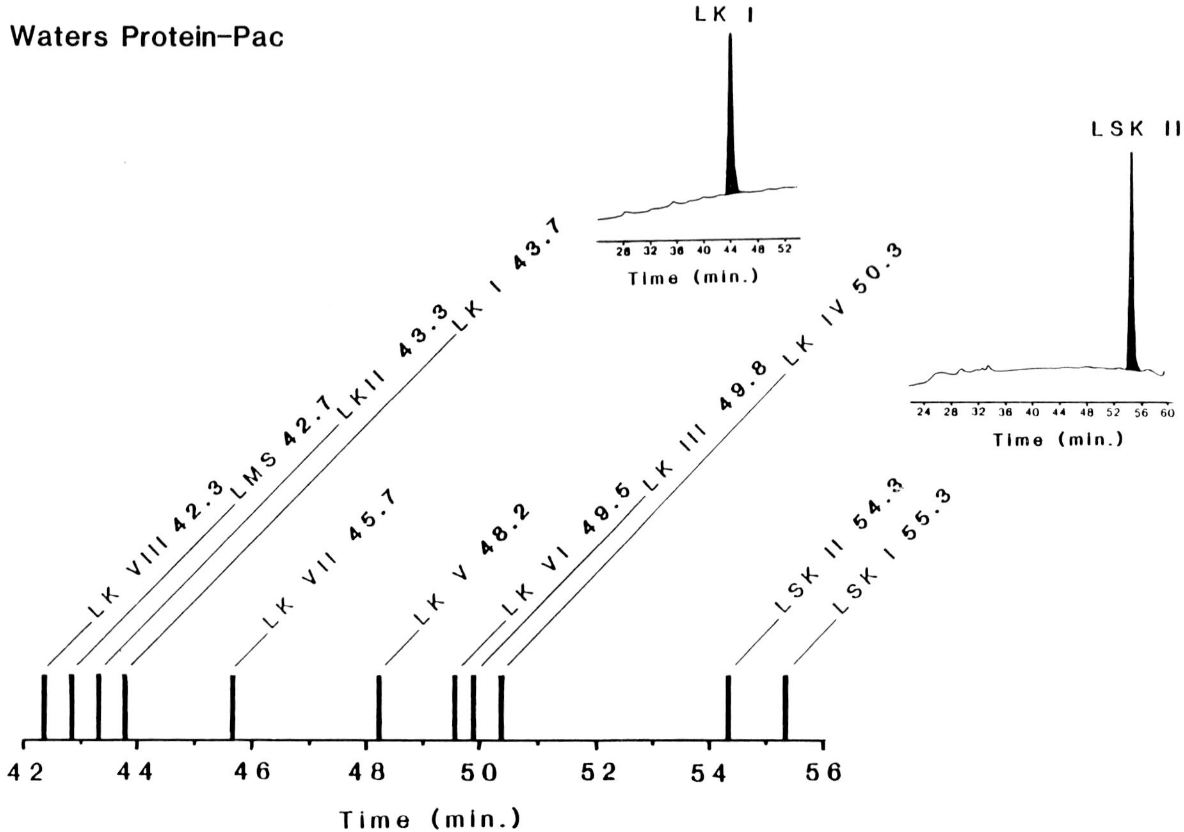

FIGURE 11. Comparative retention times for 11 cephalomyotropins and leucomyosuppressin (LMS) found in head extracts of *Leucophaea maderae* after HPLC on 125 Protein-Pac (7.8 mm × 30 cm; Waters). Peptides were eluted using a gradient from 95% acetonitrile (with 0.01% TFA) to 75% acetonitrile (with 0.01% TFA) over 40 min after 8 min at initial conditions. Bars represent fractions containing activity. Leucokinins (LK I to VIII), leucosulfakinins (LSK I and LSK II), and leucopyrokinin (LPK). Peak absorbance values of insets are 50 and 77% of 0.2 absorbance units full scale (AUFS) for LSK I and LSK II, respectively. (Chromatogram insets excerpted with permission from References 16 and 20.)

a rapid method for determining if the active component of each chromatographic peak is a peptide and/or if the amino (N-) terminus of the peptide is blocked or modified. This may be accomplished by digesting aliquots sufficient for biological activity with peptidases, which cleave the peptide bonds between amino acids. Loss of activity after enzymatic digestion indicates that a peptide is responsible for the measured biological activity. In many initial studies, aliquots of chromatographic fractions from HPLC purification were taken to dryness, reconstituted in appropriate buffers, and digested with exopeptidases, such as aminopeptidase M (APM), leucine aminopeptidase (LAP), or carboxypeptidase A, B, or Y (CPA, CPB, CPY), or with specific endopeptidases, such as trypsin or chymotrypsin. Of the 12 *Leucophaea* myotropic peptides tested, 8 are inactivated by APM[16,17,19,21,22] while all are inactivated by CPY.[15] Proctolin is also digested by APM, CPY, or LAP, but is unaffected by treatment with CPA, CPB, or chymotrypsin.[10,70] The myoactive factor I (CCI, neurohormone D, or hyperglycemic factor I) from *Periplaneta* is stable to CPA hydrolysis, but is inactivated by trypsin or chymotrypsin, while myoactive factor II (CCII or hyperglycemic factor II) from *Periplaneta* is readily inactivated by trypsin or chymotrypsin.[28,30] The *Blaberus* hypertrehalosemic hormone is digested to some extent by CPA, CPY, APM, trypsin, or chymotrypsin.[32] This hydrolysis and subsequent loss of biological activity indicates that the compound tested is a peptide.

b. Amino Acid Analysis

In general, peptides are hydrolyzed in 6 *N* HCl at 110°C for 20 to 24 h or at 150°C for 1 h,[71] and the resulting amino acids are separated by ion-exchange chromatography and quantitated by reaction with ninhydrin.[72] More recently, hydrolysis has been performed with HCl vapor and the amino acids derivatized with PITC, then separated by RP-HPLC.[73] This method has the advantage of approximately a 100-fold increase in sensitivity over the ninhydrin method. During hydrochloric acid hydrolysis tryptophan is destroyed, asparagine is converted to aspartic acid, and glutamine and pyroglutamate are converted to glutamic acid. Therefore, other hydrolysis conditions must be used to identify these residues.

For analysis of the cockroach myotropic and hypertrehalosemic peptides, samples were hydrolyzed in 6 *N* HCl, followed by either ninhydrin, orthophthalaldehyde, or PITC derivatization.[27,28,32,74-76] For the *Nauphoeta* hypertrehalosemic peptide, hydrolysis was also performed in 4 *M* methanesulfonic acid[76] in order to determine tryptophanyl residues. Likewise, a basic hydrolysis in barium hydroxide[77] was also performed on LSK I and II in order to identify acid-labile sulfate groups on the tyrosyl residues.[19,20]

The results of amino acid analyses indicated that all of the APM-refractive peptides from *Leucophaea* contained one residue of glutamate per molecule of peptide.[18,20,21,23] This suggested the presence of an N-terminal pyroglutamate moiety resistant to hydrolysis by aminopeptidase M. Subsequent digestion of several head-equivalents of each of these peptides with pyroglutamate aminopeptidase (PCA-AP) followed by APM digestion abolished activity of all four of these APM-stable peptides (LK VI, LPK, LSK II, and LMS[18,20,21,23]), thus confirming the presence of pyroglutamate as the N-terminal amino acid. Because not all of the hyperglycemic hormones were treated with APM or LAP and each of these peptides contained at least one glutamate residue, the possibility of an N-terminal pyroglutamate was not ruled out on the basis of enzymatic and amino acid analyses.

2. Sequence Analysis

For each of the peptides, a combination of amino acid analysis and amino-terminal chemical sequencing or amino acid analysis and FABMS was used to determine primary structure.

a. Sequence Determination

N-terminal sequence analysis of each peptide was accomplished by either automated

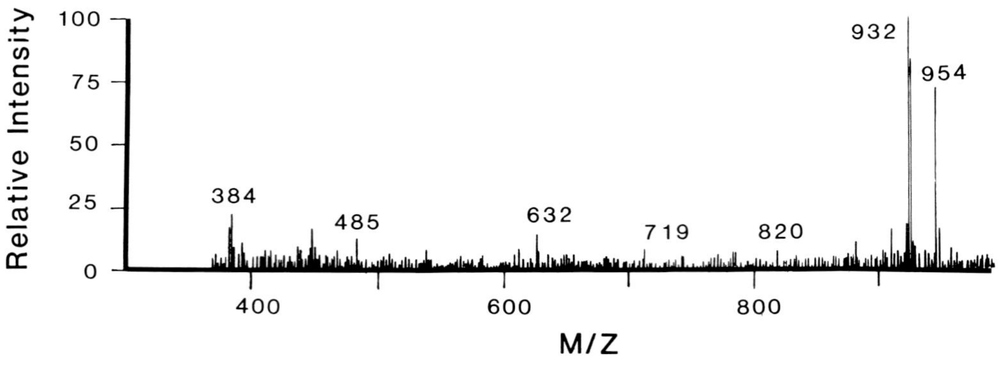

FIGURE 12. Fast atom bombardment mass spectrometry of leucopyrokinin.[83] Peptide (1 nmol) in DTT/DTE was bombarded with xenon atoms at an accelerating potential of 8 kV. M/Z ≡ mass/charge ratio.

Edman degradation[16-23,30,32,78] or by a manual dansyl-Edman technique.[11,79,80] In either case, the amino terminus of a peptide was derivatized with a chromophoric or fluorescent agent (phenylisothiocyanate or dansyl chloride, respectively) and the derivatized N-terminal amino acid cleaved from the remainder of the peptide. The cleaved amino acid was then identified chromatographically and the amino terminus of the newly generated peptide derivatized as before. Thus, a sequential N-terminal amino acid sequence was generated.

For peptides containing amino-terminal pyroglutamate (a cyclic derivative of glutamine), the pyroglutamate residue was converted to glutamate by methanolysis[81] (neurohormone D[80]) or removed from the peptide by incubation with pyroglutamate aminopeptidase[82] (*Periplaneta* HGH I and II, *Blaberus* HTH, and *Leucophaea* LK VI, LMS, LPK, and LSK II).[18,20,21,23,30,32] This technique yields a free (i.e., unblocked primary amine) amino terminus which is available for derivatization with Edman reagent (or dansyl chloride) and subsequent sequence analysis. The results of sequence analysis in conjunction with PCA-AP digestion are shown in Table 1. At this point it was not known whether the carboxy-terminal amino acids of each peptide were free acids or carboxamide derivatives.

b. Fast Atom Bombardment Mass Spectrometry

Natural products (MI and MII;[29] *Nauphoeta* hypertrehalosemic neuropeptide;[33] *Leucophaea* LSKII[20] and LPK[83]) of sufficient quantity (at least 1 nmol) were analyzed by FABMS. This technique is extremely useful for the determination of accurate nominal mass and the confirmation of predicted peptide structure, as evidenced by the structural determinations of *Periplaneta* MI and MII[28,29] and *Nauphoeta* hypertrehalosemic neuropeptide[33] based solely on amino acid analysis and FABMS.

Mass spectral data are obtained by bombardment of a sample in a liquid matrix (usually glycerol or dithiothreitol/dithioerythritol [DTT/DTE]) by an accelerated neutral atom beam. The most abundant peak in the spectrum usually corresponds to the molecular ion species (intact peptide), and fragmentation that does occur is favored at the peptide bonds between amino acids. Therefore, examination of fragmentation patterns may provide further structural details in conjunction with a known amino acid content.

An example of such a spectrum is shown for LPK[83] (Figure 12). Positive-ion mass spectral data were obtained using a Kratos® MS-50 double-focusing mass spectrometer with a high-field magnet and a postaccelerating detector (Kratos, Manchester, England). After mass calibration using a cesium iodide/glycerol reference, 1 nmol of peptide was dissolved in 0.1 *N* HCl and applied to a solid gold probe tip on which a thin film of DTT/DTE (5:1, w/w)[29] had previously been applied. Data were recorded during bombardment with a xenon atom beam accelerated at 8 kV as described elsewhere.[84]

The protonated molecular ion, $[M + H]^+$, for LPK appeared at mass ratio (m/z) 932

and an ion for the sodium adduct of LPK at m/z 954 (Figure 12). As LPK contains a pyroglutamate residue, two threonine residues, and one each of phenylalanine, leucine, proline, arginine, and serine, the protonated molecular ion at m/z 932 is consistent with the carboxamide form of this peptide, based on monoisotopic molecular weight calculations. Carboxy-terminal sequence ions were also distinguishable at m/z 820, 719, 632, 485, and 384, corresponding to fragmentations on the carboxyl side of pyroglutamate, threonine, serine, phenylalanine, and threonine. Thus, in this example, the mass spectral data together with the results of amino acid analysis could be used to determine most of the structural details of LPK, in agreement with those reported by Holman et al.[18]

3. Synthesis

a. Chemical Peptide Synthesis and Analysis

To confirm the proposed structures of peptides it is necessary to synthesize them chemically and to compare the biological activity of the synthetic materials with the natural products quantitatively. Syntheses of the hyperglycemic and myotropic peptides were carried out using routine Merrifield solid-phase protocols with *t*-butyloxycarbonyl-protected amino acids and dicyclohexylcarbodiimide coupling[16-23,85] or by solution-phase synthesis using mixed anhydride procedures.[86] After synthesis the peptides were cleaved from the solid-phase resins, and the sulfate moiety was attached to the two sulfated myotropic peptides LSK I and LSK II by treatment with concentrated sulfuric acid.[19,77]

Most of the *Leucophaea* myotropic peptides and the *Blaberus* hypertrehalosemic hormone were synthesized as both free acids and carboxamides,[16-23,32] as there were no other data available to distinguish between the two possibilities. Synthetic peptides were purified chromatographically under the same conditions as described for purification of the natural products. All synthetic peptides were subsequently analyzed for chromatographic behavior and quantitative myotropic and/or hyperglycemic activity. After HPLC separation of the various peptides and by-products formed during synthesis, the expected structures were verified by amino acid analysis and/or protein sequencing. The structures of synthetic LSK I and LSK II were also confirmed by FABMS.[19-20] With the exception of proctolin, the synthetic carboxamide form of each myotropic and/or hypertrehalosemic peptide corresponded to the structure and activity of the natural product.[16-23,27-29,32,33,71,81]

IV. CONCLUDING REMARKS

Certainly the two most revolutionary discoveries in neurobiology in the past decade have been (1) the frequent detection of more than one transmitter substance in a single neuron[87] and (2) the recognition that a large number of peptides (more than 30 to date among the vertebrates) serve as chemical signals in the CNS.[88] Although these facts have in one sense greatly complicated our understanding of nerve function, they have provided the basis for a realistic evaluation of the more pliant and spontaneous qualities of animal behavior. Just as the contact geometry of neurons represents the essence of integration in the nervous system, chemical messengers impart the properties of flexibility and resilience. Signaling agents that act through their specific membrane receptors may (1) selectively alter a neuron's channels of reception and command without disturbing the contact geometry and (2) direct the time course of the effect.

The initial studies on the chemical aspects of information processing in the nervous system of insects, as well as in other animals, have been sketched largely in terms of the release and action of small organic molecules (acetylcholine, catecholamines, serotonin, amino acids) at various types of synaptic junctions. More recently, attention has been focused on the neurosecretory cells of insects, another channel for neuronal communication. For years these cells have been known to produce hormones for release at nonsynaptic or neurohemal sites. However, the fact that many of these hormones are peptides was not

demonstrated conclusively until the structures of proctolin[11] and locust adipokinetic hormone[65] were announced. Although progress has been made since then,[28,29] it has been limited until now because of the levels of peptides found in insects and the lack of ultrasensitive methods for amino acid analysis and sequence determination. However, such difficulties no longer present an impediment to the definition of peptide structure, as this chapter illustrates.

Many of the known insect neuropeptides are structurally and functionally similar to those found in other invertebrates, and in many cases either structural or functional similarities exist between peptides isolated from insects and those isolated from vertebrate sources. For example, structural similarities between the insect adipokinetic hormones and crustacean red-pigment-concentrating hormone[89] are striking and, perhaps, anticipated because of cross-reactivity of biological actions[90] and morphological similarities of these arthropods. However, it is also of interest to note, for example, the structural similarities of insulin and the prothoracicotropic hormone of *Bombyx mori,*[91] between which there is as much as 60% homology of the amino-terminal sequences.

Of the cockroach peptides discussed within this chapter there are many structural similarities with known invertebrate and vertebrate correlates, including homologies of LPK and the hypertrehalosemic peptides with adipokinetic[18] and pigment-concentrating hormones and those of LSK I and II with cholecystokinin (CCK) and other gastrin-related peptides.[19,20] These structural homologies do not always give rise to functional similarities, and seemingly unrelated peptides may have similar biological activities. Similarities of biological action, in spite of the lack of extensive sequence homology, may instead be the result of similarities in molecular conformation, which determines receptor interaction. Also, because many physiological activities are mediated by second messenger systems, peptides with considerably different structures may activate these chemical messengers in the same way, resulting in similar physiological responses. The structural similarities of peptides which occur across many taxonomic groups are a prime example of conservation throughout nature, while extreme differences between neuropeptides of similar action prevent action of a peptide found in one species in a member of a different species.

The discovery in an insect of 12 peptides which activate the muscles of a single visceral organ is surprising, especially if one assumes that each peptide has only one site of action within an organism. However, with known insect neuropeptides like proctolin[92] and the adipokinetic hormones,[89] it has become evident that these peptides occur in diverse places and have a multiplicity of actions. Indeed, intensive studies on a variety of neuropeptides[93-95] for nearly a decade have uncovered the following functional properties that these chemical messengers seem to share, in marked contrast to the classical neurotransmitters:

1. The events activated by peptides generally have a threshold in the nanomolar (instead of the micromolar) range.
2. Neuropeptides often initiate a sustained regulation of some type of complex activity or cause a long-lasting increase in excitability.
3. Many peptides are released from neurohemal regions along axons or at neurosecretomotor endings rather than at synaptic junctions.
4. Peptides can activate receptors on cells quite remote from the point of their release.
5. The kinetics of action for peptides are frequently long lasting, with a time course of seconds or minutes.

These unique properties trace the outline of a totally different system of intercellular communication. Information transfer in this system is largely nonsynaptic. The chemical signals are transmitted through the extracellular spaces by the process of diffusion to the target cells, and the distribution of receptors for the peptide determines both the specificity and magnitude of the response. Different cellular response patterns would require additional transmitter substances. Thus, the discovery of 20 or more neuropeptides in insects provides

good evidence that a nonsynaptic system of communication exists. Certainly such findings offer useful prospects for the biomedical researcher who is interested in how peptides regulate behavioral events. The cockroach is, in fact, a particularly good model system for this kind of study because:

1. Several large identifiable peptidergic neurons have been found in the CNS.[96]
2. Neurons occur in much smaller numbers in the cockroach than they do in higher animals, consequently permitting a neuron circuit analysis that would be technically impossible in vertebrates.
3. Evidence from immunocytochemistry and radioimmunoassay strongly favors the presence of many vertebrate-type peptides in the tissues of *L. maderae*[97] and *P. americana*.[98] Moreover, the discovery of two myotropins in *L. maderae* that have close structural similarities to vertebrate cholecystokinin and gastrin is of particular interest.

In summary, both the close structural similarities of these neuropeptides with those found in vertebrates and the relative ease of the experimental preparations should present considerable appeal for research on cockroach model systems in the biomedical community.

REFERENCES

1. **Maddrell, S. H. P.,** Neurosecretion, in *Insect Neurobiology,* Treherne, J. E., Ed., Elsevier/North-Holland, Amsterdam, 1974, 307.

1a. **Scharrer, E. and Scharrer, B.,** *Neuroendocrinology,* Columbia University Press, New York, 1963.

1b. **von Euler, U. S. and Gaddum, J. H.,** An unidentified depressor substance in certain tissue extracts, *J. Physiol.,* 72, 74, 1931.

1c. **Hokfelt, T., Johansson, O., Ljungdahal, A., Lundberg, J. M., and Schultzberg, M.,** Peptidergic neurons, *Nature (London),* 284, 515, 1980.

2. **Cameron, M. L.,** Some Pharmacologically Active Substances Found in Insects, D.Sc. thesis, Cambridge University, Cambridge, England, 1953.

3. **Brown, B. E.,** Pharmacologically active constituents of the cockroach corpus cardiacum: resolution and some characteristics, *Gen. Comp. Endocrinol.,* 5, 387, 1965.

4. **Brown, B. E.,** Neuromuscular transmitter substance in insect visceral muscle, *Science,* 155, 595, 1967.

5. **Holman, G. M. and Cook, B. J.,** Pharmacological properties of excitatory neuromuscular transmission in the hindgut of the cockroach *Leucophaea maderae, J. Insect Physiol.,* 16, 1891, 1970.

6. **Davey, K. G.,** The mode of action of the corpus cardiacum on the hindgut in *Periplaneta americana, J. Exp. Biol.,* 39, 319, 1962.

7. **Holman, G. M. and Cook, B. J.,** Isolation, partial purification and characterization of a peptide which stimulates the hindgut of the cockroach, *Leucophaea maderae, Biol. Bull. (Woods Hole, Mass.),* 142, 446, 1972.

8. **Brown, B. E.,** Proctolin: a peptide transmitter candidate in insects, *Life Sci.,* 17, 1241, 1975.

9. **Brown, B. E.,** Occurrence of proctolin in six orders of insects, *J. Insect Physiol.,* 23, 861, 1977.

10. **Brown, B. E. and Starratt, A. N.,** Isolation of proctolin, a myotropic peptide, from *Periplaneta americana, J. Insect Physiol.,* 21, 1987, 1975.

11. **Starratt, A. N. and Brown, B. E.,** Structure of the pentapeptide proctolin, a proposed neurotransmitter in insects, *Life Sci.,* 17, 1253, 1975.

12. **Marks, E. P., Holman, G. M., and Borg, T. K.,** Synthesis and storage of a neurohormone in insect brains *in vitro, J. Insect Physiol.,* 19, 471, 1973.

13. **Holman, G. M. and Marks, E. P.,** Synthesis, transport, and release of a neurohormone by cultured neuroendocrine glands from the cockroach, *Leucophaea maderae, J. Insect Physiol.,* 20, 479, 1974.

14. **Holman, G. M. and Cook, B. J.,** Evidence for proctolin and a second myotropic peptide in the cockroach, *Leucophaea maderae,* determined by bioassay and HPLC analysis, *Insect Biochem.,* 9, 149, 1979.

15. **Holman, G. M., Cook, B. J., and Wagner, R. M.,** Isolation and partial characterization of five myotropic peptides present in head extracts of cockroach *Leucophaea maderae, Comp. Biochem. Physiol.,* 77C, 1, 1984.

16. **Holman, G. M., Cook, B. J., and Nachman, R. J.,** Isolation, primary structure and synthesis of two neuropeptides from *Leucophaea maderae:* members of a new family of cephalomyotropins, *Comp. Biochem. Physiol.,* 84C, 205, 1986.
17. **Holman, G. M., Cook, B. J., and Nachman, R. J.,** Primary structure and synthesis of two additional neuropeptides from *Leucophaea maderae:* members of a new family of cephalomyotropins, *Comp. Biochem. Physiol.,* 84C, 271, 1986.
18. **Holman, G. M., Cook, B. J., and Nachman, R. J.,** Primary structure and synthesis of a blocked myotropic neuropeptide isolated from the cockroach *Leucophaea maderae, Comp. Biochem. Physiol.,* 85C, 219, 1986.
19. **Nachman, R. J., Holman, G. M., Haddon, W. F., and Ling, N.,** Leucosulfakinin, a sulfated insect neuropeptide with homology to gastrin and cholecystokinin, *Science,* 234, 71, 1986.
20. **Nachman, R. J., Holman, G. M., Cook, B. J., Haddon, W. F., and Ling, N.,** Leucosulfakinin-II, a blocked sulfated insect neuropeptide with homology to cholecystokinin and gastrin, *Biochem. Biophys. Res. Commun.,* 140, 357, 1986.
21. **Holman, G. M., Cook, B. J., and Nachman, R. J.,** Isolation, primary structure, and synthesis of leucokinins V and VI: myotropic peptides of *Leucophaea maderae, Comp. Biochem. Physiol.,* 88C, 27, 1987.
22. **Holman, G. M., Cook, B. J., and Nachman, R. J.,** Isolation, primary structure and synthesis of leucokinins VII and VIII: the final members of this new family of cephalomyotropic peptides isolated from head extracts of *Leucophaea maderae, Comp. Biochem. Physiol.,* 88C, 31, 1987.
23. **Holman, G. M., Cook, B. J., and Nachman, R. J.,** Isolation, primary structure and synthesis of leucomyosuppressin, an insect neuropeptide that inhibits spontaneous contractions of the cockroach hindgut, *Comp. Biochem. Physiol.,* 85C, 329, 1986.
24. **Steele, J. E.,** Occurrence of a hyperglycaemic factor in the corpus cardiacum of an insect, *Nature (London),* 192, 680, 1961.
25. **Steele, J. E.,** The site of action of insect hyperglycaemic hormone, *Gen. Comp. Endocrinol.,* 3, 46, 1963.
26. **Traina, M. E., Bellino, M., Serpeitri, L., Massa, A., and Frontali, N.,** Heart-accelerating peptides from cockroach corpora cardiaca, *J. Insect Physiol.,* 22, 322, 1976.
27. **O'Shea, M., Witten, J., and Schaffer, M.,** Isolation and characterization of two myoactive neuropeptides: further evidence of an invertebrate peptide family, *J. Neurosci.,* 4, 521, 1984.
28. **Scarborough, R. M., Jamieson, G. C., Kalish, F., Kramer, S. J., McEnroe, G. A., Miller, C. A., and Schooley, D. A.,** Isolation and primary structure of two peptides with cardioacceleratory and hyperglycemic activity from the corpora cardiaca of *Periplaneta americana, Proc. Natl. Acad. Sci. U.S.A.,* 81, 5575, 1984.
29. **Witten, J. L., Schaffer, M. H., O'Shea, M., Cook, J. C., Hemling, M. E., and Rinehart, K. L., Jr.,** Structures of two cockroach neuropeptides assigned by fast atom bombardment mass spectrometry, *Biochem. Biophys. Res. Commun.,* 124, 350, 1984.
30. **Siegert, K. J. and Mordue, W.,** Elucidation of the primary structures of the cockroach hyperglycaemic hormones I and II using enzymatic techniques and gas-phase sequencing, *Physiol. Entomol.,* 11, 205, 1986.
31. **Siegert, K. J., Morgan, P. J., and Mordue, W.,** Isolation of hyperglycaemic peptides from the corpus cardiacum of the American cockroach, *Periplaneta americana, Insect Biochem.,* 16, 365, 1986.
32. **Hayes, T. L., Keeley, L. L., and Knight, D. W.,** Insect hypertrehalosemic hormone; isolation and primary structure from *Blaberus discoidalis* cockroaches, *Biochem. Biophys. Res. Commun.,* 140, 674, 1986.
33. **Gade, G. and Rinehart, J. L.,** Amino acid sequence of a hypertrehalosaemic neuropeptide from the corpus cardiacum of the cockroach, *Nauphoeta cinerea, Biochem. Biophys. Res. Commun.,* 141, 774, 1986.
34. **Cook, B. J. and Reinecke, J. P.,** Visceral muscles and myogenic activity in the hindgut of the cockroach, *Leucophaea maderae, J. Comp. Physiol.,* 84, 95, 1973.

34a. **Cook, B. J. and Holman, G. M.,** unpublished data, 1986.

35. **Cook, B. J. and Holman, G. M.,** The neural control of muscular activity in the hindgut of the cockroach *Leucophaea maderae:* prospects of its chemical mediation, *Comp. Biochem. Physiol.,* 50C, 137, 1975.
36. **Sullivan, R. E. and Newcomb, R. W.,** Structure function analysis of an arthropod peptide hormone: proctolin and synthetic analogues compared on the cockroach hindgut receptor, *Peptides,* 3, 337, 1982.
37. **Cook, B. J. and Holman, G. M.,** The role of proctolin and glutamate in the excitation-contraction coupling of insect visceral muscle, *Comp. Biochem. Physiol.,* 80C, 65, 1985.

37a. **Cook, B. J. and Holman, G. M.,** Pharmacological action of a new class of neuropeptides, the leucokinins I—IV on the visceral muscles of *Leucophaea maderae, Comp. Biochem. Physiol.,* in press.

38. **Wyatt, G. R.,** The biochemistry of sugars and polysaccarides in insects, *Adv. Insect Physiol.,* 4, 287, 1967.
39. **Roe, J. H.,** The determination of sugar in blood and spinal fluid with anthrone reagent, *J. Biol. Chem.,* 212, 335, 1955.
40. **Spik, G. and Montreuil, J.,** Deux causes d'erreur dans les dosages colorimetriques des oses neutres totaux, *Bull. Soc. Chim. Biol.,* 46, 5, 1964.
41. **Huggett, S. G. and Nixon, D. A.,** Enzymic determination of blood glucose, *Proc. Biochem. Soc.,* 66, 12, 1957.

42. **Hanaoka, K. and Takahashi, S. Y.,** Effect of a hyperglycemic factor on haemolymph trehalose and fat body carbohydrates in the American cockroach, *Insect Biochem.*, 6, 621, 1976.
43. **Matthews, J. R., Downer, R. G., and Morrison, P. E.,** Estimation of glucose in the haemolymph of the American cockroach, *Periplaneta americana, Comp. Biochem. Physiol.*, 53A, 165, 1976.
44. **Downer, R. G.,** Trehalose production in isolated fat body of the American cockroach, *Periplaneta americana, Comp. Biochem. Physiol.*, 62C, 31, 1978.
45. **Hayes, T. K. and Keeley, L. L.,** Properties of an *in vitro* bioassay for hypertrehalosemic hormone of *Blaberus discoidalis* cockroaches, *Gen. Comp. Endocrinol.*, 57, 246, 1985.
46. **Sutherland, E. W., Oye, I., and Butcher, R. W.,** The action of epinephrine and the role of the adenyl cyclase system in hormone action, *Recent Prog. Horm. Res.*, 21, 623, 1965.
47. **Ashida, M. and Wyatt, G. R.,** Properties and activation of phosphorylase kinase from silkmoth fat body, *Insect Biochem.*, 9, 403, 1979.
48. **Kiegert, K. and Zeigler, R.,** A hormone from the corpora cardiaca controls fat body glycogen phosphorylase during starvation in tobacco hornworm larvae, *Nature (London)*, 301, 526, 1981.
49. **Sutherland, E. W. and Wosilait, W. D.,** The relationship of epinephrine and glucagon to liver phosphorylase, *J. Biol. Chem.*, 281, 459, 1956.
50. **Ziegler, R., Masaaki, A., Fallon, A. M., Wimer, L. T., Wyatt, S. S., and Wyatt, G. R.,** Regulation of glycogen phosphorylase in fat body of *Cecropia* silkmoth pupae, *J. Comp. Physiol.*, 131, 321, 1979.
51. **Candy, D. J. and Kilby, B. A.,** The biosynthesis of trehalose in the locust fat body, *Biochem. J.*, 78, 531, 1961.
52. **Cabia, E. and Leloir, L. F.,** The biosynthesis of trehalose phosphate, *J. Biol. Chem.*, 231, 259, 1958.
53. **Murphy, T. A. and Wyatt, G. R.,** The enzymes of glycogen and trehalose synthesis in silk moth fat body, *J. Biol. Chem.*, 240, 1500, 1965.
54. **Chen, A. C. and Friedman, S.,** Hormonal regulation of trehalose metabolism in the blowfly *Phormia regina* Meig.: effects of cardiacectomy and allatectomy at the subcellular level, *Comp. Biochem. Physiol.*, 58B, 339, 1977.
55. **Childress, C. C. and Sacktor, B.,** Regulation of glycogen metabolism in insect flight muscle, *J. Biol. Chem.*, 245, 2927, 1970.
56. **Downer, R. G. H. and Steele, J. E.,** Hormonal stimulation of lipid transport in the American cockroach, *Periplaneta america, Gen. Comp. Endocrinol.*, 19, 259, 1972.
57. **Downer, R. G. H. and Steele, J. E.,** Hormonal control of lipid concentration in fat body and hemolymph of the American cockroach, *Periplaneta americana, Proc. Entomol. Soc. Ont.*, 100, 113, 1972.
58. **Goldsworthy, G. J. and Gade, G.,** The chemistry of hypertrehalosemic factors, in *Endocrinology of Insects*, Laufer, H., Ed., Alan R. Liss, New York, 1983, 109.
59. **Mutt, V.,** New approaches to the identification and isolation of hormonal polypeptides, *Trends Neurosci.*, 6, 357, 1983.
60. **Stone, J. V. and Mordue, W.,** Isolation of insect neuropeptides, *Insect Biochem.*, 10, 229, 1979.
61. **Petzel, D. H., Hagedorn, H. H., and Beyenbach, K. W.,** Peptide nature of two mosquito natriuretic factors, *Am. J. Physiol.*, 250, R328, 1986.
62. **Hayes, T. K. and Keeley, L. L.,** The isolation of insect neuropeptides using reverse-phase high performance liquid chromatography, in *Insect Neurochemistry and Neurophysiology*, Borkovec, A. B. and Kelly, T. J., Eds., Plenum Press, New York, 1984, 223.
63. **Starratt, A. N. and Steele, R. W.,** Proctolin: bioassay, isolation, and structure, in *Neurohormonal Techniques in Insects*, Miller, T. A., Ed., Springer-Verlag, New York, 1980, 1.
64. **Hayes, T., Petzel, D., Hagedorn, H., Beyenback, K., and Keeley, L.,** Studies on the purification of mosquito natriuretic peptides, *Fed. Proc.*, 46, 347, 1987.
65. **Stone, J. V., Mordue, W., Bailey, K. E., and Morris, H. R.,** Structure of locust adipokinetic hormone, a neurohormone that regulates lipid utilization during flight, *Nature (London)*, 263, 207, 1976.
66. **Steele, R. W. and Starratt, A. N.,** Rapid isolation of proctolin and other pharmacologically active constituents from cockroach tissues, in *Insect Neurochemistry and Neurophysiology*, Borkovec, A. B. and Kelly, T. H., Eds., Plenum Press, New York, 1984, 487.
67. **Baumann, E. and Gersch, M.,** Purification and identification of neurohormone D, a heart accelerating peptide from the corpora cardiaca of the cockroach, *Periplaneta americana, Insect Biochem.*, 12, 7, 1982.
68. **Bishop, C. A., O'Shea, M., and Miller, R.,** Neuropeptide proctolin (H–Arg–Tyr–Leu–Pro–Thr–OH): immunological detection and neuronal localization in insect central nervous system, *Proc. Natl. Acad. Sci. U.S.A.*, 78, 5899, 1981.
69. **Kingan, T. and Titmus, M.,** Radioimmunological detection of proctolin in arthropods, *Comp. Biochem. Physiol.*, 74C, 75, 1983.
70. **Starratt, A. N. and Brown, B. E.,** Synthesis of proctolin, a pharmacologically active pentapeptide in insects, *Can. J. Chem.*, 55, 4238, 1977.
71. **Everleigh, J. W. and Winter, G. D.,** Amino acid composition determination, in *Protein Sequence Determination*, Needleman, S. B., Ed., Springer-Verlag, New York, 1970, 91.

72. **Moore, S. and Stein, W. J.,** Chromatographic determination of amino acids by the use of automatic recording equipment, *Methods Enzymol.*, 6, 819, 1963.
73. **Heinrikson, R. L. and Meredith, S. C.,** Amino acid analysis by reverse-phase high performance liquid chromatography: precolumn derivatization with phenylisothiocyanate, *Anal. Biochem.*, 136, 65, 1984.
74. **Bohlen, R. and Schroeder, R.,** High-sensitivity amino acid analysis: methodology for the determination of amino acid compositions with less than 100 picomoles of peptide, *Anal. Biochem.*, 126, 144, 1982.
75. **Gade, G.,** Amino acid composition of cockroach hypertrehalosaemic hormones, *Z. Naturforsch.*, 40C, 42, 1985.
76. **Gade, G.,** Characterization and amino acid composition of a hypertrehalosaemic neuropeptide from the corpora cardiaca of the cockroach, *Nauphoeta cinerea, Z. Naturforsch.*, 42C, 225, 1987.
77. **Ondetti, M. A., Pluscec, J., Sabo, E. F., Sheehan, J. T., and Williams, N.,** Synthesis of cholecystokinin-pancreozymin. I. The C-terminal dodecapeptide, *J. Am. Chem. Soc.*, 92, 195, 1970.
78. **Edman, R.,** Method for the determination of the amino acid sequence in peptides, *Acta Biochem. Scand.*, 4, 283, 1950.
79. **Gray, W. R. and Smith, J. F.,** Rapid sequence analysis of small peptides, *Anal. Biochem.*, 33, 36, 1970.
80. **Baumann, E. and Penzlin, E.,** Sequence analysis of neurohormone D, a neuropeptide of an insect, *Periplaneta americana, Biomed. Biochem. Acta*, 43, 13, 1984.
81. **Kawasaki, I. and Itano, H. A.,** Methanolysis of the pyrrolidone ring of amino-terminal pyroglutamic acid in model peptides, *Anal. Biochem.*, 48, 546, 1972.
82. **Podell, D. N. and Abraham, G. N.,** A technique for the removal of pyroglutamic acid from the amino terminus of proteins using calf liver pyroglutamate aminopeptidase, *Biochem. Biophys. Res. Commun.*, 81, 176, 1978.
83. **Wagner, R. M., Pettigrew, D. W., Holman, G. M., and Fraser, B. A.,** An integrated approach to peptide structural analysis, in Proc. 1st Symp. Protein Soc., San Diego, CA, August 9—13, 1987, 56.
84. **Buko, A. M., Phillips, L. R., and Fraser, B. A.,** Peptide studies using a fast atom bombardment high field mass spectrometer and data system. I. Sample introduction, data acquisition and mass calibration, *Biomed. Mass Spectrom.*, 10, 324, 1983.
85. **Stewart, J. M. and Young, J. D.,** *Solid Phase Peptide Synthesis*, 2nd ed., Pierce Chemical Co., Rockford, IL, 1984.
86. **Anderson, G. W., Zimmerman, J. E., and Callahan, F. M.,** A reinvestigation of the mixed carbonic anhydride method of peptide synthesis, *J. Am. Chem. Soc.*, 89, 5012, 1967.
87. **Hokfelt, T., Everitt, B., Holets, V. R., Meister, B., Melander, T., Schalling, M., Staines, W., and Lundberg, J. M.,** Coexistence of peptides and other active molecules in neurons: diversity of chemical signalling potential, in *Fast and Slow Chemical Signalling in the Nervous System*, Inversen, L. L. and Goodman, E. G., Eds., Oxford University Press, Oxford, England, 1986, 204.
88. **Iversen, L. L.,** Amino acids and peptides; fast and slow chemical signals in the nervous system, *Proc. R. Soc. London*, 221, 245, 1984.
89. **Orchard, I.,** Adipokinetic hormones — an update, *J. Insect Physiol.*, 33, 451, 1987.
90. **Carlsen, J., Herman, W. S., Christensen, M., and Josefsson, L.,** Characterization of a second peptide with adipokinetic and red pigment concentrating activity from the locust corpora cardiaca, *Insect Biochem.*, 9, 497, 1979.
91. **Nagasawa, H., Kataika, H., Isogai, A., Tamura, S., Suzuki, A., Ishizaki, H., Mizoguchi, A., Fujiwara, Y., and Suzuki, A.,** Amino-terminal amino acid sequence of the silkworm prothoracicotropic hormone: homology with insulin, *Science*, 226, 1344, 1984.
92. **O'Shea, M. and Adams, M.,** Proctolin: from "gut factor" to model neuropeptide, in *Advances in Insect Physiology*, Vol. 19, Berridge, N. J., Treherne, J. E., and Wigglesworth, V. B., Eds., Academic Press, London, 1986, 1.
93. **Mayeri, E. and Rothman, B. S.,** Nonsynaptic peptidergic neurotransmission in the abdominal ganglion of *Aplysia*, in *Neurosecretion: Molecules, Cells, Systems*, Farner, D. S. and Lederis, K., Eds., Plenum Press, New York, 1981, 305.
94. **Jan, Y. N. and Jan, L. Y.,** An LHRH-like peptidergic neurotransmitter capable of action at a distance in autonomic ganglia, in *Neurotransmitters in Action*, Bousfield, D., Ed., Elsevier, New York, 1985, 94.
95. **Barker, J. L. and Smith, T. G.,** Peptides as neurohormones, in *Soc. for Neuroscience Symp.*, Vol. 2, Cowan, W. and Ferrendelli, J. A., Eds., Society for Neuroscience, Bethesda, MD, 1977, 340.
96. **O'Shea, M. and Adams, M. E.,** Pentapeptide (proctolin) associated with an identified neuron, *Science*, 213, 567, 1981.
97. **Hansen, B. L., Hansen, G. N., and Scharrer, B.,** Immunoreactive material resembling vertebrate neuropeptides in the corpus cardiacum and corpus allatum of the insect *Leucophaea maderae, Cell Tissue Res.*, 225, 319, 1982.
98. **Iwanaga, T., Fujita, T., Nishiitsutsuji-Uwo, J., and Endo, Y.,** Immunohistochemical demonstration of PP-, somatostatin-, enteroglucagon- and VIP-like immunoreactivities in the cockroach midgut, *Biomed. Res.*, 2, 202, 1981.

Chapter 16

NEUROBIOLOGY OF OPIOIDS IN *LEUCOPHAEA MADERAE*

George B. Stefano, Berta Scharrer, and Michael K. Leung

TABLE OF CONTENTS

I. INTRODUCTION

In recent years, the use of invertebrates as models in biomedical research has gained considerable attention.[1] In particular, the relevance of information gained from the study of insects has been established by the demonstration of remarkable structural, functional, and biochemical parallels between their neuroendocrine control mechanism and that of vertebrates.[2] The presence of biologically active neuropeptides, known to play a central role in these integrative systems and in neurobiology in general, has been reported in a variety of invertebrates, including insects[3-5] (see also Chapter 13 of this work). However, detailed information on a specific class of these peptides, the endogenous opioids, is largely confined to the mammalian nervous system.[6] The recent upsurge of interest in the diverse roles and modes of operation of these active principles has sparked a search for their evolutionary history. Whereas several reports on the occurrence of endogenous opioids in submammalian vertebrates have become available,[7] comparable data in invertebrates are still scarce. They pertain to the demonstration, in certain ganglia, either of opioid peptides comparable to those of vertebrates or of specific receptor sites for such active principles.[8]

Recent evidence for the presence of certain chemically identified opioids in the bivalve *Mytilus edulis*[9,10] and of opioid-like substances in several other invertebrates, including insects, makes the nervous systems of these animals promising models for the study of opioid functions.[11] In molluscs other than *Mytilus,* immunocytochemical methods have revealed enkephalin-like materials in the nervous sytems of a cephalopod, *Octopus vulgaris,*[12] and a gastropod, *Achatina fulica.*[13] Among annelids, an enkephalin-like substance has been demonstrated in a leech, *Haemopis marmorata,*[14] and β-endorphin-like products have been found in the cerebral ganglion of the earthworm *Lumbricus terrestris.*[15] Moreover, there is evidence for the presence of an α-endorphin-like peptide in the subesophageal ganglion of another lumbricid, *Dendrobaena subrubicunda.*[16]

Of particular interest in the context of this chapter are corresponding data among several insect species. Met-enkephalin-like immunoreactivity was found in the cerebral ganglion of *Locusta migratoria.*[17] α-Endorphin-like immunoreactivity has been reported in the larva of the lepidopteran *Thaumetopoea pityocampa,*[18] and Met-enkephalin- and β-endorphin-like activities were found in the midgut of *Periplaneta americana.*[19] Duve et al.[20] reported the presence of Met- and Leu-enkephalin-, β-endorphin-, and adrenocorticotropin (ACTH)-like immunoreactivities in the blowfly *Calliphora vomitoria.* Thus, the evidence available to date strongly suggests that the insect nervous system may contain a variety of opioid compounds which seem to resemble those found in mammalian systems.

In *Leucophaea maderae,* the neuroendocrine complex was shown to contain neuropeptides resembling Met-enkephalin and β-endorphin[21] as well as ACTH and melanotropin (MSH). These results will be dealt with in some detail later in this chapter.

The first demonstration of an opioid receptor mechanism in invertebrates was that in the central nervous system (CNS) of the marine mollusc *Mytilus edulis* by Stefano and colleagues.[23,24] Such a mechanism was suggested by a rise in the level of dopamine following intracardiac administration of exogenous Met- and Leu-enkephalin, an effect reversible by naloxone.[25] Similar results were obtained in the freshwater bivalve *Anodonta cygnea*[26] and the landsnail *Helix pomatia.*[27] The biomolecular properties of this high-affinity opiate binding system in *Mytilus,* analyzed in some detail by Kream et al.,[28] have been found to parallel those of mammalian systems.

Aside from brief preliminary statements by Pert and Taylor[29] and Santoro et al.[30] that membranes prepared from *Drosophila* heads avidly bind ^{3}H-Leu-enkephalin and the opioid ligand ^{3}H-diprenorphine, the currently available data on specific binding sites for opioid neuropeptides in insects are those obtained in *Leucophaea.* This information, reported in detail in Section III, includes data on the presence and putative mode of operation of opioid receptors in the CNS as well as the digestive tract.[31,32]

The search for such receptors in the midgut of *Leucophaea* was prompted by two considerations:

1. High-affinity binding sites for opioid peptides are known to be present in the myenteric plexus of the mammalian digestive tract, and their localization is closely correlated with the pharmacological effects of bioactive peptides administered to this tissue.
2. The midgut of *Leucophaea* has been shown to respond to experimental interference with its nerve supply (severance of the recurrent nerve) by tumor formation, a result suggesting a trophic function of this nerve.[33]

The view that this specific neuronal activity may well be of a peptidergic nature is supported by the immunocytochemical demonstration of several vertebrate-type neuropeptides in axons supplying the midgut and in ''diffuse neuroendocrine cells'' of a closely related insect species, *P. americana*.[34]

II. METHODOLOGY

A. THE EXPERIMENTAL ANIMAL

Insects such as the blattarian species *Leucophaea maderae* possess a neuroendocrine apparatus comparable to the hypothalamic-hypophyseal system of vertebrates. Like the posterior lobe of the pituitary gland, the corpus cardiacum stores and releases hormonal neuropeptides that are synthesized in the insect's brain. The corpus allatum, an analogue of the anterior pituitary, appears to be controlled by peptidergic (including opioid) signals. The stock colonies of *Leucophaea* supplying our experimental material have been maintained for many years under controlled conditions on a diet of dog chow and apples, at room temperature. Depending on the design of the experiments, variously aged adults and immature (nymphal) stages of either sex were selected.

B. IMMUNOCYTOCHEMICAL DEMONSTRATION OF OPIOID-TYPE NEUROPEPTIDES

The search for opioids in the neuroendocrine apparatus (brain and attached corpus cardiacum-corpus allatum complex) of *Leucophaea* was carried out in Bouin-fixed, paraffin-embedded tissues cut at 6 μm. The unlabeled-antibody enzyme method of Sternberger[35] was used. The details of the procedures were those described previously.[21,36] The antisera for endogenous opioids tested were obtained from INC (Immuno Nuclear Corporation, Stillwater, MN); they were raised in rabbits against synthetic products. These sera have been determined by the manufacturer to contain specificity against the ligands and are said to react with cells in the rat brain and pituitary, which are known to contain the peptides in question. Of the several antisera tested, two were against the endogenous opioids β-endorphin and Met-enkephalin. Immunological specificity (first-level controls) of the primary antisera was judged as detailed by Hansen et al.[21] Additional ''vertebrate neuropeptides'' demonstrated by means of region-specific immunocytochemistry in *Leucophaea* resemble ACTH and α-MSH. A detailed description and evaluation of the antisera used and of the various controls carried out is provided in the article by Hansen et al.[22]

C. HPLC ANALYSIS OF INSECT ENKEPHALINS

In order to determine possible changes in the brain's content of endogenous enkephalins with increasing age, HPLC analyses were carried out as follows. Groups of male and female adults of *Leucophaea* were separated from stock colonies shortly after their terminal molt (emergence). These pooled groups were maintained for periods of up to 16 months, the

approximate life span of these animals. At selected intervals ranging from 2 to 16 months, the brains (in sets of 50 for each group) of these "timed" adults were subjected to biochemical tests. The acid extraction of the low-molecular-weight peptides was essentially the same as described elsewhere in detail.[37] The ganglia were homogenized in a Brinkman Polytron® homogenizer with four 3-s bursts in an extraction solution (25 ganglia per millimeter) containing 1 *M* acetic acid, 20 m*M* HCl, 1 μg each of phenylmethylsulfonyl fluoride and pepstatin A per milliliter, and 1% 2-mercaptoethanol. The homogenate was clarified by centrifugation at 27,000 × g for 1 h. The supernatant was then deproteinized by the addition of 50% (w/v) trichloroacetic acid (TCA) to reach a final concentration of 10% (w/v), followed by centrifugation at 27,000 × g for 30 min. The lipids and TCA in the supernatant were removed by extracting three times with equal volumes of ether. The aqueous portion was lyophilized and stored. All these steps were carried out at 4°C; all steps subsequent to extraction were carried out at room temperature. The lyophilized extract was dissolved in a high-pressure liquid chromatography (HPLC) buffer (1 ganglion per microliter) consisting of 10 m*M* ammonium acetate, pH 4.0. The HPLC system used was a Beckman® Model 334 liquid chromatograph equipped with a Beckman®/Altex 210 sample injector and a Beckman®/Altex C-RIA integrator. Before injection, the sample was clarified by centrifugation at 10,000 × g for 10 min. An aliquot was then subjected to HPLC on a Brownlee RP-300 reverse-phase column (4.6 to 250 mm). The column was eluted at a flow rate of 2 ml/min with a solvent system consisting of HPLC ammonium acetate buffer and a linear gradient of 5 to 25% 2-propanol (30 min). The localization of Met- and Leu-enkephalin peaks was carried out by chromatographing samples with and without added enkephalin standards. The weight of the material in the acid extracts was estimated from the integrated areas under the respective peaks.

D. RECEPTOR BINDING ANALYSIS

1. Cerebral Ganglion

The specimens of *Leucophaea* used in this study were male and female adults as well as last-instar nymphs of both sexes. The adults were chosen so as to exclude freshly emerged animals.[31] The females selected had in common the presence of an ootheca in their brood pouch. Trimmed brains (without the optic lobes) from 50 adult or nymphal males and females, respectively, were pooled. For binding studies, all performed in a double-blind manner, the dissection of the neural tissue was carried out under 4°C saline. The tissues were homogenized, suspensions prepared, and binding experiments carried out as described previously. The potent synthetic enkephalin analogue ^{3}H-D-Ala$_2$-Met-enkephalinamide (15.1 Ci/mmol) was used as the exogenous opioid ligand. Aliquots of suspension (0.2 ml, 0.12 mg protein) were incubated in triplicate at 4°C for 90 min with the radiolabeled ligand in the presence of dextrorphan (10 μ*M*) or levorphanol (10 μ*M*) in 10 m*M* Tris-HCl buffer, pH 7.4, containing 0.1% bovine serum albumin (BSA) and 150 m*M* KCl. Separation of free ligand from membrane-bound ligand was done by filtration under reduced pressure through GF/B glass filters previously soaked (for 45 min at 4°C) in buffer containing 0.5% BSA. Filters were then counted in a Packard® 460 CD liquid scintillation counter (57% efficiency).

Stereospecific binding is defined as binding in the presence of 10 μ*M* levorphanol. Protein concentration was determined by the method of Lowry et al.[38] For displacement analysis, aliquots of membrane suspensions of adult male cerebral ganglia were incubated with nonradioactive opioid compounds at seven concentrations for 10 min at 22°C and then with ^{3}H-D-Ala$_2$-Met$_5$-enkephalinamide (1 n*M*) for 60 min at 4°C. Bound ^{3}H-D-Ala$_2$-Met$_5$-enkephalinamide in the presence of 10 μ*M* levorphanol represents 100% binding. IC_{50} is defined as the concentration of drug which elicits half-maximal inhibition of specific ^{3}H-D-Ala$_2$-Met$_5$-enkephalinamide binding. The mean ± SE for three experiments is reported for each compound tested.

2. Midgut

Adult *Leucophaea* were selected from stock colonies. The dissection of the delicate midgut was carried out under 4°C saline (160 m*M* NaCl, 3 m*M* KCl, 1.8 m*M* $CaCl_2$, 0.2 m*M* KCl, 1.8 m*M* $CaCl_2$, 1.8 m*M* Na_2HPO_4, 1.8 m*M* NaH_2PO_4, pH 7.2[23]). Prior to homogenization the midgut was freed of the gastric caeca, the Malpighian tubules, and the adhering fat body. The lumen was flushed clean with cold saline. The entire process took about 4 min per specimen. The net weight of the cleaned midgut was approximately 4.2 mg and the protein content approximately 0.38 mg per midgut. The tissues were homogenized, suspensions prepared, and binding experiments carried out as described previously.[32] As noted earlier, ^{3}H-D-Ala_2-Met_5-enkephalinamide (15.1 Ci/mmol) was used as the exogenous agonist opioid ligand and ^{3}H-naloxone (16.2 Ci/mmol) as the antagonist. Aliquots of suspension (0.2 ml, 0.12 mg protein) were incubated in triplicate at 4°C for 90 min with the radiolabeled ligand in the presence of dextrorphan (10 μ*M*) or levorphanol (10 μ*M* Tris-HCl buffer, pH 7.4, containing 0.1% BSA and 150 m*M* KCl for agonist and 150 m*M* NaCl for antagonist binding).[20] Separation of free ligand from membrane-bound ligand was by filtration under reduced pressure through GF/B glass filters previously soaked (for 45 min at 4°C) in buffer containing 0.5% BSA. Filters were then counted in a Packard® 460 CD liquid scintillation counter (57% efficiency).

For the *in vivo* application of drugs to the midgut prior to binding analysis, the animal was held stationary. A 100-μl syringe was positioned by a micromanipulator and the injection made laterally between the fifth and sixth tergites over a 4-min period at the rate of 25 μl/min. Following this procedure the animals were returned to their respective containers. For sustained exposure of the midgut to the opiate antagonist naloxone, four injections of the drug (0.1 to 200 μ*M*) were given daily at 6-h intervals for 8 or 11 d, respectively. This regimen was found to be effective in altering the binding characteristics under investigation, whereas two or three daily injections were not sufficient. Before the binding analysis with naloxone-treated midguts could be undertaken, numerous washings of the suspensions had to be carried out to remove residual amounts of naloxone.

For tests with the neurotoxin 6-hydroxydopamine (6-OHDA) directed against dopaminergic neurons, a single dose of 1.0, 10, or 100 μg was administered to the midgut region in the manner described above. Vehicle-treated controls received 1% ascorbic acid in distilled water. The binding properties of these specimens were determined 4, 6, or 10 d following treatment.

E. SEVERANCE OF RECURRENT NERVE

Severance of the recurrent nerve, a component of the stomatogastric (autonomic) nervous system, was performed in the neck region posterior to the corpora cardiaca-corpora allata, with which this nerve is closely connected.[33] Following this operation the insects were maintained for periods of up to 11 d before the binding assays were carried out. In none of these animals was a thickening of the midgut wall evident. Sham-operated animals served as controls.

For determining the possible effect of sustained naloxone exposure on "denervated" midguts, administrations of this opiate antagonist were scheduled in the same way as in the unoperated animals, starting on the day of surgery. The total protein content of the midgut following severance of the recurrent nerve with or without naloxone treatments was virtually the same in all of our experiments. Moreover, the ratio of specific to nonspecific binding (2.5:1) remained constant regardless of the experimental procedure, indicating that there were no significant chemical alterations related to these procedures.

F. DETERMINATION OF BIOGENIC AMINES

For analysis of the content of biogenic amines, the brains or midguts of two adult males or females, respectively, were used. These tissues were combined and assayed for dopamine,

norepinephrine, epinephrine, octopamine, and phenylethanolamine. This procedure was repeated 12 times. The tissues were homogenized by a Polytron® (setting 4 for 10 s). Tests for dopamine, norepinephrine, and epinephrine were by radioenzymatic assay according to the methods of Peuler and Johnson[39] and of Saavedra et al.[40,41] as modified by Stefano and Catapane.[42] After homogenization and centrifugation a 10-μl aliquot of the supernatant was incubated for 1 h at 37°C with a medium containing 50 μg dithiothreitol, 0.5 μM $MgCl_2$, 14 μM Tris-HCl buffer (pH 9.6), and 1 μl ^{3}H-methyl-*S*-adenosyl-1-methionine (SAM, 14.1 Ci/mmol). After incubation, the reaction vials were placed in an ice-water bath and 0.5 μl of 0.5 *M* borate buffer, pH 10, was added to stop the reaction. The samples were then processed for thin-layer chromatography (TLC) separation as described by Peuler and Johnson.[39] Internal standards consisted of 40-μl aliquots of homogenate plus 10 μl of 0.2 *N* perchloric acid containing 0.0 to 0.5 ng of dopamine, norepinephrine, or epinephrine. Blanks consisted of 40 μl of 0.2 *N* perchloric acid or tissue homogenate added to the incubation medium. The method described is able to detect 35 pg of norepinephrine and 25 pg of dopamine or epinephrine per 50-μl sample.

Octopamine and phenylethanolamine were assayed by the method of Saavedra.[40] Briefly, the cerebral ganglia were immediately frozen on dry ice and weighed. Subsequently, they were homogenized in 250 μl of ice-cold 0.02 *M* Tris-HCl buffer, pH 8.6, containing 1 m*M* nialamide. The homogenates were heated in a water bath for 1 min at 90°C. The mixture was then centrifuged for 15 min at 10,000 × g. From this mixture a 200-μl aliquot of the supernatant was removed and incubated at 37°C for 25 min in a medium containing 10 μl of phenylethanolamine-*N*-methyltransferase (PNMT), 5 μl of ^{3}H-SAM, and 35 μl of 0.02 *M* Tris-HCl buffer, pH 8.4. The reaction was terminated by the addition of 0.5 ml of 0.5 *M* borate. ^{3}H-Methyloctopamine was extracted in 6 ml of heptane containing 5% isoamyl alcohol followed by 6 ml of toluene:isoamylate (3:2). The organic phase was transferred to a mixture of 0.5 *M* borate (1 ml), which was then shaken for 1 min and centrifuged. A 4-ml portion of the organic phase was placed in counting vials and heated overnight in an oven at 80°C. The residue was redissolved in 0.1 ml of 100% ethanol; the radioactive products were then separated by TLC according to the method of Saavedra[40] and counted by liquid scintillation spectroscopy in a Packard® 460 CD counter (50% efficiency for ^{3}H). Next, 10 μg of synephrine (*N*-methyl octopamine) were added as a carrier. Blanks were a medium in which 200 μl of the buffer were substituted for the tissue. The sensitivity of the assays (two times the blank) was 2.2 ng/g for octopamine.

G. BEHAVIORAL STUDIES

The administration of drugs was achieved in the following manner. A 1-ml tuberculin syringe was positioned by hand; the needle was forced into the anterior-dorsal surface of the head in the proximity of the central suture, above the antennal suture just anterior to and between the compound eyes. The injection volume, 100 μl, was given over 1 min. The barrel of the needle was sealed in warm wax which was then pressed firmly on the head after the needle was withdrawn to prevent any backflow of solution. Dye injections confirmed that backflow was negligible — the cerebral ganglion was heavily stained. Immediately after the injection the animal was placed in an Entomex 1001 behavior monitoring chamber (Columbus Instruments, Columbus, OH) for a 15-min observation period (divided into five segments). The Entomex uses a light-sensitive grid floor to achieve the activity count. The Entomex is interfaced with an Apple® II computer. In addition to this method of data collection, organisms were placed in boxes that had checkered bottoms (0.5-cm squares), and observers, unaware of preceding treatments of insects, counted the number of squares crossed during observation periods of the same length. Experiments were also performed utilizing the Videomex behavioral monitoring equipment (Columbus Instruments) in conjunction with an enhanced Apple® IIe computer. This last method divides a surface into

small (1 cm^2) regions, and the surface is constantly evaluated for the occupation of new space. The computer program then analyzes and compares space occupancy, which is translated into locomotion. Each dose of pharmacological agent was given to at least six animals. Statistical analysis was by way of a one-tailed Student's t-test.

III. RESULTS

A. PRESENCE OF OPIOID PEPTIDES IN THE NEUROENDOCRINE SYSTEM DETERMINED BY IMMUNOCYTOCHEMISTRY

The presence of substances antigenically related to vertebrate opioid peptides was demonstrated within the corpus cardiacum of *Leucophaea*.[21] Tests with mammalian antisera to Met-enkephalin yielded substantial immunoreactive deposits, while those with anti-β-endorphin gave more moderate amounts. The reaction products of both substances were primarily located in the central release area of the organ and could therefore be judged to be of extrinsic (cerebral) origin. Their dispatch into the general circulation at this site is consistent with the view that they act in a hormonal capacity. Moreover, some of the peptidergic fibers penetrating the adjacent nonneural corpus allatum gave a positive reaction for β-endorphin. Based on electron microscopic evidence showing release sites of neuropeptides in close vicinity to the parenchymal cells of this endocrine gland, the β-endorphin-like material in this location may be presumed to exert a strictly localized control. In other words, β-endorphin seems to be capable of neurotransmitter-like mediation.

Parallel tests carried out in *Leucophaea* further demonstrated the presence of neuropeptides related to ACTH and α-MSH.[22] Region-specific immunocytochemistry revealed that the antigenic determinants present in the corpus cardiacum-corpus allatum complex recognize at least the 1—3 and the 11—17 regions of the ACTH 1—24 molecule. The distribution pattern and quantity of the reactive material vary within this organ complex. A substantial amount of the deposit is localized in the central release area of the neuroglandular corpus cardiacum. A much smaller amount is found within peptidergic fibers entering the corpus allatum by way of the corpus cardiacum. These data suggest that the differential localization and presumed modes of operation of this corticotropin-like material, demonstrated in this insect, are in line with those of the two opioids discussed above. Taken together, they support the view that mammalian ACTH molecule and related neuropeptides have a long evolutionary history.

B. PRESENCE OF OPIOIDS DETERMINED BY HPLC

HPLC revealed the presence, at detectable levels, of a peptide with a retention time corresponding to that of Met-enkephalin in the brain of *Leucophaea*.[37] However, since this peptide has not been sequenced, its identification as an opioid substance is tentative. It is of interest to note here that this Met-enkephalin-like substance showed a considerable increase with increasing "adult age" of the insects analyzed. There was a peak of 1.79 pmol per cerebral ganglion in young adults vs. a peak of 10 pmol in old adults.

C. PRESENCE OF OPIOID RECEPTORS

1. Enkephalin Receptors in the Brain

The presence of specific, high-affinity binding sites for the opioid peptide ^{3}H-D-Ala$_2$-Met$_5$-enkephalinamide was demonstrated in the cerebral ganglia of adults as well as of last-instar nymphs of *Leucophaea*. Under the conditions of our experimental design, binding was monophasic and saturable with respect to radioligand concentration in both sexes. Binding of ^{3}H-D-Ala$_2$-Met$_5$-enkephalinamide to a single class of high-affinity sites was half maximal at 9 nM for adult females and 6.4 nM for adult males. By contrast, nonspecific binding of ^{3}H-D-Ala$_2$-Met$_5$-enkephalinamide in the presence of 10 μM levorphanol increased

linearly with increasing concentrations of ^{3}H-D-Ala$_2$-Met$_5$-enkephalinamide. When the binding of ^{3}H-D-Ala$_2$-Met$_5$-enkephalinamide in brain suspensions was analyzed by a Scatchard plot, a similar pattern was observed. A class of high-affinity sites was found with affinity constants (K_ds) of 8.7 and 8.5 n*M* for male and female adults, respectively. However, the densities of the binding sites turned out to be different for the two sexes. The females had a high-affinity site receptor density (B_{max}) value of 56.0, whereas the males had a B_{max} value of 38.1 (pmol/g of protein). In other words, brain tissue from the adult females had 30% more binding sites for D-Ala$_2$-Met$_5$-enkephalinamide than an equivalent amount of brain tissue from males. It is of interest that specific binding of ^{3}H-D-Ala$_2$-Met$_5$-enkephalinamide could not be detected in suspensions of the optic lobes that had been removed from the cerebral ganglia used in this experiment and analyzed separately. This result suggests that in this insect opiate binding sites are confined to certain areas in the nervous tissue. Specific binding of D-Ala$_2$-Met$_5$-enkephalinamide to brain suspensions of last-instar nymphs was also found to be monophasic, saturable, and of high affinity. Scatchard plot analysis of D-Ala$_2$-Met$_5$-enkephalinamide binding to the cerebral ganglion revealed K_d values of 8.5 and 8.6 n*M* and B_{max} values of 38.4 and 41.2 (pmol/g of protein) for males and females, respectively. As a result, there appears to be no difference in binding site density per milligram protein between males and females in the immature state. The values for both sexes corresponded to those of the adult male specimens.

2. Enkephalin Receptors in the Midgut

The binding profile of ^{3}H-D-Ala$_2$-Met$_5$-enkephalinamide to membrane suspensions of the intact midgut of *Leucophaea* appears to be very similar to that found in the cerebral ganglion. Binding to midgut is monophasic, saturable with respect to radioligand concentration, and stereospecific. Scatchard analysis of these data revealed a single class of high-affinity binding sites with a B_{max} of 22 pmol/g protein. By contrast, with the differential binding properties of brains of adult *Leucophaea* for this opioid, no sex-related difference in binding site density was detected in the midguts examined. Moreover, binding site density in the midgut was found to be significantly lower than in the brain. The properties of this binding mechanism were analyzed further. Binding of the opiate antagonist ^{3}H-naloxone to midgut suspensions is also monophasic, saturable, and stereospecific. Maximal binding occurs at 20 n*M* to the same number of sites as in the case of ^{3}H-D-Ala$_2$-Met$_5$-enkephalinamide, and half-maximal binding results at 9 n*M* to approximately 10.1 pmol/g protein. Tests on the highly specific and differential effects of various ions on opiate binding by the midgut gave the following results. The binding of ^{3}H-D-Ala$_2$-Met$_5$-enkephalinamide is reduced by sodium, an effect that is reversed by manganese. By contrast, the binding of ^{3}H-naloxone is markedly enhanced in the presence of sodium and is unaffected by manganese. Moreover, lithium is equipotent to sodium in reducing ^{3}H-D-Ala$_2$-Met$_5$-enkephalinamide binding and enhancing ^{3}H-naloxone binding. The binding of both ligands is abolished by trypsin. Pretreatment of homogenate with soybean trypsin inhibitor prevented the inhibition of opiate binding by trypsin and confirmed that the binding site is proteinaceous.

In order to test the specific blockade effect of naloxone on the opioid binding site, repeated injections of this antagonist (30 μM/100 μl) were given, as specified in Section II, prior to tests for D-Ala$_2$-Met$_5$-enkephalinamide binding by the midgut. The result was a rise in the binding site density for D-Ala$_2$-Met$_5$-enkephalinamide which became clearly noticeable after an interval of at least 4 d. For example, Scatchard analysis of the values obtained from midguts exposed to naloxone for 8 d in this manner revealed an affinity constant of 8.2 n*M* and a B_{max} value of 29.7 pmol/g protein. This increase in binding site density was approximately 35% ($p < 0.01$). By contrast, vehicle-treated controls displayed a K_d of 8.2 n*M* and a B_{max} value of 21.6 pmol/g protein; i.e., levels closely resembled those in untreated specimens. The use of another radioligand, naloxone, in binding experiments

with membrane suspensions from animals previously treated with naloxone in the manner described also yielded an increase in binding site density of approximately 30% over that of control animals. Animals treated with a lower dose of naloxone (10 μM/100 μl) showed a more moderate increase in binding site density (19%) by day 8 of treatment. Still lower doses of naloxone (0.1 to 5.0 μM/100 μl) were ineffective in altering the B_{max} value. On the other hand, higher naloxone doses (50, 100, or 200 μM) increased the B_{max} by only a few percent above the 35% level.

It was considered of further interest to determine if the administration of naloxone to the midgut area as specified might have an effect on the binding properties of the cerebral ganglia for the same opioid demonstrated earlier.[21] Brain suspensions of adult males subjected to 30 μM naloxone revealed no change in binding site density as compared to that of untreated specimens. It should be noted here that, in accordance with comparable observations reported by other investigators[43] in other tissues, the midguts exposed to naloxone prior to binding analysis contained residual amounts of naloxone. This problem was overcome by repeated washing of the suspensions with normal buffer (see Section II), a procedure which did not alter their binding capacity for either D-Ala_2-Met_5-enkephalinamide or naloxone. Moreover, after prolonged washing, binding of agonist as well as antagonist was found to be influenced by sodium in the same manner as described above for midguts unexposed to naloxone.

3. Enkephalin Receptors in the Midgut Deprived of the Recurrent Nerve

In view of the known effect of the recurrent nerve on the midgut,[33] which appears to be neurotrophic and may be attributable to neuropeptide activity, additional binding experiments were carried out with suspensions of midguts previously subjected to severance of this autonomic nerve. Theoretically, peptidergic signals involved in the maintenance of the midgut could be addressed to it either directly or indirectly, i.e., via another (perhaps nonpeptidergic) neuron that reaches the effector organ as a component of the recurrent or possibly another afferent nerve. In either case, a neuroactive peptide would have to operate at close range, a type of activity calling for the presence of receptor sites at the level of the midgut, such as those demonstrated in the present study. If such a functional relationship actually exists, a change in opioid binding may well occur in midgut tissue that has been deprived of recurrent input for some time.

The outcome of the respective test was that the binding site density for the opioid D-Ala_2-Met_5-enkephalinamide in a midgut suspension prepared 6 to 10 d following recurrent nerve severence was decreased in comparison with that of normal and sham-operated controls. An effect of denervation also became apparent in midguts exposed to naloxone (30 μM) by repeated injections following the surgical procedure. Suspensions of such organs did not show the rise in D-Ala_2-Met_5-enkephalinamide binding site density observed in midguts from nonoperated animals following prolonged naloxone treatment. The values obtained from midguts deprived of their normal nerve supply were the same as those in the transection experiments without naloxone exposure.

Given the HPLC data noted earlier, it was of interest to determine whether the number of binding sites would change during aging. The B_{max} changed with increasing age, whereas no apparent change in the K_d was found. In old male and female *Leucophaea*, the B_{max} decreased by 25 and 40% (males, 38.4 to 29.2 pmol/g protein; females, 55.4 to 33.6 pmol/g protein), respectively, as compared to that in young adults. In fact, the B_{max} of the females approached the value for old male specimens.

D. RELATIONSHIP BETWEEN OPIOIDS AND OTHER NEUROREGULATORS

Tests for the presence of biogenic amines in the brain of *Leucophaea* were carried out to explore possible interactions between opioid-containing peptidergic neurons and conventional aminergic neurons. Of the five biogenic amines tested, two (epinephrine and phen-

ylethanolamine) seemed to be absent in the brain of *Leucophaea*, at least in concentrations detectable by the sensitivity of the methods employed. The levels of dopamine, norepinephrine, and octopamine were the same in both sexes.

The possibility that the presumed peptidergic influence of the recurrent nerve on midgut activity may occur by way of signals to another (e.g., catecholaminergic) neuron which in turn addresses the effector cell directly is suggested by a number of observations in both vertebrates and invertebrates demonstrating a close functional relationship between enkephalinergic and dopaminergic (or other nonpeptidergic) neurons. For example, in the bivalve *Mytilus*, circumstantial evidence has been obtained for the presence of opiate receptors on dopaminergic neurons.[8,37,44] In order to examine such a possible two-neuron system in *Leucophaea*, an attempt was made to influence the binding site density for the opioid D-Ala_2-Met_5-enkephalinamide by administration of 6-OHDA, a neurotoxin directed against dopaminergic neurons. However, 6-OHDA proved to be ineffective in altering the binding properties of the midgut. Moreover, tests for the presence of dopamine or norepinephrine in this organ were negative within the range of sensitivity of the assay used.[42]

E. OPIOIDS AND BEHAVIOR

Adult *Leucophaea* used in the present study were more active in the light than in the dark.[45] A probable explanation for this fluctuation of activity is that animals were routinely fed early in the mornings and, thus, were forced to leave their containers. Control organisms and those treated with vehicle all tended to have high activity in the first period of observation. This is probably associated with the trauma of being handled.

Injection of D-Ala_2-Met_5-enkephalinamide and/or morphine results in a complete and prolonged decrease in activity. This decrease is dose dependent and can be antagonized by concomitant naloxone injection. Taken together, the results appear to indicate that both D-Ala_2-Met_5-enkephalinamide and morphine exert a potent effect on locomotory control. In addition, D-Ala_2-Met_5-enkephalinamide appears to be more potent than morphine in this regard. The dose of naloxone required to antagonize this effect is rather large and will be discussed later. Injections of Met-enkephalin (10 n*M* to 100 μ*M*) do not alter the organism's locomotor activity. This lack of effect may be explained by the presence of endogenous enkephalinases or nonspecific proteolytic digestion. Interestingly, administration of D-Ala_2-Leu_5-enkephalinamide (DALA; 10 μ*M*), dynorphin A (2 μ*M*), and the benzomorphan cyclazocine (1 μ*M*) in the vicinity of cerebral ganglia results in 36, 41, and 34% increases in locomotor activity, respectively ($p < 0.05$). Concomitant naloxone and DALA injection prevents the increase in activity. Naloxone, when injected alone, causes an insignificant increase in activity. Also, given the apparent affinity of κ-ligands, it was of interest to try dynorphin A(1—13). The effects of both dynorphin A and cyclazocine are naloxone reversible (50 μ*M*).[45]

Displacement of ^{3}H-D-Ala_2-Met_5-enkephalinamide by nonradioactive D-Ala_2-Met_5-enkephalinamide was monophasic as a function of nonradioactive ligand concentration.[45] The ability of a variety of other opiates to displace specifically bound ^{3}H-D-Ala_2-Met_5-enkephalinamide was investigated. The opioid peptides (dynorphin A(1—13) > FK 33 824 > β-endorphin > D-Ala_2-Met_5-enkephalinamide) were the most potent of the ligands tested. The benzomorphans (cyclazocine > (−)ketocyclazocine) exhibited relatively high potencies as a group, and the opiate narcotics as a group (etorphine > naltrexone > morphine > naloxone) were the least potent. Surprisingly, κ-ligands appear to be quite potent in this regard.

IV. DISCUSSION

It is now a well-established fact that neuropeptides, including opioids, resembling those known in vertebrates occur in neuroendocrine centers of numerous invertebrates.[11] In insects,

such substances have been demonstrated immunocytochemically not only in several ganglia, but also in the neuroendocrine corpus cardiacum-corpus allatum complex, which receives peptidergic fibers from the brain.[21,22] Moreover, a Met-enkephalin-like peptide has been demonstrated in the brain of *Leucophaea* by HPLC.[37] What makes the result of this part of the study functionally meaningful is a considerable increase in the concentration of this compound with increasing "adult age" and a concomitant, presumably compensatory, decrease in the number of opioid receptors.[37]

The methods used for the demonstration of opioids in this insect fulfill the criteria established by investigators concerned with the mammalian nervous system. The region specificity of the antisera used in the search for ACTH-like compounds was ascertained not only by conventional liquid-phase absorption, but also by use of immunocytochemical models. The results of these immunobiological approaches seem to indicate that neuropeptides belonging to the opioid family have a long evolutionary history.[46] Based on previously obtained ultrastructural and physiological information, it can be concluded that opioid and related neuropeptides can function as hormonal or neurotransmitter-type neuroregulators, in *Leucophaea* as well as in other organisms.[11]

The demonstration of high-affinity binding sites for an enkephalin-like synthetic opioid ligand in the cerebral ganglion of the insect *Leucophaea* was the second such case among invertebrates. As in an earlier study carried out in the mollusc *Mytilus*,[28] the results strongly suggested the presence of opiate receptors in the nervous system and, concomitantly, of endogenous opioid peptides.[9,10] This reasoning is based on the body of information regarding endogenous analgesics in the mammalian nervous system and raises the following questions:

1. Where are these specific receptors localized in the insect brain?
2. How do they interact with the endogenous opioids produced by neurons of the insects' central and peripheral nervous systems?

From the fact that the optic lobes, a distinctive part of the cerebral ganglionic complex of *Leucophaea*, did not reveal high-affinity binding sites for D-Ala_2-Met_5-enkephalinamide, we may conclude that such sites are not randomly distributed, but confined to certain areas of the nervous system. As to the precise (i.e., subcellular) localization of these presumed receptors, we may, by analogy with mammalian data,[47] surmise it to be in the plasma membranes of the respective neurons. However, the possibility that glial membranes are also capable of binding neuropeptides cannot be ruled out. The second question can be discussed only in general terms with regard to the brain as well as the digestive apparatus.

To begin with a discussion of the possible role(s) of receptor-mediated opioid systems in insects and other invertebrates, the field is wide open since a variety of functions other than the control of pain can be attributed to the class of endogenous analgesics in mammals.[6]

One of the complex physiological activites of *Leucophaea*, the control of which may directly or indirectly involve enkephalin-like peptides of the brain, is the reproductive cycle of the female. In this ovoviviparous insect, periods of oocyte growth and maturation are followed by longer periods of ovarian dormancy during which the embryos develop in the mother's brood pouch. The timing of these events depends on concomitant activity cycles of the endocrine corpora allata, which in turn are known to be governed by peptidergic neurosecretory elements of the brain. It is reasonable to propose that neural directives to the corpora allata are conveyed at least in part by the opioid neuroregulators that have been demonstrated immunocytochemically in the neuroendocrine apparatus of this species. Theoretically, such an opioid could function as an allatotropic hormone or as a transmitter-like agent, since neurohemal release sites as well as synaptoid contact sites on corpus allatum cells have been demonstrated in *Leucophaea* and other insects.

The concept of receptor-mediated opioid regulation of the cyclicity in the reproductive

activity of female *Leucophaea* is supported by the observed sex-related difference in opioid binding site density.[31] Only in sexually active females, not in nymphal (immature) or very old (nonfertile) females, did this value significantly surpass that of males.

As for changes related to the aging process, it is of interest that, as stated earlier, a marked decrease in B_{max} values occurred in both sexes of *Leucophaea*. This decrease in receptor density was paralleled by an increase in the amounts of the endogenous Met-enkephalin-like substance detected by HPLC analysis. The same relationship had become apparent in studies in *Mytilus*, where the enkephalin in question was identified chemically.[11] Here, likewise, the density of high-affinity opioid binding sites decreased in older animals, while the K_d remained unchanged and the levels of endogenous opioid increased significantly. As speculated in an earlier report,[37] the aging studies carried out in *Leucophaea* and *Mytilus* seem to have revealed a compensatory mechanism between endogenous opioids and their receptors which may well operate throughout the animals' life span.

The detailed analysis of the opioid receptor mechanism in the midgut of *Leucophaea* discussed in Section III established its similarity to that of mammalian systems. The changes occurring following severance of the recurrent nerve further elucidated the known dependency of the digestive apparatus on its autonomic innervation. The interpretation of this functional relationship, considered to be based on neurotrophic influences, is now focused more directly on peptidergic intervention. The effect of "denervation" on opioid binding site density at the level of the midgut is consistent with the view that an opioid neuroregulator may be involved in the maintenance of normal midgut function.

The results of the experiments searching for possible relationships between enkephalin-ergic and conventional aminergic neurons in the brain of *Leucophaea* are of interest in comparison with data obtained in molluscs. In the bivalve *Mytilus*, exogenous opiates have been shown to influence the bioelectrical activity of certain identified neurons, to raise the dopamine level in the CNS, to depress intraganglionic levels of cyclic AMP, and to inhibit dopamine-stimulated adenylate cyclase.[8,37] Therefore, among the three biogenic amines demonstrated in the brain of *Leucophaea*, dopamine was given special attention. It was thought of interest to determine if the higher number of opioid binding sites observed in "pregnant" *Leucophaea* might be reflected in the dopamine content of their brains. However, this is not the case, since the monoamine levels measured in the brains of male and female *Leucophaea* adults were the same. Perhaps a more direct approach to learning about such relationships between monoamine levels and amount of putative endogenous opioids in the brain of this insect would be the examination of its response to exogenously applied opioids.

As to the possible participation of dopaminergic and noradrenergic factors in the maintenance of the midgut of *Leucophaea*, the search has been unsuccessful thus far. However, interaction between an enkephalin-like ligand and a "conventional" (possibly cholinergic) transmitter system seems possible based on results in mammalian systems.[48-50] Another possibility, i.e., that a neuron responsive to enkephalinergic signals may itself be peptidergic, is suggested by studies of the mammalian nervous system.[51,52] It is also suggested by immunocytochemical data in the brain of the insect *Locusta migratoria*.[17]

The data reported regarding the midgut tissues of *Leucophaea* represent the first demonstration and characterization of high-affinity binding of an opioid neuropeptide in the digestive tract of an invertebrate. In the search for the functional interpretation of this observation, two possible modes of operation of such active peptides must be considered. In the case of neurohormonal signaling, specific binding of the messenger substances can be assumed to occur at some distance from their sites of production. On the other hand, *in loco* control mechanisms call for receptors in close vicinity to sites of release of the respective peptidergic neuroregulators.

The intestinal tract of insects, like that of mammals,[53] contains candidates for both mechanisms of addressing effector cells. In analogy with the "diffuse neuroendocrine cells"

of vertebrates, those observed in the gut epithelium of some insects[34,54,55] qualify as sources of neurohormones whose receptors are in all probability not located in the midgut. Therefore, it seems reasonable to conclude that the high-affinity binding sites for D-Ala_2-Met_5-enkephalinamide in the midgut of *Leucophaea maderae* are associated with peptidergic fibers comparable to those in the closely related species *Periplaneta americana,* in which contacts between neurosecretory terminals in the recurrent nerve and the midgut wall have been demonstrated with the electron microscope[56] and three neuropeptides (vasoactive intestinal peptide, somatostatin, and pancreatic peptide) have been identified immunocytochemically.[34] Moreover, a close-range peptidergic control mechanism in the digestive tract of insects would parallel that operating in mammals,[43,50,57] with which it appears to have several characteristics in common. In both systems, the binding site density per gram of protein for the opioid under consideration varies depending on certain conditions. Thus, the ionic mechanisms characteristic of the binding capacity of the midgut of *Leucophaea* for the synthetic enkephalin analogue tested and for the opiate antagonist naloxone are in line with those reported for sodium, lithium, and manganese in mammals[58] as well as those in another invertebrate, *Mytilus.*[24] It is of interest that lithium, like sodium, reduces the binding capacity for the agonist and enhances that for the antagonist, the difference being that in the present study lithium has proved to be equipotent to sodium in this respect whereas in mammals it is less effective.[47]

The question arises whether the binding sites demonstrated in *Leucophaea* are located on cellular elements of the midgut proper or on nonenkephalinergic neurons that pass on to the gut the peptidergic directives derived via these sites. The second possibility is suggested by the fact that the information available on the digestive system of vertebrates points to neurons, rather than epithelial or muscular elements, as sites of opioid input.[43,57] More specifically, in the intestinal and nervous systems of mammals[48,49,52,59] and in the pedal ganglion of the mollusc *Mytilus,*[44] enkephalin-type peptides have been shown to exert an inhibitory control over dopaminergic and other types of neurons. Surgical intervention in the rhesus monkey[52] and the rat,[48,49,51] depriving such neurons of their input of endogenous opioids, causes a reduction in the number of the respective binding sites. Severance of the recurrent nerve in *Leucophaea* has the same effect. These results strongly indicate that axons severed from their cell bodies in *Leucophaea,* like those in vertebrates, do not survive.

The fact that sustained localized exposure of the midgut of *Leucophaea* to naloxone increases the number of binding sites for D-Ala_2-Met_5-enkephalinamide also agrees with the results obtained by Schulz et al.[43] in the guinea pig ileum. According to these authors, the blockade of opiate receptors by the antagonist naloxone leads to the formation of new binding sites for opioid peptides considered to be responsible for the resulting supersensitivity of the tissue to the inhibitory effect of the specific agonist. Because of the delicate nature of the insect tissue under investigation,[33] it was impossible to test for pharmacological activities and, thus, for supersensitivity in the same manner as was done in the guinea pig preparations. Besides, in the latter the observed supersensitivity pertains to the electrically induced contractility of the muscle in the intestinal strip preparation, whereas in the present case the possibility of a neurotrophic mechanism in whose absence the midgut undergoes aberrant growth has to be considered.

What has been established in this study is the existence of a functional relationship between opioid binding in the midgut and control over this organ by the recurrent nerve. The evidence is twofold:

1. The decrease in binding capacity observed after severance of the recurrent nerve indicates that the maintenance of the normal level of opioid binding requires the nerve to be intact.
2. The response to prolonged naloxone exposure observed in the unoperated specimen

(enhanced opioid binding) fails to occur in the midgut that has been deprived of its recurrent innervation for a sufficient period.

Available evidence supports the view that the opioid binding sites demonstrated in *Leucophaea* are located within the recurrent nerve, where enkephalin-like peptides may engage in synaptic or modulatory activities vis-à-vis nonpeptidergic and/or peptidergic components of the organ's nerve supply.

With respect to the possible participation of dopamine or noradrenergic factors in the maintenance of the insect's midgut, thus far the search has been unsuccessful. Nevertheless, interaction between an enkephalin-like ligand and another "conventional" (e.g., cholinergic) transmitter system is equally possible according to results obtained in mammalian systems.[48-50] Finally, yet another possibility, i.e., that a neuron responsive to enkephalinergic signals may itself be peptidergic, has been introduced in studies of the mammalian nervous system.[51,52] This is also suggested by immunocytochemical data on the brain of the insect *Locusta migratoria*[17] and by the demonstration of enkephalin binding in the brain of the related species *Leucophaea maderae*.[31]

The study described in this chapter also provides evidence for the presence of additional endogenous opioid mechanisms in insects, specifically in *Leucophaea*. The results indicate locomotor regulatory mechanisms which may involve a κ-like receptor population.

The demonstration of behavioral effects of opioid application to cerebral ganglia of *Leucophaea* is in agreement with previous studies.[45] D-Ala_2-Met_5-enkephalinamide was previously shown to bind with high affinity and stereoselectivity to membrane suspensions[31] of cerebral ganglia. Specific binding could not be detected in membrane suspensions of the organism's optic lobes. The earlier study suggests that in this insect opioid binding sites are confined to specific areas of the organism's nervous system. In support of our finding, Rémy and Dubois[17] demonstrated the presence of Met-enkephalin-like material in the cerebral ganglia of *Locusta*. As a result of these observations, it was of interest to determine if opioid administration would in some way alter some aspect of the organism's behavioral profile.

In mammals, opiates may either stimulate or inhibit locomotor activity.[11] These behavioral effects of opiates in mammals may have counterparts in various invertebrates. In the land snail *Helix pomatia*, morphine induces a state of immobilization and muscle rigidity (3 μ*M*), resulting in a loss of the organism's righting reflex.[11] These effects are reversible by naloxone.[8,37] This behavioral effect of morphine is reduced as treatments progress for 4 d and reoccurs if a higher dose of morphine is given on the fifth day (10 μ*M*), thus demonstrating that tolerance has occurred. Morphine also reduces locomotor activity in *Planaria*.[18] Other studies regarding opiate activities in invertebrates likewise suggest a motor-regulatory role.[60]

The *Leucophaea* study detailed here demonstrates that the endogenous opioid system which is known to exist in this organism is involved in regulating locomotor activity. It was, however, quite surprising to find that DALA stimulated locomotor activity whereas D-Ala_2-Met_5-enkephalinamide inhibited it. This interesting phenomenon may be explained by suggesting that separate opioid mechanisms exist and that separate enkephalin-like ligands modulate separate functions. However, in a recent study, the only enkephalin-like small peptide material found in the acid extract of *Leucophaea* cerebral ganglia fractionated by HPLC was Met-enkephalin-like.[37] This indicates that larger opioid molecules may be present in this tissue.

However, the "picture" may be more complex given the binding data presented in this chapter. The doyen among agonists at κ-receptors is a nonpeptide, ketocyclazocine. It has also been demonstrated that the peptide dynorphin is a potent κ-agonist in various assays.[11] These studies further note that naloxone is rather weak in displacing κ-agonists. The available data suggest that κ-like receptors may be present in the cerebral ganglia of *Leucophaea*, since dynophin and the benzomorphans are potent binding ligands whereas naloxone is weak.

In considering the displacement study in *Mytilus edulis*, peptides with more μ- and δ-affinity were found to be potent in reducing FK 33 824 binding, and the κ-agonists were the least potent.[28] In *Leucophaea*, the benzomorphans as a group were highly potent in reducing D-Ala_2-Met_5-enkephalinamide binding, whereas μ- and δ-ligands were relatively weak. These results tend to suggest the presence of multiple opioid receptor subtypes in invertebrates. Clearly, more detailed studies must be performed to evaluate these results. Interestingly, Santoro et al.[30] reported that *Drosophila* tissues appear to contain κ-like opioid receptors. Thus, in two insects a subpopulation of opioid receptors appears to be quite similar in regard to preference for binding ligands. These studies demonstrate that the opioid mechanisms present in *Leucophaea* may be extremely complex and highly diversified. This in turn strongly suggests that this signal system evolved earlier than previously believed.[46]

ACKNOWLEDGMENTS

This work was supported by NIH Grants (Minority Biomedical Support) NIMH-RR 08180 and ADAMHA-MARC 17138 (G.B.S.) and by NSF Grant BMS 74-12456 (B.S.).

REFERENCES

1. Committee on Models for Biomedical Research, *1985 Models for Biomedical Research: A New Perspective*, National Academy of Sciences, Washington, D.C., 1985.
2. **Scharrer, B.,** Insects as models in neuroendocrine research, *Annu. Rev. Entomol.*, 32, 1, 1987.
3. **Frontali, N. and Gainer, H.,** Peptides in invertebrate nervous systems, in *Peptides in Neurobiology*, Gainer, H., Ed., Plenum Press, New York, 1977, 259.
4. **Scharrer, B.,** The neurosecretory neuron in neuroendocrine regulatory mechanisms, *Am. Zool.*, 7, 161, 1967.
5. **Scharrer, B.,** Peptidergic neurons: facts and trends, *Gen. Comp. Endocrinol.*, 34, 50, 1978.
6. **Way, E. L., Ed.,** *Endogenous and Exogenous Opiate Agonists and Antagonists*, Pergamon Press, Elmsford, New York, 1980.
7. **Audigier, Y., Duprat, A. M., and Cros, J.,** Comparative study of opiate and enkephalin receptors on lower vertebrates and higher vertebrates, *Comp. Biochem. Physiol.*, 67, 191, 1980.
8. **Stefano, G. B.,** Comparative aspects of opioid-dopamine interaction, *Cell. Mol. Neurobiol.*, 2, 167, 1982.
9. **Leung, M. K. and Stefano, G. B.,** Isolation and identification of enkephalin in pedal ganglia of *Mytilus edulis* (Mollusca), *Proc. Natl. Acad. Sci. U.S.A.*, 81, 955, 1984.
10. **Stefano, G. B. and Leung, M. K.,** Presence of met-enkephalin-arg^6-phe^7 in molluscan neural tissues, *Brain Res.*, 298, 362, 1984.
11. **Leung, M. K. and Stefano, G. B.,** Comparative neurobiology of opioids in invertebrates with special attention to senescent alterations, *Prog. Neurobiol.*, 28, 131, 1987.
12. **Martin, R., Frosch, D., Weber, E., and Voigt, K. H.,** Met-enkephalin-like immunoreactivity in a cephalopod neurohemal organ, *Neurosci. Lett.*, 15, 253, 1979.
13. **Van Noorden, S., Fritsch, H. A. R., Grillo, T. A. I., Polak, J. M., and Pearse, A. G. E.,** Immunocytochemical staining for vertebrate peptides in the nervous system of a gastropod mollusc, *Gen. Comp. Endocrinol.*, 40, 375, 1980.
14. **Zipser, B.,** Identification of specific leech neurons immunoreactive to enkephalin, *Nature (London)*, 283, 857, 1980.
15. **Alumets, J., Hakanson, R., Sundler, F., and Thorell, J.,** Neuronal localization of immunoreactive enkephalin and beta-endorphin in the earthworm, *Nature (London)*, 279, 805, 1979.
16. **Rémy, C. H. and Dubois, M. P.,** Localization par immunofluorescence de peptides analogues a alpha-endorphines dans les ganglions infraoesophagiens du lombricide *Dendrobaena subrubicunda*, Eisen, *Experientia*, 35, 137, 1979.
17. **Rémy, C. H. and Dubois, M. P.,** Immunohistochemical evidence of methionine enkephalin-like material in the brain of the migratory locust, *Cell Tissue Res.*, 218, 271, 1981.
18. **Rémy, C. H., Girardie, J., and Dubois, M. P.,** Présence dans le ganglion sous-oesophagien de la Chenille processionaire du Pin (*Thaumetopoea pityocampa* Schiff) de cellules revelés en immunofluorescence par un anticorps anti-alpha-endorphine, *C.R. Acad. Sci. Ser. D*, 286, 651, 1978.

19. **Nishiitsutsuji-Uwo, J., Endo, Y. L., Takeda, M., and Saito, H.,** Brain-gut peptides in the cockroach with special reference to midgut, in *Handbook of Comparative Opioid and Related Neuropeptide Mechanisms,* Vol. 2, Stefano, G. B., Ed., CRC Press, Boca Raton, FL, 1986, 81.
20. **Duve, H., Thorpe, A., and Scott, A.,** Localization and characterization of opioid-like peptides in the nervous system of the blowfly, *Calliphora vomitoria,* in *Handbook of Comparative Opioid and Related Neuropeptide Mechanisms,* Vol. 1, Stefano, G. B., Ed., CRC Press, Boca Raton, FL, 1986, 197.
21. **Hansen, B. L., Hansen, G. N., and Scharrer, B.,** Immunoreactive material resembling vertebrate neuropeptides in the corpus cardiacum and corpus allatum of the insect, *Leucophaea maderae, Cell Tissue Res.,* 225, 319, 1982.
22. **Hansen, B. L., Hansen, G. N., and Scharrer, B.,** Immunocytochemical demonstration of a material resembling vertebrate ACTH and MSH in the corpus cardiacum-corpus allatum complex of the insect *Leucophaea maderae,* in *Handbook of Comparative Opioid and Related Neuropeptide Mechanisms,* Vol. 1, Stefano, G. B., Ed., CRC Press, Boca Raton, FL, 1986, 213.
23. **Stefano, G. B. and Catapane, E. J.,** Methionine enkephalin increases dopamine levels, *Soc. Neurosci. Abstr.,* 4, 283, 1978.
24. **Stefano, G. B., Kream, R. M., and Zukin, R. S.,** Demonstration of stereospecific opiate binding in the nervous tissue of the marine mollusc, *Mytilus edulis, Brain Res.,* 181, 440, 1980.
25. **Stefano, G. B. and Catapane, E. J.,** Enkephalins increase dopamine levels in the CNS of a marine mollusc, *Life Sci.,* 249, 1617, 1979.
26. **Stefano, G. B. and Hiripi, L.,** Methonine enkephalin and morphine after monoamine and cyclic nucleotide levels in the cerebral ganglia of the freshwater bivalve, *Anodonta cygnea, Life Sci.,* 25, 291, 1979.
27. **Osborne, N. N. and Neuhoff, V.,** Are there opiate receptors in the invertebrates?, *J. Pharm. Pharmacol.,* 31, 481, 1979.
28. **Kream, R. M., Zukin, R. S., and Stefano, G. B.,** Demonstration of two classes of opiate binding sites in the nervous tissue of the marine mollusc, *Mytilus edulis:* positive homotropic cooperativity of lower affinity binding sites, *J. Biol. Chem.,* 225, 9218, 1980.
29. **Pert, C. D. and Taylor, D.,** Type 1 and type 2 opiate receptors: a subclassification scheme based upon GTP's differential effects on binding, in *Endogenous and Exogenous Opiate Agonists and Antagonists,* Way, E. L., Ed., Pergamon Press, Elmsford, New York, 1980, 87.
30. **Santoro, C., Hall, L. H., and Zukin, R. S.,** Opioid receptor subtypes in *Drosophila melanogaster,* Abstr. O-22, Int. Narcotic Res. Conf., Falmouth, MA, June 6—10, 1985.
31. **Stefano, G. B. and Scharrer, B.,** High affinity binding of an enkephalin analog in the cerebral ganglion of the insect, *Leucophaea maderae* (Blattaria), *Brain Res.,* 225, 107, 1981.
32. **Stefano, G. B., Scharrer, B., and Assanah, P.,** Demonstration, characterization and localization of opioid binding sites in the midgut of the insect *Leucophaea maderae, Brain Res.,* 253, 205, 1982.
33. **Scharrer, B.,** Experimental tumors after nerve section in an insect, *Proc. Soc. Exp. Biol. Med.,* 60, 184, 1945.
34. **Iwanaga, T., Fujita, T., Nishiitsutsuji-Uwo, J., and Endo, Y.,** Immunohistochemical demonstration of PP-, somatostatin-, enteroglucagon- and VIP-like immunoreactivities in the cockroach midgut, *Biomed. Res.,* 2, 202, 1981.
35. **Sternberger, L. A.,** *Immunocytochemistry,* 3rd ed., John Wiley & Sons, New York, 1986, 524.
36. **Hansen, B. L., Hansen, G. N., and Hummer, L.,** The cell types in the adenohypophysis of the South American lungfish, *Lepiodosiren paradoxa,* with special reference to immunocytochemical identification of the corticotropin-containing cells, *Cell Tissue Res.,* 209, 147, 1980.
37. **Chapman, A., Gonzales, G., Burrowes, W. R., Assanah, P., Iannone, B., Leung, M. K., and Stefano, G. B.,** Alterations in high affinity binding characteristics and levels of opioids in invertebrate ganglia during aging: evidence for an opioid compensatory mechanism, *Cell. Mol. Neurobiol.,* 4, 143, 1984.
38. **Lowry, O. H., Rosebrough, N. H., Farr, A. L., and Randall, R. J.,** Protein measurement with the Folin phenol reagent, *J. Biol. Chem.,* 193, 265, 1951.
39. **Peuler, J. D. and Johnson, G. A.,** Simultaneous single isotope radioenzymatic assay of plasma norepinephrine, epinephrine and dopamine, *Life Sci.,* 21, 625, 1977.
40. **Saavedra, J. M.,** Enzymatic-isotope method for octopamine at the picogram level, *Anal. Biochem.,* 59, 628, 1974.
41. **Saavedra, J. M., Kvetnansky, R., and Kopin, I. J.,** Adrenaline, noradrenaline and dopamine levels in specific brain stem areas of acutely immobilized rats, *Brain Res.,* 160, 271, 1979.
42. **Stefano, G. B. and Catapane, E. J.,** Norepinephrine: its presence in the CNS of the bivalve mollusc, *Mytilus edulis, J. Exp. Zool.,* 214, 209, 1980.
43. **Schulz, R., Wuster, M., and Herz, A.,** Supersensitivity to opioid following the chronic blockade of endorphin action by naloxone, *Naunym-Schmiedebergs Arch. Pharmakol.,* 306, 93, 1979.
44. **Stefano, G. B., Hall, B., Makman, M. H., and Dvorkin, B.,** Opioids inhibit potassium-stimulated dopamine release in the marine mussel, *Mytilus edulis* and in the cephalopod, *Octopus bimaculatus, Science,* 213, 928, 1981.

45. **Ford, R., Jackson, D. M., Tetrault, L., Torres, J. C., Assanah, P., Harper, J., Leung, M. K., and Stefano, G. B.,** A behavioral role for enkephalins in regulating locomotor activity in the insect, *Leucophaea maderae:* evidence for high affinity kappa-like opioid binding sites, *Comp. Biochem. Physiol.,* 85C, 61, 1986.
46. **Stefano, G. B.,** Conformational matching: a determining force in maintaining signal molecules, in *Handbook of Comparative Opioid and Related Neuropeptide Mechanisms,* Vol. 2, Stefano, G. B., Ed., CRC Press, Boca Raton, FL, 1986, 271.
47. **Pert, C. D. and Snyder, S. H.,** Opiate receptor binding of agonists and antagonists affected differentially by sodium, *Mol. Pharmacol.,* 10, 808, 1974.
48. **Pollard, H., Llorens-Cortes, C., and Schwartz, J. C.,** Enkephalin receptors in dopaminergic neurons in rat striatum, *Nature (London),* 268, 745, 1977.
49. **Pollard, H., Llorens, C., Schwartz, J. C., Gros, C., and Dray, F.,** Localization of opiate receptors and enkephalins in the rat striatum in relationship with the nigrostriatal dopaminergic system: lesion studies, *Brain Res.,* 151, 392, 1978.
50. **Waterfield, A. A., Smokcum, R. W. J., Hughes, J., Kosterlitz, H. W., and Henderson, G.,** *In vitro* pharmacology of the opioid peptides, enkephalins and endorphins, *Eur. J. Pharmacol.,* 43, 107, 1977.
51. **Jessell, T. M. and Iversen, L. L.,** Opiate analgesics inhibit substance P release from rat trigeminal nucleus, *Nature (London),* 268, 549, 1977.
52. **Lamotte, C., Pert, C. D., and Snyder, S. H.,** Opiate receptor binding in primate spinal cord: distribution and changes after dorsal root section, *Brain Res.,* 112, 407, 1976.
53. **Polak, J. M., Sullivan, S. N., Bloom, S. R., Facer, P., and Pearse, A. G. E.,** Enkephalin-like immunoreactivity in the human gastrointestinal tract, *Lancet,* 1, 972, 1977.
54. **Endo, Y. and Nishiitsutsuji-Uwo, J.,** Gut endocrine cells in insects: the ultrastructure of the gut endocrine cells of the lepidopterous species, *Biomed. Res.,* 2, 270, 1981.
55. **Nishiitsutsuji-Uwo, J. and Endo, Y.,** Gut endocrine cells in insects: the ultrastructure of the endocrine cells in the cockroach midgut, *Biomed. Res.,* 1, 30, 1981.
56. **Gersch, M. and Richter, K., Eds.,** *Das Peptiderge Neuron,* Gustav Fischer Verlag, Jena, German Democratic Republic, 1981.
57. **Greese, I. and Snyder, S. H.,** Receptor binding and pharmacological activity of opiates in the guinea-pig intestine, *J. Pharmacol. Exp. Ther.,* 194, 205, 1975.
58. **Simon, E. J., Hiller, J. M., Groth, J., and Edelman, I.,** Further properties of stereospecific opiate binding in rat brain: on the nature of the sodium effect, *J. Pharmacol. Exp. Ther.,* 192, 531, 1975.
59. **Hiller, J. M., Simon, E. J., Crain, S. M., and Peterson, E. R.,** Opiate receptors in cultures of fetal mouse dorsal root ganglia (DRG) and spinal cord: predominance in DRG neurites, *Brain Res.,* 245, 396, 1978.
60. **Kavaliers, M., Hirst, M., and Teskey, G. C.,** A functional role for an opiate system in snail thermal behavior, *Science,* 220, 99, 1983.

Chapter 17

OCTOPAMINE, DOPAMINE, AND 5-HYDROXYTRYPTAMINE IN THE COCKROACH NERVOUS SYSTEM

Roger G. H. Downer

TABLE OF CONTENTS

I. INTRODUCTION

The spectrum of putative neurotransmitters, neurohormones, and neuromodulators that have been identified in the central nervous system (CNS) of cockroaches includes phenolamines, catecholamines, and indolalkylamines. Much research attention has focused on octopamine because of its proposed role as a sympathomimetic effector in insects and the consequent potential of octopaminergic systems as targets for insecticide development. However, octopamine also occurs as a "trace amine" in the CNS of vertebrates and has been implicated in certain neurological, psychiatric, and physiological disorders.[1] The relatively high concentrations of octopamine that occur in the insect nervous system suggest that these organisms offer a useful model with which to study the biochemistry, pharmacology, and physiology of octopamine. Dopamine and 5-hydroxytryptamine (serotonin), which are involved in several important neural functions in vertebrates, are also present in the insect nervous system. Thus, the cockroach also may have utility as a model for study of some aspects of dopaminergic and serotonergic function. A particular advantage of the insect nervous system as a model for neurobiological research is that the neurons are large, readily identifiable, and accessible.

The primary objective of the present chapter is to collate the information currently available on the estimation, distribution, metabolism, pharmacology, and physiological roles of octopamine, dopamine, and 5-hydroxytryptamine in the cockroach nervous system. The chemical structures of these compounds are indicated in Figure 1. Preparation of this chapter has been facilitated by the availability of several recent reviews,[1a-4] and the reader is referred to these accounts for additional background and information about other insect species.

II. ESTIMATION AND DISTRIBUTION OF MONOAMINES

A. METHODOLOGY FOR ESTIMATION

A variety of analytical procedures have been employed to estimate biogenic amine levels in insect tissues, and these have been reviewed and evaluated.[5] The most frequently used method for octopamine determination involves the transfer of a radiolabeled methyl group from methyl *S*-adenosyl-L-methionine to octopamine in the presence of phenylethanolamine *N*-methyltransferase (PNMT). The resulting synephrine is then separated and the radioactivity determined. This radioenzymatic method permits the estimation of only one compound at a time; precautions are required to ensure that substrates other than octopamine are not methylated in the presence of PNMT and that methylated products other than synephrine are not counted. Nevertheless, much useful information on octopamine and other monoamines has been generated through this technique. An alternative procedure that is gaining acceptance for octopamine estimation uses high performance liquid chromatographic (HPLC) separation of the amine with electrochemical detection. This method has been used widely to measure levels of catecholamines and indolamines in biological tissues and has the advantage of enabling simultaneous estimation of several monoamines in a single sample. However, the high oxidizing potential required to effect oxidation of phenolamines ensures that the method is difficult to use for estimation of octopamine when amperometric electrochemical detection systems are employed.[6] The development of dual- and multi-electrode coulometric detectors has overcome the problems inherent in single amperometric detectors, and coulometric detectors have been used to estimate a spectrum of monoamines in various insect tissues.[7]

In addition to the analytical procedure that is used to estimate monoamine levels, some caution must also be exercised with regard to preanalysis treatment of tissues. Evans[3] has summarized the errors that can occur as a result of freeze-thawing procedures, and the resulting decrease in monoamine levels observed under these conditions has been attributed to the activation of *N*-acetyltransferase activity within the tissue.[8]

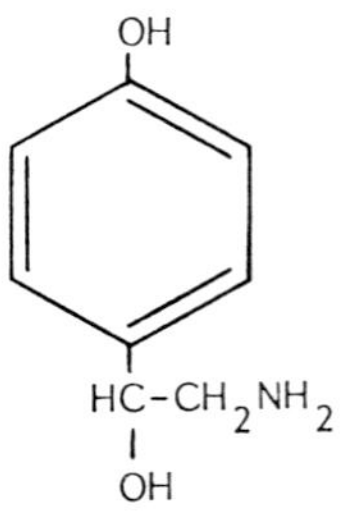

OCTOPAMINE

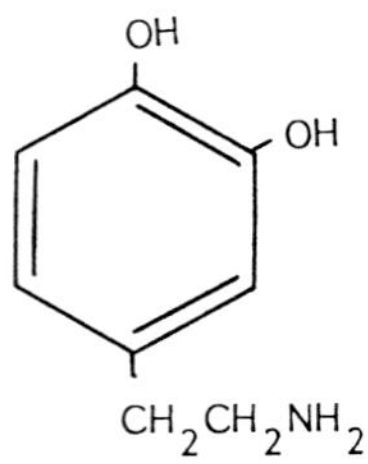

DOPAMINE

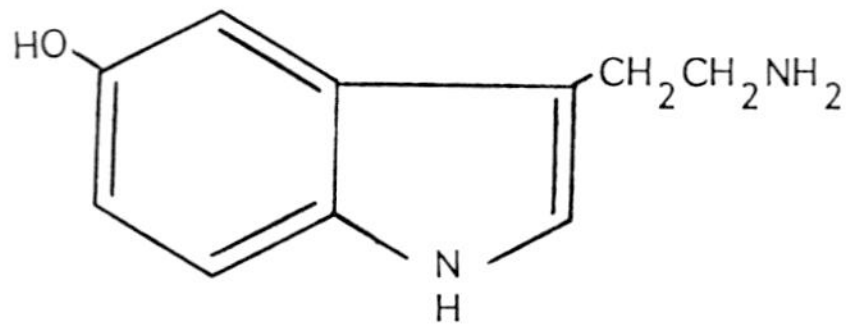

5-HYDROXYTRYPTAMINE

FIGURE 1. Chemical structures of octopamine, dopamine, and 5-hydroxytryptamine.

TABLE 1
Monoamine Levels in Various Regions of the Nervous System of Adult *Periplaneta americana*

Tissue	Monoamine (pmol/organ)			
	Octopamine[9,10]	Dopamine[10]	Norepinephrine[10]	5-Hydroxytryptamine[10]
Cerebral ganglion	14.7—21.6[9,10]	40.8	6.3	15.0
Optic lobes (pair)	7.2	—	—	—
Subesophageal ganglion	5.6[9]	—	—	—
Corpora cardiaca (pair)	1.2	—	—	—
Corpora allata (pair)	0.03[9]	—	—	—
Prothoracic ganglion	4.8—5.3[9,10]	2.4	n.d.	1.3
Mesothoracic ganglion	4.3—4.4[9,10]	2.7	n.d.	1.8
Metathoracic ganglion	4.7—5.3[9,10]	2.7	n.d.	2.3
Abdominal ganglia 1—5 (including interconnection)	4.9[10]	3.4	n.d.	3.6
Terminal abdominal ganglion	2.9—5.3[9,10]	1.4	n.d.	0.7

Note: Dash indicates tissue was not assayed; n.d. indicates that norepinephrine was not detected in the tissue in this study.

B. TISSUE LEVELS

The levels of octopamine, dopamine, norepinephrine, and 5-hydroxytryptamine, as measured by HPLC with electrochemical detection and by radioenzymatic assay, are indicated in Table 1. The reported values are consistent with those described by other workers[11-13] and demonstrate that octopamine is a major component of the total monoamine complement in the cockroach nervous system. The equivalent catecholamine, norepinephrine, is present in much lower concentrations than octopamine and in the insect may be considered a trace

amine. By contrast, another catecholamine, dopamine, is the most abundant monoamine in the cockroach brain and also occurs in appreciable amounts in the thoracic and abdominal nerve cords, while the monohydroxyphenolic equivalent, tyramine, is present only in trace amounts. Epinephrine and the monohydroxyphenolic equivalent, synephrine, have not been detected in the cockroach. 5-Hydroxytryptamine occurs in the cerebral ganglion at concentrations similar to those of octopamine and is present in the ganglia of the ventral nerve cord at concentrations similar to those of dopamine. Absolute values of 5-hydroxytryptamine have not been reported for other neuroendocrine tissues, although immunohistochemical studies (see Section II.C) indicate that this compound is localized in these regions.

The hemolymph of resting adult cockroaches contains about 33 n*M* octopamine, and this value increases to about 65 n*M* within 1 min of the onset of flight.[14] Resting levels are restored rapidly following the termination of flight. The excitation-induced elevation of hemolymph octopamine levels supports the proposal that octopamine serves a sympathomimetic role in cockroaches[15] and mediates several excitatory responses (see Section V.A). It is possible that other monoamines are released into the hemolymph and serve a neurohormonal function, but hemolymph levels have not as yet been reported.

C. DISTRIBUTION WITHIN TISSUES

Early studies with Falck-Hillarp histochemical staining indicated discrete anatomical areas in the α- and β-lobes of the corpora pedunculata that contained monoaminergic material.[16-19] However, this technique recognizes both catecholamines and indolamines; thus, it cannot be used to distinguish between specific monoaminergic neurons. More recently, the availability of specific antibodies against dopamine and 5-hydroxytryptamine has enabled the visualization and mapping of dopaminergic and serotonergic neurons in several regions of the cockroach system, using immunohistochemistry. It is beyond the scope of the present chapter to provide a detailed account of immunohistochemical studies on the cockroach nervous system (see Chapter 14 of this work), but it is useful to outline the major features in order to place the current chapter in context.

Immunocytochemical studies using a specific antibody against dopamine suggest that at least some of the aminergic activity reported by the early workers[16,17] is likely to result from the presence of dopaminergic neurons.[20] These results indicate that dopamine-like material is produced in a group of deutocerebral neurons located on either side of the brain (Figure 2). This material is transported along axons to the glomerular region of the deutocerebrum, where extensive ramification occurs. Some dopaminergic fibers leave the deutocerebral glomeruli through the tractus olfactorioglobularis and eventually enter the α- and β-lobes of the corpora pedunculata, where they produce a characteristic banding pattern (Figure 3). Axons containing Leu-enkephalin-like material have also been described in the corpora pedunculata of the cockroach, in close association with the dopaminergic-like axons.[20] In light of the proposed role of the corpora pedunculata as an association center for modulating multimodal sensory responses,[21-23] it is possible that Leu-enkephalin may modulate dopaminergic and serotonergic function (see below) in this region of the cockroach brain. Future studies using immunogold staining in conjunction with transmission electron microscopy may facilitate elucidation of the relationship between dopamine- and Leu-enkephalin-like neurons in the α- and β-lobes of the corpora pedunculata.

Cell bodies and axons that respond positively to specific antibodies against 5-hydroxytryptamine have been demonstrated in the central and peripheral nervous systems of the cockroach.[23-30] Between 110 and 140 serotonin-immunoreactive cell bodies have been reported in each half of the cockroach brain.[25] They may be arranged into 16 different groups according to their location within the neuropil, with the largest number of cells present in the cell group located in the anterior pars intercerebralis. Most of the serotonin-immunoreactive neurons remain within the brain and, therefore, appear to function as interneurons,

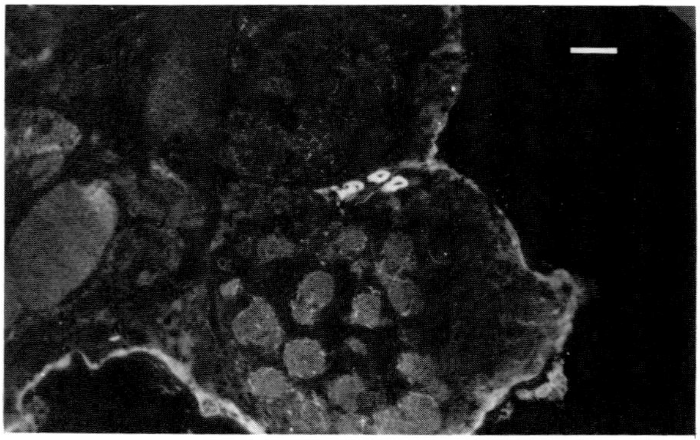

FIGURE 2. Micrograph of dorsolateral region of deutocerebrum of the cockroach, showing neuronal cell bodies containing dopamine-like material. Brain sections were fixed in 5% glutaraldehyde and successively exposed to dopamine antibody (Immunonuclear Corporation, Stillwater, MN) and a secondary antibody against the primary antibody. The secondary antibody carried a fluorescent tag which enabled the dopamine-like neurons to be visualized and photographed using a Nikon® Biophot microscope with epifluorescent attachment. (Bar = 100 μm.)

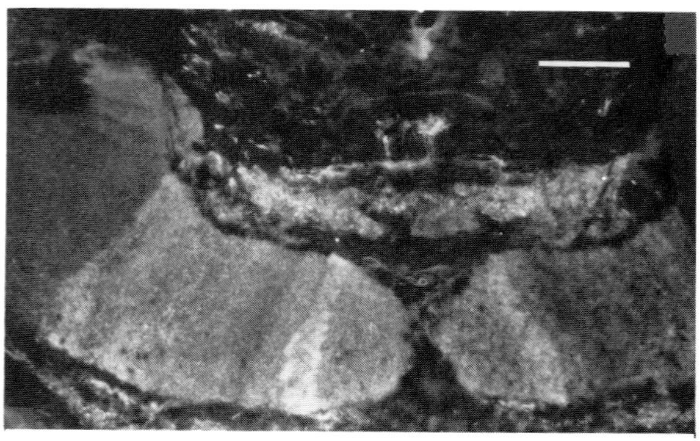

FIGURE 3. Micrograph of β-lobes of corpora pedunculata, showing banding pattern obtained with indirect immunofluorescent technique using dopamine antibody. Experimental procedure was as described in legend to Figure 2. (Bar = 100 μm.)

although some axons from the tritocerebrum may enter the frontal nerve. An extensive network of serotonin-immunoreactive fibers has been demonstrated on the surface of nerves of the subesophageal ganglion,[28] and serotonergic-like cell bodies and axons have also been described in thoracic and abdominal ganglia of the ventral nerve cord.[29] Immunohistochemical studies have also suggested serotonergic innervation of the corpora cardiaca, corpora allata, salivary glands, and visceral muscles of the stomodeum and midgut.[27-31]

The lack of a specific antibody against octopamine has prevented the localization of discrete octopaminergic cell bodies and nerve tracts in the brain. However, as octopamine does not react positively to the Falck-Hillarp stain, but stains with neutral red, it is possible to propose certain cells as octopaminergic on the basis of these two tests. Thus, a number of dorsal, unpaired cells (dorsal unpaired median [DUM] cells) that extend into the thoracic and abdominal ganglia of the locust have been identified as octopaminergic. The octopa-

minergic DUM neuron which supplies the extensor tibiae muscle (DUMeti) has been studied and characterized extensively.[32-34] Similarly, ventral unpaired median (VUM) cells have been identified on the ventral surface of the seventh abdominal ganglion of the locust and shown to contain octopamine.[35] Experimental studies on the corpus cardiacum suggest that octopaminergic neurons are present within nervi corporis cardiaci II (NCCII) and serve to modulate the release of a neurosecretory peptide from the neurohemal organ.[36,37] A similar role for octopamine has been demonstrated in the locust,[38] and it is possible that other monoamines which innervate the corpus cardiacum and other neurohemal organs may serve to modulate peptide hormone release (see Section V.B).

III. METABOLISM

A. BIOSYNTHESIS

Most reviews that impinge upon the neurochemistry of monoamines in insects assume that the biosynthetic pathways present in the insect nervous system are similar to those that have been described for vertebrates[1-4,39,40] (Figure 4). These conclusions are based, in part, upon the results of studies in which various precursors have been shown to become incorporated into specific monoamines. More definitive data are provided by studying the biochemical characteristics of the enzymes that catalyze particular steps in the biosynthetic pathway or, as has been elegantly demonstrated for *Drosophila,*[41] by studying mutant insects with a genetic lesion at the locus for a key enzyme of monoamine synthesis. The various biosynthetic enzymes that have been described in insects will be considered below, with particular emphasis on the cockroach.

1. Tryptophan Hydroxylase

Tryptophan hydroxylase catalyzes the hydroxylation of L-tryptophan to 5-hydroxytryptophan and is believed to be the rate-limiting step in the biosynthesis of 5-hydroxytryptamine in vertebrates.[42] An enzyme preparation derived from cockroach nervous tissue has been compared with the tryptophan hydroxylase of rat brain.[43] The enzyme from both sources is unstable, requires a pterin cofactor, and is inhibited by *p*-chlorophenylalanine. Indeed, the addition of this drug to cockroach nervous tissue homogenates depresses the production of 5-hydroxytryptamine.[44,45] The pH optima and enzyme kinetics of tryptophan hydroxylation differ greatly between the cockroach brain and the rat brain. The V_{max} and K_m for the cockroach enzyme, in the presence of the cofactor dimethyltetrahydropterine, are 8- and 250-fold lower, respectively, than those of the rat brain enzyme. When the presumed natural cofactor biopterin is used, the V_{max} of the insect enzyme remains low, but the K_m is sixfold greater than that of the rat brain enzyme. The estimated molecular weight of the insect enzyme (54,000 Da)[43] is similar to that reported for the mammalian enzyme (55,000 to 60,000).[46]

2. Aromatic Amino Acid Decarboxylase

Aromatic amino acid decarboxylase (AAAC) catalyzes the decarboxylation of 5-hydroxytryptophan to 5-hydroxytryptamine, L-3,4-dihydroxyphenylalanine (DOPA) to dopamine, and tyrosine to tyramine; it also has the ability to catalyze decarboxylation of phenylalanine, tryptophan, and histidine. Dopa decarboxylase plays an important role in providing dopamine as a precursor for cuticular sclerotization during the tanning process; as a result, extensive studies have examined developmental changes in activity of the enzyme.[47] These elegant studies have been conducted principally on epidermal tissue and cultured imaginal disks of Diptera and Lepidoptera, and they have elucidated molecular mechanisms underlying ecdysone-induced stimulation of *de novo* dopa decarboxylase synthesis around the time of the insect molt.

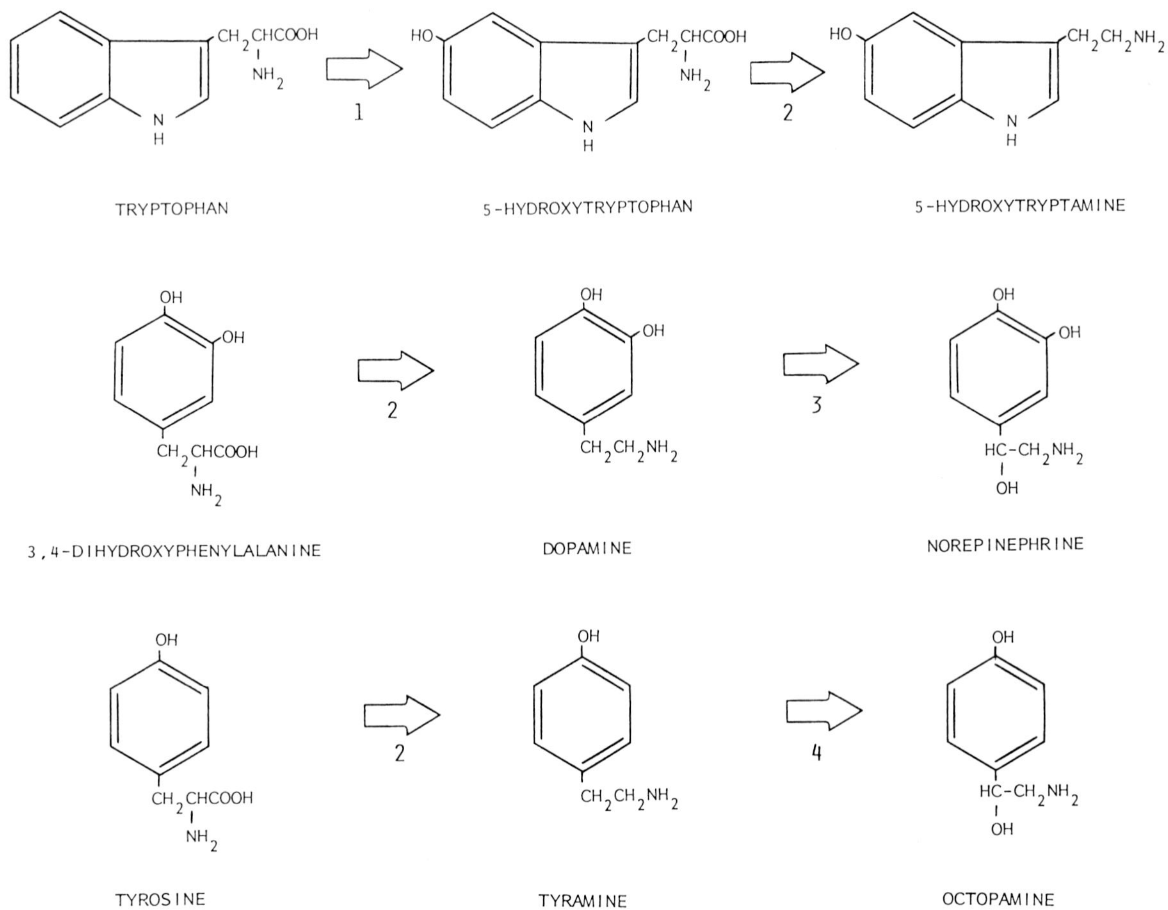

FIGURE 4. Proposed biosynthetic routes for 5-hydroxytryptamine, dopamine, and octopamine in insects. (1) Tryptophan hydroxylase; (2) aromatic amino acid decarboxylase; (3) dopamine-β-hydroxylase; (4) tyramine-β-hydroxylase.

TABLE 2
Apparent K_m and V_{max} Values for *Periplaneta americana* DOPA Decarboxylase and Tyrosine Decarboxylase[52]

	K_m (10 μM)	V_{max} (nmol/mg protein/h)
DOPA decarboxylase	2.1 ± 0.8	55.1 ± 5.1
Tyrosine decarboxylase	5.3 ± 1.8	14.8 ± 1.7

Note: For experimental details, see Reference 52.

Although there are reports of nonspecific AAAC activity against a variety of substrates in some insects,[48-50] there is strong evidence that separate molecular species of the enzyme exist. Studies on a *Drosophila* mutant with a mutation at the structural gene for AAAC indicate depressed levels of dopamine and 5-hydroxytryptamine, whereas tyramine and octopamine levels are not markedly affected.[41,51] Similarly, comparisons of AAAC activity against DOPA, 5-hydroxytryptophan, and *p*-tyrosine in sclerotized and newly ecdysed insects indicate increased enzyme activity against DOPA and 5-hydroxytryptophan, but not against *p*-tyrosine in the newly ecdysed animals.[52] These observations suggest that two enzymic species, one with affinity for tyrosine and the other with affinity for dopamine and 5-hydroxytryptophan, catalyze aromatic amino acid decarboxylation reactions in insects.

AAAC activity was demonstrated in the epidermis, fat body, hemolymph, and nervous tissue of the cockroach.[52-55] A combination of gel filtration and ion-exchange chromatography was used to separate DOPA decarboxylase from tyrosine decarboxylase activity.[52] The kinetic properties of the two enzymes are summarized in Table 2.

The existence of two molecular species of AAAC may reflect differing physiological demands of the cuticle and the nervous system for the products of the AAAC activity. As indicated above, DOPA decarboxylase activity increases markedly in newly ecdysed insects in order to satisfy cuticular demands for the large amounts of dopamine required for sclerotization. By contrast, tyrosine decarboxylase activity is not altered appreciably during the molt cycle, thereby affording a means of ensuring that monoamine levels in the nervous system remain stable.

3. Dopamine(Tyramine)-β-Hydroxylase

In mammals, dopamine(tyramine)-β-hydroxylase (DBH) catalyzes the hydroxylation of dopamine to norepinephrine, tyramine to octopamine, and other phenylethylamines to their respective β-hydroxylated derivatives. DBH activity has been reported in cockroach hemolymph,[56] and radiolabeled tyramine is converted to octopamine in the cockroach nervous system.[13] This conversion is blocked by fusaric acid, a known inhibitor of vertebrate DBH; therefore, DBH activity appears to be present in the cockroach nervous system. However, the enzyme has not been purified from cockroach tissue; therefore, no information is available on the kinetics of the enzyme or the affinity of the enzyme for various substrates.

B. CATABOLISM

A high-affinity, sodium-dependent uptake mechanism for octopamine has been described in the cockroach nerve cord and likened to the mechanism which rapidly removes norepinephrine from the synaptic gap of adrenergic nerve terminals in vertebrates.[57] A sodium-insensitive octopamine removal system associated with the cockroach nerve cord has also been reported,[57] and it has been suggested that this independent system may be located in the connective tissue sheath that surrounds the CNS.[3] This extraneuronal location could serve to stabilize monoamine levels in the nervous system during the tanning process when peripheral amine levels undergo marked fluctuations.[3]

In vertebrates the adrenergic uptake mechanism is associated with enzymic monoamine degradation, and it is probable that a catabolic reaction also occurs in insects. However, the mechanism by which this sequestered octopamine and other monoamines are degraded in the insect nervous system is not fully resolved. Several possible routes of monoamine degradation have been proposed, and these will be discussed separately below.

1. *N*-Acetylation

N-Acetylation of dopamine in the presence of *N*-acetyltransferase (NAT) is an important step in cuticular sclerotization. The *N*-acetyldopamine formed by the reaction undergoes phenol oxidation to produce *N*-acetyldopamine quinone, which links protein chains to form large, stable molecules of sclerotin. The proposal that NAT may also contribute to catecholamine degradation in the insect nervous system[58] has received increasing favor, and *N*-acetylation is now considered the primary route of monoamine degradation in several insect species.[48,59-63]

NAT activity has been demonstrated in a number of cockroach tissues, including accessory glands, fat body, salivary glands, nervous tissue, and various regions of the gut.[64] Comparison of monoamine levels in freshly dissected cerebral ganglia with those in tissues that have been frozen, thawed, and incubated for 1 h indicate marked decreases in the levels of octopamine, dopamine, and 5-hydroxytryptamine and corresponding elevations in the levels of the equivalent *N*-acetylated derivatives.[8] The NAT associated with gut, particularly that present in the Malpighian tubules, has been proposed as the major route of octopamine removal from hemolymph[65] following stress-induced elevation of hemolymph octopamine.[14] Detailed investigation of the biochemical properties of NAT from Malpighian tubules and cerebral ganglia indicates pronounced activity against octopamine, dopamine, and 5-hydroxytryptamine and differences in the sensitivity of the enzyme preparations from the two sources to various pharmacological agents. Thus, the dopamine agonists 2-amino-6,7-dihydroxy-1,2,3,4-tetrahydronaphthalene (ADTN) and epinine and the dopamine antagonist *cis*-flupenthixol inhibit the *N*-acetylation of octopamine by Malpighian tubule NAT more readily than when the reaction is catalyzed by the brain enzyme preparation.[66]

2. Oxidative Deamination

Monoamine oxidase (MAO) catalyzes an important mechanism of catecholamine and 5-hydroxytryptamine catabolism in the vertebrate nervous system; therefore, several studies have attempted to demonstrate the products of MAO activity in insect tissues. HPLC with electrochemical detection of eluted compounds affords sensitive, specific estimation of a broad spectrum of monoamines and their derivatives[5] and has been employed by several groups to investigate the monoamine complement of insect nervous tissue. With one exception, these studies have failed to demonstrate any of the anticipated products of MAO metabolism.[8,67-70] The single exception is a report of 5-hydroxyindoleacetic acid in incubated cockroach nerve cord[71] which has not yet been confirmed by other workers.

Occasional reports of MAO activity in insect tissues should be viewed with caution in light of concerns about the analytical procedures that have been employed to demonstrate enzyme activity. The classical assay for vertebrate MAO relies on toluene extraction of the product of MAO activity,[72] but *N*-acetylated derivatives are also extracted by this procedure. Therefore, this technique should not be used with insect tissues that have high NAT activity (see above). MAO activity has been reported in insect tissues on the basis of histochemical methods that indicate the presence of aldehyde moieties, although the tetrazolium salts employed in these procedures do not provide sufficient specificity for the conclusions to be definitive. Thus, there has been no conclusive demonstration of MAO activity in insects.

3. Conjugation

Amino acid, sugar, sulfate, and phosphate conjugates of monoamines have been reported

in several insect species.[2] In most cases, conjugation appears to be related to the storage or transport of monoamine precursors for cuticular sclerotization. Thus, in the cockroach, exogenous *N*-acetyldopamine is rapidly converted to *N*-acetyldopamine-3-*O*-phosphate and *N*-acetyldopamine-3-*0*-sulfate; however, only the *N*-acetyldopamine moiety is incorporated into the sclerotized cuticle.[73] Dopamine-3-*O*-sulfate (but not dopamine-4-*O*-sulfate) occurs in several tissues of newly ecdysed cockroaches, with particularly high levels in the fat body (187 ng/mg w/w) and hemolymph plasma (104 μg/μl), but the sulfate conjugate is absent or markedly decreased in mature, sclerotized insects.[74] Furthermore, the capacity to produce or metabolize dopamine-3-*O*-sulfate is greatly diminished in sclerotized cockroaches.[74]

Dopamine-3-*O*-sulfate is detected in the cerebral ganglion of adult cockroaches dissected less than 1 h after ecdysis,[74] but there is no indication that sulfate conjugation or, indeed, any other conjugation reaction contributes to monoamine inactivation in the nervous system at other developmental stages. By contrast, β-alanyl and/or sulfate conjugation appear to be important catabolic mechanisms in the nervous systems of the lobster,[75] *Aplysia,*[76] and *Manduca sexta.*[77]

IV. RECEPTOR PHARMACOLOGY

In order for monoamines to serve as effectors of physiological processes, they must bind to a specific protein receptor located in the outer cell membrane of target cells. Interaction of the monoamine with the receptor initiates a response that results in the production of an intracellular second messenger. Monoaminergic effects appear to be expressed through one of two transmembrane signaling mechanisms. In the first, the receptor is coupled to an adenylate cyclase complex which catalyzes the production of cyclic AMP when the monoamine effector binds to the receptor. The cyclic AMP serves as a second messenger to activate a specific protein kinase, which phosphorylates target proteins and leads to the physiological response. The second category of transmembrane signaling mechanism appears to be linked to the hydrolysis of phosphatidyl inositol and may involve calcium, inositol triphosphate, and/or diacylglycerol as second messengers. Again, the activation of protein kinase and protein phosphorylation are implicated in the ultimate expression of the physiological response. An understanding of the nature of the association between the monoamine and the receptor has been gained principally from studies of agonistic simulation of monoaminergic effects, antagonism of monoaminergic effects, and the capacity of various agonists and antagonists to compete with radiolabeled ligands for binding sites on the receptor. The available data on the pharmacological nature of octopamine, dopamine, and 5-hydroxytryptamine receptors in the cockroach are discussed below. The reader is referred to several reviews for comparative data on the pharmacology of aminergic receptors in other insect species.[2,3,78-80]

A. OCTOPAMINE

Pharmacological characterization of octopamine receptors in the cockroach is based upon the relative abilities of various agonists and antagonists to mimic or interfere with octopamine-mediated production of cyclic AMP. Octopamine-sensitive adenylate cyclase has been demonstrated in several cockroach tissues, including brain,[81] nerve cord,[82-84] hemocytes,[85,86] and corpus cardiacum.[87] A variety of potential agonists and antagonists have been tested against these preparations. The hemocyte preparation has been particularly useful as a model for characterizing the octopamine receptor because these cells do not elevate their cyclic AMP levels when exposed to dopamine or 5-hydroxytryptamine.[85] Thus, the properties of the octopamine receptor can be studied independently of cross-reactions with other monoaminergic receptors. Furthermore, cultured hemocytes offer an excellent source of octopamine receptor for purification and molecular characterization of the receptor protein.

TABLE 3
Effect of Receptor Antagonists on Octopamine-Mediated Increases in Cyclic AMP Production in Several Tissues of *Periplaneta americana*

	Corpus cardiacum		Nerve cord	Brain	Hemocyte
Antagonist	IC_{50} (μM)	K_i (μM)	K_i (μM)	K_i (μM)	IC_{50} (μM)
Mianserin	0.45	0.37	0.02	0.002	0.04
Gramine	2.82	2.32	0.06	—	1.12
Phentolamine	5.01	4.13	0.05	0.009	0.46
Cyproheptadine	6.92	5.70	0.02	0.016	3.98
cis-Flupenthixol	17.8	14.7	0.89	0.343	1.78
Promethazine	105.0	86.5	—	—	0.40
Propranolol	120.0	98.9	659.5	—	50.1
(+)-Butaclamol	—	—	14.1	2.038	—

Note: $K_i = IC_{50}/[1 + (S/K_a)]$, where IC_{50} = concentration of antagonist giving 50% inhibition of octopamine-mediated cyclic AMP production, S = concentration of octopamine used (5 μM), and K_a = concentration of octopamine causing 50% maximal cyclic AMP production. Not tested: —.

The hemocyte octopamine receptor is activated by synephrine (50% of octopamine response at 10 μM) and tyramine (37% of octopamine response at 10 μM), and additivity studies indicate that these effects are mediated through the octopamine receptor.[85] The adrenergic agonists naphazoline and clonodine (but not tolazoline) and the formamidine derivatives N^2-(4-chloro-*O*-tolyl)-N^1-methylformadine (DCDM) and N^1-methyl-N^2-(2,4-xylyl)formadine (BTS 27271) also increased cyclic AMP production in the cockroach hemocyte preparation.[85] A classification of octopamine receptor types based upon responses to a variety of pharmacological agents has been proposed.[78] According to this proposal, class 1 receptors are more sensitive to clonodine than to naphazoline, whereas the converse applies to class 2 receptors. Class 2A receptors can be distinguished from those of class 2B by virtue of their greater sensitivity to naphazoline than to tolazoline, whereas tolazoline is more effective than clonodine on class 2B receptors. Thus, on the basis of the proposed agonist criteria,[78] the cockroach hemocyte receptor may be classified as a class 2A receptor. However, this classification is not absolute, and the hemocyte receptor does not conform to the criteria identified for octopamine 2A receptors in terms of the order of antagonism achieved with various antagonists.[85] The effects of several potential antagonists on octopamine-mediated stimulation of cyclic AMP production in nerve cord, brain, corpus cardiacum, and hemocyte preparations of *Periplaneta americana* are summarized in Table 3. In all four tissues, the most potent inhibition is obtained with mianserin, an inhibitor of 5-hydroxytryptamine and histamine in the vertebrate nervous system. The octopamine-mediated response was also inhibited with the α-adrenergic blocking agent phentolamine, the 5-hydroxytryptamine inhibitor gramine, and another antagonist of 5-hydroxytryptamine and histamine, cyproheptadine.

It is evident from this brief description that the octopamine receptors do not fit conveniently into any recognized receptor class and, therefore, these receptors represent a new pharmacological class of aminergic receptor.[88] The identification of pharmacological agents that are specific for octopamine receptors will greatly facilitate further study and characterization of this pharmacological class. The data presented in Table 3 also demonstrate differences in the relative responsiveness of receptors from different tissues to the various agonists, and they confirm the existence of more than one class of octopamine receptor. For example, the octopamine receptor in the brain and nerve cord is only slightly less

sensitive to phentolamine than to mianserin, whereas, in hemocyte and corpus cardiacum, phentolamine is about ten times less sensitive than mianserin. Similarly, the octopamine receptors in nerve cord and brain are much more sensitive to cyproheptadine than those in the corpus cardiacum and hemocyte.

Mianserin, the most potent inhibitor of octopamine-mediated responses in all cockroach tissues tested, has been used as a ligand to further characterize the pharmacological properties of the putative octopamine receptors in cockroach nerve cord.[89] Saturable specific binding of [^{3}H]mianserin was demonstrated in nerve cord homogenates, and Scatchard analysis indicated both a high-affinity binding site (K_D = 39.6 n*M* and B_{max} = 0.8 fmol/μg) and a low-affinity binding site (K_D = 648.9 n*M* and B_{max} = 5.9 fmol/μg). Negative cooperativity between the binding sites was suggested by the slope of 0.6 determined from the Hill plot. Inclusion of the stable GTP analogue 5^1-guanylylimidodiphosphate at 10 μ*M* reduced the binding of [^{3}H]mianserin, and Scatchard analysis under these conditions indicated a single binding site with K_D = 696.2 n*M* and B_{max} = 2.0 fmol/μg. The results indicate the presence in insect nerve cord of two separate octopamine receptors, one of which is GTP dependent. The possibility that one receptor expresses octopamine effects through the mediation of adenylate cyclase and cyclic AMP, whereas the other uses phosphatidyl inositol and calcium chaneling, is strongly suggested by the demonstration of octopamine-mediated enhancement of cyclic AMP production in this tissue[89] and by octopamine-stimulated increases in intracellular calcium concentrations that have been demonstrated in insect cell culture.[90]

B. DOPAMINE

Dopamine-mediated increases in cyclic AMP production have been reported in several cockroach tissues,[84,87] and a detailed examination of the pharmacology of a putative dopamine receptor in brain membrane preparations has been completed.[91] In vertebrates, the dopamine D_1 receptor is coupled to adenylate cyclase and dopamine agonists elevate cyclic AMP levels. Thus, the demonstration that the dopamine agonists ADTN and epinine elevate cyclic AMP levels in the brain preparation suggests the presence of a D_1 receptor. However, the failure of the strongly selective D_1 agonist 1-phenyl-2,3,4,5-tetrahydro-(1H)-3-benzazepine-7,8-diol hydrochloride (SKF-38393) to elicit increases in cyclic AMP and the pronounced stimulation of cyclic AMP production obtained with the specific D_2 agonist [*trans*-(−)-4aR-4,4a,5,6,7,8,8a,9-octahydro-5-propyl-1H-pyrazolo(3,4-g)-quinoline] monohydrochloride (LY-171555) indicate that the dopamine receptor in the insect brain has pharmacological characteristics of both D_1 and D_2 receptors. This conclusion is further substantiated by the effects of various potential dopamine antagonists on the dopamine-sensitive adenylate cyclases. These data are presented for several tissues in Table 4. The most effective antagonists of the insect brain dopamine receptor are pifluthixol, *cis*-flupenthixol, and (+)-butaclamol, but as these compounds are not selective for D_1 or D_2 receptors they cannot be used to distinguish the receptor type. However, the selective D_1 antagonists R(+)-7-chloro-8-hydroxy-3-methyl-1-phenyl-2,3,4,5-tetrahydro-(1H)-3-benzazepine hydrochloride (SCH-23390) and 7-bromo-8-hydroxy-3-methyl-1-phenyl-2,3,4,5-tetrahydro-1H-3-benzazepine (SKF-83566), the D_2 antagonist spiperone, and the 5-hydroxytryptamine/histamine antagonist cyproheptadine show marked antagonism of the insect brain dopamine receptor. The inhibition obtained with spiperone contrasts with the lack of inhibition observed with two other selective D_2 antagonists, sulpiride and haloperidol. The data presented in Table 4 also indicate differences in the ratio of inhibition obtained with various antagonists in different tissues. For example, cyproheptadine is about 5 times less potent than *cis*-flupenthixol in brain tissue, but 200 times less potent than *cis*-flupenthixol in the corpus cardiacum.

The pharmacological conclusions derived from studies on dopamine-mediated elevation of cyclic AMP production are consistent with the results of binding studies using [^{3}H]pifluthixol as a ligand of dopamine receptors in the cockroach brain.[91,92] Scatchard analysis of the

TABLE 4
Effect of Receptor Antagonists on Dopamine-Mediated Increases in Cyclic AMP Production in Several Tissues of *Periplaneta americana*

	Corpus cardiacum		Nerve cord	Brain
Antagonist	IC_{50} (μM)	K_i (μM)	K_i (μM)	K_i (μM)
Pifluthixol	—	—	—	0.006
cis-Flupenthixol	1.600	0.890	0.033	0.010[a]
(+)-Butaclamol	3.160	1.760	0.004	0.015
Spiperone	—	—	—	0.045
SCH-23390	—	—	—	0.048[b]
Cyproheptadine	310.000	173.000	0.036	0.057[b]
SKF-83566	—	—	—	0.090[b]
SCH-23338	—	—	—	0.212[b]
Mianserin	10.600	5.910	0.025	0.273[a]
Phentolamine	42.200	23.500	—	1.030
Sulpiride	—	—	—	3.086[b]
SKF-82526	—	—	—	0.027[b]
Apomorphine	—	—	—	0.718[b]
Phentolamine	42.200	23.500	0.455	—
Gramine	2.140	1.190	0.103	—
Haloperidol	—	—	—	n.i.

Note: $K_i = IC_{50}/[1 + S/K_a]$, where IC_{50} = concentration of antagonist giving 50% inhibition of dopamine-mediated cyclic AMP production, S = concentration of dopamine used (normally 5 μM), and K_a = concentration of dopamine causing 50% maximal cyclic AMP production. Not tested: —; no inhibitor: n.i.

[a] Dopamine concentration = 0.5 μM.
[b] Dopamine concentration = 10 μM.

binding data indicate a B_{max} of 1.4 pmol/g protein and a K_D of 0.35 nM.[92] The results above suggest that dopamine receptors in the cockroach have pharmacological characteristics that are distinct from those of vertebrate D_1 and D_2 receptors and, therefore, comprise a new class of dopamine receptor. This proposal is supported by studies on the effects of various pharmacological agents on nervous stimulation of a putative dopamine receptor on the salivary glands of another species of cockroach, *Nauphoeta cinerea*.[93-95] Thus, phentolamine is a competitive antagonist of the hyperpolarizing response that is elicited in cockroach salivary glands by electrical stimulation and administration of dopamine, whereas the dopamine agonist apomorphine simulates the dopamine response at high concentrations.[93]

C. 5-HYDROXYTRYPTAMINE

The mechanism by which 5-hydroxytryptamine regulates fluid secretion in the blowfly salivary gland has been investigated rigorously and has contributed greatly to an understanding of transmembrane signaling mechanisms involving inositol lipids.[96-98] It is probable that some 5-hydroxytryptamine-mediated processes in the cockroach are expressed through a similar transduction mechanism, but no data are available on the pharmacology of the receptors that are likely to be involved. 5-Hydroxytryptamine causes an accelerated rate of contraction of Malpighian tubules in *P. americana*, and the effect is inhibited by 5-benzyloxygramine.[99] It has been proposed that the effect is mediated through a type D 5-hydroxytryptamine receptor, but a more rigorous pharmacological characterization is required

TABLE 5
Effect of Receptor Antagonists on 5-Hydroxytryptamine-Mediated Increases in Cyclic AMP Production in Nerve Cord and Corpus Cardiacum of *Periplaneta americana*

Antagonist	Corpus cardiacum IC_{50} (μM)	Corpus cardiacum K_i (μM)	Nerve cord K_i (μM)
Cyproheptadine	0.40	0.10	0.20
Mianserin	15.80	3.79	0.26
cis-Flupenthixol	37.20	8.93	0.01
(+)-Butaclamol	58.90	14.10	2.36
Promethazine	70.80	17.00	—
Phentolamine	321.00	77.10	0.38
Gramine	616.00	148.00	0.11

Note: $K_i = IC_{50}/[1 + S/K_a]$, where IC_{50} = concentration of antagonist giving 50% inhibition of 5-hydroxytryptamine-mediated cyclic AMP production, S = concentration of 5-hydroxytryptamine used (5 μM), and K_a = concentration of 5-hydroxytryptamine causing 50% maximal cyclic AMP production. Not tested: —.

to substantiate this claim. Contractile movements of the cockroach fore-, mid-, and hindgut are also influenced by 5-hydroxytryptamine,[100-102] but the mechanisms by which these effects are mediated have not been determined.

5-Hydroxytryptamine-sensitive adenylate cyclase activity has been indicated in cockroach nerve cord,[84,103] corpus cardiacum,[87] and heart.[104] The effects of various potential antagonists on 5-hydroxytryptamine-mediated increases in cyclic AMP levels in the corpus cardiacum and nerve cord of *P. americana* are given in Table 5. Although the data are not sufficient to permit classification of the 5-hydroxytryptamine receptors according to the scheme proposed for vertebrates, they suggest that in the cockroach at least two types of 5-hydroxytryptamine receptors are coupled to adenylate cyclase. Thus, the α-adrenergic antagonist phentolamine is a potent inhibitor of the 5-hydroxytryptamine-mediated response in the nerve cord, but not in the corpus cardiacum. Pronounced differences are also evident between the two tissues with regard to their sensitivity to gramine, a blocker of 5-hydroxytryptamine effects in vertebrates. Distinct adenylate cyclase-coupled 5-hydroxytryptamine receptor types have been proposed in vertebrates,[105] and the K_a of the insect 5-hydroxytryptamine-sensitive adenylate cyclase ($\cong 1\ \mu M$) approximates that of many vertebrate 5-hydroxytryptamine receptors.[105]

V. NEUROENDOCRINE ROLES OF MONOAMINES

The important roles of octopamine, dopamine, and 5-hydroxytryptamine in effecting and regulating various physiological processes in insects have been documented in several reviews.[1-3] The present account will consider only those studies that have used the cockroach as the experimental insect, although it is probable that many of the amine-mediated processes described for other species also occur in the cockroach.

The physiological roles of biogenic amines may be categorized under the rubric of neurotransmitter, neuromodulator, or neurohormone according to whether the immediate physiological effect is expressed at the synapse where the effector is released, in the general vicinity of a group of synapses, or at a target site that is distant from the site of release and

to which the effector is transported by the hemolymph.[106] However, the distinction between these categories is not always clear, and rigid adherence to the terms can result in oversimplication of the true physiological function. Therefore, in the present account the roles will be considered in terms of the overall physiological effect.

A. SYMPATHOMIMETIC AND BEHAVIORAL EFFECTS

Adult cockroaches rapidly respond to handling by elevating hemolymph trehalose levels at the expense of fat body glycogen.[107] This excitation-induced hypertrehalosemic effect (EXIT response) is mediated by octopamine.[108] Hemolymph levels of octopamine also increase rapidly in response to excitation.[14] Octopamine has been implicated in a number of excitatory physiological processes in other insect species, and it has been suggested that the EXIT response is part of a general excitation response, with octopamine serving as the sympathomimetic effector.[15] The demonstrated octopamine-induced increase in the rate of heartbeat[104] may be a further manifestation of this effect. 5-Hydroxytryptamine also elevates the rate of heartbeat in cockroaches, but this effect can be separated from that of octopamine by use of a specific antagonist of 5-hydroxytryptamine.[104] The proposed sympathomimetic role for octopamine is further supported by the demonstration that octopamine raises the arousal level and increases the honeybee's responsiveness to unconditioned olfactory stimuli.[109]

In addition to the increased behavioral responsiveness that occurs as a result of the sympathomimetic action of octopamine, biogenic amines appear to influence other aspects of insect behavior. The *per*^{o} mutation in *Drosophila,* which has depressed octopamine levels, shows abnormal circadian eclosion and activity rhythms as well as abnormal courtship rituals.[51] Furthermore, in locusts, iontophoresis of octopamine onto specific regions of the neuropil elicited particular patterns of response.[110,111] In light of these observations and studies of other invertebrate species,[112-116] it is reasonable to propose that the integration of much of the complex pattern of insect behavior may be effected through highly localized release of specific monoamines to trigger and modulate defined behavioral neural networks. Elucidation of the precise mechanisms by which this integration is achieved represents an exciting area for future investigation.

B. REGULATION OF NEUROSECRETION

Aminergic regulation of the release of peptidergic hormones from the hypothalamo-pituitary complex of vertebrates is well established.[117] The pars intercerebralis-corpus cardiacum complex of insects is considered functionally analogous to the hypothalamo-pituitary axis of vertebrates, and a similar involvement of monoamines in the regulation of peptidergic neurohormones has been proposed.[118] Electrical stimulation of a nerve tract supplying the corpus cardiacum results in release of a hypertrehalosemic factor[119,120] and elevated cyclic AMP levels within the neurohemal organ.[36] Application of octopamine to incubated corpora cardiaca simulates the effects of electrical stimulation and, together with studies on the effects of various pharmacological antagonists on the electrically and chemically induced responses, suggests that the release of a hypertrehalosemic peptide in the cockroach is effected by octopaminergic neurons.[36] Octopamine, dopamine, and a small amount of norepinephrine have been detected in abdominal median neurohemal organs of *P. americana,*[11] and it has been suggested that the octopamine is present in octopaminergic neurons and may regulate the release of neurosecretory material. An involvement of dopamine in the functioning of the median neurosecretory cells is also indicated by the observation that in the abdominal ganglia these cells take up label from ^{3}H-dopamine.[121]

C. REGULATION OF GUT MOTILITY

Cockroach gut is well innervated, with the foregut receiving nervous input from the stomatogastric nervous system,[122] while the hindgut is supplied by paired proctodeal nerves

from the terminal abdominal ganglion[123,124] and also may be regulated through nerve cells located peripherally on the surface of the gut.[125] Peptidergic and aminergic effectors have been implicated in the regulation of gut function, but precise roles have not been ascribed to the various factors.[125] 5-Hydroxytryptamine stimulates both the fore- and hindgut; as the effect is demonstrated in denervated preparations, the serotonergic action appears to be directly on the gut muscle.[101]

D. DOPAMINERGIC REGULATION OF SALIVARY SECRETION

The salivary glands of *N. cinerea* receive dual innervation from the stomatogastric nervous system[126] and, as the peripherally located dopaminergic synapses are readily accessible to experimental manipulation, offer an excellent model with which to study the nature and regulation of dopaminergic synapses. Electrical stimulation of the salivary duct nerve leads to fluid secretion,[127] but stimulation of the esophageal nerve does not appear to influence secretion.[128]

VI. BIOGENIC AMINES AS TARGETS FOR INSECTICIDE DEVELOPMENT

It is evident from the foregoing account that biogenic amines serve a variety of important physiological roles, as neuroendocrine effectors and as essential intermediates in the process of sclerotization. Therefore, the aminergic system must be considered a potential target against which insect-specific, efficacious insecticides might be developed. This possibility is strengthened by the demonstration of differences in the pharmacological properties of insect aminergic receptors compared with equivalent vertebrate receptors and the demonstrated importance of *N*-acetylation in both neuroamine degradation and sclerotization. One group of compounds that appear to exert their insecticidal and acaricidal influences through perturbation of aminergic function are the formamidines. These insecticides cause behavioral changes in the target insect, eventually resulting in death or loss of fecundity. The formamidine-induced effects may be expressed through any or all of the following actions: agonistic binding to octopamine receptor,[83,129] antagonistic binding to dopamine and/or 5-hydroxytryptamine receptors,[84] and inhibition of *N*-acetyltransferase.[65] Other insecticidal compounds that affect the octopamine receptor include the phenyliminoimidazolines,[130] and it is probable that other compounds targeted against the aminergic system will be developed in the future.

It is hoped that the present chapter, in addition to identifying potential models for neurobiological research, will provide impetus for the development of novel amine-targeted insecticides.

VII. IMPLICATIONS FOR BIOMEDICAL RESEARCH

Invertebrate models have contributed substantially to an increased understanding of many biological processes and have led to some important advances in biomedicine.[112] For example, basic mechanisms of neuronal function were elucidated in studies with the squid axon, the neurophysiological importance of the putative neurotransmitters γ-aminobutyric acid and glutamic acid was first recognized in arthropods, and major conceptual advances in the neuronal basis of memory and learning have resulted from studies with invertebrate models.[112] The present chapter summarizes available information on octopamine, dopamine, and 5-hydroxytryptamine in cockroaches and thereby identifies potential for using this insect as a model for monoamine research.

In particular, the cockroach (together with other arthropods) offers an excellent model with which to study the metabolism, pharmacology, and physiological role(s) of octopamine. Octopamine occurs in several vertebrate tissues[131] and has been implicated in the mediation

or modulation of some aspects of spontaneous behavior, stereotypy, locomotor activity, motivated behavior, motor control, sexual behavior, feeding, sleep, and aggression.[1] However, the low concentration and rapid turnover of octopamine have prevented detailed investigation of octopaminergic function in vertebrate tissues. By contrast, octopaminergic neurons are readily accessible in insects, and extensive data are available on the pharmacological nature of octopamine receptors.

Dopaminergic and serotonergic neurons in the cockroach are also large, easily identified, and accessible for experimental manipulation. Thus, they should be considered as possible models for studies in which specific functioning neuronal preparations are required.

ACKNOWLEDGMENTS

I am pleased to acknowledge the valued contributions of many students and associates to much of the work discussed in this chapter. I thank Mr. G. W. A. Milton for permission to include Figures 2 and 3, which were taken from his M.Sc. thesis. Financial support for studies in this laboratory was received from the National Sciences and Engineering Research Council of Canada and from American Cyanamid.

REFERENCES

1. **Boulton, A. A.,** Trace amines and the neurosciences: an overview, in *Neurobiology of the Trace Amines,* Boulton, A. A., Baker, G. B., Dewhurst, W. G., and Sandler, M., Eds., Humana Press, Clifton, NJ, 1984, 13.

1a. **Evans, P. D.,** Biogenic amines in the insect nervous system, *Adv. Insect Physiol.,* 15, 317, 1980.

2. **Brown, C. S., and Nestler, C.,** Catecholamines and indolalkylamines, in *Comprehensive Insect Physiology, Biochemistry and Pharmacology,* Vol. 11, Kerkut, G. A. and Gilbert, L. I., Eds., Pergamon Press, Oxford, 1985, 435.

3. **Evans, P. D.,** Octopamine, in *Comprehensive Insect Physiology, Biochemistry and Pharmacology,* Vol. 11, Kerkut, G. A. and Gilbert, L. I., Eds., Pergamon Press, Oxford, 1985, 499.

4. **David, J.-C. and Coulon, J.-F.,** Octopamine in invertebrates and vertebrates: a review, *Prog. Neurobiol.,* 24, 141, 1985.

5. **Downer, R. G. H., Bailey, B. A., and Martin, R. J.,** Estimation of biogenic amines in biological tissues, in *Neurobiology,* Gilles, R. and Balthazart, J., Eds., Springer-Verlag, Berlin, 1985, 248.

6. **Bailey, B. A., Martin, R. J., and Downer, R. G. H.,** Simultaneous determination of dopamine, norepinephrine, tyramine and octopamine by reverse-phase liquid chromatography with electrochemical detection, *J. Liq. Chromatogr.,* 5, 2435, 1982.

7. **Martin, R. J., Bailey, B. A., and Downer, R. G. H.,** Rapid estimation of catecholamines, octopamine and 5-hydroxytryptamine from biological tissues using high performance liquid chromatography with coulometric detection, *J. Chromatogr.,* 278, 265, 1983.

8. **Downer, R. G. H. and Martin, R. J.,** Analysis of monoamines and their metabolites by high performance liquid chromatography with coulometric electrochemical detection, *Life Sci.,* 41, 833, 1987.

9. **Evans, P. D.,** Octopamine distribution in the insect nervous system, *J. Neurochem.,* 30, 1009, 1978.

10. **Martin, R. J., Bailey, B. A., and Downer, R. G. H.,** Analysis of octopamine, dopamine, 5-hydroxytryptamine and tryptophan in the brain and nerve cord of the American cockroach, in *Neurobiology of the Trace Amines,* Boulton, A. A., Baker, G. B., Dewhurst, W. G., and Sandler, M., Eds., Humana Press, Clifton, NJ, 1984, 91.

11. **Dymond, G. R. and Evans, P. D.,** Biogenic amines in the nervous system of the cockroach, *Periplaneta americana:* association of octopamine with mushroom bodies and dorsal unpaired median (DUM) neurons, *Insect Biochem.,* 9, 535, 1979.

12. **Frontali, N. and Haggendal, J.,** Noradrenaline and dopamine content in the brain of the cockroach, *Periplaneta americana, Brain Res.,* 14, 540, 1969.

13. **Robertson, H. A. and Steele, J. E.,** Octopamine in the insect central nervous system: distribution, biosynthesis and possible physiological role, *J. Physiol. (London),* 237, 34P, 1974.

14. **Bailey, B. A., Martin, R. J., and Downer, R. G. H.,** Haemolymph octopamine levels during and following flight in the American cockroach, *Periplaneta americana* L., *Can. J. Zool.,* 62, 19, 1984.

15. **Downer, R. G. H.,** Short-term hypertrehalosemia induced by octopamine in the American cockroach, *Periplaneta americana* L., in *Insect Neurobiology and Pesticide Action (Neurotox 79),* Society of Chemical Industry, London, 1980, 335.
16. **Frontali, N. and Norberg, K.,** Catecholamine containing neurons in the cockroach brain, *Acta Physiol. Scand.,* 66, 243, 1966.
17. **Frontali, N.,** Histochemical localization of catecholamine in the brain of normal and drug treated cockroaches, *J. Insect Physiol.,* 14, 881, 1968.
18. **Mancini, G. and Frontali, N.,** On the ultrastructural localization of catecholamines in the beta lobes (corpora pedunculata) of *Periplaneta americana, Z. Zellforsch. Mikrospk. Anat.,* 103, 341, 1970.
19. **Richter, V. D. and Rutschke, E.,** Zum Vorkommen der monoamino Oxidase (MAO) in Gehirn von *Periplaneta americana* (L.), *Acta Histochem.,* 60S, 304, 1977.
20. **Milton, G. W. A.,** Immunofluorescent Localization of Neurons Containing Leucine-Enkephalin-Like and Dopamine-Like Material in the Brain of the American Cockroach, *Periplaneta americana* L., M.Sc. thesis, University of Waterloo, Waterloo, Ontario, Canada, 1988.
21. **Suzuki, H., Takeda, H., and Kuwabara, M.,** Activities of antennal and ocellar interneurones in the protocerebrum of the honeybee, *J. Exp. Biol.,* 64, 405, 1976.
22. **Erber, J. and Menzel, R.,** Visual interneurons in the median protocerebrum of the bee, *J. Comp. Physiol.,* 121, 65, 1977.
23. **Erber, J.,** Response characteristics and after effects of multimodal neurons in the mushroom body area of the honeybee, *Physiol. Entomol.,* 3, 77, 1978.
24. **Bishop, C. A. and O'Shea, M.,** Serotonin immunoreactive neurons in the central nervous system of an insect *(Periplaneta americana), J. Neurobiol.,* 14, 251, 1983.
25. **Nässl, D. R. and Klemm, N.,** Serotonin-like immunoreactivity in the optic lobes of three insect species, *Cell Tissue Res.,* 232, 129, 1983.
26. **Klemm, N., Steinbusch, H. W. M., and Sundler, F.,** Distribution of serotonin-containing neurons and their pathways in the supraoesophageal ganglion of the cockroach *Periplaneta americana* (L.) as revealed by immunocytochemistry, *J. Comp. Neurol.,* 225, 387, 1984.
27. **Nishiitsutsuji-Uwo, J., Takeda, M., and Saito, H.,** The production of an antiserum to serotonin and serotonin-like immunoreactivity in the cockroach brain-midgut system, *Biomed. Res.,* 5, 211, 1984.
28. **Davis, N. T.,** Neurosecretory neurons and their projections to the serotonin neurohaemal system of the cockroach *Periplaneta americana* (L.), and identification of mandibular and maxillary motor neurons associated with this system, *J. Comp. Neurol.,* 259, 604, 1987.
29. **Davis, N. T.,** Serotonin-immunoreactive visceral nerves and neurohemal system in the cockroach *Periplaneta americana* (L.), *Cell Tissue Res.,* 240, 593, 1985.
30. **Nässl, D. R.,** Serotonin and serotonin-immunoreactive neurons in the nervous system of insects, *Prog. Neurobiol.,* 30, 1, 1987.
31. **Sloley, B. D., Downer, R. G. H., and Gillott, C.,** Levels of tryptophan, 5-hydroxytryptamine and dopamine in some tissues of the cockroach, *Periplaneta americana, Can. J. Zool.,* 65, 797, 1987.
32. **Hoyle, G., Dagan, D., Moberly, B., and Colquhoun, W.,** Dorsal unpaired median insect neurons make neurosecretory endings on skeletal muscle, *J. Exp. Zool.,* 189, 407, 1974.
33. **Evans, P. D. and O'Shea, M.,** The identification of an octopaminergic neurone which modulates neuromuscular transmission in the locust, *Nature (London),* 270, 257, 1977.
34. **Evans, P. D. and O'Shea, M.,** The identification of an octopaminergic neurone and the modulation of a myogenic rhythm in the locust, *J. Exp. Biol.,* 73, 235, 1978.
35. **Lange, A. B. and Orchard, I.,** Ventral neurons in an abdominal ganglion of the locust *Locusta migratoria,* with properties similar to dorsal unpaired median neurons, *Can. J. Zool.,* 64, 264, 1986.
36. **Downer, R. G. H., Orr, G. L., Gole, J. W. D., and Orchard, I.,** The role of octopamine and cyclic AMP in regulating hormone release from corpora cardiaca of the American cockroach, *J. Insect Physiol.,* 30, 457, 1984.
37. **Downer, R. G. H., Bailey, B. A., Gole, G. W. D., Martin, R. J., and Orr, G. L.,** Estimation of octopamine and its role in the release of a hypertrehalosemic factor in the American cockroach, *Periplaneta americana,* in *Insect Neurochemistry and Neurophysiology,* Borkovec, A. B. and Kelly, T. J., Eds., Plenum Press, New York, 1984, 349.
38. **Orchard, I., Loughton, B. G., Gole, J. W. D., and Downer, R. G. H.,** Synaptic transmission elevates adenosine 3′-5′-monophosphate (cyclic AMP) in locust neurosecretory cells, *Brain Res.,* 259, 152, 1983.
39. **Robertson, H. A. and Juorio, A. V.,** Octopamine and some related noncatecholic amines in invertebrate nervous systems, *Int. Rev. Neurobiol.,* 19, 173, 1975.
40. **Harmer, A. J.,** Neurochemistry of octopamine, in *Noncatecholic Phenylethylamines, Part 2,* Modern Pharmacology/Toxicology Ser., Vol. 12, Mosnaim, A. D. and Wolf, M. E., Eds., Marcel Dekker, New York, 1980, 97.
41. **Livingstone, M. S. and Tempel, B. L.,** Genetic dissection of monoamine neurotransmitter synthesis in *Drosophila, Nature (London),* 303, 67, 1983.

42. **Grahame-Smith, D. G.,** The biosynthesis of 5-hydroxytryptamine in brain, *Biochem. J.,* 105, 351, 1967.
43. **Sloley, B. D. and Yu, P. H.,** Apparent differences in tryptophan hydroxylation by cockroach *(Periplaneta americana)* nervous tissue and rat brain, *Neurochem. Int.,* 11, 265, 1987.
44. **Osborne, N. N. and Neuhoff, V.,** Formation of serotonin in insect *(Periplaneta americana)* nervous tissue, *Brain Res.,* 74, 366, 1974.
45. **Pandey, A., Habibulla, M., and Singh, R.,** Tryptophan hydroxylase and 5-HTP-decarboxylase activity in cockroach brain and the effects of *p*-chlorophenylalanine and 3-hydroxybenzylhydrazine (NSD-1015), *Brain Res.,* 273, 67, 1983.
46. **Youdim, M. B. H., Hamon, M., and Bourgoin, S.,** Properties of partially purified pig brain stem tryptophan hydroxylase, *J. Neurochem.,* 25, 407, 1975.
47. **Sekeris, C. E. and Fragoulis, E. G.,** Control of DOPA-decarboxylase, in *Comprehensive Insect Physiology, Biochemistry and Pharmacology,* Vol. 8, Kerkut, G. A. and Gilbert, L. I., Eds., Pergamon Press, Oxford, 1985, 147.
48. **Dewhurst, S. A., Croker, S. G., Ikeda, K., and McCamen, R. E.,** Metabolism of biogenic amines in *Drosophila* nervous tissue, *Comp. Biochem. Physiol.,* 43B, 975, 1972.
49. **Murdock, L. L., Wirtz, R. A., and Köhler, G.,** 3,4-Dihydroxyphenylalanine (DOPA) decarboxylase activity in the arthropod nervous system, *Biochem. J.,* 132, 681, 1973.
50. **Emson, P. C., Burrows, M., and Fonnum, F.,** Levels of glutamate decarboxylase, choline acetyltransferase and acetylcholinesterase in identified motorneurones of the locusts, *J. Neurobiol.,* 5, 33, 1974.
51. **Livingstone, M. S.,** Two mutations in *Drosophila* differentially affect the synthesis of octopamine, dopamine and serotonin by altering the activities of two different amino acid decarboxylases, *Soc. Neurosci. Abstr.,* 7, 351, 1981.
52. **Yu, P. H. and Sloley, B. D.,** Some aspects on L-DOPA decarboxylase and *p*-tyrosine decarboxylase in the central nervous and peripheral tissues of the American cockroach, *Periplaneta americana, Comp. Biochem. Physiol.,* 87C, 315, 1987.
53. **Colhoun, E. H.,** Pharmacological tantalizers, in *Insects and Physiology,* Beament, J. W. L. and Treherne, J. E., Eds., Oliver and Boyd, London, 1967, 201.
54. **Whitehead, D. L.,** New evidence for the control mechanism of sclerotization in insects, *Nature (London),* 224, 721, 1969.
55. **Hopkins, T. L. and Wirtz, R. A.,** DOPA and tyrosine decarboxylase activity in tissues of *Periplaneta americana* in relation to cuticle formation and ecdysis, *J. Insect Physiol.,* 22, 1167, 1976.
56. **Lake, C. R., Mills, R. R., and Brunet, P. C. J.,** β-Hydroxylation of tyramine by cockroach hemolymph, *Biochim. Biophys. Acta,* 215, 226, 1970.
57. **Evans, P. D.,** Octopamine: a high-affinity uptake mechanism in the nervous system of the cockroach, *J. Neurochem.,* 30, 1015, 1978.
58. **Sekeris, C. E. and Karlson, P.,** Biosynthesis of catecholamines in insects, *Pharmacol. Rev.,* 18, 89, 1966.
59. **Maranda, B. and Hodgetts, R.,** A characterisation of dopamine acetyltransferase in *Drosophila melanogaster, Insect Biochem.,* 7, 33, 1977.
60. **Hayashi, S., Murdock, L. L., and Florey, E.,** Octopamine metabolism in invertebrates *(Locusta, Astacus, Helix):* evidence for *N*-acetylation in arthropod tissues, *Comp. Biochem. Physiol.,* 58C, 183, 1977.
61. **Evans, P. H., Soderlund, D. M., and Aldrich, J. R.,** *In vitro N*-acetylation of biogenic amines by tissues of the European corn borer, *Ostrinia nubilalis* Hübner, *Insect Biochem.,* 10, 375, 1980.
62. **Mir, A. K. and Vaughan, P. T. F.,** Biosynthesis of *N*-acetyldopamine and *N*-acetyloctopamine by *Schistocerca gregaria* nervous tissue, *J. Neurochem.,* 36, 441, 1981.
63. **Evans, P. H. and Soderlund, D. M.,** Biogenic amine acetylation: no detectable circadian rhythm in whole brain homogenates of the insect *Ostrinia nubilalis, Experientia,* 38, 302, 1982.
64. **Martin, R. J.,** Characterisation of *N*-Acetyltransferase Activity in *Periplaneta americana:* A Putative Mechanism for Octopamine Degradation, Ph.D. thesis, University of Waterloo, Waterloo, Ontario, Canada, 1987.
65. **Downer, R. G. H. and Martin, R. J.,** *N*-Acetylation of octopamine: a potential target for insecticide development, in *Sites of Action for Neurotoxic Pesticides,* ACS Symp. Ser. 356, Hollingworth, R. M. and Green, M. B., Eds., American Chemical Society, Washington, D.C., 1987, 202.
66. **Downer, R. G. H., Gole, J. W. D., Orr, G. L., and Martin, R. J.,** Pharmacological characteristics of octopamine-sensitive adenylate cyclase and *N*-acetyl octopamine transferase in insects, in *Trace Amines: Comparative and Clinical Neurobiology,* Boulton, A. A., Juorio, A. V., and Downer, R. G. H., Eds., Humana Press, Clifton, NJ, 1988, 3.
67. **Murdock, L. L. and Omar, D.,** *N*-Acetyldopamine in insect nervous tissue, *Insect Biochem.,* 11, 161, 1981.
68. **Omar, D., Murdock, L. L., and Hollingworth, R. M.,** Actions of pharmacological agents on 5-hydroxytryptamine and dopamine in the cockroach nervous system *(Periplaneta americana), Comp. Biochem. Physiol.,* 73C, 423, 1982.

69. **Brookhart, G. L. and Murdock, L. L.,** Endogenous *N*-acetyl-5-hydroxytryptamine (NA-5-HT) in insect cerebral ganglia, in *Insect Neurochemistry and Neurophysiology,* Borkovec, A. B. and Kelly, T. J., Eds., Plenum Press, New York, 1984, 333.
70. **Sloley, B. D. and Downer, R. G. H.,** Distribution of 5-hydroxytryptamine and indolealkylamine metabolites in the American cockroach, *Periplaneta americana* L., *Comp. Biochem. Physiol.,* 79C, 281, 1984.
71. **Scott, J. A., Johnson, T. L., and Knowles, C. O.,** Biogenic amine uptake by nerve cords from the American cockroach and the influence of amidines on amine uptake and release, *Comp. Biochem. Physiol.,* 82C, 43, 1985.
72. **Wurtman, R. J. and Axelrod, J.,** A sensitive and specific assay for the estimation of monoamine oxidase, *Biochem. Pharmacol.,* 12, 1439, 1963.
73. **Bodnaryk, R. P., Brunet, P. C. J., and Koeppe, J. K.,** On the metabolism of *N*-acetyldopamine in *Periplaneta americana, J. Insect Physiol.,* 20, 911, 1974.
74. **Sloley, B. D. and Downer, R. G. H.,** Dopamine, *N*-acetyldopamine and dopamine-3-*O*-sulphate in tissues of newly ecdysed and fully tanned adult cockroaches *(Periplaneta americana), Insect Biochem.,* 17, 591, 1987.
75. **Kennedy, M. B.,** Products of biogenic amine metabolism in the lobster: sulfate conjugates, *J. Neurochem.,* 30, 315, 1978.
76. **Goldman, J. E. and Schwartz, J. H.,** Metabolism of [^{3}H]-serotonin in the marine mollusc, *Aplysia californica, Brain Res.,* 136, 77, 1977.
77. **Maxwell, G. D., Moore, M. M., and Hildebrand, J. G.,** Metabolism of tyramine in the central nervous system of the moth, *Manduca sexta, Insect Biochem.,* 10, 657, 1980.
78. **Evans, P. D.,** Multiple receptor types for octopamine in the locust, *J. Physiol. (London),* 318, 99, 1981.
79. **Evans, P. D.,** Biogenic amine receptors and their mode of action in insects, in *Insect Neurochemistry and Neurophysiology,* Borkovec, A. B. and Gelman, D. B., Plenum Press, New York, 1984, 117.
80. **Uzzan, A. and Dudai, Y.,** Aminergic receptors in *Drosophila melanogaster:* responsiveness of adenylate cyclase to putative neurotransmitters, *J. Neurochem.,* 38, 1542, 1982.
81. **Harmar, A. J. and Horn, A. S.,** Octopamine-sensitive adenylate cyclase in cockroach brain: effects of agonists, antagonists and guanyl nucleotides, *Mol. Pharmacol.,* 13, 512, 1977.
82. **Nathanson, J. A. and Greengard, P.,** Octopamine-sensitive adenylate cyclase: evidence for a biological role of octopamine in nervous tissue, *Science,* 180, 308, 1973.
83. **Gole, J. W. D., Orr, G. L., and Downer, R. G. H.,** Interaction of formamidines with octopamine-sensitive adenylate cyclase receptor in the nerve cord of *Periplaneta americana* L., *Life Sci.,* 32, 2939, 1983.
84. **Downer, R. G. H., Gole, J. W. D., and Orr, G. L.,** Interaction of formamidines with octopamine-, dopamine- and 5-hydroxytryptamine-sensitive adenylate cyclase in the nerve cord of *Periplaneta americana, Pestic. Sci.,* 16, 472, 1985.
85. **Orr, G. L., Gole, J. W. D., and Downer, R. G. H.,** Characterisation of an octopamine-sensitive adenylate cyclase in haemocyte membrane fragments of the American cockroach, *Periplaneta americana* L., *Insect Biochem.,* 15, 695, 1985.
86. **Gole, J. W. D., Orr, G. L., and Downer, R. G. H.,** Octopamine-mediated elevation of cyclic AMP in haemocytes of the American cockroach, *Periplaneta americana, Can. J. Zool.,* 65, 1509, 1987.
87. **Gole, J. W. D., Orr, G. L., and Downer, R. G. H.,** Pharmacology of octopamine-, dopamine- and 5-hydroxytryptamine-sensitive adenylate cyclase on the corpus cardiacum of the American cockroach, *Periplaneta americana* L., *Arch. Insect Biochem. Physiol.,* 5, 119, 1987.
88. **Downer, R. G. H., Gole, J. W. D., and Orr, G. L.,** Pharmacological characterisation of octopamine-sensitive adenylate cyclase in *Periplaneta americana,* in *Neuropsychopharmacology of the Trace Amines,* Boulton, A. A., Bieck, P. R., Maitre, L., and Riederer, P., Eds., Humana Press, Clifton, NJ, 1985, 257.
89. **Minhas, N., Gole, J. W. D., Orr, G. L., and Downer, R. G. H.,** Pharmacology of [^{3}H]-mianserin binding in the nerve cord of the American cockroach, *Periplaneta americana, Arch. Insect Biochem. Physiol.,* 6, 191, 1987.
90. **Jahagirdar, A. P., Milton, G. W. A., Viswanatha, T., and Downer, R. G. H.,** Calcium involvement in mediating the action of octopamine and hypertrehalosemic peptides on insect haemocytes, *FEBS Lett.,* 219, 83, 1987.
91. **Orr, G. L., Gole, J. W. D., Notman, H. J., and Downer, R. G. H.,** Pharmacological characterisation of the dopamine-sensitive adenylate cyclase in cockroach brain: evidence for a distinct dopamine receptor, *Life Sci.,* 41, 2705, 1987.
92. **Notman, H. J. and Downer, R. G. H.,** Binding of [^{3}H]-pifluthixol, a dopamine antagonist, in the brain of the American cockroach, *Periplaneta americana, Insect Biochem.,* 17, 587, 1987.
93. **Ginsborg, B. L., House, C. R., and Silensky, E. M.,** Conductance changes associated with the secretory potential in the cockroach salivary gland, *J. Physiol. (London),* 236, 723, 1976.
94. **Bowser-Riley, F., House, C. R., and Smith, R. K.,** Competitive antagonism by phentolamine of responses to biogenic amines and the transmitter at a neuroglandular junction, *J. Physiol. (London),* 279, 473, 1978.

95. **House, C. R. and Ginsborg, B. L.,** Properties of dopamine receptors at a neuroglandular synapse, in *Neuropharmacology of Insects,* Ciba Found. Symp. 88, Pitman, London, 1982, 32.
96. **Berridge, M. J. and Heslop, J. P.,** Receptor mechanisms mediating the action of 5-hydroxytryptamine, in *Neuropharmacology of Insects,* Ciba Found. Symp. 88, Pitman, London, 1982, 260.
97. **Berridge, M. J. and Irvine, R. F.,** Inositol triphosphate, a novel second messenger in cellular signal transduction, *Nature (London),* 312, 315, 1984.
98. **Litosch, I., Fradin, M., Kasaian, M., Lee, H. S., and Fain, J. S.,** Regulation of adenylate cyclase and cyclic AMP phosphodiesterase by 5-hydroxytryptamine and calcium ions in blowfly *Calliphora erythrocephala* salivary gland homogenates, *Biochem. J.,* 204, 153, 1982.
99. **Crowder, L. A. and Shankland, D. C.,** Pharmacology of the Malpighian tubule muscle of the American cockroach, *Periplaneta americana, J. Insect Physiol.,* 18, 929, 1972.
100. **Cook, B. J. and Holman, G. M.,** Comparative pharmacological properties of muscle function in the foregut and the hindgut of the cockroach *Leucophaea maderae, Comp. Biochem. Physiol.,* 61C, 291, 1978.
101. **Brown, B. E.,** Pharmacologically active constituents of the cockroach corpus cardiacum: resolution and some characteristics, *Gen. Comp. Endocrinol.,* 5, 387, 1965.
102. **Colhoun, E. H.,** Synthesis of 5-hydroxytryptamine in the American cockroach, *Experientia,* 19, 9, 1963.
103. **Nathanson, J. A. and Greengard, P.,** Serotonin-sensitive adenylate cyclase in neural tissue and its similarity to the serotonin receptor: a possible site of action of lysergic acid diethylamide, *Proc. Natl. Acad. Sci. U.S.A.,* 71, 987, 1974.
104. **Collins, C. and Miller, T.,** Studies on the action of biogenic amines on the cockroach heart, *J. Exp. Biol.,* 67, 1, 1977.
105. **Nelson, D. L., Herbert, A., Enjalbert, A., Bockaert, A., and Hamon, A.,** Serotonin-sensitive adenylate cyclase and [^{3}H]-serotonin binding sites in the CNS of the rat, *Biochem. Pharmacol.,* 29, 2445, 1980.
106. **Hoyle, G.,** Neurotransmitters, neuromodulators and neurohormones, in *Neurobiology,* Gilles, R. and Balthazart, J., Eds., Springer-Verlag, Berlin, 1985, 264.
107. **Downer, R. G. H.,** Trehalose production in isolated fat body of the American cockroach, *Periplaneta americana, Comp. Biochem. Physiol.,* 62C, 31, 1979.
108. **Downer, R. G. H.,** Induction of hypertrehalosemia by excitation in *Periplaneta americana, J. Insect Physiol.,* 25, 59, 1979.
109. **Mercer, A. R. and Menzel, R.,** The effects of biogenic amines on conditioned and unconditioned responses to olfactory stimuli in the honeybee, *Apis mellifera, J. Comp. Physiol.,* 145, 363, 1982.
110. **Sombati, S. and Hoyle, G.,** Central nervous sensitization and dishabituation of reflex action in an insect by neuromodulator octopamine, *J. Neurobiol.,* 15, 455, 1984.
111. **Sombati, S. and Hoyle, G.,** Generation of specific behaviours in a locust by a release into neuropile of the natural neuromodulator octopamine, *J. Neurobiol.,* 15, 481, 1984.
112. **Downer, R. G. H.,** Projected research trends in invertebrate physiology and biochemistry, *Can. J. Zool.,* 65, 797, 1987.
113. **Selverston, A.,** Are central pattern generators understandable?, *Behav. Brain Sci.,* 3, 535, 1980.
114. **Nagy, F. and Dickinson, S. P.,** Control of a central pattern generator by an identified modulatory interneurone in crustacea, *J. Exp. Biol.,* 105, 33, 1983.
115. **Kandel, E. R.,** Cellular insights into behavior and learning, *Harvey Lect.,* 73, 19, 1979.
116. **Kravitz, E. A., Beltz, B., Glusman, S., Goy, M., Harris-Earwick, R., Johnston, M., Livingstone, M., Schwartz, T., and King Siwicki, K.,** The well-modulated lobster: the roles of serotonin, octopamine and proctolin in the lobster nervous system, in *Model Neural Networks and Behaviour,* Selverston, A. I., Ed., Plenum Press, New York, 1986, 339.
117. **Weiner, R. I. and Ganong, W. F.,** Role of brain monoamines and histamine in regulation of anterior pituitary secretion, *Physiol. Rev.,* 58, 905, 1978.
118. **Orchard, I.,** The role of biogenic amines in the regulation of peptidergic neurosecretory cells, in *Insect Neurochemistry and Neurophysiology,* Borkovec, A. B. and Kelly, T. J., Eds., Plenum Press, New York, 1984, 115.
119. **Gersch, M.,** Experimentelle Untersuchungen zum Freisetzungsmechanisus von Neurohormonen nach elektrischer Reizung der Corpora Cardiaca von *Periplaneta americana in vitro, J. Insect Physiol.,* 18, 2425, 1972.
120. **Gersch, M., Hentschel, E., and Ude, J.,** Aminergic Substanzen in lateralen Herznerven und im stomatogastrichen Nervensystem der Schabe *Blaberus craniifer* burm, *Zool. Jahrb. Physiol.,* 78, 1, 1974.
121. **Smalley, K. N.,** Median neurosecretory cells in the abdominal ganglia of the cockroach, *Periplaneta americana, J. Insect Physiol.,* 16, 241, 1970.
122. **Cook, B. J., Eraker, J., and Anderson, G. R.,** The effect of various biogenic amines on the activity of the foregut of the cockroach, *Blaberus giganteus, J. Insect Physiol.,* 15, 445, 1969.
123. **Cook, B. J. and Holman, G. M.,** The neural control of muscular activity in the hindgut of the cockroach, *Leucophaea maderae:* prospects for its chemical mediation, *Comp. Biochem. Physiol.,* 50C, 137, 1975.

124. **Miller, T. A.,** Neurosecretion and the control of visceral organs in insects, *Annu. Rev. Entomol.*, 20, 133, 1975.
125. **Brown, B. E.,** Proctolin: a peptide transmitter candidate in insects, *Life Sci.*, 17, 1241, 1975.
126. **Bland, K. P., House, C. R., Ginsborg, B. L., and Laszlo, I.,** Catecholamine transmitter for salivary secretion in the cockroach, *Nature (London)*, 244, 26, 1973.
127. **House, C. R. and Smith, R. K.,** On the receptors involved in the nervous control of salivary secretion by *Nauphoeta cinerea* Olivier, *J. Physiol. (London)*, 279, 457, 1978.
128. **Ginsborg, B. L. and House, C. R.,** The responses to nerve stimulation of the salivary gland of *Nauphoeta cinerea* Olivier, *J. Physiol. (London)*, 262, 477, 1976.
129. **Nathanson, J. A. and Hunnicutt, E. J.,** *N*-Dimethylchlordimeform: a potent partial agonist of octopamine-sensitive adenylate cyclase, *Mol. Pharmacol.*, 20, 68, 1981.
130. **Nathanson, J. A.,** Phenyliminoindizalidines. Characterization of a class of potent agonists of octopamine-sensitive adenylate cyclase and their use in understanding the pharmacology of octopamine receptors, *Mol. Pharmacol.*, 28, 254, 1985.
131. **Axelrod, J. and Saavedra, J. M.,** Octopamine, *Nature (London)*, 265, 501, 1977.

Chapter 18

TOXICOLOGY AND PHARMACOLOGY OF NEUROACTIVE AGENTS IN THE COCKROACH

Derek W. Gammon

TABLE OF CONTENTS

I. INTRODUCTION

The American cockroach, *Periplaneta americana*, has long provided a useful model for studies of both insect neurophysiology[1] and insect neurotoxicology.[2] As a result of studies conducted over the last 50 years, the neuropharmacology of the cockroach is better understood than that of any other insect[3-6] and, on occasion, phenomena with pharmacological relevance have been described in the cockroach which have only subsequently been found in vertebrates, e.g., the demonstration of cholinesterase in the nerve sheath (or blood-brain barrier) of the cockroach central nervous system (CNS).[7] In this chapter will be discussed the mode of action of most of the important classes of insecticide, not by chemical class, but instead from the standpoint of target site. Several sites which are apparently not insecticidal or else have not yet been exploited will also be discussed. Although some of the work to be described has not utilized the cockroach, its role as a model will nevertheless become clear. In Section IV the relative importance of the various targets in causing insect toxicity will be addressed.

II. TOXINS AND INSECTICIDES AFFECTING SODIUM AND CALCIUM CHANNELS

Both DDT and pyrethroids are considered to cause important effects on voltage-sensitive ion channels in the nerve membrane. Such effects were first shown using cockroach nerve preparations.[2,8-10]

A. *IN VITRO* EFFECTS OF PYRETHROIDS ON THE SODIUM CHANNEL

Pyrethroids have been shown to act on the sodium channels of a variety of nerve axon preparations, causing repetitive firing[9-15] as well as irreversible nerve blockage.[3,14,16] The cause of the repetitive firing appears to be an interference with the sodium channel such that it stays open longer than normal following membrane depolarization.[17-20] The maintenance of the prolonged open state could be due to slowing the closing of the inactivation (h) gate,[20] to a prolongation of the opening of the activation (m) gate,[19,21,22] or both. More recent work has been interpreted as showing that the effects of pyrethroids are principally on the activation process: batrachotoxin, which eliminates sodium channel inactivation, not only had no effect on the sodium tail current in squid giant axons, but it also had no effect on the increased and prolonged tail current caused by the pyrethroid tetramethrin.[23] It is probable that the voltage-sensitive sodium channel is affected similarly in insects,[24,25] and prolonged tail currents have been recorded in voltage-clamped cockroach giant axons in response to pyrethroids.[17]

Vijverberg and colleagues[19] and Narahashi[23,27] consider all active pyrethroids to affect sodium channels in the same way with regard to tail current prolongation, and they also believe that any apparent differences between pyrethroids are quantitative rather than qualitative. The time constant of the sodium tail current, measured in crayfish[18] and in frog[26] nerve fibers, correlates loosely with toxicity in the cockroach; i.e., the longer the time constant, the more toxic the pyrethroid. It is thus envisaged that depolarizing nerve blockage caused by prolonged sodium influx into nerve axons is the primary cause of pyrethroid toxicity to insects, regardless of the chemical structure of the pyrethroid. However, it has nonetheless been claimed that separate sites exist in the sodium channel for *cis* and *trans* pyrethroid isomers.[28]

Some pyrethroids, notably those which do not have an α-cyanophenoxybenzyl moiety, cause repetitive firing in nerve axons following stimulation. However, since in general the cyanophenoxybenzyl pyrethroids do not cause this effect and yet are more toxic to insects and other organisms, repetitive firing following stimulation in nerve axons is often regarded as being an unimportant feature of pyrethroid action.[19,29,30] It has been pointed out that

repetitive firing in a nerve as a result of sodium channel modification by a pyrethroid is a function of many variables, thus making it an unreliable measure of pyrethroid intoxication.[19] However, in Section II.C, some of the data will be presented which suggest that, at least in insects, there is evidence that the nerve axon repetitive firing caused by many pyrethroids correlates very well with the observed symptoms.

B. *IN VITRO* EFFECTS OF PYRETHROIDS ON THE CALCIUM CHANNEL

The first demonstration of an effect of an insecticide on calcium channels was the observation that very low concentrations of the pyrethroid permethrin caused bursting in segmental neurosecretory nerves of the stick insect *Carausius morosus,* in which the inward current carrier is thought to be calcium.[31] Subsequently, Narahashi[23] examined the effects of tetramethrin on two types of calcium channels in neuroblastoma cells (NIE 115). At 50 μM it reduced the fast, inactivating current by 70 to 80%, whereas it blocked the slower, noninactivating current by only 20 to 30%. The calcium channel antagonist verapamil (100 to 500 μM) specifically blocked the latter type of calcium channel.[23] The pharmacology of the NIE 115 cell line is a subject of great interest since it has also been used in ion flux studies to describe pyrethroid action.[32] Some pyrethroids, notably those with Type II properties (see Section II.C), stimulate veratridine-activated ^{22}Na influx, an effect also observed in mouse brain vesicles.[33] Veratridine is a plant alkaloid which maintains voltage-sensitive sodium channels in an open conformation, but which is also a potent blocker of calcium currents in NIE 115 cells.[34] Since it is possible that calcium channels are developmental predecessors of sodium channels,[35] particular care must be taken in the interpretation of results from immature tissue such as NIE 115 cells.

Some pyrethroids have recently been shown to be potent, noncompetitive inhibitors of the binding of the so-called peripheral benzodiazepine Ro5-4864 to rat brain membranes.[36,37] Since Ro5-4864 binding is inhibited competitively in a variety of tissues by dihydropyridines, although not by other types of calcium channel antagonist,[38] it is possible that the effect of pyrethroids on Ro5-4864 binding is a reflection of action at a type of calcium channel.

C. MODE OF ACTION OF PYRETHROIDS ON THE INSECT NERVOUS SYSTEM

Attempts to define target sites of insecticides in insects have helped to measure the relevance of *in vitro* studies and also to provide a focus for the best type of tissue to study. This has been the subject of several recent pyrethroid reviews.[39-41] The first such examination of pyrethroid target sites considered electrical activity in the nervous system of the American cockroach following topical application of a lethal dose of pyrethrin I.[42] Although this work was beset with technical difficulties (see Reference 43), it helped to move the emphasis of subsequent studies away from giant axons, which had been used for all of the earlier work.[3,8-10] In order to overcome some of the problems associated with physical stress, often found to be similar to the stressful effects of insecticides,[44-46] a preparation was developed which enabled extracellular responses to be recorded following stimulation in a free-walking cockroach.[12,13,47,48] Responses were stable for about 1 week in undosed insects, thus enabling the sequence of events from application of an insecticide to the eventual death of the cockroach to be monitored continuously in the same preparation and correlated with abnormal signs of poisoning.

The mode of action of the pyrethroid allethrin, the first synthetic analogue of pyrethrin I, was studied at 15 and 32°C using the topical LD_{95} dose at each temperature (2.1 and 20 $\mu g/g$, respectively) for the adult male American cockroach. It is perhaps worth noting that the negative temperature coefficient of toxicity of pyrethroids to insects had been shown, using the cockroach, not to be a result of metabolism or penetration differences with changing temperature.[46] By applying a pyrethroid dose which is toxic at a low (but not at a high)

temperature, it is possible to repeatedly cause signs of poisoning to disappear and reappear merely by raising and lowering the temperature. The temperatures of 15 and 32°C were chosen to be lower and higher than the critical one (26°C), above which cockroach giant axons fire repetitively *in vitro*.[9,10] The main findings were that the symptoms of poisoning at both temperatures correlated with excitatory effects on nerve, but the nerves affected and the symptoms differed with temperature. At 32°C, axons in the cercal sensory nerve as well as in the CNS discharged repetitively after stimulation, and high-frequency spike bursts were also recorded in the cercal motor nerves; these were probably direct effects of allethrin. At 15°C, however, only discharges in peripheral nerves were found to be primary effects; abnormal discharges in the CNS were concluded to be secondary. The cercal sensory nerve, affected by allethrin within a few minutes of dosing, was usually the last nerve to become blocked, suggesting that nerve blockage was probably a secondary effect of treatment.[12,13,48] Additional experiments were performed using tetrodotoxin (TTX), a potent blocker of sodium channels. Based on work showing that the sheath around the cockroach abdominal nerve cord presented a formidable barrier to penetration of this hydrophilic molecule,[49] a dose of TTX was found which, when injected into the cockroach, blocked peripheral nerves (which are less well protected[50]) while apparently leaving the CNS unaffected.[47] Over a period of about 1 week, 80% of the cockroaches made a complete recovery, regaining normal nerve responses. Incidentally, this selective and reversible nerve blockage suggests that TTX may have application as an anesthetic during surgery in mammals.

In the cockroach, an LD_{95} dose of allethrin applied at the time of TTX prostration (~4 h) became nontoxic.[48] By studying insects with implanted electrodes it was found that TTX reversibly blocked those (peripheral) nerves which normally fired repetitively after allethrin treatment, thereby demonstrating the paramount importance of repetitive firing in peripheral nerves in causing symptoms and mortality in the allethrin-dosed cockroach. It also showed that nerve blockage by allethrin, previously considered to be an important primary effect responsible for the negative temperature coefficient of toxicity of pyrethroids,[3,16,51-53] was a secondary effect of allethrin intoxication.[48]

A range of pyrethroids became available in the mid-1970s,[54] and a structure-activity study was made by Clements and May.[55] Four different types of effect were reported in a variety of physiological assays in the locust. Although some of the observations were a little puzzling (e.g., permethrin and cypermethrin seemed to cause the same effects), this work was important since it was the first indication that not all pyrethroids necessarily had the same mode of action as allethrin or pyrethrin I. A subsequent study concentrated on just two pyrethroids, permethrin and cypermethrin, which differ only in the presence of a cyano group in the α-position of the phenoxybenzyl moiety. This study was an attempt to determine whether there was a neurophysiological basis for resistance to permethrin, but not cypermethrin, in a strain of the Egyptian cotton leafworm, *Spodoptera littoralis*.[56,57] As a starting point, the temperature-dependence of permethrin resistance was established; this resistance gradually disappeared as the temperature was reduced. Since the cockroach CNS exposed to allethrin fires repetitively only at relatively high temperatures following stimulation, it was reasoned that a similar phenomenon in *S. littoralis* dosed with permethrin must implicate the CNS as the predominant site of resistance. Therefore, evoked responses were recorded from the abdominal nerve cord of *S. littoralis* exposed to permethrin *in vitro*. At a permethrin concentration of 100 n*M*, repetitive firing followed stimulation consistently until nerve blockage occurred, but only at temperatures above 19°C; below this temperature only gradual nerve blockage was observed. Nerves from resistant insects were significantly less susceptible to permethrin-induced repetitive firing. Moreover, when cypermethrin was applied to the same nerve preparation, no repetitive firing was detected — only nerve blockage. Increasing the cypermethrin concentration to 1 μ*M* and varying the temperature between 15 and 36°C failed to elicit repetitive firing or even any abnormal bursting. It was therefore concluded

that the basis for the lack of resistance to cypermethrin was qualitative rather than quantitative, i.e., that cypermethrin had a different mode of action from permethrin. It could be relevant that this strain of *S. littoralis* was cross-resistant to DDT, which has a mode of action similar to permethrin and allethrin, but not to dieldrin, which has a mode of action more like that of cypermethrin (see Section III.B.4).

A large number of pyrethroids were then studied to determine the structural basis for the different modes of action.[15] The assay system chosen was the cockroach cercal sensory nerve, since this had been shown in the allethrin-treated insect to fire repetitively across a broad temperature range. An *in vitro* assay used a suction electrode to record an electrically evoked potential, and a particular pyrethroid concentration was then applied. The same nerve was monitored *in vivo* by the insertion of a silver wire electrode into the cut end of a cercus. Various pyrethroids were assayed in this way[15,58-60] and, with the exception of fenpropathrin, no pyrethroid ester with an α-cyanophenoxybenzyl moiety caused repetitive firing in this nerve, either *in vitro* or *in vivo*, whereas their noncyano analogues did. At a range of concentrations of 1R *cis* α-*S*-cypermethrin from 0.1 p*M* to 1 μ*M* in 24 experiments using the *in vitro* assay, no repetitive firing following stimulation or abnormal discharges was detected.[37] In the *in vivo* assay, at doses exceeding the LD_{50}, pyrethroids with an α-cyanophenoxybenzyl group usually showed decreased cercal sensory nerve activity instead of repetitive firing. However, bursts of efferent potentials in the cercal motor nerve were often observed, even after sensory activity had become blocked.[15,61] The differences in nerve effects were correlated with differences in the symptoms of intoxication between the two classes of pyrethroids, and so the terms "Type I Syndrome" and "Type II Syndrome" were coined to describe the effects, the α-cyanophenoxybenzyl pyrethroids causing the latter syndrome.

Similar classifications of pyrethroids into two classes have been made based on symptoms of poisoning in the rat[62] and on the ability to cause repetitive firing in crayfish giant axons.[18] Studies on the cockroach CNS using extracellular recording techniques also suggested two types of pyrethroid action: allethrin consistently caused repetitive firing, whereas cypermethrin did not.[63] Another interesting feature of this work was the evaluation of the resistance profile of a strain of German cockroach, *Blattella germanica:* these insects were resistant to DDT and to all (three) Type I pyrethroids studied, but they were not resistant to dieldrin or the Type II pyrethroids deltamethrin or cypermethrin. However, they were resistant to fenvalerate, which causes Type II effects in other systems studied. Using a cockroach leg mechanoreceptor preparation, it was also shown that bioallethrin caused repetitive firing following stimulation, whereas deltamethrin, which caused Type II effects in an earlier study,[15] blocked the nerve without causing repetitive firing.[64] Kadethrin possessed features of both types of pyrethroid action in these studies.[15,64]

Another assay system in insects which has proven to be extremely sensitive to pyrethroids is the one developed to study the neuromuscular junction.[29,30,65] Recordings from the body-wall muscle fibers of larvae of Diptera and Lepidoptera exposed to pyrethroids at low concentrations reveal an increase in both frequency and amplitude of miniature excitatory postsynaptic potentials (mEPSPs). The effect was abolished by treatment with TTX, which indicates that the pyrethroid effect is presynaptic, but it does not rule out the possibility that the effect is at nerve terminal calcium channels rather than at sodium channels, as suggested by the authors. It is also claimed that Type I pyrethroids, which are only weakly active in this assay, nevertheless owe their toxicity to their ability to cause nerve terminal depolarization leading to transmitter release with a concomitant increase in mEPSPs.[29,30] However, the correlation with insecticidal toxicity is perhaps a little weak. It would be interesting to see the effects of pyrethroids on the preparation in the cockroach nymph which allows simultaneous, intracellular recording of both a presynaptic calcium spike and the postsynaptic excitatory potential.[66]

D. *IN VITRO* EFFECTS OF DDT ON THE SODIUM CHANNEL

DDT is often considered to have a mechanism of action on the nerve axon sodium channel similar to that of the pyrethroids,[18,67,68] causing prolonged sodium currents to flow into the cell. However, whatever the nerve preparation, DDT treatment almost invariably results in repetitive firing following stimulation, even in those where pyrethroids do not. For example, in housefly labellar sense organs, DDT, but not pyrethroids, causes multiple spikes; moreover, DDT-pyrethroid hybrid molecules structurally resembling the former the most also cause multiple firing, unlike those which most resemble pyrethroids.[69] DDT has virtually no tendency to cause nerve blockage, either *in vitro*[67] or *in vivo*.[70] In this respect, therefore, DDT action is much more similar to the Type I effects of pyrethroids than it is to Type II. The effects of DDT and some analogues on sodium tail currents in crayfish giant axons are very similar to those of pyrethroids causing the Type I syndrome.[18] Likewise, the DDT-induced repetitive firing in a cockroach cercal sensory nerve is very similar to that caused by Type I pyrethroids.[15] In more intact insect preparations, however, some differences are evident as well as some similarities.[70]

E. EFFECTS OF DDT ON THE INSECT NERVOUS SYSTEM

As with pyrethroids, DDT exhibits a negative temperature coefficient of toxicity against most insects. Therefore, many experiments to determine the mode of action of DDT in insects have made use of this feature; i.e., any effect caused by DDT which becomes more pronounced as temperature is lowered could be relevant in causing mortality. Conflicting theories had been proposed to understand the important target sites within insects,[3] and so the electrode-implanted, free-walking cockroach preparation was subjected to LD_{95} doses at three temperatures: 16, 25, and 32°C.[70] Primary, excitatory effects were recorded on cercal sensory axons and on neurons in the CNS, whether stimulated directly or presynaptically, through cercal sensory nerve stimulation. Repetitive firing in these nerves following stimulation commenced within 90 min of dosing, before the onset of overt symptoms, becoming more intense as "DDT tremors" progressed and reaching a peak during paralysis. Bursts of efferent potentials were recorded in the cerci as tremoring developed, and they continued for several days after dosing, until long after spontaneous movements by the cockroach had ceased (paralysis). The duration of the discharges in the CNS was quantified; it was found to increase as temperature was lowered, despite the fact that the dose applied (the LD_{95}) was also less as temperature was reduced.

Thus, the temperature coefficient of the central effects of DDT was opposite that shown by the pyrethroids allethrin and permethrin, where repetitive firing in the CNS disappeared rather than lengthened as the temperature was reduced.[9,10,13,57] It is therefore difficult to regard DDT as an example of a model for a peripherally acting insecticide, devoid of central effects, as suggested by Miller and Kennedy.[71] The repetitive discharges in cercal sensory axons were not quantified in the experiments using the intact cockroach, but it has been suggested that DDT-induced cercal sensory discharges in this insect increase in frequency as the temperature is raised,[72,73] thus making the central effects of even greater potential importance in causing insect mortality.

III. TOXINS AND INSECTICIDES AFFECTING NEUROTRANSMITTER SYSTEMS

A. ACETYLCHOLINE

Acetylcholine (ACh) is the best identified neurotransmitter in insects (see Reference 74). Furthermore, the enzyme responsible for its degradation, acetylcholine esterase (AChE), represents the primary target site for one of the earliest classes of neuroactive pesticides, the organophosphates. Corbett et al.[75] have ably reviewed the literature describing a diverse

range of structures in this class of insecticide, which still represents one of the most important. More recently, carbamates have also become an important group of insecticides which inhibit AChE,[75] and it seems likely that other classes of chemicals will eventually be discovered which are able to exert insecticidal effects through inhibition of AChE. The enzyme which makes ACh, choline acetyl transferase, has also been examined as a potential insecticide target. Recent efforts to make insecticides which act specifically at the ACh receptor and which are improvements over nicotine will also be discussed.

1. Acetylcholine Esterase Inhibition

The two main chemical classes of insecticides which are known to act by inhibiting AChE are the organophosphates and the carbamates.[75,76] These generic classes cover a variety of subtypes (e.g., phosphorothioates, phosphonothioates, and methyl carbamates). The long history of successful use of these compounds and the relative ease of assay of AChE combine to ensure that new chemical classes of insecticide with this basic mode of action are constantly being sought throughout the insecticide/nematicide industry.

The insect CNS is the usual source of synapses that use ACh. There is evidence for both nicotinic and muscarinic receptors for ACh in insects (see Sections III.A.3 and 6), but unlike mammals, where the skeletal neuromuscular junction has nicotinic ACh receptors, the skeletal neuromuscular junction in insects probably uses glutamate as the excitatory transmitter.[77,78] AChE inhibition leads to the accumulation of ACh in the synaptic cleft, resulting in repeated activation of receptors, abnormal nerve discharges, and hyperactivity leading to mortality. Although the housefly head is commonly used as a convenient source of AChE, the best-defined insect preparation for studying the physiological and biochemical effects of AChE inhibition is the cockroach cercal nerve-giant fiber pathway. Cercal sensory axons synapse with giant interneurons in the terminal abdominal ganglion, forming the first stage of the escape reflex of the cockroach.[1] The result of AChE inhibition in this system is that a single presynaptic spike can lead to a prolonged discharge in the giant interneurons.[3]

One of the features of organophosphates and carbamates which greatly contributes to their toxicity is that AChE inhibition is irreversible. This has led to the application of novel methods of kinetic analysis to the study of this inhibition,[79] and these methods in turn have been used with other enzymes subjected to irreversible, covalent inhibitors (see Section III.B.1). In general, the organophosphates are "more irreversible" in their action than the carbamates, and one of the main problems with both types is their high mammalian toxicity. This has been overcome at least in part for the organophosphates by using as the applied insecticide the phosphorothioate rather than the phosphonate or phosphate, which are the active inhibitors *in vitro*. This generally improves the selective toxicity considerably, as has been shown with malathion (see Reference 75); it is oxygenated to malaoxon (a potent AChE inhibitor) in insects (P = S to P = O), whereas it is preferentially hydrolyzed to inactive materials in mammals.

2. Choline Acetyl Transferase Inhibition

Since the inhibition of the breakdown of ACh results in hyperactivity and mortality in insects, Baillie et al.[80] made a series of competitive inhibitors of the enzyme which synthesizes ACh, choline acetyl transferase, as potential insecticides. The most potent of these isothiocyanates, 2-isothiocyanotoethyltrimethyl ammonium iodide, had a K_i of 60 n*M* against choline acetyl transferase *in vitro*. However, when it was applied to a desheathed sixth abdominal ganglion preparation of the cockroach, no presynaptic effects (which would be expected from inhibition of this enzyme) were observed. Instead, blockage of the excitatory postsynaptic potentials (EPSPs) was accompanied by a depolarization and fall in resistance of the postsynaptic membrane, supporting a postsynaptic mechanism of action, perhaps at the ACh receptor. Given the structural similarities between the isothiocyanate molecule and ACh, it is perhaps not too surprising that the former may act as an apparently partial ACh

agonist.[81] The above studies[80,81] represent a good example of a concerted attempt to design insecticides by a rational approach using biochemical and physiological methods, although it was concluded, since none of the compounds were toxic to insects, that choline acetyltransferase does not represent a viable target site for insecticides.

3. Nicotinic Acetylcholine Receptor (*n*AChR)

Nicotine is one of the oldest natural insecticides, but its limited insect toxicity and high mammalian toxicity have resulted in it being superseded by synthetic insecticides in most applications. However, for a variety of reasons, the nicotinic acetylcholine receptor (*n*AChR) has recently become of increasing interest to insect toxicologists seeking novel classes of insecticide. Most of our current knowledge of the insect *n*AChR has been summarized in recent reviews.[82-86] The receptor is located principally in the insect CNS, and most of the physiological experiments performed to characterize it have used nerve preparations from the cockroach.[87-89] The definition of specific binding sites for nicotinic agonists and antagonists has been achieved using receptor binding experiments with radioactive antagonists such as α-bungarotoxin (BGTX),[90,91] which was first shown to be a specific, potent *n*AChR antagonist through the use of a cockroach neurophysiological assay.[92] The ion channel which is opened (activated) by agonists acting at the *n*AChR has also been investigated as a site of action for insecticides through the use of a specific probe, tritiated histrionicotoxin ($[^3H]H_{12}$-HTX).[93]

4. Pharmacology of the Insect Nicotinic Acetylcholine Receptor

One of the most significant breakthroughs in the search for agents acting at the *n*AChR was the finding that BGTX was a potent antagonist at the cockroach receptor.[92] Among the compounds which have been shown to inhibit the binding of labeled BGTX, apart from nicotine itself, are nereistoxin (originally isolated from a marine annelid; the prototype for the insecticides cartap and thiocyclam),[94,95] toxic components of the seeds of *Delphinium* plants,[96] and certain carbamate insecticides which are known inhibitors of AChE.[97] Although BGTX is an antagonist at the *n*AChR, agonists such as nicotine will also inhibit its binding; therefore, physiological experiments are often used to distinguish between these two types of action.

One of the major differences between the insect and vertebrate *n*AChR is the lack of potency of tubocurarine on the former. Using the desheathed fast depressor motoneuron (D_f) in the cockroach metathoracic ganglion, David and Sattelle[89] showed that nicotine was a more potent agonist than ACh (when applied with the AChE inhibitor neostigmine). This in turn was better than carbamylcholine at causing postsynaptic depolarization. However, as an antagonist of ACh-induced depolarizations, BGTX was several orders better than D-tubocurarine, which was similar in potency to the muscarinic antagonists quinuclidinyl benzilate (QNB) and atropine.

Furthermore, when applied to the voltage-clamped D_f cell, it was found that whereas the block caused by BGTX was independent of voltage, as would be expected of an antagonist acting specifically at the ACh receptor component of the receptor/channel complex, the blockage caused by D-tubocurarine and atropine was strongly voltage-dependent, a sign that these compounds might be affecting the open, ACh-activated ion channel directly.[89] Similarly, amantidine, the anti-Parkinsonian and antiviral drug, caused a blockage of the ACh-induced depolarizations in cockroach D_f cell bodies which was also strongly voltage-dependent.[98]

Nereistoxin appeared to have several sites of action in the *n*AChR/ion channel complex: at low concentrations it caused a partial blockage of synaptic transmission and a voltage-dependent blockage of ACh-induced currents, suggesting an open-channel antagonist action. At higher concentrations it also caused postsynaptic depolarization in the same concentration

range as it inhibited [^{125}I]BGTX binding to cockroach nerve cord extracts.[94] Thus, there is evidence that nereistoxin acts at both the *n*AChR and the ion channel site. Since it causes rapid prostration and paralysis when injected into the cockroach, without excitation, correlated with blockage of *n*ACh transmission across the sixth abdominal ganglion,[99] it seems likely that the antagonist effects reported above are more important *in vivo* than the depolarizing effect.

Piperidine alkaloids, which make up the toxic constituents of fire ant venom, have also been shown to bind to components of the *n*AChR-ion channel complex in the cockroach.[100] *Cis* 2-methyl-6-undecyl piperidine did not inhibit BGTX binding, yet it blocked ACh-induced currents in the D_f cell body independently of voltage. This suggests an action at a site on the closed *n*AChR-ion channel site which is distinct from the BGTX site.

The drug phencyclidine and its thienylpyrolidine analogue showed a mixture of effects, not only on the *n*AChR (as judged from the inhibition of BGTX binding), but also on voltage-sensitive Na and K channels.[101] It has recently been shown that the alkaloid methyllaconitine (from *Delphinium* plants) is a potent inhibitor of BGTX binding in housefly head membranes; it also causes blockage of synaptic transmission across the sixth abdominal ganglion of the cockroach[96] and in a rat phrenic nerve/diaphragm preparation. It is possible that this information will provide useful data for molecular modeling studies of the *n*AChR, but the chemical complexity and the high mammalian toxicity probably preclude such alkaloids from being used for making analogues as commercial insecticides. Another plant alkaloid, L-lobeline, has been shown to act as an inhibitor of BGTX binding in cockroach head homogenates with a potency ~140 times higher than nicotine, although it is less potent than nicotine in housefly heads.[102]

The nitromethylenes represent a class of insecticide which has high activity against insects, but which has never been commercialized due to its low photostability. Experiments on the cockroach nervous system have suggested that the nitromethylenes act as ACh agonists. They caused bursts of spikes to arise postsynaptically in neurons of the cockroach abdominal nerve cord, and these discharges were blocked by D-tubocurarine, although (puzzlingly) not by BGTX.[103,104] The possibility of an open-channel site of D-tubocurarine blockage in the cockroach, mentioned above, suggests that the nitromethylenes may be acting at such a site. However, it is also possible that the receptors excited by the nitromethylenes represent a class of nicotinic site which is either inaccessible or insensitive to BGTX. It has also been reported that D-tubocurarine is a noncompetitive inhibitor of ^{3}H-glutamate binding in rat brain, perhaps acting, therefore, at the glutamate-activated ion channel.[85] As with some of the other chemical classes which have one or more site of action at the *n*AChR-ion channel complex, receptor binding experiments using a radioligand specific for the ion channel site, such as [^{3}H]H_{12}-HTX,[93] may help to explain some of the reported observations.

5. *n*AChR-Activated Ion Channel

This site has been thoroughly characterized using radioligand binding experiments with tritiated histrionicotoxin (HTX), combined with ion flux studies and electrophysiological measurements.[105] Its binding is enhanced in a dose-dependent manner in the presence of an *n*ACh agonist such as carbamylcholine and also with presynaptic stimulation; when considered together, these facts suggest that HTX preferentially binds to an open configuration of the *n*ACh-activated ion channel.[105] In electrophysiological studies in the cockroach, the inhibitory effect of HTX on ACh-induced currents has been shown to be strongly voltage-dependent in the D_f cell body.[106]

Aside from the potential value of HTX as a probe for a novel insecticide target, alluded to above, it has been shown that existing insecticides such as pyrethroids, which have been shown to act on voltage-sensitive Na channels (see Section II.A), also act on the HTX receptor.[107,108] The pyrethroid allethrin had no effect on the binding of [^{3}H]ACh to *Torpedo*

*n*AChR, but it reduced, in a dose-dependent manner, the stimulation by ACh of the specific binding of [^{3}H]H_{12}-HTX.[107]

A range of pyrethroids was divided into two types (very similar to those categories in Section II.C) according to their ability to inhibit the binding of [^{3}H]H_{12}-HTX. Type I pyrethroids achieved this completely within only 30 s of the start of incubation, whereas Type II pyrethroids showed a large increase in the amount of binding inhibition if they were incubated for 2 h. Furthermore, Type I pyrethroids caused a greater amount of inhibition of [^{3}H]H_{12}-HTX binding than did Type II. The binding inhibition was stereospecific (for fluvalinate isomers) and showed a negative temperature coefficient, correlating with insecticidal activity. The inhibition of [^{3}H]H_{12}-HTX binding was correlated with an inhibition of ^{45}Ca flux through the *n*ACh-activated ion channel, but, interestingly, there was no such inhibition of ^{22}Na flux.[108] Calcium ions inhibited [^{3}H]H_{12}-HTX binding competitively, whereas pyrethroids did so noncompetitively. Because Ca binding is thought to potentiate *n*AChR desensitization, it is therefore conceivable that pyrethroids block the inactivation or desensitization of the *n*ACh-activated ion channel,[84] thus leading to excitation at such synapses.

6. Muscarinic Acetylcholine Receptor

A pharmacological profile of the receptor for [^{3}H]QNB binding in the cockroach nerve cord was reported by Lummis and Sattelle.[109] The K_d was 8 n*M*, similar to that reported for [^{3}H]QNB binding to the mammalian CNS muscarinic acetylcholine receptor (*m*AChR).[110] Although the potency of [^{3}H]QNB binding inhibition by a series of ten agonists and antagonists was generally lower than that shown by these compounds against the mammalian receptor, the rank order was very similar.[110] The cockroach CNS, unlike that of vertebrates, contained seven times more *n*AChR than *m*AChR, as judged from a comparison of binding sites for BGTX with those for QNB. A very similar profile of the insect *m*AChR was obtained by Breer[111,112] using CNS tissue from the locust.

In mammals there are at least two classes of *m*ACh receptors, and the distribution and physiology are relatively well understood. However, in insect nerve only one class of binding site for [^{3}H]QNB has generally been found.[83] The intracellular transduction process resulting from *m*AChR activation in insects is unclear. Whereas in mammals there is evidence for several systems in different tissues (e.g., increased cyclic GMP, decreased cyclic AMP, changes in ion permeability, and breakdown of polyphosphoinositides), in insects there is evidence thus far only for the latter.[113] The functional role for the insect *m*AChR has not yet been established physiologically. However, Breer and Knipper[114] obtained biochemical evidence that the *m*AChR might be involved in the autoregulation of ACh release. Using locust synaptosomes, they found that a muscarinic agonist inhibited and an antagonist stimulated ACh release. This suggests that the *m*AChR may be located presynaptically, in contrast to the predominantly postsynaptic site for the *n*AChR.[115]

From a toxicological perspective, the injection of 50 μg/g of the muscarinic agonist pilocarpine into the cockroach caused tremoring lasting for about 60 min, followed by recovery, and QNB resulted in uncoordination (but not mortality) at doses up to 250 μg/g.[116] Pilocarpine was not very effective as an inhibitor of the binding of [^{3}H]QNB in the cockroach nerve cord.[109] It remains to be seen whether any commercially available insecticides inhibit the binding of [^{3}H]QNB; until then, the question of the toxicological relevance of the *m*AChR in insects is largely unanswered. Perhaps the development of a physiological assay in insects to complement the pharmacological approach will lead to a clearer answer on this point.

B. γ-AMINOBUTYRIC ACID (GABA)

γ-Aminobutyric acid (GABA) is the most abundant putative inhibitory neurotransmitter in the vertebrate brain; however, it is found in insects and other arthropods not only in the

brain and CNS, but also at the skeletal neuromuscular junction.[117] Nerve terminals of "slow" motoneurons innervating such muscle often have GABA-ergic inhibitory nerve endings, whereas no such inhibition has been found to be associated with "fast" motoneuron terminals. Many of the experiments characterizing the fundamental physiology of both neuromuscular transmission and GABA inhibition have used the opener muscle of the crayfish claw,[118] and it is thought that other arthropods, including insects, have similar mechanisms of GABA-ergic inhibition. This can be presynaptic, in which case the membrane potential of the excitatory neuron is altered so as to reduce the number of quanta of excitatory transmitter molecules released, or inhibition can be postsynaptic, in which case the conductance of the muscle fiber is increased (or input resistance reduced) through the opening of chloride channels.[118,119]

The enzyme responsible for the degradation of GABA is GABA transaminase (GABA-T), and the main enzyme which synthesizes GABA is glutamic acid decarboxylase (GAD). A study of the likely toxicological importance of inhibitors of these enzymes to insects will be described. Some agents which are insecticides and which act at the GABA receptor complex will also be discussed.

1. GABA Transaminase

This enzyme is located predominantly within mitochondria (i.e., intracellularly), and it is therefore likely that synaptically released GABA must be pumped into a presynaptic or glial cell before degradation.[120] Specific inhibitors of uptake into glial or neuronal sites are described in the vertebrate literature, but these sites have not been considered as potential targets for insecticides. However, an attempt has been made to evaluate the likely importance of GABA-T inhibition as a target for insecticides.[121,122]

GABA-T has an obligatory requirement for the cofactor pyridoxal phosphate (PLP), which forms a Schiff's base with a free amine group of the substrate or a potential inhibitor. Therefore, two types of inhibitor were considered: first, gabaculline (a cyclized analogue of GABA), which is a specific "suicide" substrate, since the enzyme induces the gabaculline ring to aromatize at the active site, forming an irreversible bond with the enzyme,[123] and second, a series of aminooxy acids, along with derivatives of both the acid and amine portion. Aminooxy acetate is a less specific inhibitor, generally affecting any enzyme requiring PLP as a cofactor; it also forms a covalent bond with PLP at the enzyme active site. Although irreversible, this type of inhibition is not considered to be a "suicide" mechanism.

Inhibitory potencies were assessed *in vitro* using a bacterial source of GABA-T. Activity was measured *in vivo* by injecting the compounds into the cockroach and recording both the symptomology and the toxicity. The findings supported the conclusion that GABA-T is not a viable target site for insecticides. First, effective inhibitors were not very toxic to the cockroach, even at doses which drastically reduced the metabolism of [^{14}C]GABA. Second, gabaculline was shown to abolish the symptoms caused by the most toxic of the aminooxy acid derivatives, which (unlike gabaculline) were demonstrated to be nonspecific, also inhibiting GAD.

2. Glutamic Acid Decarboxylase

The effectiveness of several aminooxy compounds as inhibitors of GAD was measured using cockroach flight muscle, both *in vitro* and *in vivo*.[122] It was concluded that insect GAD was probably not a good insecticide target site either. For example, there was a zero-order relationship between dose and GAD inhibition *in vivo* for the most toxic aminooxy acid derivative. Doses of the inhibitor causing either mild symptoms or paralysis resulted in identical degrees of inhibition (70 to 80%) when the flight muscle was dissected out and assayed for GAD activity 2 h after treatment. However, the lack of a specific inhibitor of GAD as a standard, analogous to gabaculline for GABA-T, prevented a definitive judgment of the toxicological importance of GAD.

3. The GABA Receptor

The inhibition of the binding of [^{3}H]GABA to the cockroach CNS by a range of chemicals has been described.[6,124] The GABA receptor is sensitive to muscimol, only moderately sensitive to isoguvacine, and insensitive to bicuculline and baclofen. Its properties are unlike those of $GABA_A$ or $GABA_B$ receptors in vertebrates, although it is perhaps more like the former than the latter. No synthetic insecticides are known which act directly at the GABA receptor, although the avermectins, a class of macrolide-like antibiotics which are extremely toxic to some insects, mites, and nematodes, are thought to cause potentiation at GABA synapses through an uncertain mechanism.[125,126] The $GABA_A$ agonist muscimol is fairly toxic to the cockroach (with an $LD_{50} \sim 40$ μg/g by injection),[122] suggesting that it may serve as a potential model, but the $GABA_A$ antagonist bicuculline, which is extremely toxic to mammals (with an LD_{50} of 0.45 μg/g in the mouse), is only weakly toxic to the cockroach, having an LD_{50} of about 150 μg/g by injection.[122] However, it has been suggested that it may act at the GABA-activated ion channel at the crayfish neuromuscular junction rather than at the $GABA_A$ receptor.[127] A similar site of action in insects thus could indicate a possible difference between arthropods and vertebrates.

The physiological effects of $GABA_A$ receptor activation involve the opening of chloride channels. This has been demonstrated at both the neuromuscular junction of the crayfish claw opener muscle[117] and the cockroach fast coxal depressor motoneuron cell body, D_f.[128] There is also evidence for accompanying changes in K permeability in cockroach neurons.[129] The GABA-activated Cl channel is blocked by picrotoxinin (PTX) in D_f[128] and at the crayfish neuromuscular junction,[119] as found for the $GABA_A$-activated channel in vertebrates. This ion channel has been found to be a target site for two major classes of insecticides, pyrethroids causing the Type II syndrome and the cyclodienes.

4. GABA-Activated Ion Channel

Lawrence and Casida[130] found that a range of cyclodiene insecticides, including dieldrin and aldrin, were potent, competitive inhibitors of the binding of [^{35}S]*t*-butylbicyclophosphorothionate ([^{35}S]TBPS, a convulsant acting at the PTX receptor) to rat brain membranes. The relative potency correlated very well with toxicity. Almost simultaneously, it was found that cyclodienes inhibited the binding of [^{3}H]dihydro-PTX to cockroach nerve homogenates.[131] Abalis et al.[132] showed that cyclodienes and lindane displaced [^{35}S]TBPS from housefly thoraces and from electroplax tissue. The presence of lindane-sensitive binding sites for [^{35}S]TBPS in electroplax tissue suggests that TBPS (and lindane) may also bind to voltage-sensitive Cl channels, since this tissue is devoid of GABA receptors.[133]

Pyrethroids causing the Type II syndrome have been found to inhibit the binding of [^{35}S]TBPS to rat brain membranes.[134] The inhibitory effect was noncompetitive and not especially potent when compared with effects reported on certain insect electrophysiological assays,[15,29,30,58,59,65] but a subsequent study[135] showed that the binding of [^{35}S]TBPS was enhanced in the presence of 1 to 10 μM GABA. Furthermore, the binding inhibition by pyrethroids became both more potent and more complete. It seems probable that GABA induces a conformational change in the receptor complex, thereby exposing more pyrethroid binding sites. This could be analogous to the opening of the Cl channel by GABA, as this (open-channel) site has been shown to be a target site for pyrethroids specifically causing the Type II syndrome in electrophysiological experiments using the crayfish neuromuscular junction.[37,136] However, it should be noted that some authors (e.g., Reference 84) find GABA to be a potent inhibitor of [^{35}S]TBPS binding in rat brain membranes.

5. Benzodiazepine Receptors

The $GABA_A$ receptor complex in vertebrates contains a benzodiazepine (BZ) receptor in addition to a GABA recognition site and a Cl channel.[137,138] Agonists at the BZ receptor

amplify the effects of GABA stereospecifically. Although insects and other invertebrates were originally thought not to have BZ receptors,[139] toxicological studies in the cockroach[61] and receptor binding experiments using housefly thoracic muscle[140] showed that insects did indeed have such receptors. Diazepam delayed the onset of signs of poisoning caused by deltamethrin and other pyrethroids bringing on the Type II syndrome, but not those causing the Type I syndrome, in both the cockroach and the mouse.[61] Diazepam also delayed the appearance of prolonged discharges in the cockroach cercal motor nerves caused by deltamethrin while not affecting the prolonged cercal sensory nerve discharges caused by permethrin, both effects recorded *in vivo* using implanted electrodes. The antagonistic effects of PTX and deltamethrin on the inhibitory action of GABA at the crayfish neuromuscular junction were also reduced by BZs.[136]

Insect BZ receptors have now been demonstrated using receptor binding techniques in several insect preparations.[124,140-143] A common finding has been that a potent displacer of the binding of [^{3}H]flunitrazepam in insect nerve and muscle was Ro5-4864 (4-Cl phenyl diazepam). This BZ is very poor at inhibiting the binding of [^{3}H]flunitrazepam binding from rat brain membranes.[144] However, there are sites in the brain (and elsewhere) in mammals which bind [^{3}H]Ro5-4864 with high affinity, and this binding is inhibited by flunitrazepam and certain other anxiolytic BZs. The receptor for [^{3}H]Ro5-4864 is often referred to as a peripheral site because (1) it is not localized specifically in the brain and spinal cord and (2) it is thought to be associated with glial tissue in the mammalian CNS.[145]

Another interesting observation reported for the insect BZ receptor is that clonazepam, which is a potent inhibitor of [^{3}H]flunitrazepam binding in rat brain, but which is very poor at displacing [^{3}H]Ro5-4864 binding in mammals, is also very poor at displacing [^{3}H]flunitrazepam from insect tissue.[140,142] It therefore seems possible that the Ro5-4864 receptor is an evolutionary predecessor of the central BZ receptor in mammals which in insects fulfills the same function. The proof of this theory must await the development of an insect receptor binding assay for [^{3}H]Ro5-4864. Meanwhile, the rat brain Ro5-4864 receptor has provided some interesting data with regard to the mode of action of insecticides, especially pyrethroids.

Lawrence et al.[36] found that pyrethroids causing the Type II syndrome displaced [^{3}H]Ro5-4864 stereospecifically and noncompetitively from rat brain membranes, but with a potency 50 times that shown for [^{35}S]TBPS binding inhibition. Furthermore, the early symptoms of dosing in the rat were identical to those of Type II pyrethroids, but not to those of Type I pyrethroids. Gammon and Sander[37] showed a similar effect using dialyzed rat brain membranes. Devaud et al.[146] studied the potency of pyrethroids as proconvulsants in the rat; the GABA-ionophore antagonist pentylene tetrazole caused convulsions soon after injection, but certain compounds (including Ro5-4864) resulted in a reduction in the dose of pentylene tetrazole needed to cause convulsions. Type II pyrethroids were very potent at reducing this threshold dose, at pyrethroid doses well below those needed to cause symptoms. However, allethrin, which causes the Type I syndrome, was ineffective at reducing the threshold for pentylene tetrazole, even at doses which themselves caused symptoms. The proconvulsant action of Ro5-4864 and of pyrethroids was blocked by PK 11195, which is a specific, competitive inhibitor of the binding of [^{3}H]Ro5-4864 to rat brain membranes. These results suggest that the receptor for Ro5-4864 may have great toxicological relevance in insects and, further, that insects could provide useful models for gaining an understanding of the function for the Ro5-4864 receptor in mammals.

C. BIOGENIC AMINE RECEPTORS

Nathanson and Greengard[147] have shown that the cockroach CNS contains receptors for octopamine, dopamine, and serotonin, each coupled positively to the enzyme adenylate cyclase. It was found that low micromolar concentrations of these biogenic amines caused

a dose-dependent elevation of cyclic AMP. The largest effect, although not the most potent one, was caused by octopamine; this effect was inhibited by phentolamine, but not by propanolol, thus indicating similarities to the α-adrenergic receptor in mammals.

Physiologically, the octopamine receptor has been studied in detail in two preparations: first, in the firefly light organ, where octopamine appears to act as a neurotransmitter,[148] and second, in the dorsal unpaired medial (DUM) neuron, with its cell body in the locust (or cockroach) metathoracic ganglion and with a (slow) motoneuron activating the extensor tibiae muscle. Here octopamine seems to act as a neuromodulator, potentiating the effect of stimulation of the motoneuron and dramatically increasing the rate of muscle relaxation.[149] Octopamine has also been shown to act as a hormone, resulting in an elevation of hemolymph trehalose levels when applied to the cockroach fat body (see Chapter 17). In all three of these roles for octopamine (i.e., neurotransmitter, neuromodulator, and hormone) there is good evidence that the insecticide/acaricide chlordimeform (CDM), after undergoing bioactivation by demethylation to desmethyl-CDM (DCDM), acts as an octopamine agonist to exert many of its toxicological effects.[150-153]

Other insecticides acting at the octopamine receptor include amitraz and chloromethiuron, both close structural analogues of CDM.[75] Perhaps through an octopaminergic action in causing the potentiation of efferent activity in thoracic motoneurons, sublethal, knockdown effects of CDM have been reported which may result in field efficacy[154] and perhaps also cause the increased effectiveness reported for certain pyrethroids when used in conjunction with CDM.[155] However, few if any commercial insecticides which are not closely related to octopamine or CDM are known to owe their toxicity to effects on the insect octopamine receptor. For example, a series of imidazolidines which were found to be potent activators *in vitro* of the octopamine receptor in the firefly light organ only caused insect mortality when used in the presence of a phosphodiesterase inhibitor to block the metabolism of the cyclic AMP formed from receptor stimulation.[156]

Serotonin and dopamine receptors remain largely untested toxicologically in insects, but it has been concluded the DCDM acts as an antagonist of the cockroach fat body serotonin receptor in addition to its agonistic action at the octopamine receptor.[153] Similarly, there is no currently available evidence that the histamine receptor in insects represents a potential toxicological target, although it has been identified biochemically in the optic lobes of the locust.[157,158]

D. GLUTAMATE RECEPTOR

The most common excitatory neurotransmitter in the mammalian brain is considered to be L-glutamate. In arthropods, it is thought that L-glutamate acts as the excitatory transmitter at the skeletal neuromuscular junction[77,78,159,160] rather than ACh, as in vertebrates. It has recently become apparent that the pentapeptide proctolin, originally isolated from the cockroach proctodeal nerve by Brown,[161] has a role as a neuromodulator at other neuromuscular junctions in insects (see Chapters 13 and 14 of this work), perhaps analogous to octopamine and serotonin.[149,162]

It is thought that the elevation of proctolin levels at the neuromuscular junction does not cause toxic effects in insects, since the inhibition of the peptidase responsible for proctolin degradation followed by the injection of a large dose of proctolin into the cockroach does not cause outward behavioral abnormalities.[163] However, domoic acid isolated from a species of red seaweed, which may act specifically at the proctolin receptor, is toxic to the cockroach and to other insects.[164] At the cockroach hindgut neuromuscular junction it enhances the sensitivity to glutamate and stimulates the calcium-dependent release of glutamate presynaptically. Both of these effects are also caused by proctolin,[164] and it is possible that domoic acid owes its toxicity to its lack of susceptibility to the peptidase which normally terminates proctolin action.

At the glutamate receptor itself, several analogues of quisqualic acid, the potent glutamate agonist, have been studied electrophysiologically,[165,166] and it is possible that a derivative of one of them will prove to have insecticidal activity worthy of commercial development. There have also been recent attempts to define the insect glutamate receptor physiologically, using neuron D_f in the cockroach CNS,[128] and pharmacologically, using housefly thoracic membranes.[85] An alternative approach, also based on natural products, is the study of wasp and spider venoms, which often paralyze insects by apparently acting as glutamate antagonists.[167,168] Considerable progress is currently being made in the elucidation of the structures of toxins obtained from orb web spiders[169] and *Philanthus* species.[167] It appears that some of the components of orb web spider toxins have relatively low molecular weights and act, at the locust neuromuscular junction, primarily as antagonists at the open, glutamate-activated ion channel.[168] The commercial insecticides which may result from this work could have two clear advantages: first, there should be little cross-resistance with current insecticides, and second, by focusing on the peripheral nervous system/neuromuscular junction, problems associated with penetration of the blood-brain barrier,[50] which may limit the toxicity of insecticides acting exclusively on the insect CNS, may be avoided.

IV. CONCLUSIONS

The range of potential target sites for insect neurotoxicants is large, but the great majority of insecticides affect only a very limited number of these targets. In summary, pyrethroids and DDT primarily interfere with voltage-sensitive Na channels in the nerve axon membrane, although there is also evidence for effects of some pyrethroids on GABA- and ACh-activated ion channels. ACh esterase is the major target site for organophosphate and carbamate insecticides, although it has also been shown that some carbamates can also bind to the nicotinic ACh receptor. It is thought that the nitromethylene insecticides primarily act as agonists at this ACh receptor, but these compounds are not used commercially. Cartap and perhaps other synthetic compounds based on nereistoxin appear to act principally as antagonists at this receptor. Cyclodiene insecticides act as antagonists at the GABA-activated Cl ionophore, and the avermectins have been shown to have several effects on the GABA-receptor complex. Formamidine pesticides appear to owe their toxicity to the fact that they act as octopamine receptor agonists, but several other effects have been claimed, and these may have importance in some instances.

Apart from these target sites, the remainder are either not viable as specific, potential insecticide targets (e.g., choline acetyl transferase and GABA transaminase) or else are largely untested (e.g., biogenic amine receptors, aside from the octopamine receptor, muscarinic ACh receptors, and the glutamate receptor). Other target sites for toxic molecules in the insect nervous system remain to be described.

The mode of action of most neurotoxicants appears to be very similar in insects and vertebrates. The pharmacological and toxicological manifestations of these effects are also similar. Thus, agents which interfere with nicotinic, cholinergic transmission can be studied readily using identified neurons in the cockroach CNS. Indeed, the sixth abdominal ganglion is as rich a source of these receptors as electroplax tissue of *Torpedo*. Likewise, many skeletal muscle fibers of the cockroach contain a GABA receptor which has a pharmacological profile similar to the vertebrate $GABA_A$ receptor. The voltage-sensitive Na channel of the cockroach can be studied both *in vitro* and *in vivo* using techniques which cannot be applied readily to the vertebrate nervous system. Pyrethroids and DDT affect this channel in similar ways in vertebrates and in the cockroach. TTX also has similar nerve blocking effects; it will be interesting to see if it will find use as an anesthetic, as in experiments described using the cockroach. There are also instances of pharmaceutical drugs having similar protective effects in mammals and in the cockroach. It is therefore suggested that

since so much knowledge has been accumulated on the pharmacology of the cockroach nervous system, including the profiling of several uniquely identifiable cells, the cockroach may provide a useful model system for comparative and vertebrate pharmacologists.

REFERENCES

1. **Pumphrey, R. J. and Rawdon-Smith, A. F.,** Synaptic transmission of nervous impulses through the last abdominal ganglion of the cockroach, *Proc. R. Soc. London Ser. B,* 122, 106, 1937.
2. **Roeder, D. D. and Weiant, E. A.,** The site of action of DDT in the cockroach, *Science,* 130, 304, 1946.
3. **Narahashi, T.,** Effects of insecticides on excitable tissues, *Adv. Insect Physiol.,* 8, 1, 1971.
4. **Pelhate, M. and Sattelle, D. B.,** Pharmacological properties of insect axons: a review, *J. Insect Physiol.,* 28, 889, 1982.
5. **Sattelle, D. B.,** Acetylcholine receptors, in *Comprehensive Insect Physiology, Biochemistry and Pharmacology,* Vol. 11, Kerkut, G. A. and Gilbert, L. I., Eds., Pergamon Press, Oxford, 1985, 395.
6. **Lummis, S. C. R., Pinnock, R. D., and Sattelle, D. B.,** GABA receptors of the insect central nervous system, in *Sites of Action for Neurotoxic Pesticides,* ACS Symp. Ser. 356, Hollingworth, R. M. and Green, M. B., Eds., American Chemical Society, Washington, D.C., 1987, 14.
7. **Smith, D. S. and Treherne, J. E.,** The electron microscopic localization of cholinesterase activity in the central nervous system of an insect, *Periplaneta americana, J. Cell Biol.,* 26, 445, 1965.
8. **Lowenstein, O.,** A method of physiological assay of pyrethrum extract, *Nature (London),* 150, 760, 1942.
9. **Narahashi, T.,** Effect of the insecticide allethrin on membrane potentials of cockroach giant axons, *J. Cell. Comp. Physiol.,* 59, 61, 1962.
10. **Narahashi, T.,** Nature of the negative after-potential increased by the insecticide allethrin in cockroach giant axons, *J. Cell. Comp. Physiol.,* 59, 67, 1962.
11. **van den Bercken, J., Akkermans, L. M. A., and van der Zalm, J. M.,** DDT-like action of allethrin in the sensory nervous system of *Xenopus laevis, Eur. J. Pharmacol.,* 21, 95, 1973.
12. **Gammon, D. W.,** Nervous effects of toxins on an intact insect: a method, *Pestic. Biochem. Physiol.,* 7, 1, 1977.
13. **Gammon, D. W.,** Neural effects of allethrin on the free walking cockroach, *Periplaneta americana:* an investigation using defined doses at 15 and 32°C, *Pestic. Sci.,* 9, 79, 1978.
14. **Narahashi, T.,** Nerve membrane as a target of pyrethroids, *Pestic. Sci.,* 7, 267, 1976.
15. **Gammon, D. W., Brown, M. A., and Casida, J. E.,** Two classes of pyrethroid action in the cockroach, *Pestic. Biochem. Physiol.,* 15, 181, 1981.
16. **Narahashi, T.,** Mode of action of pyrethroids, *Bull. W.H.O.,* 44, 337, 1971.
17. **Pelhate, M., Hue, B., and Sattelle, D. B.,** Actions of natural and synthetic toxins on the axonal sodium channels of the cockroach, in *Insect Neurobiology and Pesticide Action (Neurotox 79),* Society of Chemical Industry, London, 1980, 65.
18. **Lund, A. E. and Narahashi, T.,** Kinetics of sodium channel modification as the basis for the variation in the nerve membrane effects of pyrethroids and DDT analogs, *Pestic. Biochem. Physiol.,* 20, 203, 1983.
19. **Vijverberg, H. P. M. and de Weille, J. R.,** The interaction of pyrethroids with voltage dependent Na channels, *Neurotoxicology,* 6, 23, 1985.
20. **Lund, A. E. and Narahashi, T.,** Kinetics of sodium channel modification by the insecticide tetramethrin in squid axon membranes, *J. Pharmacol. Exp. Ther.,* 219, 464, 1981.
21. **van den Bercken, J. and Vijverberg, H. P. M.,** Voltage clamp studies on the effects of allethrin and DDT on the sodium channels in frog myelinated nerve membrane, in *Insect Neurobiology and Pesticide Action (Neurotox 79),* Society of Chemical Industry, London, 1980, 79.
22. **Vijverberg, H. P. M., van der Zalm, J. M., and van den Bercken, J.,** Similar mode of action of pyrethroids and DDT on sodium channel gating in myelinated nerves, *Nature (London),* 295, 601, 1982.
23. **Narahashi, T.,** Mechanisms of action of pyrethroids on sodium and calcium channel gating, in *Neuropharmacology and Pesticide Action, Proc. Society of Chemical Industry Symp.,* Ford, M. G., Usherwood, P. N. R., Reay, R. C., and Lunt, G. G., Eds., Ellis Horwood, Chichester, England, 1986, 36.
24. **Laufer, J., Roche, M., Pelhate, M., Elliott, M., Janes, N. F., and Sattelle, D. B.,** Pyrethroid insecticides: actions of deltamethrin and related compounds on insect axonal sodium channels, *J. Insect Physiol.,* 30, 341, 1984.
25. **Laufer, J., Pelhate, M., and Sattelle, D. B.,** Actions of pyrethroid insecticides on insect axonal sodium channels, *Pestic. Sci.,* 16, 651, 1985.

26. **Vijverberg, H. P. M., de Weille, J. R., Ruigt, G. S. F., and van den Bercken, J.,** The effect of pyrethroid structure on the interaction with the sodium channel in the nerve membrane, in *Neuropharmacology and Pesticide Action, Proc. Society of Chemical Industry Symp.,* Ford, M. G., Usherwood, P. N. R., Reay, R. C., and Lunt, G. G., Eds., Ellis Horwood, Chichester, England, 1986, 267.
27. **Narahashi, T.,** Nerve membrane ionic channels as the primary target of pyrethroids, *Neurotoxicology,* 6, 3, 1985.
28. **Lund, A. E. and Narahashi, T.,** Dose-dependent interaction of the pyrethroid isomers with sodium channels of squid axon membranes, *Neurotoxicology,* 3, 11, 1982.
29. **Salgado, V. L., Irving, S. N., and Miller, T. A.,** Depolarization of motor nerve terminals by pyrethroids in susceptible and kdr-resistant house flies, *Pestic. Biochem. Physiol.,* 20, 100, 1983.
30. **Salgado, V. L., Irving, S. N., and Miller, T. A.,** The importance of nerve terminal depolarization in pyrethroid poisoning of insects, *Pestic. Biochem. Physiol.,* 20, 169, 1983.
31. **Orchard, I. and Osborne, M. P.,** The action of insecticides on neurosecretory neurons in the stick insect *Carausius morosus, Pestic. Biochem. Physiol.,* 10, 197, 1979.
32. **Jacques, Y., Romey, G., Cavey, M. T., Kartalovski, B., and Lazdunski, M.,** Interaction of pyrethroids with the Na^+ channel in mammalian neuronal cells in culture, *Biochim. Biophys. Acta,* 600, 882, 1980.
33. **Soderlund, D. M.,** Pyrethroid-receptor interactions: stereospecific binding and effects on sodium channels in mouse brain preparations, *Neurotoxicology,* 6, 35, 1985.
34. **Romey, G. and Lazdunski, M.,** Lipid-soluble toxins thought to be specific for Na^+ channels block Ca^{2+} channels in neuronal cells, *Nature (London),* 297, 79, 1982.
35. **Bolger, G. T., Gengo, P., Klockowski, R., Luchowski, E., Siegel, H., Janis, R. A., Triggle, A. M., and Triggle, D. J.,** Characterization of binding of the Ca^{++} channel antagonist, [^{3}H]Nitrendipine, to guinea-pig ileal smooth muscle, *J. Pharmacol. Exp. Ther.,* 225, 291, 1983.
36. **Lawrence, L. J., Gee, K. W., and Yamamura, H. I.,** Interactions of pyrethroid insecticides with chloride ionophore-associated binding sites, *Neurotoxicology,* 6, 87, 1985.
37. **Gammon, D. W. and Sander, G.,** Two mechanisms of pyrethroid action: electrophysiological and pharmacological evidence, *Neurotoxicology,* 6, 63, 1985.
38. **Cantor, E. H., Kenessey, A., Semenuk, G., and Spector, S.,** Interaction of calcium channel blockers with non-neuronal benzodiazepine binding sites, *Proc. Natl. Acad. Sci. U.S.A.,* 81, 1549, 1984.
39. **Miller, T. A. and Salgado, V. L.,** The mode of action of pyrethroids on insects, in *The Pyrethroid Insecticides,* Leahey, J. P., Ed., Taylor and Francis, London, 1985, 43.
40. **Miller, T. A. and Adams, M. E.,** Mode of action of pyrethroids, in *Insecticide Mode of Action,* Coats, J. R., Ed., Academic Press, New York, 1982, 3.
41. **Casida, J. E., Gammon, D. W., Glickman, A. H., and Lawrence, L. J.,** Mechanisms of selective action of pyrethroid insecticides, *Annu. Rev. Pharmacol. Toxicol.,* 23, 413, 1983.
42. **Burt, P. E. and Goodchild, R. E.,** The site of action of pyrethrin I in the nervous system of the cockroach *Periplaneta americana, Entomol. Exp. Appl.,* 14, 179, 1971.
43. **Gammon, D. W.,** The mode of action of pyrethroids in insects — a review, in *Pyrethroid Insecticides: Chemistry and Action,* Tables Rondes Roussel Uclaf No. 37, Casida, J. E. and Elliott, M., Eds., Paris, 1980, 11.
44. **Beament, J. W. L.,** A paralyzing agent in the blood of cockroaches, *J. Insect Physiol.,* 2, 199, 1958.
45. **Sternberg, J. G. and Kearns, C. W.,** The presence of toxins other than DDT in the blood of DDT-poisoned roaches, *Science,* 116, 114, 1952.
46. **Blum, M. S. and Kearns, C. W.,** Temperature and the action of pyrethrum in the American cockroach, *J. Econ. Entomol.,* 49, 862, 1956.
47. **Gammon, D. W.,** The action of tetrodotoxin on the cockroach, *Periplaneta americana:* a toxicological and neurophysiological study, *Physiol. Entomol.,* 3, 37, 1978.
48. **Gammon, D. W.,** An analysis of the temperature-dependence of the toxicity of allethrin to the cockroach, in *Neurotoxicology of Insecticides and Pheromones,* Narahashi, T., Ed., Plenum Press, New York, 1979, 97.
49. **Thomas, M. V.,** Permeability of the Insect Blood-Brain Barrier, Ph. D. thesis, University of Cambridge, Cambridge, England, 1974.
50. **Lane, N. J. and Treherne, J. E.,** The ultrastructural organization of peripheral nerves in two insect species (*Periplaneta americana* and *Schistocerca gregaria*), *Tissue Cell,* 5, 703, 1973.
51. **Narahashi, T. and Anderson, N. C.,** Mechanism of excitation block by the insecticide allethrin applied externally and internally to squid giant axons, *Toxicol. Appl. Pharmacol.,* 10, 529, 1967.
52. **Wang, C. M., Narahashi, T., and Scuka, M.,** Mechanism of negative temperature coefficient of nerve blocking action of allethrin, *J. Pharmacol. Exp. Ther.,* 182, 442, 1972.
53. **Narahashi, T.,** Effects of insecticides on nervous conduction and synaptic transmission, in *Insecticide Biochemistry and Physiology,* Wilkinson, C. F., Ed., Plenum Press, New York, 1976, 327.
54. **Elliott, M. and Janes, N. F.,** Synthetic pyrethroids — a new class of insecticide, *Chem. Soc. Rev.,* 7, 473, 1978.

55. **Clements, A. N. and May, T. E.,** The actions of pyrethroids upon the peripheral nervous system and associated organs in the locust, *Pestic. Sci.,* 8, 661, 1977.
56. **Gammon, D. W. and Holden, J. S.,** A neural basis for pyrethroid resistance in larvae of *Spodoptera littoralis,* in *Insect Neurobiology and Pesticide Action (Neurotox 79),* Society of Chemical Industry, London, 1980, 42.
57. **Gammon, D. W.,** Pyrethroid resistance in a strain of *Spodoptera littoralis* is correlated with decreased sensitivity of the CNS *in vitro, Pestic. Biochem. Physiol.,* 13, 53, 1980.
58. **Gammon, D. W., Ruzo, L. O., and Casida, J. E.,** A new pyrethroid insecticide with remarkable potency on nerve axons, *Neurotoxicology,* 4, 165, 1983.
59. **Ruzo, L. O., Casida, J. E., and Gammon, D. W.,** Neurophysiological activity and toxicology of pyrethroids derived by addition of methylene, sulfur or oxygen to chrysanthemate 2-methyl-1-propenyl substituent, *Pestic. Biochem. Physiol.,* 21, 84, 1984.
60. **Gammon, D. W.,** Correlations between *in vitro* and *in vivo* mechanisms of pyrethroid insecticide action, *Fund. Appl. Toxicol.,* 5, 9, 1985.
61. **Gammon, D. W., Lawrence, L. J., and Casida, J. E.,** Pyrethroid toxicology: protective effects of diazepam and phenobarbital in the mouse and cockroach, *Toxicol. Appl. Pharmacol.,* 66, 290, 1982.
62. **Verschoyle, R. D. and Aldridge, W. N.,** Structure-activity relationships of some pyrethroids in rats, *Arch. Toxicol.,* 45, 325, 1980.
63. **Scott, J. G. and Matsumura, F.,** Evidence for two types of toxic actions of pyrethroids on susceptible and DDT-resistant German cockroaches, *Pestic. Biochem. Physiol.,* 19, 141, 1983.
64. **Roche, M. and Guillet, J. C.,** Effects of the pyrethroids bioallethrin, deltamethrin and RU-15525 on the electrical activity of a cuticular mechanoreceptor of the cockroach *Periplaneta americana, Pestic. Sci.,* 16, 511, 1985.
65. **Irving, S. N.,** *In vitro* activity of pyrethroids, in *British Crop Protection Conference — Pests and Diseases,* International Specialized Book Services, Beaverton, OR, 1984, 859.
66. **Blagburn, J. M. and Sattelle, D. B.,** Presynaptic depolarization mediates presynaptic inhibition at a synapse between an identified mechanosensory neurone and giant interneurone 3 in the first instar cockroach, *J. Exp. Biol.,* 127, 135, 1987.
67. **Narahashi, T. and Haas, H. G.,** Interaction of DDT with the components of lobster nerve membrane conductance, *J. Gen. Physiol.,* 51, 177, 1968.
68. **Pichon, Y., Guillet, J. C., Heilig, U., and Pelhate, M.,** Recent studies on the effect of DDT and pyrethroid insecticides on nervous activity in cockroaches, *Pestic. Sci.,* 16, 627, 1985.
69. **Holan, G., O'Keefe, D. F., Virgona, C., and Walser, R.,** Structural and biological link between pyrethroids and DDT in new insecticides, *Nature (London),* 272, 734, 1978.
70. **Gammon, D. W.,** Effects of DDT on the cockroach nervous system at three temperatures, *Pestic. Sci.,* 9, 95, 1978.
71. **Miller, T. and Kennedy, J. M.,** Flight motor activity of houseflies as affected by temperature and insecticides, *Pestic. Biochem. Physiol.,* 2, 206, 1972.
72. **Eaton, J. L. and Sternburg, J.,** Temperature and the action of DDT on the nervous system of *Periplaneta americana* (L.), *J. Insect Physiol.,* 10, 471, 1964.
73. **Eaton, J. L. and Sternburg, J. G.,** Temperature effects on nerve activity in DDT-treated American cockroaches, *J. Econ. Entomol.,* 60, 1358, 1967.
74. **Leake, L. D. and Walker, R. J.,** *Invertebrate Neuropharmacology,* Blackie, London, 1980, 358.
75. **Corbett, J. R., Wright, K., and Baillie, A. C.,** *The Biochemical Mode of Action of Pesticides,* Academic Press, London, 1984, 382.
76. **O'Brien, R. D.,** *Insecticide Action and Metabolism,* Academic Press, New York, 1967, 332.
77. **Clements, A. N. and May, T. E.,** Studies on locust neuromuscular physiology in relation to glutamic acid, *J. Exp. Biol.,* 60, 673, 1974.
78. **Usherwood, P. N. R.,** Glutamate synapses and receptors on insect muscle, in *Glutamate as a Neurotransmitter,* DiChiara, G. and Gressa, G. L., Eds., Raven Press, New York, 1981, 183.
79. **Hart, G. J. and O'Brien, R. D.,** Recording spectrophotometric method for determination of dissociation and phosphorylation constants for the inhibition of acetylcholinesterase by organophosphates in the presence of substrate, *Biochemistry,* 12, 2940, 1973.
80. **Baillie, A. C., Corbett, J. R., Dowsett, J. R., Sattelle, D. B., and Callec, J. J.,** Inhibitors of choline acetyltransferase as potential insecticides, *Pestic. Sci.,* 6, 645, 1975.
81. **Sattelle, D. B. and Callec, J. J.,** Actions of isothiocyanates on the central nervous system of *Periplaneta americana, Pestic. Sci.,* 8, 735, 1977.
82. **Sattelle, D. B.,** Acetylcholine receptors of insects, *Adv. Insect Physiol.,* 15, 215, 1980.
83. **Sattelle, D. B.,** Insect acetylcholine receptors — biochemical and physiological approaches, in *Neuropharmacology and Pesticide Action, Proc. Society Chemical Industry Symp.,* Ford, M. G., Usherwood, P. N. R., Reay, R. C., and Lunt, G. G., Eds., Ellis Horwood, Chichester, England, 1986, 445.

84. **Eldefrawi, M. E., Sherby, S. M., Abalis, I. M., and Eldefrawi, A. T.,** Interactions of pyrethroid and cyclodiene insecticides with nicotinic acetylcholine and GABA receptors, *Neurotoxicology,* 6, 47, 1985.
85. **Eldefrawi, M. E., Abalis, I. M., Sherby, S. M., and Eldefrawi, A. T.,** Neurotransmitter receptors of vertebrates and insects as targets for insecticides, in *Neuropharmacology and Pesticide Action, Proc. Chemical Industry Symp.,* Ford, M. G., Usherwood, P. N. R., Reay, R. C., and Lunt, G. G., Eds., Ellis Horwood, Chichester, England, 1986, 154.
86. **Breer, H. and Sattelle, D. B.,** Molecular properties and functions of insect acetylcholine receptors, *J. Insect Physiol.,* 33, 771, 1987.
87. **Shankland, D. L., Rose, J. A., and Donniger, C.,** The cholinergic nature of the cercal nerve-giant fibre synapse in the sixth abdominal ganglion of the American cockroach *(Periplaneta americana* L.), *J. Neurobiol.,* 2, 247, 1971.
88. **Harrow, I. D. and Sattelle, D. B.,** Acetylcholine receptors on the cell body membrane of giant interneurone 2 in the cockroach *Periplaneta americana, J. Exp. Biol.,* 105, 339, 1983.
89. **David, J. A. and Sattelle, D. B.,** Actions of cholinergic pharmacological agents on the cell body membrane of the fast coxal depressor motoneurone of the cockroach *(Periplaneta americana), J. Exp. Biol.,* 108, 119, 1984.
90. **Gepner, J. I., Hall, L. M., and Sattelle, D. B.,** Insect acetylcholine receptors as a site of insecticide action, *Nature (London),* 276, 188, 1978.
91. **Eldefrawi, M. E. and Eldefrawi, A. T.,** Coupling between the nicotinic acetylcholine receptor site and its ionic channel site, *Ann. N.Y. Acad. Sci.,* 358, 239, 1980.
92. **Harrow, I. D., Hue, B., Pelhate, M., and Sattelle, D. B.,** *Alpha*-Bungarotoxin blocks excitatory postsynaptic potentials in an identified insect interneurone, *J. Physiol.,* 295, 63P, 1979.
93. **Eldefrawi, A. T., Eldefrawi, M. E., Albuquerque, E. X., Oliveira, A. C., Mansour, N., Adler, M., Daly, J. W., Brown, G. B., Burgermeister, W., and Witkop, B.,** Perhydrohistrionicotoxin: a potential ligand for the ion conductance modulator of the acetylcholine receptor, *Proc. Natl. Acad. Sci. U.S.A.,* 74, 2172, 1977.
94. **Sattelle, D. B., Harrow, I. D., David, J. A., Pelhate, M., Callec, J. J., Gepner, J. I., and Hall, L. M.,** Nereistoxin: actions on a CNS acetylcholine receptor/ion channel in the cockroach, *Periplaneta americana, J. Exp. Biol.,* 118, 37, 1985.
95. **Sherby, S. M., Eldefrawi, A. T., David, J. A., Sattelle, D. B., and Eldefrawi, M. E.,** Interactions of charatoxins and nereistoxin with the nicotinic acetylcholine receptors of insect CNS and *Torpedo* electric organ, *Arch. Insect Biochem. Physiol.,* 3, 431, 1986.
96. **Jennings, K. R., Brown, D. G., Wright, D. P., Jr., and Chalmers, A. E.,** Methyllycaconitine: a potent natural insecticide active on the cholinergic receptor, in *Sites of Action for Neurotoxic Pesticides,* ACS Symp. Ser. 356, Hollingsworth, R. M. and Green, M. B., Eds., American Chemical Society, Washington, D.C., 1987, 274.
97. **Sherby, S. M., Eldefrawi, A. T., Albuquerque, E. X., and Eldefrawi, M. E.,** Comparison of the actions of carbamate anti-cholinesterases on the nicotinic acetylcholine receptor, *Mol. Pharmacol.,* 27, 343, 1985.
98. **Artola, A., Callec, J. J., Hue, B., David, J. A., and Sattelle, D. B.,** Actions of amantadine at synaptic and extrasynaptic cholinergic receptors in the central nervous system of the cockroach, *Periplaneta americana, J. Insect Physiol.,* 30, 185, 1984.
99. **Gammon, D. W.,** unpublished data, 1986.
100. **David, J. A., Crowley, P. J., Hall, S. G., Battersby, M., and Sattelle, D. B.,** Actions of synthetic piperidine derivatives on an insect acetylcholine receptor ion channel complex, *J. Insect Physiol.,* 30, 191, 1984.
101. **Sattelle, D. B., Hue, B., Pelhate, M., Sherby, S. M., Eldefrawi, A. T., and Eldefrawi, M. E.,** Actions of phencyclidine and its thienylpyrrolidine analogue on synaptic transmission and axonal conduction in the central nervous system of the cockroach, *Periplaneta americana, J. Insect Physiol.,* 31, 916, 1985.
102. **Battersby, M. K. and Hall, S. G.,** Lobeline, a potent *O*-acetylcholine antagonist at cockroach nicotinic receptors, may be able to distinguish between nicotinic receptor subtypes in insects, *Pestic. Sci.,* 16, 428, 1985.
103. **Schroeder, M. E. and Flattum, R. F.,** The mode of action and neurotoxic properties of the nitromethylene heterocycle insecticides, *Pestic. Biochem. Physiol.,* 22, 148, 1984.
104. **Harris, M., Price, R. N., Robinson, J., May, T. E., and Wadayama, N.,** WL108477 — a novel neurotoxic insecticide, in *British Crop Protection Conference — Pests and Diseases,* International Specialized Book Services, Beaverton, OR, 1986, 115.
105. **Aronstam, R. S., Eldefrawi, A. T., Pessah, I. N., Daly, J. W., Albuquerque, E. X., and Eldefrawi, M. E.,** Regulation of [^{3}H]-perhydrohistrionicotoxin binding to *Torpedo* electroplax by effectors of the acetylcholine receptor, *J. Biol. Chem.,* 256, 2843, 1981.
106. **Sattelle, D. B. and David, J. A.,** Voltage-dependent block by histrionicotoxin of the acetylcholine-induced current in an insect motoneurone cell body, *Neurosci. Lett.,* 43, 37, 1983.

107. **Abbassy, M. A., Eldefrawi, M. E., and Eldefrawi, A. T.,** Allethrin interactions with the nicotinic acetylcholine receptor channel, *Life Sci.*, 31, 1547, 1982.
108. **Abbassy, M. A., Eldefrawi, M. E., and Eldefrawi, A. T.,** Pyrethroid action on the nicotinic acetylcholine receptor/channel, *Pestic. Biochem. Physiol.*, 19, 299, 1983.
109. **Lummis, S. C. R. and Sattelle, D. B.,** Binding of *N*-[propionyl-^{3}H] propionylated *alpha*-bungarotoxin and L-[benzilic-4,4^1-^{3}H] quinuclidinyl benzilate to CNS extracts of the cockroach *Periplaneta americana, Comp. Biochem. Physiol.*, 80C, 75, 1985.
110. **Yamamura, H. I. and Snyder, S.,** Muscarinic cholinergic receptor binding in the longitudinal muscle of the guinea pig ileum with [^{3}H]quinuclidinyl benzilate, *Mol. Pharmacol.*, 10, 861, 1974.
111. **Breer, H.,** Characterization of synaptosomes from the central nervous system of insects, *Neurochem. Int.*, 3, 155, 1981.
112. **Breer, H.,** Properties of putative nicotinic and muscarinic cholinergic receptors in the central nervous system of *Locusta migratoria, Neurochem. Int.*, 3, 43, 1981.
113. **Trimmer, B. A. and Berridge, M. J.,** Inositol phosphates in the insect nervous system, *Insect Biochem.*, 15, 811, 1985.
114. **Breer, H. and Knipper, M.,** Characterization of acetylcholine release from insect synaptosomes, *Insect Biochem.*, 15, 337, 1984.
115. **Breer, H.,** Synaptosomes — systems for studying insect neurochemistry, in *Neuropharmacology and Pesticide Action, Proc. Society of Chemical Industry Symp.*, Ford, M. G., Usherwood, P. N. R., Reay, R. C., and Lunt, G. G., Eds., Ellis Horwood, Chichester, England, 1986, 384.
116. **Gammon, D. W.,** unpublished data, 1985.
117. **Takeuchi, A. and Takeuchi, N.,** A study of the inhibitory action of *gamma*-amino-butyric acid on neuromuscular transmission in the crayfish, *J. Physiol. (London),* 183, 418, 1966.
118. **Dudel, J. and Kuffler, S. W.,** Presynaptic inhibition at the crayfish neuromuscular junction, *J. Physiol. (London),* 155, 543, 1961.
119. **Takeuchi, A. and Takeuchi, N.,** A study of the action of picrotoxin on the inhibitory neuromuscular junction of the crayfish, *J. Physiol. (London),* 205, 377, 1969.
120. **Shepherd, D. and Tyrer, N. M.,** Inhibition of GABA uptake potentiates the effect of exogenous GABA on locust skeletal muscle, *Comp. Biochem. Physiol.*, 82C, 315, 1985.
121. **Gammon, D. W., Gingrich, H. L., Sander, G., Van Der Werf, P. A., Manly, C. J., and Crosby, G. A.,** The design and synthesis of GABAergic compounds as potential insecticides, in *British Crop Protection Conference — Pests and Diseases,* International Specialized Book Services, Beaverton, OR, 1986, 217.
122. **Gammon, D. W., Gingrich, H. L., Sander, G., Stewart, R. R., and Van Der Werf, P. A.,** GABA transaminase and glutamic acid decarboxylase as targets for insecticides: biochemical and toxicological assessment, in *Sites of Action for Neurotoxic Pesticides,* ACS Symp. Ser. 356, Hollingworth, R. M. and Green, M. B., Eds., American Chemical Society, Washington, D.C., 1987, 122.
123. **Rando, R.,** Mechanism of the irreversible inhibition of *gamma* aminobutyric acid-*alpha* ketoglutaric acid transaminase by the neurotoxin gabaculline, *Biochemistry,* 16, 4604, 1977.
124. **Sattelle, B. D. and Lummis, S. C. R.,** Acetylcholine and GABA receptors in insect CNS as sites of insecticide action, in *Membrane Receptors and Enzymes as Targets of Insecticidal Action,* Clark, J. M. and Matsumura, F., Eds., Plenum Press, New York, 1985, 51.
125. **Dybas, R. A. and Green, A. St. J.,** Avermectins: their chemistry and pesticidal activity, in *British Crop Protection Conference — Pests and Diseases,* International Specialized Book Services, Beaverton, OR, 1984, 947.
126. **Wright, D. J.,** Biological activity and mode of action of avermectins, in *Neuropharmacology and Pesticide Action, Proc. Society of Chemical Industry Symp.*, Ford, M. G., Usherwood, P. N. R., Reay, R. C., and Lunt, G. G., Eds., Ellis Horwood, Chichester, England, 1986, 174.
127. **Takeuchi, A. and Onodera, K.,** Effect of bicuculline on the GABA receptor of the crayfish neuromuscular junction, *Nature New Biol.*, 236, 55, 1972.
128. **Wafford, K. A. and Sattelle, D. B.,** Effects of amino acid neurotransmitter candidates on an identified insect motoneurone, *Neurosci. Lett.*, 63, 135, 1986.
129. **Hue, B., Pelhate, M., and Chanelet, J.,** Pre- and postsynaptic effects of tuarine and GABA in the cockroach central nervous system, *J. Can. Sci. Neurol.*, 6, 243, 1979.
130. **Lawrence, L. J. and Casida, J. E.,** Interactions of lindane, toxaphene and cyclodienes with the brain-specific *t*-butylbicyclophosphorothionate receptor, *Life Sci.*, 35, 171, 1984.
131. **Matsumura, F. and Ghiasuddin, S. M.,** Evidence for similarities between cyclodiene type insecticides and picrotoxinin in their action mechanisms, *J. Environ. Sci. Health,* B18, 1, 1983.
132. **Abalis, I. M., Eldefrawi, M. E., and Eldefrawi, A. T.,** High affinity stereospecific binding of cyclodiene insecticides and *gamma*-BHC to GABA receptors of rat brain, *Pestic. Biochem. Physiol.*, 24, 95, 1985.

133. **Eldefrawi, A. T., Abalis, I. M., and Eldefrawi, M. E.,** The GABA/benzodiazepine receptor-chloride channel: biochemical identification in insects and stereospecific binding of insecticides, in *Membrane Receptors and Enzymes as Targets of Insecticidal Action,* Clark, J. M. and Matsumura, F., Eds., Plenum Press, New York, 1985, 107.
134. **Lawrence, L. J. and Casida, J. E.,** Stereospecific action of pyrethroid insecticides on the *gamma*-aminobutyric acid receptor-ionophore complex, *Science,* 221, 1399, 1983.
135. **Seifert, J. and Casida, J. E.,** Regulation of [^{35}S]*t*-butylbicyclophosphorothionate binding sites in rat brain by GABA, pyrethroid and barbiturate, *Eur. J. Pharmacol.,* 115, 191, 1985.
136. **Gammon, D. W. and Casida, J. E.,** Pyrethroids of the most potent class antagonize GABA action at the crayfish neuromuscular junction, *Neurosci. Lett.,* 40, 163, 1983.
137. **Braestrup, C. and Squires, R. F.,** Specific benzodiazepine receptors in rat brain characterized by high-affinity [^{3}H]diazepam binding, *Proc. Natl. Acad. Sci. U.S.A.,* 74, 3805, 1977.
138. **Mohler, H. and Okada, T.,** Benzodiazepine receptor: demonstration in the central nervous system, *Science,* 198, 849, 1977.
139. **Nielsen, M., Braestrup, C., and Squires, R. F.,** Evidence for a late evolutionary appearance of brain specific benzodiazepine receptors: an investigation of 18 vertebrate and 5 invertebrate species, *Brain Res.,* 141, 342, 1978.
140. **Abalis, I. M., Eldefrawi, M. E., and Eldefrawi, A. T.,** Biochemical identification of putative GABA/benzodiazepine receptors in house fly thorax muscles, *Pestic. Biochem. Physiol.,* 20, 39, 1983.
141. **Robinson, T., MacAllan, D., Lunt, G., and Battersby, M.,** *Gamma*-Aminobutyric acid receptor complex of insect CNS: characterization of a benzodiazepine binding site, *J. Neurochem.,* 47, 1955, 1986.
142. **Lummis, S. C. R. and Sattelle, D. B.,** Binding sites for [^{3}H]GABA, [^{3}H]flunitrazepam and [^{35}S]TBPS in insect CNS, *Neurochem. Int.,* 9, 287, 1986.
143. **Ozoe, Y. and Matsumura, F.,** Effects of diazepam and chlordimeform analogs on the German and the American cockroaches, *Pestic. Biochem. Physiol.,* 26, 253, 1986.
144. **Shoemaker, H., Bliss, M., and Yamamura, H. I.,** Specific high-affinity saturable binding of [^{3}H]Ro5-4864 to benzodiazepine binding sites in the rat cerebral cortex, *Eur. J. Pharmacol.,* 71, 173, 1981.
145. **Marangos, P. J., Patel, J., Boulenger, J. P., and Clark-Rosenberg, R.,** Characterization of peripheral-type benzodiazepine binding sites in brain using [^{3}H]Ro5-4864, *Mol. Pharmacol.,* 22, 26, 1982.
146. **Devaud, L. L., Szot, P., and Murray, T. F.,** PK 11195 antagonism of pyrethroid-induced proconvulsant activity, *Eur. J. Pharmacol.,* 120, 269, 1986.
147. **Nathanson, J. A. and Greengard, P.,** Octopamine-sensitive adenylate cyclase: evidence for a biologic role of octopamine in nervous tissue, *Science,* 180, 308, 1973.
148. **Nathanson, J. A.,** Octopamine receptors, adenosine 3′,5′-monophosphate, and neural control of firefly flashing, *Science,* 203, 65, 1979.
149. **Evans, P. D. and O'Shea, M.,** An octopaminergic neurone modulates neuromuscular transmission in the locust, *Nature (London),* 270, 257, 1977.
150. **Hollingworth, R. M. and Murdock, L. L.,** Formamidine pesticides: octopamine-like actions in a firefly, *Science,* 208, 74, 1980.
151. **Evans, P. D. and Gee, J. D.,** Action of formamidine pesticides on octopamine receptors, *Nature (London),* 287, 60, 1980.
152. **Matsumura, F. and Beeman, R. W.,** Toxic and behavioral effects of chlordimeform on the American cockroach *Periplaneta americana,* in *Insecticide Mode of Action,* Coats, J. R., Ed., Academic Press, New York, 1982, 3.
153. **Downer, R. G. H., Gole, J. W. D., and Orr, G. L.,** Interaction of formamidines with octopamine-, dopamine- and 5-hydroxytryptamine-sensitive adenylate cyclase in the nerve cord of *Periplaneta americana, Pestic. Sci.,* 16, 472, 1985.
154. **Lund, A. E., Hollingworth, R. M., and Shankland, D. L.,** Chlordimeform: plant protection by a sublethal, noncholinergic action on the central nervous system, *Pestic. Biochem. Physiol.,* 11, 117, 1979.
155. **Plapp, F. W., Jr.,** Chlordimeform as a synergist for insecticides against the tobacco budworm, *J. Econ. Entomol.,* 69, 91, 1976.
156. **Nathanson, J. A.,** Phenyliminoimidazolidines: characterization of a class of potent agonists of octopamine-sensitive adenylate cyclase and their use in understanding the pharmacology of octopamine receptors, *Mol. Pharmacol.,* 28, 254, 1985.
157. **Elias, M. S. and Evans, P. D.,** Histamine in the insect nervous system: distribution, synthesis and metabolism, *J. Neurochem.,* 41, 562, 1983.
158. **Elias, M. S. and Evans, P. D.,** Autoradiographic localization of ^{3}H-histamine accumulation by the visual system of the locust, *Cell Tissue Res.,* 238, 105, 1984.
159. **Cull-Candy, S. G., Miledi, R., and Parker, I.,** Single glutamate-activated channels recorded from locust muscle fibres with perfused patch-clamp electrodes, *J. Physiol. (London),* 321, 195, 1981.
160. **Takeuchi, A. and Takeuchi, N.,** The effect on crayfish muscle of iontophoretically applied glutamate, *J. Physiol. (London),* 170, 296, 1964.

161. **Brown, B. E.,** Proctolin: a peptide transmitter candidate in insects, *Life Sci.,* 17, 1241, 1975.
162. **Breen, C. A. and Atwood, H. L.,** Octopamine — a neurohormone with presynaptic activity-dependent effects at crayfish neuromuscular junctions, *Nature (London),* 303, 716, 1983.
163. **Schooley, D. A., Quistad, G. B., Skinner, W. S., and Adams, M. E.,** Insect neuropeptides and their physiological degradation: the basis for new insecticide discovery?, Abstr. 2S-04, in *Abstr. 6th Int. Congr. Pesticide Chemistry,* International Union of Pure and Applied Chemistry, Ottawa, Ontario, Canada, 1986.
164. **Maeda, Kodama, T., Saito, M., Tanaka, T., Yoshizumi, H., Nomoto, K., and Fujita, T.,** Neuromuscular action of insecticidal domoic acid on the American cockroach, *Pestic. Biochem. Physiol.,* 28, 85, 1987.
165. **Fukami, J.,** The action of glutamate agonists at insect neuromuscular junction, in *Membrane Receptors and Enzymes as Targets of Insecticidal Action,* Clark, J. M. and Matsumura, F., Eds., Plenum Press, New York, 1985, 107.
166. **Miyamoto, T., Oda, M., Yamamoto, D., Kaneko, J., Usui, T., and Fukami, J.,** Agonistic action of synthetic analogues of quisqualic acid at the insect neuromuscular junction, *Arch. Insect Biochem. Physiol.,* 2, 65, 1985.
167. **Piek, T., Kits, K. S., Spanjer, W., Van Marle, J., and van Wilgenburg, H.,** Neurotoxic effects of wasp venoms — synaptic and behavioral aspects, *Neurotoxicology,* 6, 251, 1985.
168. **Usherwood, P. N. R.,** Noncompetitive antagonism of glutamate receptors, in *Sites of Action for Neurotoxic Pesticides,* ACS Symp. Ser. 356, Hollingsworth, R. M. and Green, M. B., Eds., American Chemical Society, Washington, D.C., 1987, 298.
169. **Quicke, D. L. J. and Usherwood, P. N. R.,** Peptides as modulators and channel blockers, *Pestic. Sci.,* 20, 315, 1987.

Section V. Metabolic Control

Chapter 19

NEUROENDOCRINE REGULATION OF FAT BODY METABOLISM

Larry L. Keeley

TABLE OF CONTENTS

I. NEUROENDOCRINE EFFECTS ON FAT BODY METABOLISM

The cockroach fat body has a metabolic role analogous to the vertebrate liver. It is the main tissue for intermediary metabolism, storage of excess metabolites, and detoxification of xenobiotic agents. Like the liver of vertebrates, the fat body of insects is the primary source for the circulating carbohydrates, lipids, and proteins found in the hemolymph (blood) of the cockroach. The production of these metabolites in insects is controlled by endocrine regulations that are similar to the regulations found in vertebrate animals. (See also Chapter 13.)

A. NEUROENDOCRINE REGULATIONS OF CARBOHYDRATE METABOLISM

The first example of neuroendocrine regulation of fat body metabolism concerned the synthesis of trehalose, the major circulatory carbohydrate in most insects. Trehalose is 1,1-glucoglucoside, a nonreducing disaccharide. It is synthesized only by the fat body tissue and constitutes about 80 to 90% of the carbohydrate in the hemolymph of cockroaches. Trehalose is present at concentrations that average about 500 mg/dl, as compared to 100 mg/dl for glucose in vertebrate blood. Trehalose levels vary widely in cockroach hemolymph, and concentrations as high as 700 mg/dl can be observed after handling stress caused by repetitive sampling of the hemolymph.[1] The ability to carry a high level of carbohydrate is an advantage to an animal with an open circulatory system, such as an insect. Hemolymph distribution is inefficient in open systems and poorly directed so that circulating metabolites must be present in high concentrations to assure that adequate supplies are available to the tissues at critical times. Trehalose, as a disaccharide, has the advantages of being carried at high concentrations in the blood and delivering the energy potential of two glucose molecules, but with only half the osmotic effect.

Stimulation of trehalose synthesis by the corpora cardiaca (CC) neuroendocrine complex was initially reported in the American cockroach, *Periplaneta americana*.[1] An identical hypertrehalosemic effect by the CC was confirmed about the same time in the tropical cockroach *Blaberus discoidalis*.[2] In both animals, the hypertrehalosemic factor stimulates glycogen phosphorylase activity in the fat body to degrade glycogen and elevate trehalose synthesis.[2,3] Phosphorylase activity and glucose-6-phosphate (G-6-P) levels double in fat bodies exposed to CC extracts.[4] The increases in phosphorylase activity and G-6-P suggest that the hormone stimulates the formation of trehalose by degrading glycogen stores to the carbohydrate intermediates that serve as trehalose precursors.

Recently, the hypertrehalosemic hormones (HTHs) of *P. americana* and *B. discoidalis* were isolated and structurally characterized (Table 1).[5,6] The *Blaberus* hormone is also found in *Nauphoeta cinerea* cockroaches.[7] The cockroach HTHs belong to a family of adipokinetic-hyperglycemic factors of which there are now 13 bioanalogues reported from 11 species of insects and 1 from crustaceans. In all cases, these neurohormones stimulate the release of either lipid or carbohydrate by the fat body of the insect species in which they were tested.

The cockroach HTHs appear to activate glycogen phosphorylase in the cockroach fat body by stimulating adenylate cyclase and cyclic AMP (cAMP) formation in the same manner that glucagon activates phosphorylase in the vertebrate liver.[3] CC extracts elevate cAMP in the fat body of *P. americana* by twofold with no effect on cGMP levels.[8] The cAMP increases within 5 min of treatment with the CC extracts and remains elevated for at least 1 h. The synthetic HTHs (M1, M2) of *P. americana* increase both cAMP and glycogen phosphorylase of the fat body in a dose-dependent manner *in vivo*.

In vitro studies have failed to confirm the activation of glycogen phosphorylase by the formation of hormone-dependent cAMP. M1 and M2 activate glycogen phosphorylase and stimulate trehalose synthesis in fat bodies from *P. americana in vitro*, but they fail to elevate cAMP.[10] However, crude CC extracts elevate both cAMP and glycogen phosphorylase in

TABLE 1
Neurohormones that Affect the Cockroach Fat Body

Hormone	Structure
Blaberus HTH	pGlu-Val-Asn-Phe-Ser-Pro-Gly-Trp-Gly-Thr-NH_2
Periplaneta M1	pGlu-Val-Asn-Phe-Ser-Pro-Asn-Trp-NH_2
Periplaneta M2	pGlu-Leu-Thr-Phe-Ser-Pro-Asn-Trp-NH_2
Locusta AKH 1	pGlu-Leu-Asn-Phe-Thr-Pro-Asn-Trp-Gly-Thr-NH_2
Red pigment-concentrating hormone	pGlu-Leu-Asn-Phe-Ser-Pro-Gly-Trp-NH_2

the fat body. Forskolin activates adenylate cyclase; it also increases glycogen phosphorylase activity in the fat body *in vitro*, but without a concomitant elevation of trehalose production.[10] These results suggest that endocrine factors in the CC stimulate adenylate cyclase, glycogen phosphorylase, and trehalose synthesis, but the elevation of cAMP and phosphorylase activity may not be related to the production of trehalose. Furthermore, M1 and M2 may not be the CC factors that account for these effects.

Neurohormone treatment elevates the pools of fat body metabolites related to trehalose formation. The levels of G-6-P and fructose-6-phosphate double in the fat body of CC extract-treated *P. americana*, but fructose-1,6-diphosphate increases only 50%.[4] In contrast, fat body citrate levels increase by fivefold over controls. The fat body does not contain measurable quantities of intracellular trehalose,[11] so the increased production of trehalose by the fat body resides at the biosynthetic level and not at the level of membrane permeability and release. Exogenous [^{14}C]glucose is incorporated readily into trehalose; however, treatment with CC extracts reduces the amount of exogenous [^{14}C]glucose incorporated. Presumably, the increase in the pool of trehalose precursors in the fat body as a result of phosphorylase activation and glycogen degradation decreases the incorporation of exogenous [^{14}C]glucose into trehalose.

The research indicates that the primary action of the HTHs is to activate glycogen phosphorylase activity and increase the pools of trehalose precursors in the fat body. The exact nature of the mechanism for this activation and the role of the second messenger cyclic nucleotides are still in question. There is no evidence to indicate that the hormones increase either the rate of trehalose synthesis directly or the permeability of adipocytes to glucose influx or trehalose efflux.

Calcium is essential for the endocrine regulation of trehalose synthesis by phosphorylase activation.[10,12,13] CC extract, along with cAMP and theophylline (a phosphodiesterase inhibitor), increases the influx of Ca^{2+} into adipocytes,[13] and the presence of Ca^{2+} increases the rate of trehalose efflux from fat body exposed to CC extracts by nearly twofold.[13] The increased trehalose efflux from hormone-stimulated adipocytes may assist in preventing intracellular feedback inhibitions imposed by the elevated rates of trehalose synthesis.[13] Mg^{2+} does not replace Ca^{2+} in phosphorylase activation.

B. NEUROENDOCRINE EFFECTS ON LIPID METABOLISM

Neurohormones affect lipid metabolism in cockroaches. Adipokinetic hormone 1 (AKH1; Table 1) from the CC of locusts stimulates the release of lipids from the locust fat body during migratory flight.[14,15] However, in cockroaches, neither locust nor cockroach CC extracts stimulate lipid release from the fat body, even though the cockroach CC extracts cause lipid release in locusts.[16] However, extracts of cockroach CC stimulate lipid uptake by the cockroach fat body[17] and the release of carbohydrates.[18] Both natural and synthetic M1 and M2 peptides elevate hemolymph concentrations and lipids in *P. americana*,[5] presumably by affecting synthesis of the metabolites by the fat body.

CC extracts affect the metabolism of both lipids and carbohydrates in the cockroach fat

body. The addition of CC extracts to *in vitro* fat body from the cockroach *Leucophaea maderae* elevated the rate of whole-body oxygen consumption by 27%.[19,20] The increased respiratory activity was accompanied by a concomitant lowering of the rate of CO_2 production and a decrease in the respiratory quotient from 0.94 to 0.84. The lowering of the respiratory quotient indicated that the CC extract caused a shift from the use of carbohydrate to lipid as the principal substrate for energy. The shift in substrates coincided with elevated phosphorylase activity and increased trehalose synthesis. Finally, the rate of conversion of [^{14}C]acetate and [^{14}C]palmitate to [^{14}C]O_2 was increased by ca. 30% after CC extract treatments.[19,20] The increased use of lipid as the primary energy source after exposure to HTH was supported by the observation that citrate increased fivefold in the fat body of CC extract-treated *P. americana*.[4] These studies indicate that AKH/HTH factors regulate fat body metabolism in cockroaches by causing a shift from carbohydrates to lipids as the source of energy during the period when carbohydrate stores are being mobilized for trehalose synthesis.

Mobile carbohydrate reserves are maintained at high levels in cockroaches, since carbohydrate is the principal source of energy for peripheral tissues such as the muscles that are essential for escape and the evasion of predators. In cockroaches, the HTHs are secreted continuously to maintain hemolymph trehalose levels, since head ligation results in a nearly 65% decrease in blood trehalose within 2 h.[21] The open circulatory system of insects does not support well-directed hemolymph circulation and is slow relative to vertebrates. Therefore, it is likely that cockroaches must at all times maintain a high level of trehalose, in excess of normal needs, so that there is a mobile reserve of carbohydrate readily available to meet emergency energy demands such as an attack by a predator. If the cockroach is successful in evading danger, then the HTH can replenish trehalose back to normal levels.

HTH maintains trehalose equilibrium in the cockroach hemolymph in much the same way that glucagon maintains blood glucose levels in vertebrate animals. It may take 30 min for HTH to elevate trehalose levels, and a maximum is reached at 5 h.[3] To cope with immediate demands for elevated trehalose the cockroach uses octopamine, which acts in a manner comparable to vertebrate epinephrine to provide a rapid response. Octopamine, an important biogenic amine of insects which appears to be relatively insignificant in vertebrates,[22] is secreted from abdominal ganglia in response to excitation and stimulates the production of trehalose by the fat body.[23] This response has been named the excitation-induced hypertrehalosemic (EXIT) response.[23] During the EXIT response, octopamine elevates hemolymph trehalose to a maximum within 15 min, with a measurable increase within 5 min. This correlates with an octopamine-induced increase in cAMP levels within 10 min in the *in vitro* fat body from *P. americana* at doses that stimulate trehalose production.[24] Injections of octopamine elevate the activity of fat body glycogen phosphorylase within 1 min.[25] Unlike the HTH response, which persists for more than 24 h,[3] the EXIT response begins to decline after 20 min and hemolymph trehalose levels return to normal after about 1 h.[23]

The cockroach, then, has two neuroendocrine factors that determine the production of circulatory carbohydrate. Octopamine causes a rapid, short-lived response by the fat body to produce a burst of trehalose in response to immediate stress stimuli in much the same way that the vertebrate liver responds to epinephrine with a burst of glucose production. Cockroach HTH serves as a slower-acting regulator for blood carbohydrate production that elevates and maintains circulating carbohydrate levels in much the same way that glucagon maintains glucose levels in vertebrate blood.

C. NEUROENDOCRINE REGULATION OF PROTEIN SYNTHESIS

Neuroendocrine regulation of fat body protein synthesis is poorly defined in insects in general and in cockroaches in particular. Although general fat body protein synthesis is not

affected when CC are implanted into decapitated female *L. maderae*,[26] the fat body may respond to neurohormones by synthesizing specific proteins. The concentrations of five out of six hemolymph proteins increase in decapitated females after implants of brain or CC.[27] Brain implants stimulate the production of four of the proteins, and three increase in response to CC implants. These findings suggest that neurohormones may regulate the synthesis of specific proteins by the fat body.

Evidence suggests that neurohormones may be important as regulators for general protein synthesis in the cockroach fat body. Treatment of decapitated female *B. discoidalis* with juvenile hormone (JH), CC extracts, and JH + CC extracts demonstrated that CC extracts alone stimulated the synthesis of proteins secreted from the fat body by only 38% over controls. Suboptimal dosages of JH increased the synthesis rate of soluble proteins by 270%, and JH + CC increased soluble protein synthesis by 400%.[28] These results indicate that the CC contain factors that stimulate fat body protein synthesis in general. This effect was only observed when protein synthesis was stimulated by a dose of JH to promote protein synthesis related to reproduction. The JH dose used was suboptimal, and the protein synthesis was below that observed for normal intact females during vitellogenesis; however, treatment with a combination of JH and CC extract caused a protein synthesis rate comparable to that in normal vitellogenic females. These findings suggest that neurohormonal factors in the CC affect general metabolic processes in the fat body and determine its level of biosynthetic activity.

II. NEUROENDOCRINE EFFECTS ON FAT BODY MITOCHONDRIOGENESIS

A. MITOCHONDRIOGENESIS IN HIGHER ANIMALS

One general physiological process of animals for which the cockroach fat body is a useful model is the endocrine regulation of mitochondriogenesis. Mitochondria contain DNA and have an intrinsic capacity for protein synthesis (see reviews by Borst[29] and by Schatz and Mason[30]). The presence of an intrinsic genetic potential along with a translational capacity suggests that mitochondria *per se* may control the synthesis of products important in the formation of new mitochondrial units. Evidence exists for the production of regulatory proteins from mitochondria that communicate with the nucleus during mitochondriogenesis and regulate the production of nucleus-directed protein products for mitochondrial structures and enzymes.[31-33] Mitochondriogenesis has been and remains a topic of intense investigation.

New mitochondria are produced in cells during times of cell growth and in response to increased energy demands. How the formation of new mitochondria is regulated in higher animals is uncertain. In simple organisms, such as yeast or the slime mold *Neurospora crassa*, mitochondriogenesis is stimulated by environmental changes such as removal of glucose from the medium and the release of glucose repression or conversion from an anaerobic to an aerobic environment. Such changes will result in rapid formation of new mitochondria. Mitochondriogenesis is not easily induced in higher animals, and regulatory controls are not well defined. The thyroid and growth hormones are implicated in the formation of new mitochondrial units in the tissues of vertebrate animals.[34-37]

It is difficult to investigate mitochondriogenesis and its regulation by hormones in higher animals. The biosynthetic events of mitochondriogenesis are investigated most easily during synchronous phases of cell growth, but tissues that exhibit both synchronous growth and defined endocrine regulation are not readily available in vertebrate animals. Synchronous growth occurs during embryogenesis, but endocrine involvement is unclear at this time. Endocrine regulation is best defined in postembryonic animals, but synchronous growth is infrequent. Rapid regeneration can be induced in the mammalian liver or kidney after partial hepatectomy or nephrectomy, but it is difficult to define normal endocrine controls when

the organs and the organism have sustained such trauma. Finally, cell cultures can be used to generate synchronous growth, but again, endocrine regulation is uncertain.

B. NEUROENDOCRINE EFFECTS ON FAT BODY RESPIRATORY MATURATION

The neuroendocrine system affects events of mitochondriogenesis in the cockroach fat body. A 25% reduction in whole body respiration is observed 30 d after removal of both the CC and corpora allata (CA) from adult male *B. discoidalis.*[38] Removal of only the CA has no effect on the respiration, indicating that the effective glands are the CC. The decrease in whole body basal metabolism was traced to a nearly 50% decrease in respiration by the intact fat body and by isolated fat body mitochondria.[39] The respiratory decrease is reflected by both the succinate-cytochrome *c* reductase and cytochrome *c* oxidase enzymes of electron transport.[40] A series of injections of CC extract into CC + CA-deficient animals returns the respiratory activity of fat body mitochondria to normal levels.[41] CA extracts have no effects. Furthermore, the direct addition of CC extracts to isolated mitochondria has no stimulative effect on the rate of oxygen consumption. The CC extracts stimulate respiratory activity of fat body mitochondria from CC + CA-deficient cockroaches only after repeated, daily injection of CC extracts over a period of several days.

In conclusion, the neuroendocrine system affects the respiratory enzymes of fat body mitochondria. After neuroendocrine deficiency by CC ablation, the respiratory activities decrease by about 50%. Hormone replacement therapy with CC extracts returns the respiratory capacity of the fat body mitochondria to normal. This is reminiscent of the situation in thyroidectomized vertebrate animals. Mitochondria from the livers of thyroidectomized rats have 50% lower succinate oxidase activity,[42] and injection of thyroid hormones returns the respiratory rate to 50% of normal in 3 h, with elevated activities for cytochromes *b* and *c*.

Fat body mitochondriogenesis was examined to determine if neuroendocrine effects on mitochondrial respiration resulted from disruption of mitochondrial formation or function. In all of the previous experiments, the CC were removed within the first several days of adult life and the effects of CC deficiency on respiratory activity were measured about 1 month later. Therefore, any mitochondriogenetic activity occurring in the fat body early in the life of the adult insect would be overlooked and only observed at 30 d as an absence of function, presumably in response to chronic endocrine deficiency.

Unlike vertebrate systems, where endocrine regulation and synchronous cell growth are difficult to study simultaneously, insect cells undergo cycles of growth and development that correlate with the molting cycles (i.e., are under endocrine control). During these growth cycles, new mitochondria are formed in a synchronous manner. The events can be studied by timing them relative to the time of molting. Thus, the cockroach fat body is a useful model for studying mitochondriogenesis and its regulation.

In the tropical cockroach *B. discoidalis*, the fat body undergoes developmental maturation during the first week of adult life. This maturation may be likened to a postemergence metamorphosis in that the adipocytes of the fat body remain intact, but the cytoplasm is degraded and reconstituted. Just prior to the cytoplasmic reorganization, mitochondria are rare and small with a poorly defined internal structure. At the time of adult emergence, the mitochondria are restricted to narrow strands of cytoplasm along the plasma membrane.[43] However, by day 4 the adipocyte mitochondria are well defined, abundant, and distributed throughout the adipocyte cytoplasm.

In agreement with the early structural development of the fat body mitochondria from adult male *B. discoidalis*, there is also respiratory development during the first 10 d of adult life.[44] The mitochondria have a low rate of O_2 consumption in newly emerged adults. The respiratory capacity of the mitochondria increases threefold during the next 10 d. After 10

d of adult age, the respiratory activity of the mitochondria remains constant through at least the next 30 d. The 10-d developmental pattern in O_2 consumption is reflected by increases in both succinate-cytochrome *c* reductase and oxidase.[44] Finally, respiratory control and ADP:O ratios also increase during the 10-d developmental period.

Removal of the CC + CA at adult emergence blocks the maturation of the respiratory enzymes beyond the 5-d level, which is 40 to 50% below the final level of mature mitochondria.[44] This finding suggests that the increase in respiratory activity that occurs between days 0 and 5 is endocrine-independent. However, the neuroendocrine system is required for the final maturation of respiratory capacity that occurs between days 5 and 10. Although respiratory control by ADP is lower in the CC-deficient animals, the ADP:O ratios remain normal. Treatment of coupled mitochondria with 2,4-dinitrophenol, an uncoupling agent, does not change the rates of state 4 (O_2 and substrate present; ADP absent) and state 3 (ADP, O_2, and substrate present).[44] This lack of response to the uncoupling agent suggests that electron transport, and not exidative phosphorylation, is the rate-limiting step for respiration in fat body mitochondria of *B. discoidalis*. Therefore, the decrease in respiration and electron transport activity that follows removal of the CC most likely results from a decrease in the cytochrome enzymes. These results suggest that a neurohormone present in the CC facilitates the formation of cytochrome enzymes for the respiratory maturation of fat body mitochondria in *B. discoidalis*.

C. NEUROENDOCRINE EFFECTS ON FAT BODY CYTOCHROME SYNTHESIS

Cytochrome synthesis was investigated using the heme-specific precursor aminolevulinic acid (ALA). ALA synthase (ALAS) is a mitochondrial enzyme that is rate limiting for heme synthesis.[45] ALAS was measured in mitochondria isolated from the fat body of adult male *B. discoidalis*.[46] A peak of ALAS activity occurs between 4 and 6 d of adult age. Thereafter, the ALAS activity declines to a sustained but minimal level that is maintained through day 30. Removal of the CC + CA at adult emergence does not affect the fat body ALAS activity at 30 d of age. These data suggest that the fat body has a single period of heme synthesis at days 4 to 6 followed by a low level of maintenance activity. The results do not indicate any neuroendocrine regulation at the level of ALAS.

The incorporation of [^{14}C]ALA into mitochondrial cytochromes was used as a means of assessing fat body heme synthesis and its neuroendocrine regulation.[47] Hemes a + b were separated from hemes c + c_1 in order to evaluate whether there were differential controls for cytochrome synthesis. Both cytohemes a + b and c + c_1 show peaks of synthesis in the fat body mitochondria between days 4 and 6 with a decline to minimal but sustained levels through 30 d.[47] Removal of the CC + CA or ligation of the head at emergence eliminates the peak of heme a + b synthesis at day 4.[47] Neurohormone replacement therapy was performed by injecting CC extract on days 2, 3, and 4 followed by measurement of [^{14}C]ALA incorporation on day 4. [^{14}C]ALA incorporation into hemes a + b returns to normal (day 4) levels in response to the CC extract injections. However, the synthesis of cytohemes c + c_1 is unaffected by the CC extracts. The results of studies on ALAS activity and [^{14}C]ALA incorporation indicate that maximum mitochondrial cytochrome synthesis occurs between days 4 and 6 and that incorporation of ALA into cytohemes a + b is neuroendocrine dependent.

The properties of the cockroach fat body make it a nearly ideal model for studies on the endocrine regulation of mitochondriogenesis in higher animals. The fat body mitochondria undergo synchronous maturation during a brief period that can be timed by the obvious event of the adult molt. In addition, there is neuroendocrine regulation of cytochrome synthesis that occurs naturally around the fourth day after the molt.

We have termed the neuroendocrine factor that regulates heme synthesis the *cytochromo-*

genic hormone (CGH).[48] We have been pursuing the isolation of CGH for several years. Early in the isolation process we found that the factor that stimulates cytochrome synthesis coisolates with the hormone that affects trehalose synthesis by the fat body.[6] We have isolated and characterized HTH and found it to be a heat-stable decapeptide (Table 1) which stimulates cytoheme a + b synthesis in the same manner as CGH. We cannot determine yet whether CGH and HTH are identical. Further research is in progress to confirm the identity of CGH.

III. CONCLUSIONS

The data suggest that although the insect fat body is not structurally comparable to the vertebrate liver, it is functionally similar. Studies on fundamental biological processes such as the mode of action of neurohormones on subcellular processes and endocrine regulation of processes related to mitochondriogenesis may be similar in both cockroaches and vertebrates, and the less expensive cockroach system may be used as a research model. The field of vertebrate biochemistry is far ahead of insect biochemistry because of its potential applications to human health. However, as we seek new biological models that can serve as illustrations for fundamental biochemical properties of systems common to all animals, the insects (and the cockroaches in particular) will become more attractive. Insects, like mammals, are highly evolved, complex animals, and there is no reason why fundamental processes such as neuroendocrine regulation and neuronal functions cannot be investigated in cockroaches and be applicable to medically related problems of vertebrate animals.

ACKNOWLEDGMENTS

Personal research cited in this review was supported in part by NSF grants PCM 7403606, PCM 8103277, and DCB 8511058, NIH grants TMP A115190 and 1 RO1 NS 20137, and by the Robert J. Kleberg, Jr. and Helen C. Kleberg Foundation.

REFERENCES

1. **Keeley, L. L.,** personal observation, 1965.

1a. **Steele, J. E.,** Occurrence of a hyperglycemic-factor in the corpus cardiacum of an insect, *Nature (London),* 192, 680, 1961.

2. **Bowers, W. S. and Friedman, S.,** Mobilization of fat body glycogen by an extract of corpus cardiacum, *Nature (London),* 198, 685, 1963.

3. **Steele, S. E.,** The site of action of insect hyperglycemic hormone, *Gen. Comp. Endocrinol.,* 3, 46, 1963.

4. **Steele, J. E., Coulthart, K. C., and McClure, J. B.,** Control of hexose phosphate and citrate in fat body of the cockroach (*Periplaneta americana*) by the corpus cardiacum, *Comp. Biochem. Physiol.,* 79B, 559, 1984.

5. **Scarborough, R. M., Jamieson, G. C., Kalish, F., Kramer, S. J., McEnroe, G. A., Miller, C. A., and Schooley, D. A.,** Isolation and primary structure of two peptides with cardioacceleratory and hyperglycemic activity from the corpora cardiaca of *Periplaneta americana, Proc. Natl. Acad. Sci. U.S.A.,* 81, 5575, 1984.

6. **Hayes, T. K., Keeley, L. L., and Knight, D. W.,** Insect hypertrehalosemic hormone: isolation and primary structure from *Blaberus discoidalis* cockroaches, *Biochem. Biophys. Res. Commun.,* 140, 674, 1986.

7. **Gade, G. and Rinehart, K. L., Jr.,** Amino acid sequence of a hypertrehalosemic neuropeptide from the corpus cardiacum of the cockroach, *Nauphoeta cinerea, Biochem. Biophys. Res. Commun.,* 141, 774, 1986.

8. **Gade, G.,** Effect of corpus cardiacum extract on cyclic AMP concentration in the fat body of *Periplaneta americana, Zool. Jahrb. Physiol.,* 81, 245, 1977.

9. **Gade, G.,** Mode of action of the hypertrehalosaemic peptides from the American cockroach, *Z. Naturforsch.,* 40c, 670, 1985.

10. **Orr, G. L., Gole, J. W. D., Jahagirdar, A. P., Downer, R. G. H., and Steele, J. E.,** Cyclic AMP does not mediate the action of synthetic hypertrehalosemic peptides from the corpus cardiacum of *Periplaneta americana, Insect Biochem.,* 15, 703, 1985.
11. **Steele, J. E. and Hall, S.,** Trehalose synthesis and glycogenolysis as sites of action for the corpus cardiacum in *Periplaneta americana, Insect Biochem.,* 15, 529, 1985.
12. **McClure, J. B. and Steele, J. W.,** The role of extracellular calcium in hormonal activation of glycogen phosphorylase in cockroach fat body, *Insect Biochem.,* 11, 605, 1981.
13. **Steele, J. E. and Paul, T.,** Corpus cardiacum stimulated trehalose efflux from cockroach *(Periplaneta americana)* fat body: control by calcium, *Can. J. Zool.,* 63, 63, 1985.
14. **Mayer, R. J. and Candy, D. J.,** Control of haemolymph lipid concentration during locust flight: an adipokinetic hormone from the corpus cardiacum, *J. Insect Physiol.,* 15, 611, 1969.
15. **Goldsworthy, G. J., Johnson, R. A., and Mordue, W.,** *In vivo* studies on the release of hormones from the corpora cardiaca of locusts, *J. Comp. Physiol.,* 79, 85, 1972.
16. **Goldsworthy, G. J., Mordue, W., and Guthkelch, J.,** Studies on insect adipokinetic hormones, *Gen. Comp. Endocrinol.,* 18, 545, 1972.
17. **Downer, R. G. H. and Steele, J. E.,** Hormonal stimulation of lipid transport in the American cockroach, *Periplaneta americana, Gen. Comp. Endocrinol.,* 19, 259, 1972.
18. **Holwerda, D. S., Weeda, E., and van Doorn, J. M.,** Separation of the hyperglycemic and adipokinetic factors from the cockroach corpus cardiacum, *Insect Biochem.,* 7, 477, 1977.
19. **Weins, A. W. and Gilbert, L. I.,** Regulation of cockroach fat-body metabolism by the corpus cardiacum *in vitro, Science,* 150, 614, 1965.
20. **Weins, A. W. and Gilbert, L. I.,** Regulation of carbohydrate mobilization and utilization in *Leucophaea maderae, J. Insect Physiol.,* 13, 779, 1967.
21. **Keeley, L. L.,** The Corpus Cardiacum as a Metabolic Regulator in *Blaberus discoidalis* Serville (Blattidae), Ph.D. dissertation, Purdue University, West Lafayette, IN, 1966.
22. **Axelrod, J. and Saavedra, J. M.,** Octopamine, *Nature (London),* 265, 501, 1977.
23. **Downer, R. G. H.,** Induction of hypertrehalosemia by excitation in *Periplaneta americana, J. Insect Physiol.,* 25, 59, 1979.
24. **Gole, J. W. D. and Downer, R. G. H.,** Elevation of adenosine 3′,5′-monophosphate by octopamine in fat body of the American cockroach, *Periplaneta americana, Comp. Biochem. Physiol.,* 64C, 223, 1979.
25. **van Marrewijk, W. J. A., van den Broek, A. Th. M., and Beenakkers, A. M. Th.,** Regulation of glycogen phosphorylase activity in fat body of *Locusta migratoria* and *Periplaneta americana, Gen. Comp. Endocrinol.,* 50, 226, 1983.
26. **Luscher, M.,** Hormonal control of respiration and protein synthesis in the fat body of the cockroach *Nauphoeta cinerea* during oocyte growth, *J. Insect Physiol.,* 14, 499, 1968.
27. **Scheurer, R.,** Endocrine control of protein synthesis during oocyte maturation in the cockroach, *Leucophaea maderae, J. Insect Physiol.,* 15, 1411, 1969.
28. **Keeley, L. L., Sowa, S. M., Hayes, T. K., and Bradfield, J. Y.,** Neuroendocrine and juvenile hormone effects on fat body protein synthesis during the reproductive cycle in female *Blaberus discoidalis* cockroaches, *Gen. Comp. Endocrinol.,* 72, 364, 1988.
29. **Borst, P.,** Mitochondrial nucleic acids, *Annu. Rev. Biochem.,* 42, 333, 1972.
30. **Schatz, G. and Mason, T. L.,** The biosynthesis of mitochondrial proteins, *Annu. Rev. Biochem.,* 43, 51, 1974.
31. **Barath, Z. and Kuntzel, H.,** Cooperation of mitochondrial and nuclear genes specifying the mitochondrial genetic apparatus in *Neurospora crassa, Proc. Natl. Acad. Sci. U.S.A.,* 69, 1371, 1972.
32. **Barath, Z. and Kuntzel, H.,** Induction of mitochondrial RNA polymerase in *Neurospora crassa, Nature New Biol.,* 240, 195, 1972.
33. **Poyton, R. O.,** Cooperative interaction between mitochondrial and nuclear genomes: cytochrome *c* oxidase assembly as a model, *Curr. Top. Cell. Regul.,* 17, 231, 1980.
34. **Maddaiah, V. T., Weston, C. L., Chen, S. Y., and Collipp, P. J.,** Growth hormone and liver mitochondria. Effects on cytochromes and some enzymes, *Arch. Biochem. Biophys.,* 173, 225, 1976.
35. **Maddaiah, V. T., Collipp, P. J., Lin, J. H., and Duffy, J. L.,** Growth hormone and liver mitochondria effect on morphology and protein turnover, *Biochem. Med.,* 16, 47, 1976.
36. **Nelson, B. D., Mutvei, A., and Joste, V.,** Regulation of biosynthesis of the rat liver inner mitochondrial membrane by thyroid hormone, *Arch. Biochem. Biophys.,* 228, 41, 1984.
37. **Booth, F. W. and Holloszy, J. O.,** Effect of thyroid hormone administration on synthesis and degradation of cytochrome c in rat liver, *Arch. Biochem. Biophys.,* 167, 674, 1975.
38. **Keeley, L. L. and Friedman, S.,** The corpus cardiacum as a metabolic regulator in *Blaberus discoidalis* Serville (Blattidae). I. Long-term effects of cardiacectomy on whole body and tissue respiration and on trophic metabolism, *Gen. Comp. Endocrinol.,* 8, 129, 1967.
39. **Keeley, L. L. and Friedman, S.,** Effects of long-term cardiacectomy-allatectomy on mitochondrial respiration in the cockroach, *Blaberus discoidalis, J. Insect Physiol.,* 15, 509, 1969.

40. **Keeley, L. L.,** Endocrine effects on the biochemical properties of fat body mitochondria from the cockroach, *Blaberus discoidalis, J. Insect Physiol.,* 17, 1501, 1971.
41. **Keeley, L. L. and Waddill, V. H.,** Insect hormones: evidence for a neuroendocrine factor affecting respiratory metabolism, *Life Sci.,* 10(2), 737, 1971.
42. **Bronk, J. R.,** Thyroid hormone: effects on electron transport, *Science,* 153, 638, 1966.
43. **Keeley, L. L.,** Neuroendocrine regulation of fat body development and function, in *Energy Metabolism and Its Regulation in Insects,* Downer, R. G. H., Ed., Plenum Press, New York, 1981, 207.
44. **Keeley, L. L.,** Biogenesis of mitochondria: neuroendocrine effects on the development of respiratory functions in fat body mitochondria of the cockroach *Blaberus discoidalis, Arch. Biochem. Biophys.,* 153, 8, 1972.
45. **Granick, S. and Urata, G.,** Increase in activity of d-aminolevulinic acid synthetase in liver mitochondria induced by feeding of 3,5-dicarbethoxy-1,4-dihydrocollidine, *J. Biol. Chem.,* 238, 821, 1963.
46. **Keeley, L. L.,** Development and endocrine regulation of mitochondrial cytochrome biosynthesis in the insect fat body. Delta-aminolevulinic acid synthase, *Arch. Biochem. Biophys.,* 187, 78, 1978.
47. **Keeley, L. L.,** Development and endocrine regulation of mitochondria cytochrome biosynthesis in the insect fat body. Delta-[^{14}C]aminolevulinic acid incorporation, *Arch. Biochem. Biophys.,* 187, 87, 1978.
48. **Hayes, T. K. and Keeley, L. L.,** Cytochromogenic factor: a newly-discovered neuroendocrine agent stimulating mitochondrial cytochrome synthesis in the insect fat body, *Gen. Comp. Endocrinol.,* 45, 115, 1981.

Chapter 20

REGULATION OF VITELLOGENESIS IN COCKROACHES

Franz Engelmann

TABLE OF CONTENTS

I. INTRODUCTION

Vitellogenesis, i.e., the production of the mature egg, involves the massive deposition of proteinaceous and lipid yolk, later utilized in embryogenesis. Many extrinsic and intrinsic factors, often species specific, influence this process. Among these are the availability of food to mature and immature animals, photoperiodic conditions, temperature, presence of mates, and ultimately hormone production, which is itself influenced by other factors.[1,2] Any one of these components may not and usually does not stand alone. Consequently, the intricate interrelationships have become fascinating research topics during the past several decades; still, few if any species are comprehensively understood today. Cockroach species have often been found amenable to in-depth and integrated studies leading to the formulation of models. As a matter of fact, many of the primary findings on various aspects of reproduction in insects were worked out on appropriate cockroach species.[1,2]

Several current symposia and books deal with specific protein synthesis under hormonal control, a central concern to endocrinologists. Hundreds of laboratories perform research on steroid hormone-affected processes in vertebrates, including the estrogen-controlled production of vitellogenin (Vg) in egg-laying species.[3] In a parallel and quite analogous fashion, juvenile hormone directs Vg synthesis in all cockroach species. Therefore, research on insect Vgs and their control has yielded key information and still is of far-reaching interest. The easy manipulation of cockroaches and their cost efficiency may make these insects the animal of choice for many questions still to be answered.

II. ENDOCRINE-REGULATED VITELLOGENESIS

A. INSECTS IN GENERAL

Wigglesworth[4] made the pioneering discovery in the bloodsucking bug *Rhodnius prolixus* that the corpus allatum (CA) secretes a hormone essential for the promotion of egg growth. The trigger for production and release of this hormone is the adequate blood meal. Since this early report, CA-controlled vitellogenesis has been documented in many more insect species of several orders.[1,2] With the exception of certain moth species in which no endocrine control is apparent and certain Diptera where ecdysone also plays a crucial role, the CA hormone has been implicated in egg production as the only hormone involved. The CA produce the juvenile hormones (JHs), and from among the four known homologues it is JH III which is identified as the gonadotropic hormone in the majority of species studied.

When one reviews the progression in research interests of insect endocrinologists, one perceives that it was first important to establish the source which produces JH and determine how these glands may be regulated in the various species. Then, with the discovery of the predominant yolk protein precursor, Vg, essential for vitellogenesis,[5] and the identification of its endocrine control in a cockroach species,[6,7] research began to center on a detailed understanding of the molecular mechanisms involved. Research on the hormonal control of an easily recognizable yolk protein precursor became the major research effort for many laboratories. Today, research ranges from the environmental influence on endocrine gland activity[2] to the molecular biology of vitellogenin production (e.g., in *Drosophila* spp.[8,9] or *Locusta migratoria*[10]).

Native Vgs of insects are glycolipophosphoproteins of a molecular mass between 200 and 600 kDa. With the exception of certain dipterans, Vgs may be produced in multiple forms and processed intracellularly or in the hemolymph to yield differently sized "subunits".[11-14] Vgs are synthesized as exportable proteins primarily by the fat bodies, the equivalent to the vertebrate liver, but in certain Diptera the ovarian follicular epithelial cells are also capable of their synthesis. Vgs are usually transported through the hemolymph and taken up by the ovaries via endocytosis. This primary yolk protein, which makes up 60 to

80% of the total yolk protein, is then termed "vitellin" (Vt) in order to distinguish it from the circulating precursor. Vts are immunologically identical to Vgs, but may differ in their physicochemical properties and subunit compositions. Vts are the major source of amino acids for the growing embryos.

B. COCKROACHES

Since Scharrer[15] first showed the suitability of the ovoviviparous cockroach *Leucophaea maderae* for endocrinological studies, several additional species of cockroaches have been the animals of choice for research on vitellogenesis. Cockroaches have served as models for many aspects of hormonal control of reproduction. Vitellogenesis depends exclusively on the CA hormone in all cockroach species studied[2] and proceeds even after decapitation or in isolated abdomina, provided JH or its analogues have been supplied (as, for example, in *L. maderae*).[16] Brain control of the synthetic activity of the CA was first suggested for *L. maderae,* in which disruption of the neural connection to the brain caused a visible increase in the CA size, denoting gland activity.[17] This was followed by vitellogenesis.[18] Neuronal inhibition of the CA during the prolonged gestation period was postulated for this species[18] as well as for the truly viviparous *Diploptera punctata.*[19] Both of these cockroach species appeared particularly suitable for research on the inhibitory control mechanisms for the CA because of their necessary inactivation during gestation, which lasts 2 months or longer. Space constraints in the abdomen would not allow the maturation of another set of eggs before parturition. Details regarding CA controls, inhibitory and stimulatory, are discussed in Chapter 21 of this work. In all cases the brain has ultimate control over the CA. It translates stimulatory events, such as mating,[19] or inhibitory influences exerted by an egg case in the brood sac of a viviparous species[18-20] into the appropriate signal to the CA. As shown here, cockroach species have served as initial models for research on certain key issues regarding control of reproduction. The principal findings have been expanded to include females from insects of other orders.

III. THE MODES OF JUVENILE HORMONE ACTION IN COCKROACHES

JH-controlled vitellogenesis in cockroaches involves numerous physiological and biochemical processes, all of which must be analyzed in detail for a comprehensive understanding of egg production. Obviously, the most prominent event is the massive synthesis of Vg by the fat bodies. This is demonstrated relatively easily, since it occurs at precise times and Vg is identified qualitatively and quantitatively by immunological procedures. Equally as important as producing a normal egg are several additional processes which are under JH control.

A. THE VITELLOGENINS AND VITELLINS

Vgs and Vts are readily available in large quantities. However, detailed characteristics of these macromolecules are known for only three species of cockroaches. Most of the pertinent findings have been published recently. Technical difficulties in handling these large proteins and the associated problems with partial degradation during the lengthy isolations may account for the limited and sometimes conflicting information.[13]

Native Vgs of the hemolymph of *Leucophaea maderae* and other cockroaches are isolated by anion exchange chromatography (DEAE or QAE), since Vgs readily bind to ion exchangers at low salt concentrations.[2,21] Elution from the resins is effected with 0.4 to 0.45 *M* NaCl. The eluted Vgs contain only minor contaminants.[22] The same procedure has been equally effective in several cockroach species.[2,13] The isolation procedure for Vts is different, but fairly simple. Eggs or ovaries containing yolky oocytes are homogenized in saline with

TABLE 1
Characterization of Cockroach Vitellogenins (Vg) and Vitellins (Vt)

Species	Native M_r (kDa)	Pro-Vg in fat bodies (kDa)	Subunits in hemolymph or oocytes (kDa)							Ref.
Blattella germanica	560	240	Vg	160		102				26
			Vt			102	95		50	26
Periplaneta americana	~600	250	Vg/Vt			123	118		57	27
	440		Vg/Vt	140	135		63		59	28
		275	Vg_1	170		105				25
			Vt_1			105	92		78	25
		266	Vg/Vt_2			105	101		60	25
Leucophaea maderae	559	260?	Vg	179?		118	96	87	57	29
			Vg	160		109	98		57	30
			Vt			105	90	85	57	30
	550	215	Vg	155		112	95	92	54	14

0.4 *M* NaCl to prevent precipitation of the Vts. Vts are virtually insoluble in low ionic strength media. Dialysis against distilled water precipitates the Vts, and they are subsequently collected by centrifugation and then redissolved in saline containing 0.4 *M* NaCl.[13] Essentially pure Vts, as determined by polyacrylamide gel electrophoresis (PAGE), can be obtained by one repetition of this procedure.

L. maderae Vg is a glycolipoprotein with a molecular mass of 550 kDa; that of *Periplaneta americana* is 600 (440) kDa, while *Blattella germanica* Vg is 560 kDa (Table 1). For *P. americana*[23,24] and several additional species of Blattidae,[25] two native forms of Vg with slightly differing M_r have been reported. In *L. maderae,* however, only one Vg species could be identified.[22,25] Lipids make up between 7 and 15% and carbohydrates 7 to 8% of the native molecular weights of the Vgs and Vts. Carbohydrates consist primarily of mannose.[13] The Vgs of *L. maderae* and *B. germanica* are phosphorylated.[14,21,26]

Subunit compositions of Vgs and Vts were determined by SDS-PAGE analysis (Table 1). Similarities in the size classes of the subunits of all three cockroach species were observed (Table 1). In both Vgs and Vts there are three to four subunits in the range of 50 to 120 kDa. For the Vts, some degree of processing within the oocytes is indicated in certain instances.[26] Differences reported for the same species may have their origin in the application of different technical procedures.[25,27,28] It is noteworthy that the hemolymph Vgs of all three cockroach species may in addition contain a slightly larger unit.[14,25,26,30] In *L. maderae* this 155-kDa unit is phosphorylated[14] and is clearly identified. It is a transitory polypeptide and not a true subunit of the mature Vg.[14] This phosphorylated protein represents a very small amount of the total Vg.[14] Probably because of the relatively small amount seen by Coomassie blue staining, it was considered to be a questionable part of Vg.[29] In contrast, a 160-kDa unit is one of two major subunits of *B. germanica* hemolymph Vg.[26] These observations suggest genuine species differences in the structure and maturation of the Vgs.

Homogenates of female vitellogenic fat bodies of the three cockroach species mentioned contain a female-specific protein with a molecular mass of between 215 and 275 kDa (Table 1). This polypeptide appears to be the pro-vitellogenin which is not exported into the hemolymph, but will give rise to the presumed "subunits" of the mature hemolymph Vgs. This has been documented for *B. germanica*[26] and *L. maderae.*[14] In *L. maderae* it has been shown that the mature "subunits" are generated by two or three specific cleavage steps which occur in part within the fat bodies and are completed in the hemolymph.[14] Data obtained from time course studies and limited proteolysis of the various "subunits" showed that these cleavages did not result in polypeptides which could stoichiometrically account for the molecular weight of the primary translation product (Figure 1). It is not known how the positions of cleavage on the primary translation product are determined. Unlike *L.*

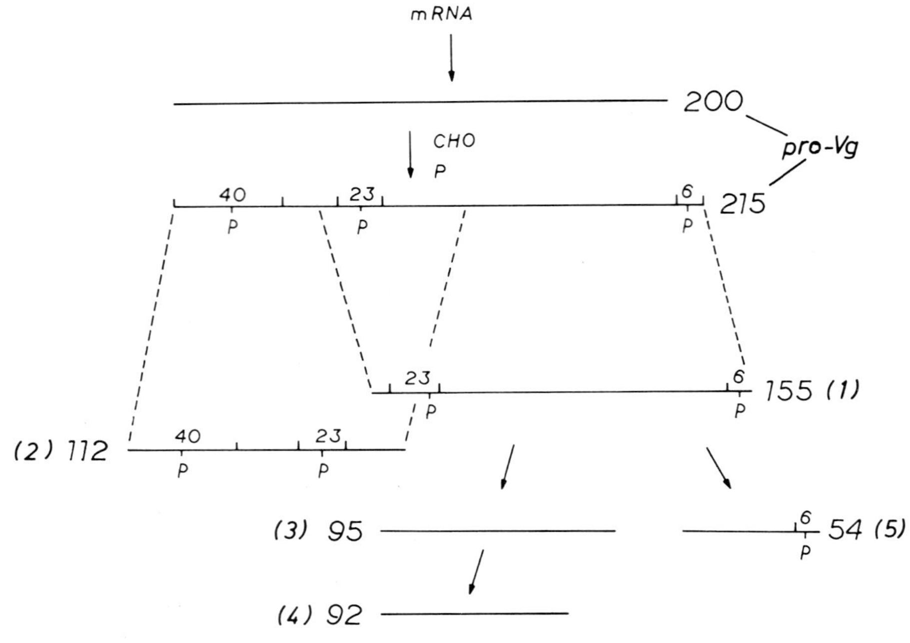

FIGURE 1. Scheme for the processing of the primary translation product, the pro-vitellogenin (pro-Vg) of *L. maderae*.[14] This conclusion was derived from data obtained from ^{35}S-methionine and ^{32}P labeling experiments by partial proteolysis of subunits and peptide mapping. The latter technique allowed the positioning of the 40-, 23-, and 6-kDa peptides. Note that the 215-kDa peptide is first clipped at two positions to yield the transitory 155-kDa peptide and a second peptide which is not processed further. CHO = carbohydrates; P = covalently attached phosphorus. Numbers in parentheses refer to the numbered "subunits" of the hemolymph Vg. (Reprinted with permission from della-Cioppa, G. and Engelmann, F., *Insect Biochem.*, 17, 401, 1987. Copyright 1987, Pergamon Journals, Ltd.)

maderae, no further processing of the hemolymph Vg occurs in *B. germanica.*[26] However, in this species the 160-kDa subunit is shown to give rise to two of the subunits of the mature Vts, which are identifiable in the oocytes before ovulation and onset of embryogenesis.

B. JUVENILE HORMONE-CONTROLLED VITELLOGENIN SYNTHESIS

Since the discovery that vitellogenesis depends on the presence of the CA (the source of JH) in *L. maderae,*[15] cockroaches have provided model systems for detailed analyses of hormone actions in insects. JH-controlled Vg synthesis was first suggested for *L. maderae*[6] and subsequently shown conclusively for this species.[7,31] Endocrine manipulations (i.e., allatectomy followed by JH application) in conjunction with the immunological identification of Vg (qualitatively and quantitatively) gave indisputable evidence for exclusive JH control. A similar experimental approach yielded comparable conclusions for *P. americana,*[23] *Byrsotria fumigata,*[32] *Nauphoeta cinerea,*[33] and, by inference, *Diploptera punctata.*[19,34] In cockroaches, no other hormones appear to be involved in the induction of Vg synthesis, since nearly complete egg growth could be achieved in isolated abdomina of *L. maderae* after treatment with a JH analogue.[16]

While the identification of the JH-induced end product, Vg, is fairly easy, knowledge concerning the details of the machinery involved in its biosynthesis requires a multifaceted research approach. Application of actinomycin D or α-amanitin limited the rate of Vg synthesis in *L. maderae,*[35] suggesting that the rate of transcription of the Vg mRNA had been reduced. In *L. maderae,* poly A+ mRNA could be extracted from vitellogenic fat bodies and translated into polypeptides precipitable by anti-Vg. Such mRNA was only

extractable from vitellogenic fat bodies and not from nonvitellogenic females or males.[36] This, together with the finding that Vg mRNA-containing polysomes are only found in vitellogenic fat bodies,[35] provides strong evidence that JH acts primarily at the transcriptional level.

C. THE JUVENILE HORMONE RECEPTOR

Hormones involved in the induction of specific proteins, such as Vg, are thought to act via hormone receptors in the cytosol and nuclei of the target tissues.[2,3] Cytosolic receptors may function in the translocation of the hormone to nuclear acceptor molecules, the latter being essential for the initiation of gene transcription.[3] This working hypothesis is thought to apply to JH and the induction of Vg in insects as well.[37] Functional JH receptors should have a high affinity for the natural hormone, exhibit saturability, and occur preferentially in the target tissue only.

Cytosol fat body preparations of *L. maderae* adults indeed contained a population of JH binding compounds with high affinity (K_d ca. 1 n*M*).[37] This compound could not be identified in nymphal tissues. Nymphs of this species normally do not make Vg. A JH binding compound with similar affinity for the hormone and identical sedimentation characteristics was extracted from nuclei of vitellogenic fat bodies of *L. maderae,* but not from nonvitellogenic tissues of nymphs or allatectomized adults.[38] Likewise, nuclei of adult male fat bodies did not contain this high-affinity JH binding compound. However, following treatment with methoprene (a JH analogue), which even induced the production of Vg in males, JH binding proteins with high affinity were extracted from those nuclei.[39] The correlation with Vg induction is striking. In addition, the natural JH enantiomer 10R-JH III had an even higher affinity for the putative receptor than the racemic mixture.[38]

The sum of these reports demonstrates that a high-affinity JH binding molecule was identified in *L. maderae* with the necessary characteristics expected of a hormone receptor (e.g., hormone saturability, tissue and stage specificity, and hormone specificity). Other species of insects have not yet been exploited to the same extent in this regard. However, it is very likely that these results will apply to other species as well. A cockroach species was most suitable for this initial search for a JH receptor because of the vast amount of background information on control of Vg production, the availability of adequate amounts of tissues, and the ability to manipulate the timing and staging of Vg production.

The significance of these findings goes beyond an interest in the biology of cockroaches and other insect species. Since we have shown that the model of hormone action worked out for vertebrate steroid hormones very likely applies to another class of hormone, the insect sesquiterpenoid JH, such an insect system may well serve as a generic experimental system for the molecular biology of hormone action. Basic principles of hormone-controlled events may well be similar in different animal groups.

D. THE JUVENILE HORMONE BINDING PROTEINS OF THE HEMOLYMPH

The sesquiterpenoid JHs are lipophilic molecules and have the tendency to adhere (bind) to lipids, lipoproteins, and even laboratory glassware. In order to reduce the degree of artifactual binding and its consequent interference with the biological information of interest, many researchers resort to coating glassware with polyethylene glycol. This precaution may help, but it does not reduce the number of potential artifacts generated by the many unknown components of any biological fluid. In other words, caution should be exercised in the interpretation of experimentally obtained JH binding data. Extensive controls are essential for a reliable evaluation of the actual data.[38,39]

JH binding proteins (JHBs) of the hemolymph had first been described for the silkmoth, *Hyalophora cecropia,*[40] and much of our knowledge regarding properties and characteristics of these proteins is derived from work with lepidopteran species.[41] The molecular weights

of the JH binding compounds of Lepidoptera are generally <100 kDa, and they usually bind JH with modest affinity (K_d = 1 to 0.1 μM).

Little information is available on hemolymph JH binding compounds in species outside the Lepidoptera. For *L. maderae,* however, a macromolecule with a molecular weight >350 kDa was identified which binds JH with relatively high affinity (K_d = 10 nM).[42] As shown, this hemolymph JHB has binding properties similar to the lower-affinity binding protein of fat body cytosol preparations of this species.[37] Since hemolymph proteins are synthesized in fat bodies, the assumption is made that these compounds are identical species of proteins. Following the earlier report,[42] more recent research identified this JHB of *L. maderae* as the lipophorin described for other species of insects.[43] It has a molecular weight of 430 kDa and is composed of one large and one small subunit (220 and 81 kDa).[43] A lipophorin with an M_r of 629 kDa was also described recently for *P. americana.*[44] The lipophorin of this species is composed of two subunits (245 and 79 kDa) and contains a lipid moiety of nearly 50%. These authors[44] assume that this hemolymph JHB is the unique JHB of this species.

JHBs of the insect hemolymph have been described primarily as JH carriers. However, additional functions of these molecules can be documented.[42] The major JHB in *L. maderae* may well function as a vehicle for distribution of JH because the concentration in the hemolymph can be higher than 10 μg/μl hemolymph during mid-vitellogenesis,[39] representing 10 to 15% of the total hemolymph protein concentration. Furthermore, this JHB of *L. maderae* has an affinity for JH III, with a K_d about 10 *n*M (i.e., approximately 1.5 to 2 orders of magnitude higher than the affinity of the JH esterases [JHEs] of the hemolymph).[45] Because of this high affinity of JH to the JHB one may conjecture that the JHB removes JH from the attack by JHEs. This is based on the assumption that the JHB covers the site of attack on the JH molecule. In *L. maderae,* at least, JHE activity rose about 1000-fold after removal of the JHB by anion exchange column chromatography, supporting the contention of substrate (JH) protection by the JHB in unfractionated hemolymph.[42,45]

The availability of pure JHB and JHE allowed an experiment to be performed which hitherto was not possible — namely, the direct identification of the effects of JHB on JHE activity *in vitro.* Pure JHB was added to a JH/JHE mixture prior to incubation. Kinetic analyses (Lineweaver-Burk plot) showed that JHB inhibited the JHE activity and functioned as an uncompetitive enzyme inhibitor.[39,42] This role of the JHB is unrelated to the high affinity for JH and the potential removal of JH from JHE attack. These reported results document a threefold action of the JHB of *L. maderae:* it functions as a JH carrier molecule, removes the substrate (JH) from the JHEs, and inhibits JHE activity. The JHB thus assures a prolonged half-life of JH in circulation. A natural half-life of more than 6 or 8 h had been determined for the JH of this species.[39] A generalization and application of these findings to other cockroach species cannot be made at this time, since detailed information is only available for *L. maderae.*

E. JUVENILE HORMONE ESTERASES

Hormone production and degradation must both be controlled efficiently in any system in order to achieve a functional hormone titer. In insects, JH degradation in circulation is generally accomplished by JHEs, whereas JH hydratases function primarily within the tissues.[46,47] Both JHE titers and JHE activities may fluctuate in some species. Consequently, rates of JH degradation may vary at different times during the animal's life cycle. It is important to mention here that it is JHE activity which is the functional measure and not the number of JHE molecules in circulation. In other words, the *in vitro* data do not necessarily reflect the situation within an intact animal. The reader is referred to Section III.D regarding the JH binding molecules of the hemolymph and their role in the determination of JHE activity in unfractionated hemolymph.

In adult females of *Diploptera punctata,* the JHE activity of diluted unfractionated

hemolymph was measured in relation to the vitellogenic cycles of the animals.[48] JHE activity toward the end of the vitellogenic period was higher than during mid-vitellogenesis; that is, lower JHE activity occurred when JH titers were very high. On the other hand, the JH analogue hydroprene increased JHE activity two- to threefold. This is an apparent paradox which was interpreted to signal that JH can, under certain circumstances, stimulate JHE production. In *L. maderae,* high JHE activity was measured throughout the vitellogenic cycle and no decline was seen during mid-vitellogenesis.[45] For this species it was concluded that the JHEs normally do not control or substantially influence the JH titers of the hemolymph. JHEs may, however, remove the remaining circulating JH efficiently once JH production by the CA has ceased.

In *L. maderae,* two JHEs have been isolated from hemolymph of adult vitellogenic females by anion exchange column chromatography.[45] By this relatively simple procedure pure JHEs were obtained, as indicated by PAGE. The two JHEs differ slightly in their K_ms and molecular weights. Both esterases have substrate-specific characteristics. Only JH is hydrolyzed, not the general esterase substrate α-naphthylacetate. In other words, genuine JHEs have been isolated. It is noteworthy that one of the esterases (JHE I) is present in the hemolymph at high levels at all times, regardless of the reproductive status of the animals. The mode of control of this esterase is unknown. The appearance of the second esterase (JHE II) is correlated with vitellogenic periods of the females, suggesting that JH may play a role in its production. Indeed, the JH analogue methoprene caused the appearance of the JHE II in allatectomized females.[45] Nymphs and adult males were never found to have JHE II and, furthermore, could not be induced to make it upon application of methoprene. The induction of a new JHE by JH or a JH analogue in *L. maderae* is the only case known in insects.[45] The detection of this JHE was possible because of the relative ease with which separation and purification were achieved.

F. JUVENILE HORMONE-STIMULATED SYNTHESIS OF THE ENDOPLASMIC RETICULUM

For several species of insects, including cockroaches, the rough endoplasmic reticulum (rER) of the fat bodies was more extensive during vitellogenesis than at other times. It is certainly more prominent during production of Vg than at other periods.[35] The implication was intuitive: JH may stimulate rER production. Synthesis of ER under JH influence was monitored through the incorporation of ^{32}P into ER phospholipids in *L. maderae*.[49,50] For this cockroach species, JH quantitatively stimulated the proliferation of the ER in the fat bodies. This could only be documented for fat bodies from adult females and not for nymphs or other tissues of adult females.[51] For no other species of insects do we have comparable experimental information available at this time. Again, a cockroach species was suitable for the initial demonstration of a novel and specific role of JH because in this species JH-controlled Vg production can be monitored with ease. There is no reason to believe that the same does not occur in other species.

Vg is synthesized on rER by polyribosomes containing the specific Vg mRNA, as shown for *L. maderae*.[52] Comparable information has been gathered in *Locusta migratoria*.[53] Vgs are exportable proteins and are synthesized by vectorial discharge, as are known exportable proteins in other animals. The availability of an extensive ER in the target tissue (i.e., the fat bodies) may facilitate the rapid and massive production of Vg. An extensive ER may provide additional binding sites for the Vg polyribosomes. Proliferation of the ER by JH stimulation is thus beneficial. This proliferation of the ER is, however, not an all-or-none event; JH only accelerates an ongoing process. In contrast to this, JH induces the transcription of the Vg mRNA *de novo*. Both of these JH-affected events synergistically lead to an efficient and massive synthesis of Vg during vitellogenesis.

JH-controlled ER proliferation in target tissues presumably is not restricted to cockroach

species, since electron micrographs from many other species allow us to speculate and expand our conclusions. Biochemical information is, however, only available for *Leucophaea maderae*.[49,50]

IV. SYNTHESIS AND OUTLOOK

The research on vitellogenesis in several species of cockroaches has been very fruitful from several points of view. This chapter serves to illustrate that conclusion. We cannot escape the notion that cockroaches are suitable insects for the biochemical and molecular biology approach, primarily because adequate amounts of material can be secured, vitellogenesis is regulated by a single hormone, and animals can be manipulated endocrinologically with ease. Details regarding JH-induced Vg production were first worked out in cockroaches, and many of the fundamental principles were found to be applicable to species of other orders. The research on cockroaches made it clear that JH regulates several additional processes associated with Vg production besides the initiation of transcription. Prominent among these are the influence on ER proliferation (occurring only in target tissues), induction of a second JHE in *Leucophaea maderae,* and the enhancement of production of the JHB.[39]

With the identification of the JH receptor in the fat bodies of *L. maderae,* researchers are now ready to explore the molecular mode of JH action at the locus of the Vg genes. How is transcription initiated by JH? We need to establish clones for the Vg gene(s) and identify regions upstream from the gene(s) which presumably recognize the hormone-occupied receptor. No such regions have been identified with certainty in either *Drosophila* or *Locusta migratoria*. While the molecular technology is available and has been exploited by researchers engaged in the study of steroid hormone actions, it is uncertain whether one can transfer these findings and conclusions to a hormone molecule of different chemistry, the sesquiterpenoid JH. Furthermore, we will have to search for the cellular localization of the JH receptor for the establishment of a model to explain how a receptor may be functional in the induction of transcription of a specific gene. Certainly, cockroach species will be profitably exploited once again. These insects may help to broaden the established concepts of steroid hormone action to encompass another type of hormone, or the findings may lead to modifications of these models.

REFERENCES

1. **Engelmann, F.,** *The Physiology of Insect Reproduction,* Pergamon Press, Oxford, 1970.
2. **Engelmann, F.,** Vitellogenesis controlled by juvenile hormone, in *Endocrinology of Insects,* Downer, R. G. H. and Laufer, H., Eds., Alan R. Liss, New York, 1983, 2.
3. **Tata, J. R., Ng, W. C., Perlman, A. J., and Wolffe, A. P.,** Activation and regulation of the vitellogenin gene family, in *Gene Regulation by Steroid Hormones,* Vol. 3, Roy, A. K. and Clark, J. H., Eds., Springer-Verlag, Berlin, 1987, 205.
4. **Wigglesworth, V. B.,** The function of the corpus allatum in the growth and reproduction of *Rhodnius prolixus* (Hemiptera), *Q. J. Microsc. Sci.,* 79, 91, 1936.
5. **Telfer, W. H.,** Immunological studies of insect metamorphosis. II. The role of a sex-limited blood protein in egg formation by the *Cecropia* silkworm, *J. Gen. Physiol.,* 37, 539, 1954.
6. **Engelmann, F. and Penney, D.,** Studies on the endocrine control of metabolism in *Leucophaea maderae* (Blattaria). I. The hemolymph proteins during egg maturation, *Gen. Comp. Endocrinol.,* 7, 314, 1966.
7. **Engelmann, F.,** Female specific protein: biosynthesis controlled by corpus allatum in *Leucophaea maderae, Science,* 165, 407, 1969.
8. **Bownes, M.,** Expression of the genes coding for vitellogenin (yolk protein), *Annu. Rev. Entomol.,* 31, 507, 1986.

9. **Postlethwait, J. H. and Kunert, C. J.,** Endocrine and genetic regulation of vitellogenesis in *Drosophila*, in *Comparative Endocrinology: Developments and Directions*, Ralph, C. L., Ed., Alan R. Liss, New York, 1986, 33.
10. **Wyatt, G. R., Dhadialla, T. S., and Roberts, P. E.,** Vitellogenin synthesis in locust fat body: juvenile hormone-stimulated gene expression, in *Biosynthesis, Metabolism and Mode of Action of Invertebrate Hormones*, Hoffmann, J. and Porchet, M., Eds., Springer-Verlag, Berlin, 1984, 475.
11. **Wyatt, G. R. and Pan, M. L.,** Insect plasma proteins, *Annu. Rev. Biochem.*, 47, 779, 1978.
12. **Hagedorn, H. H. and Kunkel, J. G.,** Vitellogenin and vitellin in insects, *Annu. Rev. Entomol.*, 24, 475, 1979.
13. **Kunkel, J. G. and Nordin, J. H.,** Yolk proteins, in *Comprehensive Insect Physiology, Biochemistry and Pharmacology*, Vol. 1, Kerkut, G. A. and Gilbert, L. I., Eds., Pergamon Press, Oxford, 1985, 83.
14. **della-Cioppa, G. and Engelmann, F.,** The vitellogenin of *Leucophaea maderae:* synthesis of a large phosphorylated precursor, *Insect Biochem.*, 17, 401, 1987.
15. **Scharrer, B.,** The relationship between corpora allata and reproductive organs in adult *Leucophaea maderae* (Orthoptera), *Endocrinology*, 38, 46, 1946.
16. **Brookes, V. J.,** The induction of yolk protein synthesis in the fat body of an insect, *Leucophaea maderae*, by an analog of the juvenile hormone, *Dev. Biol.*, 20, 459, 1969.
17. **Scharrer, B.,** Neurosecretion. XI. The effects of nerve section on the intercerebralis cardiacum-allatum system of the insect *Leucophaea maderae, Biol. Bull. (Woods Hole, Mass.)*, 102, 261, 1952.
18. **Engelmann, F.,** Die Steuerung der Ovarfunktion bei der ovoviviparen Schabe *Leucophaea maderae* (Fabr.), *J. Insect Physiol.*, 1, 257, 1957.
19. **Engelmann, F.,** The control of reproduction in *Diploptera punctata* (Blattaria), *Biol. Bull. (Woods Hole, Mass.)*, 116, 406, 1959.
20. **Stay, B. and Tobe, S. S.,** Control of the corpora allata during a reproductive cycle in a viviparous cockroach, *Am. Zool.*, 21, 663, 1981.
21. **Engelmann, F. and Friedel, T.,** Insect yolk protein precursor: a juvenile hormone induced phosphoprotein, *Life Sci.*, 14, 587, 1974.
22. **Engelmann, F., Friedel, T., and Ladduwahetty, M.,** The native vitellogenin of the cockroach, *Leucophaea maderae, Insect Biochem.*, 6, 211, 1976.
23. **Bell, W. J.,** Dual role of juvenile hormone in the control of yolk formation in *Periplaneta americana, J. Insect Physiol.*, 15, 1279, 1969.
24. **Bell, W. J.,** Demonstration and characterization of two vitellogenic blood proteins in *Periplaneta americana:* an immunological analysis, *J. Insect Physiol.*, 16, 291, 1970.
25. **Storella, J. R., Wojchowski, D. M., and Kunkel, J. G.,** Structure and embryonic degradation of two native vitellins in the cockroach, *Periplaneta americana, Insect Biochem.*, 15, 259, 1985.
26. **Wojchowski, D. M., Parsons, P., Nordin, J. H., and Kunkel, J. G.,** Processing of pro-vitellogenin in insect fat body: a role for high-mannose oligosaccharide, *Dev. Biol.*, 116, 422, 1986.
27. **Sams, G. R., Bell, W. J., and Weaver, R. F.,** Vitellogenin: its structure, synthesis and processing in the cockroach *Periplaneta americana, Biochim. Biophys. Acta*, 609, 121, 1980.
28. **Harnish, D. G. and White, B. N.,** Insect vitellins: identification, purification, and characterization from eight orders, *J. Exp. Zool.*, 220, 1, 1982.
29. **Koeppe, J. and Ofengand, J.,** Juvenile hormone-induced biosynthesis of vitellogenin in *Leucophaea maderae, Arch. Biochem. Biophys.*, 173, 100, 1976.
30. **Brookes, V. J.,** The polypeptide structure of vitellogenin and vitellin from the cockroach, *Leucophaea maderae, Arch. Insect Biochem. Physiol.*, 3, 577, 1986.
31. **Engelmann, F.,** Juvenile hormone-controlled synthesis of female-specific protein in the cockroach *Leucophaea maderae, Arch. Biochem. Biophys.*, 145, 439, 1971.
32. **Bell, W. J. and Barth, R. H., Jr.,** Quantitative effects of juvenile hormone on reproduction in the cockroach *Byrsotria fumigata, J. Insect Physiol.*, 16, 2303, 1970.
33. **Bühlmann, G.,** Haemolymph vitellogenin, juvenile hormone, and oocyte growth in the adult cockroach *Nauphoeta cinerea* during first pre-oviposition period, *J. Insect Physiol.*, 22, 1101, 1976.
34. **Mundall, E. C., Tobe, S. S., and Stay, B.,** Vitellogenin fluctuations in haemolymph and fat body and dynamics of uptake into oocytes during the reproductive cycle of *Diploptera punctata, J. Insect Physiol.*, 27, 821, 1981.
35. **Engelmann, F.,** Insect vitellogenin: identification, biosynthesis, and role in vitellogenesis, *Adv. Insect Physiol.*, 14, 49, 1979.
36. **Engelmann, F.,** unpublished data, 1987.
37. **Engelmann, F.,** Heterogeneity of juvenile hormone binding compounds in fat bodies of a cockroach, *Mol. Cell. Endocrinol.*, 24, 103, 1981.
38. **Engelmann, F., Mala, J., and Tobe, S. S.,** Cytosolic and nuclear receptors for juvenile hormone in fat bodies of *Leucophaea maderae, Insect Biochem.*, 17, 1045, 1987.

39. **Engelmann, F.,** The pleiotropic action of juvenile hormone in vitellogenin synthesis, *Mitt. Dtsch. Ges. Allg. Entomol.,* 5, 186, 1987.
40. **Whitmore, E. and Gilbert, L. I.,** Haemolymph lipoprotein transport of juvenile hormone, *J. Insect Physiol.,* 18, 1153, 1972.
41. **Goodman, W. G. and Chang, E. S.,** Juvenile hormone cellular and hemolymph binding proteins, in *Comprehensive Insect Physiology, Biochemistry and Pharmacology,* Vol. 7, Kerkut, G. A. and Gilbert, L. I., Eds., Pergamon Press, Oxford, 1985, 491.
42. **Engelmann, F.,** Juvenile hormone binding compounds in hemolymph and tissues of an insect: the functional significance, *Adv. Invert. Reprod.,* 3, 177, 1984.
43. **Engelmann, F.,** unpublished data, 1987.
44. **de Kort, C. A. D. and Koopmanschap, A. B.,** Molecular characteristics of lipophorin, the juvenile hormone-binding protein in the hemolymph of the Colorado potato beetle, *Arch. Insect Biochem. Physiol.,* 5, 255, 1987.
45. **Gunawan, S. and Engelmann, F.,** Esterolytic degradation of juvenile hormone in the haemolymph of the adult female of *Leucophaea maderae, Insect Biochem.,* 14, 601, 1984.
46. **de Kort, C. A. D. and Granger, N. A.,** Regulation of the juvenile hormone titer, *Annu. Rev. Entomol.,* 26, 1, 1981.
47. **Hammock, B. D.,** Regulation of juvenile hormone titer: degradation, in *Comprehensive Insect Physiology, Biochemistry and Pharmacology,* Vol. 7, Kerkut, G. A. and Gilbert, L. I., Eds., Pergamon Press, Oxford, 1985, 431.
48. **Rotin, D., Feyereisen, R., Koener, J., and Tobe, S. S.,** Haemolymph juvenile hormone esterase activity during the reproductive cycle of the viviparous cockroach *Diploptera punctata, Insect Biochem.,* 12, 263, 1982.
49. **della-Cioppa, G. and Engelmann, F.,** Juvenile hormone-stimulated proliferation of endoplasmic reticulum in fat body cells of a vitellogenic insect, *Leucophaea maderae* (Blattaria), *Biochem. Biophys. Res. Commun.,* 93, 825, 1980.
50. **della-Cioppa, G. and Engelmann, F.,** Juvenile hormone regulation of phospholipid synthesis in the endoplasmic reticulum of vitellogenic fat body cells from *Leucophaea maderae, Insect Biochem.,* 14, 27, 1984.
51. **della-Cioppa, G. and Engelmann, F.,** Phospholipid synthesis in the fat body endoplasmic reticulum during primary and secondary juvenile hormone stimulation of vitellogenesis in *Leucophaea maderae, Rouxs Arch. Dev. Biol.,* 193, 78, 1984.
52. **Engelmann, F. and Barajas, L.,** Ribosome-membrane association in fat body tissue from reproductively active females of *Leucophaea maderae, Exp. Cell Res.,* 92, 102, 1975.
53. **Couble, P., Chen, T. T., and Wyatt, G. R.,** Juvenile hormone-controlled vitellogenin synthesis in *Locusta migratoria* fat body: cytological development, *J. Insect Physiol.,* 25, 327, 1979.

Chapter 21

REGULATION OF CORPUS ALLATUM ACTIVITY IN ADULT FEMALES OF THE VIVIPAROUS COCKROACH DIPLOPTERA PUNCTATA

Susan M. Rankin

TABLE OF CONTENTS

I. INTRODUCTION

A. STRUCTURE AND FUNCTION

The corpora allata (CA) are a pair of insect endocrine glands closely associated with and innervated by the brain[1] (see also Chapter 4 of this book). The brain-corpora cardiaca (CC)-CA complex of insects is somewhat analogous to the hypothalamic-hypophyseal system of vertebrates.[2] Within this framework, the CA are analogous to the adenohypophysis of the vertebrate system. The CA produce juvenile hormone (JH), a sesquiterpene. Of the several identified JHs only JH III appears to be produced by the CA of the viviparous cockroach *Diploptera punctata.*[3]

In adult female *D. punctata,* JH is necessary for the production of vitellogenin, the yolk protein precursor, and for uptake of that yolk protein by the ovary. Vitellogenesis in *D. punctata* (and other insects) appears to be remarkably similar to that of vertebrates in that (1) both activities are under neuroendocrine regulation, (2) vitellogenin is produced outside the ovary in the fat body of insects or liver of lower vertebrates, and (3) the vitellogenin reaches the ovary by way of the circulatory system and is taken up and packaged into yolk granules[2] (see also Chapter 20 of this book for review).

B. METHODS (RADIOCHEMICAL ASSAY)

We use an *in vitro* radiochemical assay to measure the rate of JH synthesis. This assay for JH production, first developed by Tobe and Pratt,[4] has undergone some modifications in recent years,[5,6] but remains based on the incorporation of a labeled methyl moiety of methionine into JH in the penultimate step of hormone production. This method appears satisfactory because (1) the incorporation rates are linear for more than 24 h, (2) CA from females in different reproductive stages produce the hormone at a rate characteristic for the particular stage of oocyte development, (3) the quantity of JH in the CA is proportional to that in the medium (i.e., the hormone is not stored, but is released immediately following synthesis),[7] and (4) ultrastructural characteristics of the CA are maintained after incubation.[8,9] The assay for JH production is relatively rapid, reliable, and easy to perform. Furthermore, these calculations based on the radiochemical assay closely reflect the far more costly measurements of titers of JH in the hemolymph.[10]

C. NORMAL CYCLES OF JUVENILE HORMONE SYNTHESIS AND OVARIAN DEVELOPMENT

The cockroach *D. punctata,* like most mammals, undergoes a discrete cycle of reproductive development in which oocyte growth is followed by an extended period of pregnancy. In *D. punctata,* the reproductive stage is closely correlated with the rate of JH production.[7,11] At adult emergence, the rate of JH production is relatively low (10 to 20 pmol/h per pair). Oocytes are previtellogenic and about 0.6 mm long. By day 4 to 5, hormone production increases up to, and sometimes exceeds, 100 pmol/h per pair as the oocytes reach 1.3 to 1.4 mm in length. As oocytes near maturity (with a basal length of about 1.5 mm), JH synthesis declines and remains low (5 to 15 pmol/h per pair) for oviposition (normally on day 7) and during the 60 d of gestation which follow.[5,7,11,12]

During that first cycle in JH synthesis rates that accompanies the gonadotrophic cycle, the CA also undergo a cycle in cell number which parallels the increase and decrease in JH synthesis.[13] On the day following adult emergence, the cell number in one CA is about 5500. The number increases to about 9000 when JH production is maximal and declines to about 6000 by the time of oviposition. To date we cannot measure the activity of individual cells and, thus, cannot say with certainty whether observed changes in activity of the glands are due to cell number, with each cell an equal participant in hormone production, to recruitment of cells into the process of hormone production, or to some combination of number of active cells and level of activity.

D. ULTRASTRUCTURAL CHARACTERISTICS OF THE CORPORA ALLATA DURING THESE CYCLES

Ultrastructurally, the cells of the CA undergo striking changes during the first cycle of activity. As the rate of JH synthesis increases, the mitochondrial matrix becomes less dense, mitochondrial size increases, and mitochondrial shape becomes irregular. Similarly, smooth endoplasmic reticulum becomes more difficult to distinguish, with networks and vesicles becoming prominent. The morphology of these organelles at the end of vitellogenesis reverts to that of newly emerged females. Rough endoplasmic reticulum, present in only relatively small amounts in the CA during the cycle, typically becomes long, with a curving or circular shape at the end of vitellogenesis.[9]

II. TARGET ORGAN FEEDBACK

A. EFFECTS OF OVARIECTOMY

The CA in insects, like the adenohypophysis in vertebrates, is subject to target organ feedback. In *D. punctata* a target organ, the ovary, is required for the first cycle in the rates of JH synthesis. Ovariectomy abolishes the first cycle in JH synthesis. JH production remains low (10 to 20 mol/h per pair),[12,14] although the increase in CA cell number continues.[15] Implantation of a previtellogenic ovary into an ovariectomized female restores the cycle in rates of JH synthesis.[12] Thus, innervation of the ovary is not required for the organ to exert its effect on JH production. Furthermore, if implantation of the immature ovary into an ovariectomized host is delayed until vitellogenin titers in the hemolymph are elevated, a normal cycle in hormone production and oocyte development ensues.[12] High titers of the yolk protein precursor do not appear to reduce JH production, nor does the amount of ovarian tissue present affect the cycle. Implantation of only one half of one ovary, or as many as two pairs (four ovaries), results in a normal cycle in JH synthesis.[12] The portion of the ovary responsible for eliciting increases in JH production appears to be the basal oocytes, since implantation of basal oocytes into ovariectomized females enhances JH production whereas implantation of the apical and pedicel portions does not alter JH synthesis.[16]

B. IMPLANTATION OF OVARIES OF DIFFERENT SIZES

Basal oocytes of different stages have different capacities for eliciting increases in JH production in ovariectomized females after incubation times of 24 or 48 h. Ovaries become progressively more effective in promoting increases in JH production as vitellogenesis progresses until oocytes reach about 1.5 mm in length.[16] Thus, the ovary does not appear to merely trigger a cycle in JH synthesis, but to continually regulate activity of the CA during the gonadotrophic cycle.

JH synthesis has also been stimulated by topical application of a low (2.5 μg) dose of the JH analogue ZR 512.[17,18] The analogue appears to exert its effect on JH synthesis via the ovary, perhaps by accelerating oocyte growth, since such feedback stimulation is not observed in ovariectomized females treated with ZR 512.[18] A similar positive feedback of short duration has been reported in mammals in the estradiol-luteinizing hormone interaction.[19]

Nearly mature (>1.5-mm-long) oocytes appear to elicit decreases in JH synthesis, an inhibition which has been demonstrated in several ways. First, bilateral ovariectomy of females when CA are near-maximally active delays the decline in JH synthesis when compared with that in unilaterally ovariectomized control animals. However, differences persist for only 1 d after surgery, suggesting that while the ovary may be a source of inhibition, it is not the sole source.[20] Second, implantation into ovariectomized hosts of ovaries of two sizes, one known to stimulate increases in JH synthesis and one suspected of being inhibitory (>1.5 mm long), results in a level of JH production that is intermediate between that elicited

by either ovarian type alone.[20] Another class of nonstimulatory test ovaries, those with basal oocytes only 0.38 mm long (from pregnant females), do not elicit a decrease in JH production when implanted along with a stimulatory ovary. Thus, nearly mature ovaries appear to actively inhibit JH production, while ovaries from pregnant females do not appear to either stimulate or inhibit JH production.[20] Finally, perhaps the strongest evidence for an inhibitory effect of the ovary on JH production is derived from experiments utilizing male *D. punctata* as hosts. When CA from newly emerged females are implanted into males, hormone synthesis by the implanted glands increases and remains elevated rather than declining to low levels on days 5 to 7 (as seen in normal females). If an immature ovary is implanted into the male along with the CA from a newly emerged female, the implanted CA undergo a decrease in JH production when the implanted oocytes near maturity.[21] Furthermore, implantation of a nearly mature ovary prevents the increases in activity of implanted CA.[20]

III. NEURAL AND NEUROHORMONAL REGULATION

A. DENERVATION OF CORPORA ALLATA

Neural regulation of the CA may prove to be analogous to regulation of the anterior lobe of the hypophysis in vertebrates. Both systems may involve inhibitory and stimulatory regulation. Intact nerves from the brain to the CA have been shown to restrain JH synthesis by the CA. Virgin female CA, for example, which normally do not exhibit increases in hormone production or cell number, undergo a normal cycle of JH synthesis when denervated on day 0.[7,15] Denervation at other times, such as after ovariectomy,[12] during pregnancy,[11] or even *in vitro*,[22] also causes a significant elevation in JH synthesis. Similar documentation of an increase in hormone production following denervation of glands has been reported in vertebrates in the adrenal medulla.[23]

The concept of compensation has been demonstrated in the CA of *D. punctata* in several ways. For example, removal of one CA results in a doubling of JH production by the remaining gland,[24] even when the remaining gland is denervated.[25] Indeed, denervation of the remaining gland accelerates its increase in hormone production beyond that of innervated controls, again indicating that intact nerves can inhibit CA activity. The mechanism of compensation appears to be increased activity of cells rather than an increase in cell number.[13]

Negative feedback from high concentrations of hormone, a phenomenon common among vertebrates (as, for example, in testosterone secretion[26]), has also been documented for JH production in *D. punctata*. For this cockroach, implantation of a supernumerary pair of CA of similar age decreases the activity of innervated glands below that of the implanted ones. When host glands are denervated, supernumerary and host glands both undergo a reduced cycle and ovaries undergo accelerated development.[25] When host and implanted glands are of different ages (activities), the more active pair appears to take precedence over the less active pair, as the latter show the greater reduction in hormone synthesis.[25]

In a final example of compensation, topical application of high doses (100 μg) of the JH analogue ZR 512 inhibits JH production.[14] This feedback inhibition requires intact nerves from the brain to the CA.[18]

B. DECAPITATION AND CAUTERY

Decapitation of newly emerged females does not abolish the cycle in hormone production by implanted CA,[14] suggesting that the brain is not required for the normal cycle of CA activity that accompanies oocyte development.

Ruegg et al.[27] cauterized various regions of the brain of virgin *D. punctata* in an effort to localize the portions responsible for inhibiting JH synthesis. Radiofrequency cautery of the medial neurosecretory cell axons at the point of decussation resulted in a normal cycle of JH synthesis and oocyte growth which began within 24 h of destruction of the axons.

Unilateral cautery of the area just lateral to the pars intercerebralis (which contains axons from the pars lateralis) or of the pars lateralis activated the ipsilateral CA. This type of cautery is not sufficient to activate CA of pregnant or ovariectomized females. Based on these experiments and on nerve severance experiments,[27a] neural inhibition of CA in virgin females is hypothesized to be a result of release of inhibitory substances from the medial neurosecretory cells controlled via the pars lateralis. It is suggested that signals from the pars lateralis stimulate the medial neurosecretory cell to inhibit JH synthesis.[27]

In addition to inhibition of CA activity via intact nerves from the brain, glands in pregnant females also appear to be inhibited by humoral factors from the brain.[11] This was suspected from the observation that activation of CA by denervation was not equivalent to activation elicited by embryo removal and that the two operations had an additive effect with respect to activation of CA in pregnant females. These inhibitory centers can be removed by decapitation and restored by implantation of protocerebra from pregnant females, but not by implantation of protocerebra from first-cycle females.[11] Humoral factors carried by the hemolymph can inhibit CA activity, perhaps comparable to the action of vertebrate somatostatin,[2] and may represent an important component of the system that regulates CA activity.

C. INCUBATION OF CORPORA ALLATA IN HIGH-KCl MEDIUM

The observation that the brain inhibits CA activity in adult female *D. punctata*[7,11,12,27] coupled with the observation that CA contain numerous neurosecretory endings which might originate in the brain[9] suggested that neurosecretory material, if released from those neurosecretory endings within the CA, would inhibit JH production. The recent finding that high levels of potassium evoke a calcium-dependent release of neurosecretory material from the rat neurohypophysis[28] suggested that such treatment might evoke a similar release of neurosecretion within isolated CA. Incubation of CA *in vitro* in medium containing high (50 m*M*) levels of KCl elicits a reversible decline in JH synthesis. This effect is blocked by the calcium antagonist, magnesium.[22] The presumed release of neurosecretory material within the CA inhibits JH synthesis.

D. INCUBATION OF CORPORA ALLATA WITH BRAINS AND OTHER NEURAL TISSUE

Denervation experiments,[7,11,12] implantation of brains into decapitated pregnant females,[11] and high-KCl experiments[22] all support the hypothesis that material from the brain inhibits JH production. In further support of this hypothesis, incubation of the intact brain-CC-CA complex *in vitro* lowers JH synthesis below that of CA incubated with the brain-CC cut free at the nervi corporis allati I (NCA I), the nerves between the CC and CA[22] (see also Chapter 4 of this work). Thus, intact nerves to the CA can inhibit JH synthesis *in vitro*. The presence of the brain-CC in culture with the isolated CA results in lower JH synthesis than that of isolated CA alone.[22] One possible explanation for such a reduction in hormone production is that the brain-CC complex might release a humoral inhibitory factor such as that demonstrated from brains of pregnant females[11] and discussed in Section III.B above.

Inhibitory material can be extracted from the brain, CC, and CA, but not from optic lobes, suggesting that material which inhibits JH production is selectively distributed within the brain.[22] However, inhibitory material is not limited to the brain, but is widely distributed throughout the central nervous system, and extracts of other ganglia (such as subesophageal, thoracic, and ventral abdominal ganglia) inhibit JH synthesis *in vitro*.[29] The inhibitory activity of these extracts is trypsin sensitive, suggesting that the active factor(s) are proteins or peptides. *In vitro,* the inhibitory material is maximally effective within 1 h and maintains this effect when present. The inhibition is reversible, since it decreases when the active factor(s) are removed from the incubation medium.[29] These features are consistent with the action of a number of mammalian peptide and amine hormones.[30] (See also Chapter 13 in this work for a general discussion of peptides as chemical signals.)

Descriptions of investigations into the mechanism of action of the inhibitory neuropeptide(s) are beyond the scope of this chapter. Suffice it to say here that (1) glands of high activity are less sensitive to the inhibitory extract than those of low activity;[29] (2) the effect of the factor is exerted prior to the terminal methylation and epoxidation of the hormone;[29] and (3) the roles of calcium and cAMP, second messengers in many other systems,[30] are now being delineated with respect to the regulation of JH synthesis in *D. punctata* (see Chapter 5 of this work).

IV. CONCLUSION AND SUMMARY

CA of the viviparous cockroach *D. punctata* provide a system for investigating paradigms of regulation of hormone production. This system exhibits a hierarchy of control factors, including hormones (e.g., JH), neurohormones (e.g., the putative allatostatin [see Section III]), and target organ factors such as the ovary (see Section II), that can feed back to upper levels of control. The CA of mated females undergo a pronounced cycle of activity, cell number, and morphology during the first week of adult life. Unilateral allatectomy on the day of adult emergence results in a doubling of the rate of JH synthesis by the remaining gland. This compensatory response by the remaining CA appears to be due to an increase in activity of CA cells rather than to an increase in cell number.

The normal cycle in rates of JH synthesis by the CA requires the ovary, since ovariectomy abolishes the cycle and implantation of one half to four ovaries restores that cycle. Additionally, as shown by short-term *in vivo* tests, the basal oocytes of the ovary have an increasing capacity to stimulate increases in rates of JH synthesis while they are growing, and they become actively inhibitory when near maturity. During the 60 d of pregnancy, when embryos develop within the brood sac, the ovarian oocytes mature slowly, remaining previtellogenic and incapable of stimulating or inhibiting the sustained, low rates of JH synthesis. Thus, a target organ for JH, the ovary, appears to exert stimulatory, inhibitory, and no control over the CA, depending on the stage of the ovary.

The brain, however, is not required for a normal cycle in JH production accompanying ovarian maturation; such a cycle persists after decapitation of newly emerged females. While the brain is not necessary for the cycle in JH synthesis associated with oocyte development, it can exert an inhibitory influence over JH production. Intact nerves from the brain to the CA appear to restrain JH synthesis in virgin females, in ovariectomized females, and in pregnant females; severance of those nerves in each case results in an increase in JH production. Inhibition of hormone production via intact nerves, as occurs in virgins, is hypothesized to be a result of release of inhibitory material from the medial neurosecretory cells, which are in turn controlled by lateral neurosecretory cells.

Intact nerves from the brain to the CA appear to mediate another phenomenon common among endocrine systems, that of inhibitory feedback. Such feedback, observed in *D. punctata* after implantation of supernumerary CA or after application of high doses of the JH analogue ZR 512, is abolished by denervation. However, intact nerves from the brain to the CA are not required for feedback stimulation, i.e., when low doses of ZR 512 applied to newly emerged, mated females stimulate a temporary increase in JH production. Feedback stimulation is not abolished by denervation of CA, but is abolished by ovariectomy. Thus, this type of feedback appears to be mediated by the ovaries.

As shown by denervation of CA and removal of the batch of embryos, separately and in combination, rates of JH synthesis appear to be restrained by two centers within the brain in the pregnant female. The first one requires intact nerves to the CA, similar or identical to that observed in virgin females. The second acts via the hemolymph and responds to the presence of embryos in the brood sac. These centers can be removed effectively by decapitation, and the latter can be restored by implantation of protocerebra from pregnant females, but not by implantation of protocerebra from first-cycle females.

The role of the brain in inhibiting JH synthesis by CA is now being investigated by *in vitro* experiments. As was found *in vivo,* denervation of CA incubated *in vitro* enhances JH synthesis. Furthermore, incubation of isolated CA with high levels of potassium, a treatment known to elicit a calcium-dependent release of neurosecretion in vertebrates, elicits a decline in JH production. Additionally, factors (presumably neuropeptides extracted from the brain, other ganglia, CC, and CA) can restrain JH synthesis *in vitro.* These observations are consistent with the hypothesis that neuropeptides produced in the brain (and other ganglia) may act to inhibit JH synthesis. The precise nature of the inhibitory neuropeptide(s) and its mechanism of action are current topics of active research.

Thus, this system, the brain-CC-CA complex, considered analogous to the hypothalamic-hypophyseal system of vertebrates,[2] provides the potential for rapidly advancing our understanding of regulation of hormone production, not only in insects, but in other organisms, including vertebrates, as well.

REFERENCES

1. **Lococo, D. J. and Tobe, S. S.,** Neuroanatomy of the retrocerebral complex, in particular the pars intercerebralis and partes laterales in the cockroach *Diploptera punctata* Eschscholtz (Dictyoptera: Blaberidae), *Int. J. Insect Morphol. Embryol.,* 13, 65, 1984.
2. **Scharrer, B.,** Insects as models in neuroendocrine research, *Annu. Rev. Entomol.,* 32, 1, 1987.
3. **Tobe, S. S. and Stay, B.,** Corpus allatum activity *in vitro* during the reproductive cycle of the viviparous cockroach, *Diploptera punctata* (Eschscholtz), *Gen. Comp. Endocrinol.,* 31, 138, 1977.
4. **Tobe, S. S. and Pratt, G. E.,** The influence of substrate concentrations on the rate of insect juvenile hormone biosynthesis by corpora allata of the desert locust *in vitro, Biochem. J.,* 144, 107, 1974.
5. **Feyereisen, R., Friedel, T., and Tobe, S. S.,** Farnesoic acid stimulation of C_{16} juvenile hormone biosynthesis by corpora allata of adult female *Diploptera punctata, Insect Biochem.,* 11, 409, 1984.
6. **Tobe, S. S. and Clarke, N.,** The effect of L-methionine concentration on juvenile hormone biosynthesis by corpora allata of the cockroach *Diploptera punctata, Insect Biochem.,* 15, 175, 1984.
7. **Stay, B. and Tobe, S. S.,** Control of juvenile hormone biosynthesis during the reproductive cycle of a viviparous cockroach, *Gen. Comp. Endocrinol.,* 33, 531, 1977.
8. **Feyereisen, R., Johnson, G., Koener, J., Stay, B., and Tobe, S. S.,** Precocenes as pro-allatocidins in adult female *Diploptera punctata:* a functional and ultrastructural study, *J. Insect Physiol.,* 27, 885, 1981.
9. **Johnson, G. D., Stay, B., and Rankin, S. M.,** Ultrastructure of corpora allata of known activity during the vitellogenic cycle in the cockroach *Diploptera punctata, Cell Tissue Res.,* 239, 317, 1985.
10. **Tobe, S. S., Stay, B. A., Baker, F. C., and Schooley, D. A.,** Regulation of juvenile hormone titer in the adult female cockroach *Diploptera punctata,* in *Biosynthesis, Metabolism, and Mode of Action of Invertebrate Hormones,* Hoffmann, J. and Porchet, M., Eds., Springer-Verlag, Berlin, 1984, 397.
11. **Rankin, S. M. and Stay, B.,** Regulation of juvenile hormone synthesis during pregnancy in the cockroach, *Diploptera punctata, J. Insect Physiol.,* 31, 145, 1985.
12. **Stay, B., Tobe, S. S., Mundall, E. C., and Rankin, S.,** Ovarian stimulation of juvenile hormone biosynthesis in the viviparous cockroach, *Diploptera punctata, Gen. Comp. Endocrinol.,* 52, 341, 1983.
13. **Szibbo, C. M. and Tobe, S. S.,** The mechanism of compensation in juvenile hormone synthesis following unilateral allatectomy in *Diploptera punctata, J. Insect Physiol.,* 27, 609, 1981.
14. **Rankin, S. M. and Stay, B.,** Effects of decapitation and ovariectomy on regulation of juvenile hormone synthesis in the cockroach, *Diploptera punctata, J. Insect Physiol.,* 29, 839, 1983.
15. **Tobe, S. S., Clarke, N., Stay, B., and Ruegg, R. P.,** Changes in cell number and activity of the corpora allata of the cockroach *Diploptera punctata:* a role for mating and the ovary, *Can. J. Zool.,* 62, 2178, 1984.
16. **Rankin, S. M. and Stay, B.,** The changing effect of the ovary on rates of juvenile hormone synthesis *Diploptera punctata, Gen. Comp. Endocrinol.,* 54, 382, 1984.
17. **Tobe, S. S. and Stay, B.,** Modulation of juvenile hormone synthesis by an analogue in the cockroach, *Nature (London),* 281, 481, 1979.
18. **Stay, B. and Rankin, S. M.,** Regulation of corpus allatum activity during reproduction: the role of ovaries and nerves, in *Advances in Invertebrate Reproduction,* Vol. 4, Andries, J. C. and Dhainaut, A., Eds., Elsevier, Amsterdam, 1986, 43.

19. **Adams, T. E., Norman, R. L., and Spies, H. G.,** Gonadotropin-releasing hormone receptor binding and pituitary responsiveness in estradiol-primed monkeys, *Science,* 213, 1388, 1981.
20. **Rankin, S. M. and Stay, B.,** Ovarian inhibition of juvenile hormone synthesis in the viviparous cockroach, *Diploptera punctata, Gen. Comp. Endocrinol.,* 59, 230, 1985.
21. **Stay, B., Friedel, T., Tobe, S. S., and Mundall, E. C.,** Feedback control of juvenile hormone synthesis in cockroaches: possible role for ecdysterone, *Science,* 207, 898, 1980.
22. **Rankin, S. M., Stay, B., Aucoin, R. R., and Tobe, S. S.,** *In vitro* inhibition of juvenile hormone synthesis by corpora allata of the viviparous cockroach, *Diploptera punctata, J. Insect Physiol.,* 32, 151, 1986.
23. **Martin, C. R.,** *Endocrine Physiology,* Oxford University Press, New York, 1985, 272.
24. **Stay, B. and Tobe, S. S.,** Control of juvenile hormone biosynthesis during the reproductive cycle of a viviparous cockroach. II. Effects of unilateral allatectomy, implantation of supernumerary corpora allata, and ovariectomy, *Gen. Comp. Endocrinol.,* 34, 276, 1978.
25. **Tobe, S. S. and Stay, B.,** Control of juvenile hormone biosynthesis during the reproductive cycle of a viviparous cockroach. III. Effects of denervation and age on compensation with unilateral allatectomy and supernumerary corpora allata, *Gen. Comp. Endocrinol.,* 40, 89, 1980.
26. **Martin, C. R.,** *Endocrine Physiology,* Oxford University Press, New York, 1985, 51.
27. **Ruegg, R. P., Lococo, D. J., and Tobe, S. S.,** Control of corpus allatum activity in *Diploptera punctata:* roles of the pars intercerebralis and pars lateralis, *Experientia,* 39, 1329, 1983.
27a. **Stay, B.,** personal communication.
28. **Nordmann, J. J.,** Evidence for calcium inactivation during hormone release in the rat neurohypophysis, *J. Exp. Biol.,* 65, 669, 1976.
29. **Rankin, S. M. and Stay, B.,** Distribution of allatostatin in the adult cockroach, *Diploptera punctata* and its effects on corpora allata *in vitro, J. Insect Physiol.,* 33, 551, 1987.
30. **Rasmussen, H.,** *Calcium and cAMP as Synarchic Messengers,* John Wiley & Sons, New York, 1981.

Chapter 22

NEUROENDOCRINE REGULATION OF PHEROMONE PRODUCTION IN COCKROACHES

Coby Schal and Alan F. Smith

TABLE OF CONTENTS

I. INTRODUCTION

A central concern in endocrinology is understanding the coordination of (1) maturation of secretory and target organs, (2) synthesis of a hormone and its release into circulation, (3) synthesis of another factor (in this case, pheromone), and (4) specific behaviors associated with its release into the environment. In an ideal model system, the precise timing of developmental and behavioral events should be well defined, the inducing hormone should be known, and the product of its action should be easily assayed analytically or behaviorally. The regulation of synthesis and emission of sex pheromones in cockroaches meets these criteria.

Cockroaches have long served as a model system in studies of endocrinology and neurohormonal regulation of vitellogenesis and reproduction (see Chapters 19 to 21 in this work). Thus, a wealth of information is available about endocrine events during the gonadotrophic cycle and the roles of exogenous and endogenous factors in modulating endocrine events. The regulation of pheromone production within the insect and of its release externally may be similar to the regulation of vitellogenin synthesis in the fat body and release into the hemocoel. Like vitellogenins, pheromones may be assayed biochemically and analytically. Pheromones can also be assayed with relative ease using behavioral and electrophysiological responses of the receiving sex. Hence, mechanisms by which the corpora allata (CA) are regulated, the mode of action of juvenile hormone (JH), and the development of competence of secretory tissues may be investigated readily. The effects of exteroceptive signals (from feeding, drinking, crowding, photoperiod, and temperature) and interoceptive directives (from mating and carrying an egg case) may be studied; the interaction and hierarchical organization of these effects vis-à-vis pheromone production and release may be elucidated. As in the synthesis, release, uptake, and metabolism of other biological materials, pheromone biosynthesis, release, and catabolism must be regulated precisely and in step with other events such as oocyte maturation and sexual receptivity. Hence, different nutritional and environmental conditions are expected to modulate these events.

Cockroaches are particularly useful for studies of the regulation of chemical communication because of the variety of reproductive and oviposition tactics (oviparity, ovoviviparity, viviparity, parthenogenesis), their relatively primitive status among the insects, and their economic importance as pests; their predictable responses to pheromones are usually uncomplicated by learning and experience (compared with mammals), and their peripheral and central olfactory centers are easily accessible (see Chapter 26 in this work). In studies requiring large numbers of subjects, scarcity of insects is rarely a problem when cockroaches are used. The cockroaches, thus, may rival the white rat as a useful model in biomedical research on olfaction and chemical communication.

Cockroach pheromones may be active at a distance (volatile pheromones), at close range (most male tergal secretions), or by contact only (cuticular components). We shall confine our discussion to female-produced sex pheromones only. The production and release of aggregation pheromones, spacing (repellent) pheromones, allomones, and male-produced

sex pheromones are omitted so that we can concentrate on relationships of pheromone production and release and cyclic events associated with the gonadotrophic cycle in females.

II. ISOLATION AND ACTIVITY OF SEX PHEROMONES

A. *BLATTELLA GERMANICA*

Nishida and Fukami[1] summarized work on the isolation, identification, and behavioral activity of components of the sex pheromone of the German cockroach, *Blattella germanica*. A rigorous description of courtship behavior and ablation experiments with various sensory structures suggested that a nonvolatile pheromone contained in the cuticular wax of females elicited the wing-raising courtship response in males.[2,3] A behavioral assay developed by Roth and Willis[2] and modified by Nishida et al.[4] proved to be very effective in subsequent evaluations of fractions of female extracts and of analogues of the pheromone components. An antenna is ablated from an adult male cockroach, glued to the end of a glass rod, dipped in a solution of the test material, and used to fence with sexually mature test males. The wing-raising response of males is recorded.

Three components were isolated and identified from a hexane extract of 224,000 females. Mass spectrometry, proton and ^{13}C-nuclear magnetic resonance (^{13}C-NMR), infrared (IR) absorption, and chemical derivatization have indicated that all three possess a 3,11-dimethyl-2-nonacosanone skeleton with various functional groups at the C-29 position[1] (Figure 1B; see Reference 1 for review). Several researchers synthesized either component A or B, or both A and B.[5-13] Nishida and Fukami[1] clearly documented that for components A and B, respectively, the concentration-dependent male responses to the synthetic and natural pheromone components are identical. Compound B (29-hydroxy-3,11-dimethyl-2-nonacosanone) is more active by an order of magnitude than compound A, but each of the three components elicits the full range of behavioral responses in males. Thus, the female sex pheromone complex in the German cockroach differs from most lepidopteran pheromones, where omission of even minor components affects male response qualitatively and quantitatively.

Using stereoisomers synthesized by Mori et al.,[14,15] it has been shown that both pheromone components A and B possess 3S,11S configurations,[16] although all combinations of stereochemical isomers of the 3,11-positions in both components yield similar wing-raising activity in males,[1,14] indicating lack of stereospecificity in the pheromone receptor. Studies[1,8] of structure-activity relationships of pheromone components A and B have concluded that activity is proportional to polarity of the C-29 end, that reduction of the C-2 carbonyl to an alcohol increases activity about tenfold, that a methyl in place of the C-2 carbonyl eliminates activity, that the 3,11-dimethyl branches are important for activity, and that shortening or elongation of the alkyl chain reduces activity of the pheromone. It thus appears that this system can accommodate much greater changes in pheromone specificity than most lepidopteran systems[1] and may explain, at least in part, frequent observations of interspecific courtship in cockroaches, particularly where recognition is modulated by contact pheromones.

No work has been reported on the biosynthesis of *B. germanica* pheromones. Our research group[16a] has identified the cuticular hydrocarbons of the German cockroach, and we have shown with radiotracer studies that the female can synthesize compound A from 3,11-dimethylnonacosanone as well as from acetate and propionate,[16b] but no information is available about the pathways involved. Surprisingly, following short rinses of females with hexane and ether, the extract from a 2-d immersion in methylene chloride:methanol was inactive, indicating that little pheromone was present internally.[1] It is important to note, however, that the extracts were not fractionated and probably contained large amounts of internal fatty acids which inhibited the activity of the pheromone.[1]

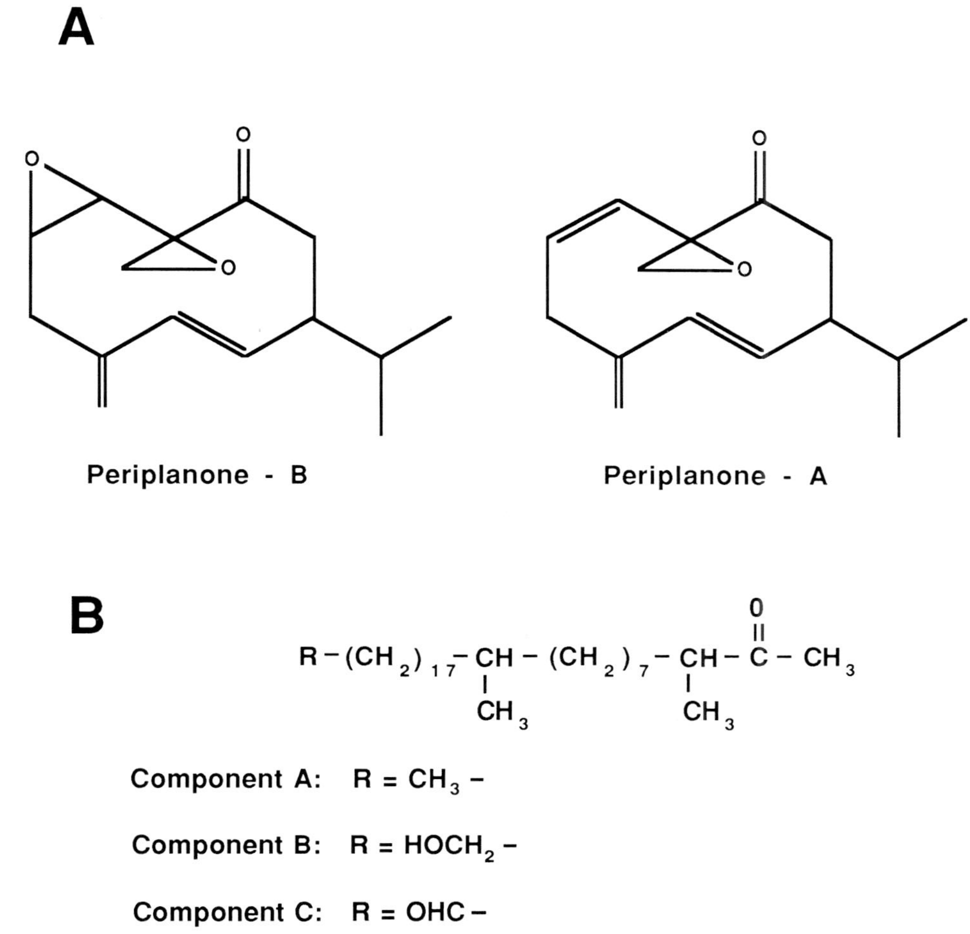

FIGURE 1. Structures of (A) the two pheromone components of *Periplaneta americana* and (B) the three components of *Blattella germanica*.

B. *PERIPLANETA AMERICANA*

1. Volatile Pheromones

Structural elucidation of the pheromone produced by the American cockroach, *Periplaneta americana,* has been hampered by the minute quantities of extremely active material produced by females and also by several misidentifications (see Reference 17 for review). Briefly, employing the behavioral assays of Roth and Willis,[2] the pheromone was identified as an aliphatic ester.[18] Jacobson et al.[19] assigned the structure of 2,2-dimethyl-3-isopropylidenecyclopropyl propionate to the active compound, but later[20] withdrew this identification when it was shown that the synthetic material was inactive.[21-23] Several groups reported attempts to identify the pheromone, but with little success.[24-26]

Persoons et al.[27] isolated two sesquiterpenoid pheromone components, periplanone-A ($C_{15}H_{20}O_2$) and periplanone-B ($C_{15}H_{20}O_3$), from 35,000 midguts and from the feces of 20,000 additional females. The active fractions were subjected to silica gel chromatography and spectrometric analyses (primarily nuclear magnetic resonance [NMR]), and a germacranoid structure was assigned to periplanone-B ([1Z,5E]-1,10[14]-diepoxy-4[15],5-germacradiene-9-one; Figure 1A). The structure was confirmed by an elegant stereoselective synthesis following a conformational analysis of intermediate germacranoid compounds,[28] and the

(−) enantiomer was shown to release behavioral responses in males at 0.1 to 1 pg.[29] Other syntheses have been reported which result in behaviorally active (+/−)-periplanone-B (for a review, see Reference 30) or the more active chiral (−)-periplanone-B.[31] Subsequent isolations of both periplanones have effectively employed the electroantennagram (EAG) technique as an initial screen prior to more time-consuming behavioral and spectroscopic assays (see Reference 32, for example).

Periplanone-A (7-methylene-4-isopropyl-12-oxa-tricyclo[4.4.2.$0^{1,5}$]-9-dodecen-2-one; stereochemistry unknown) was thought to be an unstable reduction product of periplanone-B[33,34] with no biological significance. However, electrophysiological studies have shown that some peripheral antennal cells[35] and central olfactory neurons[36] of male American cockroaches respond specifically to periplanone-A (see also Chapter 26 in this work). Seelinger[37] showed that both periplanones elicited anemotaxis in males in a wind tunnel, and both resulted in linear log dose-response (running and wing raising) relationships, but at different threshold concentrations. He further reported that when periplanones-A and -B were combined in the natural ratio (1:1) the threshold concentration for initiation of running was higher, and males oriented more slowly and stopped more frequently near the source than when periplanone-B was present alone.[38] They reasoned that periplanone-A might serve as a close-range orientation cue which dampens locomotory activity (orthokinesis) near the odor source. Sass[35] also showed, in contrast to Persoon's observations, that periplanone-A was chemically stable below 0°C, but decomposed at temperatures normally used in gas chromatography.

It is not known whether these two components comprise the entire pheromone blend in *P. americana*. Persoons and Ritter[39] made reference to four other biologically active fractions which were not investigated further, and the "large scale separation" by Sass[35] of adsorbed pheromone from 15 females showed several additional EAG and behaviorally active components, as does Seelinger's Figure 2 in Chapter 26 of this work.

Various investigators have shown that natural components of plants and monoterpenoids can elicit behavioral and/or EAG responses in males. These include, among many other compounds, germacrene-D, D-bornyl acetate, and various verbanyl analogues (for reviews, see References 40 and 41). For the most current of an 11-part series on synthesis and behavioral and physiological assays with various sex pheromone mimics, refer to the paper by Manabe et al.[42]

Simon and Barth[43,44] have concluded that males of five *Periplaneta* and one *Blatta* are attracted to and stimulated to court females of all six species. However, work in the laboratory[45] and field trapping[46] clearly showed that *P. australasiae, P. americana,* and *Blatta orientalis* males are attracted by periplanone-A, but attraction of male *P. australasiae* to periplanone-A is inhibited by periplanone-B.

2. Contact Pheromones

Before the term "pheromone" was coined, Dethier[47] stated that "no one attractant alone performs the service of guiding an insect to its proper host plant, food or mate, and that the desired end is achieved only by a complex array of stimuli, such as chemical, light, temperature and humidity, acting in harmony". Roth and Willis[2] first observed that papers conditioned by virgin females elicited wing raising in males tested in groups, but not in isolated males. This observation was repeated by Wharton et al.[48] and by Sturckow and Bodenstein,[49] and homosexual wing raising was used in part to assay sex pheromone (see Reference 50 for review). By investigating responses of isolated males to cuticular washes of *P. americana* females, Seelinger and Schuderer[51] showed that volatile pheromones (periplanone-A and -B) attract males to receptive females and that (a) contact pheromone(s) (identity unknown) mediates sex recognition and releases courtship. However, this contact pheromone is ineffective without the volatile pheromone. They also reported that virgin and

mated *P. americana* females were equally effective (85%) releasers of male courtship when placed in a plume of periplanone-B, but males elicited courtship from other males in only 15% of the cases. Interestingly, whereas males and females of other *Periplaneta* species differ in cuticular hydrocarbon composition, cuticular hydrocarbon components in *P. americana* are qualitatively and quantitatively identical in males and females.[52] Thus, it appears that if a female-specific contact pheromone occurs in *P. americana,* it is likely a component of the more polar fraction of the epicuticle. Gilby and Cox[53] reported on the fatty acid, ester, and aliphatic aldehyde composition of *Periplaneta* cast skins, but no mention was made of sex or life cycle stage of the insects used.

C. OTHER COCKROACHES

Females of other cockroach species have been shown to have either volatile pheromones or contact pheromones, or both. In the Blattidae, members of the genera *Periplaneta* and *Blatta* appear to utilize periplanone-A and/or periplanone-B (or related compounds), as evident from cross-attraction of males to congeneric females (see Section II.B.1 above). *Byrsotria fumigata,* which has served as a model for studies of the regulation of pheromone production, produces a volatile pheromone of unknown identity. Interestingly, whereas water was effective in extracting *Byrsotria* pheromone (which was adsorbed onto filter papers), methylene chloride, methanol, ethyl ether, and petroleum ether were ineffective.[54] It remains to be determined how this pheromone can adsorb to the highly hydrophobic cuticle. Other cockroaches, including both bisexual and parthenogenetic strains of *Pycnoscelus* spp.[55] as well as various tropical species,[56] produce volatile pheromones. Work is now in progress to elucidate the chemical structure of the volatile pheromone of *Supella longipalpa.*

III. TISSUES INVOLVED

In most insects, glandular modifications of epidermal cells of the integument are usually involved in pheromone production.[57] However, the location and morphology of pheromone-producing glands may be quite variable among species. The digestive tract, the tergum, the genital atrium, and the antennae have all been implicated as regions of pheromone production in different cockroach species. The morphology, secretory products, and behavioral activity of pheromone-producing glands in male cockroaches are reviewed by Sreng[58] and will not be covered here.

A. *BLATTELLA*

Nishida and Fukami[1] observed that although antennal fencing is sufficient to elicit courtship responses in males, the pheromone was not limited to female antennae in *Blattella.* Burns and Schal[58a] found that of 1.52 μg extracted from cuticles of 15-day-old females, 2% was recovered from the antennae and 3, 6, 13, 19, 24, and 33% from the ootheca, head, thorax, legs, abdomen, and wings, respectively. It is interesting that the wings contained more pheromone than the abdomen, but it is likely that pheromone was transferred during grooming activities. Clearly, assays with radiotracers and tissue culture techniques are needed to establish the site of pheromone synthesis in this cockroach.

B. *PERIPLANETA*

Using behavioral assays with various body parts of virgin and mated American cockroaches, it was erroneously concluded that the pheromone is produced in the head and that activity decreases little after mating.[49] The site of greatest pheromone concentration was later found to be the midgut,[59] which was used by various investigators to isolate active compounds for structure elucidation. Talman et al.[33] report that whereas periplanone-B is recovered from both midgut and feces, periplanone-A can be isolated only from feces.

However, Sass[35] showed that both periplanones can be extracted from female guts. Raisbeck[60] showed that a pheromone extract incubated with guts from males and non-pheromone-producing *Periplaneta americana* females rapidly lost activity and that piperonyl butoxide, a mixed function oxidase inhibitor, reduced the rate of pheromone inactivation by the gut.

Seelinger[61] has documented a specific "calling" stance (first noted by Tobin[61a]) in which *P. americana* females expose the genital chamber and anal region by lowering the seventh abdominal sternite. The behavior occurs mainly in the first 6 h of darkness, and it appears to be associated with pheromone release. Interestingly, by adsorption onto Tenax®, Sass[35] collected equal amounts of pheromone during the day and night, probably due to adsorption and desorption of pheromone in the collection apparatus. It is unknown whether specific glands are exposed during this behavior or whether pheromone produced in the midgut is simply released through the anus.

The contact pheromone of *P. americana* was shown to be present throughout the female's cuticle,[51] but 3- to 6-week-old females were used whose body parts may have been contaminated by the cuticular secretion.

C. *BYRSOTRIA*

Removal of the ovaries,[62] colleterial glands,[63] and the digestive tract does not interfere with pheromone production. Since gynandromorphs exhibiting male sexual behavior lack both female reproductive tract and female sex pheromone, Barth and Bell[64] hypothesized that the pheromone may be produced or released in the genital tract. Using wax plugs inserted into the genital atrium and electrocautery of the lining of the genital atrium, Moore and Barth[63] showed that females producing pheromone possessed active columnar epithelium along the roof of the atrium, whereas cells in cauterized nonproducers appeared smaller. A problem with proving lack of release is that pheromone production may be inhibited indirectly by the manipulation. For instance, as discussed in Section IV.G.2 below, implantation of a genital plug may mimic a spermatophore or an ootheca, which in some cockroaches may inhibit CA activity and (indirectly) pheromone production. Similarly, cautery may damage mechanoreceptors. Moore and Barth[63] stated that CA activity was unaffected, but no quantitative comparison with controls was reported.

D. *SUPELLA*

Hales and Breed[65] have described a calling posture in the brown-banded cockroach in which the female raises her wings, flexes the abdomen, and periodically exposes the genital atrium. By comparing the orientation responses of males to hexane extracts of various female body parts, our research group[65a] showed that pheromone activity was greatest on the third through fifth tergites of virgin females; the genital region, digestive tract, and sternum lacked activity. EAG responses were also greater to tergal than to genital extracts. Scanning electron and light microscopy revealed that the distribution and density of cuticular pores correlated with the activity of tergites 1 through 7 (i.e., as pheromone activity increased, so did the density of pores).

IV. REGULATION OF SEX PHEROMONES

In this section, work on the neuroendocrine regulation of pheromone synthesis and release in cockroaches is reviewed. In each subsection, selected examples from other insect models are also presented. This is not an exhaustive review of all insect studies; rather, these examples are presented to facilitate comparative discussions of regulatory mechanisms. Recent reviews of regulation of pheromone production in other insect groups are presented in a text by Prestwich and Blomquist.[66]

A. BARTH'S HYPOTHESIS

Because pheromone production and/or release are usually coordinated with specific physiological events and environmental conditions, Barth[55] proposed a hypothesis on neuroendocrine control of pheromone production, stating that "neuroendocrine control of mating behavior would occur only in those insects which are long-lived as adults and which have repeated reproductive cycles containing periods during which mating is not appropriate and perhaps not even possible . . . ". Conversely, insects with mature eggs at emergence and with a short imaginal life would not have such control. Apparent exceptions to this hypothesis are studies of short-lived Lepidoptera (moths), which have demonstrated neuroendocrine regulation,[67] and studies with long-lived flies demonstrating various degrees of ovarian control (see Reference 68 for review). The latter clearly involve endocrine regulation by the gonadotropic hormone (20-hydroxyecdysone in flies) which is produced in the ovaries (see below). Also, regulation of pheromone synthesis and release in the corn earworm moth, *Heliothis zea,* by neuropeptides[67] conforms with Barth's predictions that "in certain Lepidoptera . . . which . . . do actually feed as adults . . . endocrine or neuroendocrine control over the communication system for mating might occur independent of the endocrine events occurring during adult development".[55]

B. PHEROMONE PRODUCTION AND OVARIAN DEVELOPMENT

Female cockroaches exhibit two basic reproductive patterns: (1) a primitive pattern in which egg cases are oviposited frequently and embryogenesis proceeds away from the female, and (2) an ovoviviparous mode in which the ootheca is retracted into a brood pouch or uterus and incubated within the female until hatching (see also Chapter 1 of this work). A highly advanced viviparous condition is known in *Diploptera,* and functionally intermediate modes may occur. For example, oviparous *Blattella* females form a hard egg case which is extruded, but not deposited; young hatch after 18 to 25 d of embryogenesis during which the ootheca is carried by the female.

With a few exceptions (e.g., *Diploptera*), most oviparous and ovoviviparous females undergo a sexual maturation period preceeding the first vitellogenic cycle. Females then become behaviorally receptive to courting males. Upon mating, females become unreceptive and may cease pheromone production, but both may reappear after several oviposition cycles in oviparous species or after each protracted period of gestation in ovoviviparous species.

By employing high pressure liquid chromatography (HPLC) fractionation and EAG, Sass[35] quantified temporal changes in production of periplanone-A and -B in virgin female *Periplaneta americana* over a 60-d period. He found equal amounts of the two components in gut extracts and a 100-fold increase in both over the first 3 weeks after adult emergence. Mated females ceased production of the pheromone.[2,51,69,70] Sass[35] reported that in mated females with egg cases there was a 100-fold decrease in pheromone in the guts, whereas virgin females produced equal amounts of pheromone with or without egg cases.

When maintained at 27°C and a 12 h light:12 h darkness photoperiod, virgin *Supella longipalpa* females initiate pheromone production and release ("calling" behavior) at mean adult ages of 4 and 6 d, respectively.[70a] The onset of pheromone production correlates with the end of the previtellogenic stage of basal oocyte development and with an increase in synthesis of JH by the CA.[70b] Pheromone production and release continue in virgin females through at least 12 ovarian cycles. Pheromone production and release cease after mating and do not resume for at least 17 successive gonadotrophic cycles.

Nishida and Fukami[1] showed that teneral* *Blattella germanica* females elicit a strong wing-raising response in males (see also Reference 2). No males respond to 4-day-old females, but after day 4 the activity of female extracts increases, eliciting wing raising in 100% of the males by 7 d. However, gas-liquid chromatographic (GLC) analyses of amounts

* Term applying to recently molted, pale, soft-bodied individuals.

of 3,11-dimethyl-2-nonacosanone and 29-hydroxy-3,11-dimethyl-2-nonacosanone from cuticular extracts indicate that both components increase after the imaginal molt, with the greatest increase during the phase of most rapid oocyte growth.[70c] A slow increase in pheromone occurs during gestation (*Blattella* incubates its young in an external egg case), followed by a second phase of rapid increase in pheromone on the cuticle during maturation of the second wave of oocytes. Incorporation of [1-^{14}C]propionate into pheromone follows a similar pattern.[70d] Thus, in *B. germanica,* oocyte maturation, pheromone synthesis, and JH biosynthesis[70e] appear to be closely correlated.

C. RECEPTIVITY AND OVARIAN DEVELOPMENT

In the first ovarian cycle, receptivity of females to males, as measured either by calling behavior (pheromone release; see Section IV.D.2.b below) or by mounting of courting males, is also highly correlated with the stage of the ovarian cycle. A teneral female *B. germanica* may mount courting males, but by stilting on her hind legs or mounting the male from the side she remains unreceptive until 4 to 8 d later.[2] Whereas only 20% of individually isolated females mate on day 6, more than 74% mate on day 8. Female *Supella* and *Periplaneta* initiate calling close to the age at which they become receptive. Interestingly, virgin *Supella* females do not call while carrying an inviable ootheca, but they resume calling and mate immediately after oviposition. Thus, it appears that time courses of pheromone synthesis, pheromone release (calling), and sexual receptivity are different in the first and in subsequent gonadotrophic cycles. Unfortunately, most studies of cockroaches address only the first ovarian cycle.

D. ROLE OF THE CORPORA ALLATA

1. Pheromone Production

a. Other Insects

Roller et al.[71,72] first performed allatectomies (removal of CA) on wax moth *(Galleria mellonella)* females and determined that their ability to attract males was unaffected. Barth[55] and Steinbrecht[73] extended these observations to the giant silk moths *Antheraea pernyi* and *Bombyx mori,* respectively, and Barth showed that calling and mating were also unaffected. Riddiford and Williams[74] confirmed Barth's results, but showed that in *A. polyphemus* and *Hyalophora cecropia* removal of the CA and corpora cardiaca (CC) inhibited calling. Riddiford[75] also showed that injection of blood from calling to noncalling females induced calling behavior, but Sasaki et al.[76] showed conclusively that neither CC nor CA exerted any influence on calling in *H. cecropia.*

When female *Heliothis zea* are ligated between the head and thorax they do not produce pheromone, as determined by GLC of gland extracts.[67] Such females resumed pheromone production when injected with a brain extract. Thus, it appears that an intact brain-body connection is needed. The brain substance (a neuropeptide of subesophageal origin) was present in the brain in both photophase and scotophase, but was found in the hemolymph only in scotophase, when calling occurred. Raina and Menn[77] review regulation of pheromone production in Lepidoptera.

Allatectomy, decapitation, or removal of brains of female *Tenebrio molitor* (Coleoptera [beetles]) diminished pheromone production (the joint removal of the brain and the CA reduced pheromone activity more than the removal of either one alone), but ovariectomy did not.[78] Although reimplantation of brains or CA did not stimulate pheromone production, injection of JH analogues (JHA) did, suggesting that a brain-CA connection was needed for JH induction of pheromone production. Vanderwel and Oehlschlager[79] review endocrine regulation of pheromone production in beetles.

The CA control development of accessory sex glands, production of the maturation accelerating pheromone, and sexual behavior in males of some acridids (grasshoppers; for

a review, see Reference 80). Pheromone production in male *Schistocerca gregaria* is under CA control.[81,82]

b. Cockroaches

In cockroaches, the CA are required for synthesis and deposition of yolk into oocytes and for activation of the accessory sex glands of females. The presence of CA is necessary for mating in female *Leucophaea maderae.*[83] In *Byrsotria fumigata,* removal of the CA inhibits pheromone production and mating,[62,84] while reimplantation of CA[62] or treatment with JH and JHA restores production.[85] Onset of pheromone production in *P. americana* is also under hormonal regulation of the CA.[55] Allatectomized females resume production of sex attractant within a few days after JH injection.[86] Bowers[87] claimed that topical application of precocene II, a compound with antiallatotropic activity, terminated pheromone production in *P. americana* within 5 d.

Allatectomized *Supella* females exhibit no pheromone production, and pheromone production is restored by implantation of active CA as well as by exposure to the JHA ZR 512 (hydroprene) for as little as 24 h.[70a] High doses of ZR 512 ($>$20 μg on filter paper) fail to stimulate pheromone production in allatectomized *Supella* females, as in *Byrsotria* females.[85] Topical application of ZR 512 to intact virgin *Supella* females exhibits dose-dependent effects: low doses (0.1 and 1.0 μg) advance the onset age of pheromone production by as much as 3 d, while high doses delay or suppress pheromone production by up to 4 d.[70a] It is important to note that application of low doses (2.5 μg) of ZR 512 to intact *Diploptera* females increases JH biosynthetic rates of native CA (measured by *in vitro* incubations of CA), whereas high doses (25 or 100 μg) reduce endogenous JH synthesis.[88] It is not known whether application of high doses of JH or JHA serves to suppress pheromone production through physiological or pharmacological effects.

In vitro measurements of the rates of JH biosynthesis in *Supella* females reveal CA of relatively low activity (1 pmol/h per pair) at the age at which pheromone production is normally initiated. A 20-fold increase in JH biosynthetic rate is reached 5 d later.[70b] Thus, the initiation of pheromone production appears to be stimulated directly or indirectly by low titers of JH.

In *Blattella,* the increase in titer of the major component of the female's cuticular sex pheromone corresponds to ovarian development. In isolated females, the greatest incorporation of [1-^{14}C]propionate into pheromone occurs on day 9 — after mating (day 8), but before ovulation (day 11).[70d] Allatectomy, or inhibition of CA activity by implantation of artificial egg cases into the genital atrium of teneral adult *Blattella* females, results in low accumulation of pheromone in 15-day-old females.[70c] Exposure of intact *B. germanica* imaginal females to ZR 512 induces pheromone synthesis in a dose-dependent manner, while application of precocene II partially inhibits pheromone synthesis. Combining 600 μg precocene with 10 μg ZR 512 results in a large accumulation of pheromone on the cuticle, indicating that in *Blattella* precocene influences pheromone synthesis indirectly by inhibiting CA activity. Exogenous JH can induce pheromone synthesis in females with inhibited CA.

In *Pycnoscelus indicus,* a bisexual species, removal of the CA eliminated pheromone production, but in a parthenogenetic strain of *P. surinamensis* allatectomies had no effect on pheromone synthesis.[55,86]

2. Calling Behavior and Receptivity

a. Other Insects

Calling, or release of sex pheromone, is related to the synthesis of pheromone and may be a good indicator of onset of female receptivity. It is of interest to determine whether there is a delay between the initiation of synthesis and pheromone release and, if so, if there are different regulatory mechanisms for the two processes. For instance, in many ixodid

ticks, pheromone production coincides with the imaginal molt, but pheromone release only occurs when feeding commences (for a review, see Reference 89).

Hollander and Yin[90] have shown that calling behavior and pheromone release are separate events under different regulatory mechanisms. Experimentally, the two can be uncoupled. In the gypsy moth, *Lymantria dispar,* either removal of the brain or transection of the ventral nerve cord (VNC) anterior to the terminal abdominal ganglion (TAG) resulted in cessation of pheromone release, but not calling. Removal of the TAG or severance of nerves posterior to this ganglion eliminated calling[90] and pheromone release, suggesting that the TAG is involved in calling behavior and the brain in pheromone synthesis. However, Tang et al.[91] conclude that calling in VNC-transected females is qualitatively different and is most likely controlled by nervous input from a higher center via the VNC, a hypothesis supported by data from other moths (*Manduca sexta* and *Utetheisa ornatrix*) where transection of the VNC at any point eliminates calling.[92,93] Webster and Cardé[94] found that calling terminated and pheromone titer decreased following decapitation of virgin *Platynota stultana* moths. Applications of a JHA did not restore pheromone production in either decapitated or mated females. They did not test other secretory products of the head region (see, for example, Reference 67).

In some orthopterans (crickets, grasshoppers, locusts, katydids), female sexual receptivity to courting males is unaffected by allatectomy; in others, the CA mediate receptivity, song production, and phonotactic orientation to calling males. Koudele et al.[95] reported that, following allatectomy, phonotactic orientation of female house crickets deteriorated, but was significantly improved after topical application of JH III or ZR 512.

Ovarian development in the housefly correlates with pheromone production and mating. Females mate at a preferred stage of ovarian development. Removal of female CA, CC, or CA + CC does not affect pheromone production,[96] but allatectomy reduces female receptivity.[97] In *Calliphora vomitoria* (the blowfly), sexual receptivity is partly inhibited by ovariectomy and totally suppressed by allatectomy in newly emerged females.[98] Topical application of ZR 512 induces receptivity in previtellogenic females at low concentrations, but inhibits receptivity at high concentrations. This study, however, involved only the first gonadotrophic cycle; whether or not the blowfly shows cycles of receptivity paralleling subsequent ovarian cycles remains unknown.

b. Cockroaches

Allatectomized females of *Leucophaea maderae*[83] and *Byrsotria fumigata*[84] do not mate normally. However, topical application of female pheromone extract on allatectomized females restores mating,[99] indicating that JH controls pheromone production, but not receptivity, in females. Barth and Lester[86] state that ''the question as to whether the CA control the synthesis or merely the release of the sex pheromone in *B. fumigata* remains unresolved although the available evidence suggests that it controls synthesis of the pheromone''. Since all studies of *Byrsotria* have tested male responses to pheromone released by females and adsorbed onto filter paper, they clearly address *release,* not production. To monitor production of pheromone, extracts of females must be assayed.

The initiation of calling in *Supella longipalpa* appears to be regulated by the CA. Although low doses (0.1 and 1.0 μg) of ZR 512 accelerate pheromone production in intact females, the age of onset of calling is not altered.[70a] However, transection of the nervi corporis allati I (NCA I) within 24 h of the imaginal molt significantly accelerates the onset age of calling (4.3 d) compared with sham-operated controls (6.0 d), as well as the onset age of pheromone production. Tobe and Stay[100] review studies on the regulation of the CA via neural inhibition by the brain through the NCA I.

Supella females allatectomized within 24 h of the imaginal molt fail to call for 11 d.[70a] Allatectomized females treated topically with ZR 512 (1 μg, day 0 and 10 μg, day 5) or

exposed to ZR 512 vapors (10 or 20 μg on filter paper) resume calling as early as 24 h after treatment. Calling and pheromone production can also be restored with implantation of a pair of active CA.

In both *Supella* and *Blattella* (oviparous), unlike the ovoviviparous species, allatectomized females will not mate. It is unknown whether active CA or vitellogenic oocytes stimulate female receptivity.

E. ROLE OF THE OVARIES

1. Other Insects

In the housefly the ovaries play a key role in pheromone production. Removal of ovaries shortly after emergence inhibited production of three pheromone components. Although females did not attract males,[101] they were courted and mated normally.[97] Reimplantations of previtellogenic oocytes into females initiated pheromone production during early vitellogenesis.[96] When treated with 20-hydroxyecdysone (a hormone synthesized by the ovaries which induces vitellogenin synthesis in the fat body of flies), newly emerged ovariectomized females synthesize pheromone.[101] The ovaries only initiate synthesis and can be removed later without curtailing pheromone production.[96] Ecdysone also induces vitellogenin synthesis in Diptera (flies, mosquitos), whereas in most other insects vitellogenin synthesis is regulated by JH.

2. Cockroaches

In both oviparous (*Periplaneta americana*[102]) and ovoviviparous (*Diploptera punctata*,[103] *Nauphoeta cinerea*[104]) cockroaches, ovariectomy abolishes the cycle of JH synthesis. Implantation of previtellogenic ovaries restores the JH cycle,[103,105] but mature ovaries appear to inhibit JH synthesis[106] (see also Chapter 21 in this work). In cockroaches with protracted gestation (ovoviviparous and viviparous), presence of an ootheca in the brood sac suppresses CA activity.[107] It appears that both neural and humoral feedback from the ovaries and ootheca modulate CA activity and may influence pheromone production indirectly (see Chapter 21 in this work for a discussion of ovarian regulation of CA activity).

Barth[62] showed that removal of ovaries in newly emerged females did not affect pheromone production in *Byrsotria fumigata*. *Byrsotria* virgin females do not produce pheromone during a 12-week gestation period,[62] but in a study of the effects of mating on pheromone production it is also stated that 21 of 28 females produced pheromone 2 to 14 weeks after mating; it would appear that suppression by the ootheca should occur in mated females as well as virgins. Moreover, since JH is synthesized (albeit at low rates) during gestation, it would be of interest to quantify pheromone production during this period.

In oviparous females (e.g., *P. americana*), as described above, pheromone production is equal in virgin females with and without oothecae.[35] *Supella longipalpa* females, ovariectomized as either teneral adults or last-instar nymphs, continue to produce pheromone.[70a] Thus, it appears that the presence of ovaries is not essential to stimulate the CA at the rates necessary to initiate pheromone production. Similarly, in *Blattella germanica*, which utilizes a contact pheromone, the ovaries need not be present for pheromone synthesis.[70c]

F. ROLE OF FEEDING AND DRINKING

1. Other Insects

Topical application of JH or CA + CC implants induced the conversion of host-tree-produced myrcene to aggregation pheromones in males of the bark beetle, *Ips paraconfusus*.[108] The regulatory mechanism was complicated, however, by the interaction of feeding and JH effects. Release of JH in unfed adults was prevented by neural inhibition. Stretching of the gut during feeding (or artificially with air) removed the inhibition, resulting in release of JH, which stimulated release of brain hormone (BH) from the CC or brain neurosecretory cells.[109] It is thought that BH then stimulates pheromone production.

Observations of bark beetles and boll weevils have indicated that reduction of symbiotic microorganisms by axenic rearing or by administration of dietary antibiotics does not reduce pheromone content, and may even increase it. Gueldner et al.[110] showed that weevils free of bacteria produced more pheromone, and Conn et al.[111] hypothesized that under field conditions microbes would thus regulate pheromone levels.

In the stable fly, *Stomoxys calcitrans,* as in some bark beetles, pheromone synthesis begins only after feeding occurs.[112] However, unlike the beetles, this fly apparently does not utilize food as a pheromone precursor. In some ticks, sex pheromone production commences soon after emergence of the adult, but release of pheromones is delayed until after feeding (for a review, see Reference 89).

2. Cockroaches

Feeding is essential for mating activity in *Periplaneta americana.*[113] Whether food and water exert direct influence on pheromone synthesis and release or have indirect effects on female receptivity was not determined. Weaver[114] has shown that both food and water are essential for stimulation of CA activity and mating. Females with access to food, but not water, are unreceptive and have low rates of JH synthesis, while 75% of females with access to only water mate. In this second group of females, JH synthesis is higher and some oocyte growth occurs.[114] When newly emerged *Supella* females are deprived of food and water, pheromone production fails to occur,[70a] presumably due to relatively inactive CA, as in *Periplaneta.* It is interesting to note that in their isolation and identification of periplanone-B Persoons et al.[27] used alimentary canals of *P. americana* females starved for 10 d in order to avoid contamination. Although this cockroach is known to withstand long periods of starvation (see Chapter 1 in this work), such treatment may reduce the pheromone yield in mass-extraction procedures. Moreover, experimental neck ligations of cockroaches, performed in order to isolate the abdomen from CA influence, must be interpreted with caution because food and water may influence neuroendocrine events. For example, pheromone production, which is inhibited in neck-ligatured *Byrsotria fumigata* females, can be induced with injections of JHA,[85] but it appears that this effect may be indirectly due to starvation (although feeding is not needed to initiate vitellogenesis in *Byrsotria*).

In *Blattella germanica,* starved females (with access to water) accumulate only 55 ng of pheromone by day 5 and 103 and 110 ng by days 10 and 15, respectively.[70c] By contrast, 149, 886, and 1258 ng are recovered from 5-, 10-, and 15-day-old fed females. Topical applications of ZR 512 induce pheromone synthesis in a dose-dependent manner. Receptivity of starved females is low as measured by percent mating, although starved females commonly mount males to feed on tergal secretion, indicating that the latter is not an appropriate measure of receptivity.

G. EFFECTS OF MATING AND PREGNANCY

1. Other Insects

In the moth *Hyalophora cecropia,* deposition of sperm in the bursa copulatrix and subsequent release of a humoral factor from the bursa facilitate the termination of calling and the onset of oviposition.[115,116] Presence of sperm in the spermatheca is apparently the trigger in other moths and true bugs (Hemiptera).[117] In some flies, accessory secretion from the male is responsible for this change (for a review, see Reference 118).

Several investigators (see Reference 119) have documented stores of JH in accessory sex glands of males, and Shirk et al.[119] have reported its transfer to the female during copulation. Webster and Cardé[94] have shown that exogenous JH mediates the switch from calling to oviposition and terminates pheromone production in the moth *Platynota stultana.* Thus, it appears that in the same order of insects, endogenous and exogenous factors of humoral, endocrine, and/or neural nature can induce "mated behavior" in a virgin female.

2. Cockroaches

The mechanical stimulus of the spermatophore in the bursa copulatrix brings about mated behavior in some female cockroaches.[120] It is hypothesized that first the spermatophore and then the developing ootheca inhibit pheromone production.[121] However, since mated oviparous females stop calling for several reproductive cycles, it seems likely that the spermatophore plays a role only in the initial switch from virgin to mated behavior; in most species it is removed within several hours to several days after copulation. Furthermore, inhibition by the ootheca can only be cyclic. Hence, sperm or seminal fluid in the spermatheca may serve the same function.

The spermatophore plays a role in the initial termination of calling in *Supella*. When the spermatophore is removed within 4 min after copulation (prior to sperm transfer to the spermatheca), calling is suppressed.[70a] When a section of the ventral sternites 2 and 3 is excised within 4 min of copulation, mated females continue to call. Furthermore, implantation of an artificial spermatophore into the bursa of virgin females inhibits calling and stimulates significant basal oocyte growth, suggesting that mechanical stimulation of the bursa by insertion of the spermatophore serves in the initial termination of calling via ascending neurons of the VNC. Transection of the VNC 4 d after mating (and after sperm transfer to the spermatheca and production of an ootheca) restores the calling behavior characteristic of virgin females, indicating that maintenance of mated (i.e., noncalling) behavior may be mediated by the presence of sperm in the spermatheca.[70a] Roth and Stay[120,122] also showed that a spermless spermatophore (of a male from which the testes were removed in the last larval instar) was able to activate the CA and stimulate maturation of oocytes. A common procedure to ascertain whether females have mated recently is to expand the genital atrium with forceps and to determine whether a spermatophore has been inserted into the bursa. We caution that this procedure may stimulate the CA and accelerate the gonadotrophic cycle.

In a bisexual strain of *Pycnoscelus surinamensis,* 99% of oothecae of virgin females are aborted.[123] Of females mated to spermless males with normal spermatophores, 95% aborted oothecae, indicating that sperm are an important stimulus in termination of virgin behavior. Thus, the spermatheca may have (a) role(s) in maintaining mated behavior in cockroaches, either humorally or neurally. Implantation of sperm-filled spermathecae into the abdomens of *Supella* virgin females does not inhibit calling.[70a] Stay and Gelperin[123] have concluded that sperm-filled spermathecae lack a hormonal influence because mated females with cut spermathecal ducts behave like virgins. They showed that mated females with cut spermathecal nerves aborted oothecae, as did virgin females. However, the last abdominal ganglion, where the spermathecal nerves originate, is not sufficient to coordinate the mated state; intact connections to more anterior nervous centers (probably the brain) are required.[123]

Roth and Stay[122] showed that CA activity (as measured by oocyte maturation) in cockroaches is inhibited by natural or artificial egg cases through mechanoreceptors in the uterus. Thus, severing the VNC, removing the egg case, or denervating the CA in pregnant females initiates a second ovarian cycle. *Blattella germanica* females with implanted oothecae produced 0.5 μg of pheromone by day 15 (1.4 μg in controls), and the amount of pheromone correlated well with oocyte growth.[70c]

An interesting, as yet uninvestigated finding was that sequential injections of JH increased yolk deposition and activity of the colleterial glands, but inhibited pheromone production in *Byrsotria*.[85] Bell and Barth[85] have hypothesized that normally high JH titers occur after copulation, when oocytes are mature and sex pheromones are no longer needed. Several studies comparing peak JH biosynthetic rates during the first and second ovarian cycles of representative oviparous, ovoviviparous, and viviparous species indicate that both peaks are of similar magnitude (for a review, see Reference 100). Nonetheless, it is possible that a high (but physiological) titer of JH inhibits pheromone production, as does insertion of a spermatophore, a sperm-filled spermatheca, and an ootheca either in the uterus or held in

the bursa. Clearly, a complex neuroendocrine mechanism involving several neuronal and humoral feedbacks is implicated.

H. EXTERNAL (ENVIRONMENTAL) FACTORS REGULATING PHEROMONE RELEASE

Photoperiod is an important determinant of the timing of release of pheromones in *Periplaneta americana,*[61] *Supella longipalpa,*[65] and many other cockroaches which exhibit calling behaviors.[56] Calling in *S. longipalpa* females occurs throughout the dark phase of the photocycle.[65] When observed under conditions of continuous light or continuous dark after entrainment to a light:dark regime, the behavior free-runs, confirming the circadian nature of calling.[70a] Smith and Schal[70a] have shown that the onset of the dark phase serves as the entraining cue (Zeitgeber) by which the rhythm is kept in phase with the photocycle. Whether such exogenous cues modulate only pheromone release (calling) or also pheromone synthesis remains unknown. Current work in our laboratory addresses whether JH release and pheromone production exhibit diel periodicity, as do many physiological and behavioral activities.

I. INDUCTION IN MALES

An important question in the development of sexual competence and sexual behaviors is why sensory receptors (see, for example, Reference 124) and sex pheromones develop in a sexually dimorphic manner. In oviparous vertebrates, estrogen can induce vitellogenin synthesis in the male liver, indicating that absence of vitellogenin in males is due to the lack of an inducer. In cockroaches, JH occurs in all stages. Hence, lack of pheromones in males may be due to (1) lack of target organs (pheromone glands) or hormone receptors, (2) their inability to respond to inducing factors (i.e., competence), or (3) a lower titer of JH in circulation. JH or JHA applications to males result in vitellogenin synthesis in a dose-dependent manner in some cockroaches *(D. punctata),*[125] but not in others *(Eublaberus posticus),*[126] suggesting that different control mechanisms may operate in vitellogenin synthesis. In some flies, injection of 20-hydroxyecdysone into males induces vitellogenin synthesis;[127] female pheromones are induced in males by either ovary implants or 20-hydroxyecdysone injections.[128] Interestingly, in male flies, synthesis of pheromones decreased within several days after injection, indicating that induction of the enzymes involved in pheromone synthesis was only temporary.[129] Studies with *Blattella germanica* indicate that nymphal male fat bodies can be induced by exogenous JH to synthesize vitellogenin, but adult male fat bodies are only slightly inducible,[130] possibly indicating loss of JH receptor sites in the adult male.

Nishida and Fukami[1] observed that in *B. germanica* the male fractions corresponding to the female pheromone components did not elicit sexual responses in males. Application of 100 μg of ZR 512 induced some female pheromone (and vitellogenin[130]) synthesis in males.[70c] Control 15-day-old males contained 16 ng of female pheromone, whereas treated males accumulated 104 ng. Studies are now in progress to determine whether nymphs can be induced to a greater extent than adult males, as in Kunkel's[130] vitellogenesis model.

V. SUMMARY, CONCLUSIONS, AND A HYPOTHESIS

Much of our current knowledge about neuroendocrine regulation of reproduction in cockroaches derives from studies of *Leucophaea maderae, Diploptera punctata,* and *Nauphoeta cinerea,* all of which have protracted periods of internal incubation of embryos. Studies of positive and negative influences of feeding, drinking, mating, crowding, and enforced virginity on the CA, brain, ovaries, and other organs have been restricted largely to these species. Recently, the American cockroach has been used as an oviparous model

for studies of CA regulation, but in *Periplaneta americana* the penultimate oocyte becomes vitellogenic before the basal oocyte is ovulated, resulting in two JH biosynthetic cycles per ovulation cycle[102] (for a review, see Reference 100). Clearly, representatives of other oviparous cockroaches (e.g., *Supella*) must be studied.

A common theme in reviews of the regulation of pheromone production is that in those insects where endocrine regulation is important, behavioral regulation of release is not an available option because the pheromone is an epicuticular secretion. Here we have shown convincingly that, in cockroaches, endocrine regulation of pheromone synthesis may be coupled with behavioral regulation of its release. In all sexually reproducing cockroaches studied to date, pheromone production and release are under neuroendocrine regulation, with both events coinciding with periods of sexual competence and receptivity. Where volatile pheromones are involved, high JH titers and/or feedback from mating turn off production and release (calling) of pheromones for several ovarian cycles[70a] despite JH biosynthetic rates at each cycle reaching levels that would induce pheromone synthesis in virgin females.[70b] In *Supella*, for example, JH has a dose-dependent effect on induction of pheromone synthesis, and pheromone production is probably inhibited by neural (or possibly humoral) feedback from mating. Suppression of calling is probably a multistep process involving several organs: stretch receptors of the bursa, spermatheca, and uterus, as well as tropins and statins from the central nervous system.

Where contact pheromones are employed (e.g., *Blattella*), however, release is not mediated by specific behaviors (calling) and, thus, cannot be turned off quickly; the pheromone remains on the cuticle in its active form throughout the nonreceptive period of gestation or incubation.[70c] Moreover, production and release of contact pheromones may occur at each gonadotrophic cycle as JH biosynthesis increases. Thus, in virgin females, JH biosynthetic rates increase, stimulating both oocyte growth and contact pheromone production; there appears to be no neural or humoral feedback from mating to decrease pheromone production, as in the case of volatile pheromones. Mated and virgin females accumulate similar amounts of cuticular pheromones, since JH induces pheromone synthesis, and the time course of JH biosynthesis and oocyte growth is similar in both groups. Suppression of pheromone production is related to inhibition of CA activity, as during gestation.

Several researchers have classified the regulation of pheromone production and release in cockroaches into a "tonic release system", characteristic of females, where the CA control production and release is continual, and a "phasic release system", characteristic of male cockroaches and controlled by motoneurons during courtship (see, for example, Reference 121). From the discussion above, it is clear that both systems may be found in females, with endocrine regulation of synthesis and release coupled with motor control of calling behavior.

Scharrer[131] summarized structural and functional similarities between vertebrate and invertebrate neuroendocrine models, with special reference to the cyclic events of reproduction in females. There are remarkable similarities in the neurosecretory cells in the protocerebrum of insects and those in the hypothalamus of vertebrates. Regulation of the CA in the cockroach through a brain-CC-CA axis is similar to regulation of the hypothalamic-hypophyseal system in vertebrates. In both systems, stimulatory (allatotropins in insects) or inhibitory (allatostatins or allatinhibins in insects, somatostatin in vertebrates) directives may reach the endocrine gland via a circulatory system (hemolymph in insects) as well as by way of neurosecretory neurons innervating the gland (NCA I in insects). As in vertebrates, cockroach ovaries (particularly in viviparous and ovoviviparous species) undergo cycles of growth and yolk uptake followed by periods of dormancy subsequent to ovulation. In both systems, signals emanating from the ovary and the developing embryo (or the brood pouch) may suppress activity of the endocrine gland.

Most studies of the roles of JH in the adult cockroach examine the sex-specific induction

of vitellogenin synthesis and the endocytotic uptake of vitellogenins by the oocytes. JH-mediated induction of synthesis of specific gene products (vitellogenins) and selective uptake of yolk proteins by the oocytes resemble similar events in vertebrates. The insect, therefore, is an ideal model for the study of integration of the nervous and endocrine systems, as well as for the investigation of gene regulation through neuroendocrine directives.

In this chapter we have shown that pheromones can be induced by hormones in a sex-specific manner. Events regulating synthesis and release of the inducing factor (JH) are similar to those regulating other reproductive events (e.g., vitellogenesis). In the virgin female, JH induces pheromone synthesis and release. However, in some cockroaches, neural feedback signaling a ''mated state'' appears to suppress expression of the pheromone in spite of JH biosynthetic rates that would induce pheromone synthesis in virgin females. It is not known, however, what specific gene-encoded products are induced that permit the synthesis of pheromones.

Many questions about regulation of cockroach pheromones remain unanswered. However, as cockroaches continue to serve as subjects for neuroendocrine regulation of synthesis, release, and uptake of vitellogenin (see Chapter 20 of this work), the models generated should be tested to evaluate their heuristic value in relation to regulation of other reproductive events. Thus, pheromones may serve as easily assayable products to study neuroendocrine events in cockroaches. Because the chemical identities of most cockroach pheromones are unknown, most studies have been limited in the past by use of behavioral bioassays. Such assays offer remarkable sensitivity, but quantitation is difficult. Moreover, in cases where calling behavior was not evident, most studies could not distinguish biosynthesis from release of pheromones. Recently, work on chemical identification of cockroach pheromones has stimulated a renewed interest in studies of their regulation.

ACKNOWLEDGMENTS

We thank E. Burns, M. Gadot, R. Hamilton, and D. Liang for citations of unpublished results and E. Burns and M. Gadot for helpful suggestions. Some of the work was funded in part by grants from the U.S. Public Health Service (NIH Grant HD-21891) and the Rutgers University Research Council. This is New Jersey Agricultural Experiment Station Publication No. D-08170-06-88, supported by State and U.S. Hatch Act funds.

REFERENCES

1. **Nishida, R. and Fukami, H.,** Female sex pheromone of the German cockroach, *Blattella germanica, Mem. Coll. Agric. Kyoto Univ.,* 122, 1, 1983.
2. **Roth, L. M. and Willis, E. R.,** A study of cockroach behavior, *Am. Midl. Nat.,* 47, 66, 1952.
3. **Ishii, S.,** Sex discrimination by males of German cockroach, *Blattella germanica* (L.) (Orthoptera: Blattellidae), *Appl. Entomol. Zool.,* 7, 226, 1972.
4. **Nishida, R., Fukami, H., and Ishii, S.,** Sex pheromone of the German cockroach (*Blattella germanica* L.) responsible for male wing-raising: 3,11-dimethyl-2-nonacosanone, *Experientia,* 30, 978, 1974.
5. **Nishida, R., Fukami, H., and Ishii, S.,** Female sex pheromone of the German cockroach, *Blattella germanica* (L.) (Orthoptera: Blattellidae), responsible for male wing-raising, *Appl. Entomol. Zool.,* 10, 10, 1975.
6. **Nishida, R., Sato, T., Kuwahara, Y., Fukami, H., and Ishii, S.,** Female sex pheromone of the German cockroach, *Blatella germanica* (L.) (Orthoptera: Blattellidae), responsible for male wing-raising. II. 29-Hydroxy-3,11-dimethyl-2-nonacosanone, *J. Chem. Ecol.,* 2, 449, 1976.
7. **Nishida, R., Sato, T., Kuwahara, Y., Fukami, H., and Ishii, S.,** Female sex pheromone of the German cockroach, *Blattella germanica* (L.), responsible for male wing-raising. III. Synthesis of 29-hydroxy-3,11-dimethyl-2-nonacosanone and its biological activity, *Agric. Biol. Chem.,* 40, 1407, 1976.

8. **Sato, T., Nishida, R., Kuwahara, Y., Fukami, H., and Ishii, S.,** Syntheses of female sex pheromone analogues of the German cockroach and their biological activity, *Agric. Biol. Chem.*, 40, 391, 1976.
9. **Schwarz, M., Oliver, J. E., and Sonnet, P. E.,** Synthesis of 3,11-dimethyl-2-nonacosanone, a sex pheromone of the German cockroach, *Blattella germanica, J. Org. Chem.*, 40, 2410, 1975.
10. **Burgstahler, W. A., Weigel, L. O., Bell, W. J., and Rust, M. K.,** Synthesis of 3,11-dimethyl-2-nonacosanone, a contact courting pheromone of the German cockroach, *J. Org. Chem.*, 40, 3456, 1975.
11. **Burgstahler, W. A., Weigel, L. O., Sanders, M. E., Shaefer, C. G., Bell, W. J., and Vuturo, S. B.,** Synthesis and activity of 29-hydroxy-3,11-dimethyl-2-nonacosanone, compound B of the German cockroach sex pheromone, *J. Org. Chem.*, 42, 566, 1977.
12. **Rosenblum, L. D., Anderson, R. J., and Henrick, C. A.,** Synthesis of 3,11-dimethyl-2-nonacosanone, a sex pheromone of the German cockroach, *Blattella germanica, Tetrahedron Lett.*, 6, 419, 1976.
13. **Katsuki, T. and Yamaguchi, M.,** Syntheses of optically active insect pheromones, (2R,5S)-2-methyl-5-hexanolide, (3S,11S)-3,11-dimethyl-2-nonacosanone, and serricornin, *Tetrahedron Lett.*, 28, 651, 1987.
14. **Mori, K., Suguro, J., and Masuda, S.,** Stereocontrolled synthesis of all of the four possible stereoisomers of 3,11-dimethyl-2-nonacosanone, the female sex pheromone of the German cockroach, *Tetrahedron Lett.*, 37, 3447, 1978.
15. **Mori, K., Masuda, S., and Suguro, J.,** Stereocontrolled synthesis of all of the possible stereoisomers of 3,11-dimethynonacosan-2-one and 29-hydroxy-3,11-dimethynonacosan-2-one, the female sex pheromone of the German cockroach, *Tetrahedron*, 37, 1329, 1981.
16. **Nishida, R., Kuwahara, Y., Fukami, H., and Ishii, S.,** Female sex pheromone of the German cockroach, *Blattella germanica* (L.) (Orthoptera: Blattellidae), responsible for male wing-raising. IV. The absolute configuration of the pheromone, 3,11-dimethyl-2-nonacosanone, *J. Chem. Ecol.*, 5, 289, 1979.
16a. **Jurenka, R. A., Schal, C., Burns, E., Chase, J., and Blomquist, G. J.,** Structural correlation between the cuticular hydrocarbons and the female contact sex pheromone of the German cockroach, *Blattella germanica* (L.), *J. Chem. Ecol.*, 15, 939, 1989.
16b. **Chase, J., Jurenka, R. A., Schal, C., Halarnkar, P. P., and Blomquist, G. J.,** Biosynthesis of methyl branched hydrocarbons: precursors to the female contact sex pheromone of the German cockroach, *Blattella germanica* (L.) (Orthoptera: Blattellidae), *Insect Biochem.*, in press.
17. **Ritter, F. J., Bruggemann, I. E. M., Gut, J., and Persoons, C. J.,** Recent pheromone research in the Netherlands on muskrats and some insects, in *Insect Pheromone Technology: Chemistry and Applications*, Leonhardt, B. A. and Beroza, M., Eds., American Chemical Society, Washington, D.C., 1981, 107.
18. **Wharton, D. R. A., Black, E. D., Meritt, C., Jr., Wharton, M. L., Bazinet, M., and Walsh, J. T.,** Isolation of the sex attractant of the American cockroach, *Science*, 137, 1062, 1962.
19. **Jacobson, M., Beroza, M., and Yamamoto, R. T.,** Isolation and identification of the sex attractant of the American cockroach, *Science*, 139, 48, 1963.
20. **Jacobson, M. and Beroza, M.,** American cockroach sex attractant, *Science*, 147, 748, 1965.
21. **Day, A. C. and Whiting, M. C.,** The structure of the sex attractant of the American cockroach, *Proc. Chem. Soc. (London)*, p. 368, 1964.
22. **Day, A. C. and Whiting, M. C.,** On the structure of the sex attractant of the American cockroach, *J. Chem. Soc. (London)*, p. 464, 1966.
23. **Wakabayashi, N.,** A new synthesis of 2,2-dimethyl-3-isopropylidenecyclopropyl propionate, *J. Org. Chem.*, 32, 489, 1967.
24. **Chen, S. M. I.,** Sex Pheromone of the American Cockroach, *Periplaneta americana*, Isolation and Some Structural Features, Ph.D. thesis, Columbia University, New York, 1974.
25. **Chow, Y. S., Lin, Y. M., Lee, M. Y., and Lee, M. Y.,** Sex pheromone of the American cockroach, *Periplaneta americana* (L.). I. Isolation techniques and attraction tests for the pheromone in a heavily infested room, *Bull. Inst. Zool. Acad. Sin.*, 15, 39, 1976.
26. **Kitamura, C. and Takahashi, S.,** Isolation procedure of the sex pheromone of the American cockroach, *Periplaneta americana* L., *Appl. Entomol. Zool.*, 11, 373, 1976.
27. **Persoons, C. J., Verwiel, P. E. J., Ritter, F. J., Talman, E., Nooijen, P. J. F., and Nooijen, W. J.,** Sex pheromone of the American cockroach, *Periplaneta americana:* a tentative structure of periplanone-B, *Tetrahedron Lett.*, 24, 2055, 1976.
28. **Still, W. C.,** (+/−)-Periplanone-B. Total synthesis and structure of the sex excitant pheromone of the American cockroach, *J. Am. Chem. Soc.*, 101, 2493, 1979.
29. **Adams, M. A., Nakanishi, K., Still, W. C., Arnold, E. V., Clardy, J., and Persoons, C. J.,** Sex pheromone of the American cockroach: absolute configuration of periplanone-B, *J. Am. Chem. Soc.*, 101, 2495, 1979.
30. **Schreiber, S. L.,** [2 + 2] photocycloadditions in the synthesis of chiral molecules, *Science*, 227, 857, 1985.
31. **Kitahara, T., Mori, M., Koseki, K., and Mori, K.,** Total synthesis of (−)-periplanone-B, the sex pheromone of the American cockroach, *Tetrahedron Lett.*, 27, 1343, 1986.
32. **Nishino, C., Manabe, S., Kuwabara, K., Kimura, R., and Takayanagi, H.,** Isolation of sex pheromones of the American cockroach by monitoring with electroantennogram responses, *Insect Biochem.*, 13, 65, 1983.

33. **Talman, E., Verwiel, P. E. J., Ritter, F. J., and Persoons, C. J.,** Sex pheromones of the American cockroach, *Periplaneta americana, Isr. J. Chem.*, 17, 227, 1978.
34. **Persoons, C. J., Verwiel, P. E. J., Ritter, F. J., and Nooyen, W. J.,** Studies on sex pheromone of American cockroach, with emphasis on structure elucidation of periplanone-A, *J. Chem. Ecol.*, 8, 439, 1982.
35. **Sass, H.,** Production, release and effectiveness of two female sex pheromone components of *Periplaneta americana, J. Comp. Physiol.*, 152, 309, 1983.
36. **Burrows, M., Boeckh, J., and Esslen, J.,** Physiological and morphological properties of interneurones in the deutocerebrum of male cockroaches which respond to female pheromone, *J. Comp. Physiol.*, 145, 447, 1982.
37. **Seelinger, G.,** Behavioural responses to female sex pheromone components in *Periplaneta americana, Anim. Behav.*, 33, 591, 1985.
38. **Seelinger, G. and Gagel, S.,** On the function of sex pheromone components in *Periplaneta americana:* improved odour source localization with periplanone-A, *Physiol. Entomol.*, 10, 221, 1985.
39. **Persoons, C. J. and Ritter, F. J.,** Pheromones of cockroaches, in *Chemical Ecology: Odour Communication in Animals,* Ritter, F. J., Ed., Elsevier, Amsterdam, 1979, 225.
40. **Bowers, W. S.,** Phytochemical disruption of insect development and behavior, in *Bioregulators for Pest Control,* Hedin, P. A., Ed., American Chemical Society, Washington, D.C., 1985, 225.
41. **Takahashi, S., Kitamura, C., and Horibe, I.,** Sex stimulant activity of sesquiterpenes to the males of the American cockroach, *Agric. Biol. Chem.*, 42, 79, 1978.
42. **Manabe, S., Nishino, C., and Matsushita, K.,** Studies on relationship between activity and electron density on carbonyl oxygen in sex pheromone mimics of the American cockroach. XI, *J. Chem. Ecol.*, 9, 1275, 1985.
43. **Simon, D. and Barth, R. H., Jr.,** Sexual behavior in the cockroach genera *Periplaneta* and *Blatta.* I. Descriptive aspects, *Z. Tierpsychol.*, 44, 80, 1977.
44. **Simon, D. and Barth, R. H., Jr.,** Sexual behavior in the cockroach genera *Periplaneta* and *Blatta.* IV. Interspecific interactions, *Z. Tierpsychol.*, 45, 85, 1977.
45. **Seelinger, G.,** Interspecific attractivity of female sex pheromone components of *Periplaneta americana, J. Chem. Ecol.*, 11, 137, 1985.
46. **Waldow, U. and Sass, H.,** The attractivity of the female sex pheromone of *Periplaneta americana* and its components for conspecific males and males of *Periplaneta australasiae* in the field, *J. Chem. Ecol.*, 10, 997, 1984.
47. **Dethier, V. G.,** *Chemical Insect Attractants and Repellents,* Lewis, London, 1947.
48. **Wharton, D. R. A., Miller, G. L., and Wharton, M. L.,** The odorous attractant of the American cockroach, *Periplaneta americana* (L.). I. Quantitative aspects of the response to the attractant, *J. Gen. Physiol.*, 37, 461, 1954.
49. **Sturckow, B. and Bodenstein, W. G.,** Location of the sex pheromone in the American cockroach, *Periplaneta americana* (L.), *Experientia,* 22, 851, 1966.
50. **Bell, W. J.,** Pheromones and behavior, in *The American Cockroach,* Bell, W. J. and Adiyodi, K. G., Eds., Chapman and Hall, New York, 1982, 371.
51. **Seelinger, G. and Schuderer, B.,** Release of male courtship display in *Periplaneta americana:* evidence for female contact sex pheromone, *Anim. Behav.*, 33, 599, 1985.
52. **Jackson, L. L.,** Cuticular lipids of insects. II. Hydrocarbons of the cockroaches *Periplaneta japonica* and *Periplaneta americana* compared to other cockroach hydrocarbons, *Comp. Biochem. Physiol.*, 41B, 331, 1972.
53. **Gilby, A. R. and Cox, M. E.,** The cuticular lipids of the cockroach, *Periplaneta americana* (L.), *J. Insect Physiol.*, 9, 671, 1963.
54. **Bell, W. J., Burns, R. E., and Barth, R. H., Jr.,** Quantitative aspects of the male courting response in the cockroach *Byrsotria fumigata* (Guerin) (Blattaria), *Behav. Biol.*, 10, 419, 1974.
55. **Barth, R. H., Jr.,** Insect mating behavior: endocrine control of a chemical communication system, *Science,* 149, 882, 1965.
56. **Schal, C. and Bell, W. J.,** Calling behavior in female cockroaches (Dictyoptera: Blattaria), *J. Kan. Entomol. Soc.*, 58, 261, 1985.
57. **Percy-Cunningham, J. E. and MacDonald, J. A.,** Biology and ultrastructure of sex pheromone-producing glands, in *Pheromone Biochemistry,* Prestwich, G. D. and Blomquist, G. J., Eds., Academic Press, Orlando, FL, 1987, 27.
58. **Sreng, L.,** Morphology of the sternal and tergal glands producing the sexual pheromones and the aphrodisiacs among the cockroaches of the subfamily Oxyhaloinae, *J. Morphol.*, 182, 279, 1984.

58a. **Burns, E. and Schal, C.,** unpublished data, 1989.

59. **Bodenstein, W. G.,** Distribution of female sex pheromone in the gut of *Periplaneta americana* (Orthoptera: Blattidae), *Ann. Entomol. Soc. Am.*, 63, 336, 1970.

60. **Raisbeck, B.,** Pheromone inactivation by the gut of *Periplaneta americana, Nature (London),* 240, 107, 1972.
61. **Seelinger, G.,** Sex-specific activity patterns in *Periplaneta americana* and their relation to mate-finding, *Z. Tierpsychol.,* 65, 309, 1984.
61a. **Tobin, T.,** personal communication.
62. **Barth, R. H., Jr.,** The endocrine control of mating behavior in the cockroach *Byrsotria fumigata* (Guerin), *Gen. Comp. Endocrinol.,* 2, 53, 1962.
63. **Moore, J. K. and Barth, R. H., Jr.,** Studies on the site of sex pheromone production in the cockroach, *Byrsotria fumigata, Ann. Entomol. Soc. Am.,* 69, 911, 1976.
64. **Barth, R. H., Jr. and Bell, W. J.,** Reproductive physiology and behavior of *Byrsotria fumigata* gynandromorphs (Orthoptera (Dictyoptera): Blaberidae), *Ann. Entom. Soc. Am.,* 64, 874, 1971.
65. **Hales, R. A. and Breed, M. D.,** Female calling and reproductive behavior in the brown-banded cockroach, *Supella longipalpa* (F.) (Orthoptera: Blattellidae), *Ann. Entomol. Soc. Am.,* 76, 239, 1983.
65a. **Schal, C.,** unpublished data, 1989.
66. **Prestwich, G. D. and Blomquist, G. J., Eds.,** *Pheromone Biochemistry,* Academic Press, Orlando, FL, 1987.
67. **Raina, A. K. and Klun, J. A.,** Brain factor control of sex pheromone production in the female corn earworm moth, *Science,* 225, 531, 1984.
68. **Blomquist, G. J. and Dillwith, J. W.,** Pheromones: biochemistry and physiology, in *Endocrinology of Insects,* Downer, R. G. H. and Laufer, H., Eds., Alan R. Liss, New York, 1983, 527.
69. **Wharton, M. L. and Wharton, D. R. A.,** The production of sex attractant substance and of oothecae by the normal and irradiated American cockroach, *Periplaneta americana* L., *J. Insect Physiol.,* 1, 229, 1957.
70. **Jacobson, M.,** *Insect Sex Pheromones,* Academic Press, New York, 1972.
70a. **Smith, A. F.,** Endogenous and Exogenous Factors Regulating Pheromone Production and Release in the Adult Female Brown-Banded Cockroach, Ph.D. thesis, Rutgers, The State University of New Jersey, New Brunswick, 1988.
70b. **Smith, A. F., Yagi, K., Tobe, S. S., and Schal, C.,** *In vitro* juvenile hormone biosynthesis in adult virgin and mated female brown-banded cockroaches, *Supella longipalpa, J. Insect Physiol.,* in press.
70c. **Schal, C., Burns, E., and Blomquist, G. J.,** Endocrine regulation of female contact sex pheromone production in the German cockroach, *Blattella germanica, Physiol. Entomol.,* in press.
70d. **Schal, C., Burns, E., Jurenka, R. A., and Blomquist, G. J.,** unpublished data, 1989.
70e. **Gadot, M., Chiang, A.-S., and Schal, C.,** Farnesoic acid-stimulated rates of juvenile hormone biosynthesis during the gonotrophic cycle in *Blattella germanica, J. Insect Physiol.,* 35, 537, 1989.
71. **Roller, H., Piepho, H., and Holz, I.,** The problem of hormone dependency of copulation behavior of insects. Studies of *Galleria mellonella* (L.), *J. Insect Physiol.,* 9, 187, 1963.
72. **Roller, H., Biemann, K., Bjerke, J. S., Norgard, D. W., and McShan, W. H.,** Sex pheromones of pyralid moths. I. Isolation and identification of the sex attractant of *Galleria mellonella* (greater waxmoth), *Acta Entomol. Bohemoslov.,* 65, 208, 1968.
73. **Steinbrecht, R. A.,** Die Abhängigkeit der Lockwirkung des Sexualduftorgans weiblicher Seidenspinner *(Bombyx mori)* von Alter und Kopulation, *Z. Vgl. Physiol.,* 48, 341, 1964.
74. **Riddiford, L. M. and Williams, C. M.,** Role of the corpora cardiaca in the behavior of saturniid moths. I. Release of sex pheromone, *Biol. Bull. (Woods Hole, Mass.),* 140, 1, 1971.
75. **Riddiford, L. M.,** The role of hormones in the reproductive behavior of female wild silkmoths, in *Experimental Analysis of Insect Behavior,* Barton-Browne, L., Ed., Springer-Verlag, New York, 1974, 278.
76. **Sasaki, M., Riddiford, L. M., Truman, J. W., and Moore, J. K.,** Re-evaluation of the role of corpora cardiaca in calling and oviposition behaviour of giant silk moths, *J. Insect Physiol.,* 29, 695, 1983.
77. **Raina, A. K. and Menn, J. J.,** Endocrine regulation of pheromone production in Lepidoptera, in *Pheromone Biochemistry,* Prestwich, G. D. and Blomquist, G. J., Eds., Academic Press, Orlando, FL, 1987, 159.
78. **Menon, M.,** Hormone-pheromone relationships in the beetle, *Tenebrio molitor, J. Insect Physiol.,* 16, 1123, 1970.
79. **Vanderwel, D. and Oehlschlager, A. C.,** Biosynthesis of pheromones and endocrine regulation of pheromone production in Coleoptera, in *Pheromone Biochemistry,* Prestwich, G. D. and Blomquist, G. J., Eds., Academic Press, Orlando, FL, 1987, 175.
80. **Pener, M. P.,** Endocrine research in orthopteran insects, *Occas. Pap. Pan Am. Acridol. Soc.,* 1, 3, 1983.
81. **Loher, W.,** The chemical acceleration of the maturation process and its hormonal control in the male of the desert locust, *Proc. R. Soc. London,* 153, 380, 1960.
82. **Norris, M. J. and Pener, M. P.,** An inhibitory effect of allatectomized males and females on the sexual maturation of young male adults of *Schistocerca gregaria* (Forsk.) (Orthoptera: Acrididae), *Nature (London),* 208, 1122, 1965.
83. **Engelmann, F.,** Hormonal control of mating behavior in an insect, *Experientia,* 16, 69, 1960.
84. **Barth, R. H., Jr.,** Hormonal control of sex attractant production in the Cuban cockroach, *Science,* 133, 1598, 1961.

85. **Bell, W. J. and Barth, R. H., Jr.,** Quantitative effects of juvenile hormone on reproduction in the cockroach, *Byrsotria fumigata, J. Insect Physiol.,* 16, 2303, 1970.
86. **Barth, R. H., Jr. and Lester, L. J.,** Neuro-hormonal control of sexual behavior in insects, *Annu. Rev. Entomol.,* 18, 445, 1973.
87. **Bowers, W.,** Discovery of insect antiallatotropins, in *The Juvenile Hormones,* Gilbert, L. I., Ed., Plenum Press, New York, 1976, 394.
88. **Tobe, S. S. and Stay, B.,** Modulation of juvenile hormone synthesis by an analogue in the cockroach, *Nature (London),* 281, 481, 1979.
89. **Sonenshine, D. E.,** Neuroendocrine regulation of sex pheromone-mediated behavior in Ixodid ticks, in *Pheromone Biochemistry,* Prestwich, G. D. and Blomquist, G. J., Eds., Academic Press, Orlando, FL, 1987, 271.
90. **Hollander, A. L. and Yin, C.-M.,** Neurological influences on pheromone release and calling behaviour in the gypsy moth, *Lymantria dispar, Physiol. Entomol.,* 7, 163, 1982.
91. **Tang, J. D., Charlton, R. E., Cardé, R. T., and Yin, C.-M.,** Effect of allatectomy and ventral nerve cord transection on calling, pheromone emission and pheromone production in *Lymantria dispar, J. Insect Physiol.,* 33, 469, 1987.
92. **Itagaki, H. and Conner, W. E.,** Physiological control of pheromone release behaviour in *Manduca sexta* (L.), *J. Insect Physiol.,* 32, 657, 1986.
93. **Itagaki, H. and Conner, W. E.,** Neural control of rhythmic pheromone gland exposure in *Utetheisa ornatrix* (Lepidoptera: Arctiidae), *J. Insect Physiol.,* 33, 177, 1987.
94. **Webster, R. P. and Cardé, R. T.,** The effects of mating, exogenous juvenile hormone and a juvenile hormone analogue on pheromone titre, calling and oviposition in the omnivorous leafroller moth *(Platynota stultana), J. Insect Physiol.,* 30, 113, 1984.
95. **Koudele, K., Stout, J. F., and Reichert, D.,** Factors which influence female crickets' *(Acheta domesticus)* phonotactic and sexual responsiveness to males, *Physiol. Entomol.,* 12, 67, 1987.
96. **Dillwith, J. W., Adams, T. S., and Blomquist, G. L.,** Correlation of housefly sex pheromone production with ovarian development, *J. Insect Physiol.,* 29, 377, 1983.
97. **Adams, T. S. and Hintz, A. M.,** Relationship of age, ovarian development, and the corpus allatum to mating in the house-fly, *Musca domestica, J. Insect Physiol.,* 15, 201, 1969.
98. **Trabalon, M. and Campan, M.,** Etude de la receptivite sexuelle de la femelle *Calliphora vomitoria* (Dipteres, Calliphoridae) au cours du premier cycle gonadotrope. I. Approaches comportementale et physiologique, *Behaviour,* 90, 241, 1984.
99. **Roth, L. M. and Barth, R. H., Jr.,** The control of sexual receptivity in female cockroaches, *J. Insect Physiol.,* 10, 965, 1964.
100. **Tobe, S. S. and Stay, B.,** Structure and regulation of the corpus allatum, *Adv. Insect Physiol.,* 18, 305, 1985.
101. **Adams, T. S., Dillwith, J. W., and Blomquist, G. J.,** The role of 20-hydroxyecdysone in housefly sex pheromone biosynthesis, *J. Insect Physiol.,* 30, 287, 1984.
102. **Weaver, R. J.,** Radiochemical assays of corpus allatum activity in adult female cockroaches following ovariectomy in the last nymphal instar, *Experientia,* 37, 435, 1981.
103. **Stay, B., Tobe, S. S., Mundall, E. C., and Rankin, S.,** Ovarian stimulation of juvenile hormone biosynthesis in the viviparous cockroach, *Diploptera punctata, Gen. Comp. Endocrinol.,* 52, 341, 1983.
104. **Lanzrein, B., Wilhelm, R., and Gentinetta, V.,** On relations between corpus allatum activity and oocyte maturation in the cockroach *Nauphoeta cinerea,* in *Regulation of Insect Development and Behavior,* Sehnal, F., Zabza, B., Menn, J. J., and Cymborowski, B., Eds., Wroclaw Technical University Press, Wroclaw, Poland, 1981, 523.
105. **Lanzrein, B., Wilhelm, R., and Buschor, J.,** On the regulation of the corpora allata activity in adult females of the ovoviviparous cockroach *Nauphoeta cinerea,* in *Juvenile Hormone Biochemistry,* Pratt, G. F. and Brooks, G. T., Eds., Elsevier/North-Holland, Amsterdam, 1981, 149.
106. **Tobe, S. S. and Stay, B.,** Control of juvenile hormone biosynthesis during the reproductive cycle of a viviparous cockroach. III. Effects of denervation and age on compensation with unilateral allatectomy and supernumerary corpora allata, *Gen. Comp. Endocrinol.,* 40, 89, 1980.
107. **Tobe, S. S. and Stay, B.,** Corpus allatum activity *in vitro* during the reproductive cycle of the viviparous cockroach, *Diploptera punctata* (Eschscholtz), *Gen. Comp. Endocrinol.,* 31, 138, 1977.
108. **Hughes, P. R. and Renwick, J. A. A.,** Neural and hormonal control of pheromone biosynthesis in the bark beetle, *Ips paraconfusus, Physiol. Entomol.,* 2, 117, 1977.
109. **Hughes, P. R. and Renwick, J. A. A.,** Hormonal and host factors stimulating pheromone synthesis in female western pine beetles, *Dendroctonus brevicomis, Physiol. Entomol.,* 2, 289, 1977.
110. **Gueldner, R. C., Sikorowski, P. P., and Wyatt, J. M.,** Bacterial load and pheromone production in the boll weevil, *Anthonomus grandis, J. Invert. Pathol.,* 29, 397, 1977.

111. **Conn, J. E., Borden, J. H., Hunt, D. W. A., Holman, J., Whitney, H. S., Spanier, O. J., Pierce, H. D., Jr., and Oehlschlager, A. C.,** Pheromone production by axenically reared *Dendroctonus ponderosae* and *Ips paraconfusus* (Coleoptera: Scolytidae), *J. Chem. Ecol.,* 10, 281, 1984.
112. **Meola, R. W., Harris, R. L., Meola, S. M., and Oehler, D. D.,** Dietary-induced secretion of sex pheromone and development of sexual behavior in the stable fly, *Environ. Entomol.,* 6, 895, 1977.
113. **Brousse-Gaury, P.,** Starvation and reproduction in *Periplaneta americana* (L.): control of mating behaviour in the female, *Adv. Invert. Reprod.,* 1, 328, 1977.
114. **Weaver, R. J.,** Effects of food and water availability, and of NCA-1 section, upon juvenile hormone biosynthesis and oocyte development in adult female *Periplaneta americana, J. Insect Physiol.,* 30, 831, 1984.
115. **Truman, J. W. and Riddiford, L. M.,** Hormonal mechanisms underlying insect behavior, *Adv. Insect Physiol.,* 10, 297, 1974.
116. **Riddiford, L. M. and Ashenhurst, J. B.,** The switchover from virgin to mated behavior in female cecropia moths: the role of the bursa copulatrix, *Biol. Bull. (Woods Hole, Mass.),* 144, 162, 1973.
117. **Davey, K. G.,** Copulation and egg production in *Rhodnius prolixus:* the role of the spermathecae, *J. Exp. Biol.,* 42, 373, 1965.
118. **Leopold, R. A.,** The role of male accessory glands in insect reproduction, *Annu. Rev. Entomol.,* 21, 199, 1976.
119. **Shirk, P. D., Bhaskaran, G., and Roller, H.,** The transfer of juvenile hormone from male to female during mating in the cecropia silkmoth, *Experientia,* 36, 682, 1980.
120. **Roth, L. M. and Stay, B.,** Oocyte development in *Diploptera punctata* (Eschscholtz), *J. Insect Physiol.,* 7, 186, 1961.
121. **Hartman, H. B. and Suda, M.,** Pheromone production and mating behavior by allatectomized males of the cockroach, *Nauphoeta cinerea, J. Insect Physiol.,* 19, 1417, 1973.
122. **Roth, L. M. and Stay, B.,** A comparative study of oocyte development in false ovoviviparous cockroaches, *Psyche,* 69, 165, 1962.
123. **Stay, B. and Gelperin, A.,** Physiological basis of ovipositional behaviour in the false ovoviviparous cockroach, *Pycnoscelus surinamensis* (L.), *J. Insect Physiol.,* 12, 1217, 1966.
124. **Schafer, R. and Sanchez, T. V.,** The nature and development of sex attractant specificity in cockroaches of the genus *Periplaneta.* II. Juvenile hormone regulates sexual dimorphism in the distribution of antennal olfactory receptors, *J. Exp. Zool.,* 198, 323, 1976.
125. **Mundall, E. C., Szibbo, C. M., and Tobe, S. S.,** Vitellogenin induced in adult male *Diploptera punctata* by juvenile hormone and juvenile hormone analogue: identification and quantitative aspects, *J. Insect Physiol.,* 29, 201, 1983.
126. **Bell, W. J. and Barth, R. H., Jr.,** Initiation of yolk deposition by juvenile hormone, *Nature (London),* 230, 220, 1971.
127. **Bownes, M.,** The role of 20-hydroxy-ecdysone in yolk-polypeptide synthesis by male and female fat bodies of *Drosophila melanogaster, J. Insect Physiol.,* 28, 317, 1982.
128. **Blomquist, G. J., Adams, T. S., and Dillwith, J. W.,** Induction of female sex pheromone production in male houseflies by ovary implants or 20-hydroxyecdysone, *J. Insect Physiol.,* 30, 295, 1984.
129. **Blomquist, G. J., Dillwith, J. W., and Adams, T. S.,** Biosynthesis and endocrine regulation of sex pheromone production in Diptera, in *Pheromone Biochemistry,* Prestwich, G. D. and Blomquist, G. J., Eds., Academic Press, Orlando, FL, 1987, 217.
130. **Kunkel, J. G.,** A minimal model of metamorphosis-fat body competence to respond to juvenile hormone, in *Current Topics in Insect Endocrinology and Nutrition,* Bhaskaran, G., Friedman, M., and Rodriguez, J., Eds., Plenum Press, New York, 1981, 107.
131. **Scharrer, B.,** Insects as models in neuroendocrine research, *Annu. Rev. Entomol.,* 32, 1, 1987.

Section VI. Sense Organs, Plasticity, and Behavior

Chapter 23

STRUCTURE AND FUNCTION OF THE VISUAL SYSTEMS OF THE AMERICAN COCKROACH

Michael I. Mote

TABLE OF CONTENTS

I. THE COMPOUND EYE

A. THE ANATOMY OF THE COMPOUND EYE

1. The Retina and Ommatidia

The paired compound eyes of the American cockroach, *Periplaneta americana,* have been the subject of numerous anatomical studies, first at the level of the light microscope[1-3] and more recently with the electron microscope.[4-7] This species has been by far the most extensively studied of the cockroaches and will therefore serve as the example.

The sheet of tissue that contains the primary photoreceptors is called the retina, as in the vertebrates. The structural unit of the retina is the ommatidium. The structure of an ommatidium is presented schematically in Figure 1. The number of ommatidia per eye varies with age and species, but is about 2000 in an adult *P. americana.* Each ommatidium is composed of a number of elements which together comprise the visual unit. Distally, each ommatidium is bounded by the dioptric apparatus, which consists of the protective cornea and the corneal lens. This is equivalent to the cornea of the mammalian eye. The crystalline cones are shielded from those of adjacent ommatidia by the presence of a sleeve of dark pigment provided by a pair of primary pigment cells. The base of the cone is in contact with each of the eight photoreceptor cells, termed retinular cells, which comprise the main body of each ommatidium.

A portion of the plasmalemma of each retinular cell has been modified into a series of microvilli containing the visual pigment. This structure, called the rhabdomere, is fused with the rhabdomeres of the seven remaining retinular cells to form a central core of photosensitive membrane called the rhabdom. The retinular cell bodies have a peripheral position in the ommatidium, resembling the petals of a flower with a cross section of about 20 μm. In the long axis, the ommatidium measures between 250 to 350 μm. The ommatidium is shielded along its length by pigment contained in numerous accessory pigment cells. Figure 2 presents an artist's illustration of the general features to be described.

2. The Lamina Ganglionaris

The ommatidia are bounded centrally by a pair of basement laminae.* The two laminae are closely apposed at the edges of the eye, but are separated by some 40 μm at the center. The retinular cells produce axons which leave the ommatidium at varying depths, depending on the location of the nucleus, but which all course through the basement laminae toward the first synaptic region, the lamina ganglionaris. The structures of these projections and of the lamina were described by Ribi,[7] who used both the electron microscope and Golgi methods in his investigations. Figure 3 summarizes his findings. The eight axons from a single ommatidium pass through the first basement lamina as a bundle ensheathed by thin glial processes. They remain closely associated in an orderly fashion until the second basement lamina is traversed. At this point, the bundles take on a new organization in which they contain from 6 to 20 axons per bundle and are again enclosed by a glial sheath. The close association between axons from a single ommatidium appears to have been lost.

Ribi[7] describes the lamina ganglionaris as having two basic layers, the cell body layer and the first synaptic region. These are separated from the inner basement lamina by a region rich in glial elements, tracheal supply to both the retina and lamina, and the axon bundles. This is the fenestrated layer. The axons soon enter the cell body layer, where they pass between stacks of second order neurons called L cells. This is the outer region of the lamina ganglionaris. At this point the retinular axons are joined by axons of the L cells and enter into the first synaptic region. Here a third class of neuron is encountered which runs horizontally, in bundles, through the lamina neuropil. These are centrifugal fibers whose cell

* The terms "basement lamina" and "basement membrane" are used interchangeably in this text. The former is preferred, the latter being older and less accurate.

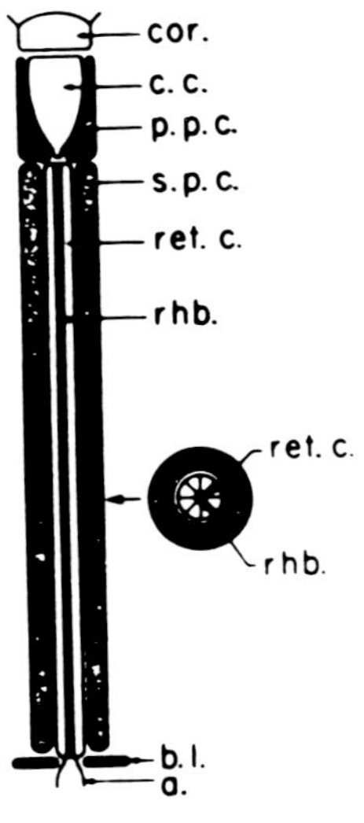

FIGURE 1. A diagramatic representation of an ommatidium. Components are described in the text and are indicated by the labels as follows: cor., cornea; c.c., crystalline cone; p.p.c., primary pigment cells; s.p.c., secondary pigment cells; ret.c., retinular cells; rhb., rhabdom; b.l., basement lamina; a., axons. A cross section at the level indicated by the arrow is shown at right.

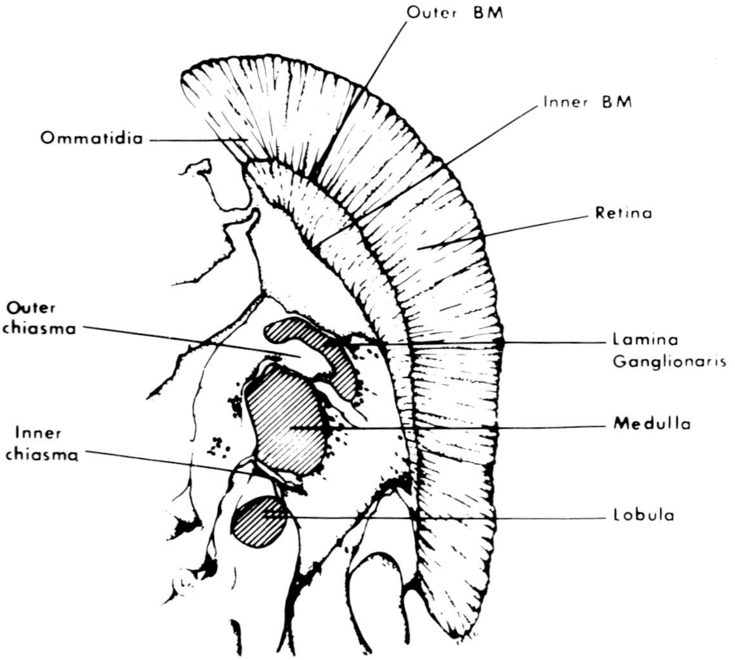

FIGURE 2. An artist's rendition of a histological cross section through the right eye and optic lobe of *P. americana,* observed from a dorsal aspect. Labels and pointers indicate the major features that have been described in the text. Individual cells and fibers have been omitted for clarity.

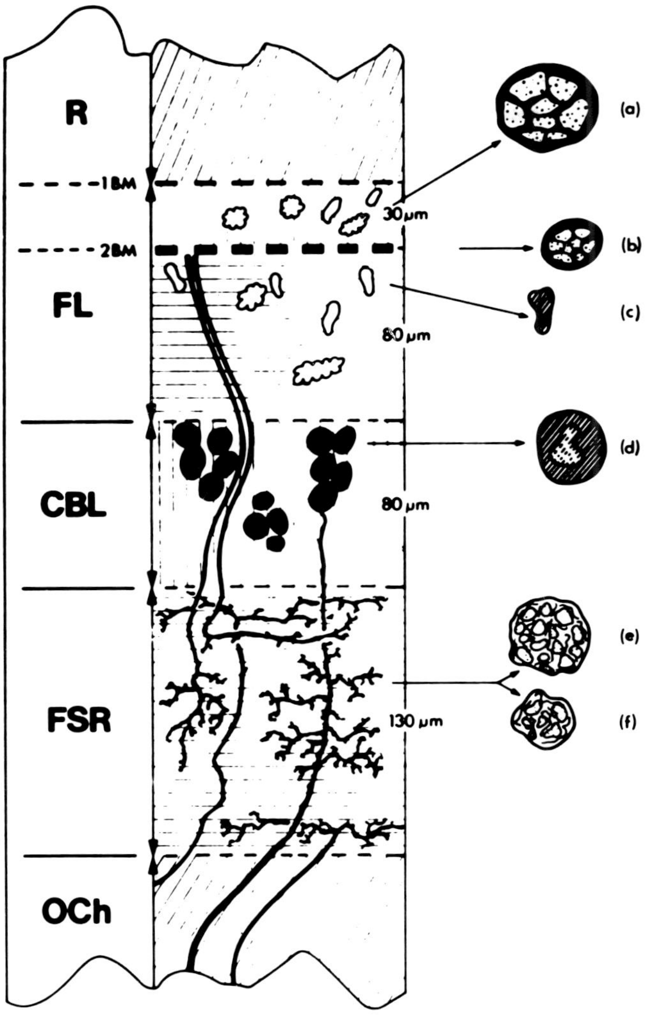

FIGURE 3. Schematic drawing of a section of the lamina showing the different layers and the structures of their components at right (the hatching distinguishes the layers from one another and does not reflect actual structures in the layers). (a) Cross section of an axon bundle between the first (outer) and second (inner) basement membranes (see Figure 2); (b) axon bundle in cross section at the level of the second basement membrane; (c) glial cell body; (d) monopolar cell body (L fiber); (e,f) bundles in the first synaptic zone contain between 8 and 20 axons. R, retina; 1 BM and 2 BM, first and second basement membranes; FL, fenestrated layer; CBL, monopolar cell body layer; FSR, first synaptic region; OCh, outer chiasma. (From Ribi, W. A., *Cell Tissue Res.*, 9, 57, 1977. With permission.)

bodies lie near deeper structures. It is in the neuropil that synaptic contacts between receptors, L cells, and horizontal cells are formed.

Based on morphology, Ribi[7] recognizes distinct classes of the various neural elements. For example, he identifies two classes of receptor axons: one type follows the pattern described and terminates in the lamina neuropil (short visual fibers [SVF]), while the second type (long visual fibers [LVF]) does not synapse in the lamina, but passes through and terminates in the second synaptic region described below. Within the SVF type, three groups

can be detected based on the depth of their termination within the neuropil and, secondarily, by the shape of their arborizations. Four types of L-cell terminations and three classes of horizontal cells can be identified by the same criteria.

3. The Medulla-Lobula Complex

The lamina is actually crescent shaped when viewed in cross section. It forms a cap over the larger, more spherical, and centrally located ganglion, called the medulla, which is the second synaptic zone of the optic lobe. The two are connected by fibers emanating from the L cells, LVF, and centrifugal horizontal fibers. These processes form the outer chiasma, crossing in the anteroposterior plane of the visual system. The rind of the medulla is made up neuron cell bodies which send their processes into the central neuropil. The LVF terminate in a distinct band in the more distal portion of the synaptic zone. Little more can be said about the possible arrangement of cells in this region. The cells of the medulla rind are, in general, of a uniform size of 10 to 15 μm. In some instances neurons have larger somata (50 to 100 μm); these appear to be at specific locations in the ganglion and can be recognized and identified in any animal.[8] Some of these cells were those identified by Sokolove[9] as being involved in a system coupling the two optic lobes on either side of the animal and generating patterns of circadian running activity. Other than that, the functional role of these identifiable cells is unknown.

The third synaptic region is the lobula. It is also spherical, but is smaller than the medulla. The two are connected by the inner chiasma which also crosses in the anteroposterior plane of the visual system. Fibers originating from neurons of the medulla and possibly the lamina comprise the inner chiasma and can either terminate in the lobula or run on into the protocerebrum and ventral nerve cord. Presumably, the inner chiasma must also contain at least some centrifugal fibers, since there are connections between the two optic lobes through the protocerebrum and projections from the ocellus. The neurons of the lobula also form the rind of the ganglion, but no more can be said about specializations at this time. More work must be done on this structure and its connections.

4. Projection Patterns

It has long been recognized that the compound eye contains two spectral types of photoreceptors, one maximally sensitive at 370 nm (UV) and one sensitive at 507 nm (green [G]; see below).[10] These receptor types reside in the same ommatidium.[11] In fact, three of the eight retinular cells are of the UV type, occupying the same position in each ommatidium as shown by morphological changes after selective adaptation of the retina.[12] We corroborated that finding using the metabolic marker nitrile blue tetrazolium and selective adaptation with 572-nm light.[8] Under these conditions only active cells take up the marker, and only the G cells are active at this wavelength. These studies show that, in distal regions where there are four or five large cell profiles, only one of them, ventrally located, is of the UV type. In deeper regions, where all eight retinular profiles are seen, a second ventral and a third dorsally located cell appear to be of the UV type. These would be cells numbered 1, 5, and 7 in the scheme of Butler.[12]

We tried to establish the degree of regularity with which the cells of the ommatidium projected to the optic lobe.[8] Receptors were impaled with electrodes filled with fluorescent dye, and their response characteristics were determined. The dye was injected iontophoretically and the site and morphology of the termination determined histologically. When a UV-type cell was impaled, it was almost always located ventrally and was maximally sensitive to light polarized along the horizontal axis (polarizational sensitivity [PS]). This result is to be expected because these cells are large and easily impaled. They correspond to cells 1 and 5 in Butler's scheme. Much less frequently, a dorsally located UV cell was encountered which was smaller (7 in Butler's scheme) and was also sensitive to the horizontal

PS vector. In 54 cells so studied, we observed no unequivocal case where a UV cell did not pass through the lamina without making terminations and then end in the medulla. When terminals were seen they were extensive in the band of LVF terminations. On the contrary, when G cells were impaled, they were only seen to terminate in the lamina. It seems that Ribi's class of LVF is equivalent to the functional class of UV receptor and the G receptor type to the morphological class of SVF. Furthermore, when G cells with a vertically oriented PS axis were impaled, all of them terminated in the deep layers of the lamina, while G cells with a horizontally oriented maximum PS axis ended in all regions of the lamina. It is possible that there is a relationship between sensitivity and termination site.

The question was also asked in a different way. In this case, fluorescent dye was injected into the medulla region and the retinular cells were backfilled into the retina.[8] With this procedure, we found that only one cell showed the marker — namely, the large ventral cell. This would corroborate Ribi's finding of only one LVF in *Periplaneta*. If this is so, then we must explain why UV terminals are never seen in the lamina. If this is not the case, then we must explain why the other cells do not take up the dye. Note that Ribi[7] leaves open the possibility that there may be more than one LVF.

B. POSTEMBRYONIC DEVELOPMENT

Periplaneta is an insect of the hemimetabolous type. In our culture conditions it goes through a series of eight instars or nymphal stages.[13] Each of these stages is terminated by a molt where the insect sheds its cuticle, grows, and enters the next instar. Except for the final molt to the adult, when the wings are fully developed, each instar is essentially a larger version of the last. The duration of the intermolt period varies from 8 to 14 d for the first and 30 to 45 d for the last. The total period of postembryonic development is roughly 6 months. During this period the number of ommatidia increases from about 50 at hatching to about 2000 in the adult. There is a concomitant increase in the size of the optic lobe. This process has been studied in detail by Stark and Mote[13] using tritiated thymidine autoradiography.

The results of these studies show that, unlike all other tissues, the compound eyes (and probably the ocelli) grow and differentiate independently of the molt cycle. A period of quiescence follows the molt; then cell division begins and continues for the duration of the instar. By contrast, cell division can be detected in other tissues only in the days just preceding the molt. The rest of the nervous system seems to lack postembryonic cell division, except for the corpora pedunculata of the protocerebrum. This could represent the proliferation of the ocellar or antennal sensory receptors and their postsynaptic elements, but this point remains to be resolved.

The regions of cell addition during these immature stages are the growth zones. Each compound eye and optic lobe system has four of these zones (Figure 4). They contribute cells to one of the developing areas of the visual system from which it takes its name. The zone of retinal growth and differentiation is easily recognized along all but the most posterior margin of the eye. Elements are provided from a thickened cluster of epidermal cells at the margin of the eye and surrounding cuticle. Cell division is not observed in this cluster. Cells move out into a region of single cells and clusters of cells which are in the process of forming ommatidia and extending axons toward the basement laminae. These cells incorporate the label and are undergoing division. Cells in ommatidia formed at earlier stages continue to grow during this period, but do not divide. Therefore, a gradation is established, with the oldest and largest ommatidia found along the posterior margin of the eye and the smallest and youngest units being found along the anterior, dorsal, and ventral margins.

1. Specificity

Since the elements that make up each ommatidium seem to do so in a very precise manner in terms of the position of each cell, the visual pigment produced, and the nature

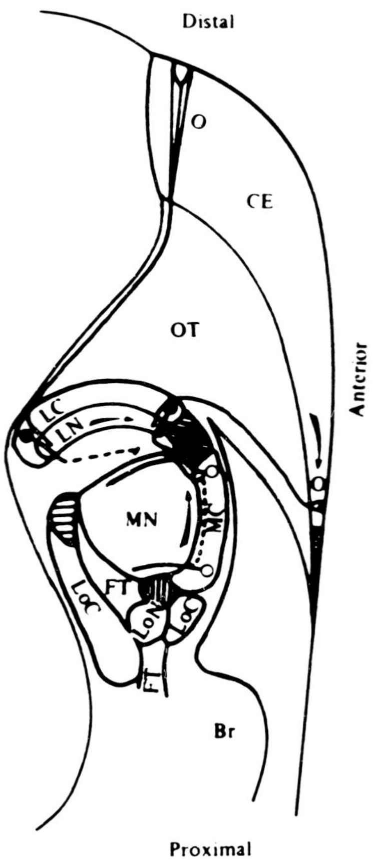

FIGURE 4. A diagrammatic section through the head of *Periplaneta americana,* representing the pattern of growth of the compound eye and the optic lobe ganglia.[13] The arrows indicate the direction of growth by addition of new cells. Symbols: Br, brain; CE, compound eye; FT, fiber tracts; LC, lamina cortex; LN, lamina neuropil; LoC, lobula cortex; LoN, lobula neuropil; MC, medulla cortex; MN, medulla neuropil; O, ommatidia; OT, optic tract. Regions with single hatching represent the growth zones of the eye, lamina, medulla, and lobula. The lamina-medulla proliferation center is represented by the darkened area between the respective growth zones. Lamina neurons are depicted by filled somata, while medulla neurons are open. The hatched line depicts the fiber path for the lamina cell in the older established region of the ganglion; it is interrupted at the growth region for clarity. Head is oriented as in Figure 2.

of the projection to the optic lobe, a very important question is raised. How is this information transmitted to a seemingly undifferentiated population of cells? Shelton et al.[14] investigated the question of the role of cell lineage in this process. They grafted cells taken from eye color mutants of one type into hosts with a different eye color. This allowed tracing the fates of transplanted cells based on the type of pigment they contained. They found that cells did not become committed to form a particular ommatidial element in response to some feature of their lineage, but rather from their position in a field of cells. The question now becomes the following: how does their position in a field regulate the expression of genes that control their fates?

The growth zones of the lamina and medulla lie in close proximity at the anterior margin of these ganglia and are separated by a single proliferative zone.[13] Thus, the neurons of the

medulla and lamina are produced by division of a few prominent cells, the neuroblast mother cells. They divide repeatedly to produce a stack of presumptive neurons called ganglion mother cells, with the oldest members at the bottom of the stack and the youngest at the top. As the cells reach the base of the stack, they become elliptically shaped and migrate away. In the case of the lamina growth zone, the migration is distally along incoming retinal fibers, coming to rest in the stacks of lamina cells of the cell body layer. At the same time, they put out fibers that enter the neuropil with retinular axons. Thus, the oldest portion of the lamina neuropil lies at the posterior margin of the ganglion, and the newest neuropil is added at the anterior margin. This is the same relationship found in the retina.

Presumptive medullar ganglion cells leave the region of proliferation and migrate centrally into the medulla growth zone. They put out processes which enter the newly formed medulla neuropil along with processes from the lamina and the LVF of the retina. A result of this is that the older established neuropil is displaced anteriorly, and new neuropil is added at the posterior face of this ganglion. This is opposite to the situation found in the retina and lamina. Therefore, the oldest established contacts between the retina-lamina elements and the medulla neurons must be crossed by the youngest lamina-medulla projections to maintain ordered relations. It is this crossing that produces the chiasma in the anteroposterior plane of the visual system. There has been much speculation as to the possible role of the chiasmata in the arthropod visual system; it may be that they are just a result of the pattern of development.

The medulla and lobula are connected by tracts that form an inner chiasma. The lobula growth zone was not studied in detail in these experiments,[13] but the presence of a chiasma indicated that the patterns of growth were similar to that of the lamina and retina and that it was the eccentricity of the morphogenesis of medulla which resulted in chiasmata being formed.

It was noted[8] that the architecture of the lamina was modified by creating lesions in the retinal growth zone. The lamina neuropil does not form without invasion of retinal fibers. Whether this represents an inductive process or just the loss of appropriate substrate remains to be investigated.

II. THE PHYSIOLOGY OF THE COMPOUND EYE

A. RETINAL ELECTRORETINOGRAPHIC (ERG) STUDIES

The early studies of the physiology of the compound eye of *Periplaneta* were carried out by the German workers Walther and Dodt.[15] They used the electroretinographic (ERG) method of recording made popular by Autrum and Gallwitz[16] in their studies of a number of insect eye types. This method involves making electrophysiological recordings across the entire eye and the optic ganglia of the animal. A complex waveform results, and this serves to indicate the sensitivities of the underlying cells. Furthermore, the contributions of certain elements to a particular feature of the ERG can be estimated. This was done for *Periplaneta* by Wolbarsht et al.[17] A second method employed selective adaptation, mentioned above, in which continuous illumination at one wavelength suppresses the contribution of one spectral type of receptor to the ERG, revealing the contribution of others. Using these methods it was shown that the retina was at least dichromatic, having peaks of sensitivity in the UV and green regions of the spectrum. Also, the cockroach ERG has the waveform altered by the process of selective adaptation, becoming prolonged in time. Walther[18,19] found these distinctions were not evenly distributed across the eye, but rather were located in the upper portion. In the ventral regions of the eye the ERG seems to be dominated by the G-type receptor, suggesting the absence of the UV type. However, in Butler's morphological study,[12] no regional difference was detected. This discrepancy remains to be clarified.

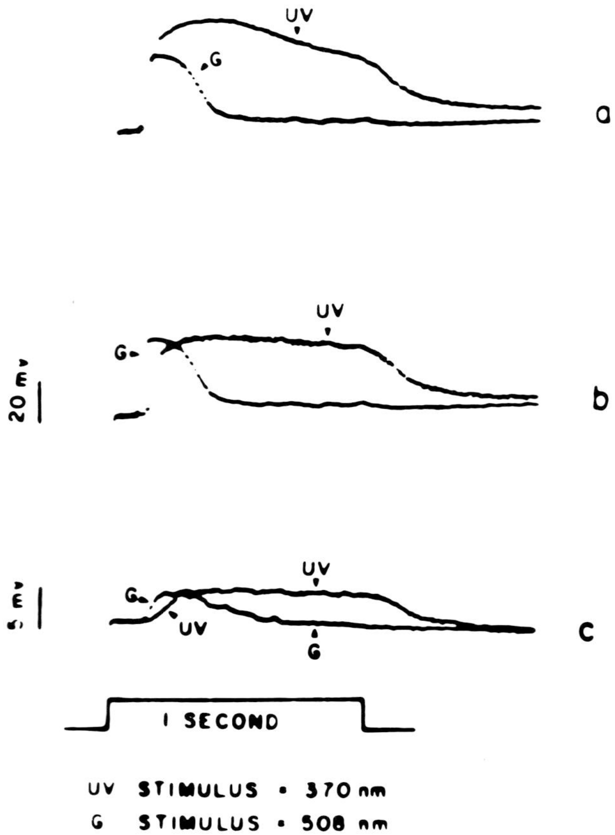

FIGURE 5. Averaged responses (n = 5) of receptor potentials recorded intracellularly from UV and G-type photoreceptors in the compound eye of *P. americana*. The responses were recorded successively in the same preparation. The stimulus consisted of a 1-s flash of either 370-nm light for the UV cell or 508-nm light for the G cell. (a) Maximum responses are compared; (b) a maximal response from the G cell is compared to a response of equal amplitude in the UV cell; (c) responses of equal amplitude to dim stimuli are compared. Labels and arrows indicate waveform for each type. Note change of scale for trace C. (From Mote, M. I., Kumar, V. S. N., and Black, K. R., *J. Comp. Physiol.*, 141, 403, 1981. With permission.)

B. INTRACELLULAR RECORDINGS

The basic findings of the ERG studies were confirmed and extended by Mote and Goldsmith,[10] who made intracellular recordings from the retinular cells. They found two spectral classes of retinular cells in the retina of *Periplaneta*. One was maximally sensitive to light of 370 nm and the second to light of 507 nm. These peaks corresponded to the UV and green peaks detected in the ERG studies. The intracellular studies also revealed a temporal difference in the time course of the response generated in the two receptor types. This was studied in greater detail by Mote et al.[20] using computer averaging to study the difference (Figure 5). The UV type shows a slower rise time, a sustained plateau, and a prolonged after-potential. In contrast, the G receptor has a more rapid rise time, little or no sustained plateau, and little or no after-depolarization. Thus, one of these receptors (UV) is "tonic" in its time course and the other (G) is "phasic". The altered time course of the ERG after selective adaptation can be explained by the increase in prominence of the slower UV component.

Butler and Horridge[21,22] extended our understanding of the retina further. They determined the angular sensitivity of cells and found that it changed from 2.4° (width at 50% sensitivity) in the light-adapted condition to 6.9° in the dark-adapted state. This suggests that during the adaptation process morphological changes alter the angle of acceptance of the ommatidium. They were able to establish morphological measures of adaptation states. However, a particular morphological state does not correlate well with electrophysiologically determined sensitivity. However, the movement of intracellular material toward or away from the rhabdom could be correlated with the state of adaptation and angle of acceptance. These workers detected two classes of PS retinular cells with axes of sensitivity orthogonal to each other. A lack of spectral information does not allow the determination of the spectral classes of these cells. The sensitivity ratio of the cells (the difference in sensitivity between the two axes) was reported to be about 5. This has been confirmed in our laboratory, although we have found the ratio to be slightly higher.[8]

III. THE PHYSIOLOGY OF THE OPTIC LOBES

A. ON-TONIC CELLS

The physiology of neurons in the optic lobes was investigated by Mote and his colleagues.[23-26] These studies are restricted to cells of the medulla which give regenerative responses to light stimuli (spikes) that can be recorded both extracellularly and intracellularly. Some recording sites are marked in the outer chiasma, and some cells that have processes in the lamina are stained. The latter are centrifugal and the former are assumed to be so. The L cells of the lamina have yet to be investigated. If the second order neurons of other insects (e.g., dragonflies and mantids) could serve as an example, we would expect these elements to be hyperpolarized by light stimuli which depolarize receptors.[27,28] The retinular-L-cell synapses are sign inverting. L cells in other insects produce only graded responses and therefore would not be responsible for the spiking response classes to be described. We assume this to be true for the cockroach. In any case, the lack of information about the response characteristics of these cells represents a large gap in our knowledge.

Studies of the optic lobe have concentrated on two broad response classes. In one, the cells are not active in the dark, but respond to a flash of light with a burst of spikes. The second, to be discussed later, contains cells which are active in the dark; their activity is modulated by a flash of light. The former are termed "on" cells. Initial studies were aimed at determining the spectral sensitivity of these units.[23] Flashes of different wavelengths produced qualitatively different responses in the cell. Responses to short-wavelength flashes produced a burst of spikes followed by a tonic discharge for the 1-s duration of the flash. As the wavelength of light increased, the tonic portion of the response disappeared, leaving only the initial burst or phasic portion in the yellow and orange regions of the spectrum. Figure 6 provides records from a cell of this type. These cells were classified as on-tonic to violet (OTV). If spike number was used to assess their spectral sensitivity, then they were most sensitive to short wavelengths and least sensitive to long ones. If, however, the intensity at threshold was used instead, the sensitivity function peaked at 508 nm and was very similar to that function measured in the G receptors of the retina. The obvious interpretation is that these cells are driven by both receptor types and that their response pattern reflects the differences in spectral sensitivity and time course of response in the primary receptors. Spike number relates directly to the level of depolarization in the receptors.

These findings were confirmed by experiments that employed selective adaptation in which the G receptor type was suppressed by long-wavelength light.[23] The results suggested interactions between receptor channels. This could occur at the cell being recorded or more peripherally in the visual system. The response of cells to UV stimulation changed in the presence of adaptation. The pattern was exaggerated in time, spiking continuing after the

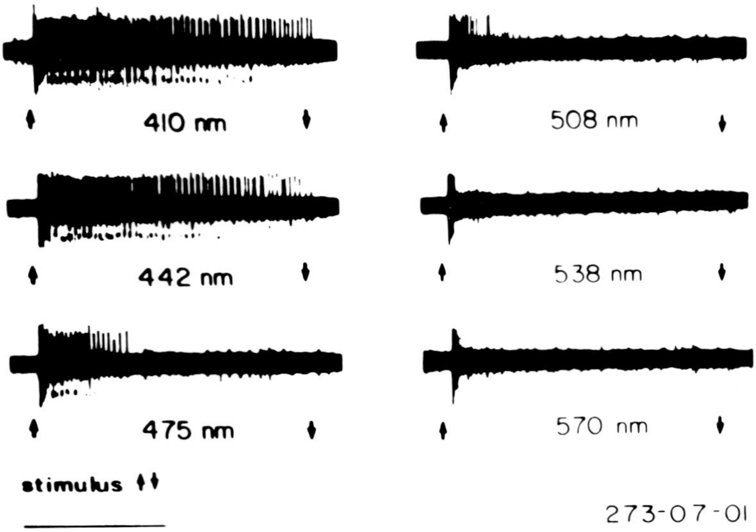

FIGURE 6. Representative "on"-type interneuron recorded extracellularly in the optic lobe of *P. americana*.[23] The eye was exposed to a 1-s stimulus (arrows indicate on and off) of approximately equal photon content at the test wavelengths indicated below each trace.

stimulus flash, and in some cases the threshold measured was lowered. This suggested that activity in the G channel suppressed or inhibited activity in the UV channel. There was no indication from these results that the opposite was true. Also, the response pattern coded for the wavelength of the stimulus because responses containing an equal number of spikes showed a different temporal distribution of those spikes at different wavelengths.

B. LATENCIES

The temporal characteristics of OTV responses were investigated further by measuring response latencies as a function of wavelength.[20] In general, latency was minimal at the highest intensities and increased as the intensity was lowered, the response approaching threshold values. Once again, a wavelength dependence was observed, since the maximum latencies to UV stimuli were much longer than to green or orange, while minimum latencies were approximately equal. The simplest interpretation of these differences is based on the differences in time course of the receptor group's responses. The prolonged latency to UV stimuli is explained, at least in part, by the slow rise time of the UV receptor. Dual input was shown by selective adaptation experiments in which a limited dynamic range and very short latencies were observed with long-wavelength stimuli, and a general increase in latency without a concomitant increase in sensitivity was seen with short-wavelength stimuli. In this case, the threshold in the G channel with adaptation was higher than the brightest UV flash that could be produced. Thus, only the UV channel was responsive. One interpretation of this result is that, in the normal condition, latency to UV stimuli is determined in part by the faster G receptor channel. With selective adaptation, this influence is removed and the latency is increased.

C. RECEPTIVE FIELDS

Receptive field studies were performed in an attempt to understand the input relations between "on" cells and the receptors in the retina. Stimulus points subtending 0.8° were flashed at various points within the visual field. Any point viewed by the eye, if made

sufficiently intense, was able to drive this type of cell. Schemes were developed to relate sensitivity at threshold to a stimulus point and its location in the visual field. This resulted in a complex pattern of field shapes and sizes which varied from cell to cell. A number of field characteristics were evaluated and compared to each other for coincidence. For instance, size and shape are positively correlated with each other, while size and spatial location are negatively correlated (see Mote et al.[20] for further details).

In about half of the cells investigated (n = 101), a wavelength dependence of the receptive field characteristics was noted. Most of these required selective adaptation, but some did not. The usual result was that the field measured with violet stimuli after adaptation was larger and spatially shifted when compared to fields measured without adaptation and with green or violet stimuli. Some of the shifts were so great that the two fields did not overlap. The shift was usually in the dorsal direction, as might be expected from the ERG data. Two obvious conclusions regarding organization are drawn from these results. The first is that in at least half the cells studied the input from retinal receptor classes is not spatially homogeneous. The most effective UV receptors are spatially distinct from the most effective G-type receptor. Second, the increased size of the light-adapted field suggests that there are spatial interactions between the receptors, as shown previously.

Points in the visual field vary in their effectiveness as stimuli when the threshold of response is used as a measure. Does their latency change accordingly? It seems it does not. We suppose that the point with the lowest threshold will generate a response with the longest latency for any given wavelength, but this is not always the case. Points requiring more intense stimulation (higher threshold) often show a longer latency. This means that the spatial location of the stimulus is also important in determining latency and that spatial interactions are involved. This fact is dramatized by the results of other experiments.[20] Here, the procedure was to measure latency to a point in the receptive field and then insert a mask that blocked part of the eye from the light of that point. In some cases, the expected result is obtained in that the cells' sensitivity to the point decreases and the latency is also decreased. This is so because a more intense stimulus is required to excite the cell, and this causes a response with a shorter latency. However, in other cases the point decreases in effectiveness, but shows a greatly increased latency, or it increases in effectiveness and decreases in latency. This phenomenon is depicted in the graphs of Figure 7. The interpretation must be that spatial interactions in the lamina and medulla are very important in determining the basic response characteristics of this class of cell. These relationships need to be understood in more detail.

D. SUSTAINING UNITS

The second functional class of unit that has been investigated is that of the sustaining, or spontaneously active, interneuron of the medulla. These structures were studied by Kelly[24] using intracellular recording methods and dye injection techniques to obtain anatomical information. These cells are always encountered in the medulla, occasionally project out to the lamina, and seem to run through the lobula without branching. When the cell body is stained, it is invariably found in the medulla neuropil. The response pattern is such that the unit is continuously active in the dark and that activity is modulated by a light flash delivered to the whole eye. These modulations are either excitatory or inhibitory and, as one might expect, they are strongly wavelength and intensity dependent. Kelly has adopted the classification scheme of Kien and Menzel[29] to describe his response types. These have to do with the spectral properties of input from the receptors.

E. BROADBAND CELLS

The broadband neurons are described as those cells which receive qualitatively similar input (i.e., excitation or inhibition) from both the receptor types. The resulting sensitivity

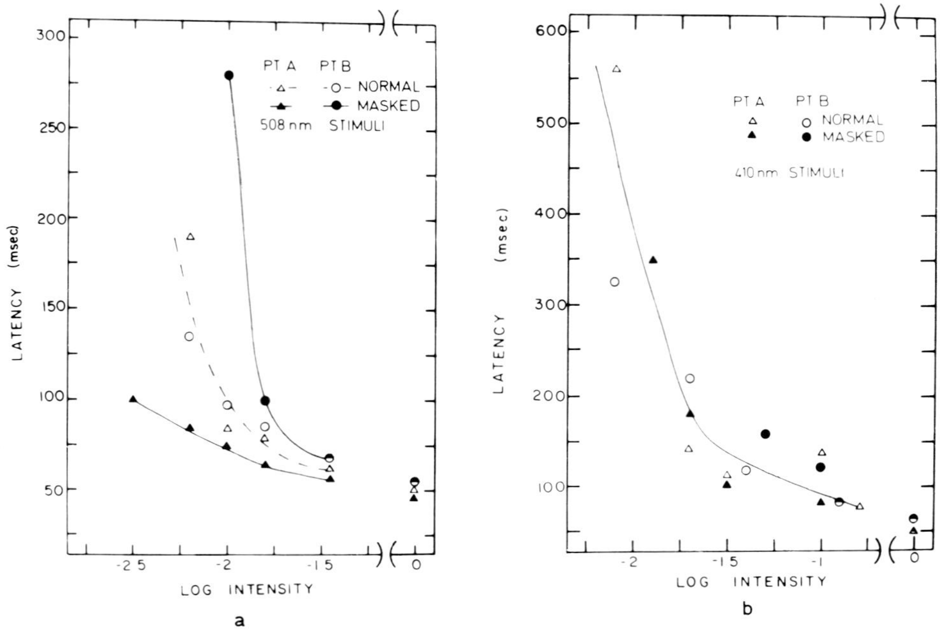

FIGURE 7. (a) Plot of the effects of masking on the latency vs. intensity function of an "on"-type interneuron in the optic lobe of *P. americana* with a pair of 508-nm stimulus points separated by 10 in the visual field. In the unmasked condition (open symbols), both points (A = triangles, B = circles) yield a curve which falls along the dashed line. Relations in the presence of the mask (filled symbols) are connected by solid lines. Note the extreme shifts of the functions to either increasing (PT B) or decreasing (PT A) latency when compared to the normal function. (b) An example taken from a different unit in a different preparation where there were no changes after masking stimulus points at 410 nm. (From Mote, M. I., Kumar, V. S. N., and Black, W. R., *J. Comp. Physiol.*, 141, 403, 1981. With permission.)

function is also broad; it is at a peak in the UV region, where the receptors maximally sum their influence, and declines above 530 nm. Figure 8 shows an example of responses from a cell of the excitatory broadband type. The exact shape of the curve is strongly dependent on the criterion selected for determining sensitivity. Broadband inhibitory units are encountered in which discharge is suppressed by light flashes at all wavelengths. Although spectral sensitivity cannot be determined precisely, the same general properties emerge as described above. Figures 9 and 10 show that such cells can show either restricted or extensive arborization. The straightforward interpretation is that the input comes from L cells for the green input and from UV receptors for the violet.[24]

Interneurons receiving either excitatory or inhibitory input from one receptor type are termed narrowband. These are present in the cockroach, and one can obtain receptor-like sensitivity functions in portions of the dynamic range. These distinctions break down for most cells when the intensity is varied from threshold to saturation. In this case, interactions with both receptor types can be detected and the propriety of calling the cells narrowband is questionable. In most cases, these cells have been dominated by the G receptor class. Dye fills of this cell type often show neurons with restricted branches in both the lamina and medulla. Such cells can represent at least one type of horizontal cell of the lamina. The cell body is located centrally between the lamina and medulla.

F. OPPONENCY

The third response class encountered is the opponent type. It receives antagonistic input from the two receptor groups. An example is shown in Figure 11. This opponency occurs

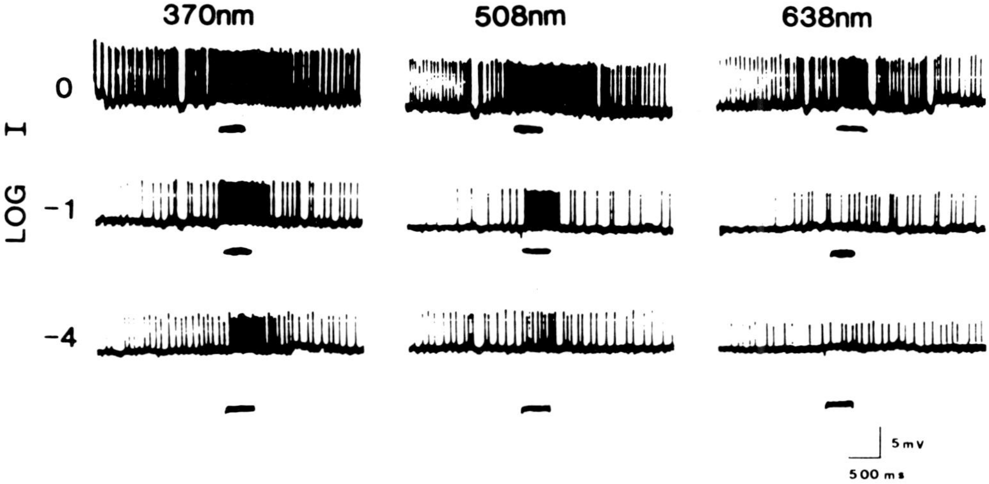

FIGURE 8. Selected spike records from an excitatory broadband unit. The records show responses to 370-, 508-, and 638-nm light ranging over four log units in intensity. The cell responds with increased firing from excitatory input provided by the UV and G receptors.

during the sustained response to the flash, but it does not occur in transients at the beginning or end of the stimulus. In these responses, it is possible to analyze slow potential or postsynaptic events. When one considers these as well as spikes, the resulting patterns are quite complex. Many of the fine details are not understood. What does seem consistent in these cells is that they are always excited by the G receptor type and inhibited by the UV receptor type. This is in contrast to the finding in the honeybee, where just the opposite is true.[27] Dye fills have revealed branching patterns in both the lamina and the terminations of LVF in the medulla. The simplest interpretation is that these cells receive excitatory drive from L cells (and possibly directly from G receptors) in the lamina and inhibitory drive from LVF in the medulla. The presence of color opponency similar to that found in vertebrates[49] suggests that this visual system is capable of processing color information delivered from a dichromatic retina. These features make this preparation useful for the study of visual integration. Not only is opponency present, but the retina is dichromatic and a less complicated system to study.

G. POLARIZATIONAL SENSITIVITY

Kelly[24] tested some sustaining cells for sensitivity to the vector of linearly polarized light. On occasion, he found cells that responded to the rotation of a polarizing filter placed in the light path. In these cases the illumination was constant. The cells were excited by rotation in one direction and inhibited by rotation in the other. Such cells were not common, and detailed analysis was not possible. It is tempting to suggest that the cockroach visual system is equipped to detect polarizational information in the environment. However, the stimulus might have created an apparent motion which could have been the effective stimulus. This question requires more attention, since motion sensitivity has not been studied in the optic lobes of this insect.

IV. DESCENDING CONTRALATERAL MOTION DETECTOR

The question of motion sensitivity was addressed at levels above the optic lobe by Edwards,[30,31] who studied the descending contralateral motion detector (DCMD) neuron that could be recorded in the ventral nerve cord. This cell seems to be analogous to the DCMD

found in locusts by Rowell.[32] In the case of the locust, this cell is driven by a giant lobular interneuron that integrates over the visual space of an entire eye. In the cockroach no such cell is known, but the receptive field is the same, so it must be inferred that either there is such a cell in the cockroach or the DCMD is itself a lobular interneuron. The neuron responds to movement of a small spot on a contrasting background. Edwards[30,31] used this response to describe the type of neural network which could act as the input.

If the intensity of the moving spot is increased, the responsiveness of the DCMD increases with a sigmoidally shaped function in the light-adapted animal. This is similar to many of the visual units described thus far. If, however, two spots are moved together (separated by about 10°), then the response of the DCMD is decreased. In addition, if the intensity function of the spot is measured in the dark-adapted state, then the function is bell shaped, showing less response at higher intensities. These results suggest that the input to the cell is through a number of identical channels that interact in a nonrecurrent lateral inhibitory scheme as described by Ratliff.[33] This scheme allows the elements of the network to maintain their acuity in restricted regions while blocking out the effects of whole field movement. Edwards[30] also uncovered a tonic inhibitory network by showing that a steadily illuminated bar of light also inhibited the response to the moving spot. A feature of the second system was that it did not adapt to the presence of the steady light. This distinguished it from the first system, which did adapt. Thus, he proposed two lateral inhibitory nets acting on the input to the cell. One adjusts the system to conditions of ambient illumination, while the other enhances acuity to moving objects and cancels out the effects of large-field moving stimuli created, for example, by movement of the animal.

Edwards[30,31] described some additional response properties of the DCMD. He found that habituation was a prominent feature of the inputs and that it occurred independently in different channels. The system did not habituate as a unit, and recovery from local habituation had a time constant of about 8.5 s. This left open the question of exactly what these channels are in terms of the elements described above. Edwards calculated that they have a receptive field of an ommatidium. Spectral sensitivity of the DCMD suggested that both receptor types must act as input because the sensitivity was constant across the spectrum. In addition, habituation to one wavelength generalizes to another, but no evidence for interaction between the two types is detected. This is an interesting point because all the optic lobe studies cited above show interaction to be a prominent feature of the response. If that is the case, what could be serving as input to the DCMD? It is clear that many interesting relationships between light-sensitive elements in the optic lobes and in the ventral nerve cord remain to be investigated.

V. VISUALLY GUIDED BEHAVIORS

A. CIRCADIAN RHYTHMS

In view of the rather extensive information on the response properties of various visual elements, the cockroach has not served as the subject of many behavioral experiments. The one exception has been the study of the role of the compound eye and optic lobe in regulating the pattern of running activity in a circadian fashion. This is discussed in detail by Page in Chapter 24 of this work. Experiments have used circadian behavior to assess visual function in this animal.[34] Since illumination of the compound eyes serves to entrain the circadian oscillator responsible for rhythms in daily running activity, it is possible to determine when an animal or a group of animals is unable to detect the presence of light. The determination of this threshold for a number of different wavelengths allows the construction of an action spectrum. This relationship shows that the entrainment mechanism is influenced mainly by the G photoreceptors. They are 1000 times more effective than the UV group, which contributes only weakly even in conditions of selective adaptation.[34] Activity in the UV

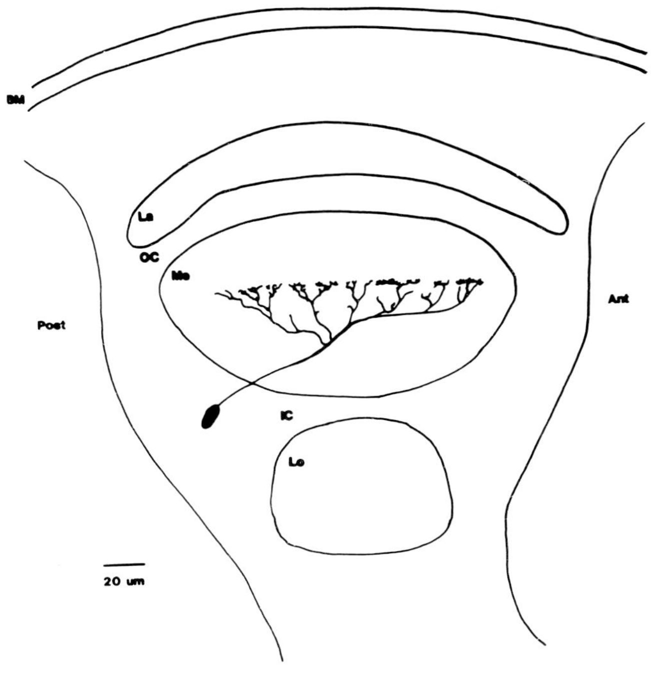

FIGURE 9. Anatomy of the cell which showed UV and G excitation in the sustained response. The major anatomical features are labeled as outlined in the text; these are the lamina (La), medulla (Me), lobula (Lo), outer chiasma (OC), inner chiasma (IC), and basement membranes (BM). Anterior (Ant) and posterior (Post) are also indicated.

channel is somehow isolated from the entrainment mechanism. An additional finding of note in these studies is the extraordinary sensitivity of the eye-clock system. The energy at threshold can only be measured indirectly with a calibrating photodiode and is 6 fJ/m^2. This value converts to a mean photon flux of $160/cm^2/s$. Each eye was estimated to be on the order of 3 mm^2, and the flux per eye was about 5 per eye per second. Estimates of how many quanta actually interact with the photoreceptor mechanism reduce this number even further. By comparison, the human visual threshold is several orders of magnitude higher.

B. CONDITIONING EXPERIMENTS

Training experiments in which animals are required to choose between a pair of lights in a Y maze based on their wavelength have been carried out in our laboratory.[8] In these tests, water-deprived cockroaches were given the choice between pairs of red, orange, and blue lights and were rewarded with water for making a certain choice. Controls for brightness consisted of changing the intensity of either light. We anticipated that the insects would only be able to learn to discriminate wavelength in the spectral region where their receptors overlap in sensitivity and not beyond. Thus, they should be able to distinguish blue from red or orange, but not red from orange. The results indicated that this is the case. With a confidence level of 99%, animals correctly chose between red-blue and orange-blue pairs, but not orange-red pairs.

C. AVOIDANCE EXPERIMENTS

An avoidance paradigm has been used to test the visual capabilities of *Periplaneta*.[34,35] In these studies an animal is placed in a darkened circular arena with a spot of light projected onto a portion of the floor adjacent to the wall. The animals tend to circle the arena with

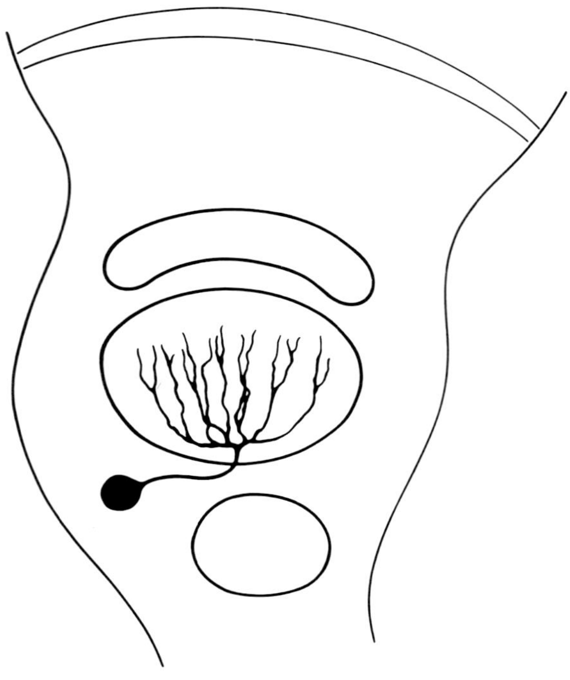

FIGURE 10. An artist's rendition of a sustaining cell's anatomy obtained by the dye injection method. This cell was of the excitatory broadband type, but in this case showed a much more extensive branching pattern. The outlines of the general anatomical landmarks are the same as in Figure 9.

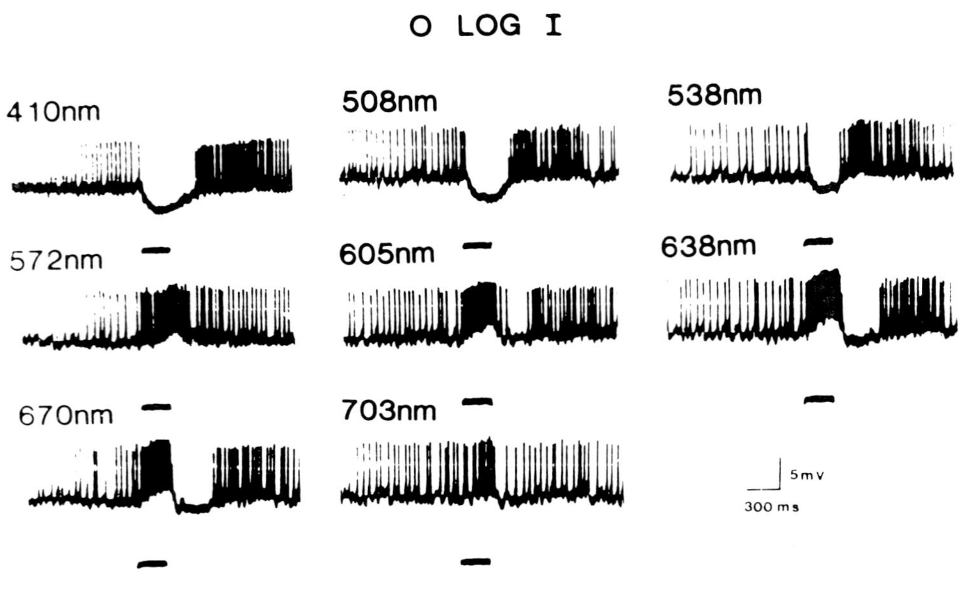

FIGURE 11. Sample records from a color opponent cell stimulated with equal photon flashes at eight different wavelengths at high intensity (0 log I). The timing of the flash is indicated in the lower trace. The wavelength is indicated above the trace.

an antenna against the wall. In a certain proportion of times when the light is encountered it is avoided either by circling or by reversal of direction. This proportion depends on the intensity of the light; the more intense the light, the higher the proportion of avoidance. We can measure the action spectrum for this behavior using this proportion. The results suggest that avoidance is driven by both photoreceptor groups, but that the UV type only exerts an influence at higher intensities. The shape of the action spectrum is intensity dependent, with G dominating at low intensities.

Another interesting result was obtained when a polarizing filter was inserted in the light path. The animals were unable to detect a stationary polarized pattern. This was judged by comparing avoidance between two stimuli of equal intensity, with one being polarized. No difference was seen. If the polarizer was rotated, at about 2 Hz, avoidance was much greater than it was with a stationary polarizer. The difference was equivalent to raising the light intensity by 1.5 log units. It is tempting to propose that polarizational detection systems are present in all insect visual systems, even in those of a crepuscular and nocturnal creature such as this species. However, the same caution is offered here as with the optic lobe findings: the results could be an artifact generated in a motion detection system.

VI. THE OCELLUS

A. ANATOMY

Cockroaches have a second, "simpler" system of photodetector organs called the ocelli. In most insects, these organs are three in number and form a characteristic triangle on the head between and dorsal to the compound eyes. An unusual situation is found in *Periplaneta* where two ocelli are located more ventrally between the eye and the antennae. Once thought to represent a degenerate receptor system in which the receptors had lost their dioptric apparatus,[1,36] the ocelli are now believed to be functioning systems and are currently receiving attention.[36-48]

The structure of the ocellar retina has been investigated in the pioneering studies of Ruck[37-39] and later by others.[40-44] These structures are elliptical in shape, measuring roughly 700 × 500 μm, and are distinctly convex. They have a characteristic white color which makes them easily distinguishable from the dark brown cuticle surrounding them. Each is bounded distally by a concave cornea of about 75 μm in thickness. An unpigmented corneal epithelium is formed by a single layer of corneagenous cells. This epithelium abuts against the ocellar retina. Each retina contains about 10,000 cells whose membranes form the microvillar rhabdomeres which in turn form fused rhabdoms. Figure 12 shows a histological section through the retina of an ocellus. Retinula cells can be seen forming clusters where the rhabdomeres combine to form rhabdoms. The number of cells forming a rhabdom is variable, ranging from two to seven.[40] The consequence of this is a variable retinular pattern across the retina, especially in terms of rhabdom shape. The microvilli are mainly parallel within the rhabdoms, which often bend and loop around. Weber and Renner[40] describe membrane-bound organelles which are stacks of from 10 to 40 microtubules. These can be found in a wide variety of cells, but not in other insect retinular cells.

1. Second Order Cells

The retinular cells extend axons from a region of the soma that contains a curious spiral of endoplasmic reticulum.[40] They course through a tapetum created by white, alcohol-soluble crystals. This tapetum gives the ocelli their white appearance. This coloration gave earlier workers the impression that the ocelli were degenerate organs, since other insects possess pigmented ocelli. Beyond the tapetum, retinular axons contact the four (usually) second order L cells in a region of synapses called the ocellar neuropil. This neuropil extends into the ocellar nerve, connecting the ocellus to the protocerebrum, and finally into the ocellar tract which traverses the protocerebrum and enters the ventral nerve cord. The ocellus and

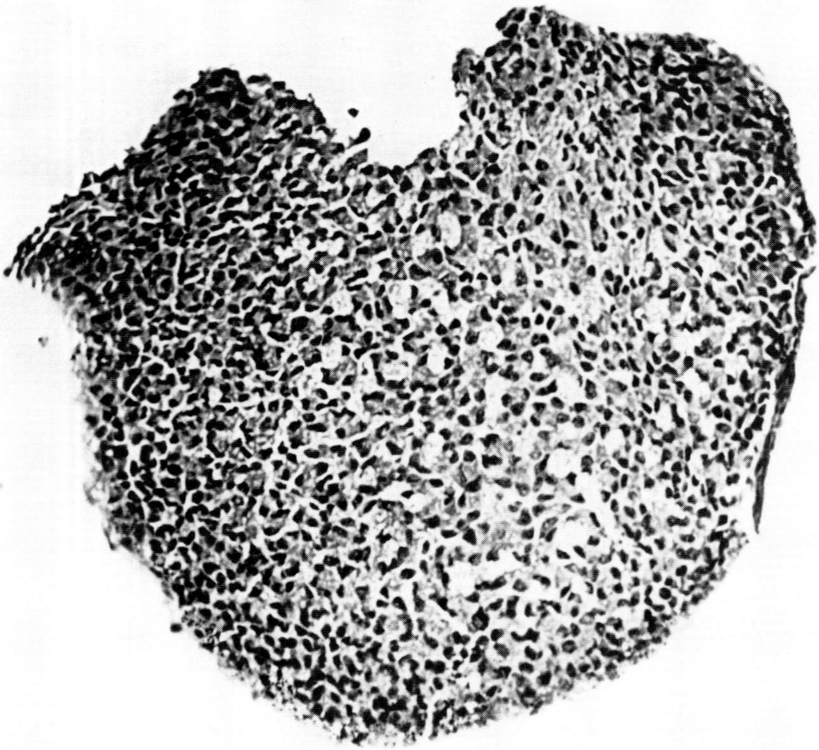

FIGURE 12. A 15-μm-thick section through the distal retina of the ocellus of *Periplaneta americana*. The clusters of retinular cells form the rhabdoms, which appear as dark spots. The rhabdoms appear to be more concentrated at the edges of the section, but this is an artifact. The ocellus has a distinctly convex shape. A section that is perpendicular to the surface at the center will become more parallel to the surface at the edges and will therefore contain the profiles of more rhabdoms. The clear oval profiles seen in the central regions are bundles of retinular cell axons running toward the ocellar neuropil.

ventral nerve cord are therefore connected by a continuous trunk of neuropil in which there are numerous contacts between neurons. Some retinular axons do not end in the ocellar neuropil, but instead enter the ocellar nerve and make synapses there or further down in the ocellar tract. Toh and Sagara[43] claim that these "long" retinular axons can be traced back to anomalous cells in the retina that contain many electron-dense vacuoles. They believe that these represent old retinular cells which formed pioneer fibers for the subsequent guidance of elements of the ocellar nerve during development.

Mizunami et al.[45] have studied the anatomy of the L cells in some detail by intracellular recording and cobalt dye injection. They find that all L-cell profiles in the ocellar nerve are centrifugal. The cell bodies lie in the protocerebrum. Distal processes in the ocellar neuropil enlarge to cover the whole neuropil region, presumably making synapses with all (or nearly all) of the retinular cells. Two morphological types of L cells can be recognized. In one type, the neurite emanating from the cell body bifurcates into distal and central branches; the central branch terminates in the ocellar tract. In the other type, a single process is directed peripherally, making contacts in the ocellar tract, nerve, and neuropil.

2. Higher Order Cells

Toh and Hara[44] identify at least six different types of fiber processes in the ocellar nerves. The origin and function of some of these have yet to be elucidated. Mizunami and Tateda[46] have described nine types of fibers that are probably third order, since they receive

their input in the ocellar nerve-tract neuropil, where the L cells terminate, and project to various locations in the nervous system. Identification was made by intracellular injection with cobalt, and classification was mainly on anatomical grounds. Five of the cells are termed posterior slope cells, since their terminations are found on the posterior slope of the protocerebrum. They are distinguished by their branching patterns. Two types of cells (OL) project into the optic lobes, and two types of cells (D) have projections which descend into the ventral nerve cord. OL cells can be either unilateral or bilateral (connecting both ocellar and compound eye systems). D cells can either project ipsilaterally or decussate and descend on the contralateral side. All of these cells have elaborate branching patterns which show extensive projection from the ocellar system to the rest of the nervous system.

B. PHYSIOLOGY ERG

Ruck[37-39] provided the pioneering work on ocellar responses by studying the ERG response in the cockroaches *Blaberus* and *Periplaneta*. The waveform of the response was analyzed by altering the electrode geometry, shifting the electrode position, and isolating the organ from its nerve and the brain. He detected components attributable to responses arising in the retina, hyperpolarizing and depolarizing, and to off discharges in the ocellar nerve tract. He visualized a depolarization of primary receptors followed by a hyperpolarizing response postsynaptically. Goldsmith and Ruck[47] studied the spectral sensitivity of this organ. They found a single receptor type with a maximum at 500 nm, not too different from the G receptor of the compound eye. This is in contrast to most other insects, in which the ocelli are dichromatic.[36] Intracellular recordings from receptor elements have not been reported for *Periplaneta*.

1. Second Order Cells

Intracellular recordings from the second order L cells have been reported.[45] The response consists of a simple hyperpolarization with phasic and tonic components at high stimulus intensities and one or a few spikes when the light is turned off. The number of spikes is greatest at intermediate intensities. There seems to be no after-depolarization as reported for other insects. More recently, the dynamics of the dorsal ocellus have been investigated using white noise analysis.[48] This is accomplished by modulating a constant stimulus and cross-correlating that modulation with the cell's output. The transfer function was entirely linear and did not change over a four log unit range in the mean luminance of the stimulus. The ocellar-L-cell system functions in a simple way, but is unique in that all other visual systems studied show dynamics that shift to higher frequency response with higher mean luminances.

2. Higher Order Cells

Mizunami and Tateda[46] studied the responses of higher order cells.[46] They described a basic response pattern from the ocellar nerve that was similar to that described above for the second order L cells, namely, an on-tonic hyperpolarization followed by depolarizing transients. There was variation in the way cells could be driven by one or both ocelli. Some respond only with transients to stimulation of one ocellus and give a complete response when the other is illuminated. When recording in the protocerebrum they could discern differences in response pattern that could identify cell types. Most notable was the OL type cell, giving spontaneous discharges in the dark. These were inhibited when the ocellus ipsilateral to the cell body was stimulated. Stimulating the contralateral organ did not elicit much of an effect.

VII. CONCLUSION

The visual system of the cockroach shares some basic similarities with the visual systems of mammals. It is in these areas that the insect will serve as a useful model for biomedical

applications. Color opponency is a strategy used by the nervous systems of both *Periplaneta*[24,25] and mammals.[49] This process is less complex in *Periplaneta* because there are only two color receptors.[10] The two types can be distinguished by their waveform,[10] and selective adaptation is an effective means of manipulation.[15] Their influence can be detected at higher levels of organization. These approaches are not available in more complex systems.[49] Note that the influence of the G receptors is pronounced at low light intensities, while the UV group is effective at higher intensities. This is similar to the relationship found between the rods and cones of the mammalian retina. Thus, this preparation could be a useful analogue.

At higher levels of organization the visual system of *P. americana* is very different than that of vertebrates. It is at these levels that the insect will not serve as a useful model for the study of visual function in humans.

The ocellar system is adapted for efficient capture of photons at low intensities. The presence of a tapetal reflecting layer and the convergence of 10,000 receptors onto four second order cells are such adaptations. This organ will serve as a useful model for the study of the basic principles of organization to enhance sensitivity. However, since the vertebrates do not possess a comparable organ, the ocellus will be of limited utility as a general model.

REFERENCES

1. **Miall, L. C. and Denny, A.,** *The Cockroach,* Lovell Reeve, London, 1886.
2. **Grenacher, H.,** *Untersuchungen über die sehorgane der Arthropoden, insbesondere der Spinnen, Insekten und Crustaceen,* Vandenhoek and Ruprecht, Göttingen, Federal Republic of Germany, 1879.
3. **Butler, R.,** The anatomy of the compound eye of *Periplaneta americana.* I. General features, *J. Comp. Physiol.,* 83, 223, 1973.
4. **Wolken, J. J. and Gupta, P. D.,** Photoreceptor structures. IV. The retinal cells of the cockroach, *J. Biophys. Biochem. Cytol.,* 9, 720, 1961.
5. **Trujillo-Cenoz, O. and Melamed, J.,** Spatial distribution of photoreceptor cells in the ommatidia of *Periplaneta americana, J. Ultrastruct. Res.,* 34, 397, 1971.
6. **Butler, R.,** The anatomy of the compound eye of *Periplaneta americana* (L.). II. Fine structure, *J. Comp. Physiol.,* 83, 239, 1973.
7. **Ribi, W. A.,** Fine structure of the first optic ganglion (lamina) of the cockroach, *Periplaneta americana, Cell Tissue Res.,* 9, 57, 1977.
8. **Mote, M. I.,** unpublished observations, 1987.
9. **Sokolove, P. G.,** Localization of the cockroach optic lobe circadian pacemaker with microlesions, *Brain Res.,* 87, 13, 1975.
10. **Mote, M. I. and Goldsmith, T. H.,** Spectral sensitivities of color receptors in the eye of the cockroach, *Periplaneta, J. Exp. Zool.,* 173, 137, 1970.
11. **Mote, M. I. and Goldsmith, T. H.,** Compound eyes: localization of two color receptors in the same ommatidium, *Science,* 171, 1254, 1971.
12. **Butler, R.,** The identification and mapping of spectral types in the retina of *Periplaneta americana, Z. Vgl. Physiol.,* 72, 67, 1971.
13. **Stark, R. J. and Mote, M. I.,** Postembryonic development of the visual system of *Periplaneta americana.* I. Patterns of growth and differentiation, *J. Embryol. Exp. Morphol.,* 66, 235, 1981.
14. **Shelton, P. M. J., Anderson, H. J., and Eley, S.,** Cell lineage and cell differentiation in the developing eye of the cockroach *Periplaneta americana, J. Embryol. Exp. Morphol.,* 39, 235, 1977.
15. **Walther, J. B. and Dodt, E.,** Electrophysiologische Untersuchungen über die ultraviolettempfindlichkeit von Insekt Augen, *Experientia,* 13, 333, 1957.
16. **Autrum, H. J. and Gallwitz, U.,** Zur Analyse der Belichtungspotential des Insektenauges, *Z. Vgl. Physiol.,* 33, 407, 1951.
17. **Wolbarsht, M. L., Wagner, H. G., and Bodenstein, D.,** Origin of electrical responses in the eye of *Periplaneta americana,* in *The Functional Organization of the Compound Eye,* Bernhard, C. G., Ed., Pergamon Press, London, 1966, 207.
18. **Walther, J. B.,** Untersuchungen am Belichtungspotential des Komplexauges von *Periplaneta,* mit farbigen Reizen und selektiver Adaptation, *Biol. Zentralbl.,* 77, 63, 1958.

19. **Walther, J. B.,** Changes in spectral sensitivity and form of retinal action potential of the cockroach eye by selective adaptation, *J. Insect Physiol.*, 2, 142, 1958.
20. **Mote, M. I., Kumar, V. S. N., and Black, K. R.,** "On" type interneurons in the optic lobes of *Periplaneta americana*. II. Receptive fields and latencies, *J. Comp. Physiol.*, 141, 403, 1981.
21. **Butler, R. and Horridge, G. A.,** The electrophysiology of the retina of *Peripianeta americana* (L.). I. Changes in receptor acuity upon light/dark adaptation, *J. Comp. Physiol.*, 83, 263, 1973.
22. **Butler, R. and Horridge, G. A.,** The electrophysiology of the retina of *Periplaneta americana* (L.): receptor sensitivity and polarized light sensitivity, *J. Comp. Physiol.*, 83, 279, 1973.
23. **Mote, M. I. and Rubin, L. J.,** "On" type interneurones in the optic lobe of *Periplaneta americana*. I. Spectral characteristics of response, *J. Comp. Physiol.*, 141, 395, 1981.
24. **Kelly, K. M.,** Electrophysiological Properties and Anatomy of Medulla Interneurones in the Optic Lobe of the Cockroach *Periplaneta americana*, Ph.D. thesis, Temple University, Philadelphia, 1987.
25. **Kelly, K. M. and Mote, M. I.,** Intracellular recordings of optic lobe interneurones in the cockroach *Periplaneta americana:* color opponency and polarizational sensitivity, *Neurosci. Abstr.*, 11, 165, 1985.
26. **Kelly, K. M. and Mote, M. I.,** in preparation.
27. **Laughlin, S. B.,** Neural integration in the first optic lobe of the dragon flies. I. Signal amplification in dark-adapted second order neurons, *J. Comp. Physiol.*, 92, 357, 1973.
28. **Barnes, S. N. and Mote, M. I.,** Lamina monopolar cells of the Praying Mantis: response pattern and receptive field, *Assoc. Res. Vision Ophthalmol. Abstr.*, p. 277, 1980.
29. **Kien, J. and Menzel, R.,** Chromatic properties of interneurones in the optic lobe of the bee, *J. Comp. Physiol.*, 113, 17, 1977.
30. **Edwards, D. H., Jr.,** The cockroach DCMD neurone. I. Lateral inhibition and the effects of dark and light adaptation, *J. Exp. Biol.*, 99, 61, 1982.
31. **Edwards, D. H., Jr.,** The cockroach DCMD neurone. II. Dynamics of response habituation and convergence of spectral inputs, *J. Exp. Biol.*, 99, 91, 1982.
32. **Rowell, C. H. F.,** The orthopteran descending motion detector (DCMD) neurones: a characterization and review, *Z. Vgl. Physiol.*, 73, 167, 1971.
33. **Ratliff, F.,** *Mach Bands: Quantitative Studies on Neural Networks in the Retina*, Holden-Day, San Francisco, 1966.
34. **Mote, M. I. and Black, K. R.,** Action spectrum and threshold of entrainment of circadian running activity in the cockroach *Periplanenta americana, Photochem. Photobiol.*, 34, 257, 1981.
35. **Kelly, K. M. and Mote, M. I.,** Visually mediated behavior in the cockroach, *Neurosci. Abstr.*, 12, 41, 1986.
36. **Goodman, L. J.,** The structure and function of the insect dorsal ocellus, *Adv. Insect Physiol.*, 7, 97, 1970.
37. **Ruck, P.,** The electrical responses of the dorsal ocellus in cockroaches and grasshoppers, *J. Insect Physiol.*, 1, 109, 1957.
38. **Ruck, P.,** Dark adaptation of the ocellus of *Periplaneta americana:* a study of the electrical response to illumination, *J. Insect Physiol.*, 2, 189, 1958.
39. **Ruck, P.,** Electrophysiology of the insect dorsal ocellus, *Gen. Physiol.*, 44, 605, 1961.
40. **Weber, G. and Renner, M.,** The ocellus of the cockroach, *Periplaneta americana* (Blattariae). Receptory area, *Cell Tissue Res.*, 168, 209, 1976.
41. **Cooter, R. J.,** Ocellus and ocellar nerves of *Periplaneta americana* (Orthopetera: Dictyoptera), *Int. J. Insect Morphol. Embryol.*, 4, 273, 1975.
42. **Bernard, A.,** Étude topographique des interneurones ocellaires et de quelques uns de leurs prolongements chez *Periplaneta americana, J. Insect Physiol.*, 22, 569, 1976.
43. **Toh, Y. and Sagara, H.,** Dorsal ocellar system of the American cockroach. I. Structure of the ocellus and ocellar nerve, *J. Ultrastruct. Res.*, 86, 119, 1984.
44. **Toh, Y. and Hara, S.,** Dorsal ocellar system of the American cockroach. II. Structure of the ocellar tract, *J. Ultrastruct. Res.*, 86, 135, 1984.
45. **Mizunami, M., Yamashita, S., and Tateda, H.,** Intracellar staining of the large ocellar second order neurons in the cockroach, *J. Comp. Physiol.*, 149, 215, 1982.
46. **Mizunami, M. and Tateda, H.,** Classification of ocellar interneurones in the cockroach brain, *J. Exp. Biol.*, 125, 57, 1986.
47. **Goldsmith, T. H. and Ruck, P.,** The spectral sensitivities of the dorsal ocelli of cockroaches and honeybees, *J. Gen. Physiol.*, 41, 1171, 1958.
48. **Mizunami, M., Tateda, H., and Naka, K.,** Dynamics of cockroach ocellar neurons, *J. Gen. Physiol.*, 88, 275, 1986.
49. **DeValois, R.,** Central mechanisms of color vision, in *Handbook of Sensory Physiology*, Vol. 7/3a, Jung, R., Ed., Springer-Verlag, Berlin, 1973, 210.

Chapter 24

CIRCADIAN ORGANIZATION IN THE COCKROACH

Terry L. Page

TABLE OF CONTENTS

I. INTRODUCTION: CIRCADIAN ORGANIZATION IN THE METAZOA

Most organisms must contend with environments that exhibit marked daily fluctuations in a variety of biotic and abiotic factors. The response of natural selection has been the evolution of regulatory systems whose primary function is to match these cyclic environmental variations with appropriate, periodic alterations in physiology, biochemistry, and behavior. These regulatory systems function as "biological clocks" generating a precise temporal program within the organism that coordinates its activities with the periodic, and thus predictable, environment in which it lives.

The two most prominent characteristics of biological clocks, which are found in most eukaryotic organisms, are that

1. They are based on endogenously generated, self-sustaining oscillations whose periods approximate those of the natural daily cycle. Even when organisms are isolated from all identifiable periodic variation in the external environment, the rhythmic changes in the biology of the organism persist with a period that is "circadian" — it is usually close to (but is never exactly) 24 h.
2. The oscillations can be synchronized (entrained) by a limited number of environmental cycles (usually light and temperature). The significance of entrainment is that it appropriately phases the oscillation to the individual's immediate environment. A secondary consequence of entrainment is that the period of the endogenous oscillation assumes the period of the environmental cycle to which it is entrained.

These two characteristics have given rise to a conceptual model of the circadian system, first proposed by Pittendrigh and Bruce,[1] which forms the basis for most experimental strategies that are used in the physiological and anatomical analysis of circadian clocks. This model, illustrated in Figure 1, consists of four functionally defined elements — a pacemaker that generates the primary timing signal, a photoreceptor for transduction of the light information used in entrainment, and two coupling pathways, one that mediates the flow of information from the photoreceptor to the pacemaker and another that couples the pacemaker to the effector mechanisms that it controls.

Since 1960 there has been a substantial effort to identify the anatomical correlates of these functionally defined components and to characterize the physiological mechanisms by which these components are integrated to form an organized regulatory system. Two general experimental approaches have been used. The first essentially treats the system as a black box, and its properties are inferred from the behavior of overtly expressed rhythms (output) in response to external stimuli (input) that affect either the phase or period of the pacemaking system. The results of this approach have led to a sophisticated understanding of the entrainment process and have provided important clues to the physiological organization of the circadian system.[2]

The second strategy has been to employ the traditional experimental approaches of physiology and endocrinology (e.g., lesions and transplantation) to identify specific structures in the organism which are important to circadian system function. One consequence of these efforts is that in the Metazoa attention has been focused on the nervous and endocrine systems. Circadian pacemakers have been localized to restricted portions of the nervous systems of a wide variety of organisms; photoreceptors for entrainment have been localized to visual organs or, more frequently, to the central nervous system (CNS), and in several cases the pathways by which pacemakers impose rhythmicity on some process have been shown to involve specific neural or endocrine pathways. Thus, the problem of understanding circadian organization in metazoans is largely a problem of neurobiology.

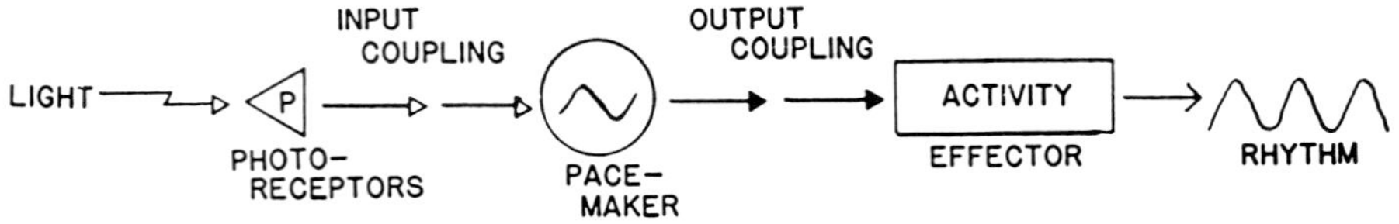

FIGURE 1. Functionally defined model of the circadian system. An entrainment pathway that consists of a photoreceptor and a coupling mechanism (input) synchronizes a self-sustaining oscillator (pacemaker) to the external light/dark cycle. The output of the pacemaker regulates the timing of various processes (e.g., activity) via coupling to effector mechanisms.

The results of both experimental approaches have led to the general proposition that circadian organization is derived from a population of anatomically discrete circadian oscillators whose temporal order is maintained by their submission to entrainment, either via direct access to temporal cues from the environment or by other oscillators within the population.[2,3]

Importantly, at both the level of the analytical (input/output) analysis and that of physiological analysis, circadian systems of metazoans appear to exhibit similar principles of organization. As a consequence, it has been possible to exploit the simpler, invertebrate systems to gain insight into the organizational principles that underly circadian systems across phylogenetic lines; additionally, several specific invertebrate models have been developed to address a variety of specific questions about the way in which circadian systems are organized. Cockroaches have proven to be particularly suitable models for investigating aspects of the physiology of circadian rhythmicity, and as a result of three decades of study the anatomical and physiological organization of the circadian system of the cockroach is one of the best understood of any organism.

II. CIRCADIAN RHYTHMICITY IN THE COCKROACH

Endogenously generated daily rhythms have been observed in a variety of behavioral, physiological, and metabolic processes in several cockroach species. The most extensively studied of these rhythms has been locomotor activity, both because of the ease with which it can be recorded for long periods of time and because it can be measured with minimal disturbance to the animal. An example of the activity rhythm of the cockroach *Leucophaea maderae* is shown in Figure 2. When exposed to a light cycle that consists of 12 h of light alternated with 12 h of darkness (LD 12:12), activity primarily occurs in the dark phase of the cycle, with activity onset occurring near the light-to-dark transition. The endogenous nature of the periodic cycle of activity and rest is most easily revealed by placing the animal in constant environmental conditions, generally constant darkness (DD) at a constant temperature. In most cases the rhythm of activity persists, or ''freeruns'' (Figure 2). In cockroaches the period of the freerunning rhythm (τ) is generally slightly $<$24 h. As a consequence, activity onset drifts to a slightly earlier time in each successive cycle. Within the adult the value of τ is relatively stable. It is unaffected by changes in the level of the constant ambient temperature[4] or by aging.[5] However, τ can be permanently modified by exposure of larval stages to different lighting conditions.

While most of the data on the locomotor activity rhythms in cockroaches have been obtained from the study of adult animals, some work has been done on the nymphal stages. The clearest evidence that early developmental stages exhibit circadian rhythms in locomotor activity comes from work done on the first through fourth instars of *L. maderae*.[5] Approximately one third of first-instar nymphs were found to express a clear circadian rhythm of activity, recorded in constant darkness, within the first few days of larval life, even in animals which had never been exposed to a periodic environment. Activity of most other

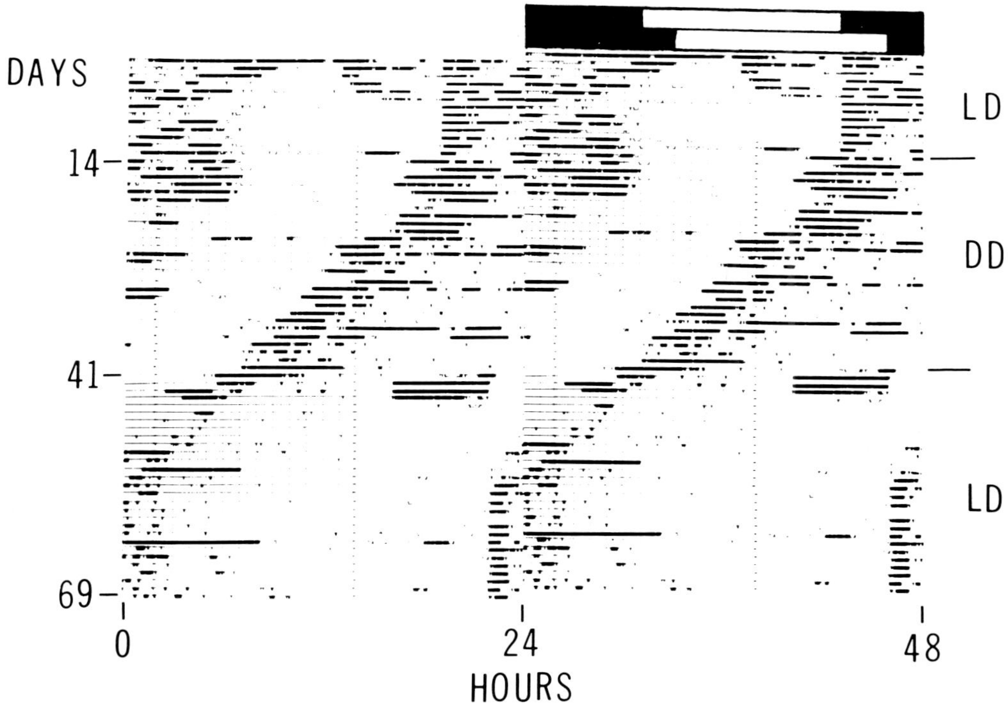

FIGURE 2. Event recording of the wheel running activity of a cockroach, *Leucophaea maderae*. Data for successive days are placed one below the other in chronological order, and the record has been duplicated to provide a 48-h time base to aid in visual inspection of the data. The bars at the top of the right-hand portion of the record represent the light cycles to which the animal was exposed, with the shaded bar corresponding to the dark portion of the LD 12:12 cycle. The record begins with the animal entrained to the LD cycle illustrated by the top bar. On day 14 the LD cycle was discontinued and the animal began to freerun in DD with a period of about 23.5 h. On day 41 the animal was exposed to the second LD cycle, and after several transient cycles the rhythm became phase locked to the new LD cycle.

animals became clearly periodic within 4 to 8 weeks of hatching. Thus, at least in some individuals, the pacemaking system that drives the circadian rhythm of activity in *L. maderae* is present and functional at the time of hatching. Although the pacemaker is functional in the larval stages, the environment to which the animal is exposed during development can have significant and permanent effects on the properties of the adult circadian system. In particular, exposure of nymphs to non-24-h light cycles or to constant darkness has a major impact on the freerunning period of the adult activity rhythm (Figure 3) and on the sensitivity of the pacemaker to phase shifting by light pulses.[5,6] Exposure of adults to these same light cycles has little or no effect. Recent data suggest that there is a critical period during approximately the first half of development when the environment can permanently modify the properties of the circadian system. The ability to manipulate the period of oscillation in the adult has proven to be useful experimentally (see Section III.A.4 below), and a more detailed knowledge of the mechanism and time course of these developmental effects may provide additional clues to the physiological and anatomical organization of the circadian system.

While the locomotor activity rhythms of cockroaches have received the most attention (particularly in *L. maderae* and *Periplaneta americana*) and for that reason are the focus of this review, a variety of other daily rhythms have been described (for review and additional citations, see References 3 and 7). These include rhythms in oxygen consumption,[8] feeding,[9]

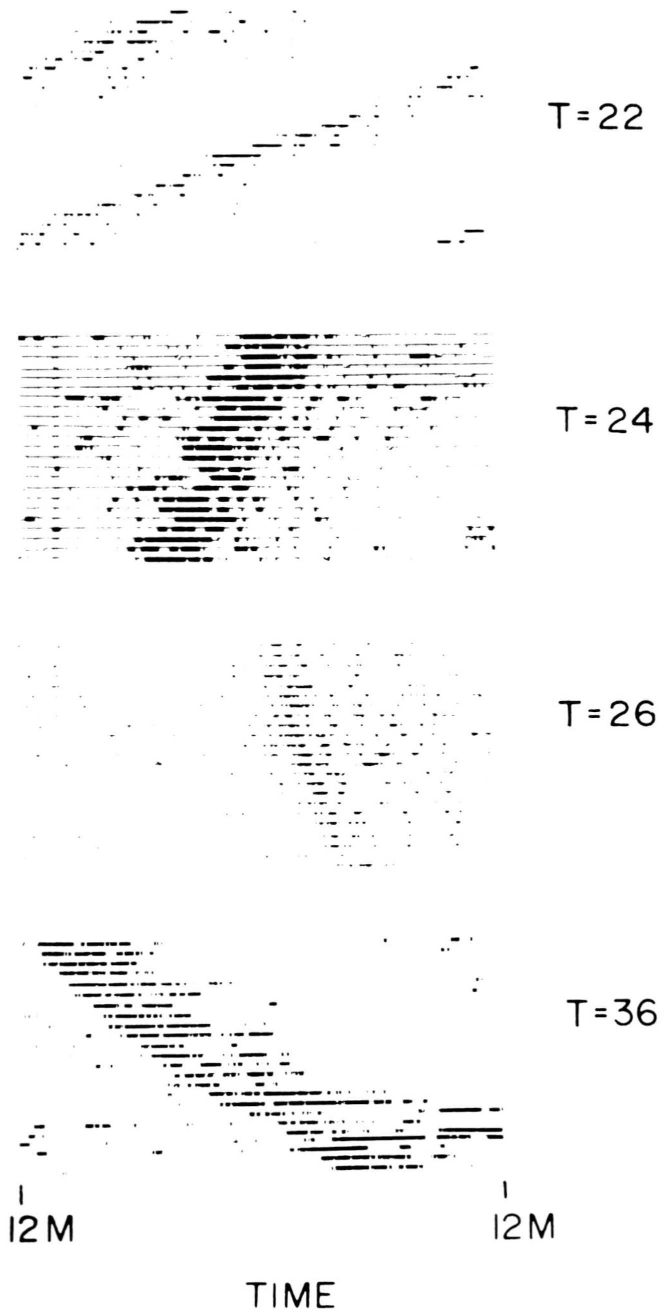

FIGURE 3. Freerunning rhythms (in DD) of *Leucophaea maderae* raised in light cycles with periods (T) of 22, 24, 26, and 36 h.

sensitivity to insecticides,[10] cuticular growth,[11] and electroretinogram (ERG) amplitude.[12] Two of these additional rhythms, the ERG and cuticular growth rhythms, are discussed in more detail in Section III.B below.

III. NEURAL ORGANIZATION OF THE CIRCADIAN SYSTEM

Interest in the neural basis of circadian rhythmicity in the cockroach was initiated by

two early and controversial studies by Harker.[13,14] On the basis of results of a series of parabiosis and transplantation experiments on *P. americana,* Harker concluded that the circadian rhythm in locomotor activity is controlled by a circadian pacemaker in the subesophageal ganglion that regulates activity via the rhythmic release of a hormone. Although evidence obtained in subsequent studies has argued convincingly against a hormonal clock in the subesophageal ganglion (reviewed in References 3 and 15), these early studies provided the impetus for an extensive effort over the next three decades to elucidate the neural organization of the circadian system in the cockroach.

A. THE LOCOMOTOR ACTIVITY RHYTHM

The focus of initial efforts was on the anatomical localization of the endogenous oscillators that generate the timing signal for the locomotor activity rhythm, the photoreceptors involved in the entrainment of these oscillators, and pathways by which the pacemaking system imposes rhythmicity on the pattern of activity. These goals have been realized to a large extent and have provided the foundation for a second level of physiological questions about the mechanisms by which these elements of the circadian system are linked into an organized regulatory system. The results of these studies are summarized in this section.

1. Photoreceptors for Entrainment

Several studies have provided unequivocal evidence that the compound eyes of cockroaches (*L. maderae* and *P. americana*) are the sole site of phototransduction for entrainment of the circadian rhythm of locomotor activity and that neither the ocelli nor the CNS photoreceptors are involved. Roberts[16] showed that when the eyes of intact roaches that were entrained to an LD cycle were painted over with black lacquer, the animals began to freerun even though the single pair of ocelli remained intact and exposed to the LD cycle. In contrast, surgical ablation of the ocelli had no effect on entrainment. Roberts' conclusion that the eyes were the necessary photoreceptive pathway for entrainment was subsequently verified by Nishiitsutsuji-Uwo and Pittendrigh,[17] who were able to repeat the observation on the effects of painting the eyes and further showed that surgical ablation of the eyes or section of the optic nerve also abolished entrainment by light. They also found no evidence that the ocelli were either necessary or sufficient for entrainment.

The photoreceptors within the compound eye that are responsible for entrainment have not been identified; however, experiments involving partial ablation of the eyes in *L. maderae* indicate that there are several photoreceptors, widely distributed over the eye, that are sufficient for entrainment.[8] Recent studies on *P. americana* have also provided information on the spectral sensitivity of the entrainment pathway.[18] The action spectrum for entrainment exhibits a peak sensitivity near 495 nm, indicating that transduction for entrainment primarily involves the green photoreceptors.

2. The Optic Lobes: Locus of the Driving Oscillation?

Nishiitsutsuji-Uwo and Pittendrigh[19] first focused attention on the importance of the optic lobes of the protocerebrum in the regulation of cockroach locomotion when they found that removal of these structures disrupted the circadian rhythm of activity in *L. maderae.* This observation has been confirmed for other cockroach species (*P. americana*[20] and *Blaberus craniifer*[21]), and it has been shown that the loss of rhythmicity caused by optic-lobe ablation persists indefinitely.[22]

Neural isolation of the optic lobes from the midbrain by bilateral section of the optic tracts also disrupts the locomotor activity rhythm.[19] However, it was discovered that when the optic lobes were left *in situ* following optic-tract section, the rhythm consistently reappeared in 3 to 5 weeks[22] (Figure 4).

Several lines of evidence indicated that the recovery of rhythmicity depended on re-

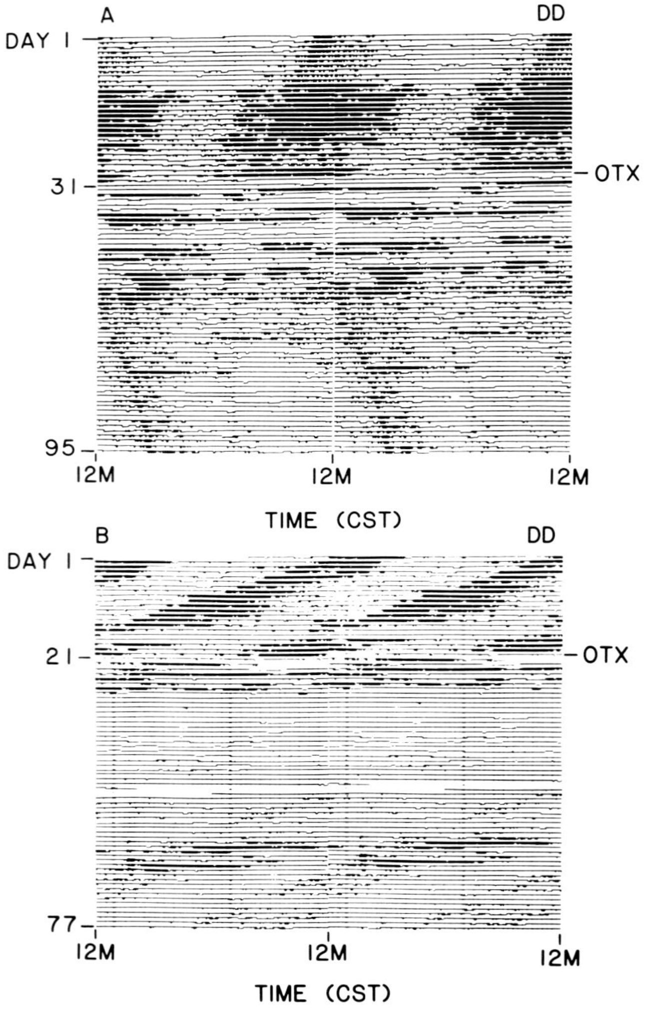

FIGURE 4. Activity records of two animals, A and B, showing the loss of the activity rhythm and its subsequent reappearance following bilateral severance of the optic tracts (OTX). CST = Central Standard Time. (From Page, T.L., *J. Comp. Physiol.*, 152, 231, 1983. With permission.)

generation of the neural connections between the optic lobes and the midbrain.[22] Histological examinations showed that structural regeneration had occurred in the brains of animals in which rhythmicity returned. Furthermore, the insertion of a glass barrier between the optic lobe and the midbrain to block regeneration prevented or slowed the recovery of rhythmicity. That functional regeneration of optic-tract pathways was occurring was also demonstrated in experiments in which light-evoked activity, driven via the compound eyes and recorded in the cervical connectives, was found to return some weeks after section of the optic tracts. The time course of the recovery of this neural activity paralleled the recovery of behavioral rhythmicity.

These results reiterated the central importance of the optic lobes in maintaining the rhythm in locomotor activity and further indicated that their effect was mediated by axons in the optic tract. The loss of rhythmicity following optic-lobe ablation or optic-tract section has raised the possibility that the optic lobes are the locus of the endogenous oscillation that

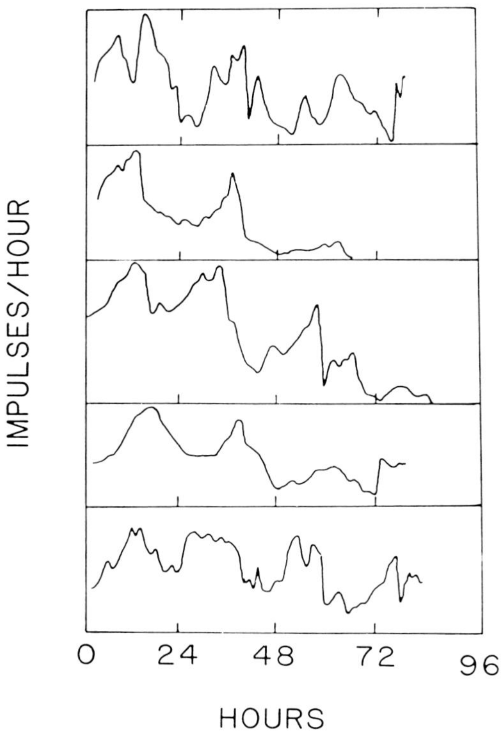

FIGURE 5. Five examples of action potential frequency (impulses per hour) plotted as a function of time. Multiunit activity was recorded with suction electrodes on the optic tracts of optic lobes maintained *in vitro* under constant conditions. (Reprinted with permission from Page, T. L., in *Trends in Chronobiology,* Hekkens, W., Kerkhof, G., and Rietveld, W., Eds., Copyright 1988, Pergamon Press, Inc.)

controls the temporal distribution of locomotor activity. This suggestion has been confirmed in a series of experiments that have shown that the optic lobes are not only necessary for the expression of the rhythm, but that they also (1) control the freerunning period of the rhythm, (2) control the phase of the freerunning rhythm, and (3) are able to generate a circadian rhythm in the absence of humoral or neural cues from the rest of the organism.

3. A Circadian Oscillation *In Vitro*: Pacemaker Activity in the Optic Lobe

The only conclusive evidence that a piece of tissue is a competent circadian pacemaker is the demonstration that the tissue can generate an oscillation in some biochemical or physiological function when completely isolated from neural or humoral signals from the rest of the animal. This demonstration has been accomplished recently for the cockroach optic lobe in experiments in which it was found that the lobe exhibited a rhythm in spontaneous impulse activity when maintained in organ culture under constant environmental conditions.[23,24] Optic lobes were removed from *L. maderae* and maintained in either Mark's M-20 medium[25] or the "5+4" medium of Chen and Levi-Montalcini,[26] and spontaneous multiunit activity was recorded from the cut end of the optic tract with a suction electrode. Under these conditions the optic lobes typically exhibited spontaneous neural activity for 3 to 5 d. Half of the optic lobes investigated (13 of 26) exhibited circadian rhythms in the number of action potentials per hour based on the criteria that the plot of activity vs. time subjectively appeared periodic (Figure 5) and that the times of peak activity on days 1 and

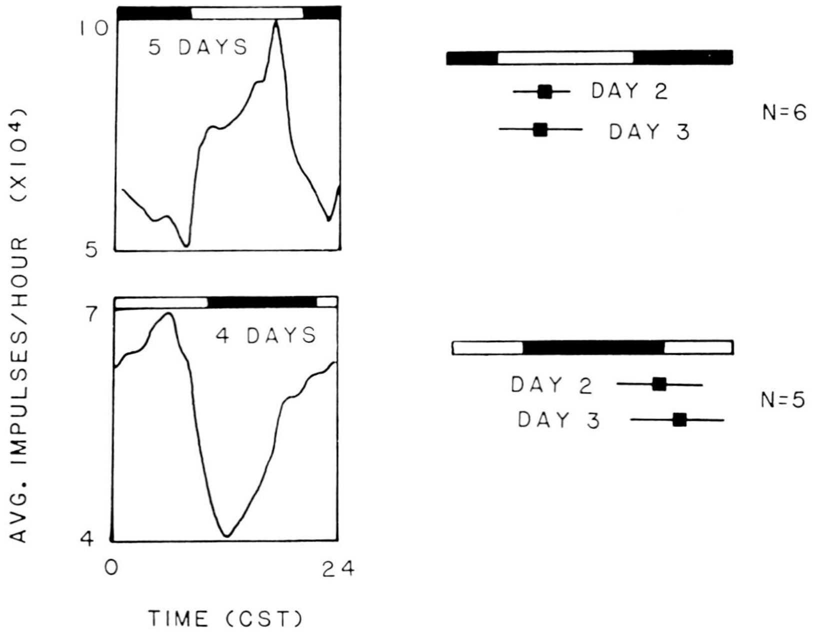

FIGURE 6. **Left:** Form estimates of impulse frequency as a function of time of day of multiunit activity recorded from the optic tracts of optic lobes *in vitro*. The top figure is generated from 5 d of data and the bottom from 4. In both figures the bar at the top of the record illustrates the light cycle to which the intact animal had been entrained prior to optic lobe removal. Peak activity occurred in the subjective day in both instances. **Right:** Average time of peak activity (± SD) on day 2 of recording for optic lobes from animals maintained in two light cycles that were 10 h out of phase. Number of optic lobes in each group (N) is also given. (Reprinted with permission from Page, T. L., in *Trends in Chronobiology*, Hekkens, W., Kerkhof, G. A., and Rietveld, W. J., Eds., Copyright 1988, Pergamon Press, Inc.)

2 *and* days 2 and 3 were approximately 24 h apart (day 0 = the day of dissection). The average period of the rhythms in impulse activity was 23.8 ± 1.90 h.

It was also shown that the time of day at which peak activity occurred was dependent on the light cycle to which the animals had been exposed prior to optic lobe removal. Interestingly, peak activity generally occurred in the early subjective day or very late subjective night (Figure 6), a time when locomotor activity is typically low.

These results provide compelling support for the proposition that the optic lobe of the cockroach brain does indeed contain a competent circadian oscillator. Additional evidence from studies on the control of the period and phase of the locomotor activity rhythm further indicates that this oscillator provides the timing signal for the activity rhythm.

4. The Optic Lobes Control the Freerunning Period

One aspect of the locomotor activity exhibited after section and regeneration of the optic tracts — the period of the freerunning rhythm — has been of particular interest. The freerunning period, after recovery of the activity rhythm, was found to be strongly correlated with the period of the preoperative rhythm (r = 0.87).[22] This result prompted a series of experiments in which the optic lobes were transplanted between individuals that had substantially different freerunning periods by virtue of their having been raised in non-24-h light cycles[22,27] (see Section II above). Of 22 animals that received optic-lobe transplants, 14 survived and exhibited a clear circadian rhythm of locomotor activity several weeks after surgery (Figures 7A and B). The preoperative period of the donor and the postoperative

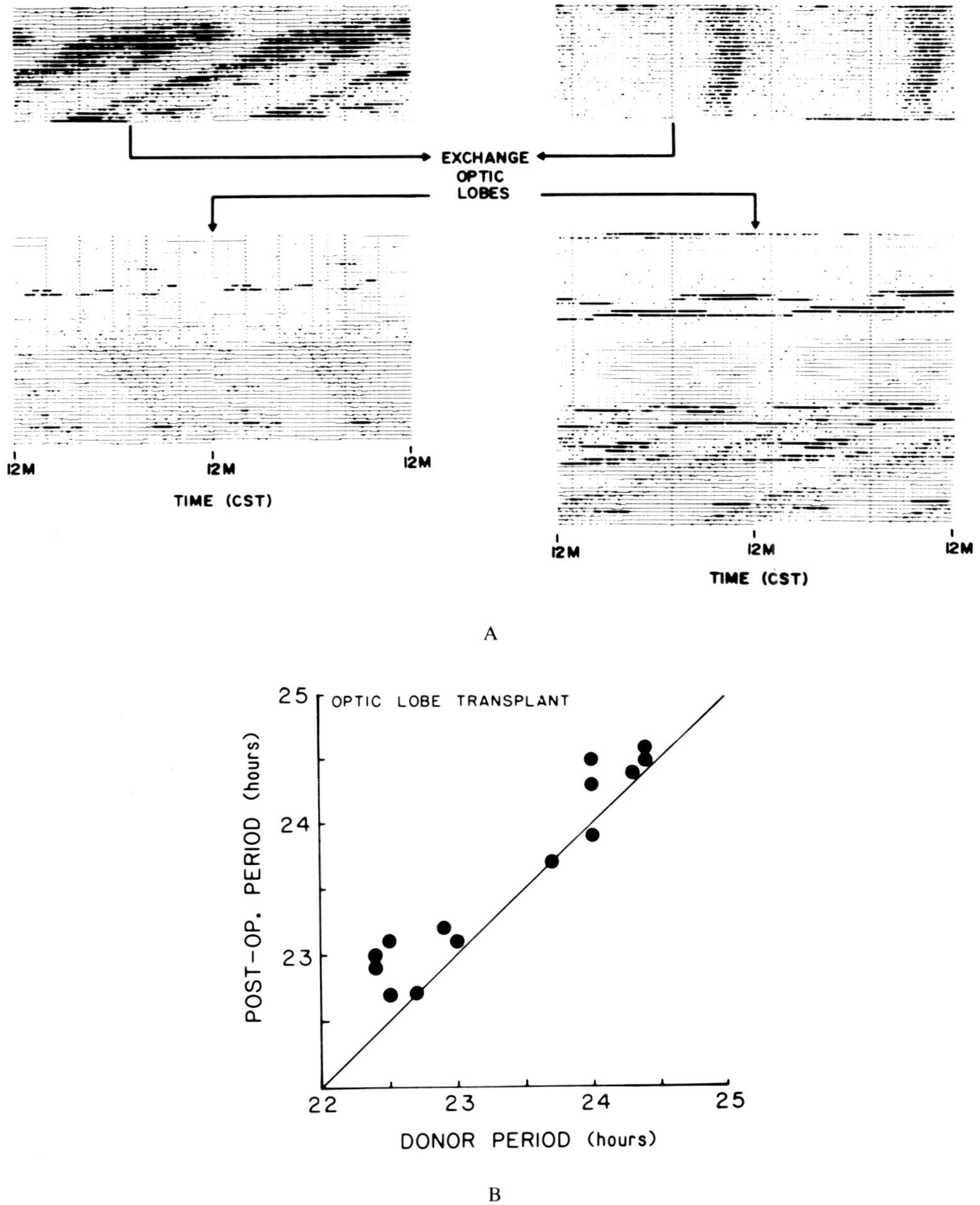

FIGURE 7. Effects on locomotor activity of optic-lobe transplantation between groups of animals raised in LD 11:11 or LD 13:13. (A) Locomotor activity records showing restoration of rhythmicity after transplantation ("exchange"). The records begin with intact animals freerunning in constant darkness. After 4 weeks the optic lobes were exchanged between the two animals and, after several weeks, rhythmicity returned in both cases. (B) Plots of the freerunning period of the activity rhythm after transplantation as a function of the period of the donor animal before surgery. Diagonal line shows the values expected if the pre- and postoperative periods had been identical.

period of the host were strongly correlated ($r = 0.96$), although τ after regeneration was slightly longer on average than τ for the donor prior to surgery ($\Delta\tau = 0.2 \pm 0.21$ h). These data showed that the transplantation of the optic lobes from one animal to another whose own optic lobes had been removed could both restore the rhythm of locomotor activity and impose the period of the donor animal's rhythm on the rhythm of the host.

It is interesting to note that similar transplantation experiments have recently been successful in mammals. Following ablation of the suprachiasmatic nuclei (SCN) of the hypothalamus in hamsters, which abolishes the circadian rhythm locomotor activity, restoration of rhythmicity was accomplished by transplantation of fetal SCN tissue into the third ventricle of the brain.[27a] Additional experiments will be necessary to determine whether or not the transplanted tissue can confer the period of the donor's rhythm on the host.

5. The Optic Lobes Control the Phase

Two independent lines of evidence demonstrate convincingly that the optic lobes control the phase, as well as the period, of the freerunning rhythm. One series of experiments, also based on the ability of the optic tracts to regenerate, showed that in animals in which the optic tracts were severed and allowed to regenerate in DD the phase of the activity rhythm was conserved during regeneration.[22] When onsets of the activity rhythm that was expressed after regeneration were projected back to the day of surgery and compared to the phase of the preoperative rhythm, the phase of activity onsets before and after surgery were correlated ($r = 0.61$, $p < 0.1$), and for the majority of animals (13 of 19) the projected phase was within 4 h of the preoperative rhythm phase on the day of surgery. In addition, it was found that the phase of the rhythm expressed after regeneration could be shifted by light cycles to which animals were exposed for several days immediately after optic-tract section.[22] These results suggested that (1) the oscillation that controls locomotor activity continued in motion in the neurally isolated optic lobe, with a freerunning period similar to the period of the rhythm after regeneration; and (2) the oscillation conserved the information on the phase of the rhythm at the time of surgery, but since the optic lobe was still attached to the compound eye after optic tract section, the oscillation in the neurally isolated optic lobe could still be phase shifted by light.[22]

Another approach to demonstrating that the optic lobes control the phase of the freerunning locomotor activity rhythm involved localized low-temperature pulses.[28] Localized cooling of a single optic lobe was accomplished by positioning a cooled insect pin near one optic lobe of animals in which one of the optic tracts had been cut. Animals that were freerunning in DD were removed from their activity monitors, and either the intact or neurally isolated optic lobe was cooled to about 7.5°C for 6 h. The animals were then returned to their activity monitors and the phases of the pre- and postoperative rhythms were compared (Figures 8A and B). Cooling the intact optic lobe consistently caused a phase shift ($\Delta\phi$) of several hours ($\Delta\phi = -7.1 \pm 1.9$ h, $n = 10$), while cooling the neurally isolated optic lobe had no effect on the phase of the activity rhythm ($\Delta\phi = -0.6 \pm 0.8$ h, $n = 5$). Cooling the midbrain also had no phase-shifting effect. The demonstration that low-temperature pulses shifted the phase of the rhythm only when an intact lobe (rather than a neurally isolated lobe or the midbrain) was cooled provided additional evidence that the optic lobes could regulate the phase of the circadian rhythm of locomotor activity.

In summary, it has been shown that (1) the optic lobe is able to generate a circadian oscillation when isolated from neural or humoral input from the rest of the animal, (2) the integrity of neural connections between the optic lobe and the midbrain is required for the maintenance of the circadian locomotor activity rhythm, (3) the optic lobe controls the period of the freerunning rhythm, and (4) the optic lobe controls the phase of the activity rhythm. The conclusion that the optic lobes contain a circadian pacemaking system that regulates the temporal distribution of daily activity via axons in the optic tract is inescapable.

6. Localization of the Oscillator Function within the Optic Lobe

The region within the optic lobe which contains the cells that generate the pacemaking oscillation for the locomotor activity rhythm has been localized further. After removal of one optic lobe, surgical or electrolytic lesions distal to the second optic chiasma or dorsal

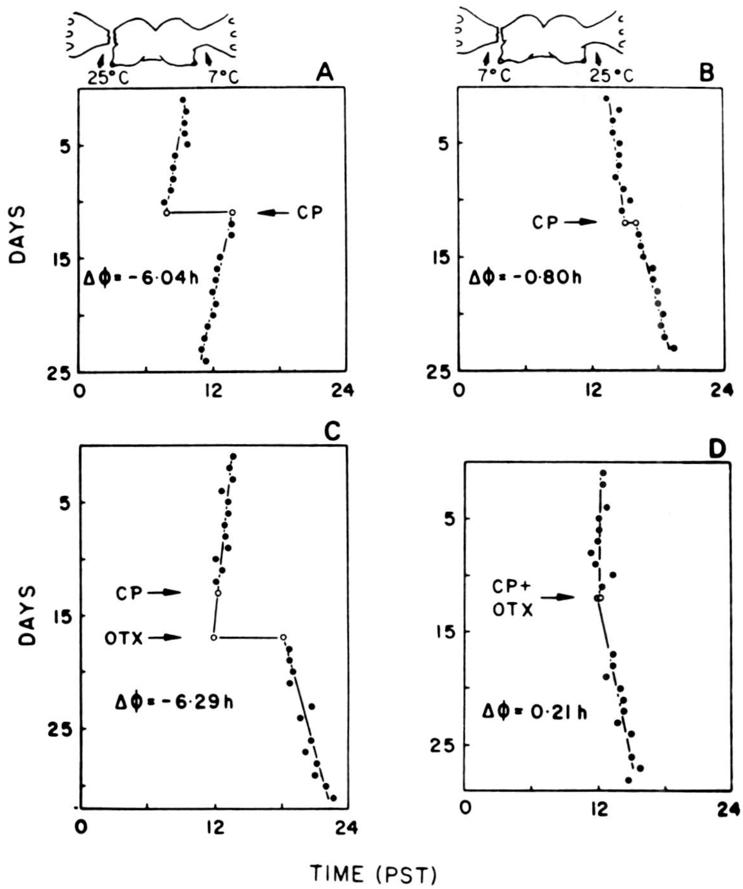

FIGURE 8. Examples of data from animals treated with localized low-temperature pulses. Closed circles are time of activity onset for each day; open circles are projected phases of the rhythms before and after the pulses; lines are linear regressions. Pulses were 6 h in duration and began at activity onset. (A) The intact optic lobe of an animal with one sectioned optic tract was cooled to 7°C, while the neurally isolated lobe was maintained at 25°C. (B) The neurally isolated optic lobe was cooled. Cooling the intact lobe caused a large phase delay (Δϕ), while cooling the neurally isolated optic lobe had little effect. (C,D) The effects of cooling one lobe on the pacemaker in the contralateral lobe of an intact animal are illustrated. (C) The optic tract of the treated lobe was cut (OTX) 4 d after the pulse (CP); the rhythm was delayed by several hours. (D) The optic tract of the treated lobe was cut 0.5 h after the pulse, thus preventing the phase shift of the rhythm. PST = Pacific Standard Time.

to the lobula have no effect on rhythmicity, while lesions near the lobula in the ventral half of the lobe frequently abolish the activity rhythm.[20,29,30] The results suggest that the cells responsible for generating the circadian signal have their somata and/or processes in this region of the optic lobe. Further work will be necessary to determine whether one, a few, or many cells within this region comprise the oscillator or whether the pacemaking signal may be generated redundantly in several cells.

7. The Pacemaker: Two Mutually Coupled Oscillators

In early studies it was shown that unilateral ablation of one optic lobe or section of a single optic tract did not abolish the rhythm of locomotor activity.[19] This suggested that each of the optic lobes could contain an oscillator sufficient to drive the activity rhythm.

This possibility raised two additional questions. First, if there are two oscillators, are they functionally equivalent? Second, is there a pathway by which the two oscillators could interact with one another? Initial investigations involved a systematic study of the effects of unilateral optic-lobe ablation on the freerunning rhythm of locomotor activity.[31] The two optic lobes were found to be functionally redundant, at least as measured by their ability to maintain rhythmicity and their average freerunning period in DD. However, ablation of either the right or the left optic lobe or section of one optic tract consistently led to a small but significant increase in τ ($\Delta\tau = 0.2 \pm 0.26$ h). This observation led to the suggestion that bilaterally distributed oscillators in the optic lobes were mutually coupled via the optic tracts and that the freerunning period of the coupled pair (ca. 23.7 h) was shorter than the period of the individual oscillators (23.9 h).

Support for this interpretation was obtained in experiments using localized low-temperature pulses[28] (see Section III.A.5 above) delivered to one optic lobe of intact animals. The treated lobes were removed at various times after the pulse to assay the phase of the rhythm driven by the contralateral oscillator. If the treated lobe was removed 4 d after the pulse, the subsequent rhythm (driven by the untreated lobe) was phase delayed by several hours (Figure 8C), but if the treated lobe was removed only 0.5 h after the pulse, the phase shift in the contralateral lobe was prevented (Figure 8D). The results indicated that the low-temperature pulse caused a phase shift in the treated lobe (without directly affecting the contralateral oscillator) which was subsequently transmitted to the oscillator in the contralateral optic lobe. It was also shown that cooling one lobe of an intact animal resulted in a phase shift that was nearly 2 h less than the phase shift obtained when the optic lobe contralateral to the pulse was isolated neurally by optic-tract section prior to treatment. This suggested that after desynchronization of the two oscillators the return of the coupled system to steady state involved a phase advance in the treated oscillator as well as a phase delay in the oscillator of the untreated optic lobe. Similar results have recently been obtained in experiments utilizing localized low-temperature pulses in another cockroach, *B. craniifer*.[32] The results indicate a similar organization of the pacemaking systems of the two species.

The notion that the two oscillators in the optic lobes were coupled was further reinforced by the finding that either one of the compound eyes was sufficient to entrain both oscillators. After isolation of one optic lobe from its (ipsilateral) compound eye by optic-nerve section, the oscillator in that optic lobe could still be phase shifted by a delay in the light cycle[31] or could be entrained by a 23-h light/dark cycle.[33] The response to light was abolished by section of the contralateral optic tract.

B. THE ELECTRORETINOGRAM (ERG) AMPLITUDE AND CUTICLE DEPOSITION RHYTHMS

The anatomical localization of the pacemaking system for the locomotor activity rhythm prompts the question of whether or not this pacemaking system also regulates circadian rhythms other than locomotor activity. In *Leucophaea* two additional rhythms have been studied — the rhythms in the amplitude of the ERG of the compound eyes[12] and the rhythm of cuticle deposition in newly molted adults.[34]

1. ERG Amplitude

A very common feature of arthropod visual systems is that the sensitivity of the eye to light exhibits a circadian variation[35,36] which is reflected in a daily change in the amplitude of the ERG recorded in response to a standard light pulse given at regular intervals throughout the day. Depending on the species, several mechanisms may contribute to producing the sensitivity change; however, it does appear invariably to be associated with daily migration of screening pigments which surround the rhabdomeres of the ommatidia.

In the cockroach *L. maderae* it was recently discovered that the amplitude of the ERG,

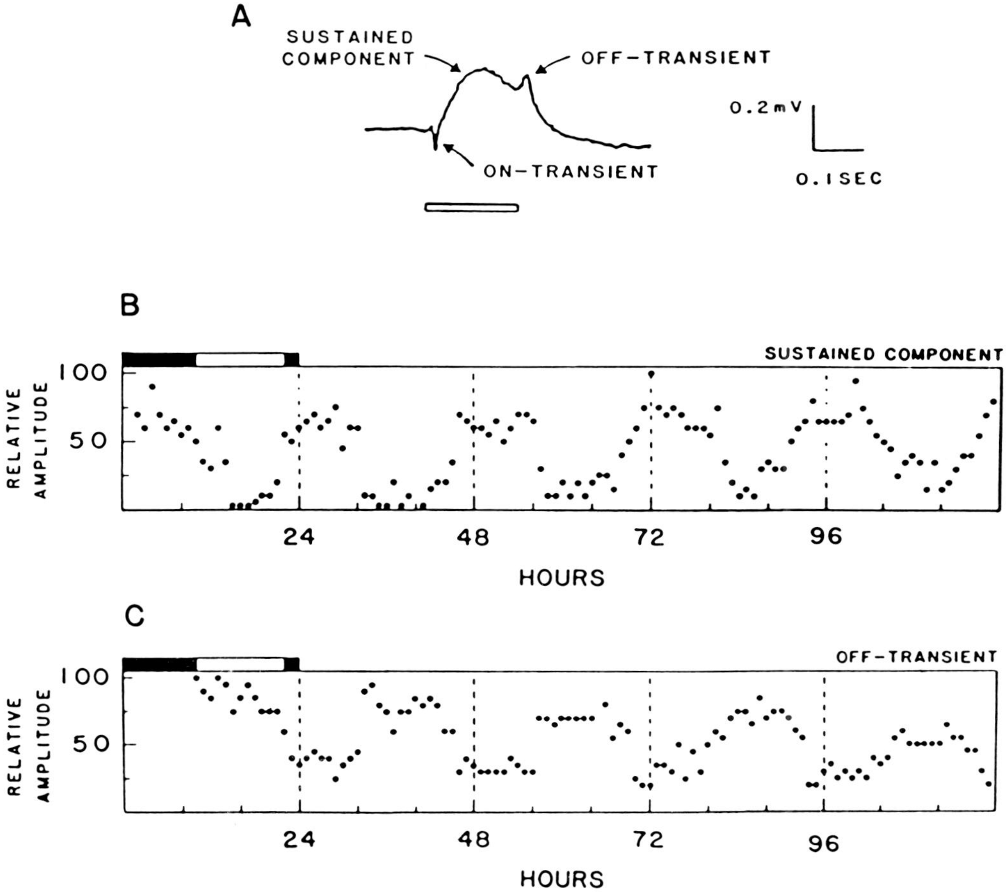

FIGURE 9. (A) The waveform of the ERG of a cockroach in response to a light pulse (bar). Below are plotted the relative amplitudes of two of the ERG components in response to a standard light pulse, given once per hour, as a function of time. (B) Sustained component. (C) Off-transient. The bar at the top of each record illustrates the light cycle to which the animal had been entrained prior to the beginning of the experiment.

in response to brief light pulses given once per hour, exhibited a circadian rhythm (Figure 9). The rhythm in the cockroach was of relatively low amplitude (the change from maximum to minimum amplitude during the day representing a sensitivity change equivalent to about 0.5 log units of intensity) compared to rhythms described in other arthropods. Two components of the ERG waveform, the "sustained component" which originates in the retina and the "off-transient" which arises in the optic lobe,[37] were found to exhibit circadian rhythms in amplitude. Interestingly, the peak amplitudes of the two components were approximately 12 h out of phase. The sustained component exhibited maximum amplitude during the subjective night while the off-transient was maximal during the subjective day.[12]

The results of a series of lesion studies investigating the pacemaker locus for the rhythms in amplitude expressed in two components of the ERG have suggested that the optic lobes also generate the driving oscillation for this rhythm. Rhythms in both ERG components persisted after section of the optic tract proximal to the lobula, neurally isolating the optic lobe/compound eye complex from the rest of the nervous system. On the other hand, the rhythms were abolished by cuts through more distal regions of the optic lobe separating the compound eye from the lobula, the site of the pacemaker for the activity rhythm. These data indicate that the oscillation that drives the rhythm in ERG amplitude is generated near

the lobula and suggest that a single pacemaker locus in the optic lobe is responsible for control of both the ERG and activity rhythms.[12]

2. Cuticle Deposition

In contrast to the results of localization experiments on the activity and ERG amplitude rhythms, studies on the cuticle deposition rhythm indicate that this rhythm is driven by an independent pacemaking system in the adult cockroach. The cuticle of the newly molted adult is secreted rhythmically in layers in both *Leucophaea* and *Blaberus*. The rhythm persists in constant conditions with a circadian period that is temperature compensated.[34] Interestingly, this rhythm's phase and period are not influenced by exposure of animals as adults to a variety of light cycles.[34] The phase appears to be set by the time of the imaginal molt. The results clearly indicate the cuticle rhythm's independence of the entrainable optic-lobe oscillator that drives the activity and ERG amplitude rhythms. Furthermore, in *Blaberus*, the rhythm in cuticle deposition persists after complete ablation of the optic lobes.[38] Thus, this rhythm appears to be driven by an independent circadian pacemaking system, and the results raise the possibility that the epidermal cells which secrete the cuticular material may be autonomously rhythmic.

C. INTEGRATION OF CIRCADIAN INFORMATION

1. Anatomy of the Coupling Pathways

The suggestion that a single pacemaker in the optic lobe drives both the ERG and locomotor activity rhythms and, in addition, sends a coupling signal to the pacemaker in the contralateral optic lobe raises the question of what anatomical pathways and physiological mechanisms are involved in the distribution of the pacemaker's output to these varied structures.

Several studies have been directed toward describing the pathway by which locomotor activity is regulated by the optic lobes. A first step in the pathway is via axons in the optic tracts.[22,27] It has also been reported that lesions of either the pars intercerebralis[39] or the central body[20] or section of the circumesophageal connectives[40] disrupt(s) the activity rhythm. Therefore, these structures also may be involved in coupling the pacemaker to activity. Beyond the suggestion that the optic tract, pars intercerebralis, central body, and circumesophageal connectives are involved in regulation of the activity rhythm, little is known about either the physiology or anatomy of this pathway. The only other available information comes from studies in which animals that were made aperiodic by optic-lobe ablation were found to exhibit a rhythm in activity when subjected to various temperature cycles. The results suggest that the optic-lobe pacemaker may control the activity rhythm via entrainment of a damped oscillator in the midbrain.[41]

The fact that the ERG amplitude rhythm persists after section of the optic tracts, but is lost following bisection of the optic lobe distal to the second optic chiasm,[12] indicates that the pathways that couple the pacemaker to locomotor activity and to the retina must diverge in the optic lobe. Those structures in the optic tract and midbrain that are required for expression of the activity rhythm do not appear to be involved in regulation of the ERG rhythm. In contrast, the maintenance of the ERG rhythm depends on distal optic-lobe structures that are not involved in driving the activity rhythm.[12,30]

The final demonstrated output of each optic-lobe oscillator is to its companion oscillator in the contralateral optic lobe.[28] This pathway also must involve axons in the optic tracts, since section of the tracts uncouples the two oscillators.[33] Interestingly, while the pathway by which the optic lobe regulates activity can regenerate after optic-tract section, the coupling pathway between optic-lobe oscillators does not (see Section III.C.2 below). Thus, the two pathways are functionally distinct.[33]

The results of the experiments summarized in the preceding sections provide a clear picture of the organization of the circadian pacemaking system of the cockroach. The

pacemaker* that regulates the activity and ERG amplitude rhythms is composed of two oscillators, each located ventrally near the lobula of one of the optic lobes of the protocerebrum (Figure 10). Each oscillator has three output pathways that are functionally and anatomically distinct: one controls the rhythm in ERG amplitude, a second regulates activity via a driven system (possibly a damped oscillator) in the midbrain, and a third couples the oscillator to its companion oscillator in the contralateral optic lobe. There are also two input pathways to each oscillator. In addition to the input from the contralateral optic lobe, there is also a light-entrainment pathway from the photoreceptors in the retina of the ipsilateral compound eye (Figure 10). Finally, it appears there is an independent pacemaking system that controls the circadian rhythm in cuticle deposition.

2. Representation and Transmission of Circadian Information

In addition to determining the anatomical organization of the internal coupling pathways of the circadian system, it is important to discover which physiological parameters and processes are involved in the representation and transmission of the circadian information generated by the optic-lobe oscillators. *A priori,* several alternative mechanisms are plausible. Phase information could be represented by the level of a circulating hormone, impulse frequency in specific neural circuits, changes in general levels of neural excitability (perhaps mediated by a neuromodulator), or by some combination of these mechanisms. Furthermore, for any specific rhythm, the output of the pacemaker could regulate the effector via a phase-dependent excitation, inhibition, or both. At present there is very little information that directly bears on this issue in the cockroach, although recently it has been speculated on the basis of electrophysiological data that the function of the oscillator in the optic lobe is to modulate the level of excitability of the CNS globally rather than to activate or inactivate a specific and limited group of cells.[24] It has further been suggested that the regulation of locomotor activity involves, in part, a mechanism by which each individual optic-lobe oscillator inhibits activity from occurring at an inappropriate phase in the circadian cycle.[24,33] This latter conclusion is based on experiments designed to investigate the question of whether or not the pathway that couples the two optic-lobe oscillators is capable of regeneration.

Since the connections between the oscillator in the optic lobe and the midbrain structures that regulate activity can regenerate after optic-tract section or optic-lobe transplantation, an obvious question is whether or not the pathway that couples the two optic-lobe oscillators can also regenerate. The data suggested that

1. Although the output pathway by which each oscillator contributes to the regulation of activity consistently regenerates after section of the optic tracts, the coupling pathway between the bilaterally paired oscillators does not.
2. The oscillator of an intact optic lobe that has not been forced to regenerate its output pathway can suppress the expression (as a component of activity) of its companion oscillator in a contralateral optic lobe in which the optic tract has been cut and regenerated.

One study involved transplantation of the optic lobe.[27,33] In these experiments a single optic lobe was exchanged between animals whose freerunning periods differed by over 1 h. Regeneration of the coupling pathway between the host optic lobe and the transplanted pacemaker would be expected to cause a significant change in the freerunning period of the activity rhythm of one or both of the optic-lobe pacemakers. No significant change in period was detected during several weeks of monitoring after surgery; thus, there was no indication

* It is useful to distinguish between *pacemakers* and *oscillators*. Here, pacemaker refers to the self-sustained oscillatory unit that determines the period of the freerunning rhythm. The pacemaker may itself be composed of more than one oscillator, or it may control effectors by entrainment of secondary oscillators.

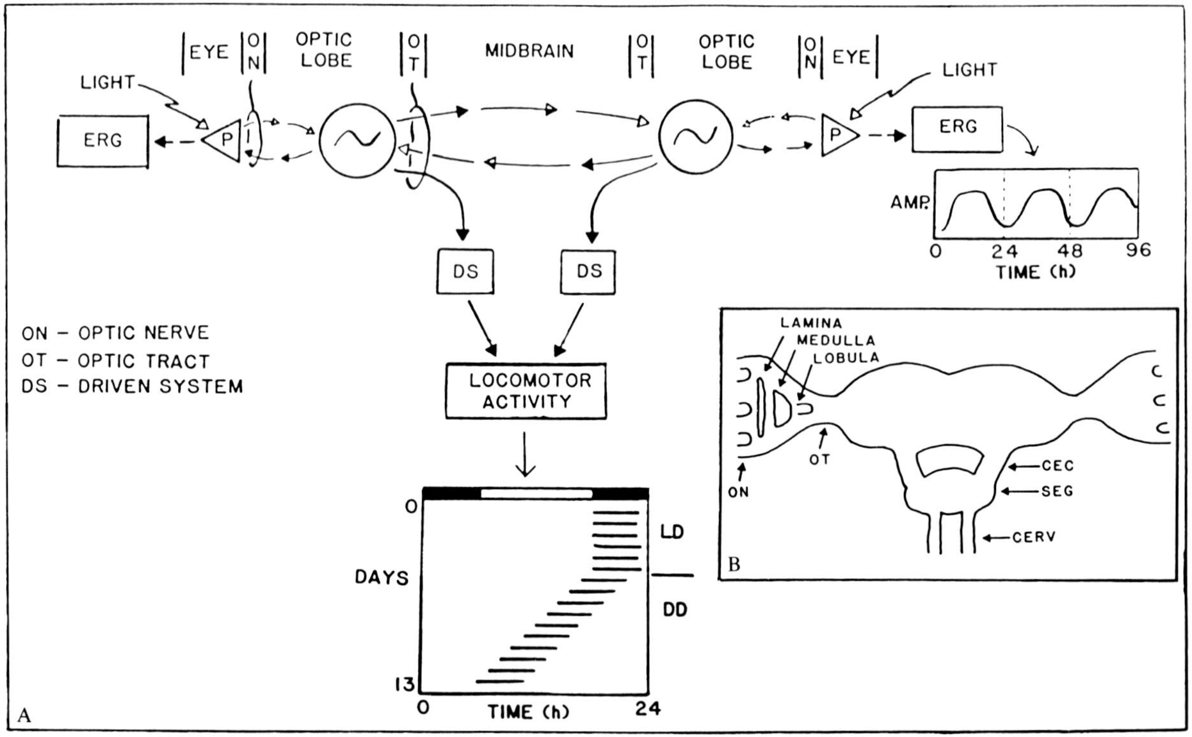

FIGURE 10. (A) Schematic model of the circadian timing system of *Leucophaea maderae*. There are two, bilaterally paired driving oscillators, each located in one of the optic lobes of the protocerebrum. Each has three output pathways; one controls a rhythm in ERG amplitude, a second regulates activity via a driven system in the midbrain (ERG amplitude and activity rhythm records are shown schematically), and a third couples the oscillator to its companion oscillator in the contralateral optic lobe. There are also two input pathways to each oscillator. In addition to the input from the contralateral optic lobe, there is also a light-entrainment pathway from photoreceptors in the retina of the ipsilateral compound eye. (B) Schematic of the anterior of the cockroach CNS, showing midbrain, the three regions of neuropil of the optic lobe (lamina, medulla, lobula), the optic nerve (ON), optic tract (OT), circumesophageal connectives (CEC), subesophageal ganglion (SEG), and the cervical connectives (CERV). (Reprinted with permission from Page, T. L., in *Trends in Chronobiology*, Hekken, W., Kerkut, G. A., and Rietven, W. J., Eds., Copyright 1988, Pergamon Press, Inc.)

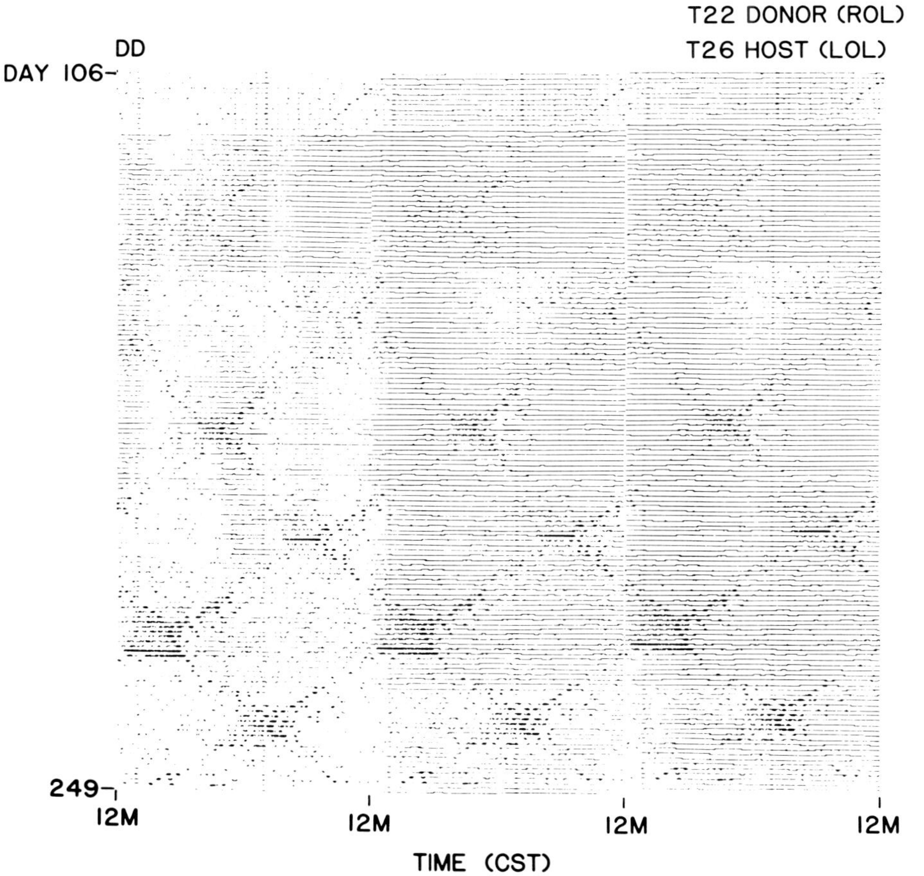

FIGURE 11. Activity record of an animal with a unilateral optic-lobe transplant. The host animal had been raised in LD 13:13 (T26) and the donor in LD 11:11 (T22). The right optic lobe (ROL) of the donor was transplanted. The host retained its left optic lobe (LOL). The record begins 106 d after transplantation and 33 d after section of the host's left optic tract; the record is ''triple plotted'' and shows 72 h across. Two clear components to the activity rhythm appear to be freerunning independently. An increase in activity level is evident when the two components come into phase, and activity is nearly eliminated when the components are in antiphase. (From Page, T. L., *J. Comp. Physiol.*, 152, 231, 1983. With permission.)

of regeneration of a coupling pathway between the host and donor oscillators. In addition, there was no evidence of the appearance of a second freerunning component, which might have been expected if the transplanted lobe's oscillator had regenerated connections with midbrain structures involved in regulating activity.

The absence of a component of activity driven by the regenerated lobe suggested that the intact (host) optic lobe had suppressed the expression of the transplanted oscillator. That regeneration had occurred was demonstrated by section of the optic tract of the host optic lobe. Most animals exhibited a clear circadian rhythm in activity beginning within 3 d of the surgery, with a period near that expected for a rhythm driven by the donor lobe.

In these experiments, there was frequently evidence that the host optic lobe eventually regenerated its connections with the midbrain after its optic tract was cut. This regeneration was reflected in the appearance of two activity components, one driven by the regenerated host oscillator and the other driven by the regenerated donor oscillator (Figure 11). Interestingly, the daily amount of activity was modulated with a periodicity that corresponded

to the beat frequency of the two oscillators. Activity was relatively intense on those days that the two oscillators were in phase, while activity was nearly completely absent on days during which the oscillators were in antiphase. The results were suggestive of a mutual suppression of activity on those days when the oscillators were in an inappropriate phase relationship.

The data from these experiments and others involving optic-tract regeneration[33] suggest that the regulation of activity by the circadian system may involve, in part, a mechanism by which the individual oscillator inhibits activity from occurring at an inappropriate phase in its circadian cycle. The strength of the inhibition from an intact lobe is sufficient to completely suppress any independent expression of activity driven by an optic lobe that has been forced to regenerate its connections. Interestingly, a similar suppression of activity driven from one optic lobe by the other lobe has been observed in crickets following spontaneous desynchronization of the bilaterally paired optic lobe oscillators;[42] hence, this suppression may be a common mechanism of bilateral integration, at least in insects. This is a novel mode of integration in the circadian system and deserves further investigation.

IV. THE COCKROACH: A TIMELY MODEL SYSTEM

The physiological organization of circadian systems that regulate behavioral processes has been investigated in a wide variety of organisms, including several species of vertebrates and invertebrates. Initial questions were primarily ones of localization — could the anatomical correlates of the functionally defined components of the circadian system illustrated in Figure 1 be identified? Substantial progress has been made on this front, and although there are no instances in which pacemakers, photoreceptors for entrainment, or coupling pathways have been linked to specific cells, much evidence indicates that restricted regions of the nervous and neuroendocrine systems can fulfill the functions of the various components. Pacemakers, photoreceptors, and steps in the coupling pathways have been localized in various organisms. However, in the course of these experiments, evidence has surfaced which indicates that the model is incomplete as a representation of the circadian system in at least one significant way. Within the individual there may be several circadian oscillators, and any one overt rhythm may be under the control of a pacemaking system composed of two or more oscillatory structures; furthermore, different overt rhythms may be under the control of separate pacemaking systems. Thus, temporal organization appears to be derived from a population of discrete oscillators that in some cases may be independent or in other instances interact through internal coupling pathways that connect them.

Research on the physiology of circadian rhythmicity in the cockroach has contributed to this current understanding of the organization of circadian systems. The data from the studies on the activity, ERG, and cuticle growth rhythms of cockroaches not only provide one of the most precise localizations (in the optic lobes) of a circadian pacemaker, they also represent one of the clearest demonstrations that circadian systems in multicellular organisms may contain more than one pacemaker, that a single pacemaker can control more than one rhythm, and that pacemaking systems may be comprised of multiple oscillators. Furthermore, it is one of the few systems in which two anatomically localized circadian oscillators have been shown to be coupled.

In addition to the contribution these studies have made to our current understanding of the neural organization of circadian systems in the Metazoa, a number of the observations on the cockroach have helped define and focus attention on several immediate, key questions and have provided a solid foundation on which to base future research efforts. Specifically, the clear anatomical picture of the cockroach circadian system that has been developed has opened the way to a detailed analysis of the mechanisms by which these elements are physiologically linked into an organized regulatory system — e.g., how is the circadian

information that originates in these anatomically discrete oscillators distributed and integrated within the individual to generate a coherent temporal program in its metabolism, physiology, and behavior, and what role does the immediate and developmental environment play in shaping the properties and behavior of the system?

In many respects the cockroach represents a unique model system for an experimental attack on these questions — the wealth of background information on the physiological and anatomical organization of its circadian system, the demonstrated potential of cockroaches for neurophysiological analysis, and the new promise that *in vitro* studies of the pacemaking mechanism in the optic lobe are feasible suggest that the cockroach will continue to serve as an excellent model for studies on the neural mechanisms of the biological clock.

ACKNOWLEDGMENTS

This work was supported by U.S. Public Health Service-NIH Grant GM 30039 and National Science Foundation Grant DCB 87-00555 to the author and Biomedical Research Support Grant 2S07 RR07201 to Vanderbilt University.

REFERENCES

1. **Pittendrigh, C. S. and Bruce, V.,** An oscillator model for biological clocks, in *Rhythmic and Synthetic Processes in Growth,* Rudnik, D., Ed., Princeton University Press, Princeton, NJ, 1957, 75.
2. **Pittendrigh, C. S.,** Circadian organization and the photoperiodic phenomena, in *Biological Clocks and Reproductive Cycles,* Follett, B. and Follett, D., Eds., Wright, Bristol, U.K., 1981, 1.
3. **Page, T. L.,** Clocks and circadian rhythms, in *Comprehensive Insect Physiology, Biochemistry and Pharmacology,* Vol. 6, Kerkut, G. and Gilbert, L., Eds., Pergamon Press, Oxford, 1985, 577.
4. **Pittendrigh, C. S. and Caldarola, P.,** General homeostasis of the frequency of circadian oscillations, *Proc. Natl. Acad. Sci. U.S.A.,* 70, 2697, 1973.
5. **Page, T. L. and Block, G. D.,** Circadian rhythmicity in cockroaches: effects of post-embryonic development and aging, *Physiol. Entomol.,* 5, 271, 1980.
6. **Page, T. L. and Barrett, R. K.,** unpublished data, 1987.
7. **Sutherland, D. J.,** Rhythms, in *The American Cockroach,* Bell, W. and Adiyodi, K., Eds., Chapman and Hall, London, 1982, 247.
8. **Richards, A. G.,** Oxygen consumption of the American cockroach under complete inanition prolonged to death, *Ann. Entomol. Soc. Am.,* 62, 1313, 1969.
9. **Lipton, G. R. and Sutherland, D. J.,** Activity rhythms in the American cockroach, *J. Insect Physiol.,* 16, 1555, 1970.
10. **Beck, S.,** Physiology and ecology of photoperiodism, *Bull. Entomol. Soc. Am.,* 9, 8, 1963.
11. **Lukat, R.,** Circadian growth layers in the cuticle of behaviorally arrhythmic cockroaches, *Experientia,* 34, 477, 1978.
12. **Wills, S. A., Page, T. L., and Colwell, C. S.,** Circadian rhythms in the electroretinogram of the cockroach, *J. Biol. Rhythms,* 1, 25, 1986.
13. **Harker, J.,** Diurnal rhythms in *Periplaneta americana* L., *Nature (London),* 173, 689, 1954.
14. **Harker, J.,** Factors controlling the diurnal rhythm of activity in *Periplaneta americana, J. Exp. Biol.,* 33, 224, 1956.
15. **Brady, J.,** The search for the insect clock, in *Biochronometry,* Menaker, M., Ed., National Academy of Sciences, Washington, D.C., 1971, 517.
16. **Roberts, S.,** Photoreception and entrainment of cockroach activity rhythms, *Science,* 148, 958, 1965.
17. **Nishiitsutsuji-Uwo, J. and Pittendrigh, C. S.,** Central nervous system control of circadian rhythmicity in the cockroach. II. The pathway of light signals that entrain the rhythms, *Z. Vgl. Physiol.,* 58, 1, 1968.
18. **Mote, M. I. and Black, K. R.,** Action spectrum and threshold sensitivity in entrainment of circadian running activity in the cockroach *Periplaneta americana, Photochem. Photobiol.,* 34, 257, 1981.
19. **Nishiitsutsuji-Uwo, J. and Pittendrigh, C. S.,** Central nervous system control of circadian rhythmicity in the cockroach. III. The optic lobes, locus of the driving oscillation, *Z. Vgl. Physiol.,* 58, 14, 1968.
20. **Roberts, S.,** Circadian rhythms in cockroaches: effects of optic lobe lesions, *J. Comp. Physiol.,* 88, 21, 1974.

21. **Lukat, R. and Weber, F.,** The structure of locomotor activity in bilobectomized cockroaches, *Experientia,* 35, 38, 1979.
22. **Page, T. L.,** Regeneration of the optic tracts and pacemaker activity in the cockroach *Leucophaea maderae, J. Comp. Physiol.,* 152, 231, 1983.
23. **Page, T. L.,** A circadian rhythm *in vitro* recorded from the optic lobe of the cockroach, *Neurosci. Abstr.,* 13, 49, 1987.
24. **Page, T. L.,** Circadian organization and the representation of circadian information in the nervous systems of invertebrates, in *Trends in Chronobiology,* Hekkens, W., Kerkhof, G. A., and Reitveld, W. J., Eds., Pergamon Press, Oxford, 1988, 67.
25. **Marks, E. P., Ittycheriah, P. I., and Leloup, A. M.,** The effect of beta-ecdysterone on insect neurosecretion *in vitro, J. Insect Physiol.,* 18, 847, 1972.
26. **Chen, J. S. and Levi-Montalcini, R.,** Axonal outgrowth and cell migration *in vitro* from nervous system of cockroach embryos, *Science,* 166, 631, 1970.
27. **Page, T. L.,** Transplantation of the cockroach circadian pacemaker, *Science,* 216, 73, 1982.
27a. **Lehman, M. N., Silver, R., Gladstone, W., Kahn, R., Gibson, M., and Bittman, E.,** Circadian rhythmicity restored by neural transplant. Immunocytochemical characterization of the graft and its integration with the host brain, *J. Neurosci.,* 7, 1626, 1987.
28. **Page, T. L.,** Effects of localized low-temperature pulses on the cockroach circadian pacemaker, *Am. J. Physiol.,* 240, R144, 1981.
29. **Sokolove, P. G.,** Localization of the cockroach optic lobe circadian pacemaker with microlesions, *Brain Res.,* 87, 13, 1975.
30. **Page, T. L.,** Interactions between bilaterally paired components of the cockroach circadian system, *J. Comp. Physiol.,* 124, 225, 1978.
31. **Page, T. L., Caldarola, P., and Pittendrigh, C. S.,** Mutual entrainment of bilateraly distributed circadian pacemakers, *Proc. Natl. Acad. Sci. U.S.A.,* 74, 1277, 1977.
32. **Zimmerman, E. C. and Page, T. L.,** unpublished data, 1987.
33. **Page, T. L.,** Effects of optic tract regeneration on internal coupling in the circadian system of the cockroach, *J. Comp. Physiol.,* 153, 353, 1983.
34. **Weidenmann, G., Lukat, R., and Weber, F.,** Cyclic layer deposition in the cockroach endocuticle: a circadian rhythm?, *J. Insect Physiol.,* 32, 1019, 1986.
35. **Barlow, R. B., Chamberlain, S. C., and Kass, L.,** Circadian rhythms in retinal function, in *Molecular and Cellular Basis of Visual Acuity,* Hilfer, S. and Sheffield, J., Eds., Springer-Verlag, New York, 1983, 31.
36. **Fleissner, G. and Fleissner, G.,** Neurobiology of a circadian clock in the visual system of scorpions, in *Neurobiology of Arachnids,* Barth, F., Ed., Springer-Verlag, Berlin, 1985, 351.
37. **Colwell, C. S. and Page, T. L.,** unpublished data, 1987.
38. **Weber, F.,** Postmolt cuticle growth in a cockroach: *in vitro* deposition of multilamellate and circadian-like layered endocuticle, *Experientia,* 41, 398, 1985.
39. **Nishiitsutsuji-Uwo, J., Petropulos, S., and Pittendrigh, C. S.,** Central nervous system control of circadian rhythmicity in the cockroach. I. Role of the pars intercerebralis, *Biol. Bull. (Woods Hole, Mass.),* 133, 679, 1967.
40. **Roberts, S., Skopik, S., and Driskell, R.,** Circadian rhythmicity in cockroaches: does brain hormone mediate the locomotor cycle?, in *Biochronometry,* Menaker, M., Ed., National Academy of Sciences, Washington, D.C., 1971, 505.
41. **Page, T. L.,** Circadian organization in cockroaches: effects of temperature cycles on locomotor activity, *J. Insect Physiol.,* 31, 235, 1985.
42. **Weidenmann, G.,** Splitting in a circadian activity rhythm: the expression of bilaterally paired oscillators, *J. Comp. Physiol.,* 150, 51, 1983.

Chapter 25

MECHANORECEPTORS: EXTEROCEPTORS AND PROPRIOCEPTORS

Sasha N. Zill

TABLE OF CONTENTS

I. INTRODUCTION: PROBLEMS IN THE STUDY OF PROPRIOCEPTION

The life of every animal requires the detection and regulation of mechanical forces. These forces can be caused by factors in the environment or may be generated by an animal's own movements and muscular contractions. In these processes, multicellular animals are aided invaluably by information provided by sensory neurons known as mechanoreceptors, which respond specifically to mechanical forces. The responses of mechanoreceptors are often limited to a single modality, such as touch, pressure, or detection of muscle tension, based upon their location in the body and specializations of their receptive endings. Sherrington[1,2] proposed that different types of mechanoreceptors could be grouped according to the source of the forces to which they respond: he termed "exteroceptors" those sense organs that respond mainly to forces generated by the environment and "proprioceptors" those afferents that respond predominantly to forces produced by the animal itself. In mammals, touch receptors in the dermis of the skin are examples of exteroceptors while receptors in muscles joints and tendons are exemplary proprioceptors.

In the study of exteroceptors and proprioceptors contemporary neurobiologists confront two major problems: to understand what information about mechanical forces is encoded by particular sense organs and to determine how these inputs are incorporated into specific behaviors. The resolution of these problems requires specific knowledge in three areas: (1) the anatomical and physiological mechanisms by which mechanical forces are detected and encoded as electrical signals, (2) the ways in which these signals are processed by the central nervous system (CNS), and (3) the effects that sensory inputs have in modifying motor outputs. While much research has been done in these areas in vertebrates, progress has often been hampered by the large numbers of mechanoreceptors present in vertebrate systems.[3,4] There are, for example, thousands of muscle receptors found in a single vertebrate limb.[5] As a consequence, it has not been possible to determine which types of exteroceptors or proprioceptors are specifically responsible for modifications of motor output in vertebrates. In contrast, invertebrates, particularly insects, not only have far fewer receptors, but individual mechanoreceptive sense organs are identifiable in that they are identical from one animal to another.[6] This numerical simplification has greatly aided research into how mechanoreceptors function and influence behavior.

Much progress has been made recently on these fundamental problems using mechanoreceptive systems in cockroaches. In this chapter a brief overview of the types of exteroceptors and proprioceptors found in cockroaches and a summary of their morphology and response properties will first be given as a general introduction to the reader. Next, recent work on the reflex effects of these receptors and their functions in behavior will be reviewed and summarized. Finally, experiments will be discussed which have examined some newly discovered mechanisms by which these afferent inputs and reflex effects can be modulated and processed by the CNS. This chapter is not intended to be comprehensive or to encompass all earlier literature on these subjects; for this the reader is referred to the extensive reviews of Dethier,[7] Wright,[8] and Seelinger and Tobin.[9] Instead, this chapter particularly focuses upon mechanoreceptors found in the legs of cockroaches and attempts to summarize what is known about how these receptors function as a system. Recent findings in the literature that are felt to contribute significantly to our knowledge about mechanoreceptors and the basic mechanisms of sensory-motor integration are also included.

II. MORPHOLOGY AND RESPONSE PROPERTIES OF COCKROACH MECHANORECEPTORS

The bodies and appendages of cockroaches are richly endowed with a variety of mechanoreceptive sense organs. In cockroaches, the main types of mechanoreceptors found on

the leg are isolated hair sensilla, hair plates, campaniform sensilla, chordotonal organs, and multipolar receptors. The responses of mechanoreceptors are generally classified as follows: the response is considered to be *phasic* if the sensory discharge occurs only at the onset of applied mechanical forces or is due to changes in force levels; receptors are considered to show *tonic* responses if their discharge persists throughout the duration of applied force and the sensory frequency encodes the level of force; and responses are *phasicotonic* if a high level of discharge occurs at the onset of force application, followed by adaptation to lower tonic frequencies. Each of these responses can be found in mechanoreceptors of cockroaches.

A. ISOLATED HAIR SENSILLA

Cuticular hairs cover much of the body and most of the appendages of cockroaches.[7] The simplest hairs, known as trichoid sensilla, are found on all leg segments and consist of a rigid cone of cuticle attached to a flexible membrane. Each hair is thought to be innervated by a single bipolar neuron. The dendrite of the sensory neuron is linked to the hair shaft and is distorted by shaft displacement (Figure 1A). Pringle[10] recorded from isolated hairs on the maxillary palps and trochanter and found that all receptors responded phasically to imposed movements of hair shafts. These receptors could therefore signal the onset of hair displacement, as when an appendage touched an external object, but they would not encode maintained contact.

B. HAIR PLATES

Pringle[10] also described groups of cuticular hairs located close to limb joints and termed them "hair plates" (Figure 1B). Hair plates differ from isolated hairs in that they are deflected by changes in joint angles and, thus, they respond to the animal's own postures and movements.

Response properties of the cockroach trochanteral hair plate have been studied extensively by Wong and Pearson.[11] This hair plate consists of approximately 30 hairs that are deflected by flexion of the coxotrochanteral joint. Receptors associated with individual hairs were classified according to their response characteristics: type I receptors respond only phasically to sustained hair displacement, while type II receptors exhibit a phasicotonic response that reflects both the rate and degree of hair bending. Both types of receptors also show responses to hair movements following release of bending, although these discharge frequencies are lower than those occurring during displacement away from the joint. Thus, the trochanteral hair plate clearly can provide information about both the degree of flexion and rate of movement of the coxotrochanteral joint.

C. CAMPANIFORM SENSILLA

Campaniform sensilla are receptors which respond to mechanical forces that produce strains (changes in length) in the exoskeleton.[12] Each sensillum consists of a single bipolar neuron whose cell body is located in the epidermis and whose dendrite inserts into a thin, ovoid cuticular cap at the surface of the exoskeleton[13] (Figure 1C). As elegantly demonstrated by Spinola and Chapman,[14] strains in the exoskeleton produce an indentation of the cuticular cap that deforms the dendrite and evokes a phasicotonic discharge in the sensory neuron. Campaniform sensilla are directionally sensitive and respond best to strains that produce compressions perpendicular to the major (long) axis of the ovoid cuticular cap.[15] Pringle[12] was the first to note that campaniform sensilla are located in discrete groups on the leg of the cockroach: four groups can be found on the trochanter, one on the proximal femur (group 5), one on the proximal tibia (group 6), and one on each of the five tarsal segments. Many of these groups are located near joints and close to muscle insertions; group 5, for example, overlies the insertion of the muscle that causes leg autotomy (active breaking off of a leg). Pringle also noted that the sensilla in a group have consistent orientations of the major axis of their cuticular caps.

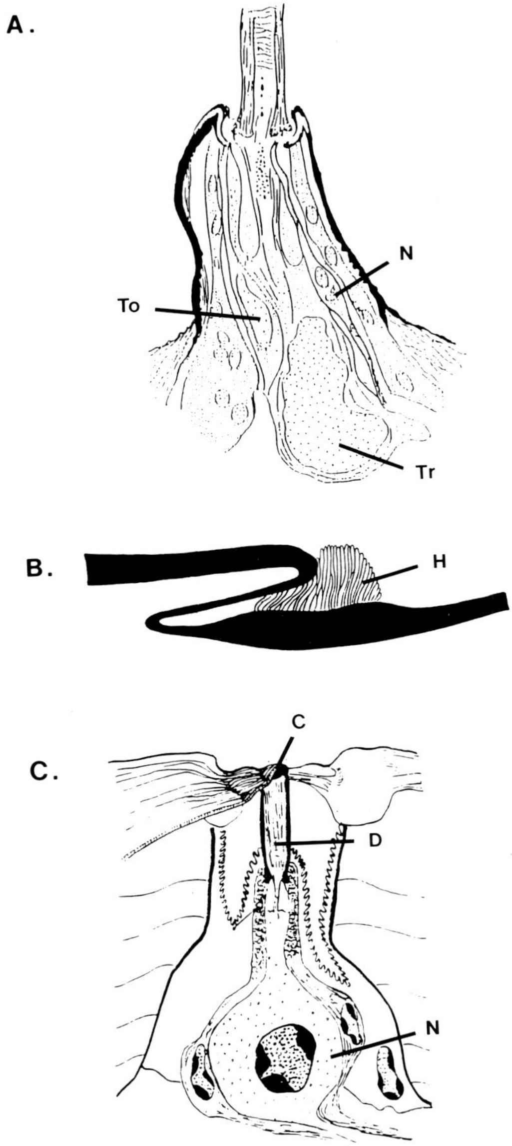

FIGURE 1. Morphology of mechanoreceptors of cockroaches. (A) Isolated hairs. Tactile hairs (trichoid sensilla) are innervated by a single sensory neuron (N), whose dendrite inserts at the base of the hair shaft. Supporting cells (Tr = trichogen cell, To = tormogen cell) are also associated with the receptor. (B) Hair plates. Hair plates (H) are concentrations of hairs located near joints. These hairs are displaced by the joint membrane when one leg segment moves relative to another. (C) Campaniform sensilla.[15] Each campaniform sensillum (C) is innervated by a single sensory neuron (N) whose dendrite (D) is linked to a cuticular cap (C) embedded in the exoskeleton. (A redrawn from Wigglesworth, V. B., *Principles of Insect Physiology,* Neuthen, London, 1965. With permission. B redrawn from Pringle, J. W. S., *J. Exp. Biol.,* 15, 467, 1938; C from Zill, S. N. and Moran, D. T., *J. Exp. Biol.,* 91, 1, 1981. With permission of Company of Biologists Ltd.)

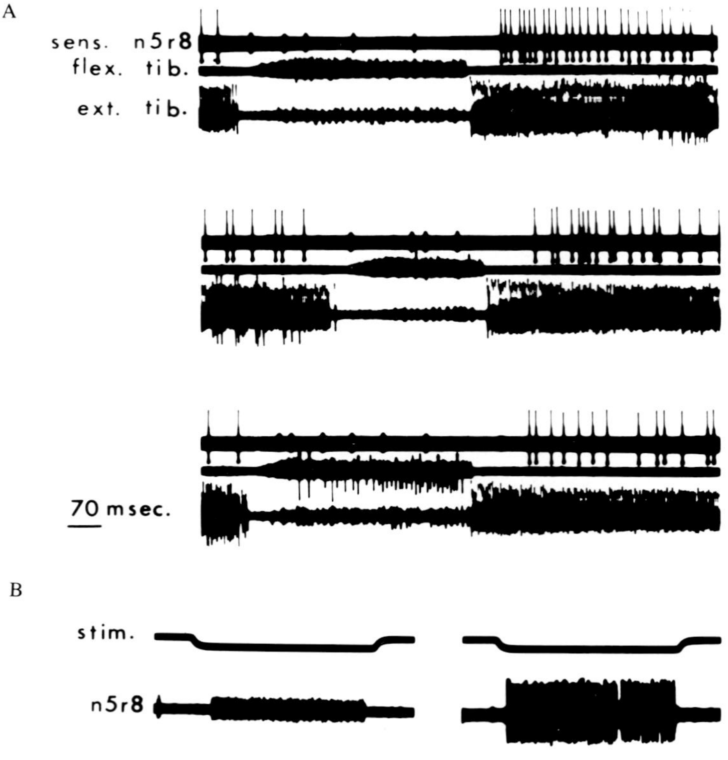

FIGURE 2. Responses of tibial campaniform sensilla to resisted muscle contractions. (A) Rapid alternating bursting was recorded in myograms of the flexor (flex. tib.) and extensor (ext. tib.) tibiae muscles (middle and lower traces). The resulting muscle contractions were resisted by placing a staple over the tibia. These resisted contractions induced alternating bursts of activity in two different campaniform sensilla (three sets of continuous traces). sens. = sensory. (B) Punctate stimulation (stim.) identifies these as proximal (small size action potential at left) and distal campaniform sensilla. These receptors are therefore proprioceptors that respond directionally to cuticular strains resulting from muscle contractions. (From Moran, D. T., Rowley, J. C., III, and Varela, F. G., *Cell Tissue Res.*, 161, 445, 1975. With permission.)

The response properties of the tibial group (group 6) of campaniform sensilla have been characterized in several studies.[14,15] This group is unique in its spatial arrangement in that the sensilla may be divided into two groups according to the orientation of their ovoid cuticular caps: a distal group consists of three to five receptors whose caps have their major axis oriented parallel to the long axis of the tibia, and a proximal group has six to ten receptors with caps oriented perpendicular to the tibial axis. The proximal and distal groups respond differentially to externally imposed mechanical forces and those generated by contractions of leg muscles.[15] Of the various types of externally imposed forces, group 6 campaniform sensilla respond best to bending forces — that is, mechanical forces that act perpendicularly to the long axis of the tibia. Responses to bending are directional: the proximal sensilla respond only to dorsal bending, while the distal sensilla discharge only to ventral bending. Both groups show a maximal response when bending is applied in the plane of joint movement. Strong orientation-dependent responses also occur to resisted contractions of tibial muscles: the proximal sensilla respond only to tibial flexor muscle contractions, while the distal sensilla discharge only in response to strains produced by tibial extensor muscle contractions (Figure 2). Thus, the tibial campaniform sensilla clearly have been

identified as proprioceptive sense organs, since they respond both to external mechanical forces and to strains generated by contraction of the cockroach's own muscles.

Many other groups of campaniform sensilla can also function as proprioceptors, although their response characteristics have not been examined systematically. As noted previously, group 5 is located on the proximal femur and overlies the point of insertion of the muscle that can produce leg autotomy.[12] Group 5 sensilla are positioned so as to be able to monitor the tension in the autotomy muscle by registering the strains its contractions produce in the exoskeleton. However, it is important to note that many campaniform sensilla are also associated with cuticular structures that are clearly exteroceptive. Chapman[16] first demonstrated that the tibial and femoral spines, which are large cuticular cones projecting from the leg, possess campaniform sensilla at their bases. Displacement of the spine in its socket produces a compression and discharge of the campaniform sensillum.

Recently, considerable progress has been made in understanding the specific electrical and ionic events underlying the phasic component of the discharge of the campaniform sensilla associated with the distal femoral spine. French[17] measured the receptor potential (i.e., the change in membrane potential of the receptor) during sinusoidal mechanical stimulation of the femoral spine by recording extracellularly from the afferent axons about 0.5 mm from the sensory neuron. He found that the receptor potential was a linear function of the position of the spine and that the frequency response was flat in a range from 0.1 to 100 Hz. The receptor potential did not exhibit any measure of adaptation, unlike the frequency of action potential discharge.[18] Furthermore, phasic responses were also obtained when the sense organ was stimulated electrically near the point where its axon left the cell body.[19] Both of these experiments indicate that the adaptation shown by the discharge of the campaniform sensillum at the base of the femoral spine occurs at the level of the action potential initiating region.

French[20] has also investigated the ionic mechanisms underlying the response adaptation in order to determine which types of membrane conductances are its cause. Adaptation is not affected by changing calcium concentrations in the bathing medium or by calcium channel blocking agents. Similar experiments ruled out potassium or chloride conductance as a major factor in adaptation. However, French[21] has recently found that oxidizing agents, which in other systems have been shown to affect sodium channel inactivation, alter adaptation in the femoral spine substantially. These oxidizing agents are thought to alter the structure of amino acids in the protein which forms the sodium channel. The effect of such oxidizing agents upon the discharge of the femoral spine is to greatly reduce adaptation to sustained stimulation. These experiments demonstrate that inactivation of sodium channels is an important component in the adaptation of the receptor, a unique new finding in the field of sensory physiology.

D. "CONNECTIVE" CHORDOTONAL ORGANS

Several types of insect mechanoreceptors possess a sensory cell that is anchored inside the body of the animal with a dendrite which terminates in a specialized, rod-like structure known as a scolopale.[7] These scolopidial organs differ in their response modality according to the cuticular structures to which the scolopale is linked mechanically. In auditory organs,[22] the scolopalia attach to an external cuticular tympanic membrane which vibrates and stimulates the sensory neurons in response to sounds. In subgenual organs,[23,24] the scolopidial cells span the tibial segment of the leg transversely and detect external vibrations transmitted through the leg. In "connective" chordotonal organs,[25] the scolopidial cells span joints between adjacent leg segments (Figure 3). Those receptors, therefore, encode joint position and its rate of change. "Connective" chordotonal organs are clearly analogous to vertebrate joint receptors.

Nijenhuis and Dresden[26,27] described seven chordotonal organs in the cockroach leg: one

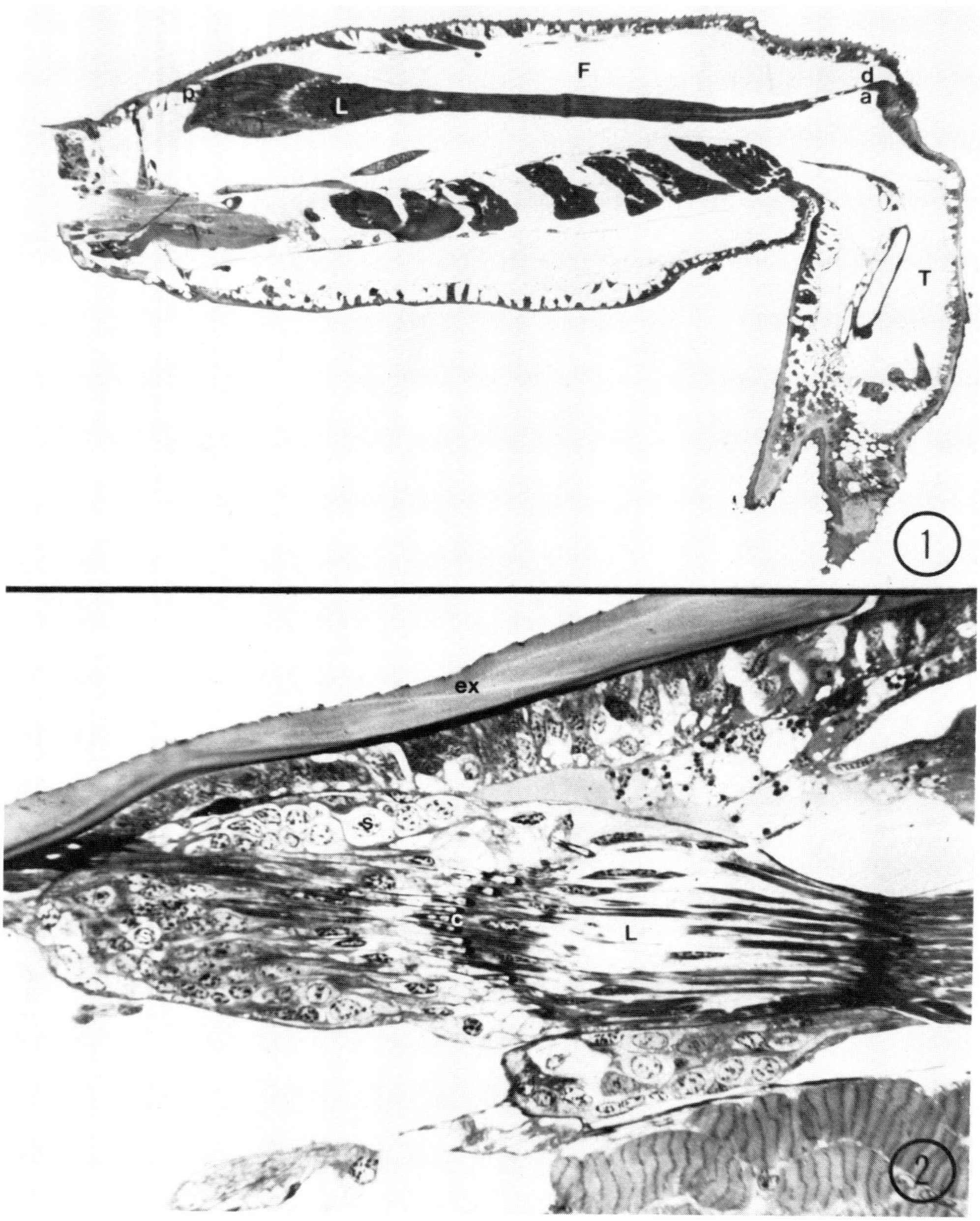

FIGURE 3. Morphology of a connective chordotonal organ. (1) A sagittal section through the femur (F) and tibia (T) demonstrates the ligament (L), apodeme (a), and proximal (p) and distal (d) attachments of the femoral chordotonal organ. This receptor thus spans the femorotibial joint. (Magnification × 194.) (2) Cells of the femoral chordotonal organ at higher magnification. Cell bodies (S) of sensory neurons have dendrites that insert into scolopidial caps (C) which are attached to the ligament (L). (Magnification × 620.) (From Zill, S. N., Moran, D. T., and Varela, F. G., *J. Exp. Biol.*, 94, 43, 1981. With permission of Company of Biologists Ltd.)

coxotrochantinal, three coxal, one femoral, one tibiotarsal, and one tarsal-pretarsal. Young[28] has characterized the morphology and response properties of the chordotonal organ of the tibiotarsal joint extensively. This joint is of interest as it is a ball-and-socket structure with considerable freedom of movement as compared to most other joints of the leg, which allow only flexion/extension movements. The tibiotarsal chordotonal organ consists of 26 neurons

that Young divided into three groups: a proximal group of large cells, an intermediate group with smaller cell bodies, and a distal group of small cells. Units of the organ also may be divided into two types according to their response properties: large fibers (presumably from the proximal group) show phasicotonic discharges and encode both velocity of joint movement and position, while small fibers (from the intermediate group) are tonic and encode only position. The responses of all units are directional, increasing as the tarsus is depressed and moved posteriorly. This response range is of interest from the standpoint of motor control of the joint: the two muscles that act directly upon the tarsus produce tarsal depression when they act in concert. Thus, while the chordotonal organ responds only in a limited range of joint angle, its area of response is matched to the range of movement produced by local muscle activity.

E. MULTIPOLAR RECEPTORS

Guthrie[29] has described several sensory neurons with multiple dendrites that are located near joints of the cockroach leg. Two such receptors are found at the femorotrochanteral joint and three at the femorotibial joint. Recordings from the dorsal sensory nerve (n5r8) in the femur show small units, presumed to be the femorotibial multipolar receptors, that respond phasically to joint movement and tonically encode the joint angle. The femorotibial receptors thus provide information that presumably complements that provided by the femoral chordotonal organ. The functions of the femorotrochanteral receptors, however, are uncertain, as little movement occurs at the femorotrochanteral joint.

III. REFLEX EFFECTS OF COCKROACH MECHANORECEPTORS

The functions of exteroceptive and proprioceptive sense organs in behavior have classically been evaluated according to their reflex effects — that is, the direct effects that stimulation of these sense organs has upon activity in motoneurons. Sherrington[1] first characterized a number of reflexes in vertebrates and termed the reflex "the unit reaction of nervous integration". While a number of reflexes have subsequently been characterized in vertebrates, it has rarely been possible to identify the particular type of sense organ that generates a specific reflex, mostly due to technical constraints.[3,4] In contrast, the accessibility and identifiable nature of cockroach mechanoreceptors have permitted a number of studies of reflexes using precisely controlled stimulation of single sense organs. In this section, a brief description of the motor innervation of those leg muscles in which reflexes have been studied will be given; then studies of sense organs that have been demonstrated to produce exteroceptive and proprioceptive reflexes will be reviewed. Finally, general principles of proprioception will be presented, and a discussion on how particular reflexes can function in behavior will follow.

A. MUSCLES AND MOTONEURONS IN WHICH REFLEXES HAVE BEEN STUDIED

Most studies of reflexes in cockroaches have recorded activities in the proximal segments of the middle and hindlegs. The muscles that have been examined most carefully are the paired antagonist trochanteral extensor and flexor muscles and the homologous tibial extensor and flexor[6] (Figure 4). The extensor muscles of each joint are innervated by a single slow excitatory motoneuron, at least one fast excitor, and a number of peripheral inhibitory neurons.[30,31] Inhibitory neurons produce hyperpolarizations in muscle cells and relaxation of tension.[32] The flexor muscles are both multiply innervated.[32,33] The specific innervation of the trochanteral flexor muscle has been characterized in detail:[32] it receives axons of the common inhibitory neuron (axon 3), at least three slow excitatory neurons (axons 4, 5, and

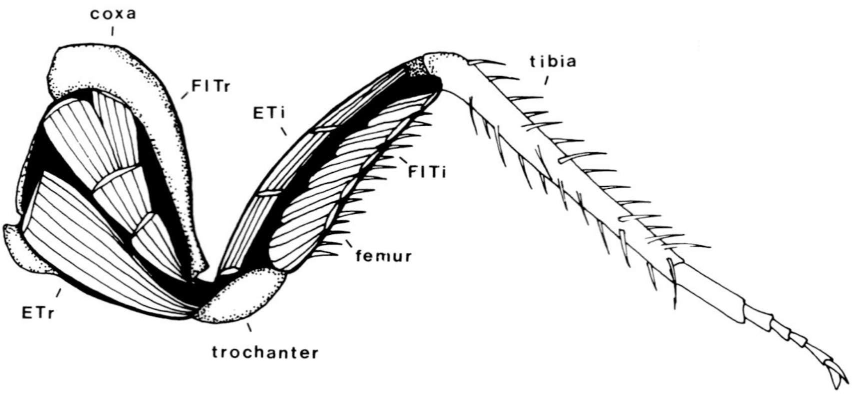

FIGURE 4. Intrinsic muscles of the cockroach leg (see text for description). ETr = trochanteral extensor; FlTr = trochanteral flexor; ETi = tibial extensor; FlTi = tibial flexor.

6), and a number of fast excitatory motoneurons. The activities of these muscles have been recorded myographically in posture and locomotion;[31,34,35] in upright posture, the extensors are active to support the animal. During walking, the extensors fire during the stance phase, when the leg is pressed against the walking surface, to support and propel the animal. The flexors are active during the swing phase to lift the leg and move it forward.

B. EXTEROCEPTIVE REFLEXES

Pringle[36] first described exteroceptive reflexes in cockroaches. He found that touching the surface of the tarsus or applying tactile stimuli to the large spines produced a momentary lifting of the leg. These findings were expanded recently by Meyer and Siegler.[37] They systematically studied the responses of decapitated cockroaches to light localized touching of the leg with a fine paintbrush while monitoring muscle activity myographically. This type of stimulation presumably excited trichoid hairs on the leg. The responses they obtained generally consisted of directional avoidance reflexes which tended to move the leg away from the site of stimulation. For example, touching the dorsal tibia produced tibial flexion, while tactile stimuli applied to the ventral tibia elicited tibial extension. These findings are of interest as they imply that the cockroach nervous system is able to respond selectively to hairs that are at different locations on the leg. A potential anatomical basis for such differential responses has also been demonstrated in the organization of afferents in the main leg nerve:[38] degeneration studies have shown that afferents from the leg are organized systematically in the main leg nerve when it enters the CNS. This organization reflects a general somatotopic map of the different leg segments. Further research is necessary to determine whether this mapping is preserved in the responses of central interneurons that encode sensory inputs.

C. PROPRIOCEPTIVE REFLEXES

1. Hair Plates

Pearson and colleagues[11,39] have demonstrated that displacement of sensilla of the trochanteral hair plate excites the slow trochanteral extensor motoneuron (D_s) and inhibits activity of all trochanteral flexor motoneurons (including the common inhibitory neuron, axon 3). The latency between the time of arrival of the sensory action potential to the CNS and the onset of the excitatory postsynaptic potential in the extensor motoneuron is in the range of 0.9 to 1.6 ms, sufficiently brief for this connection to be monosynaptic.[11] Tests using injected current to change the membrane potential of the motoneuron supported this hypothesis and suggested that this connection was also mediated chemically.[39] Similar mea-

surements of the latency of the inhibitory postsynaptic potential in flexor motoneurons are in the range of 2.5 to 4 ms, and those connections were thought to be polysynaptic and mediated by interneurons. A number of nonspiking interneurons have also been demonstrated to receive short-latency postsynaptic potentials following electrical stimulation of hair plate afferents.[39]

These reflex connections are postulated to function in walking in the following way: as the trochanter is flexed during the swing phase, sensilla of the hair plate are increasingly bent. Bending should produce an afferent discharge that would inhibit the flexor burst reflexively and help to terminate the swing phase of walking. The discharge of the hairs should also aid in the initiation of the extensor burst in the succeeding stance phase. These postulated functions are supported by ablation studies:[11] removal of the hair plate produces overstepping during walking, and the operated leg often collides with the leg anterior to it as the result of exaggerated leg flexions. Another postulated function of the hair plate, however, is to be of potential aid in load compensation. Unexpected shifts in body position that produce flexions at the coxotrochanteral joint should excite the trochanteral extensor motoneuron reflexively. This hypothesis has not been examined in behavioral studies.

2. Campaniform Sensilla

The reflex effects of both the tibial and trochanteral campaniform sensilla have been characterized.[33,35] Mechanical stimulation of the cuticular caps of individual tibial campaniform sensilla with a fine-etched tungsten wire produces reflexes in motoneurons to leg muscles that depend upon the cap orientation.[33] Stimulation of the proximal sensilla produces short-latency excitation of the trochanteral and tibial extensor motoneurons and longer-latency inhibition of trochanteral and tibial flexor motoneurons. Stimulation of the distal sensilla has the opposite effects, inhibiting the extensors and exciting the flexors at both joints. However, among the trochanteral flexor motoneurons only one is excited reflexively (axon 4).[40] These reflex connections clearly form a negative-feedback system, in that individual sensilla are excited by strains resulting from muscle contractions generated by motoneurons that the sensilla inhibit reflexively. This system could readily function to limit the amplitude of muscle contractions in posture and locomotion. Tibial campaniform sensilla also can aid in load compensation, as these receptors respond to external bending forces such as those that occur when the animal places its weight upon the leg. Recordings of the activities of tibial campaniform sensilla support both hypotheses about their functions: the proximal sensilla fire early in the stance phase, when the animal places its weight upon the leg, while the distal sensilla fire late in stance, when the firing rate of the tibial extensor motoneurons is highest.[34] Thus, the tibial campaniform sensilla apparently function both in load compensation and in the limitation of muscle tension.

Reflex effects have also been produced by mechanical stimulation of the trochanteral campaniform sensilla. Pearson[35] produced short-latency excitation in the D_s by applying pressure to the trochanter. While this method of stimulation is quite nonspecific and presumably excites a number of groups of campaniform sensilla, similar effects can be produced by stimulation of individual receptors.[68] Pearson postulated that this reflex connection generates positive feedback, as these receptors should be maximally excited when the leg is pressed upon the walking surface by the trochanteral extensor muscle. While positive feedback could occur, it should be noted that the actual response characteristics of the trochanteral campaniform sensilla have not yet been determined.

3. Chordotonal Organs

Rijlant[41] first demonstrated that imposed movement of the joints of insect legs produces reflex discharges in motoneurons to leg muscles to resist limb displacement. Wilson[42] examined such responses in the cockroach leg. He recorded reflexes in the slow and fast tibial

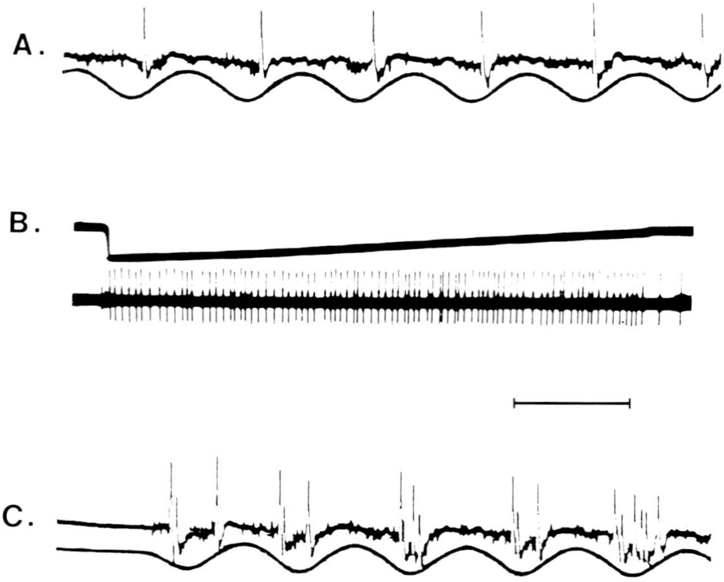

FIGURE 5. Reflex responses to joint movement and chordotonal organ stimulation. (A) Resistance reflex responses. Activity of the slow tibial extensor motoneuron (upper trace) was recorded while the femorotibial joint was moved sinusoidally. The motoneuron fired to resist joint flexion (up on lower trace). (Calibration = 125 ms.) (B) Response to stimulation of the femoral chordotonal organ. Firing in the slow tibial extensor (lower trace) was recorded during a ramp release and stretch of the ligament of the femoral chordotonal organ, mimicking joint flexion. The motoneuron again was active to resist the apparent joint movement. (Calibration = 2.2 s.) (C) Reflex reversal. Experimental setup and traces as in A. During the first series of cycles of sinusoidal movement of the tibia, the extensor motoneuron fired during both joint flexion and extension. (Calibration = 125 ms.) (A and C from Wilson, D. M., *J. Exp. Biol.*, 43, 397, 1965. With permission of Company of Biologists Ltd.)

extensor motoneurons elicited by sinusoidal movements of the leg (Figure 5A). These reflex discharges followed leg movements at frequencies of >20 c/s and, thus, could be active at the highest reported frequencies of leg movement during running. Wilson also reported some "anomalous" responses in which the phase of the reflex response shifted and motoneurons fired at two different times during a single cycle of movement[42] (Figure 5C). This finding was the earliest report of a reflex "reversal" in insects, but its function remains unclear. Subsequent experiments by Zill and Moran[40] demonstrated that resistance responses could also be elicited in the trochanteral flexor muscle to impose movements of the tibia. However, only one flexor motoneuron is excited regularly (axon 4). Experiments by Brodfuehrer and Fourtner[43] demonstrated that similar resistance reflexes also could be elicited by direct mechanical stimulation of the coxal and femoral chordotonal organs (Figure 5B). These resistance reflexes could function in load compensation variations if leg loading or environmental perturbations produced changes in joint angle.

4. Multipolar Receptors

Guthrie[29] demonstrated that multipolar receptors located in the cockroach limb could also generate resistance reflexes. He attached the apodeme of the tibial flexor muscle to a force transducer after severing the ligament of the femoral chordotonal organ, eliminating chordotonal input. Imposed extensions of the femorotibial joint still produced flexor muscle contractions. Thus, discharges of multipolar receptors have also been demonstrated to be potential mediators of load-compensatory reactions.

Sense Organs that Monitor Joint Angles and Movements

adequate stimuli	receptor	reflexes	functions
ct joint			
flexion	TrHP	+ → ETr – → FTr	limit protraction, initiate retraction, load compensation
extension	3 CxCO's	+ → FTr	load compensation
ft joint			
flexion	FCO	+ → ETi – → FTi	load compensation
extension	MR	+ → FTi	load compensation

Receptors that Monitor Muscle Tensions and Leg Load

adequate stimuli	receptor	reflexes	functions
ct joint			
extension?	TrCS	+ → ETi	increase force in stance?
ft joint			
extension, dorsal bend	ProxTiCS	+ → ETi + ETr – → FTi + FTr	decrease flexor muscle contractions, increase extensor activity during stance
flexion, ventral bend	DistTiCS	– → ETi + ETr + → FTi + FTr	decrease extensor muscle contractions, terminate stance, initiate swing

FIGURE 6. Summary of adequate stimuli, demonstrated reflex effects, and probable functions in behavior of cockroach proprioceptors (see text for discussion). Abbreviations: ct = coxotrochanteral; ft = femorotibial; TrHP = trochanteral hair plate; CxCO = coxal chorodotonal organ; FCO = femoral chordotonal organ; MR = multipolar receptor; TrCS = trochanteral campaniform sensilla; ProxTiCS and DistTiCS = proximal and distal tibial campaniform sensilla; ETr and FTr = extensor and flexor trochanteris muscles; ETi and FTi = extensor and flexor tibialis muscles; + = excitation; – = inhibition.

IV. FUNCTIONS OF PROPRIOCEPTIVE INPUTS IN POSTURE AND LOCOMOTION

Figure 6 summarizes the adequate stimuli, reflex effects, and potential functions that have been demonstrated for identified proprioceptive sense organs of the cockroach leg. There are a number of aspects of these reflexes that warrant useful consideration and comparison with other proprioceptive systems.

A. PROPRIOCEPTIVE SENSE ORGANS AT ALL JOINTS MONITOR JOINT ANGLES AND CUTICULAR STRAINS RESULTING FROM MUSCLE FORCES

The trochanteral hair plate and the three chordotonal organs monitor the angle of the

coxotrochanteral joint, while one chordotonal organ and three multipolar receptors respond similarly at the femorotibial joint. It is of interest to note that there are a greater number of neurons in the sense organs monitoring the angle of the more proximal coxotrochanteral joint.[26,27] This may reflect the fact that changes in the angle of the proximal joint produce larger changes in the actual position of the leg.

There are also groups of campaniform sensilla associated with both joints.[12] While the tibial campaniform sensilla have been shown to respond to resisted muscle contractions and externally applied forces, the response characteristics of the trochanteral groups have not been established. However, if these groups respond in a fashion similar to the tibial group, at least some sensilla should be excited by resisted leg extension, such as occurs when the leg is pressed against the walking surface. The overall arrangement of these sense organs implies that monitoring of both joint position and muscle forces at each joint is essential for motor control in the cockroach leg, despite its simplicity in comparison to vertebrate systems.

B. RECEPTORS THAT MONITOR CHANGES IN JOINT ANGLE MEDIATE LOAD-COMPENSATORY REFLEXES

Both the coxal and femoral chordotonal organs and the femorotibial multipolar receptors detect changes in joint angle and produce reflexes that excite motoneurons to muscles that oppose the apparent movement. These reflexes are similar to resistance responses that have been demonstrated in a number of arthropods,[43-45] and they could function readily in load compensation when changes in the environment produce changes in joint angle. While the actual changes in joint angles have not been measured as yet during load-compensatory responses in cockroaches, swaying movements around joints commonly accompany such reactions in vertebrates.[46] In humans, for example, sudden displacements of the substrate upon which a subject is standing produce changes in the angles of the ankle and hip joints.[46] However, the specific sense organs mediating these responses have not been identified. Sense organs of several modalities, including the visual and vestibular systems as well as muscle spindles, are thought to mediate load-compensatory responses. A similar role for vertebrate joint receptors,[47] the sense organs most similar in function to cockroach chordotonal receptors, has neither been established nor denied.

C. RECEPTORS THAT MONITOR FORCES REFLEXIVELY MODULATE MUSCLE TENSIONS

The tibial campaniform sensilla have been shown to respond to external and muscle-generated forces that produce strains in the exoskeleton.[15] These receptors also reflexively excite motoneurons to limit bending of the leg, and they inhibit motoneurons to prevent excessive muscle contractions.[33] Receptors that respond to forces and that have similar reflex effects have been found in vertebrates and in other invertebrates. In crabs, force-sensitive mechanoreceptors[48,49] respond to muscle-generated and externally imposed strains in the exoskeleton. In vertebrates, Golgi tendon organs[50] located in muscle tendons have also been shown to directly monitor the magnitude of forces produced by muscle contractions. In each of these systems, force-sensitive proprioceptors produce reflex effects that limit the magnitude of muscle contractions. The activities of all three types of receptors have also been recorded in freely walking animals;[34,49,51] each type of receptor is maximally active during the stance phase. Thus, in many systems, receptors that monitor muscle forces are active in posture and those phases of locomotion when loading and muscle contractions are greatest.

D. MOST PROPRIOCEPTIVE REFLEXES MUST BE MODULATED AND ADAPTED BY THE CENTRAL NERVOUS SYSTEM

While the reflex effects of proprioceptors of the cockroach leg provide useful insights into their functions in behavior, there are a number of factors which strongly suggest that

these reflexes are not constant and that they can be changed and adapted by the CNS. First, all receptors that monitor changes in joint angles produce reflexes to oppose those displacements.[29,42] However, it is apparent that these reflexes do not occur in response to limb movements initiated by the animal itself. If that were the case, any leg movement by the cockroach would invariably generate reflexes in antagonist muscles and all movements would be uncoordinated. Second, the reflex effects of some receptors have been shown to occur as positive-feedback loops. The femoral chordotonal organ[42] and possibly the trochanteral campaniform sensilla[35] can excite the extensor muscle reflexively, which in turn increases afferent inputs from those receptors. If these reflexes were invariably active, the leg would be caught in a self-perpetuating cycle that would lead to complete leg extension every time weight was placed on the leg or extension movements were made. Finally, studies of the activities of the tibial campaniform sensilla during rapid running[6,34] have shown that the phase of sensory activity can shift relative to motor output.[52,53] This implies that the demonstrated reflex effects of these receptors could produce inappropriate motor activities during rapid locomotion if they were still active.[34] All of these findings imply that there are mechanisms within the nervous system to modulate proprioceptive reflexes.

Similar malleability of reflexes has been demonstrated in a number of systems. Many reflexes show changes in gain[54] (intensity of motor output for a given sensory input) depending upon the state or phase[55] of behavior of an animal. Reflexes of both vertebrates[56] and invertebrates[57] can be changed by training or instruction. In addition, a number of proprioceptive reflexes can show complete changes in "sign",[43,58] i.e., the specific populations of motoneurons excited. In most cases, the mechanisms underlying these changes in reflexes or their functions in behavior are unknown.

Modulation of neuronal activity could occur readily at a number of levels in a reflex arc. Reflex could be changed by (1) direct efferent control of sensory responsiveness, as occurs in the vertebrate muscle spindle;[4] (2) neurohormonal control of afferent sensitivity; (3) presynaptic inhibition of sensory terminals in the CNS;[59] (4) changes in the activities of interneurons mediating proprioceptive reflexes;[60] (5) changes in the baseline levels of activity of motoneurons; and (6) modulation of reflexively evoked neuronal activity at the neuromuscular junction.[40] A major problem remaining in the study of proprioception is to determine at which of these levels modulation of reflexes occurs.

V. RECENT DEVELOPMENTS IN RESEARCH ON MODULATION AND PROCESSING OF PROPRIOCEPTIVE INPUTS AND REFLEXES

In addition to eliciting reflexes which can be modulated, proprioceptive inputs may also be used as cues to indicate how the legs are being used in behavior. Several recent studies have examined both of these problems in the cockroach. The first papers[40,61] to be reviewed have demonstrated novel mechanisms by which afferent transmission and reflex effects can be modulated. The last paper[62] provides insight into how proprioceptive inputs are processed by the CNS.

A. NONSYNAPTIC REGULATION OF CERCAL AFFERENTS

Libersat et al.[61] have recently demonstrated a new mechanism for regulation of self-generated afferent inputs in the cercal system of the cockroach. The cerci are abdominal appendages that bear wind-sensitive, filiform hairs which are excited by air movements generated by approaching predators.[63] These afferents, in turn, synapse upon giant interneurons whose axons ascend the ventral nerve cord[64] and excite other interneurons that mediate escape turning and running. However, cockroaches also fly and produce considerable air turbulence during flight. If these self-generated air currents excited the cercal afferents and giant fibers, they could produce inappropriate escape movements of the legs at the onset

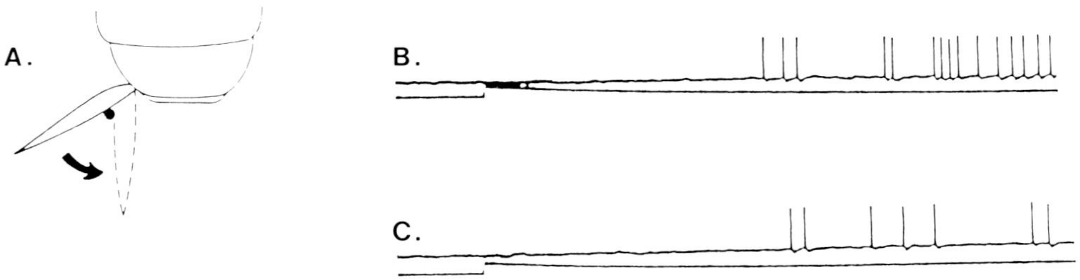

FIGURE 7. Regulation of inputs from cercal afferents when the cercus is in the flight position. In normal posture and walking the cerci (abdominal appendages) are held laterally (solid line in A). These appendages bear wind-sensitive hairs. Single hairs show bursts of action potentials (intracellular recording on upper trace) to puffs of air (displacement on lower trace) when the cerci are in the lateral position (B). The cercus is held in a medial position during flight (dashed line in A). When the cercus is held in the medial position mechanically by a pin the response of the hair is reduced (C). This effect is probably due to a partial pressure block on the cercal nerve that prevents self-generated air currents from exciting the giant fibers during flight. Duration of each set of traces = 0.21 s. (From Libersat, F., Goldstein, R. S., and Camhi, J. M., *Proc. Natl. Acad. Sci. U.S.A.*, 84, 8150, 1987. With permission.)

of flight. Furthermore, continued excitation during flight would produce habituation of the escape system, making the animal vulnerable to prey when it landed.

In order to examine how these potential problems are resolved, Libersat et al.[61,65] recorded wind puff-evoked sensory discharges in the medial cercal nerve. They noted that when a cockroach begins flying, the cerci are moved medially by 45 to 60° and are held in that position for the duration of flight (Figure 7A). In order to study the effect of cercal displacement, cerci in a restrained animal were moved and held in the medial position with a pin. The recorded wind puff-evoked sensory discharge in the medial cercal nerve was then decreased by a mean of 34%, on both extra- and intracellular recordings (Figures 7B and C). A number of tests were applied to establish the source of this reduction in afferent activity. Cutting the nerves near the ganglion did not affect the reduction, eliminating the possibility of active inhibition by the CNS. The reduction also persisted when the preparation was bathed in zero-calcium saline, blocking synaptic activity. The decrease in afferent activity was also not attributable to a change in the angle at which the wind was delivered. These investigators concluded that movement of the cercus to the medial position actually produces mechanical pressure on the cercal nerve, effectively blocking conduction of action potentials in sensory axons. This finding is supported by micrographs of the cerci that show that the cercal nerve touches the cuticle in the medial position. This pressure block thus reduces the amount of self-generated sensory activity that reaches the CNS and helps to prevent giant fiber activity during flight.

B. REGULATION OF REFLEXIVELY GENERATED MUSCLE TENSIONS

Modulation of reflexes by interaction of excitatory motoneurons and peripheral inhibitory neurons has been demonstrated in the cockroach trochanteral flexor muscle. As noted previously in Section III.A, the neurons that innervate this muscle can be identified individually according to the height of their action potential in extracellular recordings from the flexor nerve.[32] Studies using stimulation of the tibial campaniform sensilla[33] (Figure 8B) and the coxal chordotonal organs[43] (Figure 8A) have shown that only one single slow excitatory motoneuron (axon 4) to this muscle is activated reflexively at moderate rates of stimulation. These findings are of interest for several reasons. First, Pearson and Fourtner[60] have shown that nonspiking premotor interneurons that excite flexor motoneurons affect only three of these axons: excitatory neurons 5 and 6 and the common inhibitory neuron, axon 3. Activity in axon 4 is not affected by these interneurons. Second, in their study of the distribution of

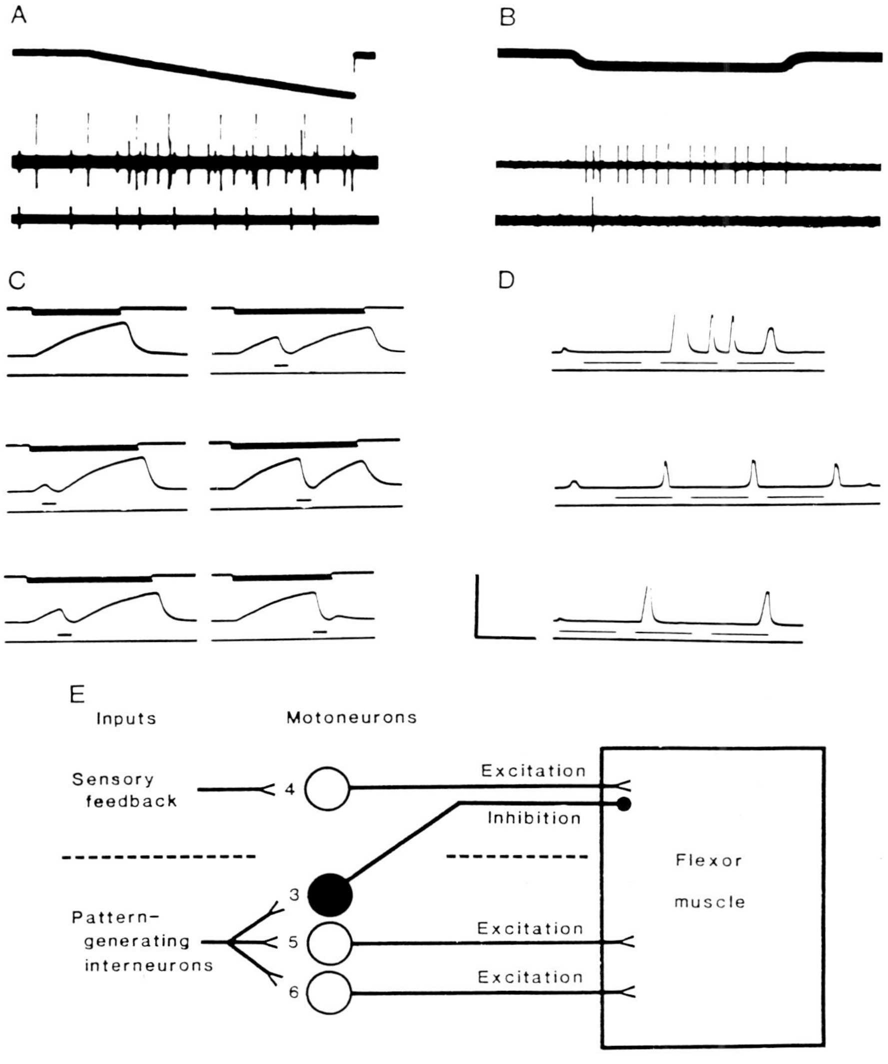

FIGURE 8. Specific suppression of reflexively generated muscle tensions by a peripheral inhibitory neuron. (A) Resistance reflex responses in the trochanteral flexor muscle. Extension of the coxotrochanteral joint at a moderate rate (upper trace) reflexively excites one axon (axon 4) in the flexor nerve (middle trace). This is not the common inhibitory neuron, which innervates the flexor and other muscles, whose activity is also monitored in the lower trace and shows 1:1 spiking. (B) Reflex effects of campaniform sensilla. Punctate stimulation of the cap of a single distal tibial campaniform sensillum (upper trace) excites only the same neuron in the flexor nerve (axon 4, middle trace) and not the common inhibitor (lower trace). (C) Suppression of reflex tensions by inhibitory neuron stimulation. Repetitive extension of the coxotrochanteral joint (thick bars in upper trace) causes large tensions in the flexor muscle (middle trace) as resistance reflex responses. Stimulation of the common inhibitory neuron (bars over lower trace) rapidly suppresses these tensions at all levels. (D) Stimulation of the common inhibitor does not affect many spontaneous tensions. Spontaneous movements that were monitored as tensions developed in the flexor muscle (upper trace) apparently were not affected by stimulation of the inhibitor. (E) Model of interactions in the flexor muscle. The common inhibitor, which is driven in concert with some flexor excitatory neurons, is specifically matched and suppresses tensions produced by activity in the flexor motoneuron that is activated by leg proprioceptors. Calibration: vertical C,D = 3.0 g; horizontal A,B = 150 ms; C = 4.5 s; D = 1.2 s. (From Zill, S. N. and Moran, D. T., *Science*, 216, 751, 1982. Copyright 1982 by the AAAS. With permission.)

axons in flexor muscle cells, Pearson and Bergman[32] observed that excitatory axon 4 and inhibitory axon 3 invariably accompany each other when they innervate flexor muscle cells. The other flexor excitatory axons innervate many muscle cells unaccompanied by the inhibitor. This compartmentalization of innervation suggests that the peripheral inhibitory neuron is matched with axon 4 and potentially can specifically modulate tension developed by proprioceptive reflexes. To test this hypothesis, the trochanter was extended repeatedly, exciting the coxal chordotonal organs, while tensions were monitored in the trochanteral flexor muscle.[40] Tensions that developed from resistance reflexes could be modulated or completely inhibited by concurrent stimulation of the common inhibitory neuron[40] (Figure 8C). In contrast, spontaneous, ballistic muscle contractions were unaffected by the inhibitor (Figure 8D). This peripheral mechanism could serve to specifically modulate or override reflexively developed muscle tensions (Figure 8E). It could also eliminate tensions from postural reflexes prior to rapid movements, such as in escape running.

C. PROPRIOCEPTIVE INPUTS TO INTERNEURONS MEDIATING ESCAPE REACTIONS

Little is known about how the cockroach CNS processes proprioceptive information and incorporates it into behaviors that are more complex than reflex reactions. However, afferent inputs into second-order interneurons that mediate escape reactions have been characterized recently.[62] These interneurons are defined as part of the escape system because they receive monosynaptic inputs from the abdominal giant fibers.[65] As is known from behavioral studies, cockroaches react to strong wind puffs to the cerci by initially turning away from the source of air turbulence and then running rapidly.[66] These reactions pose problems to the CNS. Prior to escape, any individual leg may have been used in postural support or have engaged in other behaviors such as grooming or the swing phase of walking. To execute successful escape reactions, cockroaches must initiate coordinated turns rapidly while also maintaining postural stability. It would therefore seem essential that the interneurons generating these movements should receive sensory information about the particular state and use of the leg at the onset of escape. Murrain and Ritzmann[62] have recently examined this possibility by stimulating the trochanteral campaniform sensilla, trochanteral hair plate, and femoral chordotonal organ while recording intracellularly from escape-mediating interneurons. Their results have demonstrated that all interneurons which receive input from the giant fibers also receive synaptic input from proprioceptive sense organs of the leg. In one interneuron, the lambda cell[67] (Figure 9), proprioceptors give a particular behavioral bias to the cell. The lambda cell receives strong depolarizing input from the trochanteral campaniform sensilla of the ipsilateral leg (Figure 9B1) and the trochanteral hair plate of the contralateral leg (Figure 9A2). This interneuron therefore would be maximally depolarized during walking when the ipsilateral leg was in stance and the contralateral leg was in the swing phase, as normally occurs in the alternating tripod gait.[52] Excitatory inputs from the giant fibers during this specific phase of walking could excite the lambda cell preferentially relative to other interneurons that receive different sensory inputs. While considerable work is needed to clarify the escape circuitry, these initial findings provide a unique insight into how proprioceptive inputs may function in complex, coordinated behaviors.

VI. COCKROACHES AS MODELS FOR RESEARCH IN PROPRIOCEPTION

In conclusion, identified proprioceptive sense organs of cockroaches offer considerable promise as model systems for understanding basic problems in sensory-motor integration. This review has attempted to demonstrate that research in problems of mechanoreception in cockroaches offers the following advantages:

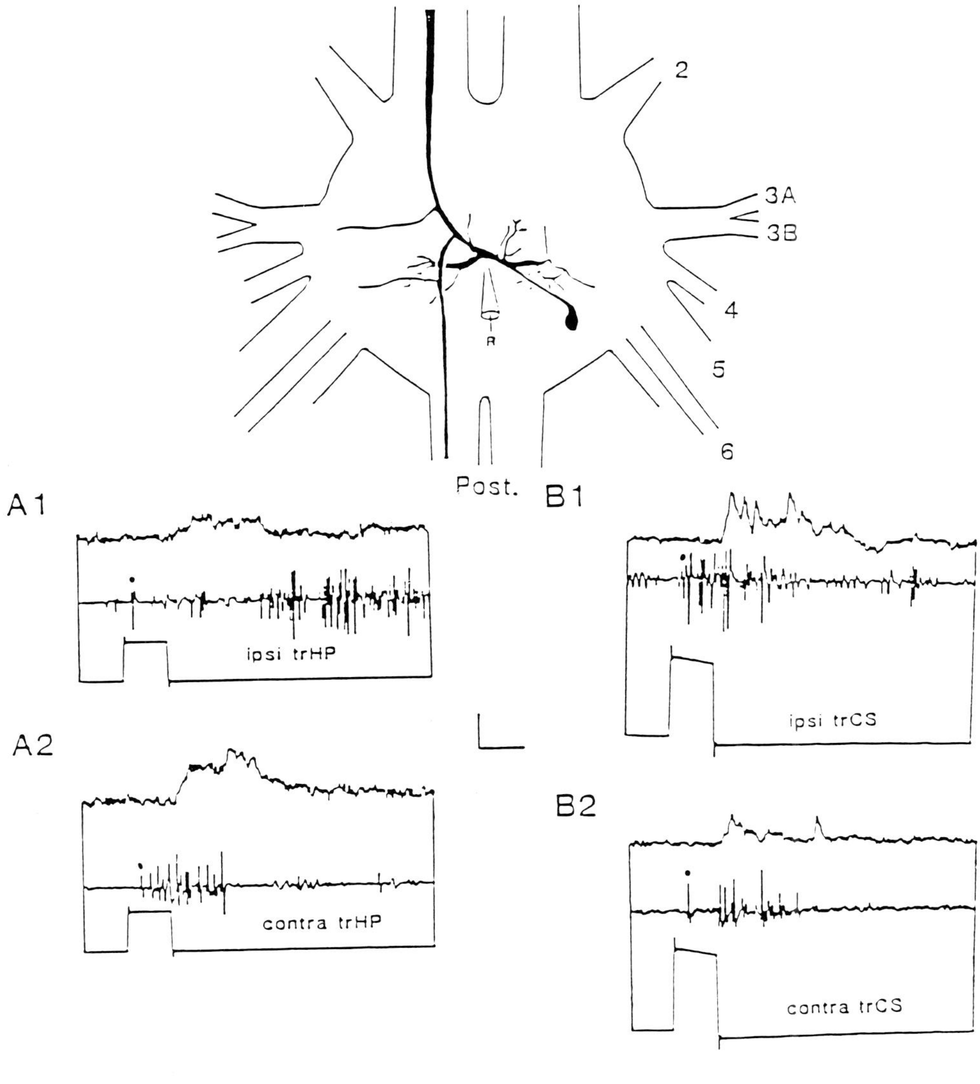

FIGURE 9. Differential responses of an escape-mediating interneuron to leg proprioceptive sense organs. The lambda cell is an interneuron that receives monosynaptic inputs from the abdominal giant fibers and presumably mediates escape reactions (not shown). Its morphology, as demonstrated by intracellular dye filling, is shown in the upper part of the figure. The lower part of this figure demonstrates the responses of the lambda cell (intracellular recording in upper trace) to stimulation of various leg sense organs (stimuli indicated in lower trace). Sensory activity is recorded in the middle trace. Responses are greatest to stimulation of the trochanteral campaniform sensilla (trCS) of the ipsilateral leg (B1) and the trochanteral hair plate (trHP) of the contralateral leg (A2). Responses are smaller to stimulation of the ipsilateral hair plate (B2) and the contralateral campaniform sensilla (A1). As the insect walks, this interneuron is depolarized most strongly when the ipsilateral leg is in stance and the contralateral leg in swing. Calibration: horizontal = 10 ms; vertical, A = 5 mV and B = 4 mV. (From Murrain, M. P., Ph.D. thesis, Case Western Reserve University, Cleveland, OH, 1987. With permission.)

1. Individual sense organs are often identifiable, and their exact morphology and response properties may be characterized reliably in different animals.
2. The motor innervation of most muscles is sufficiently simplified numerically so as to permit extensive characterization of the reflex effects of individual types of mechanoreceptors.

3. Adequate information can be compiled to allow for the consideration of mechanoreceptive inputs from a single appendage as a complete system, functioning in a sensory-motor integration in a variety of behaviors.

It is perhaps surprising that mechanoreceptive sense organs which monitor joint angles and muscles forces and elicit reflexes in leg muscles can be found in animals as divergent as arthropods and mammals. However, it is important to note that the basic mechanisms of sensory responsiveness and motor output are quite similar in both groups. Also, as more fundamentally noted by Dethier,[7] "the gross characteristics of the planet earth are the same for all living things inhabiting it." In resolving common problems such as the adaptation and coordination of posture and locomotion, many animals may utilize the information provided by exteroceptive and proprioceptive sense organs to similar advantage.

ACKNOWLEDGMENTS

The author wishes to thank Andrew French, Jeff Camhi, Frederic Libersat, Michelle Murrain, Roy Ritzmann, and Melody Siegler for providing preprints of manuscripts. This work was supported by NIH NINCDS grant NS22682 and by a grant from the Whitehall Foundation.

REFERENCES

1. **Sherrington, C.,** *The Integrative Action of the Nervous System,* Charles Scribner's Sons, New York, 1906.
2. **Sherrington, C. S.,** On the proprioceptive system, especially in its reflex aspect, *Brain,* 29, 467, 1906.
3. **McCloskey, D. I.,** Kinesthetic sensibility, *Physiol. Rev.,* 58, 763, 1978.
4. **Loeb, G. E.,** The control and response of muscle spindles during normally executed motor tasks, *Exercise Sport Sci. Rev.,* 12, 197, 1984.
5. **Mountcastle, V. B.,** *Medical Physiology,* C. V. Mosby, St. Louis, 1974.
6. **Zill, S. N.,** Proprioceptive feedback and the control of cockroach walking, in *Feedback and Motor Control in Invertebrates and Vertebrates,* Barnes, W. J. P. and Gladden, M. H., Eds., Croom Helm, London, 1985, 187.
7. **Dethier, V. G.,** *The Physiology of Insect Senses,* John Wiley & Sons, New York, 1964.
8. **Wright, B. R.,** *Structure and Function of Proprioception in the Invertebrates,* Mill, P. J., Ed., John Wiley & Sons, New York, 1976, 323.
9. **Seelinger, G. and Tobin, T. R.,** Sense organs, in *The American Cockroach,* Bell, W. J. and Adiyodi, K. G., Eds., Chapman and Hall, London, 1981, 218.
10. **Pringle, J. W. S.,** Proprioception in insects. III. The function of the hair sensilla at the joints, *J. Exp. Biol.,* 14, 467, 1938.
11. **Wong, R. K. S. and Pearson, K. G.,** Properties of the trochanteral hair plate and its function in the control of walking in the cockroach, *J. Exp. Biol.,* 64, 233, 1976.
12. **Pringle, J. W. S.,** Proprioception in insects. II. The action of the campaniform sensilla on the legs, *J. Exp. Biol.,* 15, 114, 1938.
13. **Moran, D. T., Chapman, K. M., and Ellis, R. A.,** The fine structure of cockroach campaniform sensilla, *J. Cell Biol.,* 48, 155, 1971.
14. **Spinola, S. M. and Chapman, K. M.,** Proprioceptive indentation of the campaniform sensilla of cockroach legs, *J. Comp. Physiol.,* 96, 257, 1975.
15. **Zill, S. N. and Moran, D. T.,** The exoskeleton and insect proprioception. I. Responses of tibial campaniform sensilla to external and muscle-generated forces in the American cockroach, *Periplaneta americana, J. Exp. Biol.,* 91, 1, 1981.
16. **Chapman, K. M.,** Campaniform sensilla on the tactile spines of the legs of the cockroach, *J. Exp. Biol.,* 42, 191, 1965.

17. **French, A. S.,** The receptor potential and adaptation in the cockroach tactile spine, *J. Neurosci.*, 4, 2063, 1984.
18. **French, A. S.,** Action potential adaptation in the femoral tactile spine of the cockroach, *Periplaneta americana, J. Comp. Physiol.*, 155, 803, 1984.
19. **French, A. S.,** Strength-duration properties of a rapidly adapting insect sensory neuron, *J. Comp. Physiol.*, 159, 757, 1986.
20. **French, A. S.,** The role of calcium in the rapid adaptation of an insect mechanoreceptor, *J. Neurosci.*, 6, 2322, 1986.
21. **French, A. S.,** Removal of rapid sensory adaptation from an insect mechanoreceptor neuron by oxidizing agents which affect sodium channel inactivation, *J. Comp. Physiol.*, 161, 275, 1987.
22. **Oldfield, B. P., Kleindienst, H. U., and Huber, F.,** Physiology and tonotopic organization of auditory receptors in the cricket *Gryllus bimaculatus* DeGeer, *J. Comp. Physiol. A*, 159, 457, 1986.
23. **Schnorbus, H.,** Die subgenualen Sinnesorgane von Periplaneta americana Histologie und Vibrationsschewellen, *Z. Vgl. Physiol.*, 71, 14, 1971.
24. **Moran, D. T. and Rowley, J. C., III,** The fine structure of the cockroach subgenual organ, *Tissue Cell*, 7, 91, 1975.
25. **Debaisieux, P.,** Organes scolopidiaux des pattes d'insectes, *Cellule*, 47, 78, 1938.
26. **Nijenhuis, E. D. and Dresden, D.,** A micro-morphological study on the sensory supply of the mesothoracic leg of the American cockroach *Periplaneta americana, Proc. K. Ned. Akad. Wet.*, C55, 300, 1952.
27. **Nijenhuis, E. D. and Dresden, D.,** On the topographical anatomy of the nervous system of the mesothoracic leg of the American cockroach, *Periplaneta americana, Proc. K. Ned. Akad. Wet.*, C58, 121, 1956.
28. **Young, D.,** The structure and function of a connective chordotonal organ in the cockroach leg, *Philos. Trans. R. Soc. London Ser. B*, 256, 401, 1970.
29. **Guthrie, D. M.,** Multipolar stretch receptors and the insect leg reflex, *J. Insect Physiol.*, 13, 1637, 1967.
30. **Pearson, K. G. and Iles, J. F.,** Innervation of coxal depressor muscles in the cockroach *Periplaneta americana, J. Exp. Biol.*, 54, 215, 1971.
31. **Krauthamer, V. and Fourtner, C. R.,** Locomotory activity in the extensor and flexor tibiae of the cockroach *Periplaneta americana, J. Insect Physiol.*, 24, 913, 1978.
32. **Pearson, K. G. and Bergman, S. J.,** Common inhibitory motoneurones in insects, *J. Exp. Biol.*, 50, 445, 1969.
33. **Zill, S. N., Moran, D. T., and Varela, F. G.,** The exoskeleton and insect proprioception. II. Reflex effects of tibial campaniform sensilla in the American cockroach, *Periplaneta americana, J. Exp. Biol.*, 94, 43, 1981.
34. **Zill, S. N. and Moran, D. T.,** The exoskeleton and insect proprioception. III. Activity of tibial campaniform sensilla during walking in the American cockroach, *Periplaneta americana, J. Exp. Biol.*, 94, 57, 1981.
35. **Pearson, K. G.,** Central programming and reflex control of walking in the cockroach, *J. Exp. Biol.*, 56, 173, 1972.
36. **Pringle, J. W. S.,** The reflex mechanism of the insect leg, *J. Exp. Biol.*, 16, 220, 1939.
37. **Meyer, C. A. and Siegler, M. V. S.,** Tactile reflexes of cockroach leg, *Neurosci. Abstr.*, 13, 619, 1987.
38. **Zill, S. N., Underwood, M. A., Rowley, J. C., and Moran, D. T.,** A somatotopic organization of afferents in insect peripheral nerves, *Brain Res.*, 198, 253, 1980.
39. **Pearson, K. G., Wong, R. K. S., and Fourtner, C. R.,** Connexions between hairplate afferents and motoneurones in the cockroach leg, *J. Exp. Biol.*, 64, 251, 1976.
40. **Zill, S. N. and Moran, D. T.,** Suppression of reflex postural tonus: a role of peripheral inhibition in insects, *Science*, 216, 751, 1982.
41. **Rijlant, P.,** Les manifestation electrique du tonus et des contractions volontaires et reflex chez les arthropodes, *C.R. Soc. Biol.*, 111, 631, 1932.
42. **Wilson, D. M.,** Proprioceptive leg reflexes in cockroaches, *J. Exp. Biol.*, 43, 397, 1965.
43. **Brodfuehrer, P. and Fourtner, C. R.,** Reflexes evoked by the femoral and coxal chordotonal organs in the cockroach, *Periplaneta americana, Comp. Biochem. Physiol.*, 74A, 169, 1983.
44. **DiCaprio, R. A. and Clarac, F.,** Reversal of a walking leg reflex elicited by a muscle receptor, *J. Exp. Biol.*, 90, 197, 1984.
45. **Barnes, W. J. P., Spirito, C. P., and Evoy, W. H.,** Nervous control of walking in the crab, *Cardisoma quanhumi*. II. Role of resistance reflexes in walking, *Z. Vgl. Physiol.*, 76, 16, 1972.
46. **Nashner, L. M.,** Adaptive reflexes controlling the human posture, *Exp. Brain Res.*, 26, 59, 1976.
47. **Ferrell, W. R.,** The discharge of mechanoreceptors in the cat knee joint at intermediate angles, *J. Physiol. (London)*, 268, 23P, 1977.
48. **Libersat, F., Clarac, F., and Zill, S.,** Force-sensitive mechanoreceptors of the dactyl of the crab: single-unit responses during walking and evaluation of function, *J. Neurophysiol.*, 57, 1618, 1987.
49. **Libersat, F., Zill, S., and Clarac, F.,** Single-unit responses and reflex effects of force-sensitive mechanoreceptors of the dactyl of the crab, *J. Neurophysiol.*, 57, 1601, 1987.

50. **Schoultze, T. W. and Swett, J. E.,** The fine structure of the Golgi tendon organ, *J. Neurocytol.,* 1, 1, 1972.
51. **Loeb, G. E.,** Somatosensory input to the spinal cord during normal walking, *Can. J. Physiol. Pharmacol.,* 59, 627, 1981.
52. **Delcomyn, F.,** The locomotion of the cockroach *Periplaneta americana, J. Exp. Biol.,* 54, 443, 1971.
53. **Delcomyn, F. and Usherwood, P. N. R.,** Motor activity during walking in the cockroach *Periplaneta americana.* I. Free walking, *J. Exp. Biol.,* 59, 629, 1973.
54. **Bassler, U.,** *Neural Basis of Elementary Behavior in Stick Insects,* Springer-Verlag, Berlin, 1983.
55. **Forssberg, H., Grillner, S., and Rossignol, S.,** Phasic gain control of reflexes from the dorsum of the paw during spinal locomotion, *Brain Res.,* 132, 121, 1977.
56. **Melvill Jones, G. and Watt, D. G. D.,** Observations on the control of stepping and hopping movements in man, *J. Physiol.,* 219, 709, 1971.
57. **Zill, S. N. and Forman, R. R.,** Proprioceptive reflexes change when an insect assumes an active, learned posture, *J. Exp. Biol.,* 107, 385, 1983.
58. **Bassler, U.,** Reversal of a reflex to a single motoneuron in the stick insect *Carausius morosus, Biol. Cybern.,* 24, 47, 1976.
59. **Eccles, J. C., Eccles, R. M., and Magni, F.,** Central inhibitory action attributable to presynaptic depolarization produced by muscle afferent volleys, *J. Physiol.,* 147, 1961.
60. **Pearson, K. G. and Fourtner, C. R.,** Nonspiking interneurons in walking system of the cockroach, *J. Neurophysiol.,* 38, 33, 1975.
61. **Libersat, F., Goldstein, R. S., and Camhi, J. M.,** Non-synaptic regulation of sensory activity during movement in cockroaches, *Proc. Natl. Acad. Sci. U.S.A.,* 84, 8150, 1987.
62. **Murrain, M. P. and Ritzmann, R. E.,** Analysis of proprioceptive inputs to DPG interneurons in the cockroach, *J. Neurobiol.,* 19, 552, 1988.
63. **Westin, J.,** Responses to wind recorded from the cercal nerve of the cockroach *Periplaneta americana, J. Comp. Physiol.,* 133, 97, 1979.
64. **Westin, J., Langberg, J. J., and Camhi, J. M.,** Response of giant interneurons of the cockroach, *Periplaneta americana,* to wind puffs of different directions and velocities, *J. Comp. Physiol.,* 121, 307, 1977.
65. **Libersat, F., Goldstein, R. S., and Camhi, J. M.,** Mechanism of non-synaptic regulation of sensory activity during movement, *Neurosci. Abstr.,* 13, 398, 1987.
66. **Camhi, J. M., Tom, W., and Volman, S.,** The escape behavior of the cockroach *Periplaneta americana.* I. Turning responses to wind puffs, *J. Comp. Physiol.,* 128, 193, 1978.
67. **Ritzmann, R. E. and Pollack, A. J.,** Identification of thoracic interneurons that mediate giant interneuron-to-motor pathways in the cockroach, *J. Comp. Physiol.,* 159, 639, 1986.
68. **Zill, S. N.,** unpublished data.

Chapter 26

CHEMORECEPTION

Günter Seelinger

TABLE OF CONTENTS

I. INTRODUCTION

The task of a chemical sense organ is to detect and recognize specific molecules. However, the number of organic compounds which may function as odorants is virtually endless, and natural flavors may consist of hundreds of different substances. Olfactory receptors, on the other hand, show a puzzling variety of individual specificities. Odor qualities cannot be arranged along a linear scale like colors are according to wavelength. Multiple independent variations in the structure of a molecule result in a multidimensional ''space'' of odor qualities. Therefore, it is often difficult for an investigator to find the adequate natural stimulus for a chemoreceptor, or even to search for it systematically without knowing the biological context.

Ideally, an investigation of odor recognition might proceed as follows. Ecological and ethological studies reveal biologically important odor sources and characteristic behavioral responses. Chemical analysis then determines the qualitative and quantitative composition of a stimulus, and perhaps its natural variability. A behavioral assay next must show which of the identified components or their combinations elicit the observed responses in the animal. Finally, learning experiments quantify the animal's ability to discriminate between the natural odor and similar stimulus constellations.

Having found biologically relevant odors in this (or a simpler) way, the investigator may ask some questions concerning the morphological substrate and the physiological mechanisms of olfaction — namely, which are the sense organs perceiving the stimuli? How do stimulus molecules find their way from the environment to the receptor cell? Are there mechanisms to adjust the sensitivity for a certain odor to the natural concentration range? What are the cellular mechanisms of odor recognition and of sensory transduction? How specific is a receptor cell? Are there physiological types of receptors, or is there individual variation of specificity? Are behavioral responses to complex odors released by characteristic components specifically stimulating one receptor type (''labeled line'') or by a bouquet creating a graded response pattern in many different receptors (''across fiber pattern'')? With the latter questions, we come to the problem of central processing of olfactory information.

No single organism is, of course, the best one for studying all of these problems. Cockroaches appear to be a reasonable compromise, as will be discussed at the end of this chapter. In the following account, emphasis will be placed on the olfactory system of the American cockroach, *Periplaneta americana,* but results from other insects are also cited wherever necessary.

II. THE COCKROACH WORLD — A WORLD OF ODORS

Typically, cockroaches are nocturnal insects hiding in the dark during the day, so their use of optical signals is limited. Odors play a major role in guiding a cockroach to its food, mate, and shelter.

Only a few cockroaches are food specialists (for example, feeding on wood). Most species feed on all kinds of plant materials such as fruits, flowers, decaying leaves, and other debris, but also on fungi and dead animals.[1] Species associated with man are attracted to bread, cheese, and other groceries. Preference tests show that some food odors (e.g., cheesecake, beer) are more attractive than others (e.g., bread, lemon) to hungry American cockroaches. All of these materials emit complex blends of various alcohols, esters, acids, aldehydes, ketones, amines, terpenes, etc., which are effective stimuli for many antennal receptor cells.[2,3] One might expect that cockroaches are able to discriminate between a great number of olfactory stimuli, but to date this has been shown for only two odors in learning experiments.[4]

The most spectacular behavioral responses are elicited from males of the Blattidae by the female sex pheromone.[5] *P. americana* males are attracted from a distance of over 30 m when the female sex pheromone is carried by the wind.[6] While flying upwind in the odor plume, males experience a wide range of concentrations. Two components of the pheromone ''share'' the role as an olfactory guideline to the female: periplanone-B attracts males from far away, while periplanone-A improves orientation near the female.[7] Sexually excited males will court the female upon contact and expose a tergal gland with an attractive odor for the female.[8] Male tergal glands are widespread among cockroaches and may have been their original way of chemical communication during courtship.[9,10] In addition, components of epicuticular waxes function as female contact sex pheromones.[11]

Some epidermal glands of cockroaches produce allomones of mostly unknown chemical composition and biological function. Among those investigated are defensive secretions containing caustic substances, e.g., quinones, phenols, and hexenal from the sternal glands of Blattidae[12], or sticky proteins which disable attacking arthropods mechanically.[13] In only one case have the chemistry, production site, and behavioral effect of a nonsexual pheromone been elucidated: the mandibular glands of some blaberid species produce a mixture of carbohydrates, alcohols, and ketones which leads to aggregation in the natural environment.[14] Many species, especially their larvae, aggregate at the feces of their own or different species. Chemical analysis of fecal extracts suggests, however, that this response is due to a number of ubiquitous constituents which also act as feeding stimulants rather than to specific pheromones.[14,15]

III. THE MAIN CHEMOSENSORY ORGANS

Chemoreceptive sensilla are found on various parts of the insect body. Usually located on antennae and mouthparts, they also may occur on the tarsi and wings of flies and on the ovipositor of many other insect orders. In cockroaches, however, most of the chemoreceptors have only been reported on the head and its appendages, although some will probably be found elsewhere. The main chemosensory organs are the antennae, each one bearing about 70,000 sensory hairs in the male American cockroach. Most of these sensilla are olfactory, but the long sensilla chaetica are also gustatory and tactile hairs. They represent about 30% of the antennal sensory input to the brain.[16]

Sensilla on the maxillary palps are morphologically and physiologically similar to those on the antennae.[17] In addition, there is a specialized region of densely packed gustatory hairs on the ventral side of the last palpal segment. Contact of these hairs with certain sugars[18,19] and amino acids[20] elicits feeding movements. Salt-sensitive receptors located in the oral cavity seem to be involved in a final control so that food may be imbibed first and rejected afterward.[18] Numerous organic substances have been reported to stimulate[21] or inhibit[22] feeding by influencing antennal as well as mouthpart receptors.

IV. STRUCTURE AND FUNCTION OF CHEMOSENSILLA

A. GENERAL ORGANIZATION

Chemosensilla resemble mechanoreceptive hairs in their general organization (see Chapter 25 in this work). As a rule, they contain a few bipolar receptor cells with ciliary dendrites ascending into the hair lumen, while the axons project directly into the brain or ganglion. The receptor cell bodies are encircled concentrically by three to four auxiliary enveloping cells. During ontogenesis or a molt, the auxiliary cells build the cuticular elements of the hair. In the fully developed state they produce the receptor lymph, which fills the hair lumen and plays an important role in sensory transduction. In contrast to mechanoreceptive sensilla, chemosensilla have pores in their walls.[23,24,25] Number, distribution, size, and fine structure of the pores vary among sensilla and thus provide valuable criteria for classification.[26]

B. TYPES OF CHEMOSENSILLA IN *PERIPLANETA*

The structure of the antennal sensilla of *Periplaneta* was described by several authors[16,27-29] using different terminologies.[30] The classification of Altner[26] is used in the present discussion.

1. Tip Pore Sensilla

A tip pore (tp) sensillum (Figure 1a) either has a single apical pore or its very tip is perforated by a number of small pores. This category includes the largest sensilla on the antennae, the long bristles or sensilla chaetica, which are up to 170 μm long and have a deeply grooved surface. They are thick walled with a flexible socket; a dendrite ending in the socket region with a tubular body indicates a mechanoreceptive function. Up to six other dendrites ascend through the hair lumen to the apical pore; they are contact chemoreceptors responding to various food stimuli[31] and to female contact sex pheromone.[32,33] A second type of tp sensillum is found on the maxillary palps in high density: a field of about 2600 taste hairs within an area of 0.1 mm^2. The tip region of each hair is perforated by about ten narrow slits of 10 nm diameter.[24] The functional significance of the different pore types is unknown. In both cases the pores are filled with electron-dense material. For tp sensilla with a single pore, closing mechanisms have been discussed as possible regulatory processes.[24]

2. Wall Pore Sensilla

Wall pore (wp) sensilla are characterized by numerous small wall pores more or less regularly distributed over their surfaces. They always contain olfactory receptors. Two main subtypes are distinguished on the basis of their cuticular architecture: double wall (dw) and single wall (sw) hairs.

The dw hair has a grooved surface. The inner wall is continuous with the dendritic sheath so that the unbranched dendrites are separated from the receptor lymph (Figure 1b). Cuticular bridges connect the inner and outer walls; they enclose radial spoke channels of 150 to 200 nm diameter filled with an electron-dense secretion of the tormogen cell which may influence the specificity of the sensillum.[34] Olfactory cells responding to fatty acids or to amines have been found in this sensillum type, as has a thermoreceptive cell.[34] Several subtypes of dw sensilla are distinguished on the basis of hair length and cell number.[16]

An sw sensillum has a smooth surface with structurally specialized pores from which finger-like "pore tubules" project into the hair lumen. Three subtypes of sw sensilla are found on the *Periplaneta* antenna.[16] One type, the swB sensillum of the male, may serve for a more detailed description of structure and function of a typical olfactory hair (Figure 1c). swB hairs are blunt tipped, smooth, and slightly curved, 20 μm long and 3 to 4 μm thick at the base. They are found exclusively on the antenna of the adult male, but correspond to a shorter type on larval and female antennae. Some of the male's long hairs are derived from the shorter larval type, but others are formed *de novo* during adult ecdysis.[16] Together they number 70,000 per male, which is about 50% of all antennal sensilla.

The wall of an swB sensillum is perforated by about 3000 pores in a fairly regular pattern. Each pore consists of an outer pore funnel narrowing into a pore canal of 100 Å diameter which then widens to form a pore kettle of 400 Å (Figure 1d). Proximally, two to five tubules of 150 to 200 Å diameter penetrate the cuticle of the hair wall within a channel filled with receptor lymph. The length of the pore tubules varies with the thickness of the cuticle along the hair shaft so that they always project some distance into the lumen.[35] The pore system is filled with an electron-lucent epicuticular material. Studies on the cockroach *Arenivaga* suggest that this material consists of lipid and nonlipid (protein?) components.[36] In some insects, tracer substances were shown to penetrate into the pore tubules from outside.[36,37] The dense outer layer of the tubules becomes more and more diffuse toward

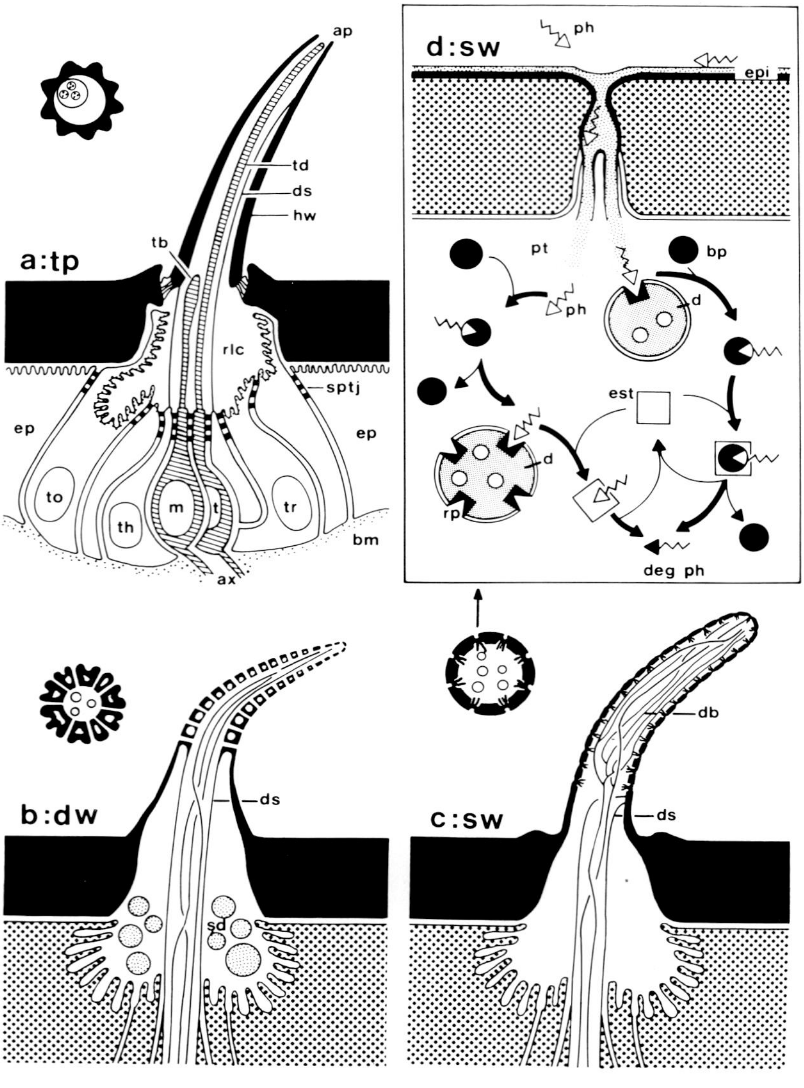

FIGURE 1. Functional organization of cockroach chemosensilla. (**a**) Cellular organization of a tp sensillum.[24] Only one chemoreceptor cell is shown for the sake of clarity, and the structural specializations of the hair base for mechanical stimulation are neglected. Abbreviations: ap = apical pore; ax = axons of receptor cells; bm = basement membrane; ds = dendritic sheath; ep = epidermal cell; hw = hair wall; m = mechanoreceptor; rlc = receptor lymph cavity; sptj = septate junctions; t = taste receptor; tb = tubular body of the mechanoreceptive dendrite; td = dendrite of taste receptor; th, to, and tr = thecogen, tormogen, and trichogen auxiliary cells, respectively. Note the granules at the folded apical membranes representing potential ion pumps. (**b,c**) Cuticular architecture of double- (dw) and single-wall (sw) sensilla, respectively.[24] Abbreviations: db = dendritic branches; ds = dendritic sheath; sd = droplets of a secretion produced by the tormogen cell which is then transported to the spoke channels. Insets in **a**, **b**, and **c** show cross sections through the hair wall. (**d**) Fine structure of a wall pore in a pheromone-sensitive swB hair, and two hypothesized ways in which stimulus molecules may interact with binding, receptor, and degrading proteins in the sensillum (bold arrows), assuming direct contact between pore tubules and dendritic membrane to occur (right) or not to occur (left).[35,40] Abbreviations: bp = binding protein; d = dendrite; deg ph = degraded (inactivated) pheromone molecule; epi = three-layered epicuticle; est = pheromone-degrading esterase; ph = pheromone molecules; pt = pore tubules; rp = receptor protein.

the hair lumen and is unlikely to provide a barrier to substances moving along the core strand.[35]

The outer dendritic segments of the four receptor cells leave their sheaths via a number of openings before they enter the hair lumen; thus, they are in contact with the receptor lymph. They divide into a total of about 25 branches, enlarging their surface area considerably. The branches run up to the hair tip, curling and crossing the hair lumen repeatedly. Wherever they come close to the wall they may come in contact with the pore tubules. Many tubules, however, end freely in the receptor lymph without touching a dendrite.[35] The number of contacts can vary with the fixation technique, but it is not clear whether contact or noncontact (or both) represents a natural condition.[38] Possibly, the dendrites may even move and touch different tubules at different times.

Although there are thousands of pores in an swB sensillum, the total cross-sectional area of the pore canals amounts to only 0.2% of the hair surface.[35] A stimulus molecule will usually hit the hair somewhere between the pores so that it must move along the epicuticle before penetrating into the hair. It appears difficult for hydrophobic molecules to reach the receptor membrane unless the dendrite is in direct contact with the pore tubules. However, studies on the chemical composition of the receptor lymph have shown that it is rich in hyaluronic acid, a substance which can form macromolecular complexes with cations and changes the physicochemical properties of the receptor lymph.[39] In moths, a highly concentrated pheromone binding protein was demonstrated in the receptor lymph of pheromone-sensitive sensilla. It could act as a carrier protein for stimulus molecules, while a second pheromone-degrading protein might inactivate the molecule after its interaction with the receptor membrane[40] (Figure 1d). A detailed survey of molecule adsorption, transport, and inactivation as well as of electrical events in the receptor cell is given by Kaissling.[41]

C. SENSORY TRANSDUCTION

Molecules penetrating into the sensillum have to interact with the receptor membrane in a primary process and induce a chain of electrical events which finally leads to the formation of action potentials, which are the cell's messages to the brain. Except for the primary process, the chain of events is basically the same in chemosensilla as in other insect sensilla.

A characteristic feature of insect sensilla is the existence of a transepithelial potential (TEP) of some +30 mV between receptor lymph and hemolymph. The TEP is due to potassium transport into the receptor lymph by an electrogenic ion pump across the folded membranes of enveloping cells (Figure 1a). As a consequence, the receptor lymph is rich in potassium.[42] The primary sensory process causes an increase in conductivity at the dendritic membrane, and an inward current flow is driven by the TEP.[43] The resulting drop in TEP amplitude can be recorded as a "receptor potential", the intensity of which depends on the number of simultaneously opened ion channels in the dendritic membrane and, thus, on the concentration of stimulus molecules. The changes in electrical circuitry finally cause depolarization at the inner dendritic or soma region, where the spike-generating membrane is suspected. The complicated structure of insect sensilla, however, makes direct evidence for most of these steps difficult to obtain.[41,44]

The primary process is usually believed to involve receptor proteins in the dendritic membrane, but the small amount of material is a handicap to direct biochemical investigation. The protein hypothesis is partly based on the high specificity of the response, which reminds us of molecular interactions between enzymes or hormone receptors and their substrates.[45] On the cockroach maxillary palp, receptor-stimulating disaccharides are hydrolyzed *in vivo* by gustatory sensilla, whereas ineffective sugars are not.[19] α-Glucosidases as receptor proteins might explain the observed specificity. Protein-like membrane particles are found more frequently on the distal portion of the dendrites of these sensilla than on the proximal portion,

as was demonstrated by freeze-fracturing.[24] It is the very tip of the dendrite where the transduction process takes place in tp sensilla.[45] Nevertheless, specific acceptor sites with ion channels also may occur on the somata of chemoreceptors, as shown in the crayfish with patch-clamp electrodes.[46]

D. CLASSICAL RECORDING TECHNIQUES

A simple method used to record physiological responses from antennal chemoreceptors is the electroantennogram (EAG).[47] The antenna is cut at its tip and its base, and two electrodes are inserted at each end. Stimulation with effective odors results in voltage changes between both electrodes, probably by summation of receptor potentials from individual sensilla. The amplitude of the signal depends on the number of receptor cells responding and on stimulus concentration. Thus, biologically important odors may not be detected if only a few cells respond to them. The EAG technique also provides little information about receptor specificity, even when differential adaptation is used for discrimination between different receptors. Nevertheless, it is a fast and sensitive bioassay for tracking highly effective components of, for example, crude pheromone or food extracts during the chemical fractionation process.

Signals from individual sensilla are obtained by inserting a finely tapered tungsten electrode at the hair base while a reference electrode is in the hemolymph.[48] A receptor potential and spikes may be seen upon stimulation. Spikes produced by different receptor cells usually differ in amplitude and shape (Figure 2). Although the reasons for this phenomenon are poorly understood, it can be used to determine the specificity of single olfactory cells. Cockroach sensilla are well suited for this purpose because they harbor only a few receptors, while up to several dozen are found in some other insect olfactory hairs.

Kaissling[49] recorded from the long pheromone sensilla of moths by cutting their tips and carefully attaching a capillary electrode to the open end so that the hair shafts were still exposed to airborne stimuli. In gustatory hairs, the natural tip opening can be used in the same way to gain electrical contact with receptors.[50] The recording electrode also must contain the stimulus in this case, and the electrolytes necessary for electrical contact may interfere with the receptors' responses to the test substances. Lateral penetration of the hair shaft with a recording electrode and separate stimulation with a second capillary not only avoids this problem, but also allows for studying electrical events before and after stimulation.[51]

E. SENSITIVITY AND RESPONSE CHARACTERISTICS

In cockroaches as well as in moths, female sex pheromones are the most effective odor stimuli in terms of threshold concentrations. As little as 10^3 molecules per cubic centimeter of air are sufficient to elicit wing fluttering in male *Bombyx*.[52] Estimates indicate that male cockroaches are just as sensitive.[53] This is due to (1) the high number of receptor cells (there are about 4×10^4 specific receptors for each pheromone component per antenna in a male cockroach, as compared to only about 10^3 cells of other types); (2) the convergence of many receptor fibers onto one central neuron (see below); and (3) differences in sensitivity of the individual receptor cells between pheromone and other olfactory cells.[54]

A single pheromone molecule can elicit a spike in moth receptors, probably by opening a single ion channel.[55] Slight changes in molecular structure, however, reduce effectiveness by several orders of magnitude. This may be due to different steps in the transduction process: lower affinity of the molecule to the receptor, lower probability of opening an ion channel when bound, or yet other processes. As a result, dose-response curves for the receptor potential can be shifted to higher threshold concentrations, rise more slowly with concentration, and have lower saturation levels for less effective stimuli.[41]

The relationship between receptor potential and nerve impulses is complex. Highly

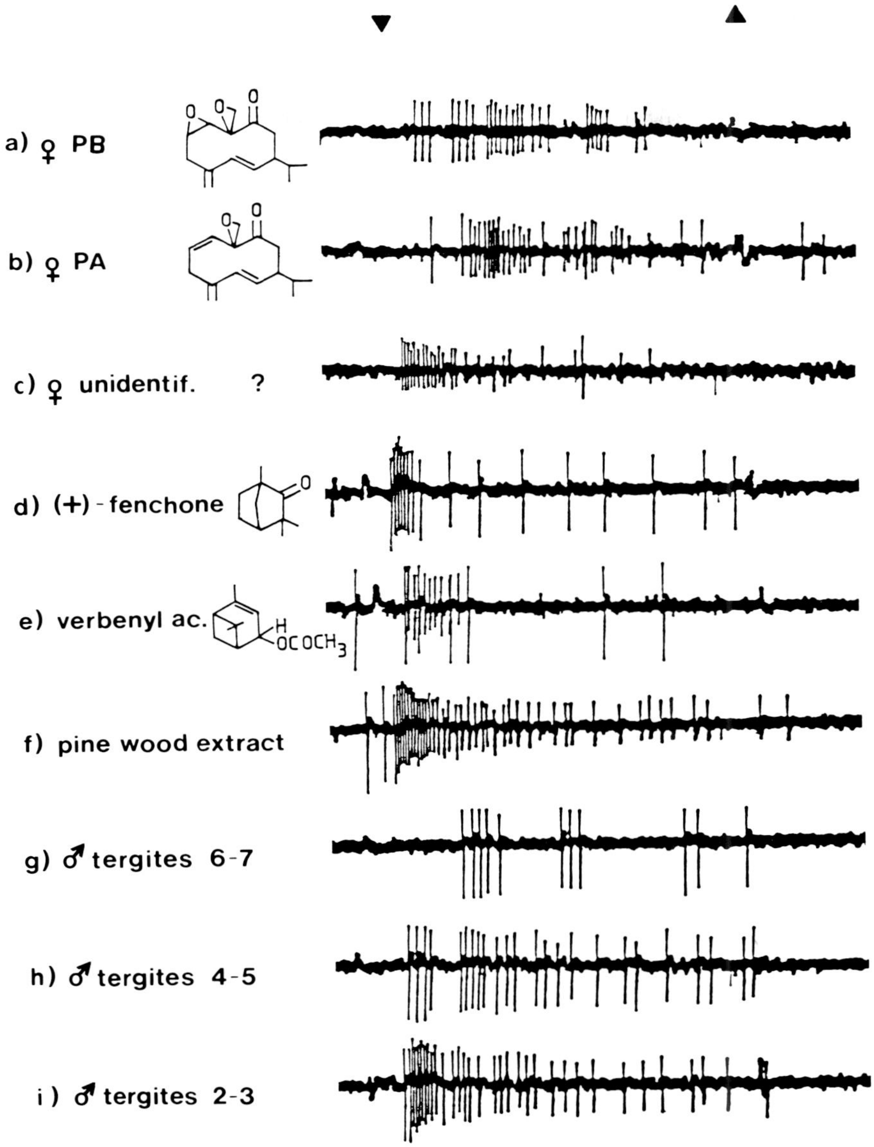

FIGURE 2. Responses of olfactory receptors in an swB sensillum of a male American cockroach, recorded with a tungsten electrode. AC filtering eliminates the slow receptor potential. Spike amplitudes are characteristic for different receptor cells, but may change during stronger excitation (e.g., in **c, f,** and **i**). Three spike types can be recognized in these examples: small-amplitude spikes from the PA cell in **c**, medium-sized spikes from the PB cell in **a**, and large spikes from a "monoterpene" cell in **d** and in **f** to **i**. Two cells respond to periplanone-A: the PB cell and, with some delay, the PA cell (**b**). Responding to verbenyl acetate are the "monoterpene" cell and the PB cell (**e**). Stimulus in **c** was an unidentified minor component of the female extract. Extracts used in **g** and **h** were obtained by washing male tergites in acetone; the male sexual glands are situated on tergites 2 to 4. Arrowheads at top indicate beginning and end of a stimulus airflow lasting 600 ms. (Original data courtesy of N. Hartmann.)

effective odor substances elicit an irregular impulse pattern near threshold concentration due to stochastic arrival of molecules at the membrane. Each spike is preceded by a small drop in TEP which reflects the opening of an ion channel. With rising concentrations, the response becomes smoother and is usually phasicotonic. Strong stimuli often yield transient excitation and off-responses.[41] Repetitive stimulation reduces the effect of the individual odor puffs, but moth pheromone receptors can follow a flicker stimulus of up to three per second with a time-coupled single spike.[56] This is of biological importance, since the natural odor plume emitted by a female is broken up into small puffs by wind turbulence.[57]

V. OLFACTORY CODING

A. RECEPTOR SPECIFICITY

The olfactory system of the male American cockroach is sensitive to a great variety of naturally occurring odors.[2,3,58,59] Each receptor cell responds to only a limited number of substances. The physicochemical parameters determining the efficiency of a molecule in stimulating a given receptor cell are difficult to characterize. A cell sensitive to aliphatic alcohols may, for example, have a lower threshold for hexanol than for pentanol or heptanol. The loss of efficiency with shorter or longer chain length can, however, be compensated for by adding a side chain, by introducing a double bond, or by some other manipulation of the molecular structure. The cell may even respond to substances quite unrelated to the alcohols. It is speculated that different types of receptor proteins might be present on one dendritic membrane in a qualitatively uneven distribution among different cells.[59] Attempts have been made to group receptor cells into physiological categories (''reaction groups'') by statistical analysis of their responses.[3] About 25 reaction groups have been described.[54] Cells responding best to the same stimulus substance are usually also similar in their responses to other odors. However, there is considerable variation among individual cells within a group. Response spectra of different groups often overlap, and some cells cannot be classified at all.

Cells from different reaction groups seem to occur in fixed combinations within sensilla, thus defining physiological subtypes of the morphological sensillum types.[2] A very distinct sensillum type, clearly defined physiologically as well as morphologically, is the swB sensillum. Each of the 70,000 swB sensilla of a male contains the same combination of four receptor cells belonging to four physiological types. Responses from a certain receptor type are recognized by their specific shape and size of nerve impulses, but it is not possible to attribute these differences to morphological characters of the cells. At least three of the four types seem to play a distinct role in eliciting certain behavioral responses. Two cells in the swB sensillum respond to the main components of the female attractant odor,[60] namely, the sesquiterpenes periplanone-A[61] and periplanone-B[62] (Figure 2a,b). These cells are referred to below as the PA and PB cells. Periplanone-B exclusively stimulates the PB cell and elicits sexual excitement, search behavior, and chemoanemotaxis, while periplanone-A primarily stimulates the PA cell and modulates chemo-orientation near the female.[7] At higher dosages, periplanone-A also elicits sexual excitement and search behavior; however, this is due to the fact that the PB cell also responds to periplanone-A, but with a higher threshold than to periplanone-B. Substances which exclusively stimulate the PA cell (as shown in Figure 2c) are not behaviorally active by themselves, but modulate locomotory activity elicited by periplanone-B in much the same way as periplanone-A does. Thus, sexual excitation and long-distance attraction generally are elicited by stimulation of the PB cell, while modulation of search behavior is accomplished via stimulation of the PA cell. The behavioral response to any active compound can be explained by its relative stimulatory effect on both receptor types.

It may be mentioned that olfactory cells in corresponding sensilla of the sympatric *P.*

australasiae males also respond to the *P. americana* sex pheromone components.[65] Periplanone-A is a potent attractant for this species, while periplanone-B inhibits the behavioral response.[66,67]

The two remaining cells of the swB sensillum are stimulated by a number of monoterpenes (e.g., cineol, fenchone, and terpineol).[63] One of them also responds to extracts of pine wood and of the male tergal gland[68] (Figure 2d through i). Females have a corresponding receptor type on their antennae; they are attracted to fresh pine wood and feed on it. It appears that the males imitate this odor and make the females feed on their tergal gland during courtship, bringing them into proper position for copulatory attempts.[8] Interestingly, some monoterpenes such as bornyl acetate occur in pine resin[69] and stimulate the same receptor cell as the male tergal gland extract does. Bornyl acetate had been reported as a mimic of the female sex pheromone,[70] but in fact it imitates a male pheromone which also sexually excites other males and elicits feeding on the tergal gland (''pseudofemale behavior''[8]).

While the receptors for periplanone respond very uniformly, the monoterpene cells show more variation, overlapping in specificity with cells from other sensilla. Verbenyl acetate closes the gap between the periplanone receptors and the monoterpene cells in that it stimulates the PB cell as well as the cell responding to male tergal gland extract (Figure 2e). Terpenes may represent a class of odorants which have developed from food odors (the biological importance of which must still be elucidated) to specialized signals in intraspecific communication, and receptor specificity may reflect various steps in this evolutionary process.

Olfactory cells not responding to pheromones are often classified as receptors for ''general odors''. This should be done with care, since there still may be unknown specific chemical signals in nature which are adequate stimuli. However, many receptor types of *Periplaneta* are stimulated by fruits, meat, bread, etc., and none of these complex odors has been found to stimulate only one receptor type.[2] Food odors seem to be encoded peripherally by a graded ''across fiber'' excitation pattern in many different olfactory cells rather than by detection of one or a few characteristic compounds. This is so not only in the polyphagous cockroach, but also in specialized monophagous insects.[71]

B. CENTRAL PROCESSING

Antennal olfactory fibers project into spherical, well-defined neuropil regions in the deutocerebrum. These ''glomeruli'' are of identical number and arrangement in different individuals of the same species.[72,73] In *P. americana* there are about 125 glomeruli on each side of the brain, receiving axons from ca. 200,000 olfactory receptors. Each glomerulus is innervated by the dendritic arborizations of one single projection neuron, which in turn innervates only this one glomerulus. The projection neuron sends its axon via the medial tractus olfactorioglobularis (tog) to the lateral lobe of the protocerebrum and to the calyces of the mushroom bodies. There are other types of deutocerebral neurons, the best known of which are local spiking interneurons without axons, but with dendritic arborizations in numerous glomeruli[74] (Figure 3).

Fine structural studies, with double-labeling of receptor axons (by degeneration) on one side and of physiologically identified projection or local interneurons (by intracellular cobalt staining) on the other, reveal that only a few synaptic contacts exist between receptors and projection neurons, while the majority of synaptic inputs on local interneurons come from the receptors.[75] In addition, receptor axons are postsynaptic to unidentified central neurons.[76] Such a neuronal circuitry, in which the peripheral excitation from about 2×10^5 receptor fibers converges onto only about 10^2 fibers in the tog, is suggestive of the performance of intensive quality processing. The local interneurons, linking input and output of the deutocerebrum vertically and different glomeruli horizontally, should play a major role.[54] However, physiological investigations have not yet revealed any simple principles explaining

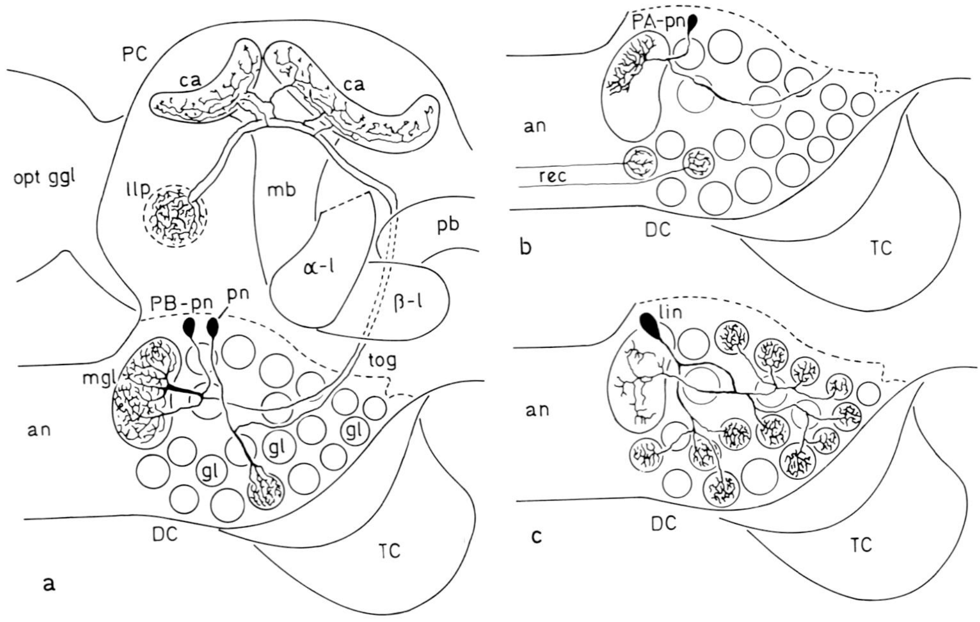

FIGURE 3. Neuronal elements in the olfactory pathway of the cockroach brain.[74,79] (**a**) Schematic drawing of the right brain side in frontal view. Two characteristic central neurons are shown: a typical projection neuron (pn) innervating one glomerulus and a pheromone projection neuron responding best to periplanone-B (PB-pn). (**b,c**) The deutocerebral region with a pheromone projection neuron for periplanone-A (PA-pn), a typical local interneuron (lin), and two olfactory receptor fibers (rec). Abbreviations: an = antennal nerve; DC = deutocerebrum; gl = glomerulus; lip = lateral lobe of the protocerebrum; mb = mushroom body with calyces (ca), α-lobe (αl), and β-lobe (βl); mgl = macroglomerulus; opt ggl = optic ganglia; pb = protocerebral bridge; PC = protocerebrum; TC = tritocerebrum; tog = tractus olfactorioglobularis.

how complex odor qualities are analyzed centrally. A topographical representation of different food odors in different glomeruli was not found. Morphologically corresponding identified glomeruli in two individuals may be quite dissimilar physiologically. Projection neurons are no more specific to certain food odors than are receptor cells, and they are less specific to single compounds. Thus, it appears that the across fiber pattern in the antennal nerve is transformed into a different, but no less complex, pattern in the tog.[73,75] To complicate the matter, olfactory and mechanical inputs converge in the deutocerebrum.[77]

Female sex pheromones are processed separately from other odors. A large "macroglomerulus" is situated near the antennal nerve within the deutocerebrum of the male (Figure 3). It is the only glomerulus innervated by several projection neurons (about 15). Ontogenetically, the macroglomerulus is formed by fusion of a number of glomerular regions.[78] Physiological recordings from projection neurons and cobalt stainings of individual swB sensilla indicate that receptors for female sex pheromone project exclusively into this neuropil.[54] Two morphologically and physiologically distinct types of macroglomerular projection neurons were found.[79,80] Their responses to female sex pheromones are not dramatically different from those of the receptor types, but there are two properties of pheromone neurons with functional significance for odor processing. Each macroglomerular projection neuron receives inputs from 10^3 to 10^4 receptor axons and can detect pheromones at very low concentrations, as when only a small proportion of the receptors are hit by a stimulus molecule.[54] This convergence from the entire antennal length is certainly a prerequisite for the extreme sensitivity of the behavioral response. Some other odors also stimulate the macroglomerular projection neurons, but the response (e.g., to octanol) is phasic in contrast

to the tonic pheromone response.[81] Lateral inhibition by nonpheromone neurons may enhance specificity with time delay.

Olfactory information is widely distributed by different neuron types within the protocerebrum, but systematic studies on further steps of information processing are just beginning.[54]

VI. SUMMARY AND CONCLUSIONS

Cockroaches have a highly developed sense of smell. On one hand, there are specific pheromones which allow species and sex recognition and trigger stereotypical behavioral patterns. Numerous adaptations on the sensillum, receptor, and central levels provide the highest sensitivity and specificity of pheromone perception. In many ways the cockroach pheromone system resembles that of moths. Both insect groups use several components perceived by specific receptor types in distinct sensilla and have a distinct macroglomerulus for female odor in the brain.[82] On the other hand, the olfactory sense of the cockroach seems to be designed for detection of a huge variety of odors. Thus, it also should provide an opportunity to study a nonspecialized sensory system similar to our own and probably characterized by plasticity through learning and imprinting. Unfortunately, we know little about odor discrimination in cockroaches. Learning experiments are urgently needed for a deeper understanding of complex odor processing.

Modern neurobiological techniques were recently introduced into the field of insect olfaction, including simultaneous intracellular recording from two neurons with reciprocal electrical stimulation and differential labeling,[83] radioactive deoxyglucose mapping of neuronal activity upon stimulation with different odors,[84] and immunocytochemistry of neurotransmitters.[85] Such techniques will allow a finer morphological analysis of the neuronal network, but also may provide insight into short- and long-time regulatory processes such as the timing of behavioral responsiveness to the correct activity period or reproductive state. With these tools applied to the study of cockroaches and moths, it will be interesting to see to what extent convergent evolution has shaped similar neuronal designs for similar functions.

There are also some remarkable parallels between the olfactory systems of insects and those of vertebrates. As with insects, vertebrate receptors project into glomerular regions of neuropil. About 1000 receptor axons per glomerulus converge onto a single mitral cell, which is the primary output neuron type of the olfactory bulb. The mitral cells distribute olfactory information to different projection areas in the cortex by axon collaterals.

There are important differences between both groups of animals as well. The number of olfactory cells in the mucosa may be as high as 10^8 in macrosmatic vertebrates, compared to only about 10^5 in the "macrosmatic" cockroach. Consequently, the whole neuronal mass of the olfactory brain is bigger and more complex in vertebrates. Receptor spectra of olfactory receptors are broad; they vary individually and over time with aging. Receptors are not connected to the brain for a considerable part of their limited life span. It is therefore very difficult to estimate the information transmitted to the brain when a certain odor is presented. Instead of physiological reaction groups, topographical representation of odor qualities on the mucosa is discussed as a peripheral coding mechanism. Different odor qualities elicit their strongest responses in different regions of the mucosa. The spatial pattern of excitation is projected onto the olfactory bulb in a fairly complicated manner. Divergence and convergence of odor qualities onto "vertical" mitral cells and "horizontal" periglomerular cells create a basis for intensive lateral interactions in the bulb. Intrabulbar recurrent loops and efferent inhibition also contribute to the complicated temporal excitation patterns of the mitral cells. Just as in the case of insects, it is not known how complex odors are encoded in the neuronal output of the first olfactory ganglion. For further information about vertebrate olfaction the reader is referred to recent review articles.[86-89]

Until recently, the study of olfactory receptor physiology was a domain of insect research. Although rapid progress has been made in the study of vertebrate olfaction, insects (especially cockroaches) are still valuable models in this field. Their advantages are the overall simpler organization, better accessibility of receptors to physiological investigation, and the existence of rather simple biological stimuli eliciting well-defined behavioral responses. More efforts should be made to analyze meaningful nonpheromone stimuli for both groups of organisms. The importance of temporal patterns created by sniffing in vertebrates, by antennal flicking in insects, and, generally, by uneven distribution of odors in the environment was probably also underestimated when temporal excitation patterns were discussed as a quality code for odors.

ACKNOWLEDGMENTS

I thank Drs. J. Boeckh, P. Selsam, and N. Hartmann for helpful comments and for their permission to include unpublished results.

REFERENCES

1. **Schal, C., Gautier, J.-Y., and Bell, W. J.,** Behavioural ecology of cockroaches, *Biol. Rev.*, 59, 209, 1984.
2. **Sass, H.,** Olfactory receptors on the antenna of *Periplaneta:* response constellations that encode food odours, *J. Comp. Physiol.*, 128, 227, 1978.
3. **Selzer, R.,** The processing of a complex food odor by antennal olfactory receptors of *Periplaneta americana, J. Comp. Physiol.*, 144, 509, 1981.
4. **Balderrama, N.,** One trial learning in the American cockroach, *Periplaneta americana, J. Insect Physiol.*, 26, 499, 1980.
5. **Roth, L. M. and Willis, E. R.,** A study of cockroach behaviour, *Am. Midl. Nat.*, 47, 66, 1952.
6. **Seellinger, G.,** Sex-specific activity patterns in *Periplaneta americana* and their relation to mate-finding, *Z. Tierpsychol.*, 65, 309, 1984.
7. **Seelinger, G. and Gagel, S.,** On the function of sex pheromone components in *Periplaneta americana:* improved odour source localisation with periplanone-A, *Physiol. Entomol.*, 10, 221, 1985.
8. **Simon, D. and Barth, R. H.,** Sexual behaviour in the cockroach genera *Periplaneta* and *Blatta*. I. Descriptive aspects, *Z. Tierpsychol.*, 44, 80, 1977.
9. **Roth, L. M.,** The evolution of male tergal glands in the Blattaria, *Ann. Entomol. Soc. Am.*, 60, 928, 1969.
10. **Sreng, L.,** Comportement Sexuel et Communication Chimique chez les Blattes, Ph.D. thesis, University of Dijon, Dijon, France, 1983.
11. **Nishida, R., Fukami, H., and Ishii, S.,** Female sex pheromone of the German cockroach, *Blattella germanica,* responsible for male wing-raising: 3,11-dimethyl-2-nonacosanone, *Experientia,* 30, 978, 1974.
12. **Brossut, R.,** Allomonal secretions in cockroaches, *J. Chem. Ecol.*, 9, 143, 1983.
13. **Plattner, H., Salpeter, M., Carre, J. E., and Eisner, T.,** Struktur und Funktion des Drüsenepithels der abdominalen Tergite von *Blatta orientalis, Z. Zellforsch.*, 125, 45, 1972.
14. **Brossut, R.,** Gregarism in cockroaches and in *Eublaberus* in particular, in *Chemical Ecology: Odour Communication in Animals,* Ritter, F. J., Ed., Elsevier/North-Holland, Amsterdam, 1979, 237.
15. **Fuchs, M. E. A., Franke, S., and Francke, W.,** Carbonsäuren im Kot von *Blattella germanica* und ihre mögliche Rolle als Teil des Aggregationspheromons, *Z. Angew. Entomol.*, 99, 499, 1986.
16. **Schaller, D.,** Antennal sensory system of *Periplaneta americana, Cell Tissue Res.*, 191, 121, 1978.
17. **Altner, H. and Stetter, H.,** Das sensorische system des Maxillartasters der Schabe *Periplaneta americana, Verh. Dtsch. Zool. Ges.*, p. 247, 1982.
18. **Frings, H.,** Gustatory thresholds for sucrose and electrolytes for the cockroach, *Periplaneta americana, J. Exp. Zool.*, 102, 23, 1946.
19. **Wieczorek, H.,** Biochemical and behavioural studies of sugar reception in the cockroach, *J. Comp. Physiol.*, 124, 353, 1978.
20. **Sugarman, D. and Jakinovich, W.,** Behavioural gustatory responses of adult cockroaches, *Periplaneta americana,* to D and L amino acids, *J. Insect Physiol.*, 32, 23, 1986.

21. **Tsuji, H.,** Attractive and feeding stimulative effect of some fatty acids and related compounds on three species of cockroaches, *Jpn. J. Sanit. Zool.,* 17, 89, 1966.
22. **Norris, D. M.,** Anti-feeding compounds, in *Chemistry of Plant Protection,* Vol. 1, Hoffmann, H. and Haug, G., Eds., Springer-Verlag, Berlin, 1985, 97.
23. **Zacharuk, R. Y.,** Ultrastructure and function of insect chemosensilla, *Annu. Rev. Entomol.,* 25, 27, 1980.
24. **Altner, H. and Prillinger, L.,** Ultrastructure of invertebrate chemo-, thermo- and hygroreceptors and its functional significance, *Int. Rev. Cytol.,* 67, 69, 1980.
25. **Keil, T. A. and Steinbrecht, R. A.,** Mechanosensitive and olfactory sensilla of insects, in *Insect Ultrastructure,* Vol. 2, King, R. C. and Akai, H., Eds., Plenum Press, New York, 1980, 477.
26. **Altner, H.,** Insektensensillen: Bau und Funktionsprizipien, *Verh. Dtsch. Zool. Ges.,* 1977, 139, 1977.
27. **Toh, Y.,** Fine structure of antennal sense organs of the male cockroach, *Periplaneta americana, J. Ultrastruct. Res.,* 60, 373, 1977.
28. **Norris, D. M. and Chu, H.,** Morphology and ultrastructure of male *Periplaneta americana* as related to chemoreception, *Cell Tissue Res.,* 150, 1, 1974.
29. **Schafer, R. and Sanchez, T. V.,** Antennal sensory system of the cockroach *Periplaneta americana:* postembryonic development and morphology of the sense organs, *J. Comp. Neurol.,* 149, 335, 1973.
30. **Seelinger, G. and Tobin, T. R.,** Sense organs, in *The American Cockroach,* Bell, W. J. and Adiyodi, K. G., Eds., Chapman and Hall, London, 1982, 217.
31. **Rüth, E.,** Elektrophysiologie der Sensilla chaetica auf den Antennen von *Periplaneta americana, J. Comp. Physiol.,* 105, 55, 1976.
32. **Seelinger, G. and Schuderer, B.,** Release of male courtship response in *Periplaneta americana:* evidence for female contact sex pheromone, *Anim. Behav.,* 33, 599, 1985.
33. **Krempien, W.,** Responses of antennal taste hairs to female contact sex pheromone in *Periplaneta americana,* in preparation.
34. **Altner, H., Sass, H., and Altner, I.,** Relationship between structure and function of antennal chemo-, thermo- and hygroreceptive sensilla in *Periplaneta americana, Cell Tissue Res.,* 176, 389, 1977.
35. **Heine, K.,** Sinnesorgane für Weiblichen Sexuallockstoff bei Verschiedenen Arten der Gattung *Periplaneta,* Thesis, University of Regensburg, Regensburg, Federal Republic of Germany, 1983.
36. **Hawke, S. D. and Farley, R. D.,** The role of pore tubules in the selective permeability of antennal sensilla of the desert burrowing cockroach, *Arenivaga* sp., *Tissue Cell,* 3, 665, 1971.
37. **Ernst, K. D.,** Die Feinstruktur von Riechsensillen auf der Antenne des Aaskäfers *Necrophorus, Z. Zellforsch. Mikroskop. Anat.,* 94, 72, 1969.
38. **Keil, T.,** Contacts of pore tubules and sensory dendrites in antennal chemosensilla of a silkmoth: demonstration of a possible pathway for olfactory molecules, *Tissue Cell,* 14, 451, 1982.
39. **Gnatzy, W. and Weber, K. M.,** Tormogen cell and receptor lymph space in insect olfactory sensilla: fine structure and histochemical properties in *Calliphora, Cell Tissue Res.,* 189, 549, 1978.
40. **Vogt, R. G. and Riddiford, L. M.,** Pheromone perception: a kinetic equilibrium, in *Mechanisms in Insect Olfaction,* Payne, T. L., Birch, M. C., and Kennedy, C. E. J., Eds., Clarendon Press, Oxford, 1986, 201.
41. **Kaissling, K. E.,** Chemo-electrical transduction in insect olfactory receptors, *Annu. Rev. Neurosci.,* 9, 121, 1986.
42. **Thurm, U. and Küppers, J.,** Epithelial physiology of insect sensilla, in *Insect Biology in the Future,* Locke, M. and Smith, D. S., Eds., Academic Press, New York, 1980, 735.
43. **Thurm, U.,** Mechanisms of electrical membrane responses in sensory receptors, illustrated by mechanoreceptors, in *Biochemistry of Sensory Functions,* Jaenicke, L., Ed., Springer-Verlag, Heidelberg, 1974, 367.
44. **de Kramer, J. J., Kaissling, K. E., and Keil, T.,** Passive electrical properties of insect olfactory sensilla may produce the biphasic shape of spikes, *Chem. Senses,* 8, 289, 1984.
45. **Hansen, K.,** Insect chemoreception, in *Taxis and Behaviour,* Hazelbauer, G. L., Ed., Chapman and Hall, London, 1978, 221.
46. **Hatt, H.,** Wechselwirkung von Reizmolekülen mit Rezeptoren: einzelkanalmessungen an chemorezeptiven Sinneszellen auf den Laufbeinen des Flusskrebses, *Verh. Dtsch. Zool. Ges.,* p. 211, 1986.
47. **Schneider, D.,** Elektrophysiologische Untersuchungen von Chemo- und Mechanorezeptoren der Antenne des Seidenspinners *Bombyx mori, Z. Vgl. Physiol.,* 40, 8, 1957.
48. **Boeckh, J.,** Elektrophysiologische Untersuchungen an einzelnen Geruchsrezeptoren auf den Antennen des Totengräbers, *Z. Vgl. Physiol.,* 46, 212, 1962.
49. **Kaissling, K. E.,** Sensory transduction in insect olfactory receptors, in *Biochemistry of Sensory Function,* Jaenicke, L., Ed., Springer-Verlag, Heidelberg, 1974, 243.
50. **Hodgson, E. S. and Roeder, K. D.,** Electrophysiological studies of arthropod chemoreception. I. General properties of the labellar chemoreceptor of diptera, *J. Cell. Comp. Physiol.,* 48, 51, 1956.
51. **Morita, H. and Yamashita, S.,** Generator potential of insect chemoreceptor, *Science,* 130, 922, 1959.
52. **Kaissling, K. E. and Priesner, E.,** Die Riechschwelle des Seidenspinners, *Naturwissenschaften,* 57, 23, 1970.

53. **Beck, K.,** Quantitative Bestimmung der Lockstoffreize bei *Periplaneta americana* als Grundlage der Reizmetrik in elektrophysiologischen Versuchen und Verhaltenstests, Thesis, University of Regensburg, Regensberg, Federal Republic of Germany, 1981.
54. **Boeckh, J. and Ernst, K. D.,** Contribution of single unit analysis in insects to an understanding of olfactory function, *J. Comp. Physiol. A.,* 161, 549, 1987.
55. **Kaissling, K. E. and Thorson, J.,** Insect olfactory sensilla: structural, chemical and electrical aspects of the functional organization, in *Receptors for Neurotransmitters, Hormones and Pheromones in Insects,* Sattelle, D. B., Hall, L. M., and Hildebrand, J. G., Eds., Elsevier/North-Holland, Amsterdam, 1980, 261.
56. **Kaissling, K. E.,** Temporal characteristics of pheromone receptor cell responses in relation to orientation behaviour of moths, in *Mechanisms in Insect Olfaction,* Payne, T. L., Birch, M., and Kennedy, C. E. J., Eds., Clarendon Press, Oxford, 1986, 193.
57. **Murlis, J. and Jones, C. D.,** Fine scale structure of odour plumes in relation to insect orientation to distant pheromone and other attractant sources, *Physiol. Entomol.,* 6, 71, 1981.
58. **Sass, H.,** Zur nervösen Codierung von Geruchsreizen bei *Periplaneta americana, J. Comp. Physiol.,* 107, 49, 1976.
59. **Selzer, R.,** On the specificities of antennal olfactory receptor cells of *Periplaneta americana, Chem. Senses,* 8, 375, 1984.
60. **Sass, H.,** Production, release and effectiveness of two female sex pheromone components of *Periplaneta americana, J. Comp. Physiol.,* 125, 309, 1983.
61. **Hauptmann, H., Mühlbauer, G., and Sass, H.,** Identifizierung und Synthese von Periplanon-A, *Tetrahedron Lett.,* 27, 6189, 1986.
62. **Persoons, C. J., Verwiel, P. E. J., Ritter, F. J., Talman, E., Nooijen, P. J. F., and Nooijen, W. F.,** Sex pheromones of the American cockroach, *Periplaneta americana:* a tentative structure of periplanone-B, *Tetrahedron Lett.,* 24, 2055, 1976.
63. **Hartmann, N.,** Das "Sexsensillum" von *Periplaneta americana* -Männchen: qualitative und quantitative Charakterisierung der einzelnen Zellen, Thesis, University of Regensburg, Regensberg, Federal Republic of Germany, 1984.
64. **Seelinger, G.,** Behavioural responses to female sex pheromone components in *Periplaneta americana, Anim. Behav.,* 33, 591, 1985.
65. **Beckmann, G.,** Rezeptoren für Periplanon-A und Periplanon-B auf der Antenne von *Periplaneta australasiae,* Thesis, University of Regensburg, Regensburg, Federal Republic of Germany, 1984.
66. **Waldow, U. and Sass, H.,** The attractivity of the female sex pheromone of *Periplaneta americana* and its components for conspecific males and males of *Periplaneta australasiae* in the field, *J. Chem. Ecol.,* 10, 997, 1984.
67. **Seelinger, G.,** Interspecific attractivity of female sex pheromone components of *Periplaneta americana, J. Chem. Ecol.,* 11, 137, 1985.
68. **Hartmann, N., Seelinger, G., and Sass, H.,** Correlating pheromone receptor types with specific behavioural patterns in the cockroach, in *Sense Organs, Proc. 16th Göttingen Neurobiology Conf.,* Elsner, N. and Barth, F. G., Eds., Thieme Verlag, Stüttgart, 1988, abstr. 51.
69. **Sturm, U.,** Untersuchung der Inhaltsstoffe von Kiefernharzen, Ph.D. thesis, University of Hamburg, Hamburg, Federal Republic of Germany, 1982.
69. **Sturm, U.,** Untersuchung der Inhaltsstoffe von Kiefernharzen, Ph.D., thesis, University of Hamburg, Hamburg, Federal Republic of Germany, 1982.
70. **Bowers, W. S. and Bodenstein, W. G.,** Sex pheromone mimics of the American cockroach, *Nature (London),* 232, 259, 1971.
71. **Visser, J. H.,** Host odor perception in phytophagous insects, *Annu. Rev. Entomol.,* 31, 121, 1986.
72. **Rospars, J. P. and Chambille, I.,** Deutocerebrum of the cockroach, *Blaberus craniifer.* Qualitative study and automated identification of the glomeruli, *J. Neurobiol.,* 12, 221, 1981.
73. **Selsam, P.,** Duftcodierung, Morphologie und synaptische Verschaltung identifizierter Neurone im Deutocerebrum der Schabe, Ph.D. thesis, University of Regensburg, Regensberg, Federal Republic of Germany, 1987.
74. **Ernst, K. D. and Boeckh, J.,** A neuroanatomical study on the organisation of the central antennal pathways in insects, *Cell Tissue Res.,* 229, 1, 1983.
75. **Selsam, P., Seelinger, U., and Boeckh, J.,** Contacts between antennal receptor cell axons and identified central neurons in the glomeruli of the deutocerebrum, in preparation.
76. **Seelinger, U.,** Elektronenmikroskopieshe Untersuchungen zur Organisation des Deutocerebrum und erste Schritte zur Aufklärung der Verschaltung identifizierter Neurone bei der Amerikanischen Schabe *Periplaneta americana,* Ph.D. thesis, University of Regensburg, Regensberg, Federal Republic of Germany, 1985.
77. **Waldow, U.,** Multimodal Neurone im Deutocerebrum von *Periplaneta americana, J. Comp. Physiol.,* 101, 329, 1975.
78. **Prillinger, L.,** Postembryonic development of the antennal lobes in *Periplaneta americana, Cell Tissue Res.,* 215, 563, 1981.

79. **Burrows, M., Boeckh, J., and Esslen, J.,** Physiological and morphological properties of interneurons in the deutocerebrum of male cockroaches which respond to female pheromone, *J. Comp. Physiol.,* 145, 447, 1982.
80. **Boechk, J. and Selsam, P.,** Quantitative investigations of the odour specificity of central olfactory neurons in the American cockroach, *Chem. Senses,* 9, 369, 1984.
81. **Waldow, U.,** CNS units in the cockroach *(Periplaneta americana)*: specificity of response to pheromones and other odor stimuli, *J. Comp. Physiol.,* 116, 1, 1977.
82. **Matsumoto, S. G. and Hildebrand, J. G.,** Olfactory mechanisms in the moth *Manduca sexta:* response characteristics and morphology of central neurons in the antennal lobes, *Proc. R. Soc. London Ser. B,* 213, 249, 1981.
83. **Christensen, T. A. and Hildebrand, J. G.,** Functions, organisation, and physiology of the olfactory pathways in the lepidopteran brain, in *Arthropod Brain: Its Evolution, Development, Structure and Functions,* Gupta, A. P., Ed., John Wiley & Sons, New York, 1987, 457.
84. **Rodrigues, V. and Buchner, E.,** (^{3}H)2-Deoxyglucose mapping of odor-induced neuronal activity in the antennal lobes of *Drosophila melanogaster, Brain Res.,* 324, 374, 1984.
85. **Hoskins, S. G., Homberg, U., Kingan, T. G., Christensen, T. A., and Hildebrand, J. G.,** Immunocytochemistry of GABA in the antennal lobes of the sphinx moth *Manduca sexta, Cell Tissue Res.,* 244, 243, 1986.
86. **Doty, R. L.,** Odor-guided behavior in mammals, *Experientia,* 42, 257, 1986.
87. **Getchell, T. V.,** Functional properties of vertebrate olfactory receptor neurons, *Physiol. Rev.,* 66, 772, 1986.
88. **Scott, J. W.,** The olfactory bulb and central pathways, *Experientia,* 42, 223, 1986.
89. **Kauer, J. S.,** Coding in the olfactory system, in *Neurobiology of Taste and Smell,* Finger, T. E., Ed., John Wiley & Sons, New York, 1987, 205.

Chapter 27

LEARNING AND MEMORY IN COCKROACHES: METHODS AND ANALYSES

Stephen Zawistowski and Ivan Huber

TABLE OF CONTENTS

I. INTRODUCTION

The analysis of the biological correlates of learned behavior has seen an important resurgence with the introduction of molecular technology. Continued study of learning may allow us to extend our understanding of this important process to the molecular level. Historical, theoretical, and methodological reasons have often presented certain systems as models for the analysis of fundamental problems. *Drosophila,* the fruit fly, has long been recognized as a practical system for the study of problems in genetics. The study of learning has resulted in similarly useful preparations. As our questions evolve, the available models may or may not continue to serve our needs. *Drosophila* were an important tool in the early study of transmission genetics and have continued to be a significant resource in the analysis of molecular genetic problems. Studies of learning at the molecular and cellular levels will require equally flexible investigators and preparations.

Analyses of the biological correlates of learning have, at times, been thwarted by the complex neural networks of the organisms typically used in the study of learning. Kandel and colleagues[1,2] have enjoyed considerable success elaborating the cellular processes associated with simple learning in the sea slug, *Aplysia.* Quinn and co-workers[3-5] have shown that hypotheses formulated using the *Aplysia* system can be verified using *Drosophila* with induced genetic lesions (i.e., mutations).

This chapter is designed to serve as a brief introduction to the basic terminology and procedures associated with the study of learning, a survey of how these procedures have been applied in the study of cockroaches, a consideration of studies focused on the study of the biological correlates of learning in cockroaches, and a description of additional roles for cockroach learning procedures. Given the rising cost of laboratory animal care and increasingly vociferous protests by animal welfare groups, cockroaches are an economical and practical alternative as a research subject. Cockroaches offer small schools with limited facilities the opportunity to provide their students with laboratory experience in animal learning and behavior.

Unless otherwise indicated, the species referred to throughout the chapter will be *Periplaneta americana.*

II. THE STUDY OF LEARNING

The study of learning has always seemed to fascinate individuals outside the traditional bounds of psychology. The obvious importance of what we intuitively feel is learned no doubt fans this interest. Unfortunately, a century of meticulous research and analysis has spawned a bewildering array of procedures and associated arcane terminology which more often intimidate rather than encourage interdisciplinary contributions. This chapter will assume a simple definition of learning: any relatively permanent change in behavior as a result of practice.[6] This excludes temporary changes due to fatigue or more permanent maturational effects.

Most people are aware that there is a distinction between classical and instrumental conditioning. This distinction is procedural in nature and does not necessarily imply different learning processes.[7] In classical conditioning a contingency is arranged between a stimulus and an outcome (e.g., Pavlov signaled the delivery of food with a tone), whereas instrumental procedures utilize a contingency between a response and an outcome (e.g., a rat presses a bar to get food). Instrumental and classical conditioning procedures are both assumed to promote associative learning; that is, the observed change in behavior is a result of an association between either the stimulus and outcome or response and outcome. This assumption is bolstered by a variety of control procedures which reduce or eliminate the required contingency.[8,9] The observed change in behavior should attenuate or disappear when the contingency is absent.

Changes in behavior that result simply from repeated presentation of a stimulus are called nonassociative learning.[7] An example of this would be habituation, where repeated presentation of a stimulus results in a subsequent reduction in response to that stimulus. This attenuation of response is not a result of fatigue or deterioration of the sensory response. Sensitization results when the presentation of a stimulus results in either increased responsiveness to that stimulus or some other stimulus.

III. LEARNING IN COCKROACHES

A. EARLY STUDIES

Two early (circa 1912) methods used for the study of conditioning in cockroaches were mazes and light/dark avoidance. The light/dark avoidance task has probably received the most use because it is a relatively simple procedure. The procedure utilizes a box or similar structure, part dark and part lighted. Most cockroaches are photonegative and will enter the darkened area and remain there. If the surface of the dark area is electrified, animals will first escape from the shock and then avoid it.[10] Performance is measured as a change in the amount of time spent in the dark and the number of shocks received. Turner[11] showed that cockroaches forced to run through a maze to avoid light took less time and made fewer errors after training. Subjects successfully completing the maze were allowed to enter their home container. Turner's maze was suspended over water, and he did have some trouble with the subjects falling off the apparatus.

In 1946, Minami and Dallenbach[12] used an avoidance procedure in their classic study of learning and memory in the cockroach. Their study, often cited in psychology texts, demonstrated that cockroaches restrained between training and testing performed better than individuals allowed normal levels of activity or those forced into continuous activity by being placed on a moving belt. They interpreted their results within the context of interference theory.[7] Experiences interpolated between training and testing would reduce performance during the test period. The acceptance of these data was a high point in this first era of cockroach learning research.

These early studies represented a time when the study of learning involved workers from many different backgrounds using a wide variety of different species. As the study of learning became increasingly specialized, the choice of subjects began to narrow, with white rats and pigeons dominating the list.[12a] As new methods are exploited to elucidate the biological correlates of learning, the cockroach may once again play an important role.

B. NONASSOCIATIVE LEARNING

1. Habituation

Habituation is the waning of a response as a result of repeated stimulations. This waning is not the result of fatigue or sensory adaptation. Zilber-Gachelin and Chartier[13,14] have demonstrated that repeated air puffs directed at a cercus result in habituation of response by the cercal nerve and connectives anterior to the sixth abdominal ganglion. This habituation is specific to the cercus stimulated; that is, the loss of responsiveness is not generalized to the other cercus. In addition, habituation is facilitated by successive series of stimulations.

2. Sensitization

Zilber-Gachelin and Paupardin[15-17] studied the facilitation of a response as a result of stimulation. This process, termed sensitization, was demonstrated by presenting mechanical shock to the leg of a cockroach, resulting in a generalized increase in motor reactivity to stimuli of any modality. This is in contrast to the high specificity of the habituation response mentioned above. These authors also reported that stronger sensitizing stimuli resulted in longer lasting periods of sensitization and stronger responses. This process appears to be

similar to the central excitatory state (CES) described in the blowfly *Phormia regina*, where sucrose stimulation sensitizes individuals to water stimulation.[18] Exploration of this state may merit further consideration, since individual differences in CES are associated with conditioning performance in the blowfly.[19-20] Exploration of this relationship in the cockroach might lead to a better understanding of interactions between associative and nonassociative learning processes.

C. LEG FLEXION LEARNING

1. Behavioral Analyses

Horridge's[21] experiments on leg flexion learning mark the opening of a second era in the study of learning and conditioning using cockroaches. He demonstrated that cockroach leg position could be conditioned to avoid shock. Individual animals were suspended over a saline solution, and shock was contingent upon contact with that solution when or if the leg was dipped. Animals receiving dip-contingent shocks made fewer dips than yoked controls that received shocks unrelated to their own behavior and not contingent upon leg dips. This simple avoidance procedure has been used by a number of investigators to probe the role of the nervous system during learning in the cockroach. Some of these findings will be discussed in a later section (III.C.2) of this chapter.

Subsequent analyses by Eisenstein and Cohen[22] confirmed Horridge's initial observations and extended the work to the study of isolated ganglia. Pritchatt[23,23a] provided further confirmation and showed that if contingencies were reversed, and shock was delivered upon leg lift and avoided by a leg dip, cockroaches would dip more often. This indicates that the lifting behavior is not an artifact resulting from the shock. Rather, it demonstrates that leg position can be made contingent upon punishment schedules, although subjects have been shown to learn more slowly to avoid shock by lowering a leg.[24] If individuals are presented with a series of reversals (shock-on-dip, then shock-on-lift), they adjust to the new contingencies, but do not show progressive improvement in reversal learning. That is, they will alter their behavior to receive the fewest shocks, but they do not make this adjustment more rapidly with each successive reversal or change of contingency.

Attempts to use light as a conditioned stimulus (CS) to signal shock were unsuccessful.[23a] However, Chen et al.[24a] reported that they were successful in using a low-intensity shock as a CS to signal a high-intensity-shock unconditioned stimulus (US) when working with *Blatta orientalis*. This may be an example of α-conditioning, where CS and US are similar in nature and so are the responses.[25] This differs from β-conditioning, where the CS and US are different. During α-conditioning, the stimulus used as the CS is initially subthreshold — that is, when first presented to the cockroaches, it does not result in an avoidance or escape response. Only after pairing with the more intense US shock does it stimulate a response. MacMillan[26] reported that *P. americana* are also capable of learning an association between a mild shock CS and a more intense shock US.

Chen et al.[24a] also reported that while intact animals were capable of learning such a task, headless animals were not. Memory of the intact animals was good and remained high if the head was removed after training. This suggests a unique role for the cephalic region in the more complex learning situations for cockroaches.

Disterhoft's[27] analysis of the leg flexion task for intact animals suggested that experimental animals learn to avoid shock, not escape it. By comparing the average number of shocks received during the first five dips and the last five dips of the first 10 min of training, he showed that the experimental subjects received fewer shocks than controls because they avoided lowering the leg to a position where shock would be received rather than removing it (the leg) more rapidly from the lowered position when shock was received. Eisenstein et al.[27a] pointed out, however, that in their experiments they observed that most escape learning occurred by the second dip and that Disterhoft's analysis may have obscured an escape effect through averaging.

Church and Lerner[28] and Willner[29] questioned whether learning does indeed occur in the leg flexion preparation and whether the yoked control is adequate. Using computer simulation, Church and Lerner indicated that the differences observed between experimental and yoked control subjects could be explained by differences in reactivity and not by learning, and they suggested that investigators not rely on a single-control procedure, such as the yoked control. Willner concluded that headless cockroaches can learn, but interpretation of the data is complicated by generally short-lived memory in headless animals and a variety of confounding factors, including the height of the subject over the saline solution. In response to Church and Lerner, Buerger et al.[30] pointed out inadequacies in the simulation model employed by the former. Buerger et al. assumed a continuous, rather than discrete, model of responding and determined that while the data presented are not definitive proof of learning, they are consistent with the interpretation that learning accounts for the differences observed between experimental and yoked control groups.

2. Neurophysiological Analyses

Because neurophysiological study usually requires limiting subject mobility, these analyses have generally been confined to studies of leg flexion to avoid electric shock. This preparation is well suited for detailed study of the transfer of information at the neuron/ganglion level because the animals are restrained and lesions can be made at several stages of the procedure.

In his original experiments, Horridge[21] showed that headless cockroaches could learn to lift their legs to avoid shock. Additional studies[22,31] have confirmed that the cephalic region is not needed for this learning to occur and that it can, in fact, be mediated by a ganglion isolated from the rest of the nervous system. However, there are some important differences between intact and headless cockroaches.[24a] Whereas intact cockroaches are capable of learning a signaled shock avoidance, in which a mild shock serves as a CS and a more intense shock as a US, headless animals are generally poor at such learning, with retention being better in intact animals as well.

Reep et al.[32] considered a different question concerning ganglionic function and leg position learning. If a prothoracic leg is trained to avoid shock and a mesothoracic leg is tested, cockroaches which received shock contingent upon prothoracic leg extension will receive fewer shocks to the mesothoracic leg than animals which serve as yoked controls. By severing connections between the pro- and mesothoracic ganglia, Reep et al. demonstrated that "the association between leg position and shock is made separately in the prothoracic and mesothoracic ganglia." A preformed association is not transferred; rather, sensory and motor information is transferred independently and an association is formed in the mesothoracic ganglion. No transfer to the mesothoracic leg occurs if the neural connectives between the ganglia are cut before training of the prothoracic leg. Both ipsi- and contralateral connectives must be cut, since information carried by the contralateral connectives is sufficient to effect a transfer.

3. Biochemical analyses

Kerkut and colleagues,[33-35] among others, have investigated the nature of the biochemical changes associated with cockroach leg flexion learning. Training results in an increased RNA turnover in the ganglion that undergoes training, when compared to yoked controls,[36] suggesting an increased rate of protein synthesis. Kerkut et al.[37] later indicated that they observed an increased uptake of radioactively labeled amino acids as a result of training, an observation consistent with the hypothesis that training results in enhanced protein synthesis. Reduced cholinesterase (ChE) activity is also observed after training.[36] This would apparently facilitate the excitatory synaptic activity mediated by the neurotransmitter acetylcholine (ACh). ChE activity returns to normal levels in about 3 d, and this corresponds

to the time course for memory attenuation. ChE activity is inversely correlated with the degree of performance,[34] and behavioral training alters the K_m of ChE.[35]

Training also results in lowered levels of glutamate decarboxylase (GAD) and γ-aminobutyric acid (GABA) in thoracic ganglia.[33] Considered as a whole, these data suggest that during behavioral training there is a reduction in ChE levels and ChE activity; this will result in enhanced ACh activity, facilitating excitatory responses in the ganglion. Reduced levels of GABA, an inhibitory neurotransmitter, also would facilitate excitatory responses mediated by the ganglion. It should be noted, however, that other studies contradict these findings, dictating caution in their interpretation.[38,39]

When a metabolic precursor to RNA, orotic acid, was injected into subjects, it increased extinction times in isolated metathoracic ganglia as well as in headless and intact animal preparations.[40] Extinction refers to the attenuation of a conditioned response (CR) when the CS is no longer paired with the US. In this case, subjects were trained to flex their legs to avoid shock (CR), and then this CR was extinguished by no longer shocking leg dips. These data are difficult to interpret since orotic acid had no effect on original acquisition or on reacquisition of the behavioral response.

D. LIGHT/DARK AVOIDANCE

Lovell and Eisenstein[41] examined the disruption of information transfer from short- to long-term memory. Cockroaches were trained to avoid the dark side of a box. Subjects showed retention or memory for this task for up to 2 h after training. CO_2-induced narcosis immediately after training resulted in no retention, and CO_2 administration 1 h after training attenuated but did not eliminate retention, supporting the results of earlier reports on the effects of CO_2 on learning and memory.[42] These data have been interpreted as indicating that CO_2 disrupts the transfer or consolidation of information from short- to long-term stages. These findings are consistent with other analyses showing disruption of memory in rats using CO_2 and electroconvulsive shock.[43] While the time course for memory disruption varies across different species, the generality of the disruptive effect is impressive.

Whereas shock is the aversive stimulus typically used to promote the avoidance of the dark, Ebeling et al.[44] report successfully using insecticide dusts as aversive stimuli to train cockroaches to avoid darkened areas. Their results have important implications for pest control. Insecticides which repel without rapid killing will result in animals avoiding those areas where the insecticide is typically placed, thus avoiding accumulation of a toxic dose. They found that boric acid powder, with a relatively weak toxicity, was the most effective insecticide application because it had such a low level of repellency that cockroaches continued to reenter a treated area until they died. This suggests the value of adding a pheromonal attractant to an insecticide preparation to reduce its repellency. Rust and Reierson[44a,44b] have, in fact, demonstrated that pheromonal extracts added to blatticides do reduce repellency and increase efficacy.

E. TWO-CHOICE MAZES

Longo[45] used a simple Y maze to study probability learning and habit reversal. The entrance to the maze was prepared to deliver shock, and the cockroach could escape shock by moving through the maze and entering the "correct" arm of the maze. If one arm (right or left) was consistently designated as correct, animals would go to that arm nearly 100% of the time. If one arm was designated correct 70% of the time and the other designated correct the remaining 30% (70:30), the animals would match these probabilities. That is, if the right arm was correct on 70% of the trials and the left arm correct on 30% of them, the subjects tended to go to the right side in 70% of the trials and to the left side in 30%. Longo interpreted these results within the context of Bitterman's[46] comparative psychology of learning. This concept suggests that analysis of performance by different species on similar

learning tasks could facilitate a phylogenetic interpretation. Longo indicates that while rats show maximization during probability learning, fish show matching. An individual that maximizes continues to direct nearly all responses to the side which is most frequently correct. This results in the greatest probability of a correct response on any particular trial. Matching results in fewer correct responses. Matching may be an earlier phylogenetic development than maximization, and Longo notes that the random probability matching observed in the cockroach is similar to that shown in fish and pigeons and contrasts with the maximization strategy generally observed in mammals. Longo's analysis of reversal patterns is generally consistent with Pritchatt's[23a] observation that evidence for improved performance during successive reversals is weak. In this way, cockroaches are also more similar to fish than to rats. Wilson and Fowler[47] indicate that alternation behavior of cockroaches *(B. orientalis)* at a choice point is similar to that of other organisms, with subjects tending to choose an alternative associated with a novel cue.

As with the other learning preparations, subsequent analyses of T-maze behavior in cockroaches emphasized the elucidation of consolidation processes. Shim and Dixon,[48] for example, reported that subjects *(B. orientalis)* cooled to 0°C immediately after each learning trial, when compared to control subjects that were not cooled, showed longer latencies (i.e., time needed to make a choice) during training and more errors and longer latencies during extinction. This would indicate that cooling impaired the consolidation process.

Eisenstein and colleagues[49-50a] report that the antibiotics cyclohexamide (CXM) and puromycin (PURO) inhibit protein synthesis, but only PURO produces retention deficits in T-maze learning. This suggests that new protein synthesis *per se* is not required for the consolidation and retrieval of long-term memory. This is true in mammals, birds, and insects. PURO apparently will block both storage and retrieval if administration is delayed up to 24 h. They point out other similarities of amnesiac agent effects in insects and mammals. Given the previously described data on biochemical changes occurring during learning in cockroaches (i.e., RNA turnover), it would appear that further analysis of the consolidation process using biochemical probes is called for.

Mote, in Chapter 23 of this publication, presents an interesting use of a Y maze to study the ability of *P. americana* to discriminate light of different wavelengths. By depriving subjects of water and using different wavelengths of light at the choice point, he showed that they could learn to discriminate between different wavelengths (i.e., blue vs. red, but not red vs. orange) if water was used as a positive reinforcer. Similar adaptations of learning procedures could be useful for the study of other cockroach sensory systems.

F. MISCELLANEOUS STUDIES

1. Grooming Posture

Luco and Aranda[51] report an intriguing procedure used with *B. orientalis*. Cockroaches normally hold an antenna with the opposite foreleg while grooming. These authors have studied individuals forced to acquire a new motor pattern. Cockroaches which have had both forelegs amputated attempt to hold their antennae with the middle legs. Shortly after amputation, an operated individual is unable to retain its balance and stand on the remaining legs. Within 4 to 5 d, the animal is able to use three legs as a tripod and to groom with one of the middle legs. This newly acquired motor behavior was observed unchanged up to 40 d after amputation. Analysis of the central nervous system of operated cockroaches revealed correlated changes in electrical activity. Subjects which had undergone amputation and subsequently adopted the new motor pattern showed a facilitated nervous pathway not observed in 90% of unoperated individuals. One example of this facilitation was a reduced delay of transmission of an impulse across the third paired ganglion. This change was not observed in subjects which had undergone amputation of the forelegs, but were given insufficient time to acquire the new motor response; neither was it observed in subjects

having the middle legs amputated, which were not reported to have learned a new motor pattern.

Luco[52] reported that animals loaded with a piece of lead weighing approximately as much as their body showed an increase in synaptic efficacy similar to that induced by amputation of the forelegs. This increase in synaptic efficacy also facilitated learning of the new grooming posture. Animals loaded with a weight before amputation of their forelegs learned to groom with their middle legs in less time than nonloaded subjects.

2. Olfactory Conditioning

Studies of olfactory conditioning in cockroaches are interesting since they involve free-moving subjects and include appetitive rather than avoidance responses. Balderama[53] associated menthol and vanilla odors with either food (a sugar solution) or saline. In a preference test, subjects generally showed a preference for vanilla. Subjects were then trained against this preference; that is, saline was paired with vanilla, and menthol with sugar solution. A subsequent preference test showed that individuals now spent more time in the presence of the odor (menthol) that had been paired with the sugar.

Pougalan and Masson[54] demonstrated that the gregarious cockroach *Blaberus craniifer* showed olfactory learning associated with odors from their home cage which would include conspecific pheromones. They paired coumarin with the conspecific odor by introducing coumarin to the air flow in a group home cage. Later tests of these subjects in a two-choice maze showed an attraction to the coumarin.

3. Social Effects

Gates and Allee[55] investigated learning in *P. americana* tested singly and in groups of two or three. They used a maze similar to that employed by Turner[11] and worked only with females of a specified size (3 cm). All subjects were initially trained individually on the maze. Subjects were subsequently trained again in pairs and then in triads. The data indicated that individual roaches learned to negotiate the maze whether trained alone or as members of a group (i.e., the same subjects were used individually and in groups). However, the presence of other individuals during training tended to reduce overall activity levels and, therefore, to increase the time required for conditioning in the grouped conditions. This contrasts with the report of Zajonc et al.[55a] that the presence of spectator roaches enhances escape from a light. An important difference was that Zajonc et al. separated subject roaches from the spectators with a Plexiglas® screen. Gates and Allee suggested that the presence of additional roaches primarily serves to interfere with their activity. In the Zajonc et al. study the Plexiglas® would prevent this behavioral interference. It would, of course, be interesting to replicate these experiments comparing cockroaches of gregarious and nongregarious species, separating the effects of olfactory and visual stimuli from physical presence.

4. Transfer of Learning

Transfer-of-learning experiments seek to demonstrate that, when an individual undergoes training, a physical/chemical trace is produced and that, if this product is isolated and given to a recipient, the recipient will either exhibit the learned behavior or learn it more rapidly. Successful transfer has been claimed for several species, including planarians[56] and rats,[57] with conflicting reports concerning failure to replicate.[58,59] Kumar and Muhar[60] and Cheney et al.[61] reported transfer of learning in cockroaches. Kumar and Muhar trained naive cockroaches in a dark-avoidance procedure. At the completion of training, these cockroaches were beheaded and their heads were fed to other naive roaches. A control group was fed the heads of cockroaches that did not undergo training. Both of the fed groups were then trained. The group which had consumed the heads of previously trained roaches learned significantly faster than those which had eaten the heads of naive roaches. A difficulty with

this experiment is that the control group had been fed the heads of naive "donors". Exposure to shock may cause biochemical changes unrelated to learning. For this reason, the control group should have been fed the heads of individuals exposed to shock not contingent with dark — that is, shocked but not conditioned. A second criticism of this study is the mechanism of transfer. It is difficult to conceive of a macromolecule, whether protein or nucleic acid, retaining its conformation and being capable of facilitating learning after passing through the digestive process.

The study by Cheney et al.[61] was similar to the Kumar and Muhar experiment. In the former case, however, neural homogenates were prepared from the donors and injected into the recipients. A control group was injected with neural homogenate provided by naive donors. When both the experimental and control recipient groups were trained, the group which had received the homogenate derived from the trained donors performed significantly better. Similar criticisms apply to this study as well. The control group should have been given homogenate derived from donors that had been exposed to shock but not conditioned. In addition, the homogenization process probably would have caused shearing of molecules of significant size. The validity of claims of transfer of learning between trained and untrained cockroaches remains to be confirmed by independent replication employing adequate control procedures.

IV. SUMMARY AND CONCLUSIONS

This brief survey of learning in cockroaches should make it abundantly clear that an enormous wealth of methods and data are available for investigators who choose to work with cockroaches. Our review has not been exhaustive; rather, we have tried to choose a variety of examples that emphasize the versatility of cockroaches as models for biomedical research. Eisenstein and Reep's[62] review is a more comprehensive analysis of some specific topics. Their "model systems" approach represents a tremendously successful attempt to study the learning process in a simplified preparation. This approach has traditionally focused on typological analyses of physiological processes. The great power of this method is demonstrated in Byrne's[63] massive and informative review of cellular analyses of learning. While Byrne does articulate the great advances made in the elucidation of these processes, he also reports that no single, complete mechanism is known yet for any species.

Methods in this area have emphasized similarities between individuals and species. We would like to suggest the value of analyzing individual differences (IDs). If we are able to discern the origins of these differences, it may provide further insight into proposed mechanisms. The work with *Drosophila* provides ample evidence of the power of this approach.[64] The identification of genetic variations in cockroaches that influence learning could open this area to the techniques of molecular biology.

An example of where this approach could prove valuable would be in studies on signaled shock avoidance.[24a,26] Differences in rate of acquisition could be probed with genetic analyses. Zawistowski and Hirsch[65] have demonstrated that IDs in an unselected population of blowflies can be measured. These flies can be bred selectively for various levels of learning performance. While their original population showed "modest" performance in a conditioned discrimination task, a selected high-performance line showed dramatic improvement on the conditioning task. The protocols for typological experiments attempt to reduce individual variation. We suggest that procedures be developed to analyze specifically, reliably measured IDs. This is, in fact, the approach taken in our own research. We are currently developing several methods to study behavioral variation in cockroaches. We hope to assess the level of variation occurring in different populations. Through selective breeding and other methods we can estimate the level of genetic variation correlated with the behavioral variation. This could provide important information on the role that learning might play in the ethogram,

or natural behavior, of cockroaches. For example, the work of Ebeling et al.[44] suggests that learning may play a role in the avoidance of pesticides. Olfactory conditioning to pheromonal cues[54] may help to regulate courtship and mating of cockroaches, as it does in *Drosophila*[66-68] and sweat bees.[69] Our understanding of the learning process certainly would be enhanced by a greater appreciation for the context in which this behavior normally occurs.

ACKNOWLEDGMENT

We would like to acknowledge the helpful comments of E. M. Eisenstein and A. Powers on earlier drafts of this manuscript.

REFERENCES

1. **Kandel, E. R.,** *Behavioral Biology of Aplysia,* Freeman, San Francisco, 1979.
2. **Kandel, E. R., Abrams, T., Bernier, L., Carew, T. J., Hawkins, R. D., and Schwartz, J. H.,** Classical conditioning and sensitization share aspects of the same molecular cascade in *Aplysia, Cold Spring Harbor Symp. Quant. Biol.,* 48, 821, 1983.
3. **Quinn, W. G., Harris, W. A., and Benzer, S.,** Conditioned behavior in *Drosophila melanogaster, Proc. Natl. Acad. Sci. U.S.A.,* 71, 708, 1974.
4. **Aceves-Piña, E. O., Booker, R., Duerr, J. S., Livingstone, M. S., Quinn, W. G., Smith, R. F., Sziber, P. P., Tempel, B. L., and Tully, T. P.,** Learning and memory in *Drosophila,* studied with mutants, *Cold Spring Harbor Symp. Quant. Biol.,* 48, 831, 1983.
5. **Tully, T. and Quinn, W. G.,** Classical conditioning and retention in normal and mutant *Drosophila melanogaster, J. Comp. Physiol. A,* 157, 263, 1985.
6. **Kimble, G. A.,** *Conditioning and Learning,* Appleton-Century-Crofts, New York, 1961.
7. **Mackintosh, N. J.,** *The Psychology of Animal Learning,* Academic Press, New York, 1974.
8. **Rescorla, R.,** Pavlovian conditioning and its proper control procedures, *Psychol. Rev.,* 74, 71, 1967.
9. **Schneiderman, N.,** *Classical (Pavlovian) Conditioning,* General Learning Press, Morristown, NJ, 1973.
10. **Szymanski, J. S.,** Modification of the innate behavior of cockroaches, *J. Anim. Behav.,* 2, 81, 1912.
11. **Turner, C. H.,** Behavior of the common roach (*Periplaneta orientalis* L.) on an open maze, *Biol. Bull.,* 25, 348, 1913.
12. **Minami, H. and Dallenbach, K. M.,** The effect of activity upon learning and retention in the cockroach, *Periplaneta americana, Am. J. Psychol.,* 59, 1, 1946.

12a. **Beach, F. A.,** The snark was a boojum, *Am. Psychol.,* 5, 115, 1950.

13. **Zilber-Gachelin, N. F. and Chartier, M. P.,** Modification of the motor reflex responses due to repetition of the peripheral stimulus in the cockroach. I. Habituation at the level of an isolated abdominal ganglion, *J. Exp. Biol.,* 59, 359, 1973.
14. **Zilber-Gachelin, N. F. and Chartier, M. P.,** Modification of the motor reflex responses due to repetition of the peripheral stimulus in the cockroach. II. Conditions of the activation of the motorneurons, *J. Exp. Biol.,* 59, 383, 1973.
15. **Zilber-Gachelin, N. F. and Paupardin, D.,** Sensitization and dishabituation in the cockroach. Main characteristics and localization of the changes in reactivity, *Comp. Biochem. Physiol.,* 49A, 441, 1974.
16. **Zilber-Gachelin, N. F. and Paupardin, D.,** Desensitization by leg contact in the cockroach: localization of the reactivity changes and a study of their possible sensory origin, *Comp. Biochem. Physiol.,* 49A, 471, 1974.
17. **Zilber-Gachelin, N. F. and Paupardin, D.,** Main characteristics of an induced decrease in leg motor reactivity (''desensitization'') in the cockroach, *Comp. Biochem. Physiol.,* 49A, 491, 1974.
18. **Dethier, V. G., Solomon, R. L., and Turner, L. H.,** Sensory input and central excitation and inhibition in the blowfly, *J. Comp. Physiol. Psychol.,* 606, 303, 1965.
19. **Tully, T., Zawistowski, S., and Hirsch, J.,** Behavior-genetic analysis of *Phormia regina.* III. A phenotypic correlation between the central excitatory state (CES) and conditioning remains in replicated F_2 generations of hybrid crosses, *Behav. Genet.,* 12, 181, 1982.
20. **McGuire, T. R.,** Further evidence for a relationship between the central excitatory state and classical conditioning in the blowfly *Phormia regina, Behav. Genet.,* 13, 509, 1983.
21. **Horridge, G. A.,** Learning of leg position by the ventral nerve cord in headless insects, *Proc. R. Soc. London Ser. B,* 157, 33, 1962.

22. **Eisenstein, E. M. and Cohen, M. J.,** Learning in an isolated prothoracic insect ganglion, *Anim. Behav.*, 13, 104, 1965.
23. **Pritchatt, D.,** Avoidance of electric shock by the cockroach *Periplaneta americana, Anim. Behav.*, 16, 178, 1968.

23a. **Pritchatt, D.,** Further studies on the avoidance behaviour of *Periplaneta americana* to electric shock, *Anim. Behav.*, 18, 485, 1970.

24. **Weiss, A. and Penzlin, H.,** A comparison of shock avoidance learning in headless cockroaches *Periplaneta americana* in leg-lifting and lowering task, *Physiol. Behav.*, 34, 697, 1985.

24a. **Chen, W. Y., Aranda, L. C., and Luco, J. V.,** Learning and long and short-term memory in cockroaches, *Anim. Behav.*, 18, 725, 1970.

25. **Razran, G.,** *The Evolution of Mind,* Houghton Mifflin, Boston, 1971.
26. **MacMillan, D. L.,** A classical conditioning paradigm for the study of learning in a ganglion of the cockroach *(Periplaneta americana), Anim. Behav.*, 21, 492, 1973.
27. **Disterhoft, J. F.,** Learning in the intact cockroach *(Periplaneta americana)* when placed in a punishment situation, *J. Comp. Physiol. Psychol.*, 79, 1, 1972.

27a. **Eisenstein, E. M., Reep, R. L., and Lovell, K. L.,** Avoidance and escape components of leg position learning in the prothoracic and mesothoracic ganglia of the cockroach, *P. americana, Physiol. Behav.*, 34, 129, 1985.

28. **Church, R. M. and Lerner, N. D.,** Does the headless cockroach learn to avoid?, *Physiol. Psychol.*, 4, 39, 1976.
29. **Willner, P.,** What does the headless cockroach remember?, *Anim. Learn. Behav.*, 6, 249, 1978.
30. **Buerger, A. A., Eisenstein, E. M., and Reep, R. L.,** The yoked control in instrumental avoidance conditioning: an empirical and methodological analysis, *Physiol. Psychol.*, 9, 351, 1981.
31. **Aranda, L. C. and Luco, J. V.,** Further studies on an electric correlate to learning. Experiments in an isolated insect ganglion, *Physiol. Behav.*, 4, 133, 1969.
32. **Reep, R. L., Eisenstein, E. M., and Tweedle, C. D.,** Neuronal pathways involved in transfer of information related to leg position learning in the cockroach, *Periplaneta americana, Physiol. Behav.*, 24, 501, 1980.
33. **Oliver, G. W. O., Taberner, P. V., Rick, J. T., and Kerkut, G. A.,** Changes in GABA level, GAD and ChE activity in CNS of an insect during learning, *Comp. Biochem. Physiol.*, 38B, 529, 1970.
34. **Kerkut, G. A., Beesley, P., Emson, P., Oliver, G., and Walker, R. J.,** Reduction in ChE during shock avoidance learning in the cockroach CNS, *Comp. Biochem. Physiol.*, 39B, 423, 1971.
35. **Beesley, P., Emson, P. C., and Kerkut, G. A.,** Change in K_m of insect ChE after behavioral training, *J. Physiol.*, 221, 26P, 1972.
36. **Kerkut, G. A., Oliver, G., Rick, J. T., and Walker, R. J.,** Biochemical changes during learning in an insect ganglion, *Nature (London)*, 227, 722, 1970.
37. **Kerkut, G. A., Emson, P. C., and Beesley, P. W.,** Effect of leg-raising learning on protein synthesis and ChE activity in the cockroach CNS, *Comp. Biochem. Physiol.*, 41B, 635, 1972.
38. **Woodson, P. B., Schlapfer, W. T., and Barondes, S. H.,** Postural avoidance learning in the headless cockroach without detectable changes in ganglionic cholinesterase, *Brain Res.*, 37, 348, 1972.
39. **Willner, P. and Mellanby, J.,** Cholinesterase activity in the cockroach CNS does not change with training, *Brain Res.*, 66, 481, 1974.
40. **Rick, J. T., Oliver, G. W., and Kerkut, G. A.,** Acquisition, extinction and reacquistion of a conditioned response in the cockroach: the effects of orotic acid, *Q.J. Exp. Psychol.*, 24, 282, 1972.
41. **Lovell, K. L. and Eisenstein, E. M.,** Effects of central nervous system lesions on leg lift learning in the cockroach *Periplaneta americana, Physiol. Behav.*, 28, 265, 1982.
42. **Freckleton, W. C. and Wahlsten, D.,** Carbon dioxide-induced amnesia in the cockroach *(Periplaneta americana), Psychon. Sci.*, 12, 179, 1968.
43. **Paolino, R. M., Quartermain, D., and Miller, N. F.,** Different temporal gradients of retrograde amnesia produced by carbon dioxide anesthesia and electroconvulsive shock, *J. Comp. Physiol. Psychol.*, 62, 270, 1966.
44. **Ebeling, W. Wagner, R. E., and Reierson, D. A.,** Influence of repellency on the efficacy of blatticides. I. Learned modification of behavior of the German cockroach, *J. Econ. Entomol.*, 59, 1374, 1966.

44a. **Rust, M. K. and Reierson, D. A.,** Using pheromone extract to reduce repellency of blatticides, *J. Econ. Entomol.*, 70, 34, 1977.

44b. **Rust, M. K. and Reierson, D. A.,** Increasing blatticidal efficacy with aggregation pheromone, *J. Econ. Entomol.*, 70, 693, 1977.

45. **Longo, N.,** Probability-learning and habit-reversal in the cockroach, *Am. J. Psychol.*, 77, 29, 1964.
46. **Bitterman, M. E.,** Toward a comparative psychology of learning, *Am. Psychol.*, 15, 704, 1960.
47. **Wilson, M. M. and Fowler, H.,** Variables affecting alternation behavior in the cockroach *Blatta orientalis, Anim. Learn. Behav.*, 4, 490, 1976.
48. **Shim, E. and Dixon, P. W.,** The effect of cold temperature exposure on avoidance learning in *Blatta orientalis, J. Biol. Psychol.*, 21, 8, 1979.

49. **Eisenstein, E. M., Lovell, K. L., Reep, R. L., Barraco, D. A., and Brunder, D. G.,** Cellular and behavioral studies of learning and memory in simpler systems, in *Cellular Analogues of Conditioning and Neural Plasticity, Proc. 28th Int. Congr. Physiological Sciences, Szeged, Hungary, 1980,* Vol. 36, Feher, O. and Joo, F., Eds., Pergamon Press, Oxford, 1981, 263.
50. **Barraco, D. A., Lovell, K. L., and Eisenstein, E. M.,** Effects of cyclo-heximide and puromycin on learning and retention in the cockroach *Periplaneta americana, Pharmacol. Biochem. Behav.,* 15, 489, 1981.
50a. **Eisenstein, E. M., Altman, H. J., Barraco, D. A., Barraco, R. A., and Lovell, K. L.,** Brain protein synthesis and memory: the use of antibiotic probes. Symposium presented at the Federated American Societies of Experimental Biology Annual Meeting, New Orleans, La., April 1982, *Fed. Proc.,* 42(14), 3080, 1983.
51. **Luco, J. V. and Aranda, L. C.,** An electrical correlate to the process of learning. Experiments in *Blatta orientalis, Nature (London),* 201, 1330, 1964.
52. **Luco, J. V.,** Increase of synaptic efficacy as a correlate to learning in *Blatta orientalis, Physiol. Behav.,* 21, 743, 1978.
53. **Balderama, N.,** One trial learning in the American cockroach *Periplaneta americana, J. Insect Physiol.,* 26, 499, 1980.
54. **Pougalan, M. and Masson, C.,** Plasticité des résponses olfactives chez la blatte grégaire *Blaberus craniifer* Burm., *Biol. Behav.,* 6, 1, 1981.
55. **Gates, M. F. and Allee, W. C.,** Conditioned behavior of isolated and grouped cockroaches on a simple maze, *J. Comp. Psychol.,* 15, 331, 1933.
55a. **Zajonc, R. B., Heingartner, A., and Herman, E. M.,** Social enhancement and impairment of performance in the cockroach, *J. Pers. Soc. Psychol.,* 13, 83, 1969.
56. **McConnell, J. V.,** Memory transfer through cannibalism in planarians, *J. Neuropsychiatry,* 3, 42, 1962.
57. **Babich, F. R., Jacobson, A. L., Berbash, S., and Jacobson, A.,** Transfer of a response to naive rats by injection of ribonucleic acid extracted from trained rats, *Science,* 149, 656, 1965.
58. **Hartry, A. L., Keith-Lee, P., and Morton, W. D.,** Planaria: memory transfer through cannibalism reexamined, *Science,* 146, 274, 1964.
59. **Gross, C. G. and Carey, F. M.,** Transfer of learned response by RNA injection: failure of attempts to replicate, *Science,* 150, 1749, 1965.
60. **Kumar, P. J. and Muhar, I. S.,** An experimental study of the chemical transfer of learning by ingestion, *Indian J. Psychol.,* 50, 215, 1975.
61. **Cheney, C. D., Klein, A., and Snyder, R. L.,** Transfer of learned dark aversion between cockroaches, *J. Biol. Psychol.,* 16, 16, 1974.
62. **Eisenstein, E. M. and Reep, R. L.,** Behavioral and cellular studies of learning and memory in insects, in *Comprehensive Insect Physiology, Biochemistry and Pharmacology,* Vol. 9, Kerkut, G. A. and Gilbert, L. I., Eds., Pergamon Press, Oxford, 1985, chap. 11.
63. **Byrne, J. H.,** Cellular analysis of associative learning, *Physiol. Rev.,* 67, 329, 1987.
64. **Tully, T.,** *Drosophila* learning: behavior and biochemistry, *Behav. Genet.,* 14, 527, 1984.
65. **Zawistowski, S. and Hirsch, J.,** Conditioned discrimination in the blow fly, *Phormia regina:* controls and bidirectional selection, *Anim. Learn. Behav.,* 12, 402, 1984.
66. **Siegel, R. W. and Hall, J. C.,** Conditioned responses in courtship of normal and mutant *Drosophila, Proc. Natl. Acad. Sci. U.S.A.,* 76, 3430, 1979.
67. **Zawistowski, S. and Richmond, R. C.,** Experience-mediated courtship reduction and competition for mates by male *Drosophila melanogaster, Behav. Genet.,* 15, 561, 1985.
68. **McRobert, S. and Tompkins, L.,** Two consequences of homosexual courtship performed by *Drosophila melanogaster and Drosophila simulans, Evolution,* 42, 1093, 1988.
69. **Wcislo, W. T.,** The role of learning in the mating biology of a sweat bee *Lasioglossum zephyrum* (Hymenoptera: Halictidae), *Behav. Ecol. Sociobiol.,* 20, 179, 1987.

Chapter 28

THE COCKROACH NERVOUS SYSTEM AS A MODEL FOR AGING

Manfred J. Kern and Rajindar S. Sohal

TABLE OF CONTENTS

I. INTRODUCTION

The process of aging is an integral component of the ontogeny of organisms. Unlike the earlier maturation phase, it reduces the adaptive ability of the organism to maintain homeostasis, which progressively increases the likelihood of disease and death.[1] It is widely believed that deteriorative changes in the nerve cells play a major role not only in the alteration of the function of the nervous system, but also in the aging of the whole organism.[2-6] Metabolic disturbances in the brain during aging have multiple physiological effects involving other organs.

A review of the gerontological literature indicates that insects are increasingly being used as model systems to analyze the underlying causes of the aging processes.[7-11] Previous reviews of aging in insects have provided relatively little coverage of aging in the nervous system, partly due to the paucity of information. For example, in a detailed review of age-related structural, physiological, and biochemical changes in the insect brain, there was hardly a mention of cockroaches because no data were available on brain aging in these species.[12] It is both unfortunate and surprising that such an excellent model system, which is used successfully in electrophysiology, neuroendocrinology, toxicology, and pharmacology, has not merited similar attention in the field of experimental gerontology. In general, short-lived insects such as the fruit fly *(Drosophila melanogaster),*[13] the housefly *(Musca domestica),*[14] the blowfly *(Calliphora erythrocephala),*[15,16] and the silk moth *(Bombyx mori),* with a maximal life span of a few weeks, have been preferred over long-lived species of the order Blattodea, with maximal life spans of over 1 year. The following are examples of the relatively long life span of Blattodea; *Blaberus giganteus,* 20 months;[17] *Periplaneta americana,* 20 months[18,19] or 30 months;[20] and *Blattella germanica,* 13 months.[21] It is hoped that in the future more attention will be paid to longer-lived insect species, of which the cockroach offers a very suitable model for the study of the age-related changes in the structure and function of the nervous system.

A photomicrograph and a schematic view of the brain of the cockroach *Blaberus craniifer* are presented in Figures 1 and 2, respectively. The anatomy of the cockroach brain is described in detail in Chapter 4 of the work.

II. ADVANTAGES OF THE COCKROACH NERVOUS SYSTEM AS A MODEL FOR AGING

The following features of the cockroach brain indicate its versatility for the investigation of a variety of biological questions:

1. In general, nerve cells in the cockroach are postmitotic; thus, structural and functional changes can be related to the age of the individual.
2. The cockroach brain lacks blood vessels. The nervous system is bathed directly in the hemolymph, and nutrients and waste products are exchanged directly between the brain and the hemolymph. Therefore, secondary changes resulting from pathology of vasculature, which are superimposed on age-related changes, can be avoided.
3. Oxygen is supplied directly to the brain. The gaseous exchange is achieved by an extensive network of tracheae and tracheoles within the brain. No neuron is more than about 9 μm away from a tracheole.[22] Thus, effects of experimentally altered ambient atmosphere can be studied advantageously.
4. The blood-brain barrier is very efficient and well developed.[23,24]
5. Glycogen is the most important stored nutrient in the insect brain.[25,26] Glycogen levels in the cockroach brain are ten times higher than the levels reported for mammalian brains,[27] an advantageous feature for *in vitro* studies.

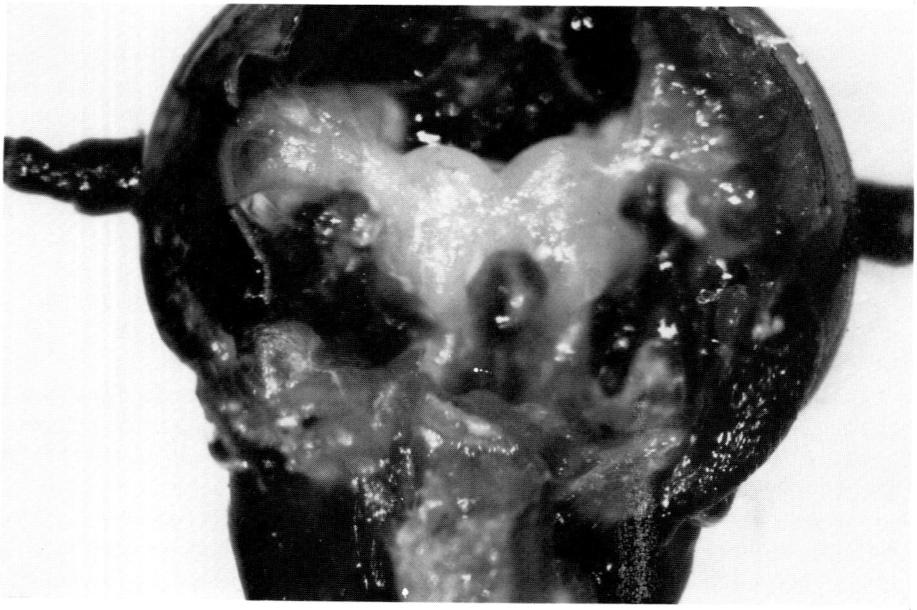

FIGURE 1. Photomicrograph of the frontal view of the brain of a female cockroach, *Blaberus craniifer*.

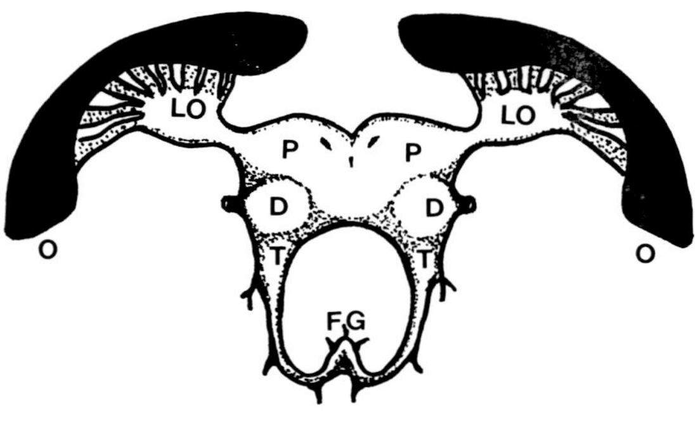

FIGURE 2. Diagram of the frontal view of the brain of the cockroach *Blaberus craniifer*. LO, optic lobes; O, eyes; P, protocerebrum; D, deutocerebrum; T, tritocerebrum; FG, frontal ganglion.

6. Besides glucose, some insect species oxidize amino acids such as proline[28] or even fatty acids such as oleic acid and lauric acid.[16] Brains of Blattodea exhibit little proline utilization.[28]
7. The insect brain exhibits remarkable stability *in vitro*.[16,29]
8. The cockroach brain consumes 3.5 to 6.6 μl O_2 per milligram dry weight brain per hour (Table 1). In contrast, brains from highly evolved insects such as some Lepidoptera or Diptera have a metabolic rate seven or eight times higher.[16,30,31]
9. The cockroach brain is highly resistant to anoxia.[32,33]
10. In general, brains of insects are small in size; they usually have <1 mg dry weight[31] and are composed of only a few hundred thousand cells.[34-37] The small size permits serial sectioning of the entire brain, which reduces sampling errors.[14] Furthermore, physiological and biochemical experiments can be repeated rapidly because the isolation of insect brain is quite easy.

TABLE 1
Rate of Oxygen Consumption in Brains of Different Cockroaches[31]

Insect	Dry weight of brain (mg)		Oxygen consumption ($\mu l\ O_2/\mu g$ dry weight/h)	
	Female	Male	Female	Male
Blaptica dubia	0.34	0.42	3.5	4.6
Blaberus craniifer	0.52	0.47	3.7	6.6
Gromphadorhina portentosa	0.65	0.56	4.2	4.7
Periplaneta americana	0.31	—	4.5	—
Leucophaea maderae	0.31	0.26	5.6	6.1

III. AGE-RELATED CHANGES IN THE COCKROACH NERVOUS SYSTEM

A. STRUCTURAL CHANGES

Shortly after eclosion, brains from young adult *Blaberus craniifer* seem to have more cell fluid than brains from old animals. Brains from old animals seem to be reduced in size. This is in agreement with changes reported in aged bees *(Apis mellifera)*,[38] blowflies *(C. erythrocephala)*,[16] and fruit flies *(D. melanogaster)*.[39]

Sharma and Bahadur[40] found that a significant increase in brain wet weight and protein content occurs in *P. americana* during the first 5 months of adult life, followed by a considerable decline (−20%) during old age. These authors suggest that changes in neurosecretory cell number and protein concentration are related to aging. In female *Blaberus craniifer* there is no decrease in brain wet weight and protein content until day 575 of adult life. In female *C. erythrocephala* there is an increase in brain dry weight until day 7 of adult life, but no significant decrease occurs after this age.[16] Overall, it seems that loss of brain mass or protein content does not occur during aging. Any age-related impairment in function of the nervous system is probably due to intraneuronal alterations rather than to the loss of nerve mass or nerve cells. Similarly, in humans, where brain weight decreases about 11% and the number of nerve cells in the brain decreases 20 to 25% during aging, the depletion of nerve cells is not considered to be an explanation for dysfunction and death.[41-43]

The most prominent age-related ultrastructural changes in nerve cells of both mammals and insects are loss of ribosomal content, mitochondrial swelling, and accumulation of age pigment or lipofuscin.[12,14] Swelling of mitochondria has been reported in the brain of *Periplaneta*,[44] the ant *Formica lugubris*,[45] the sphingid moth *Manduca sexta*,[46] and *Calliphora*.[16] Large clusters of mitochondria or "giant mitochondria" are seen in the brains of 18- to 39-day-old *C. erythrocephala*[16,47] as well as in the flight muscles of 31-day-old *Musca*.[48] One of the most consistent manifestations of aging in human and insect[11] neurons is the accumulation of lipofuscin within the cytoplasm.[49] During maturation, neurons of the pars intercerebralis of the cockroach *Blaberus craniifer* contain numerous lamellar bodies.[50] The authors suggested a relationship between degenerative mitochondria and lipofuscin. Unfortunately, they did not examine the nervous tissue in aged animals. An accumulation of lipofuscin or concentrically lamellated structures has also been reported in the brain of aged *Musca*,[11,51] scorpion flies *(Panorpa vulgaris)*,[52] *Drosophila*,[13] and *Calliphora*.[47] It is generally believed that lipofuscin itself has no deleterious effect on cell function.[51,53] A considerable number of studies have been conducted on the composition and genesis of lipofuscin. These studies were reviewed recently by Sohal and Wolfe.[54]

The current hypotheses concerning the mechanism of origin of lipofuscin can be divided into two schools of thought. One view is that lipofuscin accumulates due to the inability of the cells to degrade substances such as dolichol phosphates and polymerized lipids. The

other view is that lipofuscin is formed as a result of oxygen-free-radical-induced damage and polymerization of lipids and proteins.

B. PHYSIOLOGICAL AND BIOCHEMICAL CHANGES

Oxygen consumption by mammalian nerve cells decreases with age.[5,55,56] However, information concerning age-related changes in the metabolic rate of brains from cockroaches is presently unavailable. Oxygen consumption in the brain of *Calliphora* varies, with a maximal rate attained at about the middle of the maximal life span.[15,16,47] Within 2 to 3 d after this point the rate of oxygen consumption is reduced to a lower plateau until death. It is noteworthy that rates of oxygen consumption of brains run parallel to those of the whole resting animal.[16] Overall, it seems that the metabolic rate in the insect nervous system decreases with age.

A reduced capacity for glucose turnover has been reported in the aging human brain.[57,58] In insects, the oxidation of glucose is markedly reduced in the brain of the aged *Calliphora*.[16] This is not due to a shortage of available glucose, but results from decreased ability to utilize glucose during the aging process.

Acetylcholinesterase (AChE) activity in the brain of *A. mellifera*[59] workers and in the head of *Musca*[60] increases during the first week after emergence of the adult and stays constant thereafter until death. A similar pattern in AChE activity was found in the brain of *Calliphora*.[68] In *Blaberus craniifer,* the AChE activity increases at the beginning of the adult life span and is not reduced until day 575, suggesting that AChE activity of the brain is not a contributing factor in aging of the cockroach.[69]

The activity of acid and alkaline phosphatases increases immediately after eclosion in the brain of female *B. craniifer*. However, there is no reduction in activity of the phosphatases in the central nervous system of *B. craniifer* during old age.[69]

Transamination is one of the chief mechanisms whereby the balance between the amino acid pool and protein synthesis is regulated.[60a] The transaminases play key roles in the formation of nonessential amino acids, gluconeogenesis, and metabolism of waste nitrogen products.[60b] After eclosion there is an increase in alanine aminotransferase and aspartate aminotransferase activities in the brain of the female *B. craniifer* which is retained until day 575.[69]

Glucose metabolism and glycolysis are affected appreciably in nerve cells of aged humans.[61-64] Hexokinase[65] and phosphofructokinase[63] activities decrease in old age. There are no data available about the age-related changes in the activity of glycolytic enzymes in the brain of cockroaches. However, in the brain of *Calliphora,* the activities of glycogen phosphorylase, hexokinase, and phosphofructokinase decline markedly during aging. The activities of these enzymes increase from eclosion until the middle of the life span and then gradually decrease until death.[12] Apparently, the efficiency of glycolysis is reduced during aging in the insect brain.

There are no available data concerning age-related changes in the activity of cytochrome *c* oxidase in the brain of cockroaches. Nevertheless, in the brain of *Calliphora,* cytochrome *c* oxidase activity tripled until the third week of adult life, followed by a rapid decrease.[12] This reduction is a further indication that the reduced function of the mitochondria causes disturbances in the energetic processes that occur in the insect nervous system during aging.

The activities of Na^+-K^+-ATPase and Mg^{2+}-ATPase increase in the brains of *A. mellifera* immediately after eclosion.[66] In the case of Na^+-K^+-ATPase, activity decreases after 6 weeks of adult life. In the brain of female and male *Calliphora,* Na^+-K^+-ATPase and Mg^{2+}-ATPase activities increase until the seventh day after eclosion and remain unchanged thereafter.[67] Presumably, neither of these enzymes plays a key role in the aging of the insect nervous system.

IV. CONCLUSIONS

Age-related changes in the brain function of insects are probably due to intraneuronal alterations rather than to the loss of nerve cells or nerve mass. Swelling of mitochondria and accumulation of lipofuscin are the main structural indicators of neuronal aging. The metabolic rate, the capacity for glucose utilization, and the activities of some key enzymes in the glycolytic pathway decline in the brain during aging.

Activities of enzymes such as acetylcholinesterase, acid phosphatase, alkaline phosphatase, aspartate aminotransferase, alanine aminotransferase, Na^+-K^+-ATPase, and Mg^{2+}-ATPase are not affected by the aging process.

ACKNOWLEDGMENT

The help of Dr. H. Grötsch, Pharmaforschung-Biochemie, Hoechst AG for the determination of the activities of acetylcholinesterase, acid phosphatase, alkaline phosphatase, asparatate aminotransferase, and alanine aminotransferase is gratefully acknowledged.

REFERENCES

1. **Comfort, A.,** *The Biology of Senescence,* 3rd ed., Elsevier, New York, 1979.
2. **Ordy, J. M.,** *Neurobiology of Aging,* Plenum Press, New York, 1975.
3. **Samorajski, T.,** Neurochemical changes in the aging human and nonhuman primate brain, in *Psychopharmacology and Aging,* Eisdorfer, C. and Fann, W. E., Eds., Spectrum Publication, Jamaica, NY, 1980, 145.
4. **Buschmann, M. B. T.,** Brain structure and its implication in metabolism in aging: a review, *Clin. Nutr.,* 36, 759, 1982.
5. **Hoyer, S.,** The aging brain, *Exp. Brain Res.,* Suppl. 5, 67, 1982.
6. **Frolkis, V. V., Tanin, S. A., Martynenko, O. A., Bogatskaya, L. N., and Bezrukov, V. V.,** Aging of the neurons, *Interdiscip. Top. Gerontol.,* 18, 1, 1984.
7. **Clark, A. M. and Rockstein, M.,** Aging in insects, in *The Physiology of Insecta,* Vol. 1, Academic Press, London, 1964, 227.
8. **Rockstein, M. and Miquel, J.,** Aging in insects, in *The Physiology of Insecta,* Vol. 1, Academic Press, London, 1973, 371.
9. **Stoffolano, J. G.,** Insects as model systems for aging studies, in *Special Review of Experimental Aging Research,* Elias, M. F., Ed., EAR, Bar Harbor, ME, 1976, 407.
10. **Lamb, M. J.,** Ageing, in *The Genetics and Biology of Drosophila,* Ashburner, M. and Wright, T. R. F., Eds., Academic Press, London, 1978, 43.
11. **Sohal, R. S.,** Aging in insects, in *Comprehensive Insect Physiology, Biochemistry and Pharmacology,* Vol. 10, Kerkut, G. A. and Gilbert, L. I., Eds., Pergamon Press, Oxford, 1985, 595.
12. **Kern, M. J.,** Brain aging in insects, in *Insect Aging,* Collatz, K. G. and Sohal, R. S., Eds., Springer-Verlag, Berlin, 1986, 90.
13. **Herman, M. H., Miquel, J., and Johnson, M.,** Insect brain as a model for the study of aging, *Acta Neuropathol.,* 19, 167, 1971.
14. **Sohal, R. S. and Sharma, S. P.,** Age-related changes in the fine structure and number of neurons in the brain of the housefly, *Musca domestica, Exp. Gerontol.,* 7, 243, 1972.
15. **Kern, M. and Wegener, G.,** Age dependent changes in the metabolism of insect brains, presented at the 13th Meet. Eur. Biochem. Soc., Jerusalem, August 24 to 29, 1980.
16. **Kern, M.,** Das Insekt als Modell für Altersstudien. Altersabhängige Untersuchungen zum Gehirnstoffwechsel von *Calliphora erythrocephala* and *Bombyx mori,* Ph.D. thesis, Johannes Gutenberg-Universität, Mainz, Federal Republic of Germany, 1982.
17. **Piquett, P. G. and Fales, J. H.,** Life history of *Blaberus giganteus* (L.), *J. Econ. Entomol.,* 46, 1089, 1953.
18. **Klein, H. Z.,** Zur Biologie der amerikanischen Schabe (*Periplaneta americana* L.), *Z. Wiss. Zool. Abt. A,* 144, 102, 1933.

19. **Edmunds, L. R.,** Observations on the biology and life history of the brown cockroach *Periplaneta brunnea* Burmeister, *Proc. Entomol. Soc. Wash.*, 59, 283, 1957.
20. **Gould, G. E. and Deay, H.,** The biology of the American cockroach, *Ann. Entomol. Soc. Am.*, 31, 489, 1938.
21. **Pope, P.,** Studies of the life histories of some Queensland Blattidae (Orthoptera), *Proc. R. Soc. Queensl.*, 63, 23, 1953.
22. **Burrows, M.,** Principles of organization of insect central nervous systems, in *Insect Neurobiology and Pesticide Action (Neurotox 79),* Society for Chemical Industry, London, 1980, 5.
23. **Treherne, J. E. and Pichon, Y.,** The insect blood-brain barrier, *Adv. Insect Physiol.*, 9, 257, 1972.
24. **Treherne, J. E. and Schofield, P. K.,** Ionic homeostasis of the brain microenvironment in insects, *Trends Neurosci.*, 2, 227, 1979.
25. **Wigglesworth, V. B.,** The nutrition of the central nervous system in the cockroach *Periplaneta americana* L. The role of perineurium and glial cells in the mobilization of reserves, *J. Exp. Biol.*, 37, 500, 1960.
26. **Clement, E. M. and Strang, R. H. C.,** A comparison of some aspects of the physiology and metabolism of the nervous system of the locust *Schistocerca gregaria in vitro* with those *in vivo, J. Neurochem.*, 31, 135, 1978.
27. **Walter, D. C.,** Bioenergetic processes of the cockroach nervous system, *Diss. Abstr. B,* 40, 5, 1979.
28. **Kern, M. J.,** Utilization of glucose and proline in the brain of adult insects, *Insect Biochem.*, 16, 567, 1986.
29. **Strang, R. H. C.,** Energy metabolism in the insect nervous system, in *Energy Metabolism in Insects,* Downer, R. G. H., Ed., Plenum Press, New York, 1981, 169.
30. **Kern, M.,** Relation of insect life to body weight and energy metabolism and the problem of brain weight, metabolic rate, and life span, presented at the 17th Int. Congr. Entomology, Hamburg, Federal Republic of Germany, August 20 to 26, 1984.
31. **Kern, M.,** Metabolic rate of the insect brain in relation to body size and phylogeny, *Comp. Biochem. Physiol,* 81A, 501, 1985.
32. **Walter, D. C. and Nelson, S. R.,** Energy metabolism and nerve function in cockroaches *(Periplaneta americana), Brain Res.*, 94, 485, 1975.
33. **Wegener, G.,** Comparative aspects of energy metabolism in nonmammalian brains under normoxic and hypoxic conditions, in *Animal Models and Hypoxia,* Stefanovich, V. and Kriegelstein, J., Eds., Pergamon Press, Oxford, 1981, 87.
34. **Farrell, S. and Kuhlenbeck, H.,** Preliminary computation of the number of cellular elements in some insect brains, *Anat. Rec.*, 148, 369, 1964.
35. **Becker, H. W.,** The number of neurons, glial and perineurium cells in an insect ganglion, *Experientia,* 21, 719, 1965.
36. **Witthöft, W.,** Absolute Anzahl und Verteilung der Zellen im Hirn der Honigbiene, *Z. Morphol. Tiere,* 61, 160, 1967.
37. **Weidner, H.,** Morphologie, Anatomie und Histologie, in *Handbuch der Zoologie,* Vol. 4, 2nd ed., Helmcke, J. G., Starck, D., and Wermuth, H., Eds., Walter de Gruyter, Berlin, 1982, chap. 11.
38. **Lucht-Bertram, E.,** Degenerative Erscheinungen am Gehirn alternder Bienen-Königinnen (*Apis mellifera* L.), *Z. Bienenforsch.*, 6, 169, 1962.
39. **Miquel, J.,** Aging of male *Drosophila melanogaster:* histological, histochemical, and ultrastructural observations, *Adv. Gerontol. Res.*, 5, 39, 1979.
40. **Sharma, P. K. and Bahadur, J.,** Age-related changes in the total protein in the brain of *Periplaneta americana* (L.), *Mech. Ageing Dev.*, 20, 49, 1982.
41. **Sandoz, P. and Meier-Ruge, W.,** Age-related loss of nerve cells from the human inferior olive, and unchanged volume of its gray matter, *IRCS Libr. Compend.*, 5, 376, 1977.
42. **Tomlinson, B. E.,** The ageing brain, in *Recent Advances in Neuropathology,* Smith, Th. and Cavanagh, J. B., Eds., Churchill-Livingstone, Edinburgh, 1979, 129.
43. **Brody, H. and Vijayashankar, N.,** Anatomical changes in the nervous system, in *Handbook of the Biology of Aging,* Finch, C. and Hayflick, L., Eds., D Van Nostrand, New York, 1977, 241.
44. **Hess, A.,** The fine structure of young and old spinal ganglia, *Anat. Rec.*, 123, 399, 1955.
45. **Lampareter, H. E., Akert, K., and Sandri, C.,** Wallersche Degeneration im Zentralnervensystem der Ameise. Elektronenmikroskopische Untersuchungen am Prothorakalganglion von *Formica lugubris* Zett., *Schweiz. Arch. Neurol. Neurochir. Psychiatr.*, 100, 337, 1967.
46. **Stocker, R. F., Edwards, J. S., and Truman, J. W.,** Fine structure of degenerating abdominal neurons after eclosion in the sphingid moth, *Manduca sexta, Cell Tissue Res.*, 191, 317, 1978.
47. **Kern, M. and Wegener, G.,** Age affects the metabolic rate of insect brain, *Mech. Ageing Dev.*, 28, 237, 1984.
48. **Sohal, R. S. and Allison, V. F.,** Age-related changes in the fine structure of the flight muscle of the housefly, *Exp. Gerontol.*, 6, 167, 1971.

49. **Brizzee, K. R. and Ordy, J. M.,** Cellular features regional accumulation, and prospects of modification of age pigments in mammals, in *Age Pigments,* Sohal, R. S., Ed., Elsevier/North-Holland, Amsterdam, 1981, 101.
50. **Willey, R. B. and Chapman, G. B.,** Fine structure of neurons within the pars intercerebralis of the cockroach, *Blaberus craniifer, Gen. Comp. Endocrinol.,* 2, 31, 1962.
51. **Sohal, R. S.,** Metabolic rate, aging and lipofuscin accumulation, in *Age Pigments,* Sohal, R. S., Ed., Elsevier/North-Holland, Amsterdam, 1981, 303.
52. **Collatz, K.-G. and Collatz, S.,** Age-dependent ultrastructural changes in different organs of the mecopteran fly, *Panorpa vulgaris, Exp. Gerontol.,* 26, 183, 1981.
53. **Sohal, R. S. and McArthur, M. C.,** Cellular aspects of aging in insects, in *Cell Biology of Aging,* Cristofalo, V. J., Ed., CRC Press, Boca Raton, FL, 1985, 497.

53a. **Hendley, D. D., Mildvan, A. S., Reporter, M. C., and Strehler, B. L.,** The properties of isolated human cardiac age pigment. I. Preparation and Physical Properties, *J. Gerontol.,* 18, 144, 1963.

53b. **Hendley, D. D., Mildvan, A. S., Reporter, M. C., and Strehler, B. L.,** The properties of isolated human cardiac age pigment. II. Chemical and enzymatic properties, *J. Gerontol.,* 18, 250, 1963.

53c. **Wolfe, L. S., Ng Ying Kin, M. M. K., and Baker, R. R.,** Batten disease and related disorders: new findings on the chemistry of the storage material, in *Lysosomes and Lysosomal Storage Diseases,* Callahan, J. W. and Lowden, J. A., Eds., Raven Press, New York, 1981, 315.

53d. **Hasan, M. and Glees, P.,** Genesis and possible dissolution of neuronal lipofuscin, *Gerontologica,* 18, 217, 1972.

54. **Sohal, R. S. and Wolfe, L. S.,** Lipofuscin: characteristics and significance, *Progr. Brain Res.,* 70, 171, 1986.
55. **Patel, M. S.,** Age-dependent changes in oxidative metabolism in rat brain, *J. Gerontol.,* 32, 643, 1977.
56. **Parmacek, M. S., Fox, J. H., Harrison, W. H., Garron, D. C., and Swenie, D.,** Effect of aging on brain respiration and carbohydrate metabolism of CBF mice, *Gerontology,* 25, 185, 1979.
57. **Gottstein, U., Bernsmeier, A., and Sedlmeier, I.,** Der Kohlenhydratstoffwechsel des menschlichen Gehirns. II. Untersuchungen mit substratspezifischen enzymatischen Methoden bei Kranken mit verminderter Hirndurchblutung auf dem Boden einer Arteriosklerose der Hirngefaße, *Klin. Wochenschr.,* 42, 310, 1964.
58. **Sokoloff, L.,** Cerebral circulatory and metabolic changes associated with aging, *Res. Publ. Assoc. Res. Nerv. Ment. Dis.,* 41, 237, 1966.
59. **Rockstein, M.,** The relation of cholinesterase activity to change in cell number with age in the brain of the adult honeybee, *J. Cell. Comp. Physiol.,* 35, 11, 1950.
60. **Babers, F. H. and Pratt, J. J.,** Studies on the resistance of insects to insecticides. I. Cholinesterase in houseflies *(Musca domestica)* resistant to DDT, *Physiol. Zool.,* 23, 58, 1950.

60a. **Kaur, S. P., Sidhu, D. S., Dhillon, S. S., and Kumar, N.,** Transaminases during development of the bruchid, *Zabrotes subfasciatus* (Boh.) (Coleoptera: Bruchidae), *Insect Sci. Applic.,* 6, 585, 1985.

60b. **Cohen, P. P.,** in *The Enzymes,* Vol. 1, Summer, J. B. and Myrback, K., Academic Press, New York, 1951, 1040.

61. **Sokoloff, L.,** Relation between physiological function and energy metabolism in the central nervous system, *J. Neurochem.,* 29, 13, 1977.
62. **Bowen, D. M., White, J. A., Spillane, M. J., Goodhardt, G., Curzon, G., Iwangoff, P., Meier-Ruge, W., and Davison, A. N.,** Accelerated ageing or selective neuronal loss as an important cause of dementia?, *Lancet,* 1, 11, 1979.
63. **Iwangoff, P., Armbruster, R., Enz, A., Meier-Ruge, W., and Sandoz, P.,** Glycolytic enzymes from human autoptic brain cortex: normally aged and demented cases, in *Biochemistry of Dementia,* Roberts, P. J., Ed., John Wiley & Sons, New York, 1980, 258.
64. **Meier-Ruge, W., Iwangoff, P., Reichlmeier, K., and Sandoz, P.,** Neurochemical findings in the aging brain, in *Ergot Compounds and Brain Function,* Goldstein, M., Calne, D. B., Lieberman, A., and Thornes, M. O., Eds., Raven Press, New York, 1980, 323.
65. **Potapenko, R. I.,** Age specifics of energy metabolism of various parts of the brain, author's abstract of the dissertation for a candidate's degree (medicine), Medical Institute, Kiev, U.S.S.R., 1974. In Russian.
66. **Cheng, E. Y. and Cutkomp, L. K.,** Aging in the honeybee *Apis mellifera,* as related to brain ATPases and their DDT sensitivity, *J. Insect Physiol.,* 18, 2285, 1972.
67. **Rivera, M. E. and Langer, H.,** Effect of light on ATPases in eyes and brain of the blowfly, *Calliphora, J. Comp. Physiol.,* 123, 245, 1978.
68. **Rottman, M.,** personal communication.
69. **Kern, M. J.,** unpublished data.

Index

INDEX

VOLUMES I AND II

A

B

C

D

E

F

G

H

I

J

K

L

M

N

Q

R

S

T